An Introduction to Earth-Space Science

3rd Edition

Brent Zaprowski
Darren Parnell
Matthew Bailey
Daniel Harris
Vanessa Smullen

Kendall Hunt
publishing company

www.kendallhunt.com
Send all inquiries to:
4050 Westmark Drive
Dubuque, IA 52004-1840

ISBN 978-1-4652-7249-2

Printed in the United States of America

TABLE OF CONTENTS

ACKNOWLEDGMENTS

The authors would like to acknowledge all the people who have contributed to the writing of this book. Good books need good editors, and we are lucky to have two excellent ones who proofread this work, Dr. Frank Watson and Dr. Brent Skeeter. Other people who also contributed their proofreading skills were Heather Bailey and Dr. Andrew Pica. Thank you!

We also want to acknowledge our colleagues, former students and others who graciously contributed photos for the book: Kelly Baccala, Zack Baccala, Sarah Burton, Greg Burzynski, Tom Cawthern, Casey Condon, Cãtãlin Fãtu, Stuart Hamilton, Heather Hyre Hansen, Mike Peel, Ben Reed, Michelle Saunders, Dave Votta, Amy Zaprowski and Caylin Zaprowski.

THE EARTH AS A SYSTEM: IT'S ALL CONNECTED

Scientists have begun to realize that we cannot consider the different parts of the natural world separately, compartmentalized into individual "boxes". Why? Because seemingly different disciplines of science (such as geology and biology) are too interconnected to draw any meaningful conclusions about their place in the world if only considered in isolation. Thus, we use the term **Earth Science** to describe the Earth as one complete system. Earth Science is really a synthesis of all the major sciences. Everything in the universe is bound by the laws of physics, chemistry and mathematics. **Geoscientists** use this body of knowledge along with mapping science techniques to study the nature of the Earth and the Universe that surrounds it.

The Earth System has five major components. The **Geosphere** includes all of the rocks that make up the solid Earth. The **Hydrosphere** encompasses all of the water which is found in the Earth System. The **Atmosphere** is the layer of gasses found above the geosphere and hydrosphere. The **Biosphere** encompasses all of the flora and fauna which inhabit the Earth. The **Universe** is everything that surrounds the planet Earth. The processes which occur in the air, in the ocean, in the core of the Earth and in space overlap in a symbiotic manner to produce the world we see around us every day. Although the other parts of the Earth System have a very strong control over where and when different types of plants and animals live, living organisms also alter the other parts of the Earth System in important ways. Humans in particular have fundamentally and permanently changed the Earth System. Thus, the body of knowledge generated by geoscientists could really be thought of as a user's manual for Planet Earth.

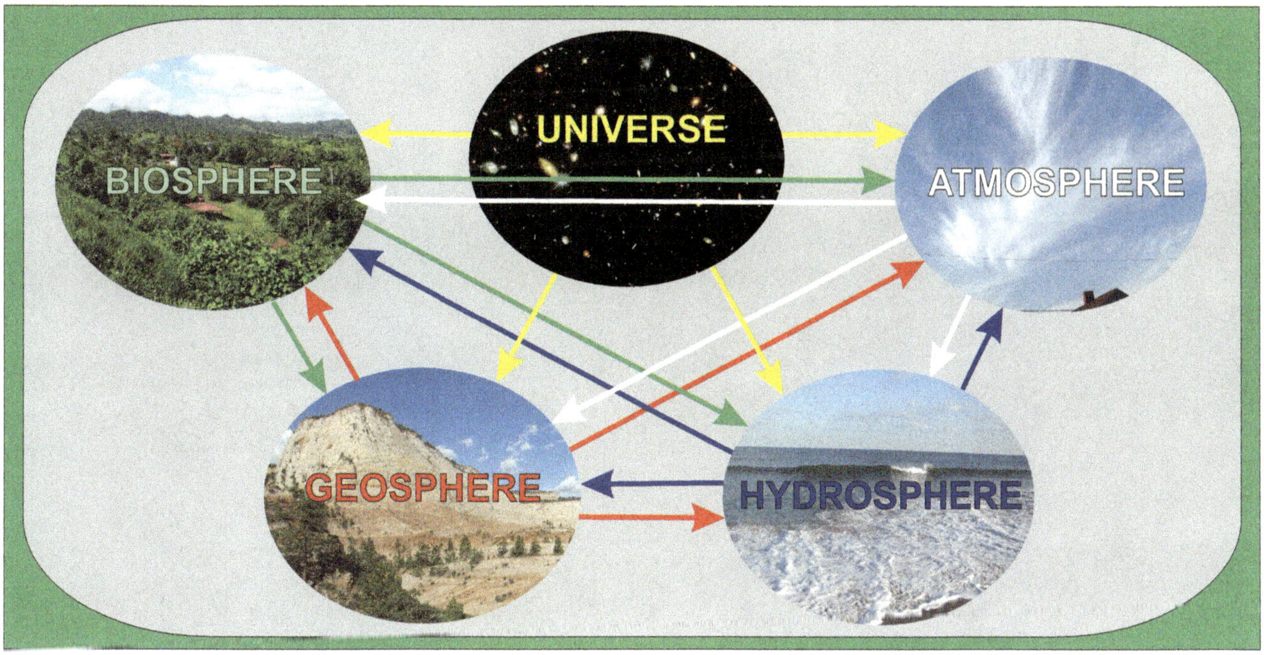

And do we ever need one. Every year hundreds of thousands of people are injured or killed by a wide range of natural hazards caused by changes in the Earth System. Some of these hazards are catastrophic and completely unpredictable. The sudden shaking of the Earth leading to a deadly rock slide is a good example of this. Other hazards are predictable, and with proper precautions, property damage and the loss of life can be minimized. For example, the analysis of past flood events allows us to predict the likelihood of a flood event occurring in the future. Some hazards are more subtle, long-term phenomena, like changes in climate or sea-level. Unfortunately, many people fail to use proper judgement when natural disasters strike because they don't understand the basic principles of the Earth System. One of the goals of this book is to inform the reader about how different aspects of the Earth System can impact human activities, and how they can use this knowledge to protect themselves and their property.

As you can imagine, Earth Science is also a broad subject. Earth Science is typically broken into three major components: Meteorology, Geology and Astronomy. That's a lot of information! As such, another goal of this book is breadth. We will cover a wide range of different topics from across all parts of the Earth System, but in many cases we will only be scratching the surface of knowledge. Those who are interested in learning more about the geosciences are encouraged to enroll in other courses which cover these topics in more depth.

USGS

NASA

NOAA

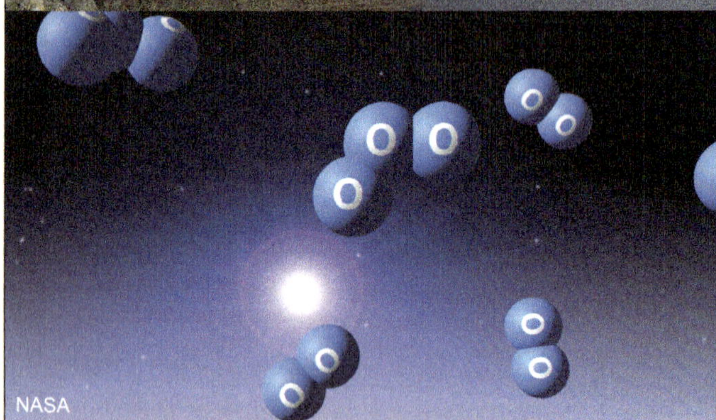
NASA

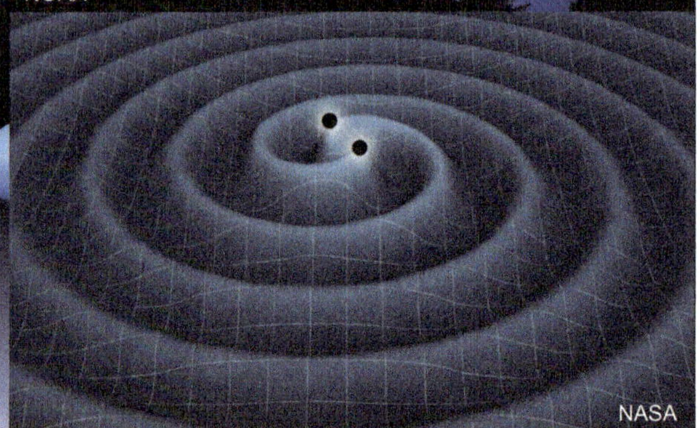
NASA

Chapter 1 - Introduction to Basic Science Concepts

In this book we are going to explore all of the major facets of the Earth and the Universe we live in. In order to understand the physical processes which govern how the natural world behaves, it is important that we first review the basic principles of chemistry and physics. In this chapter, we will look at the properties of matter, the properties of energy, and the physics of motion along with some examples of how they apply to earth science. First, let's review the basic units used to make scientific measurements.

Section 1-1: Scientific Quantities, Units and Conversion Factors

Scientific inquiry is based on measuring and collecting data. Numerical data communicate how much of something there is, and units are used to communicate how much, or how many, of *what* has been measured. A **unit** is a qualitative descriptor of a quantity. Why are units necessary? Let's assume your friend told you that she ran 15. Would you know how far she ran? Did she run 15 meters, or 15 miles, or 15 feet? Maybe she ran for 15 minutes? Without units, it is impossible to communicate about quantitative variables. To effectively communicate, scientists have developed standard units of measurement. That is, units of measurement were defined and used universally. Currently, two sets of standard units are used; the British Imperial System (also referred to as English, Standard or SI) and the Metric System (Table 1-1).

Property	English unit	Metric unit
Length	inch, foot	meter
Mass	ounce, pound	gram
Volume	quart, gallon	liter
Temperature	Fahrenheit	Celsius

Table 1-1: Comparing the English System with the Metric System for scientific measurements.

$$\frac{12\text{ eggs}}{1\text{ dozen}} = \frac{1\text{ dozen}}{12\text{ eggs}}$$

$$36\text{ eggs} \times \frac{1\text{ dozen}}{12\text{ eggs}} = \_\_\text{ dozen}$$

$$36\text{ eggs} \times \frac{1\text{ dozen}}{12\text{ eggs}} = 3\text{ dozen}$$

Figure 1.1: An example of conversion factors and how to use them

We often need to convert from one unit to another. To do so, conversion factors are used. A **conversion factor** is a ratio of two units that equals one. Since 12 eggs are in one dozen, we can write 12 eggs = one dozen. We can also write this as a fraction (Figure 1.1). Notice that the numerator equals the denominator and thus the fraction is equal to one. The ratio can also be written as the reciprocal of that, which still is equal to one. A good way to begin converting from one unit to another is to first write what you are given or what you already know. Next, multiply by a conversion factor that allows the unneeded unit to "cancel out." Let's do an example. Say we need to determine how many dozen eggs comprise 36 eggs. In this example, we will begin by writing 36 eggs. Next, we need to choose a conversion factor that includes eggs and dozen, but the conversion factor chosen should have eggs in the denominator. This will allow us to multiply the conversion factor by the given unit of eggs so that the unit of eggs is cancelled in both the first and second parts of the equation, leaving dozen as the unit of the answer. The next step is to cross out units. Finally, complete the math to determine the correct answer. Remember, a conversion factor can be made from any two quantities with different units that equal each other.

Section 1-2: The Properties of Matter

All matter is made up of tiny particles, the basic unit of which is the **atom**. Each atom is composed of even smaller sub-atomic particles; protons, neutrons and electrons (Figure 1.2). Protons are positively-charged particles found in the nucleus of an atom. Neutrons are also found in the nucleus of an atom, but they have no electrical charge. Protons and neutrons are relatively large and make up almost all of an atom's mass. Electrons on the other hand are extremely small,

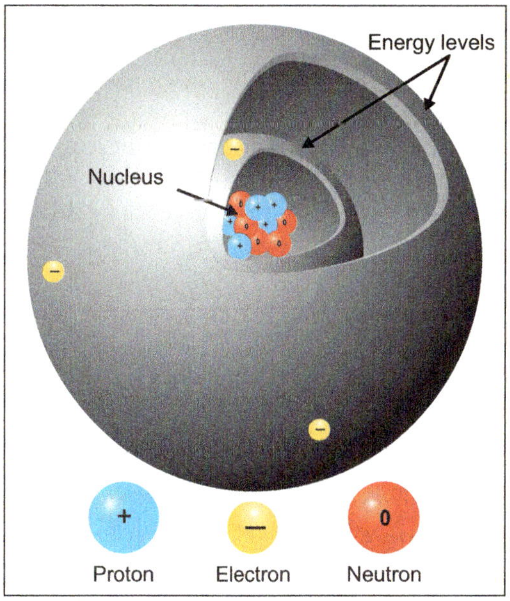

Figure 1.2: The parts of an atom. This particular example is the element boron which has an atomic number of 5.

PERIODIC TABLE OF THE ELEMENTS

Legend: Alkali metals; Alkali earth metals; Transition metals; Poor metals; Non-metals; Noble gasses

Period \ Group	1	2	3	4	5	6	7	8	9	10	11	12	13	14	15	16	17	18
1	1 H																	2 He
2	3 Li	4 Be											5 B	6 C	7 N	8 O	9 F	10 Ne
3	11 Na	12 Mg											13 Al	14 Si	15 P	16 S	17 Cl	18 Ar
4	19 K	20 Ca	21 Sc	22 Ti	23 V	24 Cr	25 Mn	26 Fe	27 Co	28 Ni	29 Cu	30 Zn	31 Ga	32 Ge	33 As	34 Se	35 Br	36 Kr
5	37 Rb	38 Sr	39 Y	40 Zr	41 Nb	42 Mo	43 Tc	44 Ru	45 Rh	46 Pd	47 Ag	48 Cd	49 In	50 Sn	51 Sb	52 Te	53 I	54 Xe
6	55 Cs	56 Ba		72 Hf	73 Ta	74 W	75 Re	76 Os	77 Ir	78 Pt	79 Au	80 Hg	81 Tl	82 Pb	83 Bi	84 Po	85 At	86 Rn
7	87 Fr	88 Ra		104 Rf	105 Db	106 Sg	107 Bh	108 Hs	109 Mt	110 Ds	111 Rg	112 Cn	113 Uut					

Lanthanides	57 La	58 Ce	59 Pr	60 Nd	61 Pm	62 Sm	63 Eu	64 Gd	65 Tb	66 Dy	67 Ho	68 Er	69 Tm	70 Yb	71 Lu
Actinides	89 Ac	90 Th	91 Pa	92 U	93 Np	94 Pu	95 Am	96 Cm	97 Bk	98 Cf	99 Es	100 Fm	101 Md	102 No	103 Lr

Figure 1.3: The periodic table of the elements.

negatively-charged particles which orbit around the atom's nucleus at high speeds. The electrons do not orbit around the nucleus randomly; instead they occupy discrete paths (atomic orbitals) around the nucleus. The orbitals represent different energy levels of the electrons.

Chemical elements are commonly arranged in the **Periodic Table** and each has its own unique atomic number based on the number of protons in the nucleus (Figure 1.3). Elements are organized on the Periodic Table based on their atomic number and characteristics. The elements increase in atomic number (and thus size and mass) as they progress from the left to the right, and down each row (or **Period**). Further, the elements are arranged such that elements with similar characteristics are in the same column (or **Group**).

A **molecule** is *a combination of atoms* and is the *smallest particle* of a compound or a gaseous element that can exist and still retain the characteristic properties of that substance. Molecules interact with each other; sometimes they are attracted to each other and other times they repel each other. It just depends on the type of molecule. An example of a molecule is the water molecule. A water molecule has two atoms of hydrogen and one atom of oxygen (Figure 1.4).

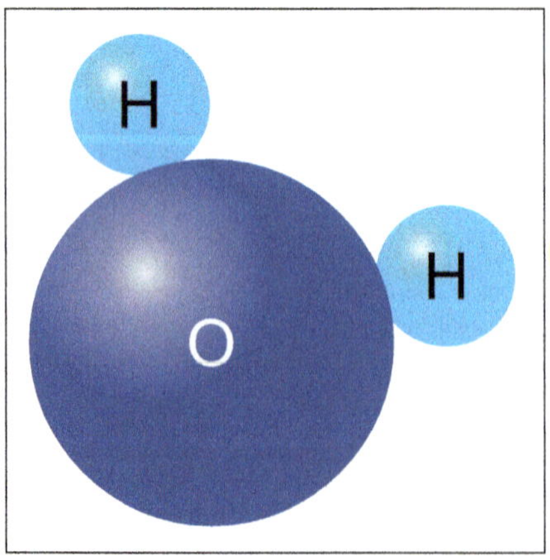

Figure 1.4: The humble water molecule, probably the most important substance on Earth!

Mass describes the total amount of matter in an object. Mass is the just the sum of all the atoms in a body of matter. Mass is a fixed value, regardless of how much space the matter occupies. **Weight** is not mass, but rather the measure of the force of gravity on a mass of matter. For example, you would "weigh" less on the moon than you do on the Earth because the moon's surface gravity is less than that of Earth's (it is about 5/6th of the Earth's surface gravity). However, your mass would remain the same. More on gravity later in the chapter.

Volume is the amount of three-dimensional space an object occupies. **Density** is the ratio of the mass of an object to the volume of an object. Density is really a measure of how many atoms are packed into a given space. The density of two objects made of the same material, but having different volumes, will be the same because density is the ratio of mass to volume (Figure 1.5). Density is important because matter, especially fluid matter, tends to spontaneously sort itself into layers based on density. For example, the world's oceans are stratified by density. The warm water at the surface of the oceans has a lower density than the cold waters at the bottom of the ocean. Although the actual difference in density between the surface and bottom of the ocean is relatively small, this density difference drives deep ocean circulation patterns on a global scale (More on this in Chapter 14). **Buoyancy** is an upward force exerted on an object, typically by a fluid. Buoyancy is driven by differences in density. For example, a rubber duck floats because its density is lower than water's. In this case, the water is literally "pushing" the duck upwards. Hot air balloons also rise because of buoyancy, but in this example, the buoyant force is due to the difference in temperature between the air in the balloon and the air outside the balloon. Hot air is less dense than cold air, so the warm air in the balloon is less dense than the surrounding air in the atmosphere. This causes the denser atmosphere to push the balloon upwards.....and uplift is achieved!

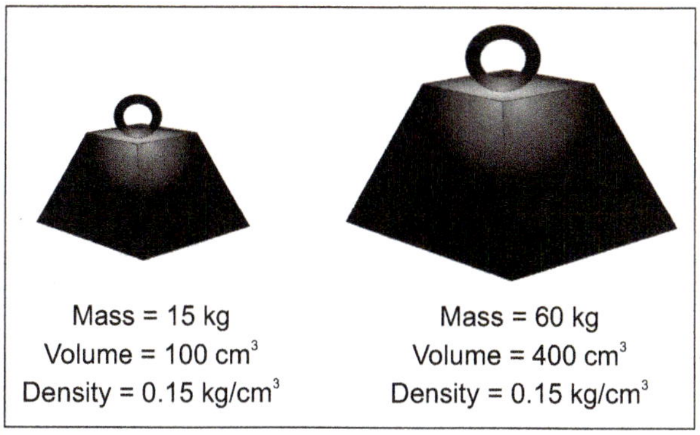

Figure 1.5: The relationship between mass, volume and density.

Pressure is defined as the force per unit area exerted by a group of molecules upon some object or surface. In equation form, Pressure (P) = Force (F) / Area (A). For gasses, the pressure of a gas is inherently linked to the volume and temperature of the gas; following **the Ideal Gas Law, which is:**

$$PV = nRT$$

Where P is pressure , V is volume, T is temperature, n = the number of molecules (in moles) and R is the universal gas constant (If you are curious, R = 8.314 J/mol-K). This formula is important for understanding many aspects of weather and climate. Since the atmosphere is made of gasses, its behavior is largely explained by the ideal gas law. To correctly and accurately apply the ideal gas law to the atmosphere, we must look at air parcels as closed systems; that is, no heat is entering or exiting the parcel of air. Since an air parcel is a **closed system**, that is no air (or gas or matter or heat or anything) will be entering or leaving it, n, the number of molecules in it will not change and is assumed to be constant. Since R is also a constant, this allows us to see how changes in P, V and T are related to each other (Figure 1.6). If we hold the volume constant, and raise the temperature, the pressure will increase. If we hold the pressure of the gas parcel constant and raise the temperature, the volume of the gas will increase. Thus, temperature has a linear relationship with both pressure and volume. However, if we hold the temperature constant and increase the pressure, the volume decreases. In the case of volume and pressure, the relationship is an inverse one. Note that during these processes temperature and pressure experience the same change, while volume experiences the opposite change.

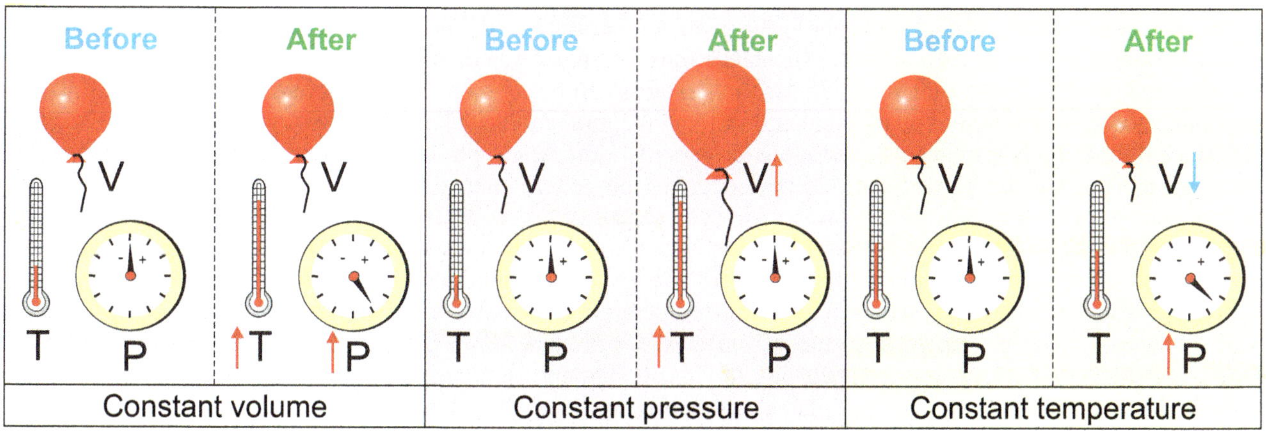

Figure 1.6: Visualizing the ideal gas law.

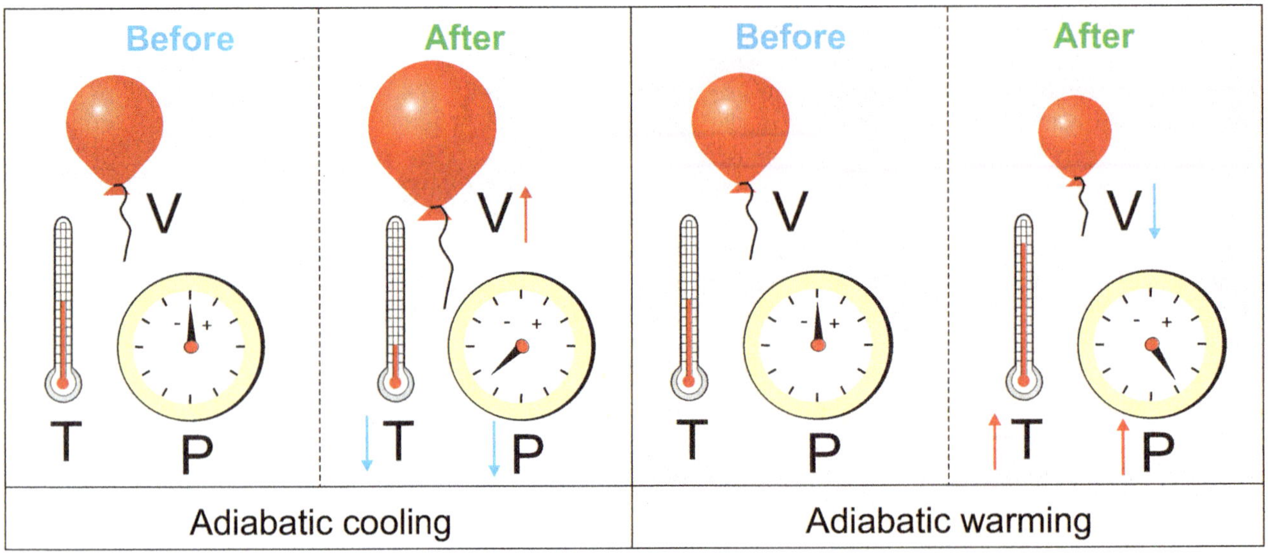

Figure 1.7: Adiabatic changes in temperature explained. In both examples, a change in pressure leads to a simultaneous change in both temperature and volume.

The term **adiabatic** describes a thermodynamic process that occurs **without the loss or gain of heat (or matter) between a system and its surroundings (thus in a closed system).** We use this "adiabatic" term to coin the terms "adiabatic cooling" and "adiabatic warming." Such adiabatic cooling and warming processes occur in the atmosphere. Adiabatic cooling and warming occur in the atmosphere due to a change in volume or pressure within a closed air parcel. For example, if the pressure exerted on an air parcel decreases in the atmosphere, its volume will increase *and its temperature will decrease* (Figure 1.7). So the decrease in pressure ultimately led to a decrease in temperature. Adiabatic warming occurs when the pressure exerted on an air parcel increases causing a decrease in its volume and an increase in the temperature of the air. Adiabatic processes play a vital role in the formation of clouds and precipitation as you will see in Chapter 3.

The Phases of Matter

Matter exists in three common phases; solid, liquid and gas (or vapor), and each is characterized by different molecular arrangements. **Solids** have molecules that are fixed distances apart. Strong cohesive forces bond the molecules together, and the spacing between the molecules determines the density, which does not change. **Liquids** on the other hand have molecules that are not confined by a rigid structure and are therefore free to move around. The cohesive forces in liquids are not as strong. Liquids have a definite volume but obviously do not have a definite shape; they are free to be poured from one shape to another. The molecules in **gasses** have a much higher amount of energy and are moving fast. The molecules in gases have weak cohesive forces and are free to move around, bounce off of each other, and are spaced much further apart than the molecules in liquids and solids.

Every type of molecule exists in some phase at a given temperature, and that phase is dictated by the energy state of the substance. The molecules in a cooler substance move more slowly than the molecules in a warmer substance, and the motion of the molecules in any of the three phases can be changed by adding or subtracting heat. When a substance cools, the motion of its molecules slows and the molecules are pulled closer together. We can also state that since the solid phase of a substance is cooler than its liquid or gas phases, its molecules have less energy and therefore exist at a lower energy state. Likewise, gases have more energy and thus we say gases are at a higher energy state than its liquid or solid phases. However, the addition or removal of heat from a substance *does not always result in a change in temperature*. This is because energy is needed to change substances between their different phases. If you were to put ice and water in a cup, along with a thermometer, and stir the ice and water while watching the ice melt, would you expect to see the temperature increase on the thermometer as the ice melted? You probably would. However, the temperature *will not increase* until all the ice has melted and *additional heat from the surrounding environment then* heats up the liquid. This phenomenon is also experienced when trying to thaw frozen meat in the microwave. Have you ever placed something frozen in the microwave and when you check it after an allotted amount of time, it is still frozen solid? So you put it back in for additional time, and when you check it again, it is still frozen. You again put it back in for more time, and when you check it again, it is still frozen. You start the microwave again for more time, and when you check it again not only has it thawed but it has partially cooked! All the heat that was flowing into the food during the times when it still felt frozen was actually going into *changing the phase of those food molecules.*

When a solid, liquid, or gas changes from one phase to another, the transition is called a **phase change.** A phase change either absorbs or releases a quantity of heat ***that does not lead to a temperature change***. That heat is called **latent heat** (It is also sometimes referred to as "hidden heat" but that is not the official definition). Latent heat is energy or heat that goes into or comes out of *internal potential energy* and does *not change the temperature of the substance during the phase change.* Note that latent heat is a constant. There are three basic types of phase changes (Figure 1.8):

1. A **Solid-liquid phase change** is a change from the solid to the liquid phase (**melting**), or vice versa (**freezing** – changing from liquid to solid). Freezing and melting occur at freezing or melting point temperatures. Note that the melting and freezing points are the same temperature for any given substance.

2. A **Liquid-gas phase change** is a change from the liquid to the gas phase (**evaporation**) or a change from the gas to the liquid phase (**condensation**). These two occur at what we call the boiling point and the condensation point. The boiling and condensation points are the same temperature for any given substance.

3. A **Solid-gas phase change** is a change from the solid phase directly to the gas phase (**sublimation**) or a change from gas directly to a solid (**deposition**). An example of sublimation is the visible vapor that comes off dry ice (CO_2). An example of deposition is frost on a window. Sublimation and deposition occur at the boiling point and freezing point, respectively. The boiling and freezing points are the same temperature for any given substance.

Notice that changes which proceed from the left of the diagram to the right require or absorb energy. Such processes are **endothermic.** Notice also that changes that proceed from the right to the left of the diagram release or give off energy. These processes are **exothermic. Endothermic reactions** require or absorb energy *because the substance is changing from a lower energy state to a higher one.* These changes proceed from left to right on Figure 1.8. **exothermic reactions** give off or release energy *because the substance is changing from a higher energy state to a lower one.* These changes proceed from right to left on Figure 1.8. Speaking of energy, let's take a closer look at that.

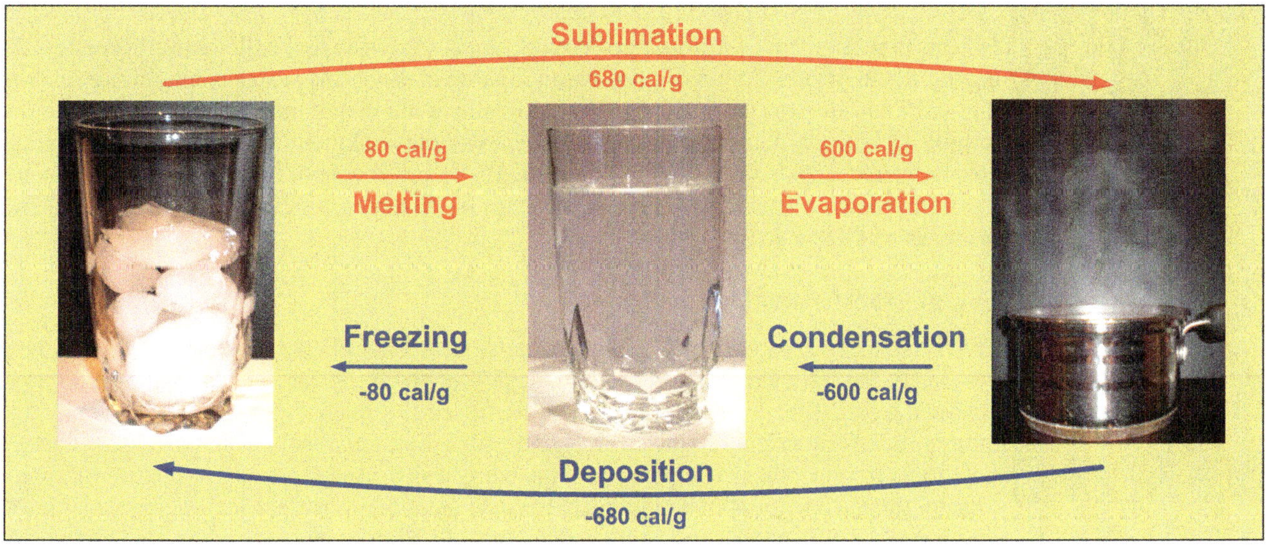

Figure 1.8: The three states of water and the change in latent heat needed to convert water between its different states. **cal/g = calories per gram**

Section 1-3: The Properties of Energy

Energy is defined as the ability to do work. Is work what we know of as the 9-5 workday? No. **Work** refers to the application of a force to an object and the distance the object moves as a result. Thus, when energy is exerted on an object, and the object moves, work has been done on the object. The equation for work is:

Work (W) = Force (F) x distance (d), or more simply **W = F·d**

The metric units for work are Joule (J), and the English units for work are foot-lb ("Foot-pound").

Figure 1.9: The common forms of energy.

Energy exists in many different forms such as **mechanical, chemical, electrical, nuclear, thermal, and radiant or electromagnetic** (Figure 1.9). **Mechanical energy** is the energy of motion and usually is comprised of simple machines (i.e., a lever, a pulley, etc.) working together. **Chemical energy** is stored energy, such as that found in food or fuel. **Electrical energy** harnesses the potential of a difference in voltage to conduct a charge in conductors. **Nuclear energy** is the energy contained within the nucleus of an atom which holds the nucleus together. Nuclear energy is released by nuclear **fusion**, which is the fusing together of two atoms to make a heavier one, or by nuclear **fission**, which is the breaking of heavier atoms into lighter ones (this is what happens in a nuclear power plant). **Thermal energy** is the internal energy within a substance due to the vibration of the atoms it contains. Thermal energy is the formal term for **heat. Radiant or electromagnetic energy** is energy which travels through space in the form of electrical and magnetic waves. We will delve into more details about mechanical energy, thermal energy and radiant energy in the next three sections.

Energy is also categorized as **potential or kinetic. Potential energy** is stored energy. It is also called the energy due to position, structure or composition. For example, gravitational energy is energy due to position. The water stored behind a dam demonstrates this concept well. The water molecules were first lifted up from the Earth's surface by evaporation and they gained gravitational potential energy. The gravitational potential energy stored in the water now has the potential to do work. A lump of coal would be a good example of energy due to structure or composition of a substance (chemical energy). When coal is burned, the potential energy in the organic molecules is released. The equation for potential energy (PE) is **PE = mgh** where m is mass, g is gravity, and h is height.

Objects in motion have the ability to do work, so they have energy called **kinetic energy,** or energy due to motion. So from one example above, when the water behind a dam flows through a turbine, the potential energy it contains will change into kinetic energy and the turbine will spin, producing electricity. Other examples of kinetic energy include a flowing electrical current, sound waves moving through the air, a beam of light, and the vibration of the atoms inside matter (heat). The equation for kinetic energy (KE) is **KE = ½ mv^2**, where m is mass and v is velocity.

Energy Transfer and Thermodynamics

The **Law of Conservation of Energy** states that energy can neither be created nor destroyed, but it may be transformed into different types. So for the example of the water behind the dam, the potenital energy the water loses as it passes through a turbine matches the amount of kinetic energy the turbine gains (plus the energy lost to friction and heat loss). Let's look at some other examples of energy being transformed from one type to another. In a flashlight, the chemical energy in the battery is converted into electrical energy and finally into light (radiant energy) and some heat energy. Another example is when coal burns in a power plant. In that process, the chemical energy stored in the coal is converted into heat energy. The heat energy is then used to produce steam, which then uses mechanical energy to turn a turbine. The spinning turbine turns a generator which then converts the mechanical energy into electrical energy. How about an example that is independent of us humans? When sunlight shines on plants, some of that radiant energy is ultimately converted to chemical (stored) energy through the process of photosynthesis.

In each of these cases, the total energy of the system *will remain the same*, even though energy may be transferred from one form or type to another *within the system*. The total energy in a closed system (one in which heat and matter do not enter or exit the system) will always equal the total potential energy in the system plus the total kinetic energy in the system. This is represented by the equation: **Potential Energy (PE) + Kinetic Energy (KE) = Total Energy (TE)**. This equation allows us to calculate the total energy of an object in a system.

For example, let us look at the total energy in a ball as it falls off a table (Figure 1.10). In the initial setting, the ball is on top of a table which is 10 feet tall. It is not moving, so its kinetic energy is zero. The ball does however have energy due to position, which is potential. Let us calculate potential and kinetic energy in the initial setting. Remember that PE = mgh and KE = ½ mv^2. Also remember that: TE = PE + KE. So initially, the ball has zero kinetic energy and 98 kg·m^2/s^2 of potential gravity (which it gained when it was lifted up to the table). Now suppose the ball has fallen half-way to the ground such that it is 5 m off the floor. Now, the ball has 49 kg·m^2/s^2 kinetic energy and 49 kg·m^2/s^2 of potential energy (Figure 1.10). Immediately before the ball touches the floor, let us assume that at this final point, it has no more distance to travel and that height = 0 m. It is moving as fast as it will be before it hits the floor. Now the potential energy is 0 kg·m^2/s^2 is the kinetic energy is 98 kg·m^2/s^2. Thus, the total energy of the ball before it falls is equal to the energy of the ball in motion and the energy of the ball when it reaches the floor.

Ball on table	Ball halfway to floor	Ball on floor
PE = mgh	PE = mgh	PE = mgh
m = 1 kg	m = 1 kg	m = 1 kg
g = 9.8 m/s^2	g = 9.8 m/s^2	g = 9.8 m/s^2
h = 10 m	h = 5 m	h = 0 m
PE = 98 kg·m^2/s^2	PE = 49 kg·m^2/s^2	PE = 0 kg·m^2/s^2
KE = ½ mv^2	KE = ½ mv^2	KE = ½ mv^2
m = 1 kg	m = 1 kg	m = 1 kg
v = 0 m/s	v = ½ (1) (9.89 m/s)^2	v = 14 m/s
KE = 0 kg·m^2/s^2	KE = 49 kg·m^2/s^2	KE = 98 kg·m^2/s^2
TE = 98 kg·m^2/s^2	TE = 98 kg·m^2/s^2	TE = 98 kg·m^2/s^2

Figure 1.10: Calculating the total energy of a ball falling from a table.

Of course this is an example from a perfect world. In the real world, some of the energy would have been lost from the ball due to air resistance as the ball fell to the ground, but the total energy of the *system* would still be the same. This is explained by the First Law of Thermodynamics (**Thermodynamics** is the study of heat transfer and mechanical energy). The **First Law of Thermodynamics** states that energy cannot be created or destroyed, but it may be transformed to different types of energy, and the *total energy of a closed system is constant*. One thing to keep in mind is that during the energy transfer, the energy will also change from a higher energy state to a lower energy state. So in the case of the ball, it is at a higher-energy state when it is sitting on the table, and at a lower-energy state once it is sitting *still* on the floor.

In many cases, the transfer of energy is spontaneous. This happens due to energy gradients. A **gradient** is simply a range of difference. If the temperature in Tampa, Florida is 90°F and the temperature in Portland, Maine (on the same day) is 60°F, then the temperature gradient between the two cities would be 30 degrees. More importantly, there is an energy gradient between the two cities. One aspect of thermodynamics is the concept that *nature will always work to even out energy gradients*. So not only will energy change from a state of higher energy to a state of lower energy, but it will also naturally dissipate in nature over time.

Figure 1.11: A calm, still pool of water is a good example of a system which is exhibiting a high level of uniformity. Photo credit: B. Zaprowski.

For example, imagine in your mind a perfectly flat, still pool of water, one that you can see your reflection in (Figure 1.11). At this point, the energy in the system is perfectly and evenly distributed throughout all of the water molecules (which is why the water is still). Now imagine someone throws a large rock in the water. When the rock hits the water, the total energy of the system will increase as kinetic energy is transferred from the rock to the water, and ripples are produced. However, the energy in the water will eventually dissipate and the surface of the pool will become calm once again. What you have just observed is the **Second Law of Thermodynamics**, which states that closed systems always evolve toward thermodynamic equilibrium, a state with the maximum amount of uniformity or randomness possible (also called **entropy**).

A good example of this concept is the spontaneous movement of air from areas of higher pressure to areas of lower pressure (a pressure gradient). Consider Figure 1.12A, which shows two jars with a closed valve connecting them. As shown, the molecules in vessels A and B are moving around, pushing against each other and the sides of the container, that is, exerting a pressure. But the pressure is higher in A because there are more molecules pushing than in B. Likewise, because there are more molecules, A has a higher energy level than B. If we open up the valve between A and B, molecules of gas will then flow into B. Why? There are more molecules pushing in A than in B, so the molecules flow into B because it is the path of least resistance. Eventually there will be the same number of molecules in A and B. Now there is a balance because the force being exerted in each vessel is the same. Thus, the pressure and energy level in A will decrease while the pressure and energy level in B increases until the pressure and energy levels are the same in both containers.

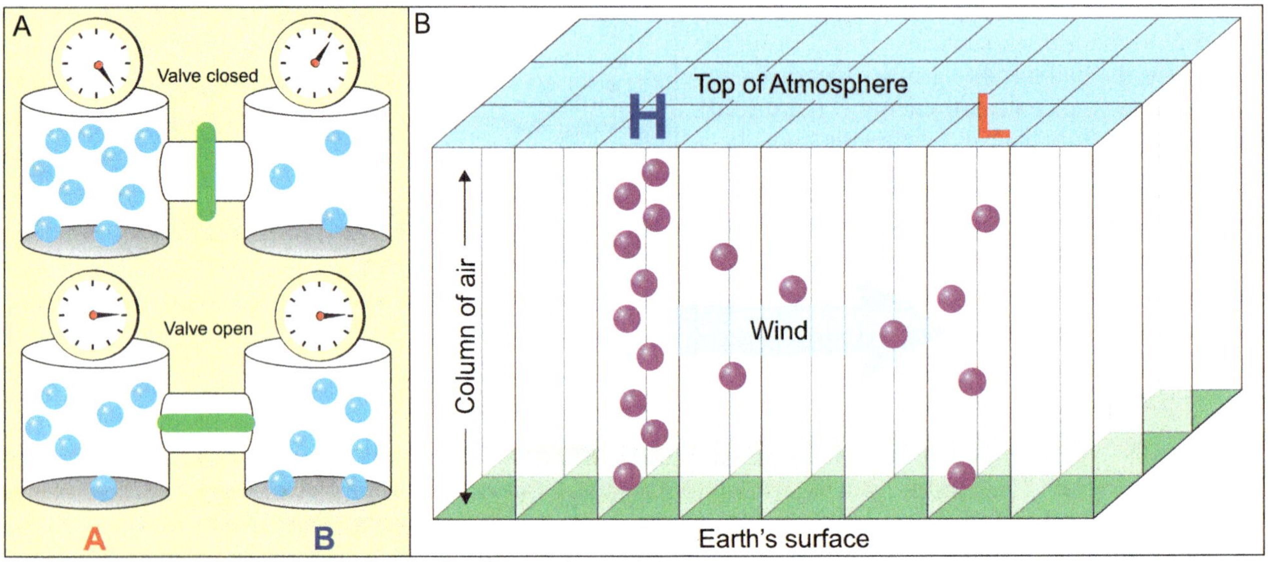

Figure 1.12: A. The second law of thermodynamics applied to pressure gradients. Before the valve is opened, the pressure is higher in jar A than jar B. When the valve is opened, some of the gas spontaneously moves from jar A to jar B until the pressure is equal in both jars. B. The second law of thermodynamics also explains why wind forms. The atmosphere naturally has areas where the pressure is higher and lower than average. These differences in pressure lead to the formation of pressure gradients. These result in a spontaneous transfer of air molecules from one part of the atmosphere to another part of the atmosphere, and wind is produced as pressure is equalized.

This relatively simplistic example will help you to understand many facets of weather. For example, wind is simply the movement of air from areas of high pressure to areas of low pressure within the atmosphere. However, the gasses of the atmosphere are not distributed equally. In some places, the number of molecules is higher while other areas have a lower number of molecules (Figure 1.12B). Just like in the example above, the gas molecules in the areas of higher pressure will spontaneously migrate towards the areas of lower pressure and wind forms as a result (we will discuss wind in more detail in Chapter 3). Once again, the second law of thermodynamics has been obeyed. Now that we've looked at the basic properties of energy, let's look at some of these forms of energy in more detail, starting with mechanical energy.

Section 1-4: Mechanical Energy and the Physics of Motion

As we discussed at the beginning of the previous section, energy is the ability to do work, and work is the transfer of energy to an object, resulting in the movement of the object. For example, when you ride a bicycle, you are doing work as you transfer energy from your leg muscles to the pedals, causing them to turn. Another good example of work is the transfer of wind energy to a body of water, producing waves (more on this in Chapter 14). A **force** is the application of energy to an object. Whether or not the application of a force will result in the object's movement (work) is another matter. For example, if you push against a building, you are applying a force, but that force will not cause the building to move, *so no work is being done*. Let's explore forces more.

Forces

There are two basic types of forces, contact forces and action-at-a-distance forces. **Contact forces** are forces applied to an object through direct contact with that object. Kicking a football is a good example of a contact force. **Action-at-a-distance forces** are forces applied to an object through open space. Gravity, electrical charges and magnetism are all examples of action-at-a-distance forces. A compass needle points toward North because the Earth's magnetic field (a force) causes the needle to move *from a distance*.

In order to understand forces better, we need to discuss the terms speed, velocity and acceleration. **Speed** is simply the change in the location of an object over some amount of time (and thus equals distance/time). An example of speed is a car's speed which is commonly reported as miles per hour (mph). **Velocity (v)** is a two-part concept that includes speed and *also direction*. Any change in velocity means that *either* the speed *and/or* the direction of an object changed. So if a car is moving and changes direction but not speed, its velocity has still changed because of the change in its direction of travel. The change in velocity over some period of time is called **acceleration (a)** such that:

$$a = \frac{v_f - v_i}{t}$$

where v_f is final velocity and v_i is initial velocity. The acceleration of an object is proportional to the forces acting upon it and inversely proportional to its mass. This brings us to the basic equation for force:

$$F = ma$$

Where m = mass. So the amount of energy that can be transferred to an object, that is, the force, is a function of both its mass and its acceleration (Figure 1.13). Massive objects, like a large boulder, can impose a large amount of force on other objects, even if they are moving slowly. Conversely, objects which are moving at very high rates of speed can also impact other objects with a large amount of force, even if their mass is small. A bullet is a good example of this. A bullet exerts a high amount of force on a target because it is moving very fast. When an object is both massive *and* moving a high speeds (like a meteorite), the amount of force generated is spectacular (just ask the dinosaurs how that went for them).

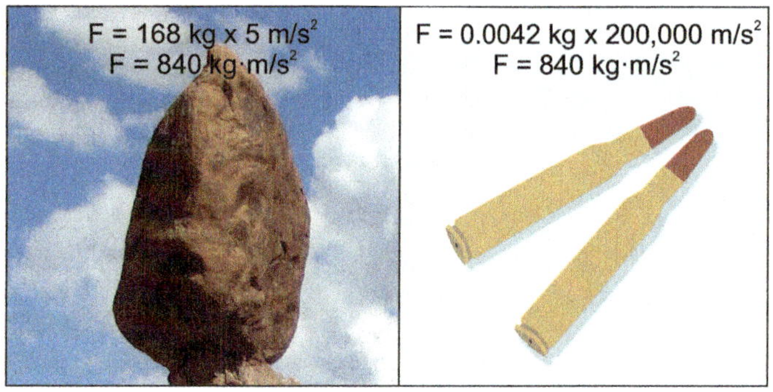

Figure 1.13: The relationship between the force, mass and acceleration of an object. A large boulder in motion will exert a large force because of its large mass, even if it is moving slowly. A bullet will exert a large force because of its high acceleration, even though its mass is small. Photo credit: B. Zaprowski.

Forces also help explain the concept of weight (Section 1-2). Remember that weight is the force of gravity acting on a mass. **Gravity** is the attraction of two bodies to each other and occurs through acceleration. Gravitational forces can be described by the formula:

$$F_{gravity} = \frac{G(m_1 \cdot m_2)}{d^2}$$

where $F_{gravity}$ is gravitational force, m_1 is the mass of the first object, m_2 is the mass of the second object, d is the distance between the two masses and G is the gravitational constant (6.674×10^{-11} N·m^2/kg^2). So as the distance between two objects grows, the force of gravity between the two objects diminishes. For example, the Moon and Sun both exert a gravitational force on the Earth (Figure 1.14). Because the Sun is many, many times more massive than the Moon, its overall gravitational force is many, many times greater than that of the Moon. But....the moon is only 384,400 km away from the Earth and the Sun is 152 million km away, so the moon's gravitational pull on the Earth is greater than that of the Sun (it's about 2:1). Here on Earth, the gravitational pull (or the acceleration due to gravity) that the Earth exerts on objects is a constant 9.8 m/s^2 because the Earth's mass isn't changing and because the distance between an object and the Earth's surface at sea level (10's of kilometers) is negligible relative to the distance between the Earth's surface and the center of the Earth (~6,400 km).

One more final important concept of motion is that of inertia. **Inertia** is the tendency of an object in motion to

Figure 1.14: An example of how mass and distance both play a role in the strength a gravitational force exerts on an object. The Sun has a very large mass as compared to the Moon, but the Moon is very close to Earth, so it has a stronger gravitational pull on the Earth's surface. This difference can be seen in the pattern of the tides. The lunar tides are about twice as strong as the solar tides.

continue moving in a straight line at a continuous speed unless a force acts upon it. Conversely, an object at rest will remain at rest unless a force acts upon it. The mass of an object has a large influence over the amount of inertia an object may have. An object with a large mass will resist acceleration more than an object with a small mass. For example, compare two balls that are the same size. One ball is made of rubber while the other ball is made of lead. The rubber ball is easier to put into motion than the lead ball, and the rubber ball is easier to stop than the lead ball. Thus, the lead ball is said to have more inertia than the rubber ball.

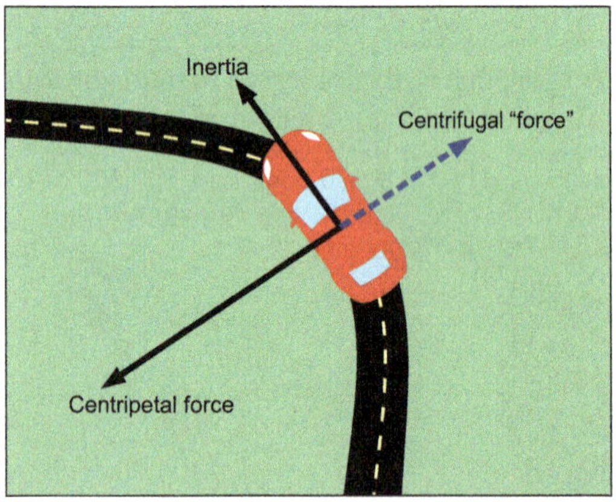

Figure 1.15: When you drive your car around a curve, it experiences an inward acceleration caused by centripetal forces, but you feel centrifugal "forces" pulling you outwards. What you are really feeling is your body's inertia resisting the inward forces.

Apparent Forces

Many of the "forces" we experience in everyday life are not real forces. These forces are called **apparent forces**. With apparent forces, it may seem like a force is being applied to an object, but no energy is actually being transferred to that object, so no force actually exists. Centrifugal forces are an excellent example of an apparent force. Let's say you are going for a ride in your car. As you are driving down the road, you come to a bend in the road. In order for you to successfully negotiate the curve and not continue forward in a straight line off the road, your turn the steering wheel. When you do that, the car continuously accelerates *toward* the center of the curve (Figure 1.15). This force, a **centripetal force**, is the force that pulls your car toward the center of rotation when you turn the steering wheel. But as your car accelerates around the bend, it feels as though your body is being "forced" outwards away from the bend in the road. The apparent force you are experiencing is called a **centrifugal force**. Centrifugal "forces" are the phenomena you experience when you rotate around a common point (especially if you go really fast). You also experience this phenomenon when you ride on a roller coaster. But centrifugal forces aren't real. Your body is actually accelerating inwards, towards the center of the curve as you are driving around it. There is no force trying to throw you outwards. The centrifugal force you "feel" is really just your body's inertia resisting the change in direction imposed on your body by your car. This is what we mean by an apparent force.

The **Coriolis force** (or effect) is another very important apparent force. It is a "force" that "appears" to interfere with the movement of free-moving objects across the surface of the Earth. Free-moving objects subject to the Coriolis force include planes, missiles, the atmosphere and large bodies of water, like the ocean. The Coriolis force causes all free-moving objects to *appear* to be deflected from their intended direction of travel, but no transfer of energy is occurring, so no force actually exists. So what does that mean? Let's say you live in Florida and you are going fly a plane to Maine to visit your friend. You take off and program the autopilot to fly in a straight line directly towards the airport near your friend's house. A few hours later your friend radios you and says the plane missed the airport by many miles to the East. You are now out over the ocean. So what happened? What happened is that you failed to account for the change in the speed of the rotation of the Earth beneath the plane as it flew through the air (Figure 1.16A). When your plane was sitting on the ground on the runway in Florida, it had zero northward velocity, but it did have an eastward velocity of ~950 mph due to the rotation of the Earth. When you took off in your plane, it had some northward velocity, say 500 mph, and it will still have the eastward velocity due to the rotation of the Earth of

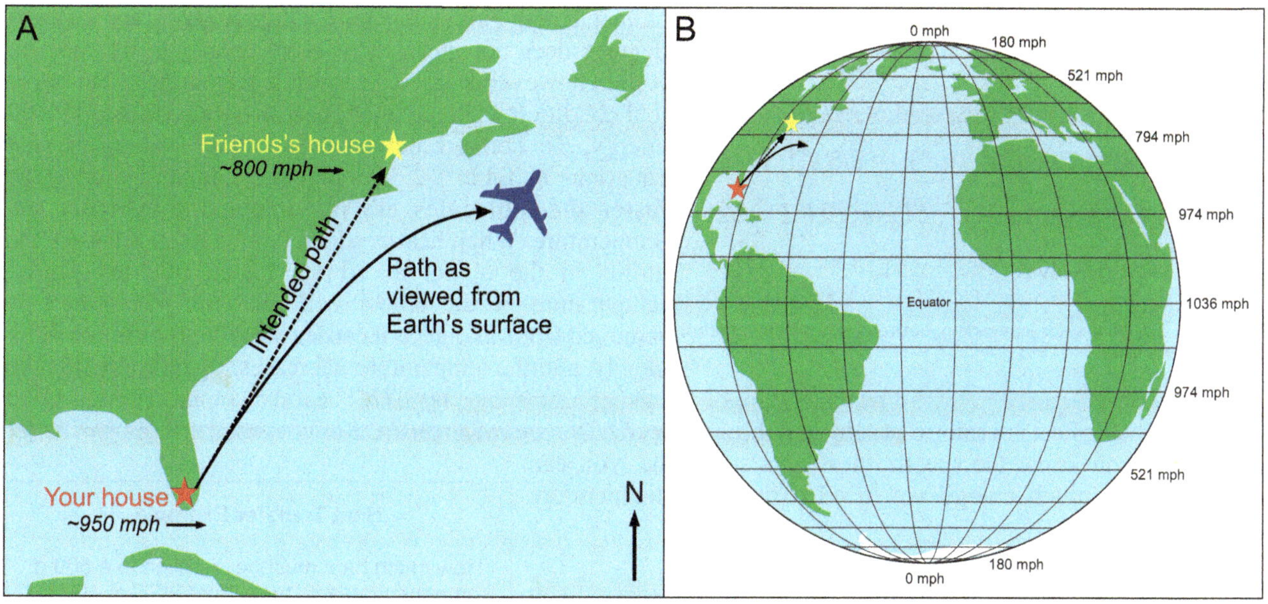

Figure 1.16: Coriolis explained. A. If you fly from your house in Florida to your friend's house in Maine and you don't account for the difference in eastward velocities of your plane and the Earth, you will miss your target to the right (east). B. The variation of the Earth's rotational velocity with latitude. Note that the velocity goes to zero at the poles.

~950 mph. But beneath you, the eastward speed of the Earth is decreasing! So when you get to Maine, you are still traveling at an eastward velocity of 950 mph, but the eastward velocity of the Earth beneath you is only ~800 mph now. Because your target (the airport) was moving at a slower eastward velocity than your plane, you ended up to the east of it. When viewed from space, the plane's path is a straight line and the Earth is simply turning beneath it (This phenomenon was also experienced by the Navy in WWII during long range missile practices). From the earthbound view, it "appears" there is some "force" which caused the plane to veer off its intended course. This is why the Coriolis force is really just an apparent force. It is simply a phenomenon caused by the rotation of the spherical Earth and the resulting variation of the Earth's rotational speed with latitude (Figure 1.16B). In the Northern Hemisphere all objects or fluids in motion appear to be deflected to the *right* of their intended path and in the Southern Hemisphere all objects or fluids in motion appear to be deflected to the *left* of their intended path.

"Rules" of the Coriolis Force:

1. The apparent deflection caused by Coriolis is at a 90° angle (left or right) relative to the direction of motion.
2. Coriolis only effects the direction of motion, not the velocity of the object.
3. The faster an object moves, the greater the deflection will appear.
4. Coriolis is strongest at the poles and weakens towards the equator, where it is non-existent.

Section 1-5: Thermal Energy

When the temperature of an atom is above absolute zero (0 K or -273°C), the atom is constantly vibrating. This vibration is referred to as **translational kinetic energy** or thermal energy. **Heat** is the total translational kinetic energy of all the atoms and molecules that make up a substance. **Temperature** is a measure of the *average* translational kinetic energy of the atoms in a substance. Thus, temperature is a measure of heat. We measure temperature with one of three different scales: Fahrenheit, Celsius and Kelvin (Figure 1.17).

Fahrenheit (°F)	Celsius (°C)	Kelvin (K)
Based on thinking that coldest day was 0 °F and warmest was 100 °F	**Based on freezing and boiling points of water**	**Based on absolute temperature and is the SI unit for temperature**
32 °F - freezing point of water	0 °C - freezing point of water	0 K - lowest known limit of temperature (i.e., absolute zero)
212 °F - boiling point of water	100 °C - boiling point of water	

Figure 1.17: The three different temperature scales.

Formula	Conversion
Fahrenheit to Celsius	$°T_F = 9/5\ (°T_C) + 32$
Celsius to Fahrenheit	$°T_C = 5/9\ (°T_F - 32)$
Celsius to Kelvin	$°T_K = °T_C + 273$

Table 1-2: Equations for converting between the three different temperature scales.

Note that both Fahrenheit and Celcius are arbitrary scales; 0 degrees does not mean there is no energy in the system. Kelvin however is based on absolute temperature and refers to a condition where an atom has zero translational kinetic energy. To convert between the three scales, use the equations in Table 1-2. The higher the kinetic energy, the faster the molecules are moving and the higher the temperature. When heat is added to an object, it causes the atoms in the object to vibrate faster, and hence, the temperature of the object will increase. When heat is removed from an object, it causes the atoms to vibrate more slowly, and the temperature decreases. **Specific heat** is the amount of heat required to raise the temperature of 1 gram of a substance by 1°C. The most common term used to describe specific heat is a calorie. A **calorie** is the amount of heat required to raise the temperature of 1 gram *of water* by 1°C. If you know the specific heat of a substance, you can calculate how much energy you need to raise its temperature (Figure1.18)

Heat Transfer Example

How much heat must be supplied to a 500 g iron pan to raise its temp from 20°C to 100°C?

The equation for heat flow is: $\mathbf{Q = m \cdot c\Delta T}$
where Q = Flow of Heat
m = mass
c = specific heat
ΔT = change in temperature

The specific heat of iron is 0.11 cal/g°C
Q = (500 g)(0.11 cal/g°C)(80°C)
Q = 4,400 calories

Figure 1.18: An example of heat flow and how to calculate it.

Heat flow is the transfer of thermal energy. There are three main types of heat transfer: conduction, convection and radiation. **Conduction** is the transfer of heat between molecules that are touching each other. When the molecules in two different objects come in contact with each other, the molecules in the hotter object will pass on some of their energy to the molecules in a colder object. This causes the colder molecules to vibrate faster and their temperatures to rise. A good example is that of a metal pan on a stove. When the bottom of the pan is heated, the molecules in the pan vibrate faster and bump into their neighboring molecules. This causes the neighboring molecules to move faster. Thus the heat, or thermal energy, moves throughout the pan in a domino-like way. If there is food in the pan, the heated molecules in the pan will then transfer their energy to the food molecules, thus heating them as well. So conduction is the transfer of heat through the collision of molecules. Don't forget that heat *always* spontaneously flows from higher temperature objects to lower temperature objects as per the second law of thermodynamics.

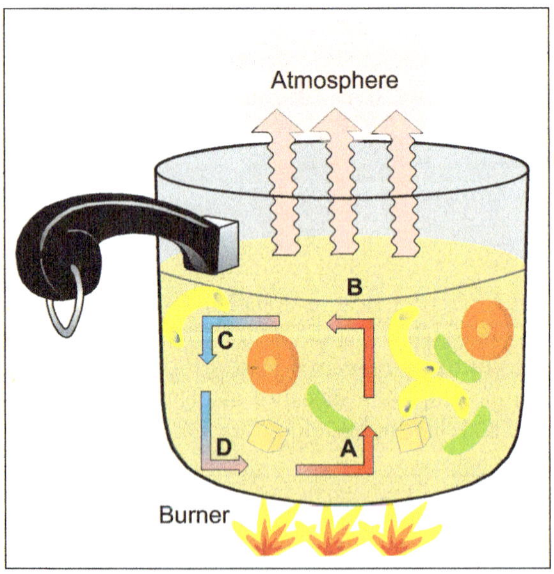

Figure 1.19: Convection explained using a pot of soup. A. The hot soup rises. B. The soup releases some heat to the atmosphere above the pot. C. Now cooled, the soup sinks back down. D. The soup is heated and repeats the cycle. In this example, heat is transferred from the burner to the atmosphere via the circulation of a fluid (the soup).

Convection is the transfer of heat because of the motion of a fluid (gasses are considered fluids too). A pot of simmering soup is a good example of this (Figure 1.19). When heat is transferred from the burner to the soup at the bottom of the pot, the soup is heated and its density decreases as a result. The hotter, lower density soup now spontaneously rises towards the surface. Once on the surface, the soup cools by transferring heat to the atmosphere above the pot. The now cooler and denser soup sinks back down to the bottom of the pot to be heated once again. In this example, heat is spontaneously transferred from the burner to the air above the pot via the circulation of the soup (a fluid).

Thermal radiation is the transfer of heat (thermal energy) through a vacuum or a transparent medium in the form of electromagnetic waves: actual waves of heat! All objects above absolute zero emit some form of radiant energy, including you! Let's look at this process in more detail.

Section 1-6: Radiant Energy

Radiant energy is also known as **electromagnetic radiation** or **electromagnetic energy** or simply as **radiation.** Radio waves, microwaves, X-rays, UV-rays and visible light are all common forms of electromagnetic energy. Everything radiates heat in the form of electromagnetic energy of some kind *at all times;* animals, rocks, trees, water, and even people emit radiation (in the form of body heat). When radiation travels through space or a transparent material (e.g. air or water) it behaves like a wave but when radiation comes in contact with matter, it acts more like a particle. All waves are simply a form of energy moving through a substance (air, water, rocks, etc.), and all waves share similar properties (Figure 1.20). The highest part of a wave is called the *crest,* while the lowest part of a wave is the *trough.* Each wave has a unique *wavelength,* that is, the distance between two consecutive crests. Waves also have a *wave height,* that is, the height between the crest and an adjacent trough. We will discuss many types of waves in this book, and we use wave properties to describe the different types of radiant energy.

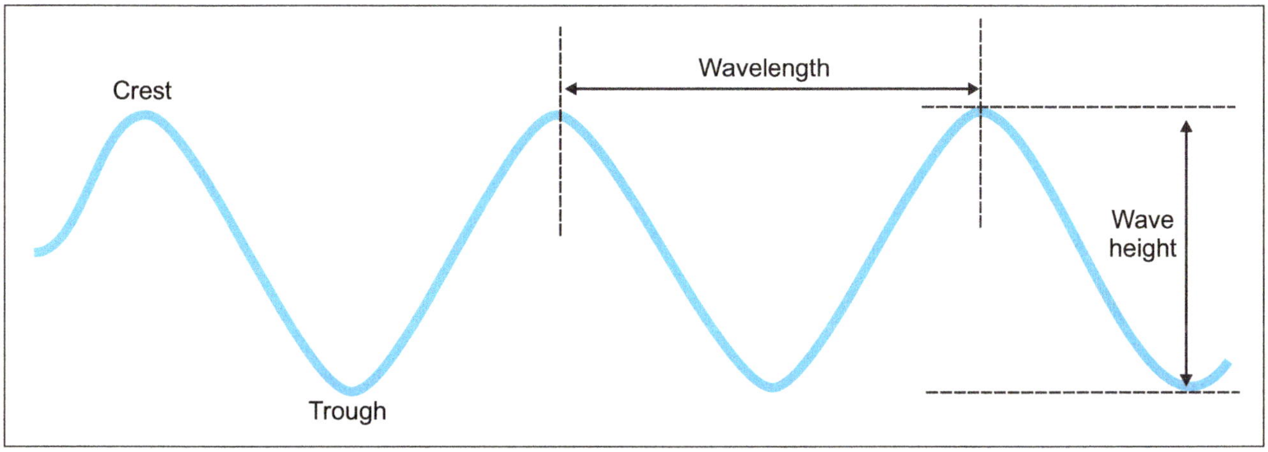

Figure 1.20: The basic properties of a wave.

The wavelengths of the different types of electromagnetic energy are determined by the temperature of the object which is emitting it. As can be seen in Figure 1.21, the **electromagnetic spectrum** includes many forms of radiation with which you may be familiar; X-rays, light, UV-rays, radio waves, and gamma rays to name a few. The energy being carried by an electromagnetic wave is contained within particle-like features called **photons**. Photons are packets of energy. Photons are unlike conventional particles, such as specks of dust or marbles in that they are not limited to a specific volume in space or time, they have no mass and they always move at a constant speed of 300,000 km/sec (186,000 mi/sec) when they are moving through a vacuum (they slow down in air or water). The amount of

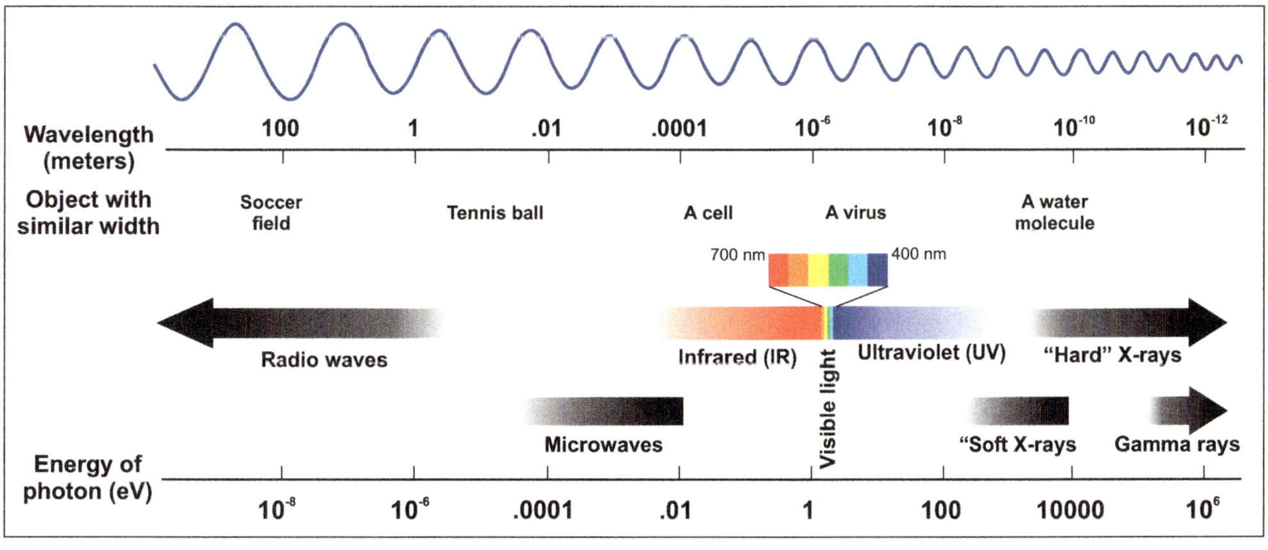

Figure 1.21: The electromagnetic spectrum. Electromagnetic waves are also known as radiation. nm = nanometer = 1 x 10^{-6} mm

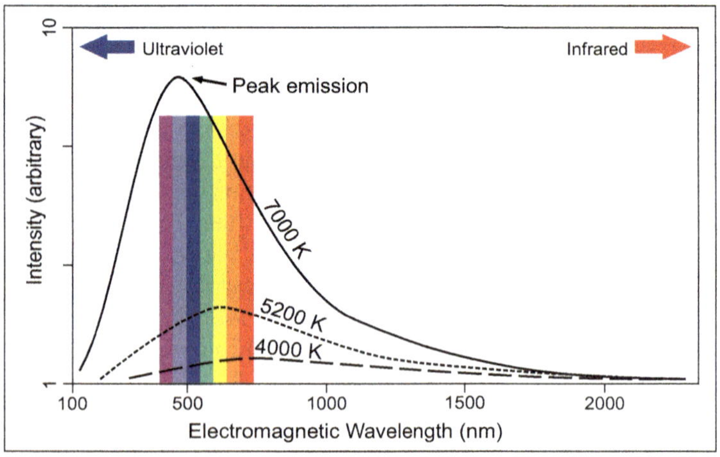

Figure 1.22: The emission spectrum for different temperature objects. Observe that when the temperature increases, the wavelength of peak emission gets shorter and bluer.

energy a photon contains is a function of its wavelength (Figure 1.21). When a photon hits an object, the energy is absorbed by the object, and the temperature of the object increases. For example, when you place some food in a microwave to heat it, microwaves and the photons they contain are absorbed by the water molecules, heating them up.

Where the wavelengths of radiation emitted from an object falls in the electromagnetic spectrum is strictly based on its temperature (Figure 1.22). The warmer the temperature of the object, the shorter the wavelengths that are emitted. Conversely, the cooler the temperature of an abject, the longer the wavelengths of the radiation that are emitted. So the sun emits shorter wavelengths of electromagnetic energy as compared to the Earth, which is considerable cooler (thankfully!). Figure 1.21 also shows the colors of visible light and their range of wavelengths. Note that the colors at the red end of the spectrum have longer wavelengths than the colors at the blue end of the spectrum.

How Radiation Interacts with Matter

When electromagnetic energy strikes an object, several things can happen to the radiation (Figure 1.23). First, some of the radiation may be **transmitted** through the object without interacting with the object's molecules and atoms at all. In that case, no energy is transferred to the object. Transmission is what makes it possible to see underwater.

Second, some of the radiation may **reflected** back towards the source. Another way to think about this is in terms of the object's reflectivity or **albedo**. **Albedo** is a measure of the amount of light that is reflected off a surface relative to the total amount of light that initially hit the surface. It can also be thought of as a percentage; the percentage of incident light that is reflected from a surface. The darker the object, the less energy it reflects (think about the amount of light reflecting from a black parking lot versus from snow).

Third, the radiation may enter an object and be broken into multiple paths and **scattered** outwards in several different directions. Scattered radiation that emerges on the other (far) side of an object still qualifies as transmission. If the radiation enters an object and emerges from the object back toward the source, the scattering is known (appropriately) as **backscattering.**

Finally, the object could **absorb** some of the incident radiation, causing the object's temperature to change or causing the object's state to change (i.e., the evaporation of water). The amount of energy absorbed by an object depends on the object's absorptivity. In the visible range of wavelengths, absorptivity is, in large part, dependent upon an object's color. Darker-colored surfaces absorb more visible radiation than do lighter-colored surfaces. Absorptivity for other wavelengths of radiation is dependent on factors such as the chemical composition of the object. For example, the atmosphere doesn't readily absorb shorter wavelengths of radiation, like UV-rays, but it does readily absorb longer wavelengths like infrared radiation.

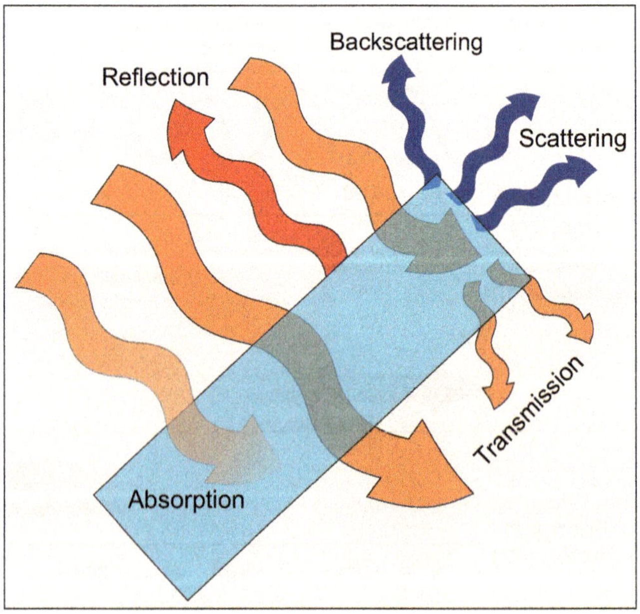

Figure 1.23: The fates of radiation when it strikes an object.

How much radiation an object absorbs is also a function of the **intensity** of the radiation. There are three things that control the intensity of the radiation that hits an object. First is the intensity of the energy emitted by the source. The hotter the source is, the higher the intensity of the radiation will be (Refer back to Figure 1.22). Second, the distance between the radiation source and an object also governs the intensity of radiation absorbed by an object. The further you are from the source, the less intense the radiation. You experience this phenomenon when you get close to a fire to warm yourself. Finally the intensity of incident radiation depends on the angle at which the radiation strikes an object.

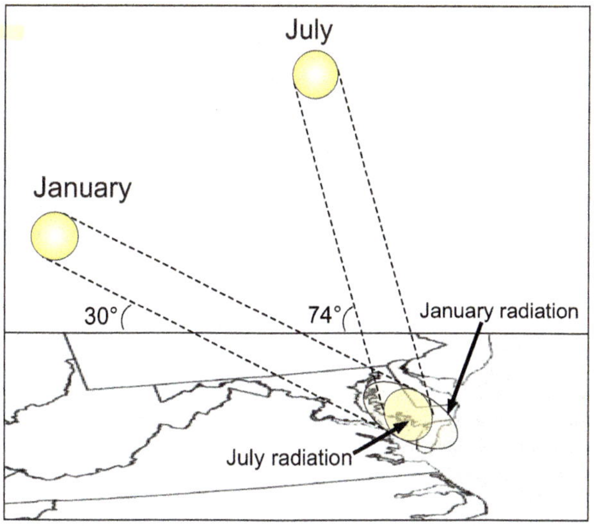

Figure 1.24: A comparison of solar intensity in January and July for 38° N latitude. Notice how the solar radiation is more concentrated in July when the sun is higher in the sky.

Consider a ray of solar radiation shining on the surface of the Earth in early July at a location near 38°N latitude (Figure 1.24). At that time of the year, the angle between the Earth's surface and the incoming solar radiation is about 74°. Now consider a ray of solar radiation hitting the same place in early January. At that time of the year, the angle between the Earth's surface and the incoming solar radiation is only 30°, less that half of what it was in July. It should be no surprise then that the intensity of the incoming radiation in January also less than half of what it is in July. In January, the same amount of incoming radiation is spread over a larger area, so the energy is diffused and less concentrated as compared to July. Conclusion: radiation is more intense when it strikes an object at a higher angle. This is why the winter months are so much colder than the summer months, even on a sunny day. The radiation is spread out over a wider area in the winter, decreasing the radiation's intensity.

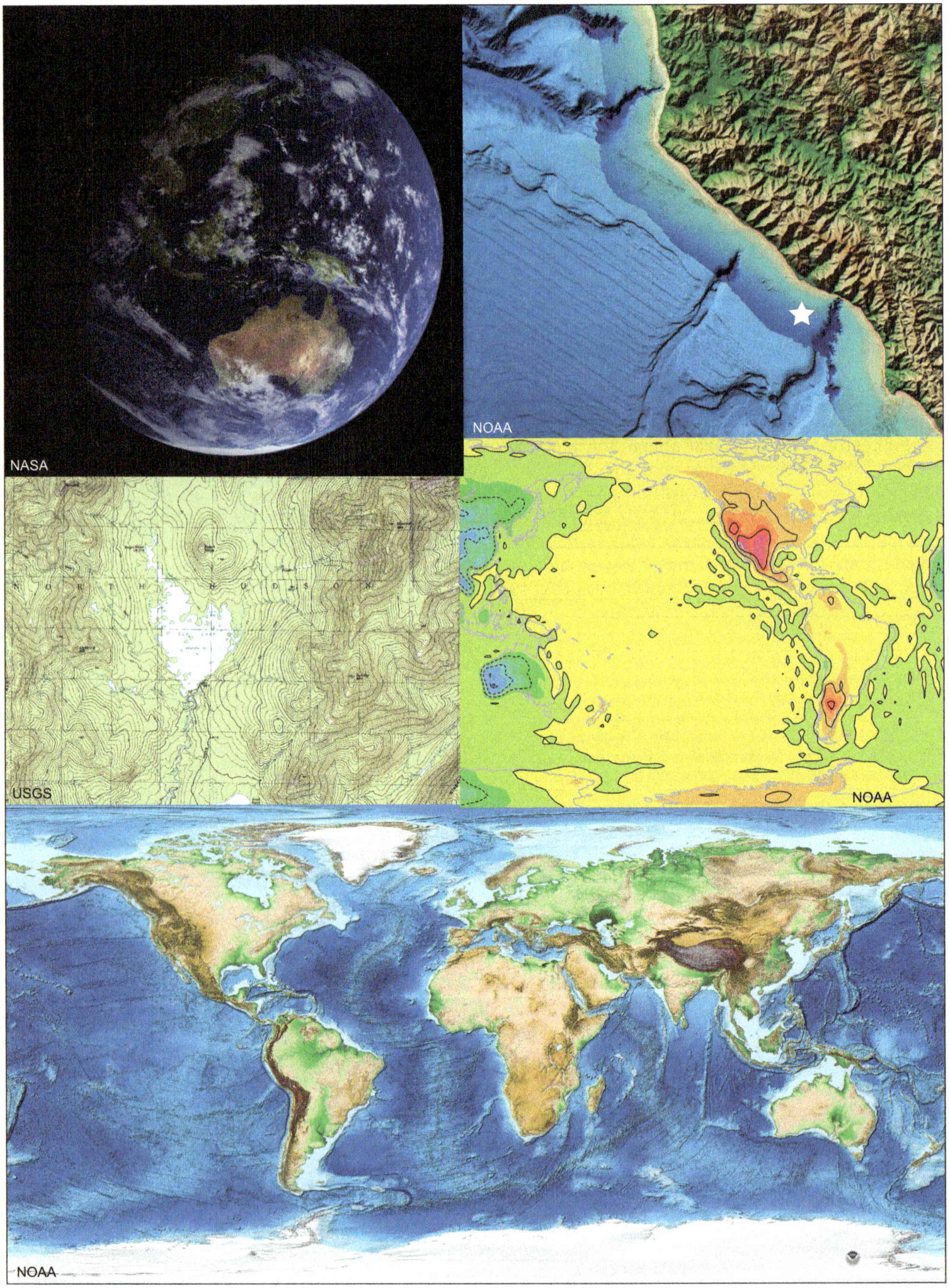
NASA
NOAA
USGS
NOAA
NOAA

Chapter 2 - Earth Science: Mapping Fundamentals and Applications

Earth's surface and atmospheric processes, fundamental components of Earth science, vary spatially and temporally resulting in significantly different landscapes and conditions across the Earth. To understand the nature and dynamics of these spatiotemporal processes and the resulting impact in shaping the Earth's surface, representations (or models) of the Earth or areas of the Earth are necessary. These representations are termed **maps** and are created by specialized Earth scientists called **cartographers**. **Maps**, which can be defined as two-dimensional representations of the natural and human features on the Earth's surface, have been and continue to be constructed to understand how Earth science processes have operated in the past and how these processes are operating in the present. **Cartography,** the science and art of developing representations of the Earth's surface or more simply the study and practice of making maps, includes both the production of traditional paper maps, as well as the development of 21st century dynamic web and mobile-based maps and geographic information science (**GIS**) databases. This introductory chapter examines the basic characteristics of the Earth, its size and shape, and describes how maps of the Earth are constructed, and their resulting limitations in representing the basic relationships on the Earth. Maps, and their primary components, are described in the following section with a special emphasis on large scale topographic maps produced by the United States Geological Survey (USGS). The chapter concludes with fundamental concepts and elements in map making and map reading and presents examples of their use and application in the Earth sciences.

Section 2-1: Earth's Shape and Selected Characteristics

To develop accurate representations of the Earth for the production of maps, Earth's physical dimensions must first be precisely understood and measured, an ongoing process improving with technological advances each year! While you may have learned that the Earth's shape is a sphere in your earlier educational experiences, in reality, the Earth is not a perfect sphere, an important distinction when cartographers attempt to produce accurate Earth representations. Because the Earth rotates on its axis from west to east, centrifugal force, an apparent force resulting from the inertia of the body away from the center of rotation, produces an outward force maximized at the equator given its higher rotational velocity and distance from the rotational axis. This outward force produces an equatorial bulge altering Earth's shape from a perfect sphere to a shape termed an **oblate spheroid** (Figure 2.1). The Earth's mean radius is 6,371 km (3,959 mi.) and varies whether measuring along the equator or between the poles due to the equatorial bulge. Earth's equatorial diameter is 7,962 mi. (12,756 km) while its polar diameter is 7,900 mi. (12,720 km). The circumference of the Earth measured along the equator is 24,902 mi. (40,075 km) while the circumference measured through the North and South Pole is 24,860 mi. (40,008 km). This deviation from a perfect sphere is small, less than 0.17% of the Earth's circumference, although important when constructing maps depicting small areas of the Earth's surface. However, when producing globes or a single map of the entire Earth, this difference is not accounted for in the map-making process, a subject examined in more detail in Section 2-3.

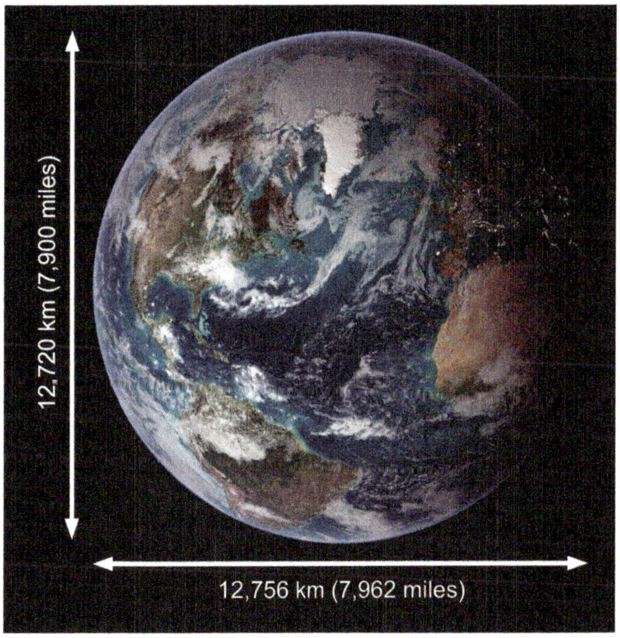

Figure 2.1: Although the Earth may look like a perfect sphere, it is actually an oblate spheroid. Image credit: Reto Stöckli/NASA.

A second consideration required when constructing maps of small areas are variations in Earth's surface topography. At the most general level, the Earth's crust can be divided into continental landmasses and ocean basins, with 71% of the surface considered oceanic and 29% continental. Both types contain significant variations in landforms and elevations/depths. The highest feature on the Earth's surface is Mt. Everest, in the Himalayan mountain chain, at an elevation of 8,848 m (29,029 ft.) above mean sea level while the lowest elevation (or depth), termed the Challenger Deep in the Mariana Trench, has been measured at 11,022 m (36,161 ft.) below mean sea level (Figure 2.2). This difference in elevation is termed **relief**, therefore the Earth's surface relief is calculated to be 65,091 ft. or 12.33 mi. (19,840 m or 19.84 km). While this value seems large, especially from the human perspective when the average height of a human is about six feet, when compared with the circumference of the Earth, this elevation difference is less than 0.05% of the Earth's circumference. Thus, if the Earth was reduced to the size of the bowling ball with a diameter of 22 cm (8.5 in.), the surface of the globe would be as smooth as a bowling bowl if the vertical scale was maintained. Because of this reality, maps portraying the Earth's relief are often exaggerated in the vertical scale to match human perceptions of the landscape. **Vertical exaggeration** is a topic examined later section in this section.

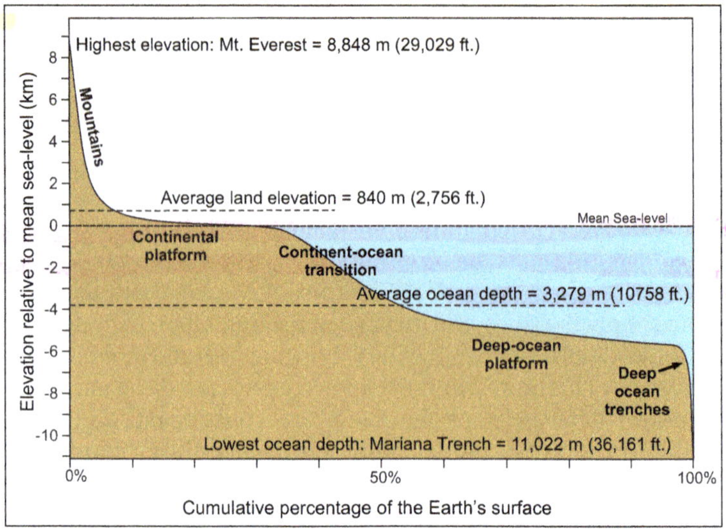

Figure 2.2: A hypsographic profile of the Earth's surface. Cumulatively, 71% of the Earth is covered by the oceans and 29% is exposed as land.

Section 2-2: Defining Earth Locations

After determining the size and shape of the Earth, a method to describe *where* features were located on the Earth was developed. If the Earth was a non-rotating, perfect sphere, identifying locations would be very difficult. Fortunately because of the Earth's rotational axis and its surface variations, we are able to describe locations that can be understood by any human with a basic understanding of the Earth. The first and easiest method of locational description is termed **relative location**, in which a feature's location is described in comparison with the location of some other known feature. For example, you can describe the location of your favorite fishing hole as "the right bank of Rockawalkin' Pond where Rockwalkin' Creek enters the pond" or the location of your favorite sunset spot as "the rocky outcrop overlooking the valley near my mountain cabin." Both of these descriptions reference known features and relate another feature's location to these known features. We often navigate our local environment using relative locations as the basis of our **mental map**, the mental construct or perception of the spatial arrangement of features in the world around us. Within this cognitive framework, we reference movements, locations and/or descriptions of places to readily observable **landmarks** obvious to others who navigate within the same area. These **landmarks** often include well-known buildings and monuments, physical features (e.g., rivers ponds, or mountain peaks) and street intersections (Figure 2.3). While this method of locational description is effective for most everyday purposes, for example providing walking or driving directions in a small area, it is not sufficient in the construction of most maps. In this case, precise locational definitions and repeatable distance and area measurements are necessary. For example, surveying public forest boundaries requires an abstracted description of the location that is uncoupled from landmark identification.

Figure 2.3: The Washington Monument is a very obvious visual landmark. Photo credit: B. Zaprowski.

This second type of locational description is termed **absolute location** which requires a mathematical construct referenced to the Earth's physical dimensions and often takes the form of a set of intersecting north-south, east-west lines resulting in a **grid** (Figure 2.4). The most commonly used grid system is the **geographic grid**, in which any location on the Earth is mathematically referenced by measured angular distance to known locations based on Earth's rotational axis and astronomical observations. The two known

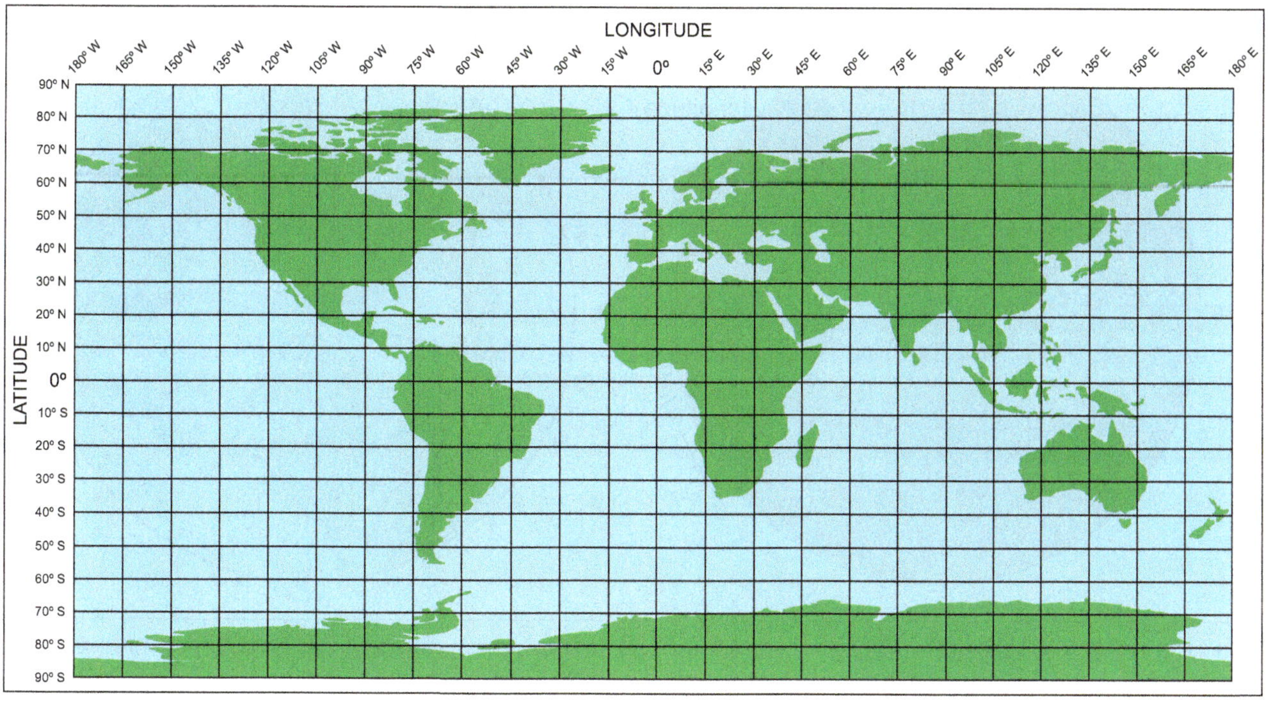

Figure 2.4: The Geographic Coordinate System is the most commonly used geographic grid system.

points obtained from Earth's rotational axis are the **North Pole** and **South Pole**, and these points enable the determination of a plane equidistant between these points and passing through the center of the Earth. The equidistant plane that bisects the Earth between the poles is the **equator,** which we define as 0°, or the basis for the north-south angular measurements for any location on the Earth (Figure 2.5). Angular distances measured north or south of the equator are called **latitudes** and when depicted on a globe, form parallel lines extending from the equator to the North Pole and South Pole. The poles are not lines but points and are at the maximum angular measurement from the equatorial plane, 90° N and 90° S. Because lines of latitude cannot intersect, these lines are also called **parallels**. One source of confusion when examining lines of latitude is that the lines themselves trace east-west across the globe while they measure north-south angular distance. When these measurements are used to

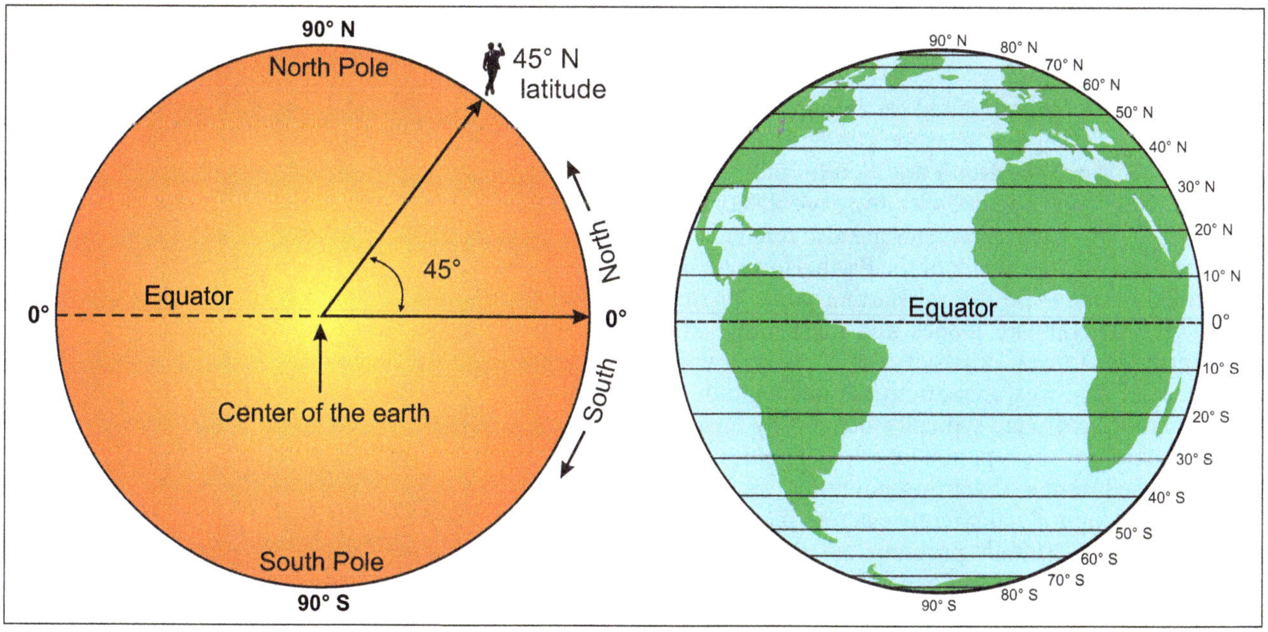

Figure 2.5: Lines of latitude explained. The equator is the arbitrary zero line and all locations are measured in degrees north or south of that line.

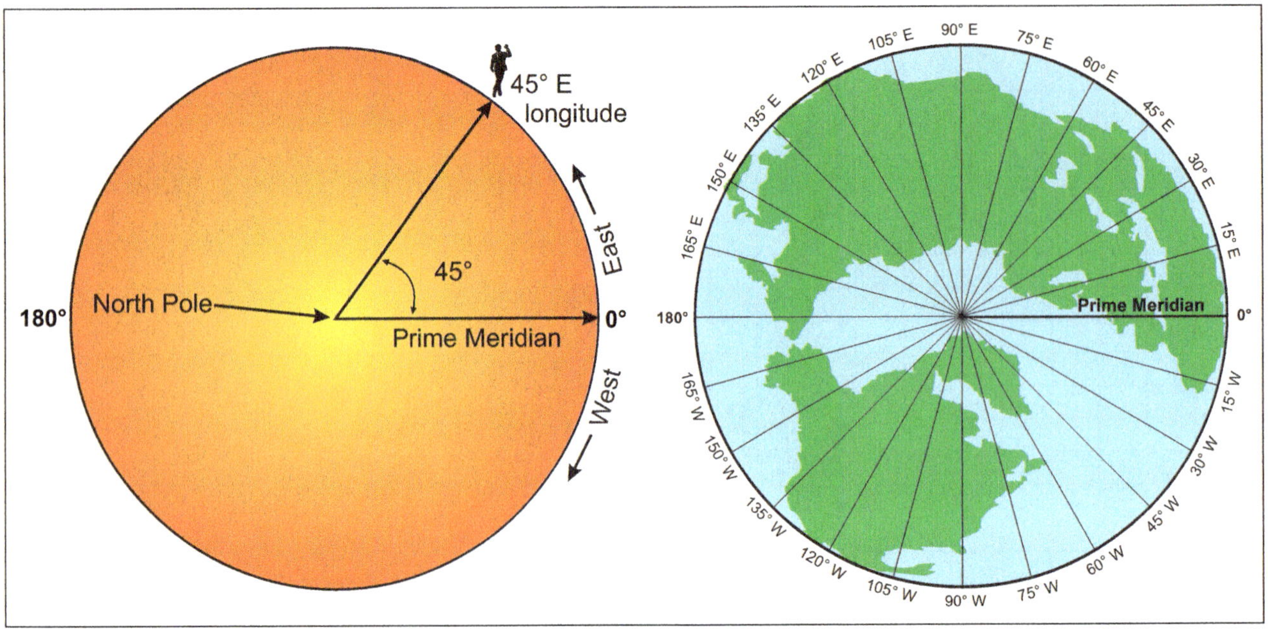

Figure 2.6: Lines of longitude explained. The prime meridian is the arbitrary zero line and all locations are measured in degrees east or west of that line.

describe a location, a capital N or S must be assigned to identify the location's angular position north or south of the equator. Because the Earth is not a perfect sphere, the distance between lines of latitude vary slightly depending on their location. Equatorial bulging results in a slightly shorter distance between a degree of latitude measured from the equator, 110.567 km (68.703 mi.), while polar flattening increases the distance between a degree of latitude measured from the Poles to 111.699 km (69.407 mi.).

The **equator** forms a plane that divides the Earth into two equal halves or **hemispheres** and is the largest circle that can be drawn on the surface of the Earth (Figure 2.7). This type of circle is called a **great circle** and is important for distance measurement and navigation because the arc of a great circle between any two points on the Earth's surface is the shortest distance between those locations. For those of you who have traveled by plane between the continents, this explains the curving north-south arc drawn on in-flight maps. Any other circle drawn on the Earth's surface, for example all lines of latitude with the exception of the equator, are termed **small circles**.

Figure 2.7: A colleague of the authors' stands astride the equator in Ecuador. Photo credit: Stuart Hamilton.

The second component required for the construction of the geographic grid are east-west angular measures. When a series of north-south lines are drawn between the North and South Poles to measure east-west angular distances on a geographic grid, we call these lines of **longitude** (Figure 2.6). These measurements are also referenced to a point on the equatorial plane in the center of the Earth. However unlike the equator for latitude, there is no natural starting point for longitude. For this reason, if you compare globes and maps made 300 years ago by different nations or on different continents, you will see that the 0° reference for east-west measurements of longitude varies with each representation and were typically defined by an important nation's capital. Therefore, the reported longitude of cities or physical features differed based on the 0° starting point that was used! This lack of a universal 0° longitudinal reference across the Earth's surface became an increasingly difficult problem as global communication and transportation improved and expanded, necessitating the designation of a global **prime meridian,** or east-west reference measure for longitude, 0° . The global economic powers of the late 19th century convened in October of 1884 at the International Meridian Conference in Washington, D.C. and defined the location for the international prime meridian as the Greenwich meridian, site of England's Royal Observatory. This conference standardized international measurements of longitude and formalized international time keeping, a concept directly related to longitudinal definitions and examined later in this chapter.

The formal designation of the international prime meridian enabled a constant mathematical definition of a central point in the Earth resulting from the intersection of the equatorial plane and Greenwich meridional plane. East-west angular measurements for all locations on the Earth are made from this point. Given the near spherical shape of the Earth and a total angular distance of 360° around the sphere, a maximum of 180° can be measured east or west from the **prime meridian.** Lines of longitude, also called **meridians,** are reported with a capital E or W to define the hemisphere of the location or feature of interest. If you carefully examine a globe, you will notice that all lines of longitude converge on the North and South Poles making the distance between lines on longitude decrease with increasing latitude. Therefore, the maximum distance of a degree of longitude is measured on the equator, 69.169 mi (111.319 km), while the minimum distance, 0 miles (0 km) is measured at the Poles. Since all lines of longitude converge on the Poles, the total length of each is exactly one-half of a great circle and taken together with their **antipode**, the line of longitude directly opposite, form great circles.

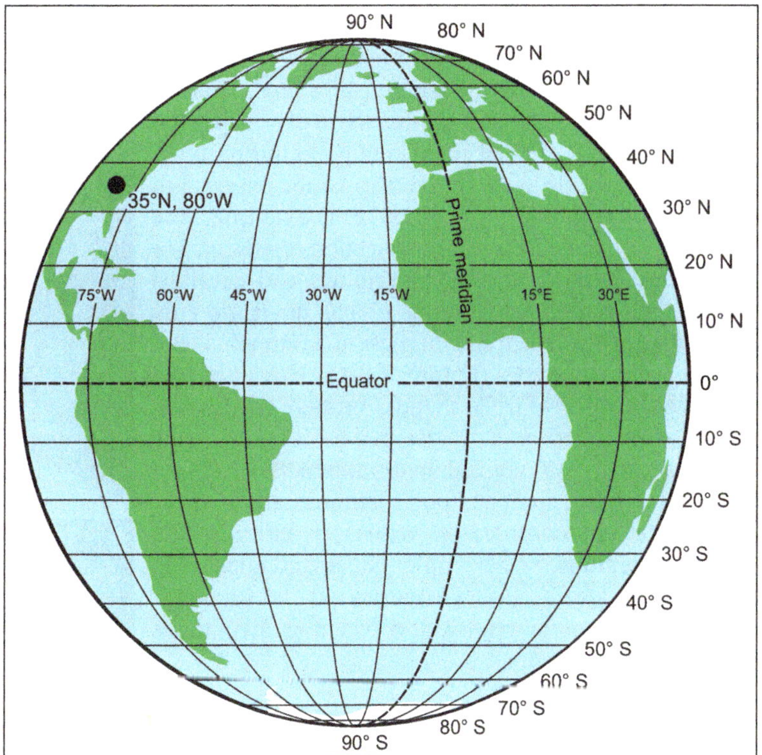

Figure 2.8: The Geographic Coordinate System projected on a globe.

The adoption of a universally-recognized geographic grid enabled the location of any feature to be defined by two coordinates, latitude and longitude, which could be communicated to anyone on Earth with an understanding of the system (Figure 2.8). Moreover, the precision with which a feature's location is defined on the Earth's surface is dictated by the units, and subdivisions of the units used to define a feature's position. For example, if the coordinates are reported at the level of a whole degree (i.e., integer), this means that the feature is defined within 111.319 km (69.169 mi.) of its location for longitude, and much closer as the location nears the Poles, and within 111.699 km (69.407 mi.) for latitude.

With the advent of computer-based mapping, geographic information systems (GIS) and global positioning systems (GPS), feature locations often need to be more and more precisely defined, especially when the coordinates are used for defining property boundaries and navigating vehicles in urban settings. To provide increased positional precision, we can report a location's geographic coordinates by subdividing a degree into minutes [1 degree (°) = 60 minutes (')] and a minute into seconds [1 minute (') = 60 seconds (")]. For a feature in our area, 38° N latitude, one second of latitude equals about 101 ft. (1 degree = 69 mi., 1 minute = 6,068 ft.) while one second of longitude equals about 80 feet (1 degree = 54.6 mi., 1 minute = 4,800 ft.). Therefore if we report the location of our Henson Science lab room as 38° 20' 45" N, 75° 36' 19" W, we are now within about 30 m (100 ft.) of its true Earth position. By adding decimal places to the seconds measure, our precision improves even further, enabling modern mapping and surveying of features to within inches or centimeters of their true position. Because we use geospatial technologies to represent and navigate our world at this level of detail, we must continue to determine the precise size and shape of the Earth and the locations of its topographic features to avoid catastrophic consequences, such as GPS guided airplanes or drones crashing into less precisely mapped surface features like mountains.

Section 2-3: Fundamentals of Maps

The preceding sections described the size and shape of the Earth and the method developed to define locations on the Earth's surface. These concepts are critically important in the construction of **maps**, two-dimensional representations of selected natural and human features on the Earth's surface, integral tools used by Earth scientists to depict surface and atmospheric phenomena. Because the Earth is an oblate spheroid, three-dimensional models (i.e., globes) provide the most accurate depictions of the Earth and its features maintaining true properties such as distance, direction, size and shape. However globes are expensive to produce, costly to purchase, unwieldy to carry and necessarily reduce the size of the Earth eliminating surface detail. To address these limitations, **cartographers** and mathematicians developed and continue to develop mathematical methods to transform the near spherical surface of the Earth to planar representations (i.e., paper maps).

Map Projections

Converting a spherical model of the Earth's surface to a plane is a mathematical transformation called **projection**. In all map projections, the parallels and meridians are systematically arranged on a plane to form a **graticule**. In this transformation process, surface properties and relationships, such as distance, direction, size and shape are modified from their true relationships on the Earth's surface resulting in distortions. However, specific projections were created to maintain the property of interest for specific areas of the map, at the expense of the remaining map properties. For example, the commonly observed Mercator projection represents the entire Earth with a series of meridians and parallels intersecting at 90°, facilitating global navigation by preserving the shapes and angles of continents and oceans in small areas (Figure 2.9A). However, the preservation of feature's shapes, termed **conformal** when referring to a map projection, exists to the detriment of areal relationships on the map. Land and ocean areas in the high and polar latitudes are severely exaggerated in comparison with low latitude and equatorial regions. For example, Greenland appears to be the same size as South America on the Mercator projection, when in reality, the island is five times smaller than the continent!

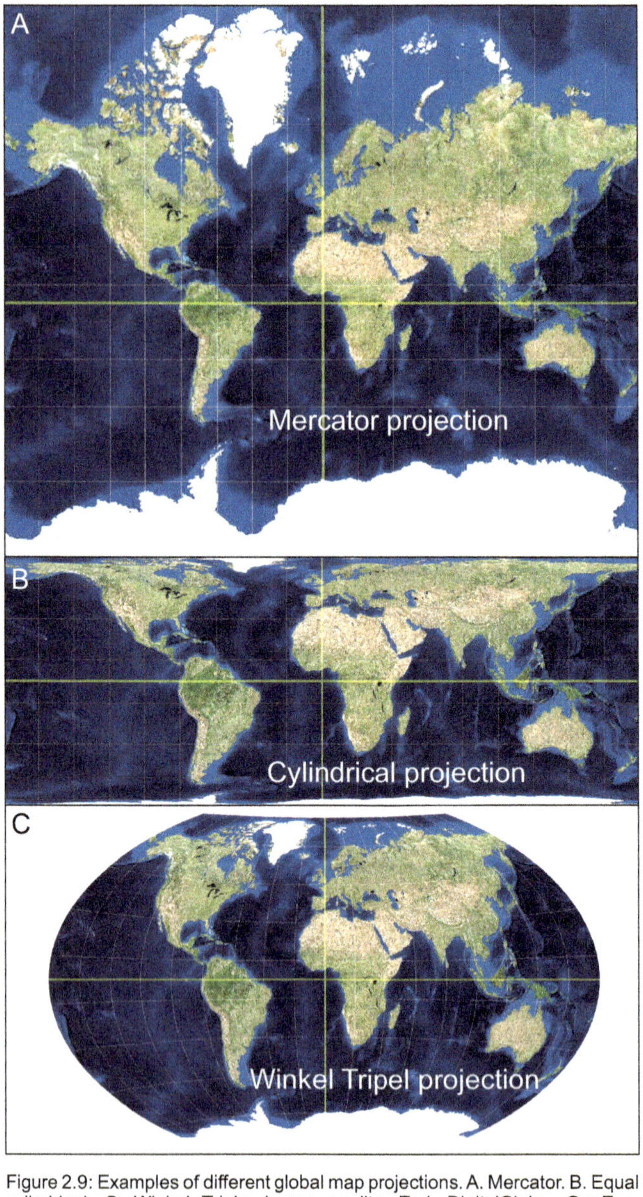

Figure 2.9: Examples of different global map projections. A. Mercator. B. Equal cylindrical. C. Winkel Triple. Image credits: Esri, DigitalGlobe, GeoEye, Earthstar Geographics, CNES/Airbus DS, USDA, USGS, AEX, Getmapping, Aerogrid, IGN, IGP, swisstopo, and the GIS User Community.

Map projections designed to maintain areal relationships are termed **equivalent** and the equal cylindrical projection is an example of an equivalent global map projection (Figure 2.9B). However, while the map enables accurate comparisons of continent, ocean and country sizes, notice the severe distortion in shape associated with this transformation in Greenland and Alaska! Additional map projections have been developed to maintain desired Earth properties including distances (i.e., **equidistant** projections). Projections have also been developed to portray the Earth on a flat plane with the perception of its appearance on a globe. While these projections do not maintain any of the previously mentioned map properties, they attempt to "look" correct by minimizing the major distortions of area, distance and direction. The

Winkel Tripel projection is an example of these projections, often termed **compromise** projections, since compromises have been made across all map properties (Figure 2.9C).

Map Scale

Another important consideration when obtaining information from a projected map is that all maps are produced at a reduced **map scale**, the relationship between a distance on Earth's surface and the equivalent distance measured on a map. **Map scale** dictates the amount of area portrayed on the map and the information or level of detail that can be portrayed on the resulting projection surface. **Map scale** can be reported on maps in three different ways although maps rarely contain all three scales. Map scale reported as the mathematical ratio of map distance to Earth distance is termed a **representative fraction** (Figure 2.10). For example, if a measured distance of one inch on the map is equal to 2,000 feet on the Earth's surface, the **representative fraction** would be 1:24,000 since the units on either side of the ratio must be the same. To calculate the representative fraction map scale, 2,000 feet must be converted to its equivalent measure in inches, 24,000 inches (12 in/ft x 2,000 ft.). The numerator or map distance in a representative fraction is always given as the number one. The more commonly recognizable method in which map scale is portrayed is called a **graphic scale** or measured distance. In this case, a graphic is drawn on the map using common increments of measurement, 1 or 2 inches or centimeters, and its equivalent Earth unit is labeled at measured distances on the graphic (Figure 2.10). This map scale is more intuitively understood and easily used by map readers with the added advantage that if an area of the map is enlarged or reduced with the scale, the graphic scale undergoes the same change and remains correct. Following reduction or enlargement, the representative fraction would need to be re-calculated. A third type of map scale often appears on continental or world maps, a **verbal scale**. In this method, the relationship between map distances and Earth distances are presented in a written statement, for example "one inch equals 43 miles" or "one centimeter equals 63 kilometers." However this type of scale is rarely included on maps which depict small areas of the Earth's surface, nor is it included on most professional maps Earth scientists construct to present surface or atmospheric data.

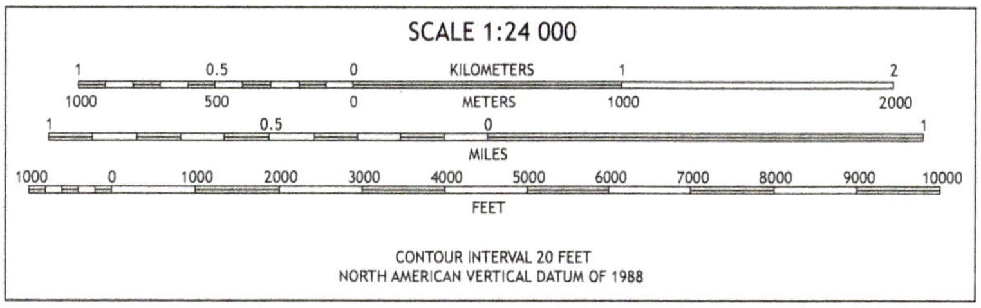

Figure 2.10: Map scales presented both as a representative fraction and as a graphic scale.

All maps can be generally categorized based on their scale and the resulting level of detail contained on the map as either **small scale** or **large scale**. **Small scale** maps portray a large portion of the Earth's surface, however at the expense of detail. For example, a 1:16,000,000 map scale is commonly used for wall-sized maps of the world; however at that scale, political and administrative boundaries must be limited to the national level therefore individual states of the United States cannot be differentiated. These maps are considered small scale maps and generally include any maps smaller than 1:250,000. **Large scale** maps depict smaller areas of the Earth's surface and therefore contain more detailed information found within the map area. For example, at a scale of 1:24,000, individual residential structures in non urban environments can be symbolized and easily read (Figure 2.11). The difference between whether a map is small scale or large scale is often confused; however the answer is easily obtained by simply dividing the numerator (i.e., map distance) by the denominator (i.e., ground distance) in a representative fraction. The larger scale map will have a larger value than the small scale map (1:24,000 = 0.00004 while 1:16,000,000 = 6.3×10^{-8}).

Map Elements

Once a cartographer selects a map projection and scale based on the map's purpose, spatial information is symbolized and map elements are added to aide understanding for the map user. Major elements contained on most scientific maps include the **mapped area**, **locational reference**, **title**, **legend**, **scale**, **orientation** and **data source(s)** (Figure 2.11). The **mapped area** contains symbolization selected to portray natural and human phenomena based on the map's purpose. Symbols, often universally recognized and standardized by long-term use, include a combination of points, lines and shaded or hatched polygons used to show features including administrative boundaries, roads, land cover, temperatures, geologic units and a multitude of additional spatial data supporting the map's purpose. Map symbolization takes many forms depending upon the type of data being

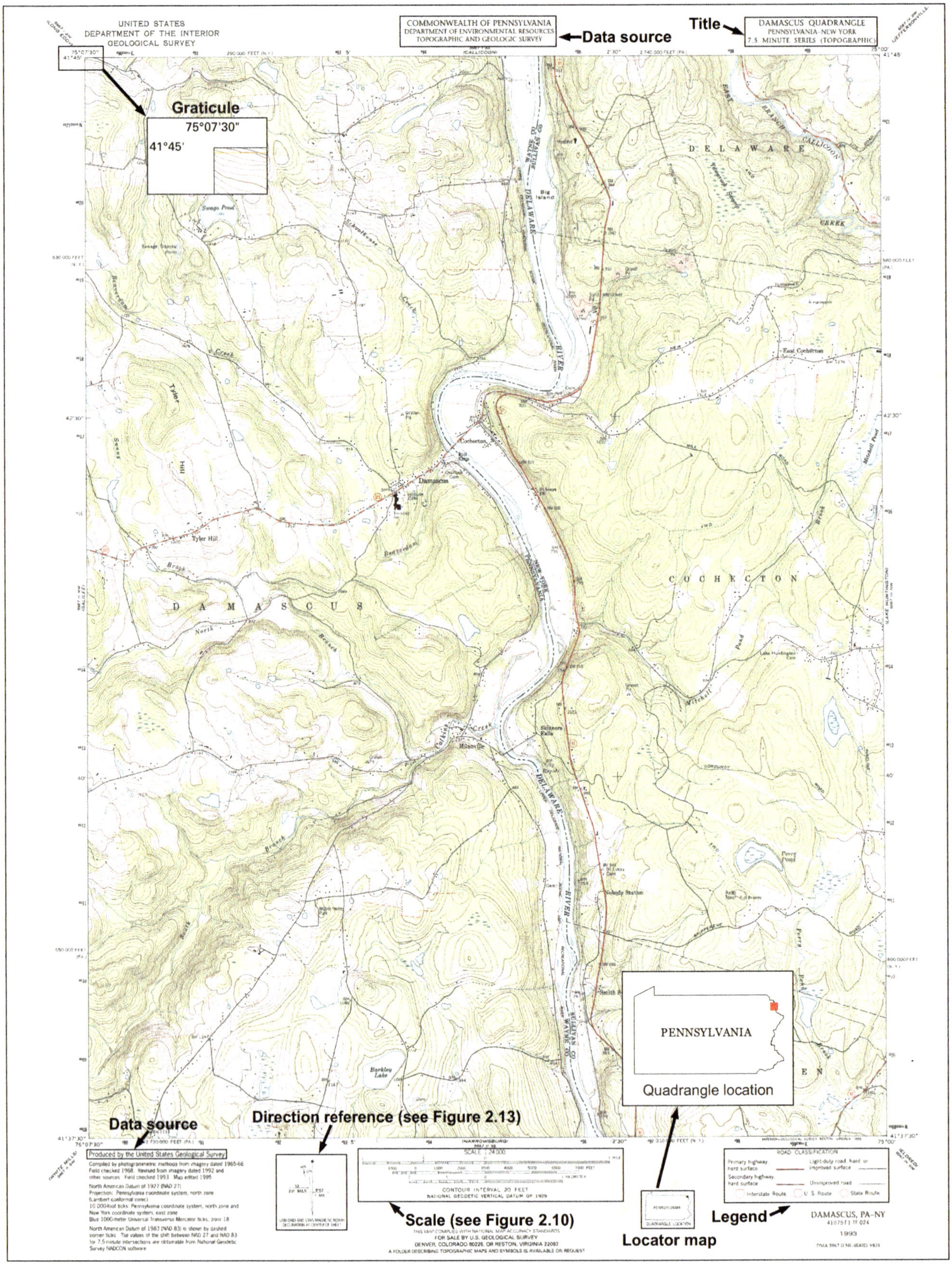

Figure 2.11: The elements of a map, in this case, a standard topographic map. Map published by the USGS.

depicted. Examples of these methods are discussed in the following section. The **mapped area** is often locationally referenced to the geographic grid by a **graticule** labeled on the map margins enabling the identification of absolute locations relative to latitude and longitude. Additionally, some maps, especially USGS topographic maps include a **locator map**, providing a graphic representation of the relative location of the map area in relation to its state's boundaries (Figure 2.11). A description of the mapped area and purpose or theme of the map is also provided in the map **title**. The title is the most important text added to the map. It has the largest font size and is placed in a prominent position on the page. The title should quickly let the map reader know the area of the map's focus and the spatial data being focused upon. Often, the title includes a date for temporal reference and a subtitle if the mapped information is especially complex and requires further description. The complexity of the mapped information and a key decoding the symbology used is contained within the map's **legend** (Figure 2.11). Any symbol which is not easily understood must be contained in the legend to facilitate reader understanding. In a **thematic map**, a map focused on a specific spatial data subject, for example, population density in Maryland, the legend presents patches containing different hues (i.e., colors), or gradations of hues based on saturation or lightness, to represent the varying nature of the map subject in space (i.e., geographic variation). In contrast, **general reference maps**, for example, most wall maps in your lecture class like the United States map, contain spatial data including boundaries, major urban centers, geographic names, landforms and major physical features to provide base understanding of the area portrayed on the map. General reference map legends are typically more complex and contain point, line and area symbology used to represent these mapped features (Figure 2.12).

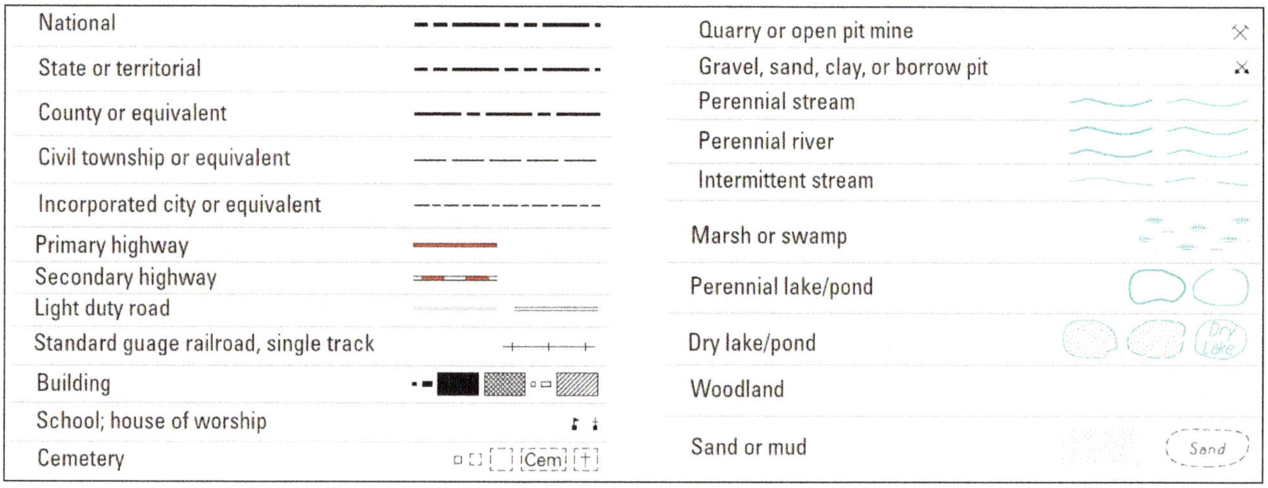

Figure 2.12: An example of the common symbols used on USGS topographic maps. Images from USGS.

All scientifically produced maps contain one or more of the **scales** discussed in the preceding section. If a graphic and/or verbal scale is used, the scales often portray measured distances in both the English and Metric systems. Some maps also include an **orientation** symbol, or north arrow/compass rose, to provide the map reader with a quick reference to north, although thematic maps tend to ignore the symbol since map convention is to orient the map such that north is toward the top. In the case of large scale maps used for navigation in small areas, for example, USGS 1:24,000 or 1:100,000 topographic maps, orientation symbols also identifies multiple commonly-used north references (Figure 2.13). These north references include true polar north, the direction to the geographic pole, magnetic north, the direction a magnetic compass's north arrow points in the center of the mapped area, and grid north (i.e., GN), the direction the central meridian of the projection (the Universal Transverse Mercator or UTM projection) runs in the center of the map. If all three north references are present, angular differences between polar north and magnetic north and grid north are also provided. The final and most important map element is the spatial information **data source** (Figure 2.11) This information is located in the margin of the map, outside of the mapped area, and provides the reader with the name of the agency or agencies that produced the map and a description

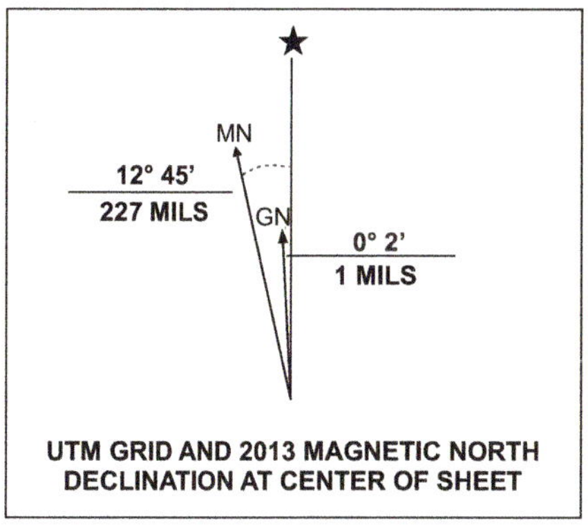

Figure 2.13: The standard orientation symbol for USGS topographic maps. ★ = True north, MN = Magnetic north, GN = Grid north.

of how and when the spatial data were obtained. The inclusion of the data source (or sources) is necessary for map readers' interested in obtaining the spatial data used to make the map and to report any erroneous information identified within the mapped area for correction on future editions of the map. All map elements combine to represent a particular area of the Earth, or the entire Earth, and selected spatial information to form a map.

Mapping Techniques

Specific techniques to portray spatial data have been developed and standardized in the Earth and atmospheric sciences especially when portraying quantitative data that are continuous in distribution (i.e., a phenomenon that can be measured at any location on the Earth's surface, in the oceans, or in the atmosphere) and relatively smooth in variation (i.e., the values of the measured phenomenon only change over short geographic distances and often form recognizable patterns). Common examples include atmospheric temperature or surface elevation. One important and commonly used technique that will be examined in many different applications in this course is the **isarithmic mapping** technique, whereby lines are used to connect locations of equal value, resulting in the construction of a series of **isolines** tracing out the pattern of variation in a given geographic area. In some cases, the areas between these lines are shaded by hue (i.e., color) to represent greater and lesser amounts of the phenomenon that is being measured and mapped (Figure 2.14). Most likely, you've seen this technique applied on the evening news or a web page to portray national average temperature patterns or local precipitation amounts flowing the passage of a storm. Thus the **isolines** technique can be utilized to portray locations of equal temperatures, **isotherms**, or precipitation amounts, **isohyets,** in a given time period. **Isolines** are also used for additional smooth and continuous phenomena that you may not have been exposed to like winds (**isotachs**), water depth (**bathymetry**) and surface elevation (**contour lines**). In all cases, **isolines** are constructed in the same manner. Instruments obtain measurements of a phenomenon of interest at a series of locations simultaneously or over a specific time period. These data values are then plotted on a map and isolines are constructed at selected intervals by drawing lines connecting values defined by the interval (Figure 2.14). In many instances, the chosen isoline value is not measured exactly by an instrument at all locations and the position of the isoline must be approximated by interpolating the location of the values between adjacent stations recording actual values above and below the desired isoline value (Figure 2.15). Following the construction of the isolines, the data values are removed from the map to produce a smooth and easily interpretable map. The technique is especially effective when a large number of measurement locations and resulting data values are used to generate the isolines, for example, in surface elevation surveying. Rather than mapping hundreds or thousands of individual data values at specific locations in a map area, isolines create a continuous surface facilitating the estimation of a data value at a location, even if an instrument did not directly measure the variable of interest at that place.

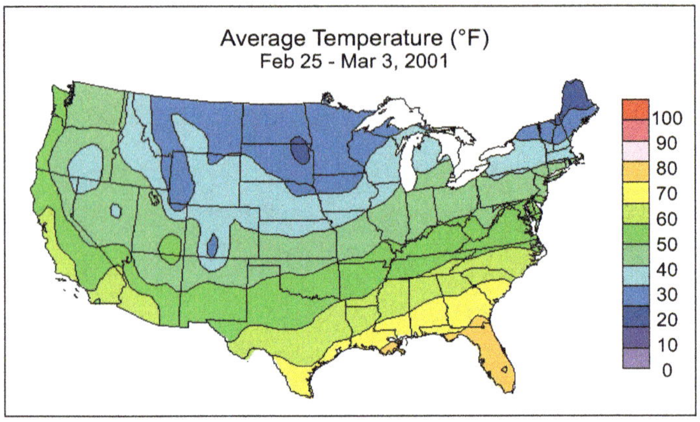

Figure 2.14: An example of a shaded isoline map, in this case, an isothermal map. This map shows average U.S. temperatures for Feb 25-Mar 3, 2001. Data credit: NOAA.

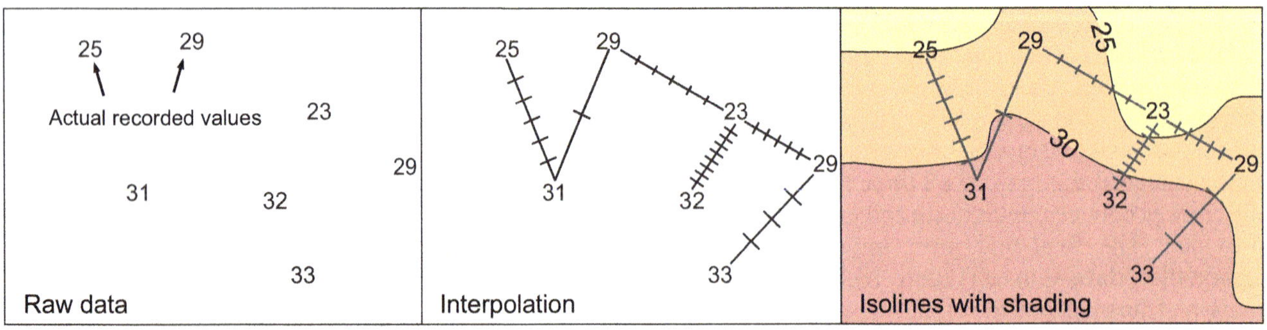

Figure 2.15: Interpolation is used to draw isoline maps, in this case, surface temperatures in °C. This is the same technique the cartographer used to draw in the isotherms on Figure 2.14 before shading the map.

Section 2-4: Topographic Maps - Reading Contour Lines

Topographic maps are an example of a commonly used isoline map in the Earth sciences requiring a tremendous number of data values. Topographic maps are considered general reference maps although these maps include **contour lines** to depict the elevation of the ground surface in the mapped area in relation to **mean sea level (msl)**, the level between average high water and low water marks. Thus contour lines enable the two dimensional portrayal of a phenomenon that is in reality, three-dimensional. Figure 2.16 exemplifies the relationship between a selected area of the three dimensional surface of the Earth and the resulting depiction using contour lines. Several patterns noted in the figure's contour lines exemplify rules related to their construction and interpretation. First, contour lines cannot cross one another since no location on the Earth's surface can have more than one elevation relative to mean sea level. In some areas where elevations change quickly in a short distance, the lines may nearly converge or touch one another, however each line will always be drawn separately.

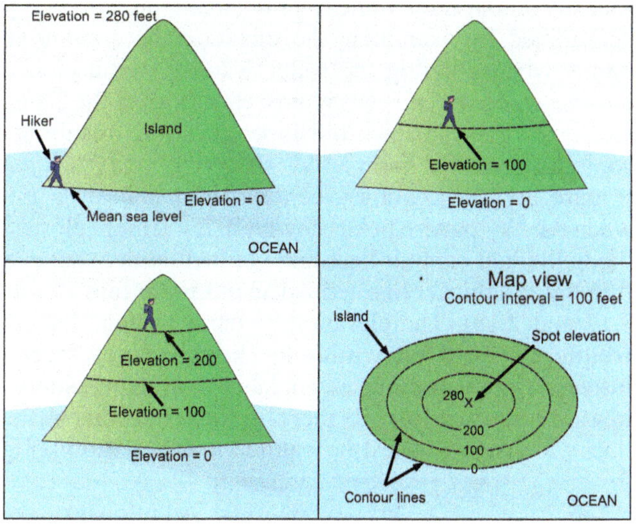

Figure 2.16: Contour lines explained using a simple island.

The difference in elevation between adjacent contour lines is called the **contour interval** and varies based on the elevation differences in the mapped area (see Figure 2.11). For example, a topographic map for a lower-lying region like the one in Figure 2.17 would likely utilize a contour interval of 10 or 20 feet. However, a topographic map representing a portion of Colorado's Rocky Mountains might use a contour interval of 50 or 100 feet given the region's higher relief. From Figure 2.16, also note that a hill or mountain peak is represented by a series of concentric, enclosed contour lines, with the higher elevation lines completely contained within the lower elevation contour lines (e.g., Diamond Hill in Figure 2.17). Recognizing this pattern in contour lines makes the identification of such features on a topographic map relatively

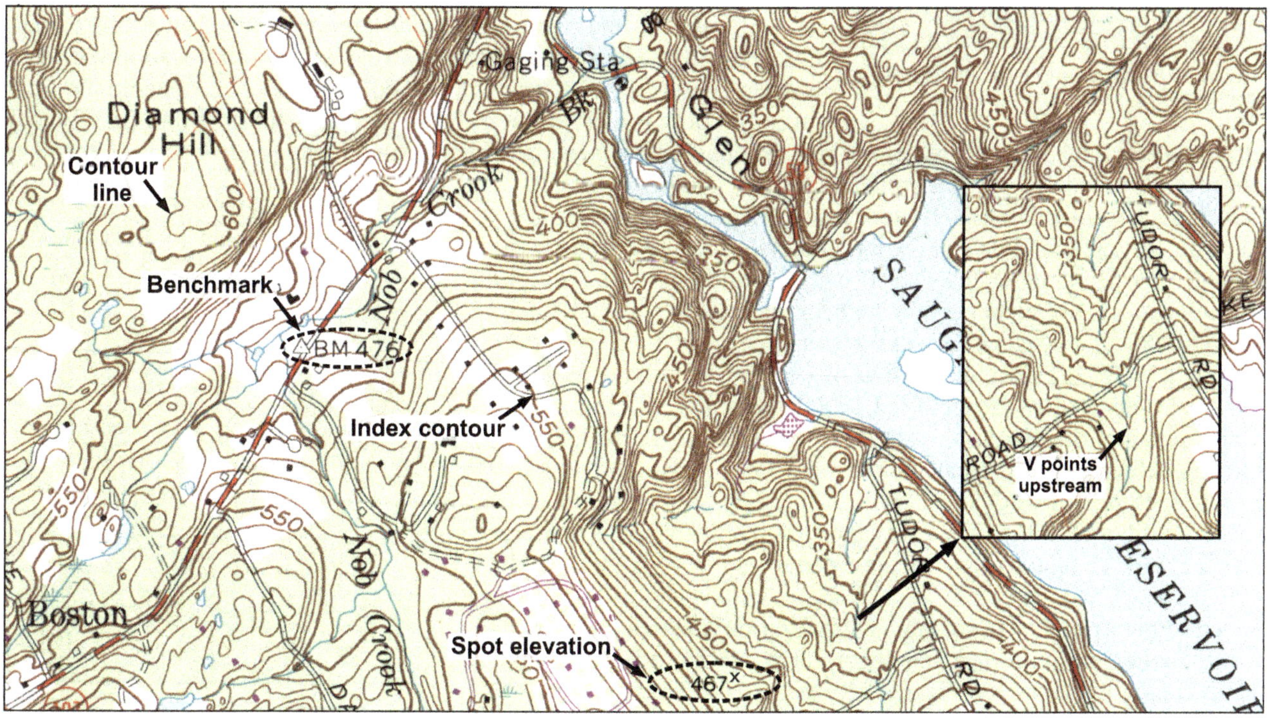

Figure 2.17: A close-up view of a topographic map with the key elements labeled. Note how the "V"s in the contour lines point upstream. Image credit: USGS.

easy. When examining USGS topographic maps, the complexity of the Earth's surface relief as portrayed by contour lines is immediately apparent (Figure 2.17). Areas where elevation does not change rapidly result in widely spaced contour lines while river or stream direction can be determined by examining the direction the contour line points when crossing the water feature. Because of the erosive action of rivers or streams, elevations of the channel are lower than adjacent hillslope elevations resulting in contours lines that form a "V" shape and point upstream (Figure 2.17). In contrast, contour lines indicate a ridge when the "V" pattern points downhill. Look more closely at the individual contour lines and you will notice that every fifth line is a thicker line, termed an **index contour** (Figures 2.17 and 2.18). The elevation of these contour lines are frequently labeled while the lighter lines, termed **intermediate contours**, are not labeled. In **depressions**, or low areas surrounded by higher elevation (like mines), the contour line or lines are represented by hachured contour lines which point toward the feature's lower elevation (Figure 2.18).

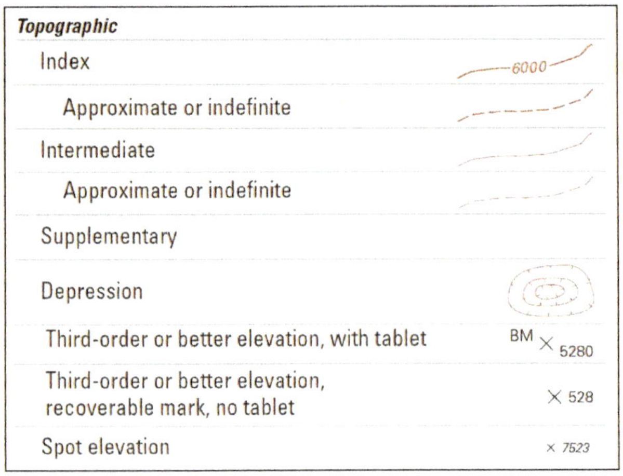

Figure 2.18: The common symbols used to denote elevation on a topographic map. Images from USGS.

Figure 2.19: A typical benchmark. Photo credit: J. Schmidt/USGS.

Additional symbols are also used to represent surface elevations in specific locations on the map. In some locations, **spot elevations** are provided to indicate the precise elevation of a feature and these locations are often symbolized with a small "x" and nearby elevation (Figures 2.17 and 2.18). Spot elevations are often found on small hills and the intersections of transportation features. All USGS topographic maps contain a special type of spot elevation called a **benchmark**. Benchmarks are symbolized with a capital "BM" alongside a small "x" or triangle filled with a dot, and a precise elevation value, and form the network of physical monuments used to measure distance, direction and elevation for mapping projects completed in the United States (Figures 2.17 and 2.18). These monuments can be located in the environment and are constructed of a brass plate engraved with the latitude, longitude and elevation of the location and set in concrete (Figure 2.19). Benchmarks and spot elevations enable the map reader to know the elevation of the indicated feature within one foot of mean sea level, the most precise elevations available in the map area. All additional feature elevations on the map can be determined using the intermediate and index contour lines. If the feature lies directly on the contour line, the feature's elevation is equal to the contour line's elevation. The elevations of features located between contour lines are usually reported by a range of possible values. For example, if a landmark is located between the 400 and 450-foot contour lines (a contour interval of 50 feet), the landmark's elevation can reported as 401 to 449 feet above mean sea level. This is often the preferred method because U.S. National Map Accuracy Standards for USGS 1:24,000 topographic maps state that elevations represented by contour lines on are correct to within one-half the contour interval of the map. By providing a range, it is likely that the correct elevation lies within the range (e.g., within at least 25 feet of the actual elevation).

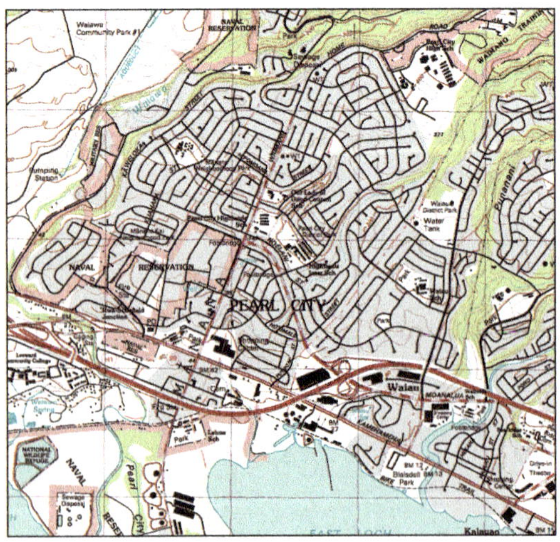

Figure 2.20: An example of how an urban area is portrayed on a topographic map. Image credit: USGS.

Topographic maps also contain symbolization associated with general reference maps to place the surface configuration of the mapped area into context (Figures 2.17 and 2.20). Blue lines and area symbols indicate hydrographic features, including rivers, streams, lakes and reservoirs, while green point and area symbols represent vegetation including forest cover, vegetation types and even certain types of agriculture. Human constructed features on the map are symbolized by red, black and gray point, line and area symbols (Figure 2.20) and at the 1:24,000 scale, are even used to represent single family residential structures in non-urban environments! Additionally, transportation features like railroads and highways are depicted using these colors as are political boundaries and settlement areas. Today, USGS topographic maps constructed at multiple scales and publication dates are available for download free of charge from several government web sites, readily enabling their use for scientific research, recreation and general interest. The National Map (http://nationalmap.gov/ustopo) and USGS Store (http://store.usgs.gov) are just two of many web sites developed to provide an interface to download topographic maps of interest.

With frequent use, map readers can begin to visualize the landscape or "lay of the land" as depicted by the patterns evident in the contour lines. However, in certain cases, for example planning a hike in mountainous terrain, the perpendicular perspective of the landscape can be difficult to understand and prepare for, necessitating the desire to view the planned path from the side or in profile (Figure 2.21). Topographic map users can use contour lines intersecting the desired path to construct a **cross sectional profile,** or side view of the landscape. Rather than determining the straight line distance between your origin (i.e., starting point) and destination (i.e., ending point), the profile enables you to determine the amount of elevation change that occurs along the path and where that change will occur. Cross sectional profiles are often used by Earth scientists to examine river valleys to determine flood plain dimensions and the locations of natural levees, for example, or to determine the slopes of hillsides and mountains. The difference between maximum elevation and minimum elevation on the route can also be calculated, a measure termed **relief**, which provides a quantitative measure of the elevation variability in the geographic area for comparison to other landscapes.

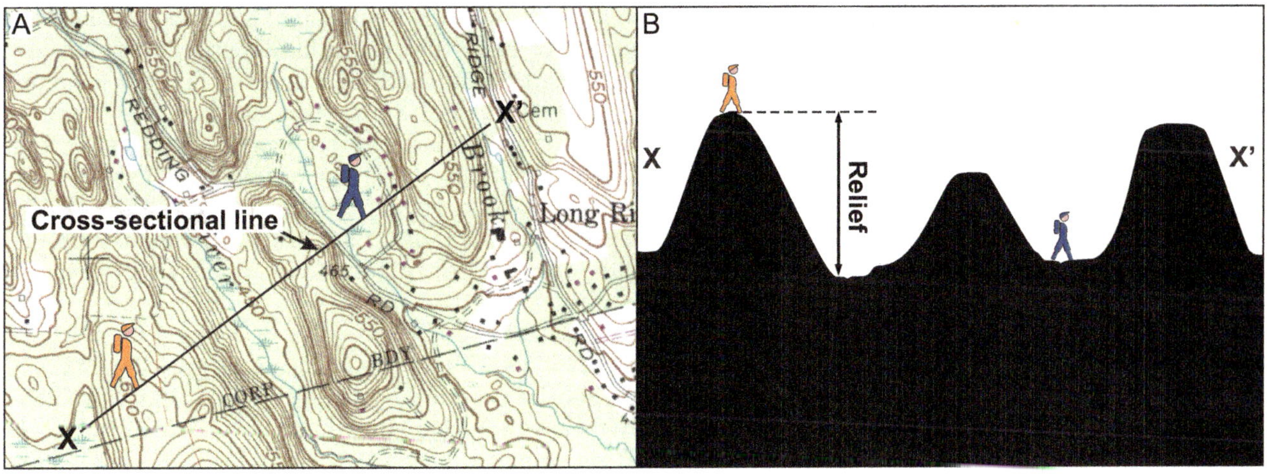

Figure 2.21: Cross-sectional profiles explained. A. A line X-X' is drawn across the map. This line is the basis for a cross-sectional profile. B. Elevation data are collected along line X-X' and drawn as shown. This cross sectional profile gives us a better sense of what the land surface looks like. Image credit: USGS.

Section 2-5: The Basis for Time and the Seasons on Earth

Throughout history, humans have observed and measured the relationship between the Earth and Sun over both short and long periods and developed methods to systematically monitor these cyclical relationships given their importance to daily life, agriculture and other activities. These Earth-Sun relationships, which form our foundation for measuring time, are based on the Earth's rotation about its axis, the basis for a day and its subdivisions, and Earth's revolution about the Sun, the basis for a year and its subdivisions. The concept of time is the result of Earth's rotation about it axis and the observed relationship between locations on the Earth and the Sun's position above the horizon. Moreover, when monitoring longer periods during Earth's revolution or orbit around the Sun, such as months, seasons and years, the Earth's locational relationships with our moon, and the stars and constellations of the celestial sphere enable the specific definitions of such temporal units. For example, months based on the time needed for the Moon to complete a cycle through the lunar phases (More on the lunar phases in Chapter 16).

At any given point in the Earth's orbit, one half of the Earth is illuminated by solar radiation, and if the Earth were viewed from space, a great circle defining the transition from light to dark could be seen at all times. This great circle which defines the boundary between day and night is termed the **circle of illumination** and if we continued to view the Earth from space, we would notice that the circle of illumination sweeps from east to west across the Earth's surface due to the direction of Earth's rotation, counter-clockwise in relation to the North Pole (Figure 2.22). This simple Earth-Sun geometry explains why the Sun rises in the east and sets in the west for all locations on Earth ***if*** the Sun rises above the horizon. This is an important and overlooked consideration for Polar Regions during certain periods of the year, a topic examined in the following section describing seasons. Archeological evidence in widely varying locations and across many ancient cultures indicates that our ancestors had an intimate knowledge of important Earth-Sun relationships, and developed calendars, aligned urban structures, and built ceremonial observatories based on the angles defined by these celestial relationships.

Figure 2.22: This photo taken by space shuttle astronauts shows the nature of the circle of illumination. Photo credit: Image Science and Analysis Laboratory, NASA-Johnson Space Center.

Humans were able to easily determine when the sun reached its maximum elevation above the horizon between sunrise and sunset, the **solar zenith**, and used the time period between consecutive solar zeniths to specifically define a **day**, a period which averages 24 hours over the course of a year. **Solar zenith** was used to define a day because the timing of sunrise and sunset was observed to fluctuate, especially in middle and high latitude locations. Solar zenith can also be termed **local noon** and the designation of local noon enabled the definition of **local time** which human settlements employed all over the world to define the day prior to the invention of pendulum clocks in the 17th century. However, because **local noon** occurs at different times for different locations based on a location's relationship to the Sun, or its longitude, every settlement, even those relatively nearby, observed local time slightly differently. Moreover, because of variations in Earth's orbital speed as its distance from the Sun changes due to its elliptical orbit, a relationship Kepler quantified, the timing of solar zenith varies by as much as 16 minutes for all locations if monitored closely. Therefore the time period between solar zeniths is not exactly 24 hours each day necessitating the idea of a **mean solar day** which we continue to maintain.

Figure 2.23: The Royal Observatory, Greenwich, England is the home of the Prime Meridian and Greenwich Mean Time. Main photo credit: Mike Peel. Inset credit: © lapas77/Shutterstock.com

Before our ability to travel and communicate at great longitudinal distances, discrepencies between local times was a minor problem. Each city based their local time on their mean solar zenith resulting in over 300 local times in the United States alone in the late 1800s. However, economic transactions began to expand by hundreds and even thousands of miles away as a result of railway construction and telegraph communication. Thus, a standard method for consistently defining time nationally and then worldwide was needed and desired. In 1884, representatives from 27 nations met in Washington, DC at the International Meridian Conference to define a global prime meridian (0°) and set standard time in the relation to this new prime meridian for all locations on Earth. Because longitude measures east-west angular distances on the Earth and the Earth rotates east-west about its axis, lines of longitude were used to define zones of time across the Earth's surface. The word meridian is derived from the Latin *meridies* and means *meri* – midpoint and *dies* – day or the midpoint of the day. From this nomenclature come A.M. or *ante meridian*, before noon, and P.M. or *post meridian,* after noon. Following 1884 and the designation of the prime meridian at the respected Greenwich Observatory (Figure 2.23) in the outskirts of London, England, standard time for longitudinal zones on the Earth were related to mean solar time in Greenwich defined as Greenwich Mean Time (GMT). The concept of **standard time** is relatively simple although its implementation required modifications that make the reality a bit more complex. We'll start with the theoretical basis for standard time and then examine today's current implementation.

Standard Time

One Earth rotation, or a mean solar day, occurs over 24 hours and 360° of longitude as measured on Earth, 180° E and 180° W from the prime meridian. Therefore, simple division allows us to determine that the Earth rotates 15° per hour (360°/24 hours = 15°/hr) and this angular distance is used to define the boundaries between **time zones,** of which there are a total of 24 to equal a day (Figure 2.24). Following the selection of the Greenwich Observatory as the prime meridian (0°), standard time and time zones for the Earth have been based on Earth-Sun time measured at the Greenwich Observatory, or Greenwich Mean Time (GMT). The prime meridian forms the **central meridian** for the Greenwich Time zone; this time zone extends 7.5° degrees east and 7.5° degrees west of the prime meridian by definition. Therefore any location within 7.5° of the prime meridian observes the time defined at the Greenwich Observatory. For example, if it is noon in Greenwich, England (0°), it is also noon in Edinburgh, Scotland (3° W), even though the sun has yet to reach its zenith in Edinburgh given its westerly longitude and the direction of Earth's rotation. World time zones are defined by the number of degrees of longitude, or hours for every 15° degree increment, that separate their position from the prime meridian and the hemisphere of their location. For example, Salisbury, Maryland's longitudinal position, 76° W, means that the time zone of its location (i.e., Eastern Standard Times or EST) is five hours different than Greenwich Mean Time (76°/15° = 5.07 or 5 hours). And because the Earth rotates from west to east, Salisbury's westerly longitude relative to the prime meridian means that the sun has not reached its zenith, therefore Salisbury's time zone is five hours ***EARLIER*** than GMT, therefore Salisbury's time, and the time of all westerly longitudes will be earlier than GMT. Thus, if it is 8 P.M. in Greenwich, England, the standard time in Salisbury, Maryland is 3 P.M. Standard times for locations east of the prime meridian can be

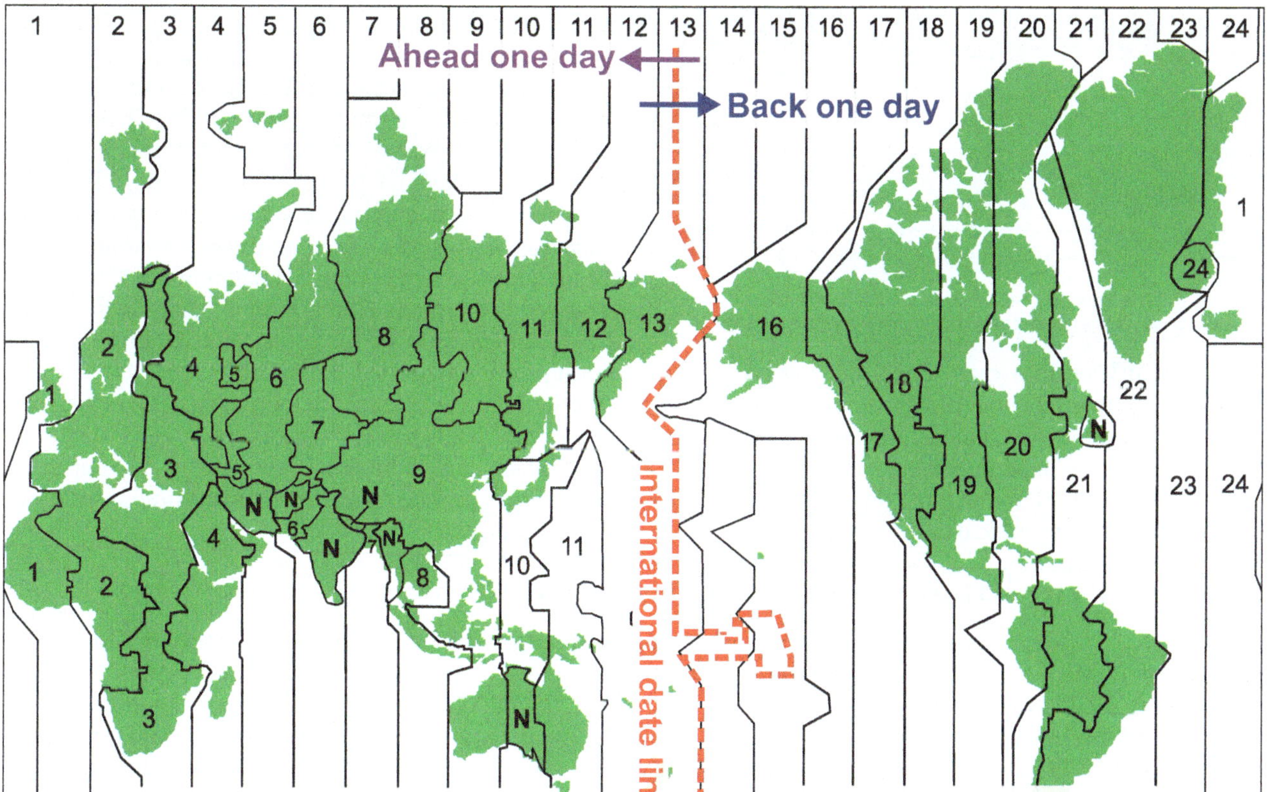

Figure 2.24: The world time zones and the International Date Line. **N = non-standard time.**

determined using the same technique, however, after calculating the longitudinal difference between the location and the prime meridian, each 15° interval of difference results in a one hour ***increase*** in time. For example, if it is 10 A.M. GMT and you are interested in determining standard time in Jakarta, Indonesia (107° E), the seven hour difference (107°/15° ≈ 7 hours) means that it is 5 P.M. Logically given a 24 hour day and 180° of east and west longitude, standard times can only vary by 12 hours from GMT (180° /15° = 12). Therefore, once we reach 180° E or W (i.e., it's the same location on the Earth's surface), we must change the calendar date to maintain our 24 hour definition of a day. This is why 180° E or W is termed the **International Date Line (IDL)**, and this is why we change the calendar date when we cross this line of longitude, or theoretically we could travel back in time if we traveled west faster than the Earth's rotational velocity (Figure 2.24)! This is also why its location in the middle of the Pacific Ocean, the **antipode** of the prime meridian and away from continents and large landmasses, made the Greenwich observatory a natural choice to define standard time. How we change the calendar date as we cross the IDL can be exemplified in the following example. If we travel across the International Date Line from Tonga (175° W), in the Western Hemisphere, to Fiji (179° E) in the Eastern Hemisphere, we advance our calendar date forward by one day. Therefore if we leave Tonga on September 16th, we will arrive in Fiji on September 17th. If we travel in the opposite direction, from Fiji to Tonga, and leave Fiji on the September 17th and we will arrive in Tonga on the September 16th. Thus locations immediately west of the IDL are the first locations on Earth to observe a new calendar day!

Figure 2.25: The boundary between the Central and Eastern time zones follows political boundaries and avoids large cities. Image credit: USGS.

The preceding section assumes that time zones and the International Date Line perfectly correspond to the defining lines of longitude in relation to the prime meridian. However an examination of the world time zone map makes it clear that the boundaries between time zones, and even the International Date Line, have been modified to accommodate political and economic realities (Figure 2.24). In the United States, notice how the boundary between Eastern Standard Time (EST) and Central Standard Time (CST) follows the political boundary between Alabama and Georgia rather than a specific line of longitude (Figure 2.25). Even more apparent is how the boundary between these time zones has been shifted eastward in northern Indiana to encompass the economic linkages of this region with Chicago, Illinois, the major midwestern urban center. On the global scale, examine the number of time zones observed in mainland China, a total of one! In a country that spans over 60°of longitude and therefore should recognize at least four time zones, the central government keeps the citizens of the entire country on a single countrywide time. You may also notice that some locations vary their time differences in increments of less than one hour. For example Newfoundland, Canada observes time 30 minutes later than its maritime provincial neighbor Nova Scotia. And in southern Asia, countries including Iran, Afghanistan, India and others, apply the time zone differences in half hour and three quarter hour increments. As can be seen, the reality of standard time is far more complex than the theoretical basis for standard time.

Coordinated Universal Time or Universal Time Coordinated (UTC)

Before 1972, world clocks were regulated based on measured time at the Greenwich Observatory or Greenwich Mean Time (GMT). With the widespread development and use of atomic clocks by standards agencies in many nations after 1972, worldwide standard time has been measured and coordinated by the International Bureau of Weights and Measures (BIPM) in France and is termed **Coordinated Universal Time** or **Universal Time Coordinated (UTC).** This current standard, which precisely defines seconds, minutes, and hours based on the vibrational frequency of atoms, is distributed to scientists worldwide and used to maintain consistent time for such purposes as the Internet, the cellular industry and global positioning systems (GPS). These atomic clocks are so precise that as the Earth's rotational time continues to slow, due to tidal friction with the Moon, the re-arrangement of crust relative to the core due to plate tectonics and changes in the distribution of the Earth's mass due to convection in the mantle, a **leap second** must be added at irregular intervals to maintain the clock's relationship to the mean solar day. For most purposes, UTC and GMT can be used interchangeably, and the same method is employed to define worldwide time zones. However, UTC is the worldwide standard and its use is observed on meteorological maps, which you will use later this semester.

Figure 2.26: Not a fan of daylight saving time? You have Ben Franklin to thank for that. Image credit: Duplessis, Joseph-Siffrede, artist. "Benjamin Franklin, head-and-shoulders portrait, facing slightly right, wearing fur collar." 1907. Prints and Photographs Division of the Library of Congress.

Daylight Saving Time

Daylight Saving Time (DST) is another important concept related to time especially for countries in the middle and high latitudes who adopted this program in an effort to save energy, although measurements of the intended effect has produced mixed results. The original idea is attributed to Benjamin Franklin and was proposed as urban populations grew and livelihoods became less reliant upon ambient light from the sun (Figure 2.26). Its basic premise was and remains to increase daylight hours in the late afternoon/early evening by advancing clocks one hour forward in the spring. It was thought that most of the population slept through the added early morning sunlight and by shifting a daylight hour to the evening in the spring, summer and fall, there would be a decrease in energy use for artificial lighting. The program continues for most areas in the United States, with the exception of Hawaii and most of Arizona, and dictates that clocks are adjusted forward one hour on the second Sunday of March at 2:00 a.m. (i.e., spring forward) and returned to local standard time on the first Sunday in November at 2:00 a.m. (i.e., fall back). Note, however, that UTC does not adjust to Daylight Saving Time so care must be taken when determining local time. For example, if you view a meteorological map in July, when we observe daylight saving time, with a time stamp of 8:00 a.m. UTC (or 8 Zulu as these times are equivalent), the local time for Salisbury, Maryland would be four hours earlier, or 4:00 a.m. However, in December, if we examine another meteorological map stamped at 8:00 a.m. UTC or 8 Zulu, the local time would be 3:00 a.m. or the standard five hour difference described above.

Earth's Revolution (Orbit) and Seasonality

Earth revolves around (or orbits) the Sun in a time period of 365.242 days, or **tropical year**, with the fractional day responsible for a leap year every four years when we add February 29th to our calendar. If we viewed our solar system from above the Earth's North Pole, Earth would revolve around the Sun in a counter-clockwise direction and form an ellipse resulting in variations in Earth's distance to the Sun. At one point in orbit, on or near January 3rd, the distance between the Earth and Sun reaches its shortest distance, or **perihelion**, ~147,095,000 km (91,400,000 mi.), while on or near July 5th, the distance between the Earth and Sun reaches its maximum distance, or **aphelion**, at ~152,083,000 km (94,500,000 mi.) (Figure 2.27A). Therefore we can see that the distance varies some 5,000,000 km (3,000,000 mi.) in a year, however, this seemingly large value is only about 3% of the overall distance, meaning that our orbit is very nearly circular. Another important insight given the timing of perihelion and aphelion, especially for Northern Hemisphere residents, is that the Earth is closer to the Sun in January, one of the coldest months of our winter!

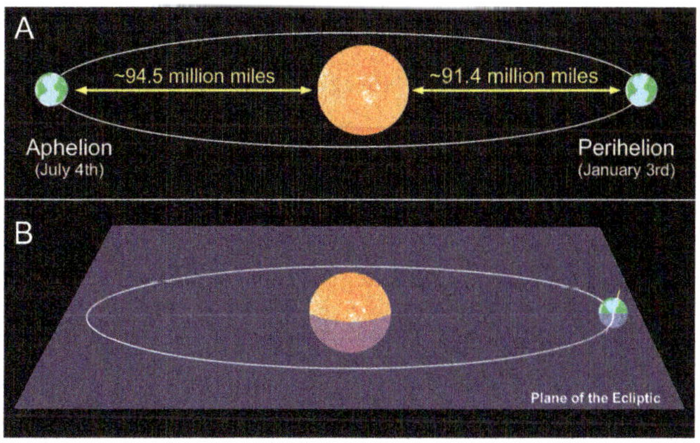

Figure 2.27: Describing the Earth's orbit around the Sun. A. The distance between the Earth and Sun varies during the year. B. The two-dimensional plane the Earth draws out around the Sun is called the plane of the ecliptic.

Obviously, seasonality is **NOT** caused by Earth's distance from the Sun! Seasonality is due to changes in Earth-Sun geometry as the Earth revolves around the Sun and the resultant changes in the amount and location of incoming solar radiation on Earth's surface. These changes in geometry occur because Earth's rotational axis is tilted in relation to the **plane of the ecliptic** (Figure 2.27B). The **plane of the ecliptic** is the two-dimensional plane in our solar system that is constructed by bisecting the Earth and Sun at every point in the Earth's revolution around the Sun. In short time spans, for example the span of a human lifetime, Earth's rotational axis maintains a consistent 23.5° tilt relative to the plane of the ecliptic (Figure 2.28). In longer time spans, for example 41,000 years, the angle of the rotational axis relative to the plane of the ecliptic actually varies between 22.1° and 24.5° degrees due perturbations of Earth's orbital plane caused by the gravity of our neighboring planets. The change in the angle of the Earth's axial tilt is termed **obliquity**. **Obliquity** is an important topic when examining long –term astronomical cycles, a component of Milankovitch cycles, which have been found to correlate with the onset of Earth's past ice ages, a topic discussed in further detail in Chapter 6. For our purposes in describing seasonality, we'll assume that Earth maintains an axial tilt of 23.5° (Figure 2.28).

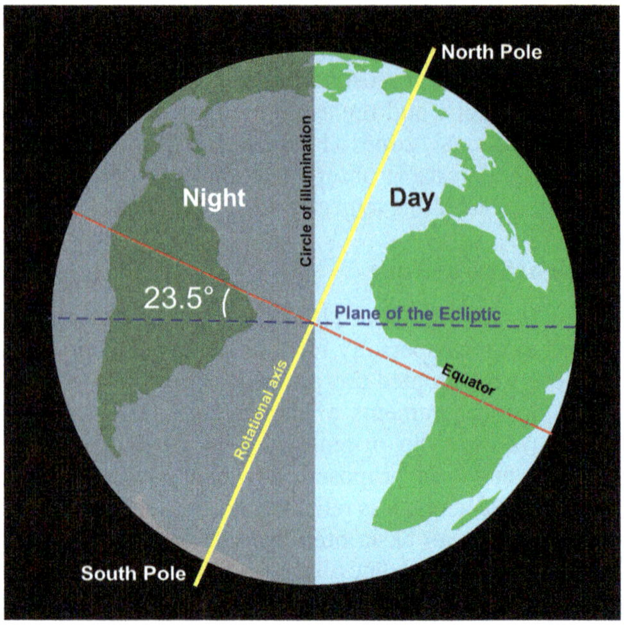

Figure 2.28: The Earth is tilted 23.5° relative to the plane of the ecliptic. This tilt contributes to the seasons.

Additionally, Earth's rotational axis maintains the same alignment relative to the background stars throughout its orbit, the reason we sometimes call Polaris the North Star given its celestial position above the Northern Hemisphere's rotational axis (Figure 2.29). This consistent orientation relative to the celestial background is termed **parallelism** and means that, as the Earth revolves around the Sun, the **subsolar point**, the location on Earth where the solar zenith is 90°, or the Sun is directly overhead, varies during the tropical year (Figure 2.30). The variation in the subsolar point is important because this location, a line of latitude also called **solar declination**, is the area on the

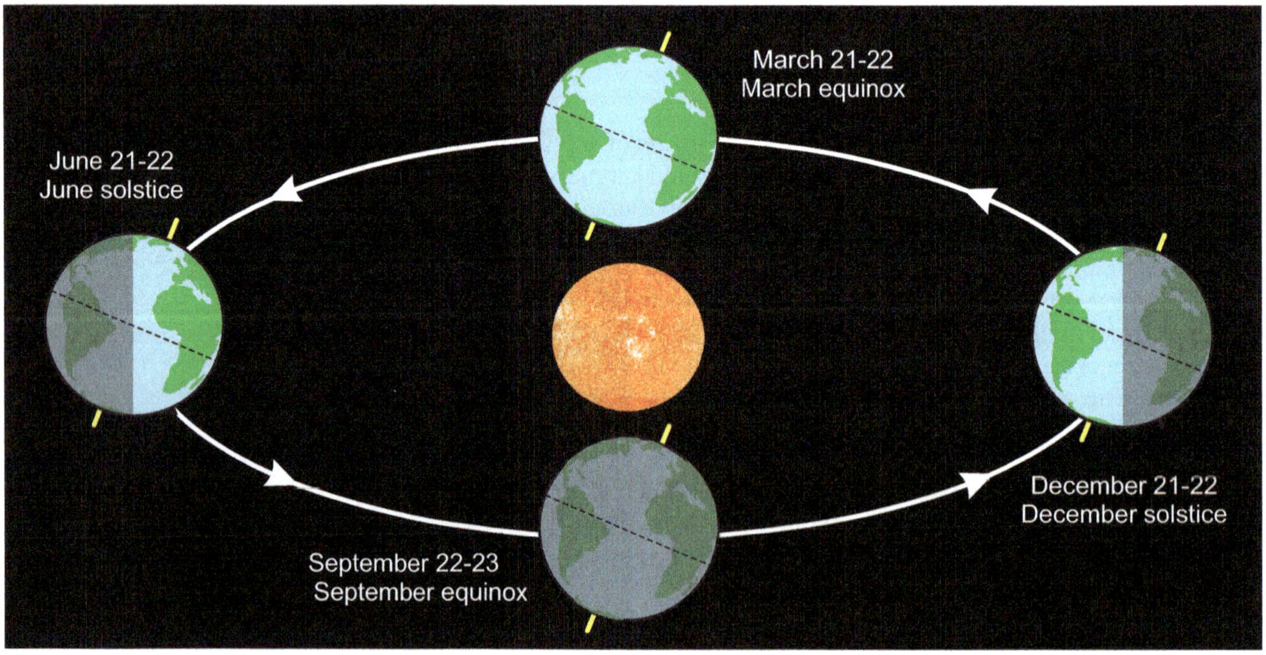

Figure 2.29: The revolution of the Earth around the Sun during the course of one year. Note how the Earth is always tilted in the same direction, pointing towards the North Star. This phenomenon is called parallelism.

Earth that receives the greatest intensity of incoming solar radiation. Because of parallelism and the 23.5° tilt, the subsolar point moves north and south on the Earth's surface between 23.5° N and 23.5° S throughout the course if the year. This is the reason that we have named these two lines of latitude, the Tropic of Cancer, 23.5° N and the Tropic of Capricorn, 23.5° S as they mark the maximum northern and southern extent of the subsolar point Figure 2.30). The daily movement of the subsolar point enables us to identify four important days in Earth-Sun geometry, and these dates are used to define our seasons given the relationship between the subsolar point and the intensity of incoming solar radiation.

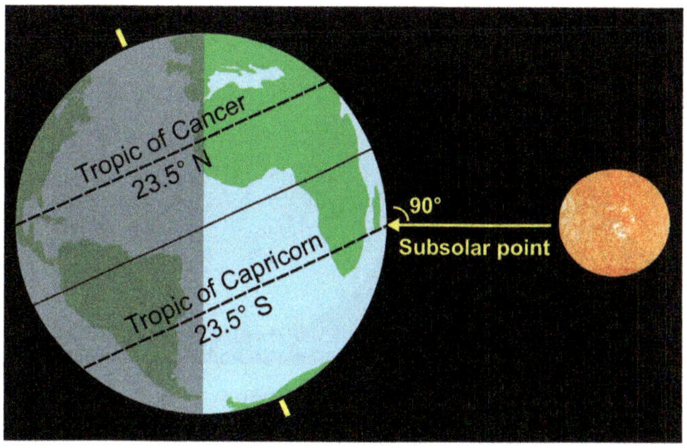

Figure 2.30: This diagram shows how there is only one latitude on any given day where the Sun's rays strike the Earth at a 90° angle. That place is known as the subsolar point.

Solstices and Equinoxes

On or near June 21st, the subsolar point reaches its maximum northern latitude, the Tropic of Cancer (23.5° N) and we call this date the **June** or **summer solstice** (Figure 2.31). Examine the figure closely and note that the circle of illumination extends from 66.5° S to 66.5° N due to the 23.5° tilt of the Earth's axis relative to the plane of the ecliptic. Also note that all locations between 66.5°N and 90° N are entirely within the daylight portion of the Earth's surface while all locations between 66.5°S and 90° S are within the nighttime section. As the Earth rotates on its axis on this date, areas north of 66.5° N, the Arctic Circle, experience 24 hours of daylight (i.e., the Sun never sets), while areas south of 66.5° S remain in darkness as the Sun never rises above the horizon. Therefore, we can see that Earth's tilt leads to variation in the number of daylight hours experienced in a 24 hour period by latitude (Table 2-1). Northern Hemisphere locations experience more daylight hours on (or near) June 21st than Southern Hemisphere locations thereby increasing the amount of solar radiation reaching the surface and leading to warmer temperatures overall. This is why we call this date the **summer solstice** in the Northern Hemisphere and use it to mark the beginning of our summer season.

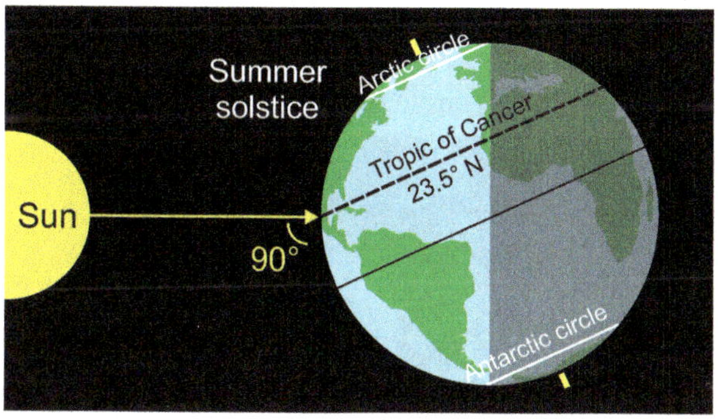

Figure 2.31: This diagram illustrates the Summer solstice.

Warmer temperatures experienced in most Northern Hemisphere locations on this date are also explained by the Sun's altitude above the horizon or its **angle of incidence**. On the June solstice, the angle of incidence reaches its maximum for all Northern Hemisphere locations. The Tropic of Cancer is equal to solar declination, therefore by definition, the angle of incidence at 23.5° N is 90° and the location that experiences the greatest intensity of solar radiation. We can easily calculate the angle of incidence for additional Northern Hemisphere locations (and really any location on Earth) if we know solar declination and the latitude of the location of interest. For example, to calculate the angle of incidence for Salisbury, Maryland (38°) on June 21st, we must first determine the total number of degrees that separate Salisbury from the latitude of solar declination. In this case, since both locations are in the Northern Hemisphere, we would subtract, 38° – 23.5° resulting in a 14.5° difference. To determine the angle of incidence, we would then subtract 14.5° from 90° and find that the altitude of the Sun on June 21st for Salisbury, Maryland is 75.5°. Compare this value to the angle of incidence for a city at 38° S, for example

	Equator	30°	40°	60°	North Pole
Feb	12:00	11:10	10:42	9:11	0:00
Mar	12:00	11:57	11:53	11:41	0:00
May	12:00	13:38	14:22	17:04	24:00
Jun	12:00	14:04	15:00	18:49	24:00
Jul	12:00	13:56	14:49	17:31	24:00
Sept	12:00	12:23	12:31	13:00	24:00
Oct	12:00	11:28	11:10	10:11	0:00
Dec	12:00	10:14	9:20	5:54	0:00

Table 2-1: The variation of the day length on the 15th day of select months throughout the year.

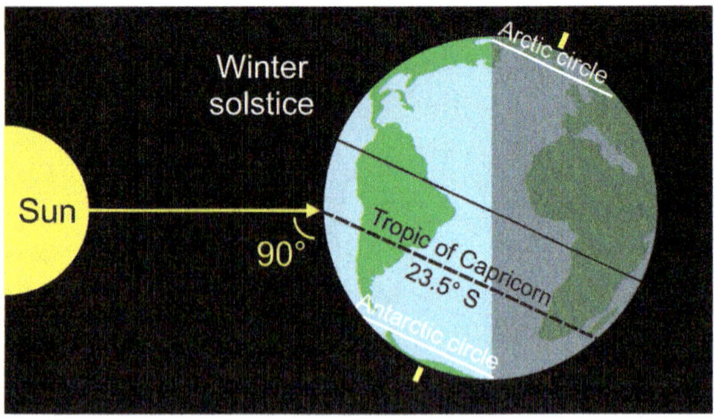

Figure 2.32 This diagram illustrates the Winter solstice.

Melbourne, Australia. Using the same steps, we would determine that the angle of incidence is 28.5° (90° - (38°+ 23.5°) = 28.5°), an indication of much lower solar radiation intensity for this Southern Hemisphere location on this date. Because lower intensity solar radiation and fewer daylight hours are received in the Southern Hemisphere locations, this hemisphere would be cooler than the Northern Hemisphere and recognize June 21st as the beginning of winter.

Six months later, on or near December 21st, the **December** or **winter solstice**, Earth-Sun geometry reverses as solar declination migrates to 23.5° S (Figure 2.32). The illuminated portion of Earth defined by the circle of illumination now encompasses all areas from the Antarctic Circle, 66.5° S, to the South Pole, 90° S, while locations north of the Arctic Circle (66.5°) remain in darkness. Angles of incidence for Southern Hemisphere locations reach their maximum and therefore so does the amount and intensity of solar radiation. The angles of incidence calculated for Salisbury, Maryland and Melbourne, Australia reverse, as does the amount and intensity of radiation for these two locations. Therefore, summer is welcomed south of the equator while the Northern Hemisphere begins the long, dark winter season.

Three months following the December and June solstices, on or near March 21 and September 23, the subsolar point migrates to a position directly above Earth's equator (Figure 2.33) and we call these dates in Earth's orbit, the **equinoxes. Equinox** or "equal night" is so named because if you examine the circle of illumination in the figure, you will notice that each line of latitude is bisected, meaning that there are 12 hours of darkness and 12 hours of daylight for all location on Earth. While this occurs each day for the equator (Figure 2.33, Table 2-1), and results in consistent climatic conditions for the region in comparison to the variability experienced in middle and high latitudes of both hemispheres, the equinoxes are the only days of the year in which this occurs for all locations on Earth. Note that the angle of incidence is at its maximum on the equator, 90° and decreases equivalently in both hemispheres with increasing latitude. In Salisbury, Maryland, our angle of incidence on the equinoxes is 52°. The equinoxes are used to define the beginning of spring and fall. In the Northern Hemisphere, March 21st is termed the **vernal equinox** or **spring equinox** while September 23rd marks the beginning of **fall** or the **autumnal equinox**. The equinox seasonal names are reversed for locations in the Southern Hemisphere.

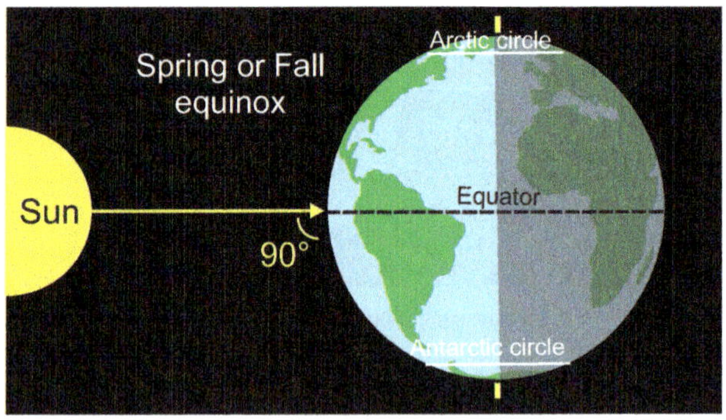

Figure 2.33: This diagram illustrates the spring and fall equinoxes.

An important reality when examining seasonal conditions on the Earth are the annual conditions experienced in the Polar Regions. While the Arctic experiences 24 hours of daylight in June, and the Antarctic experiences 24 hours of daylight in December, the angle of incidence for both locations when the Sun reaches its zenith is very low year round. The maximum solar altitude for both the North and South Pole only reaches 23.5°, an angle lower than the minimum value for Salisbury, Maryland, even though the Sun is above the horizon for six months! Moreover, both locations also experience six months of darkness. Therefore, annually, very little solar energy is received in the Polar Regions and as a consequence, year round temperatures are cold leading to a landscape dominated by snow and ice.

Annual March of Seasons

We can now examine the dynamic nature of Earth-Sun geometry for the entire year using the solstices and equinoxes as waypoints in what is termed the **annual march of the seasons**. Ancient civilizations were aware of annual variations in sun angles and day lengths and constructed observatories and monuments to mark the Sun's location on the solstices and equinoxes to monitor this annual march. Observations made with these structures enabled our ancestors to develop agricultural calendars to provide optimal planting and harvest times that ensured stable food supplies while city designers also utilized this knowledge to orient streets and buildings along important solar bearings. If we begin our year on March 21st, the vernal equinox, and observe changes for a mid-latitude Northern Hemisphere city, we will notice that as the subsolar point moves north from the equator, and daylight hours increase as does the Sun's altitude. Both continue to increase each day until they reach their maximum on June 21st, the summer solstice. Following the solstice, each begin to steadily decrease through the autumnal equinox on September 23rd as the subsolar point moves south. Daylight hours and the Sun's altitude reach their minimum on December 21st, the winter solstice. From this date onward, solar altitude and daylight hours increase through the vernal equinox and the cycle repeats again.

Figure 2.34: This simulated analemma shows the path of the sun in the sky during the course of one year at approximately 40° N latitude. Photo credit: B. Zaprowski.

The location of sunrise and sunset on the horizon also varies throughout the year and for our city on the vernal equinox, these events occur along an east-west orientation. As we march toward the summer solstice, the sun shifts on the horizon and begins to rise in the NE and set in the NW (Figure 2.34). This pattern continues through the autumnal equinox when sunrise and sunset returns to the east-west orientation. From this point in the year until the subsequent equinox, the sun rises in the SE and sets in the SW. An understanding of Earth-Sun orientations and their relationships to the cardinal compass directions enabled our ancestors to navigate the lands and seas and provides us with a built in compass when we are aware of these seasonal variations.

NOAA

Chapter 3 – The Structure and Behavior of the Atmosphere

The **atmosphere** is a very thin layer of gases that surrounds the planet. To put it into perspective, if the Earth was the size of an apple, the atmosphere would be thinner than the apple's skin. The atmosphere is closely linked to the ocean because it creates surface waves and drives the ocean's circulation. Weathering and erosion processes are also connected to the atmosphere since they rely on the atmosphere for water and heat.

Section 3-1: The Composition of the Atmosphere

The temperature and composition of today's atmosphere is much different than it used to be. As the Earth was initially forming gases were released from superheated rock material. Vast amounts of carbon dioxide and nitrogen were released into the atmosphere much like you would see today with an active volcano. The atmosphere's composition at this time would make its temperature near 85°C (185°F). Temperatures of only 60°C (140°F) would be enough to make all liquid water evaporate. The earlier atmosphere is estimated to have been made up of roughly 60 to 70 percent water vapor, 10 to 15 percent carbon dioxide and 8 to 10 percent nitrogen. Atmospheric scientists are concerned about levels of carbon dioxide in the atmosphere although today's concentrations are only measured in parts per million.

So how did the atmosphere change so drastically over time? A large portion of the water vapor would have condensed and formed raindrops as the temperature was cooling down. These raindrops would eventually accumulate in the oceans and reduce the concentration of water vapor in the atmosphere. Levels of carbon dioxide were lowered as it was absorbed by raindrops and slowly stored in the ocean and newly formed rock deposits. Photosynthesis also played a key role in changing the Earth's atmosphere (More about this in Chapter 12). Fossil evidence from ancient sediments indicates that the Earth supported life as far back as 3.5 billion years ago (Figure 3.1). Photosynthesis is valuable for our atmosphere today as it takes carbon dioxide out of the atmosphere and emits oxygen. There was a sharp increase in atmospheric oxygen levels between 1.9 and 2.2 billion years ago when photosynthetic organisms began to expand in number. Oxygen levels gradually continued to increase and carbon dioxide levels slowly decreased. Nitrogen levels also gradually increased during this time through a variety of natural processes.

Figure 3.1: The diorama at the Smithsonian Museum of Natural History depicts what the early Earth was probably like. Photo credit: B. Zaprowski.

Life could not exist on Earth without its current atmosphere. The atmosphere supplies most of the oxygen that aerobic organisms (animals, fungi, etc.) need to survive, provides carbon dioxide needed by photosythetic organisms (plants, algae, etc.), maintains an essential water supply and protects us from some of the sun's harmful rays. Nitrogen, oxygen and argon make up over 99.9% of the volume of dry air within the lower 16 kilometers (10 miles) of the atmosphere (Table 3-1). Although many assume that oxygen is the primary component of the atmosphere because it is critical to our survival, nitrogen is almost four times more abundant than oxygen. It is also interesting to note that the three most abundant components of the atmosphere have no significant influence on the weather. Water vapor determines the humidity of the atmosphere, is the source of all clouds and precipitation, and is a significant factor in the storage, movement and release of thermal energy. Water vapor is most abundant in air that overlies warm, moist surfaces and can account for up to about four percent of the total volume of the atmosphere. The air can have less than one percent water vapor in extremely dry locations such as deserts and Polar Regions.

Gas		Volume (%)
Nitrogen	N_2	78.08
Oxygen	O_2	20.95
Argon	Ar	0.93
Water Vapor	H_2O	0.01-4.0
Carbon Dioxide	CO_2	0.0375
Methane	CH_4	0.0002
Ozone	O_3	trace

Table 3-1: Important components of the atmosphere. Source: NOAA.

Carbon dioxide is found in small concentrations everywhere but varies both seasonally and geographically. Levels of some gases in the atmosphere, such as surface ozone, carbon monoxide and sulfur dioxide are primarily due to human activities. These gases are much more abundant in the Northern Hemisphere, where about ninety percent of the human population lives, and are found in significant quantities in industrialized and urbanized areas. The atmosphere is made up of more than just gases. Small solid particles, cloud drops or ice crystals suspended within the air are called **aerosols**. Aerosol concentrations influence precipitation because they act as condensation nuclei on which larger cloud droplets can form. Aerosols absorb and scatter light which can reduce local visibility and the amount of sunlight that reaches the surface. These solid particles come from natural processes and human activities and usually only spend a few days to a few weeks in the atmosphere.

Section 3-2: Atmospheric Pressure

Atmospheric pressure can simply be thought of as the weight of the overlying air. Air pressure amounts to about 14.7 lb/in^2 at sea level; we don't feel this pressure because of our body's internal pressure. A **barometer** is a device used to measure atmospheric pressure (Figure 3.2). Atmospheric pressure has been measured for hundreds of years since the barometer was invented around 1643. Atmospheric pressure is considered one of the most difficult weather elements to sense without the use of weather instruments. Meteorologists use **inches or mercury** or **millibars (mb)** to measure atmospheric pressure. The term "inches of mercury" refers to how high a column of mercury will rise in a vacuum. Mercury is used to make the device more compact since it is more than thirteen times denser than water. Aneroid barometers are also used to measure atmospheric pressure. These circular shaped instruments are commonly seen in households and do not make use of a liquid. Most meteorologists use millibars to describe atmospheric pressure and create maps of sea-level pressure. For example, at sea-level 29.92 inches of mercury = 1013.25 millibars.

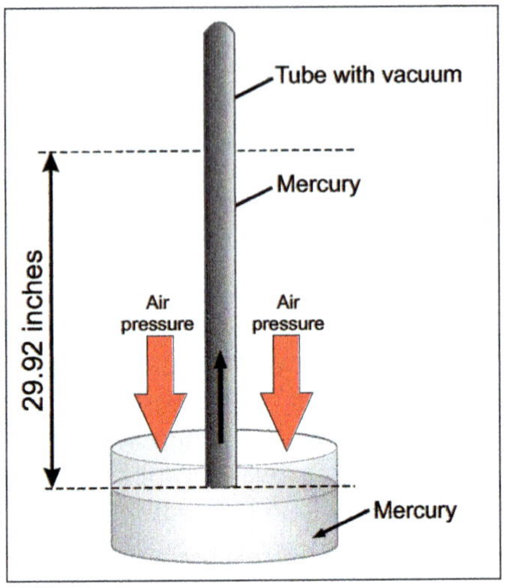

Figure 3.2: A simple mercury barometer. At sea level, mercury will rise an average of 29.92 inches.

The atmosphere is a compressible gas, much like a propane or oxygen tank. Held in place by Earth's gravity, atmospheric pressure decreases rapidly as you increase in elevation (Figure 3.3). Sometimes your ears pop when flying in an airplane or taking an elevator to the top floor of a tall building. This is caused by a rapid change in pressure due to the lower layers of the atmosphere being highly compressed by the weight of the overlying air. As a result, half of the mass of the atmosphere is found in the lowest 5.6 km (3.5 mi.). The atmosphere's oxygen levels also decrease with height, this why high altitude climbers have to set up base camps in order to allow their bodies' time to adjust to decreased levels of oxygen. Conditions are so severe at the top of Mount Everest, the Earth's highest elevation, that there is only about a two week period each year when most climbers attempt to climb to the summit.

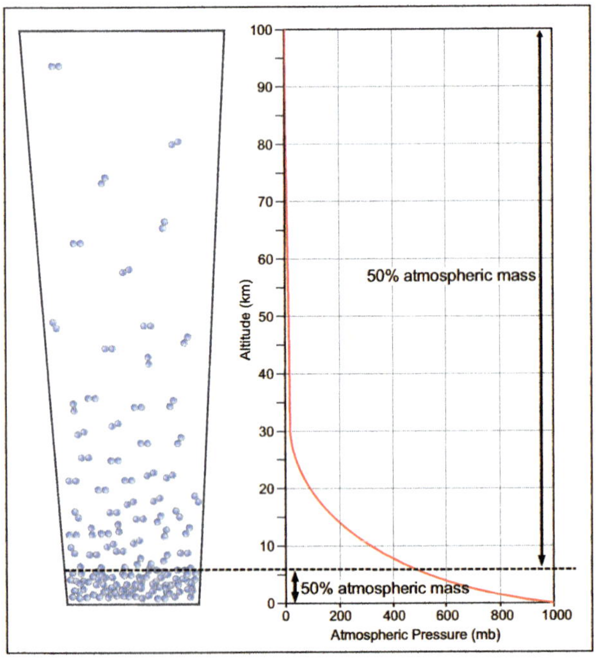

Figure 3.3: The change in atmospheric pressure with altitude. About 50% of the atmosphere's mass lies below ~5.6 km (~3.5 mi.) altitude. Also note how a column of air decreases in volume as you go towards the earth's surface.

Section 3-3: Vertical Temperature Structure of the Atmosphere

Most of us have experienced changes in temperature due to increased elevation. It used to be assumed that temperature decreased with altitude through the entire atmosphere, we know now that this is not the case. Distinct layers of the atmosphere exist due to changes in temperature with altitude. The lowest layer of the atmosphere is called the **troposphere**. Temperatures decrease with height in this layer due to the atmosphere being heated from the Earth's surface. The troposphere comprises roughly eighty percent of the mass of the atmosphere, contains practically all of the atmospheric water vapor and is where most all of our weather occurs. The troposphere is elliptical in shape and thins towards the poles due to differences in density (Figure 3.4). The thickness of the troposphere is a good example of the relationship between temperature and density. The troposphere is thicker near the equator because the atmosphere is warmer and therefore less dense. The Earth's surface is heated by incoming shortwave solar radiation received from the Sun. This energy travels through the atmosphere like a pane of glass and is absorbed by the Earth's surface. The troposphere is an unstable layer because it is heated from the ground up; the atmosphere would be very stable if it was heated from the top down. How energy is absorbed at the surface is often compared to a greenhouse (Figure 3.5). Solar energy from the Sun reaches the surface as short-wave radiation due to the Sun's high temperature. The atmosphere transmits most of this incoming short-wave radiation which is then absorbed by the Earth's surface. The Earth's surface (being much cooler than the surface of the sun) then re-readiates this absorbed energy as long-wave radiation. A number of gases, called **greenhouse gases**, in the atmosphere (CO_2, water vapor and other trace gases) absorb the vast majority of long-wave radiation that is emitted upward from the surface. The **greenhouse effect** keeps the Earth warm by absorbing this radiation emitted from the surface and then re-radiates it downward to warm the lower atmosphere. Similar to how a greenhouse was initially believed to work, the short-wave radiation easily passes through the glass while the long-wave radiation can't, resulting in an increase in temperature. It is important to understand that the greenhouse effect is a natural process and life would not be possible without it. If the atmosphere had none of the greenhouse gases that absorb outgoing long-wave radiation the Earth would be much colder and uninhabitable. The Earth's current average temperature is approximately 15°C (59°F), without the greenhouse effect the average temperature would be so low that all water would freeze. Although the greenhouse effect is a natural process human activities are enhancing it by adding more greenhouse gases to the atmosphere.

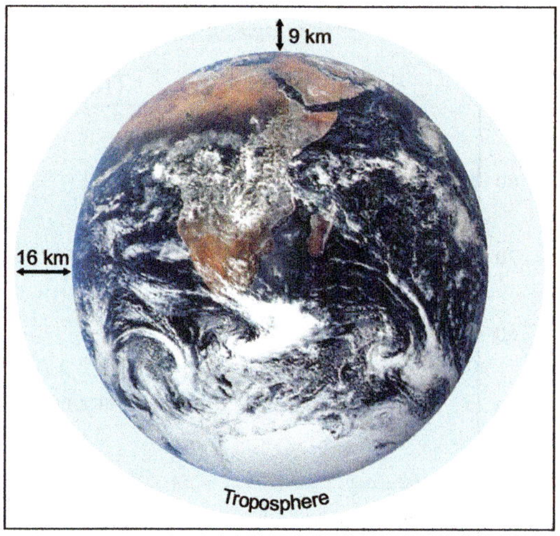

Figure 3.4: The troposphere is not evenly distributed around the Earth but thins towards the poles. This occurs because the air near the poles is colder, and therefore, occupies less volume. Photo credit: NASA/JSC.

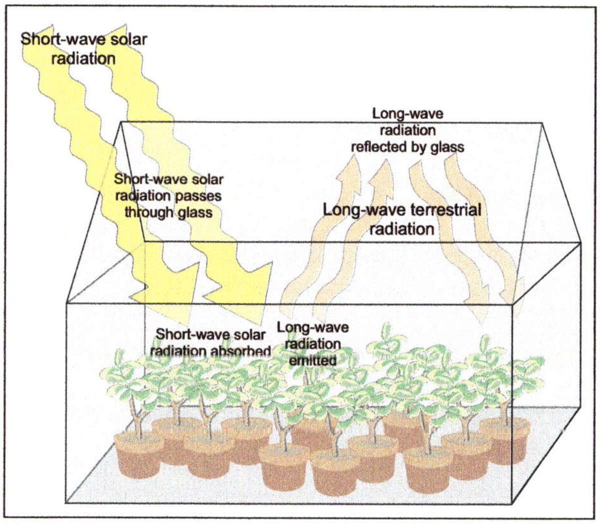

Figure 3.5: The earth's atmosphere acts much like a greenhouse. Short-wave solar radiation passes through the glass and is absorbed by objects in the greenhouse (like the plants). Long-wave radiation is emitted but cannot easily pass through the glass. Instead, the energy is absorbed by the air in the greenhouse, heating it up.

Temperatures in the troposphere are usually highest at the surface and decrease with increasing elevation. The lower atmosphere is warmest because it is closest to its heat source, the Earth's surface. This is similar to how you warm up standing next to a fire, how much you warm up depends on how close you are to the fire. Although the rate of temperature change with altitude varies across time and space, the long-term global average is approximately 6.5°C/1,000 m (3.6°F/1,000 ft.). This average rate of temperature change is called the **environmental lapse rate** (Figure 3.6). The actual environmental lapse rate for a specific area is dependent on atmospheric conditions. The relationship between temperature and elevation has a significant effect on the temperature of a given location. There are several countries in tropical latitudes that have much lower temperatures than expected due to their elevation. The lapse rate can be increased due to strong surface heating under clear and sunny conditions. Clear skies and calm winds at night can quickly decrease the lapse rate by reducing surface temperatures. A temperature inversion exists when temperatures increase with height and can form through intense surface cooling or a warm layer of air above the surface. Temperature inversions can create problems with air quality, especially in large cities where high amounts of automobile and industrial emissions

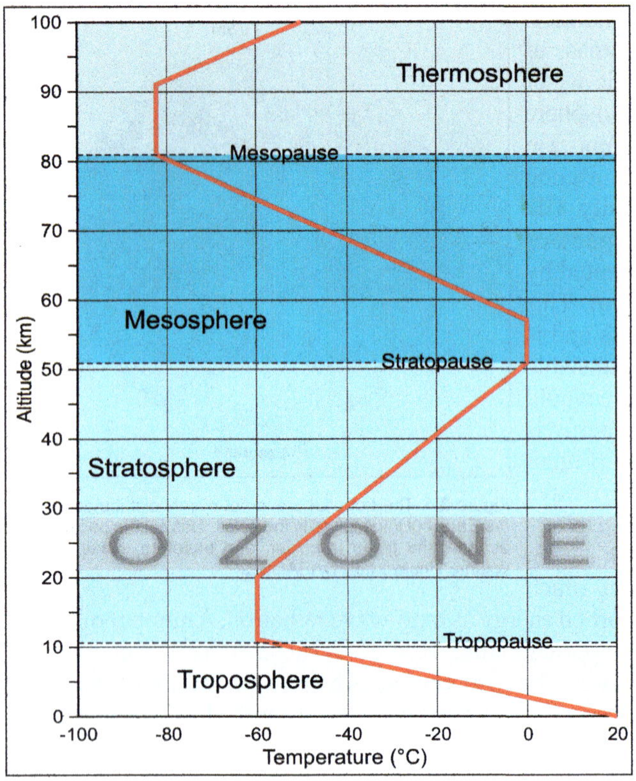

Figure 3.6: The change in atmospheric temperature with altitude. The four major layers of the atmosphere are defined based on the temperature profile.

exist. The warm air above the surface prevents the pollutants from rising through the atmosphere and traps them near the ground. Local topography can also play a role in the formation of temperature inversions. The **tropopause** defines the upper limit of the troposphere and is identified as the level of the lower atmosphere where temperatures stop decreasing with height. The average temperature of the tropopause is about -60°C (-76°F).

The **stratosphere** is the atmospheric layer directly above the troposphere (Figure 3.6). Temperatures remain consistent at about -59°C (-74°F) in the lower stratosphere before they begin to increase. Temperatures gradually rise to about 0°C (32°F) at the **stratopause**, the upper limit of the stratosphere. The stratopause is found just over 50 km (30 mi.) above the Earth's surface. The stratosphere is important because it contains the **ozone layer**. Concentration of stratospheric ozone are small but extremely important because they absorb harmful ultraviolet radiation. Stratospheric ozone (O_3) is formed naturally through chemical reactions between ultraviolet radiation and molecular oxygen (O_2). The ultraviolet radiation from the sun splits the molecular oxygen into free oxygen atoms. Atomic oxygen (O) is unstable and combines with the remaining molecular oxygen (O_2) to form ozone (O_3). Temperatures increase with height in the stratosphere because ozone adds heat to the atmosphere through its ability to absorb ultraviolet (UV) radiation. The concentration of ozone is greatest around 25 km (~15 mi.) which explains why the temperatures are warmer here than in the upper troposphere.

The atmospheric layer above the stratosphere is known as the **mesosphere**. The mesosphere extends outward to about 80 km (50 mi.). Temperatures again decrease with height as they do in the troposphere. The mesosphere is a very cold atmospheric layer because it doesn't have a heat source. The coldest temperatures are typically found at the **mesopause**, the transition zone at the top of the mesosphere, where temperatures are less than -80°C (-112°F).

The **thermosphere** is found beyond the mesosphere and is the highest recognized thermal layer of the atmosphere. Temperatures remain relatively uniform for several kilometers and then increase continually with height as the thermosphere absorbs solar energy. The heat source for the thermosphere is from charged particles ejected from the sun. Temperatures can increase in this layer if solar activity is higher than usual.

The Ozone Layer

There has been a net increase of several trace gases since the Industrial Revolution began in the late 1700s. The concentrations of these gases is very low, but they are very important. The ozone layer absorbs harmful ultraviolet radiation and prevents it from reaching the Earth's surface. The ozone layer is critical to our health and survival; it would be very difficult for anything to survive on the Earth's surface without it. Most atmospheric scientists agree that several anthropogenic gases being transported into the stratosphere release ozone destroying chlorine (or bromine). Human activities have reduced the concentrations of stratospheric ozone and created an "ozone hole" over Antarctica.

The ozone layer has been getting thinner since the 1960s, when chlorofluorocarbons (CFCs) came into major use as refrigeration and cooling agents. **CFCs** are a combination of chlorine (Cl_2), fluorine (F_2) and carbon (C). This human-made substance was once thought to be harmless and was used in a variety of applications. Practically every household used spray cans that employed CFCs as a propellant during the 1970s. Researchers at the University of California determined in 1974 that intense ultraviolet radiation can split CFCs up into atoms. This discovery led the researchers to be concerned about the health of the ozone layer as CFCs had been steadily accumulating in the atmosphere. If the CFCs are carried into the stratosphere they can destroy the ozone layer where ultraviolet radiation is intense. Under these conditions, the freed CFC atoms, especially the chlorine atom (Cl), can undergo a series of chemical reactions with ozone (O_3) to produce molecular oxygen (O_2) (Figure 3.7). This sequence of reactions decreases the amount of ozone present and also decreases the number of free oxygen atoms available to form more ozone. The CFCs are extremely harmful to the ozone layer due to their ability to continually and repeatedly destroy ozone molecules. Although CFCs were banned from aerosol spray cans in the United States in 1978 they will continue to be a problem for several decades as they persist in the atmosphere. Ozone levels measured near the South Pole in 1987 were found to be almost 50% less than their 1979 levels (Figure 3.8). These results provided strong evidence of the impact that CFCs can have on the ozone layer. The hole in the ozone layer over Antarctica is the result of a unique combination of sunlight, cold temperatures and strong winds. It is estimated that there is a 2% increase in ultraviolet radiation that reaches the Earth's surface for every 1% decrease in stratospheric ozone. There are approximately 500,000 new reports of skin cancer each year in the United States alone. Reports of skin cancer, cataracts and threats to unprotected organisms are expected to increase as more harmful ultraviolet radiation reaches the surface. Ozone depletion is also known to cause many materials to degrade faster, especially plastics and rubber. The good news is the ozone layer is projected to recover to 1980 benchmark levels by 2050 thanks to global efforts to combat this serious issue.

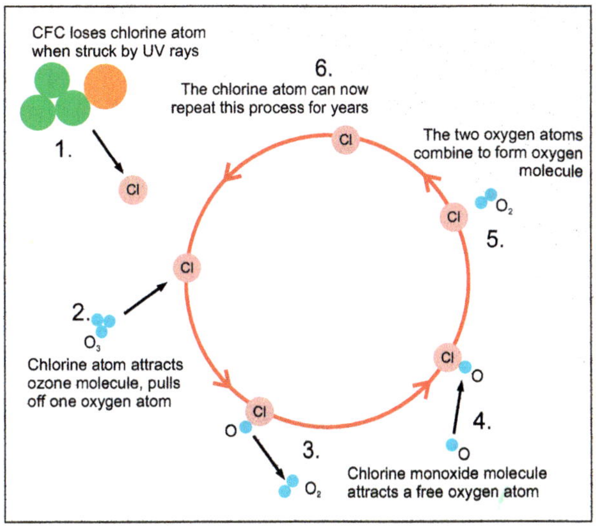

Figure 3.7: This diagram shows how CFC's effectively and repeatedly can destroy ozone molecules.

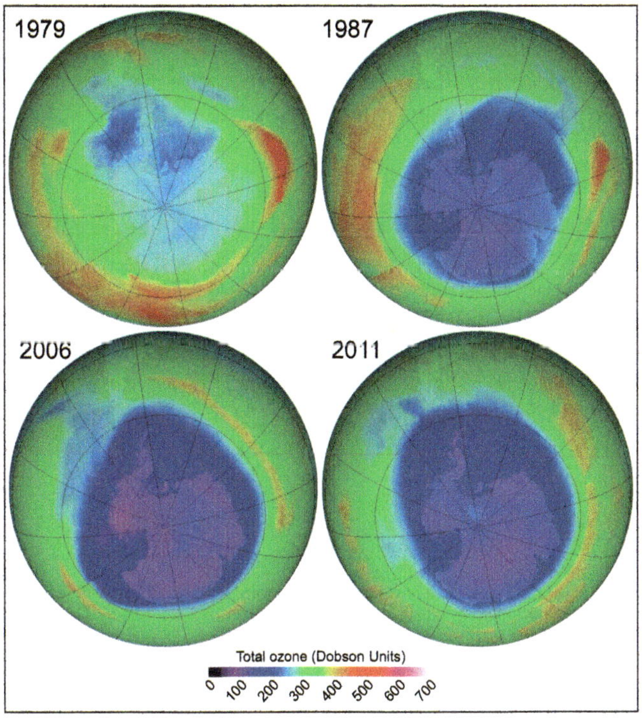

Figure 3.8: The images above show the Antarctic ozone hole on September 16 in the years 1979, 1987, 2006, and 2011. The first two maps are based on data from the Total Ozone Mapping Spectrometer (TOMS) on the Nimbus-7 satellite. The other two maps are made with data from the Ozone Monitoring Instrument on the Aura satellite. Image credit: Robert Simmons/NASA.

Section 3-4: The Solar Energy Budget: Transferring Heat Around the Globe

There is a complex set of processes between the atmosphere and the Earth's surface that occurs as solar radiation passes through the atmosphere. Not all insolation stays in the atmosphere; some is reflected off the atmosphere and bounces back to space. The Earth intercepts only a small fraction, less than one two-billionth, of the Sun's total energy output. The **solar constant** refers to the fairly constant supply of solar radiation received at the top of the atmosphere. The solar constant is equal to about 1,370 Watts per square meter (W/m^2), similar to about fourteen 100-watt light bulbs shining directly over a 1 m^2 tabletop. This tiny portion of solar energy heats the Earth's surface and atmosphere enough to make life on our planet possible.

The annual balance between incoming and outgoing radiation in the solar energy budget is usually illustrated by 100 units that represent the total insolation received at the outer edge of the atmosphere. These 100 units are used so the distribution of solar energy can be expressed as percentages (Figure 3.9). Most of the insolation that

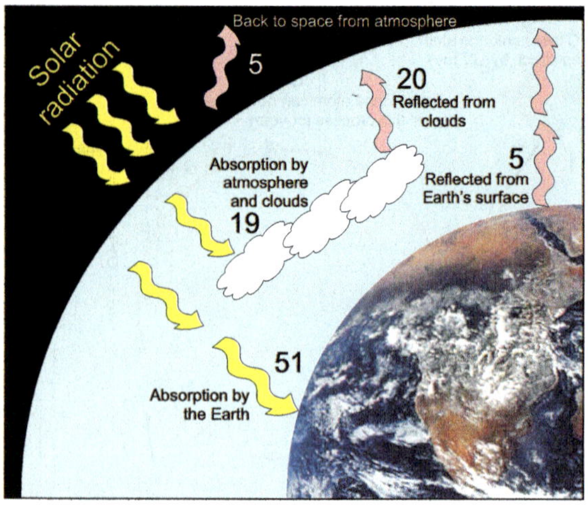

Figure 3.9: The distribution of solar energy.

enters the atmosphere does not heat it directly, 30 of these 100 units are reflected back into space, primarily by air molecules and clouds. As discussed in Chapter 1, **albedo** is the ability of an object to reflect radiation; in this case it refers to the amount of solar radiation that the Earth reflects back into space. The Earth and atmosphere combined reflect 30 units of energy back to space and therefore the Earth's planetary albedo equals 30 percent. The albedo of an object is strongly a function of its color; lighter colors reflect more energy and absorb less as compared to darker colored objects. You may notice how bright it is outside after a recent snowfall due to the snow's highly reflective nature. This is also the reason why dark surfaces like an asphalt driveway are much warmer than surrounding objects due to absorbing much more energy than it reflects. The Earth's planetary albedo of 30% means that there are only 70 of the 100 units of energy available to heat the Earth and atmosphere. Of the 70 units remaining only 19 units are absorbed in the atmosphere by clouds and air molecules. The atmosphere must also be heated from another source if only 19 percent of the solar radiation that reaches the top of the atmosphere is used to heat the air directly. There are 51 units of energy that pass through the atmosphere and are absorbed by the Earth. You can see from Figure 3.9 that the atmosphere doesn't absorb as much solar radiation as the Earth's surface. The solar radiation absorbed by the ground is the primary heat source for the atmosphere as the ground re-radiates this energy as long-wave radiation and almost all of it is absorbed by the atmosphere. This shows that the atmosphere is mostly indirectly heated by the Sun; the Sun heats the ground and the ground heats the air. The atmosphere absorbs the Earth's long-wave radiation much more effectively than solar radiation, primarily due to the presence of water vapor and carbon dioxide. Both of these greenhouse gases are effective absorbers of long-wave radiation.

A radiation deficit develops at the Earth's surface when the ground cools at night. The lower atmosphere cools after the sun sets as the ground emits more radiation than it absorbs. It is important to understand that the solar energy budget described above is a global average, local conditions such as cloud cover and ground conditions can greatly affect the distribution of solar energy and therefore influence local temperature patterns. Besides radiation, there are other ways for energy to be transferred between the Earth's surface and the overlying atmosphere.

Conduction

Energy can be transferred between the ground and the overlying air through **conduction**. The atmosphere is a much better insulator than a conductor of energy. As a result, conduction between the ground and the overlying air occurs slowly. The ground usually warms up quickly in the morning as it begins to absorb solar radiation. The air in contact with the ground is heated but can be dramatically cooler than the ground. Conduction only heats a relatively thin layer of air in contact with the ground. This explains why a runner on a hot summer afternoon may experience temperatures over 55°C (~130°F) at their feet and only around 29°C (85°F) at eye level. Be careful not to burn your feet through conduction if you walk across hot sand at the beach on a hot summer afternoon. Once the sun sets the ground begins to cool as it loses more radiation than it gains. Dramatic cooling of the surface can develop when the

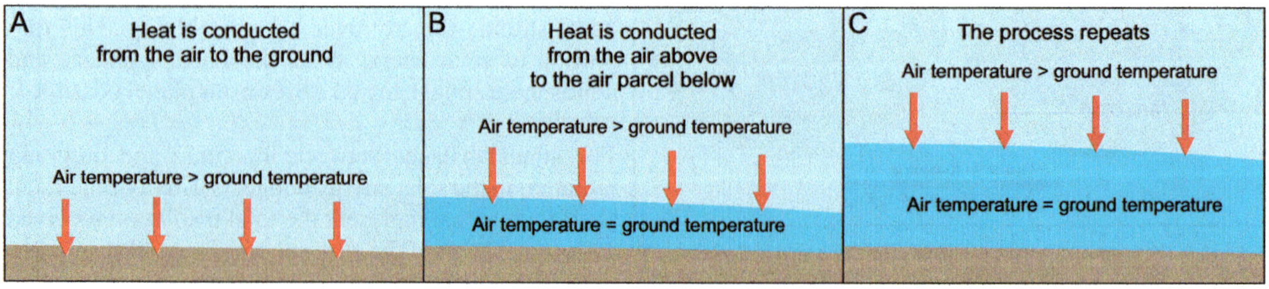

Figure 3.10: Cooling of the night air via conduction. A. Heat is transferred from the air to the ground below. B. Once the air temperature above the ground is equal to the ground temperature, heat is drawn from the air above, causing more cooling. C. Over time, the parcel of cooled air grows in size.

skies are clear and the wind is calm (Figure 3.10). The overlying air continues to cool and thicken through conduction under these conditions. This process continues through the overnight hours and often leads to cool mornings well after sunrise. As we will see, conduction is also important to the development of fog, which can have a significant impact on transportation.

Convection

Convection is another way to transfer energy between the Earth's surface and atmosphere. Convection transfers energy away from the Earth's surface through the vertical movement of air. It is often thought of as the opposite of conduction since it requires molecular movement. You are taking advantage of heat convection if you have ever warmed your hands by holding them over a campfire. Warm air can rise, just like a hot air balloon, from the ground as heat causes the air to expand and become less dense. The Earth's surface can warm significantly on a hot summer day due to a long day length and a strong sun angle. Uneven heating of the ground causes some locations to be warmer than others (Figure 3.11). You can often notice this effect by looking at the current air temperature as

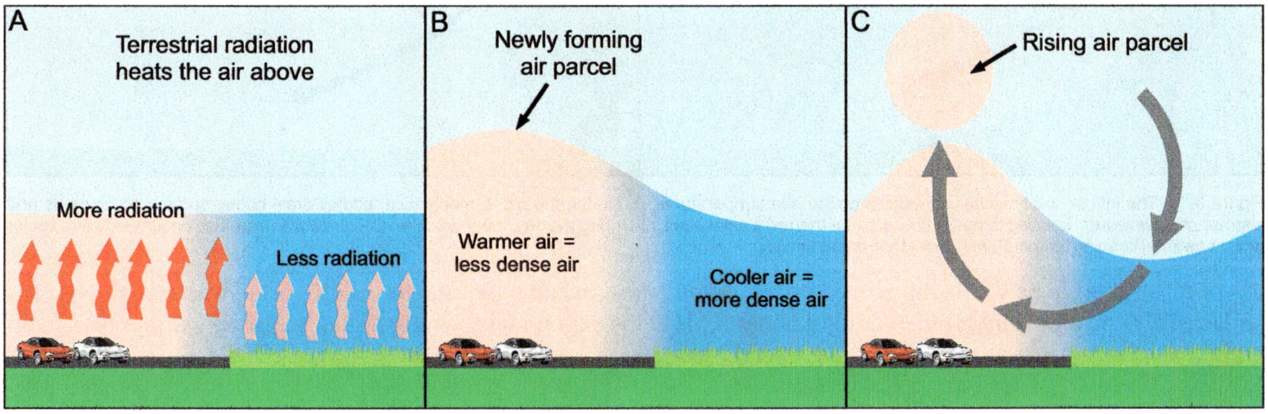

Figure 3.11: Thermally-driven convection. A. Terrestrial radiation heats the layer of air just above the Earth's surface. The heating is not even, however, as different surfaces emit different amounts of radiation. B. As the air warms, it becomes less dense and begins to rise. C. As this warm air continues to rise, it forms air parcels. These air parcels carry heat upwards while cooler air is transported downwards.

reported by a bank thermometer placed over a parking lot. Since temperature is the primary controller of air density near the Earth's surface, isolated volumes of air called **air parcels** transport heat energy upward as they become warm and less dense. Soaring birds and hang gliders are often seen using these thermal convection currents as a source of lift. Convection is very important for cloud formation and is effective at redistributing heat. Intense surface heating can often form convective clouds that produce brief bursts of rainfall from afternoon thunderstorms.

Eddies and Mechanical Convection

A different form of convection can develop when wind blows over a rough surface. **Eddies** are circular movements of air that mix the lower atmosphere and help minimize large temperature gradients near the ground. Eddies are a different form of convection because it develops from differences in wind speed, not surface heating. This type of convection is often referred to as **mechanical convection** due to the way the wind creates a rolling motion in the atmosphere (Figure 3.12). You can relate this to placing a pencil in the palm of your hand and moving your top hand over the pencil. The top hand is similar to the faster wind speeds above the surface that create the horizontal rotation. Wind speeds increase with height as we move up from the surface due to a decrease in friction. As a result, eddies can often have a large influence on surface temperatures during both the day and night (Figure 3.13). Temperatures can be cooler than expected on a hot afternoon if conditions are windy. Eddies reduce surface temperatures by transporting cooler air to the surface and warmer air upwards during the

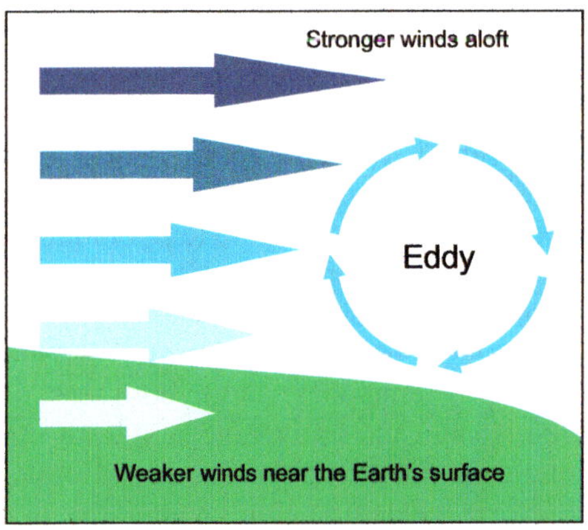

Figure 3.12: An idealized mechanical eddy. Mechanical eddies form because winds blow stronger aloft than they do near the Earth's surface.

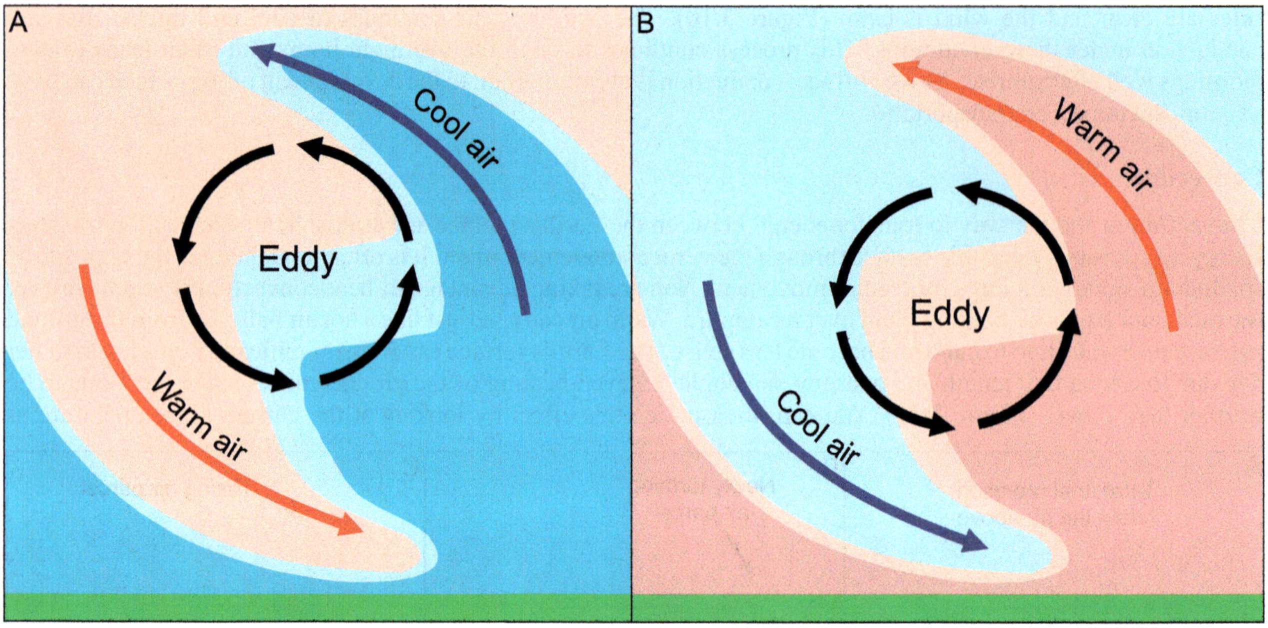

Figure 3.13: The influence of mechanical eddies on surface temperatures. A. During the night, mechanical eddies draw cooler surface air upwards and warmer air downwards, keeping temperatures warmer than on a calm night. B. During the day, mechanical eddies draw warm surface air upwards and cooler air downwards, keeping temperatures cooler than on a calm day.

daytime, this cooling effect is influenced by the air temperature and wind speeds. Eddies bring warm air downward and cool air upwards during the overnight hours. This helps explain why windy nights are often warmer than calm nights.

Evapotranspiration

Heat can also be transferred from the Earth's surface to the atmosphere in a process known as **evapotranspiration**. Evapotranspiration refers to the combined influence of evaporation and transpiration. Transpiration is how water is returned to the atmosphere from the stems and leaves of vegetation through evaporation. Since it takes energy to evaporate water, the energy is moved from the surface to the atmosphere. This cooling process helps cool plants and moves water through their root systems. It is estimated that a large oak tree can transpire around 200 gallons of water to the atmosphere in a single day! You can often feel the difference in air temperature when you drive through a wooded area like a State Park. As you would expect, massive amounts of heat are transferred from the open oceans to the atmosphere through the evaporation of water.

Section 3-5: Global Temperature Patterns

Average surface temperatures for January and July 2000-2014 are shown in Figure 3.14. You can identify which hemisphere is facing towards the Sun by locating the warmest temperatures. Note how the temperature bands are generally aligned with latitude. It is clearly summertime in the Southern Hemisphere in January as the warmest temperatures are found south of the equator. Remember that the Southern Hemisphere faces the Sun in January and therefore is warmer due to a stronger sun angle and longer day length. Latitude is the most basic factor causing the distribution of hot and cold temperatures over the Earth's surface. Due to the curvature of the Earth, solar radiation will always be more direct over the tropical latitudes compared to the Polar Regions. Temperatures are much warmer in the Northern Hemisphere in July as the direct rays of the Sun are concentrated north of the equator. Notice in Figure 3.14 how the temperature pattern is much more complicated in the Northern Hemisphere compared to the Southern Hemisphere during both seasons. This is because 68% of the Earth's landmass is located in the Northern Hemisphere and on average land heats up and cools down three times faster than water. The temperature pattern is much smoother over the Southern Hemisphere because it contains only 32% of the Earth's landmass, resulting in more gradual temperature changes throughout the year.

Weather is defined as the state of the atmosphere at a given time and place. The contrast in temperatures between the

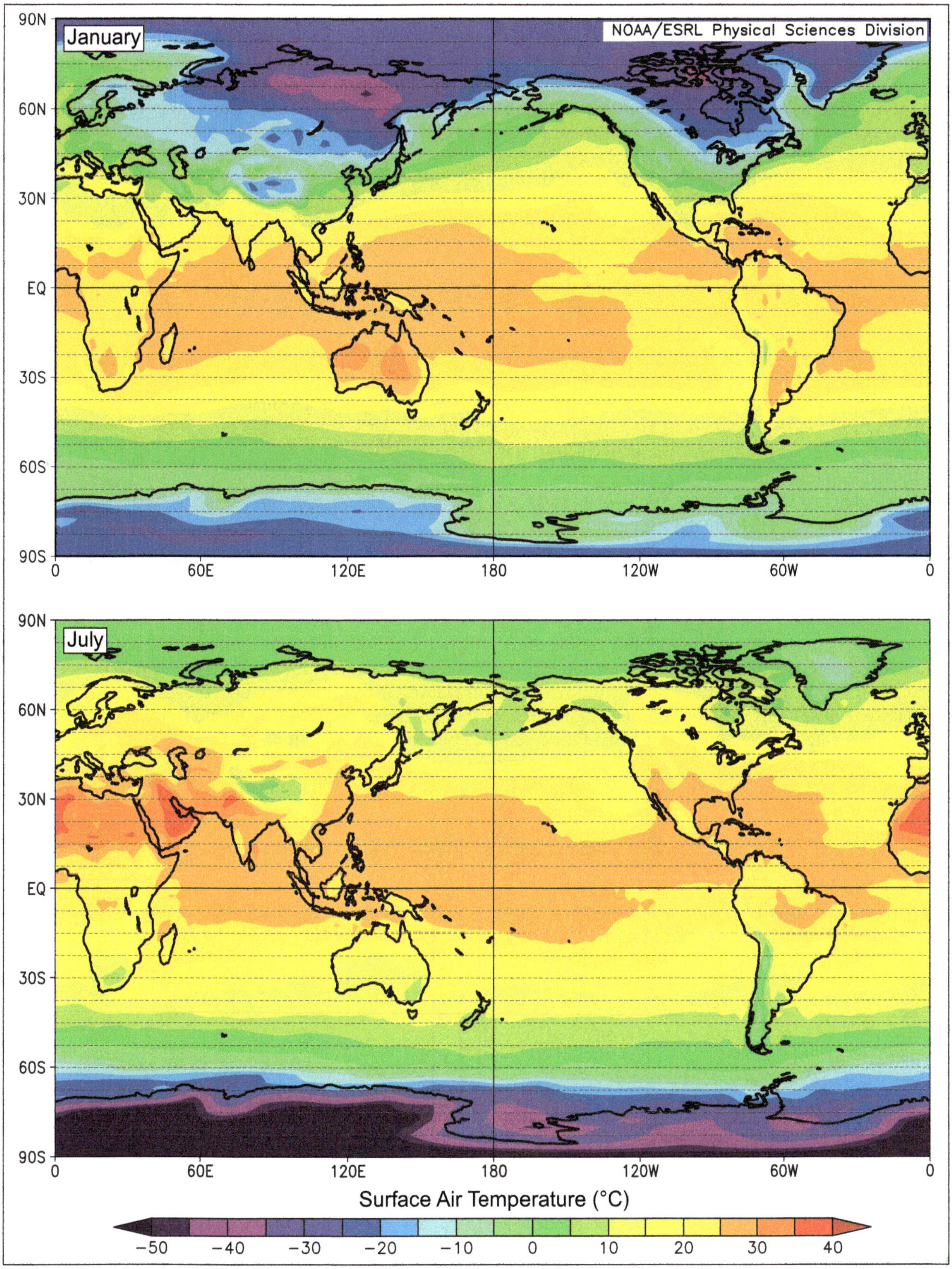

Figure 3.14: Worldwide average monthly surface temperatures for January and July, 2014. Image credits: NOAA/ESRL.

Equatorial and Polar Regions exists due to uneven heating of the Earth's surface. Remember we also have a decrease in temperature with height in the troposphere because it is heated from the ground up, making this an unstable layer of air. We have active weather patterns because the atmosphere always strives to minimize temperature and pressure differences on a global scale. Clashes between warm and cold air often lead to clouds and precipitation. As we will see, storm systems are very effective at redistributing heat and moisture. There are factors other than latitude and weather systems that control temperature variations, they are discussed next.

Difference in Heating Between Land and Water

Water bodies have a large influence on local temperature and rainfall patterns due to water's high heat capacity compared to land. Considering that 71% of the Earth's surface is covered by the oceans, coastal locations have a more moderate climate throughout the year (Figure 3.15). This maritime effect results in a lower range of surface air temperatures over both daily and annual time scales. This effect also occurs for areas near the Great Lakes and other large water bodies. Because the oceans warm and cool much slower than land, their warmest temperatures are well after the summer solstice. The warmest ocean temperatures along the East Coast of the United States are usually in late summer or early fall even though the sun's rays are the strongest in late June. Conversely, the coolest water temperatures are generally found in February, several weeks past the winter solstice.

Figure 3.15: Areas of land near large bodies of water have a moderate climate relative to areas further inland.

Figure 3.16A compares average monthly temperatures for two locations at similar latitudes but at very different continental positions. The temperatures are much more moderate throughout the year in San Francisco, CA due to its coastal location. For its latitude many would expect temperatures to be much warmer in the summer and cooler in the winter. Kansas City, MO experiences more extreme temperatures because it is located in the interior of the contiguous United States. The water temperature can also play a role in the climate of a coastal location. Figure 3.16B compares the average monthly temperatures for Charleston, SC and San Diego, CA. You can see that temperatures are warmer in Charleston, SC than San Diego, CA for eight months of the year. Charleston, SC is located near a warm ocean current that transports heat from the equator to higher latitudes. As a result, the air is warm and humid for most of the year. Temperatures in San Diego, CA are more moderate because it is located next to a cold ocean current that flows from high latitudes to the tropics. Since the winds are what drive surface ocean

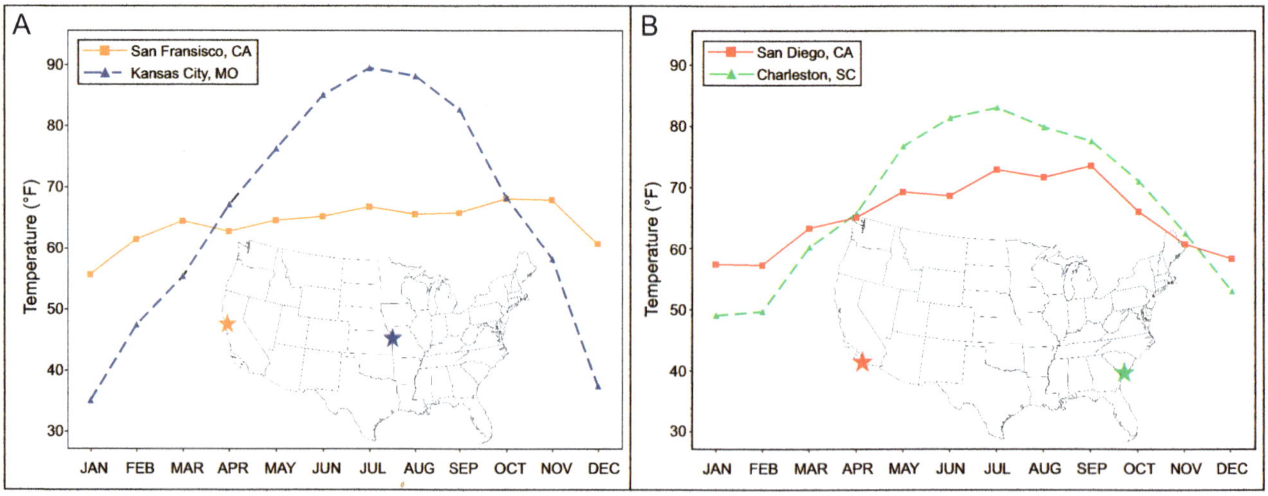

Figure 3.16: Average monthly temperatures for select cities in the United States. A. San Francisco, CA vs. Kansas City, MO for 2005. B. San Diego, CA vs. Charleston, SC for 2004.

currents, the prevailing wind direction also influences the temperature patterns at these two locations. Due to prevailing westerly winds in the mid-latitudes, temperatures along the West Coast of the United States are more strongly influenced by ocean currents than the East Coast. The prevailing westerly winds transport relatively cool, moist air eastward to the Pacific coastline while the westerly winds transport continental air masses to coastal locations in the east.

Temperature and Elevation

Elevation has a tremendous influence on local climates by affecting air temperature, annual precipitation and wind patterns. As you move away from the troposphere's heat source, the Earth's surface, the air temperature generally decreases with height. It's no surprise to climbers that temperatures in the mountains will be lower than in the valleys. Climbers can sometimes start hiking on a warm summer day and encounter much cooler temperatures at the top of the mountain (Figure 3.17). You can use the environmental lapse rate to estimate your temperature change with elevation if ever you go on a ski vacation. Remember that the environmental lapse rate is approximately 6.5°C/1,000 m (3.6°F/1,000 ft.) but can vary depending on atmospheric conditions.

Figure 3.17: The tallest peak in the world Mt. Everest, is always covered in ice and snow at its lofty elevation of just over 29,000 feet (almost 5.5 miles!). Credit: NASA/JSC.

For example, let's say that you travel to Taos Ski Valley, New Mexico with an elevation of around 3,048 m (10,000 ft.). You should expect temperatures to be about 20°C (36°F) colder than they would be if this location was at sea level. This relationship also explains how we can find tropical glaciers located high in the equatorial mountain ranges. The atmosphere's vertical temperature structure also varies with latitude. Figure 3.18 depicts the average lapse rate of the troposphere at three different latitudes. The height of the tropopause varies significantly between the poles and the tropics due to the relationship between temperature and density. Notice how the height of the tropopause varies by almost 8 km (5 mi.) while the rate of temperature change by latitude is nearly identical.

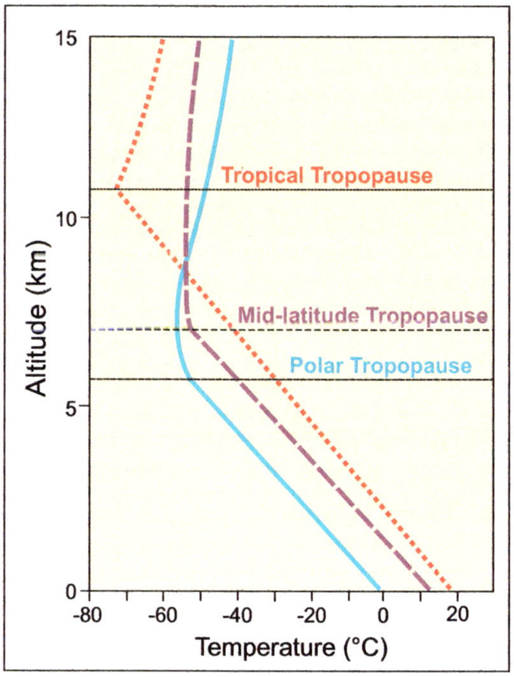

Figure 13.18: Average temperature profiles over the tropics, mid-latitudes and the poles. The tropical tropopause reaches a towering 11 km, while the polar tropopause only reaches to 6 km or so.

Ocean Circulation

When winds blow across the surface of the ocean, energy is transferred to the ocean and generates ocean currents. These currents help moderate temperatures by circulating large volumes of water and thermal energy across the oceans. The northward transport of heat through ocean circulation is illustrated in Figure 3.19. The relatively warm ocean current that flows northward is called the North Atlantic Drift (Figure 3.19A). The North Atlantic Current is fed by the Gulf Stream and has a significant impact on the winter climate of Western Europe's higher latitudes. The Gulf Stream is depicted by the plume of warm water flowing between Florida and the Bahamas that flows northeastward into the North Atlantic (Figure 3.19B). Winters in London, England would be about 5.5°C (10°F) colder without the influence of the Gulf Stream. Winters are considerably harsher in New York City compared to London despite being located more than 1,200 km (~750 mi.) closer to the equator. The Gulf Stream is one of several warm ocean currents that transfer heat to higher latitudes (more on this in Chapter 14). These currents account for almost 40 percent of the heat transfer from the tropics toward 30°N latitude. The ocean and atmosphere work together to transport heat and reduce large temperature differences between the lower and higher latitudes.

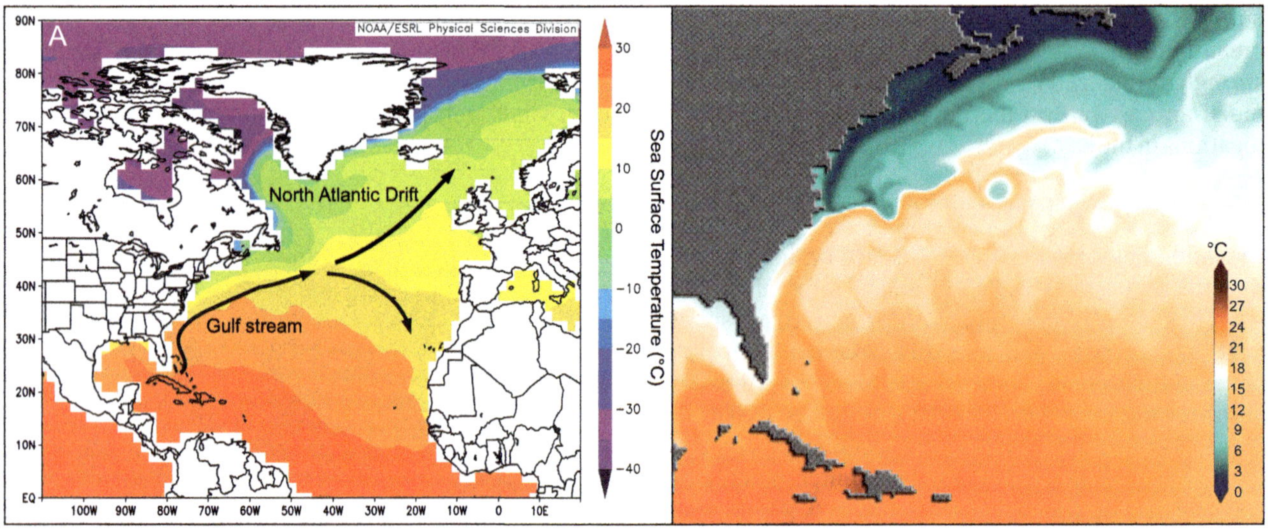

Figure 3.19: A. Sea surface temperatures for the North Atlantic in February 2015. The effects of the North Atlantic Drift are apparent. Image credit: NOAA/ESRL. B. This image of sea surface temperatures shows the trace of the Gulf Stream as it flows north and away from the North Carolina coast. Image credit: GFDL/NOAA.

Proximity to Water Bodies

Cities and towns near a large body of water can have a much different climate compared to more inland locations. Large bodies of water have both a daily and seasonal influence on moderating temperatures. On a daily basis the temperature range is, on average, smaller for coastal locations compared to more continental locations. Coastal locations typically don't get as cold at night or warm up as much during the day as inland locations due to the moderating influence of the water. The change in temperature between the warmest and coolest parts of the day is known as the diurnal temperature range. For example, locations along the Pacific Coast typically have a diurnal temperature range of less than 10°C (18°F) while during the winter and spring a continental location like the Oklahoma Panhandle can experience temperature swings of 28°C (50°F) or more in a single calendar day. Seasonally, areas near large bodies of water typically do not get as warm during the summer or as cool during the early fall compared with locations farther inland. This effect is especially felt for cities like Erie, PA and Cleveland, OH, situated south of Lake Erie (Figure 3.20). Temperatures typically stay cool during the early spring months for places like Erie, PA and Cleveland, OH due to the frequent passage of cold northwest winds blowing across Lake Erie. Spring often feels like it gets off to a late start in these locations because the waters of Lake Erie, on average, don't warm to 10°C (50°F) until late May. The good news is these locations benefit from the relatively warm water during the early fall. Lake Erie's water temperatures, on average, don't cool to below 21°C (70°F) until early September, resulting in an extended frost free period in the fall compared to locations farther from the lake. There are many vineyards along the southern shores of Lake Erie that take advantage of an extended growing season not experienced further inland.

Figure 3.20: Ice on Lake Erie keeps the towns along the eastern shore of the lake cool during the spring months. Credit: NASA/JSC.

Topography

In addition to the direct effect of elevation, topography also influences temperatures. The effect of topography is often considered when making decisions on where to plant crops for maximum yields and how to correctly design a ski resort to ensure ideal ski conditions at a minimal cost. Topography can have a large influence on overnight temperatures, especially when conditions are clear and calm. As the air cools after sunset the colder and denser air from the higher elevations slowly drains down to the valleys, making the valley floor colder than the surrounding hillsides. North facing slopes are critical for ski resorts because they tend to avoid the sun and collect more snow. In the Northern Hemisphere, the North slopes are often shaded during the winter months due to a low sun angle. This lack of direct sunlight causes cooler temperatures and a slower rate of melting on North-facing slopes compared to South-facing ones. Ski resorts can greatly benefit when topography is considered into their design due to the high cost of artificial snow-making.

Urbanization

Temperatures are generally warmer in urbanized areas compared to surrounding rural locations due to human activities. This phenomenon is known as the **urban heat island** and its effect is more pronounced in large cities due to the high density of buildings and asphalt and concrete surfaces compared to natural vegetation. The buildings and asphalt in urban areas absorb large amounts of heat during the day and then emit it at slowly at night. This can cause cities to be significantly warmer than the surrounding rural areas, especially at night. The average overnight summer temperatures for large U.S. cities are currently about 2°C (4°F) hotter than surrounding rural areas. Under the right conditions, urban temperatures in some metro areas can experience temperature increases of up to 11°C (20°F) or more compared to nearby rural areas. The temperature differences found between urban and suburban areas depend on time of the day, location, time of the year and other factors. The magnitude of the urban heat island effect for Atlanta, GA can be visualized by comparing maximum air temperatures on an average summer day (Figure 3.21). Residents of large cities, especially the elderly, are becoming increasingly vulnerable to heat waves as the population continues to expand in urban areas.

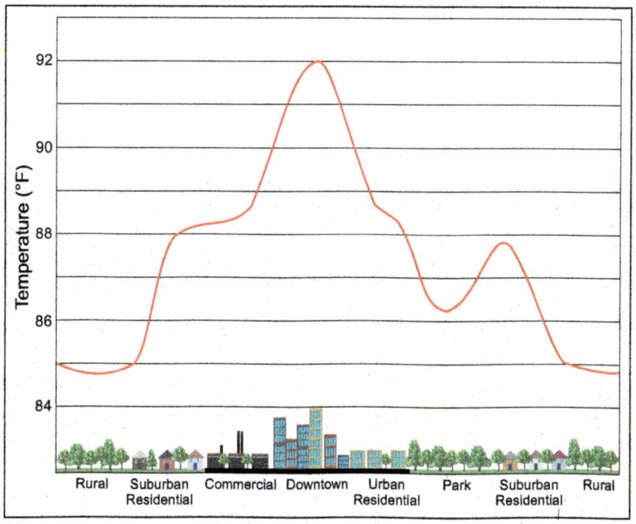

Figure 3.21: Data collected from Atlanta, GA shows the urban heat island effect. Temperatures are generally higher in the city than in the surrounding rural areas. Data courtesy NASA/Global Hydrology and Climate Center.

Section 3-6: Atmospheric Effects on Daily Temperature Variations

It usually takes the ground several hours to heat up during the day from the absorption of incoming solar radiation. The atmosphere then warms up by absorbing the re-radiated long-wave terrestrial radiation. It takes time for this transfer of energy to occur and is why the warmest temperatures don't usually coincide with solar noon, the time when the Sun is the highest in the sky. The warmest surface temperatures usually occur in the mid to late afternoon, well after solar noon. It also takes several hours for the ground and overlying air to cool off at night. The ground usually begins to slowly cool in the late afternoon and it often takes until right around sunrise for the air to reach its coldest temperature (Figure 3.22). There are times when the highest or lowest daily temperatures occur at very unusual times due to a variety of factors that are discussed next.

Wind

Winds can significantly impact surface temperatures during both the day and night. As discussed earlier, eddies can influence surface temperatures by mixing the air. Due to the Earth's topography, winds that blow across its rough surface produces eddies that allow the air to mix vertically. The amount of vertical mixing that takes place is a function of the wind speed, faster wind speeds produce greater vertical mixing. The atmosphere can also be mixed during the day from uneven surface heating. As eddies develop they can reduce local temperatures by transporting cooler air to the surface. Temperature inversions often develop near the

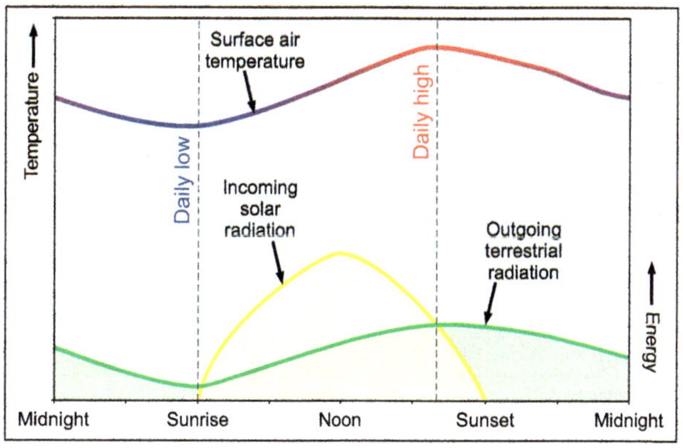

Figure 3.22: The relationship between incoming solar radiation, outgoing terrestrial radiation and surface air temperatures.

ground at night under clear and calm conditions. This allows the surface and overlying atmosphere to quickly cool. Because winds create mixing, calm or light winds at the surface are ideal for maximum cooling. Windy conditions at night result in vertical mixing through a relatively deep layer of air that prevents the coldest air from being trapped next to the ground. For this reason surface temperatures are generally warmer at night when it is windy.

Clouds and Water Vapor

Clouds influence surface temperatures through their ability to reflect sunlight and emit absorbed infrared radiation. Clouds can reduce surface temperatures during the day by reflecting visible and short-wave radiation that would otherwise be absorbed by the Earth's surface. This effect is especially noticeable during the summer months when solar radiation is intense. With all else being equal, cloudy days tend to be cooler than clear days. Clouds can elevate nighttime temperatures by absorbing and emitting infrared radiation downward (Figure 3.23). Sometimes a broadcast meteorologist will say that clouds keep nights warmer because they act like a blanket. This statement is technically incorrect because the clouds act more like space heaters and keep nights warmer due to the additional long-wave energy they radiate toward the ground. This is why, with all else being equal, cloudy nights tend to be warmer than clear nights.

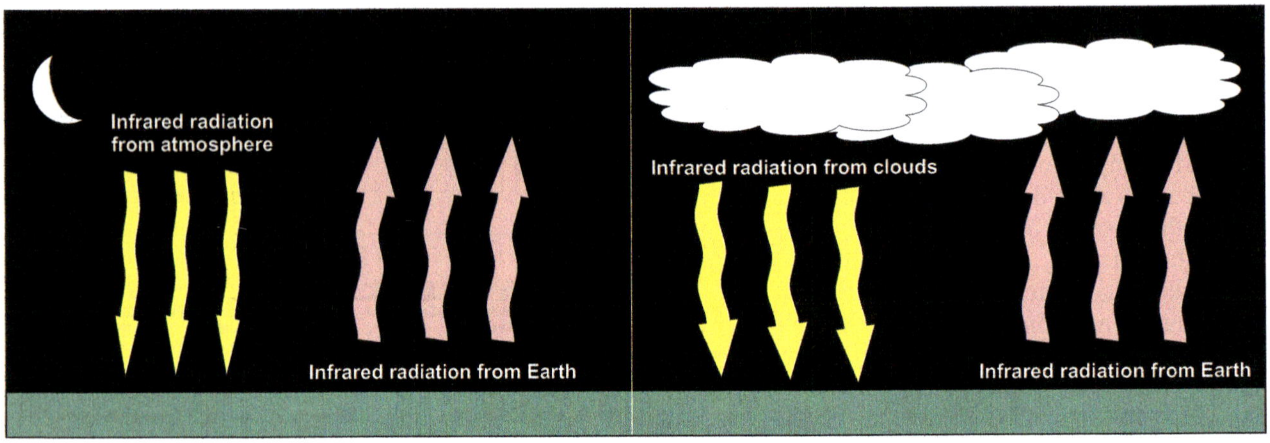

Figure 3.23: This schematic diagram shows why cloudy nights are warmer than clear nights. On a clear night, the Earth's surface loses more radiation to the atmosphere than it gains. On cloudy nights, clouds emit infrared radiation back towards the Earth's surface, keeping surface temperatures warmer.

Water vapor is an important greenhouse gas that effectively absorbs and emits longwave radiation. Water vapor influences local and regional temperature patterns, even without clouds, because it is not evenly distributed in the atmosphere. Due to larger amounts of atmospheric water vapor, humid locations tend to have a smaller diurnal temperature range and warmer nighttime temperatures compared with arid locations (Figure 3.24). For example, the humid atmosphere of the Bahamas generally keeps nighttime temperatures only 5-7°C (9-13°F) cooler than the daytime during the summer months. Extreme humidity from June through August can sometimes cause the

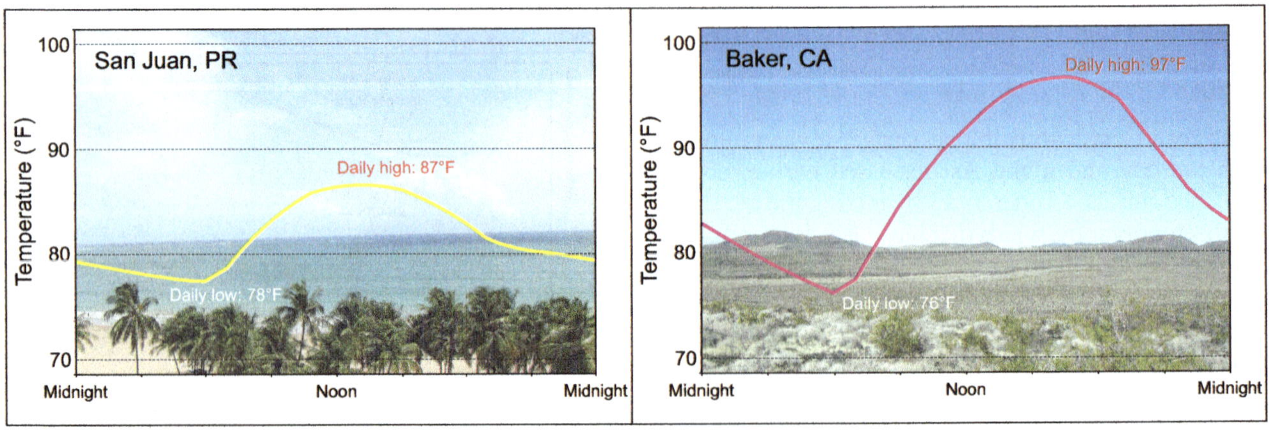

Figure 3.24: Comparing the mean daily temperature ranges on July 15th for San Juan, Puerto Rico in the Caribbean and Baker, CA in the Mojave Desert. Warm, humid tropical locales have a smaller diurnal temperature range and warmer nighttime temperatures when compared with arid locations.

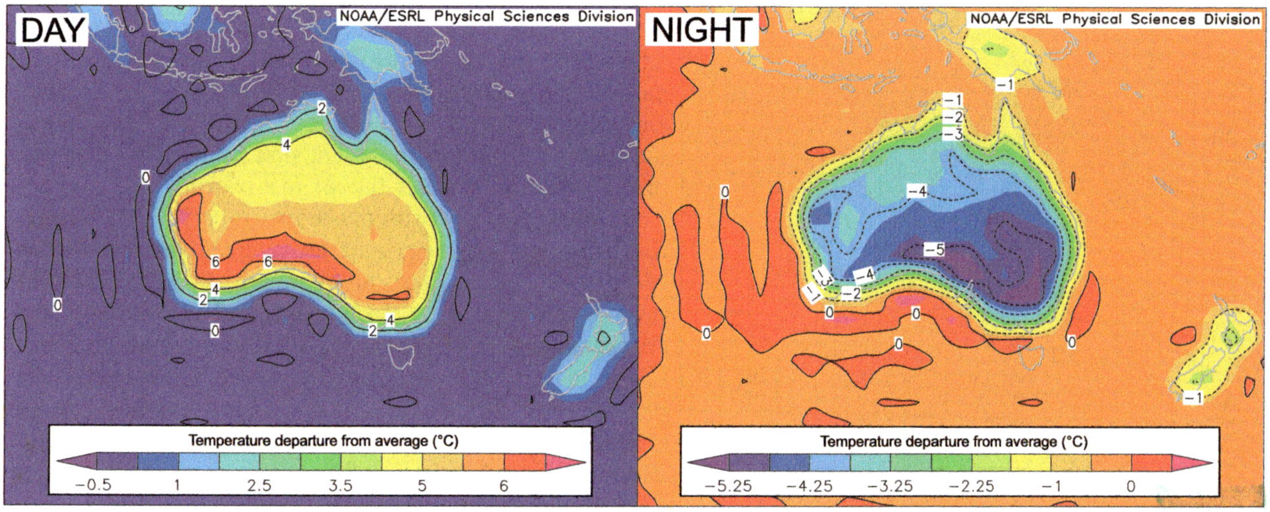

Figure 3.25: These diagrams show the dramatic difference between day and night in a desert region. Australia is one of the warmest places on Earth during the day but can become quite cool at night due to a lack of water vapor and clouds. Data courtesy of NOAA-CIRES/Climate Diagnostics Center.

minimum temperatures in Shreveport, LA to stay above 26°C (~80°F). In contrast, El Paso, Texas experiences a large diurnal temperature range due to its unique geographic setting. The mountains surrounding El Paso keep the air relatively dry by preventing the influx of moist air from the Pacific Ocean and the Gulf of Mexico. Afternoon temperatures in the 27°C to 30°C (80°F to 90°F) range can drop to 10°C (50°F) overnight. Extremely dry locations, like an arid desert, have some of the largest diurnal temperature ranges on Earth. Clear skies allow for intense surface heating during the day and maximum loss of infrared radiation at night. For example, temperatures can sometimes change as much as 28°C (50°F) in a single day in Alice Springs, Australia (Figure 3.25).

Precipitation

Precipitation can influence surface temperatures through a process known as **evaporational cooling**. Evaporation is a cooling process because it takes energy to change the state of water from a liquid to a gas; this removal of heat energy causes cooling. For clouds that produce precipitation, some of the small raindrops or snowflakes will evaporate before they reach the Earth's surface. Streaks or wisps of precipitation that evaporate before reaching the ground are known as **virga** (Figure 3.26). Virga is most easily produced when precipitation falls through a dry and warm atmosphere. This is why it sometimes appears like it is raining on a radar image well before the raindrops hit the ground. Virga is also common on radar in winter when light snow is first detected aloft and the lower troposphere is still dry. How much evaporational cooling occurs depends primarily on the amount of moisture in the air. You have experienced the effects of evaporational cooling if you jump out of a pool on a windy day or feel a rush of evaporationally cooled air right before the raindrops hit the ground in a thunderstorm.

Figure 3.26: Virga falls over the prairie. Photo credit: NOAA.

Air Masses and Fronts

Large volumes of air near the ground will acquire the temperature and moisture properties of the surface below if they remain over a uniform region for an extended period. These large volumes of air are formally called **air masses**. An air mass is a large body or air with relatively uniform temperature and moisture properties; they can control the weather for a relatively long period of time. Air masses are usually several thousand kilometers wide and typically form over large flat areas with light winds and similar surface characteristics. The areas where air masses originate are called air mass **source regions** (Figure 3.27). Air masses are grouped into one of four categories based on their source region. Ideal source regions include the oceans, large deserts, extensive snow covered areas and large forested regions. The longer an air mass remains over a source region, the more likely it will acquire the properties of the surface below.

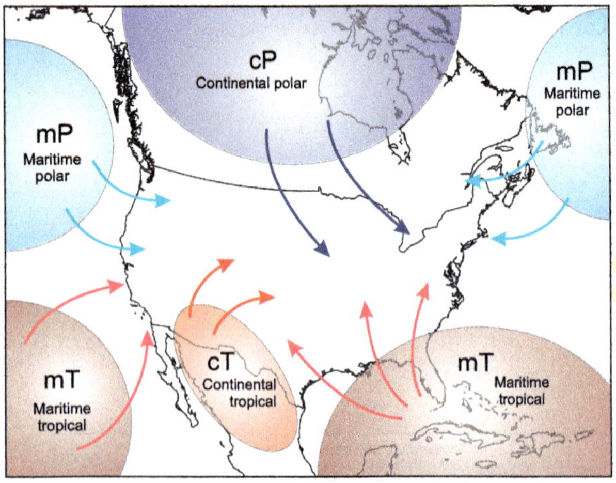

Figure 3.27: Source regions of air masses that affect the contiguous United States.

Air masses are classified using two letters according to their temperature and moisture properties, the first letter relates to moisture characteristics while the second letter indicates the temperature characteristics. These letters are combined to indicate the type of air mass. The first letter of an air mass (c or m) indicates whether the air mass formed over land or water. **Continental** (c) air masses are relatively dry because they form over large land areas like central Canada. **Maritime** (m) air masses are relatively moist because they form over large bodies of water such as the Pacific Ocean or the Gulf of Mexico. The second letter of an air mass (T or P) relates to the temperature characteristics of the air mass and is based on latitude of the source region. **Tropical** (T) air masses form within about 25 degrees of the equator and **Polar** (P) air masses are located from poleward of 60 degrees latitude. For example, continental polar (cP) air masses often bring cold, dry conditions in the winter to much of the United States as they move south from Canada. The Gulf Coast states and the eastern third of the United States commonly experience warm, humid conditions as this region is commonly situated under a maritime tropical (mT) air mass. The shading in Figure 3.28 indicates two different air masses influencing North America on an early January day.

Once formed, air masses are modified as they move out of their source region and encounter different surface conditions. Air masses are slowly modified as they acquire the temperature and moisture characteristics of the surface they travel over. For example, a continental polar (cP) air mass will be heated by the ground below as it moves south from Canada into the southern United States. How much the air mass is modified depends partly on how fast it travels from its source region. The most significant cold air outbreaks in the United States usually result from the rapid southward movement of a continental polar (cP) air mass from Canada that doesn't have enough time to be significantly warmed as it moves southward into the United States. Air masses continually move out of their source region and bring changes in weather conditions across the globe. Weather patterns continually change in the mid-latitudes due to the constant interaction between polar and tropical air masses.

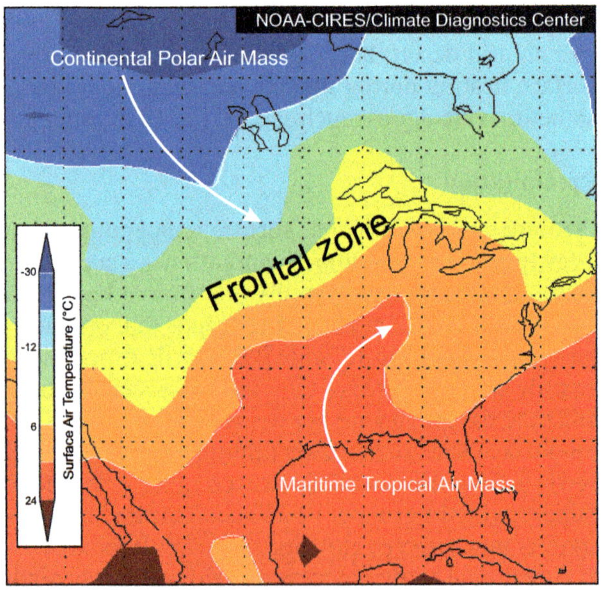

Figure 3.28: Air surface temperatures for January 3, 2004. Image credit: NOAA-CIRES/Climate Diagnostics Center.

A **front** is a boundary between two different air masses. Weather forecasters monitor the movement of air masses and the fronts that separate them. Although fronts are zones of transition, sometimes the transition zone, called a frontal zone, can be abrupt (Figure 3.28). Fronts often bring large changes in temperature, winds, clouds and precipitation as they pass over an area. Fronts extend not only in the horizontal direction, but in the vertical direction as well. A front is classified by which type of air mass (cold or warm) is replacing the other. A front is called a **cold front** if a cold air mass is replacing a warm air mass (Figure 3.29A). Cold fronts tend to move faster than all other types of fronts and generally move from northwest to southeast. They bring noticeably colder and drier conditions when they pass over an area. Cold fronts are represented on a weather map by a solid blue line with triangles pointing towards the warmer air and in the direction of its movement (Figure 3.29C). A front is called a **warm front** if a warm air mass is replacing a cold air mass (Figure 3.29B). It is important to understand that when unlike air masses clash, the warmer air is always forced to rise due to its lower density as a result of its warmer temperature. This means that the warmer air cannot advance unless the colder air retreats. Warm fronts favor the development of widespread clouds and precipitation as the

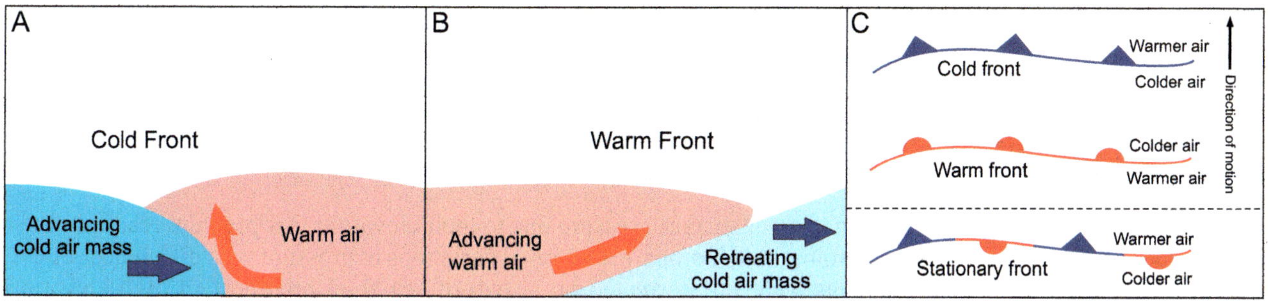

Figure 3.29: A. A cold front. B. A warm front. C. The symbols used to represent fronts on a weather map.

warm air gradually rises over the cold air. After the passage of a warm front the air becomes noticeably warmer and more humid. Warm fronts are represented on a weather map by a solid red line with semi-circles pointing towards the colder air and in the direction of its movement (Figure 3.29C). When a warm front or cold front moves very slowly or stops moving, it becomes a **stationary front**. Stationary fronts can produce significant weather and are often associated with flooding. There is usually a noticeable temperature change and a shift in wind direction when moving across a stationary front. Stationary fronts generally form from polar air masses that become significantly modified as they move over a warm surface. Stationary fronts are represented on a weather map by alternating red and blue lines with blue triangles pointing towards the warmer air and red semicircles pointing towards the colder air (Figure 3.26C).

Section 3-7: Pressure Systems and Wind

Atmospheric pressure is usually the most difficult climate element to comprehend; it is easy to understand elements we can feel like temperature, wind and moisture. We are usually unaware of pressure variations and don't notice changes without the use of a barometer. Atmospheric pressure is directly related to wind as pressure variations are responsible for air movement. On the millibar (mb) scale, sea-level pressure on the Earth usually ranges between 950 mb and 1050 mb. Tonsontengel, Mongolia has the record for the highest sea-level pressure ever recorded on Earth. A pressure of 1085.7 mb (32.06 inches of mercury) was recorded in 2001. Due to the potential biases in pressure recordings at high elevations, pressure records also exist for stations that are below 750 meters (2,460 feet). This record was set in Agata, Russia in 1968 with a pressure reading of 1083.8 mb (32.00 inches of mercury). The lowest sea-level pressure ever recorded on Earth (excluding tornadoes) was 870 mb. This record was set in 1979 by Typhoon Tip in the western Pacific Ocean. Fluctuations in sea-level pressure are much smaller on a daily basis and typically range from 980 mb to 1020 mb. For example, a sea-level pressure change of 20 mb over a horizontal distance of 800 kilometers (500 miles) only produces a pressure gradient of 0.025 mb per kilometer. Although these changes don't seem that significant, they are strong enough to generate breezes and cause tree limbs to sway.

Air pressure at the Earth's surface can be closely approximated by the weight of a column of air that extends from the ground to the "top" of the atmosphere. Remember that air pressure can be thought of as the weight of the overlying atmosphere. A column of air above a surface low-pressure system is the lightest column in this region, while the heaviest column of air is found above a high-pressure system (Figure 3.30). Columns of higher and lower atmospheric pressure are represented on weather maps with the letter H and L. For reasons to be examined later, low-pressure systems often produce clouds and rainfall while high-pressure systems are associated with dry, sunny weather. Maps need to be created to visualize how sea-level pressure changes across time and space. **Isobaric maps** use

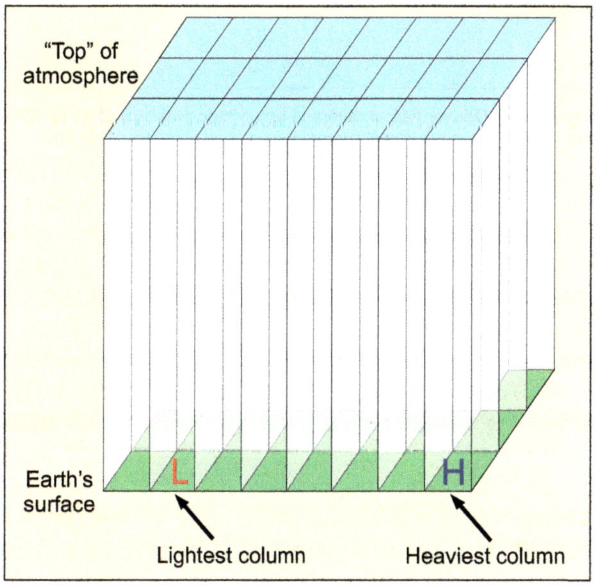

Figure 3.30: If the atmosphere was broken into an infinite number of columns stacked side by side, the lightest column would be a center of low pressure (L), while the heaviest column would be the center of high pressure (H).

isolines to show equal barometric pressures along a surface. Isobaric maps are created for several atmospheric pressure levels. A detailed analysis of isobaric maps can help understand characteristics of weather patterns and the conditions they produce. Maps of sea-level pressure allow one to see how air masses and pressure systems, driven by temperature gradients, work together to produce weather.

Isobaric Maps

Locations above sea-level naturally have lower surface pressure due to their elevation. Air pressure readings must be adjusted to sea-level to allow meteorologists to compare horizontal variations in pressure with differences due to elevation removed. If pressure readings were not corrected on a map of sea-level pressure, all high elevation locations would have much lower pressure readings compared to places near sea-level. Why is Denver's sea-level pressure adjusted to sea level? Ocean City, Maryland has one more mile of air above it than Denver. The additional pressure resulting from the extra mile of air must be added to Denver's surface pressure to adjust it to sea-level. For locations above sea level, the station pressure is adjusted to compensate for the difference between the station elevation and sea level (Figure 3.31). Making these adjustments to sea level is not perfect, errors are more likely in mountainous regions like the Rockies.

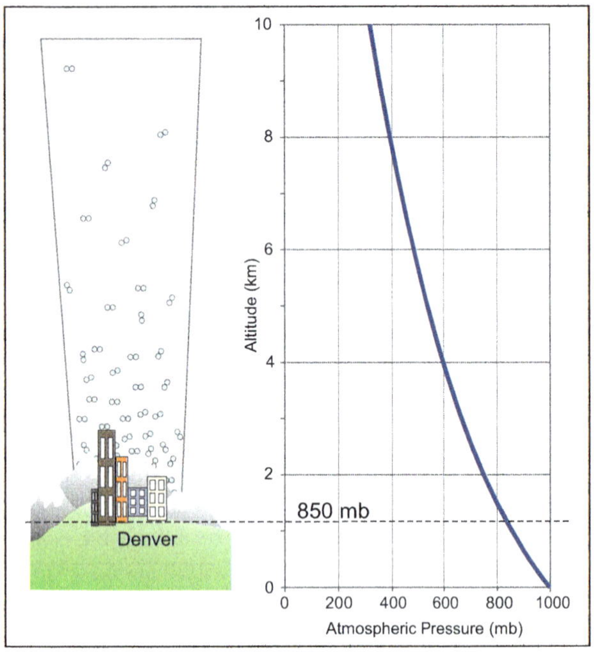

Figure 3.31: Denver has an average barometric pressure of about 850 mb, a sharp decrease from the average sea-level pressure of 1013 mb.

The fairly rapid decrease of air pressure with increasing elevation is shown in Figure 3.3. At an altitude of only about 5.5 km (3.4 mi.), air pressure decreases to 500 mb. This means that roughly half of the total weight of the atmosphere is found in the lowest 5,500 m (18,000 ft.). Figure 3.31 clearly shows that the atmosphere is a compressible gas that is held in place by gravity. Changes in air pressure, especially rapid changes, can affect your body. Gaining or losing altitude too quickly can cause a painful feeling in your ears and cause them to "pop" as they adjust to pressure changes. These symptoms commonly occur during take-off or landing in an airplane or by riding an elevator to the top floor in a tall building. Direct comparisons of pressure readings can only be made once variations due to elevation are removed.

Station models that display surface weather conditions report atmospheric pressure readings that have been adjusted to sea level. The adjusted pressure readings are organized to create an isobaric map by drawing **isobars**, which are lines of constant pressure. Large scale processes are much easier to interpret using isobaric maps than looking at raw data from individual weather stations. Isobaric maps using station model data are very useful for locating strong pressure gradients and identifying the strength and location of high and low pressure centers. The pressure pattern can also indicate the wind strength and wind direction. Isobars are normally drawn at 4 mb intervals, using 1000 mb as a base value. Figure 3.32 shows an example of how pressure readings are displayed on an isobaric map. The pressure field is displayed using isobars and the centers of the high and low pressure systems are marked. The strong low-pressure system over the upper Midwest, by convention, is depicted with a red L while the two areas of high pressure are marked with blue H's. Notice how the isobars encircle around the pressure centers and strong winds are marked where the isobars are close together.

Areas of high and low pressure that extend outward from a pressure center are also indicated on isobaric maps. Elongated areas of higher pressure, called **ridges**, typically extend from north to south away from a high-pressure center. The ridge in Figure 3.32 is marked by a serrated, or sawtooth, line that extends from Oregon to Mexico. Notice how the pressure decreases as you move away from the ridge line towards points A and B. Similar to a high-pressure center, ridges are associated with dry and sunny weather. Elongated areas of lower pressure, called **troughs**, typically extend from north to south away from a low-pressure center. The trough in Figure 3.29 is marked

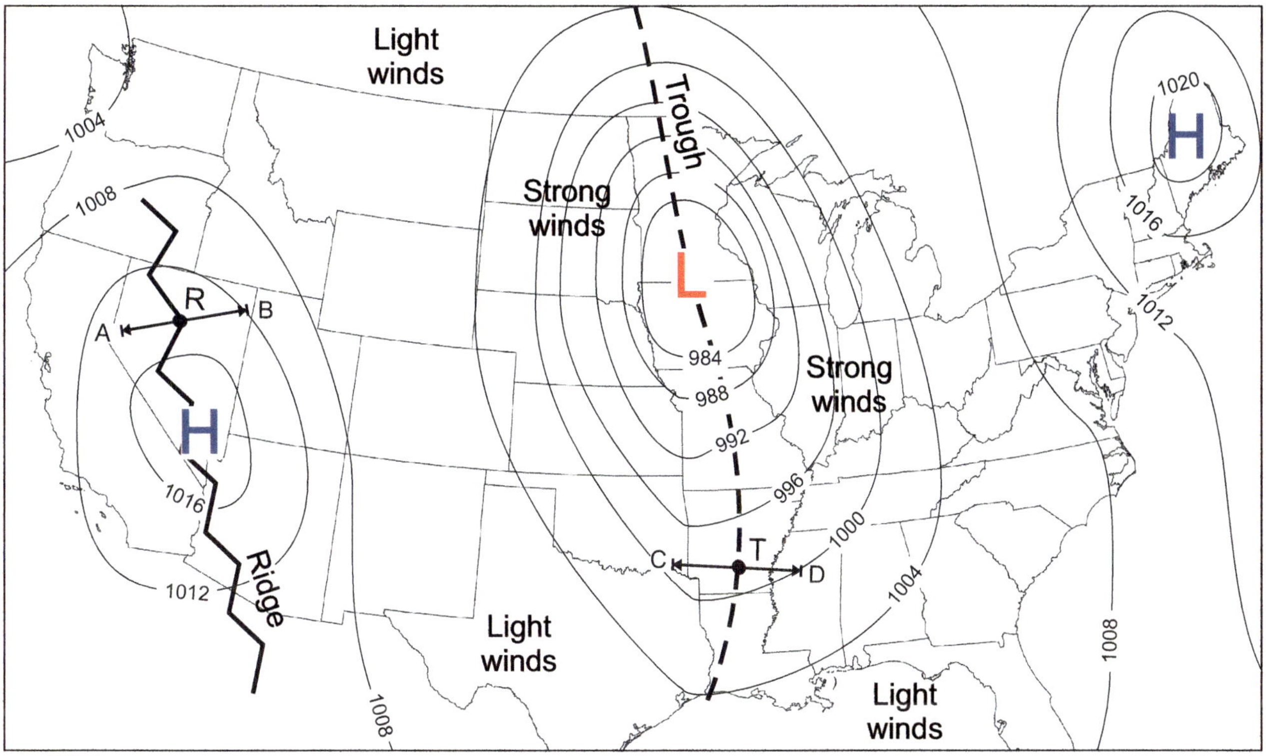

Figure 3.32: An "average" isobaric map showing the location of ridges and troughs with respect to high and low pressure systems. Barometric pressures will be lowest in a trough and highest on top of a ridge.

by a dashed line that extends from Canada to the Gulf of Mexico. Notice how the pressure increases as you move away from the trough line towards points C and D. Comparable to a low-pressure center, troughs are associated with clouds and rainfall.

Atmospheric Circulation Around Highs and Lows

The wind speed and direction can be determined by analyzing the pressure pattern on an isobaric map. According to Sir Isaac Newton (1643-1727) and his first law of motion, in order for air (or any object) to be set in motion, there must be a net force acting on it. There are three primary forces at the Earth's surface that influence wind speed and/or direction: the **pressure gradient force**, **friction** and the **Coriolis force**. These three forces all work together at the same time.

The Pressure Gradient Force

The pressure gradient force is the change in pressure over a horizontal area. The pressure gradient force tries to equalize pressure differences and is the main driving force of atmospheric motion. This force always causes air to move, although not directly, from areas of high pressure to areas of low pressure. Wind speeds are directly proportional to this force; areas that have the highest wind speeds are where the pressure gradient is the greatest. The pressure gradient force can be observed by pumping up a bicycle tire (Figure 3.33). When the tire's valve is opened, the air will automatically deflate the tire due to the pressure gradient force. Air will flow from high pressure toward lower pressure until the pressures are equalized.

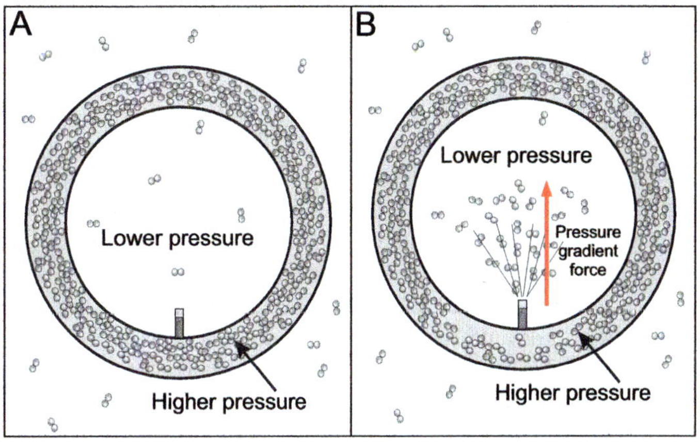

Figure 3.33: A bicycle tire can be used to demonstrate the pressure gradient force. A. An inflated tire is under higher pressure than the atmosphere around it, creating a pressure gradient. B. When the tire valve is opened, air rushes out from the tire. The force causing the air to leave the tire is the pressure gradient force.

Horizontal pressure gradients always exist at the Earth's surface due to the presence of high and low pressure systems. Horizontal pressure gradients are what force the air to move, producing **wind.** The wind attempts to minimize pressure differences by transporting air from the heavier air columns to the lighter air columns. Air would always flow directly from high to low pressure if the pressure gradient force was the only force acting on it. To understand why the strongest pressure gradients produce the strongest winds, think about pumping up a bicycle tire again. The air will flow out of the tire very fast if the tire is fully inflated because there is a strong pressure gradient between the air in the tire and the surrounding air. Wind speeds vary significantly at the Earth's surface as a result of a continually changing horizontal pressure gradient.

Identifying pressure gradients on an isobaric map is relatively easy by observing the pattern and spacing of the isobars. Regions with stronger winds are identified by large pressure gradients, where the isobars are closely spaced together. Notice how the strong pressure gradient in eastern South Dakota and Nebraska in Figure 3.32 is directly related to faster wind speeds. Light winds would be expected over Colorado and Wyoming due to a weak pressure gradient that is noted by the spacing of the isobars. However, air does flow in a direct path from areas of high pressure to areas of low pressure. There are two other primary forces that influence the winds, friction and Coriolis

Friction

The Earth's surface exerts a frictional drag on air in contact with the Earth's surface. Friction is a force that restricts motion, an object experiences friction anytime it comes in contact with a surface. Meteors heat up enough to glow and vaporize when they enter the Earth's atmosphere due to friction. The effect of friction can certainly be felt by sliding down a sliding board in a pair of shorts. The magnitude of the force of friction increases with speed. This means that winds are subject to a greater force of friction as their speed increases.

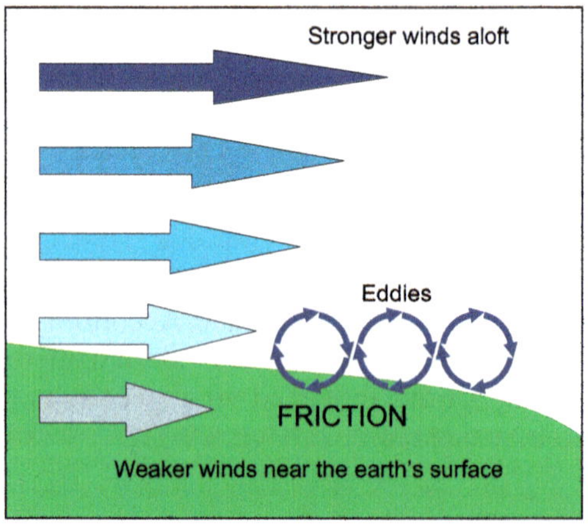

Figure 3.34: Mechanical eddies help to slow down winds near the Earth's surface thanks to friction between the ground and the air above.

Friction can act to modify the wind's direction and reduce its speed. Local topography directly affects how much friction is exerted on the overlying air. For example, winds are generally faster when they travel over a smooth surface, like a large body of water, due to the reduced force of friction. The effect of friction is increased over mountainous or rugged terrain and heavily forested areas, causing the winds to slow down. Friction reduces wind speed by causing mechanical eddies to form. Remember that mechanical eddies form when winds blow across the Earth's rough surfaces. These eddies mix the air and reduce wind speeds, especially close to the ground (Figure 3.34). Wind speeds generally increase with height as the effect of friction is reduced with increasing distance from the Earth's surface. The effect of friction becomes insignificant at roughly 0.8 km (0.5 mi.) above the Earth's surface. There is one more force that must be considered to understand why air doesn't move directly from high to low pressure.

The Coriolis Force

The pressure gradient force would create a single-cell circulation of wind in each hemisphere if the Earth did not rotate. Due to the Earth's rotation, there is a second force that affects the direction of wind and helps explain the path of winds in the atmosphere. As discussed in Chapter 1, the **Coriolis force** affects the direction of wind by producing an apparent deflection in its path and helps creates multiple circulation cells of wind on Earth. As the Earth spins on its axis, all free-moving objects appear to leave their straight-line paths. This force is named after Gustave Coriolis, a French scientist who first explained it mathematically in 1835. The Coriolis force does not actually cause the wind to deviate from its path; it appears to deviate due to the motion of the planet. Air travels in a straight line when viewed from space but appears deflected when viewed from the Earth. As air moves from high to low pressure in the Northern Hemisphere, it is deflected to the right by the Coriolis force. Air moving from high pressure to low pressure is deflected to the left by the Coriolis force in the Southern Hemisphere. This effect creates pressure systems that rotate in the opposite directions in each hemisphere. The effect of the Coriolis force can be visualized by riding a merry-go-round with someone. If you throw a ball to someone sitting directly across from you, the ball

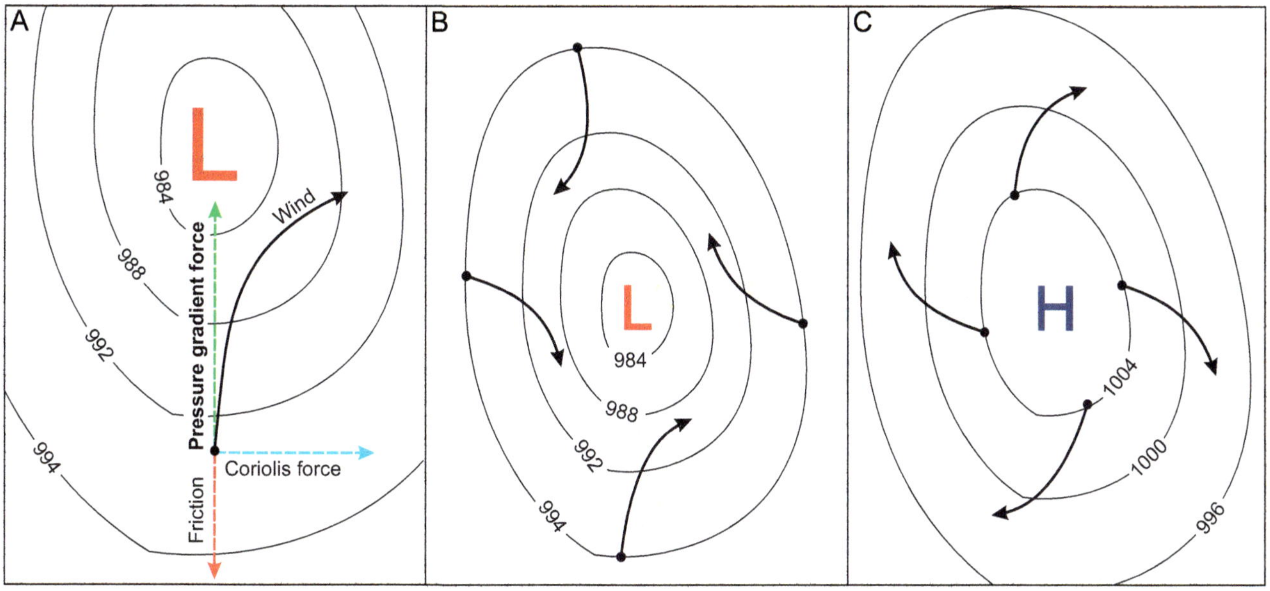

Figure 3.35: The movement of air parcels around low and high pressure systems in the northern hemisphere. A. An air parcel moving towards a center of low pressure will be deflected to the right by the Coriolis effect. B. Collectively, the Coriolis effect causes air to move counterclockwise around a low pressure center. C. The situation is reversed for high pressure systems. Air moving away from the high pressure center will be deflected to the right, producing clockwise air flow around the high.

will go in a straight line if the merry-go-round was not spinning. If the merry-go-round is spinning counterclockwise, the ball will appear to be deflected to the right even though it is travelling in a straight line.

The magnitude of the Coriolis force can be calculated if you know the travelling object's speed and latitude. Objects, like the wind, get deflected more if they are travelling at a higher speed. The Coriolis force is also a function of latitude; it doesn't exist at the equator and is a maximum at the poles. Pilots account for this deflective force based on the speed and latitude of their flight path to ensure the most efficient route possible. Figure 3.35A shows a map of sea-level pressure in the Northern Hemisphere with a low-pressure center. The air parcel initially moves in a straight path towards the low-pressure center in response to the pressure gradient force. As the parcel accelerates, the Coriolis force kicks in and deflects the parcel to the right of its initial route. There is an endless battle between the pressure gradient force and the Coriolis force. Pressure gradient force attempts to move the air directly toward the low-pressure center while the Coriolis force deflects the air to the right. As friction reduces wind speed in the lower troposphere the effect of the Coriolis force is reduced. The effect of friction is greatest near the surface and its influence decreases vertically. This results in winds crossing the isobars at an angle due to the pressure gradient force exceeding the Coriolis force (Figure 3.35B). The circulation of air around a surface low-pressure system is counterclockwise in the Northern Hemisphere with air converging toward the low. Figure 3.35C shows the clockwise circulation of air around a high-pressure system in the Northern Hemisphere. Notice how the air diverges outward and away from the high.

Arrows were added to Figure 3.32 to depict the general circulation of air around the high and low pressure systems (Figure 3.36). The wind speeds are directly proportional to the strength of the pressure gradient, indicated by the spacing of the isobars. Notice how the wind crosses the isobars at an angle and converges inward toward the low-pressure center. Also

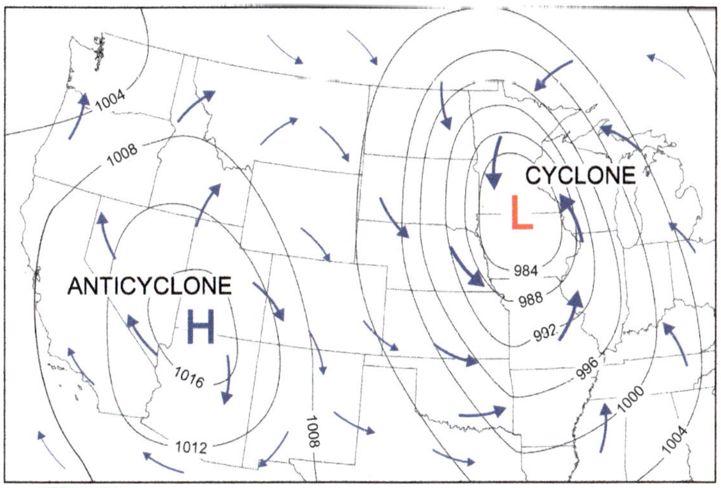

Figure 3.36: A typical isobaric map showing the wind flow. The wind moves inward towards low pressure centers and moves away from high pressure centers. The general wind speed is shown with different sized arrows.

note how the wind diverges outward away from the high-pressure center. Closed low-pressure centers are called **cyclones** while closed high-pressure centers are called **anticyclones**. These pressure centers are identified by the isobars that surround them in a roughly concentric pattern. The distinct and predictable flow patterns that develop around high and low pressure centers can be determined once the effects of the pressure gradient force, friction and the Coriolis force are understood.

Fronts And Atmospheric Pressure

Weather systems are produced as the atmosphere attempts to minimize pressure gradients. There is **convergence** of air in the vicinity of low pressure and **divergence** of air in the vicinity of high pressure that gives each its unique weather characteristics. According to the law of conservation of mass, mass cannot be created or destroyed in a given area. This means that converging air cannot "pile up" around a low-pressure system at the surface and is therefore forced to rise (Figure 3.37). Rising air cools which often results in the formation of clouds and precipitation. This is why there is often stormy weather near low-pressure areas. Air from above must sink to replace diverging air moving away from a high-pressure system (Figure 3.37). The sinking air warms up and dries out as it descends. This is why fair weather and sunny skies are often associated with high-pressure systems. It is the vertical motions generated by upper-level convergence and divergence that is responsible for most of the weather associated with high and low pressure systems. It is important to understand that low and high pressure systems are not just a surface feature; they have a vertical structure and are directly connected to the upper atmosphere.

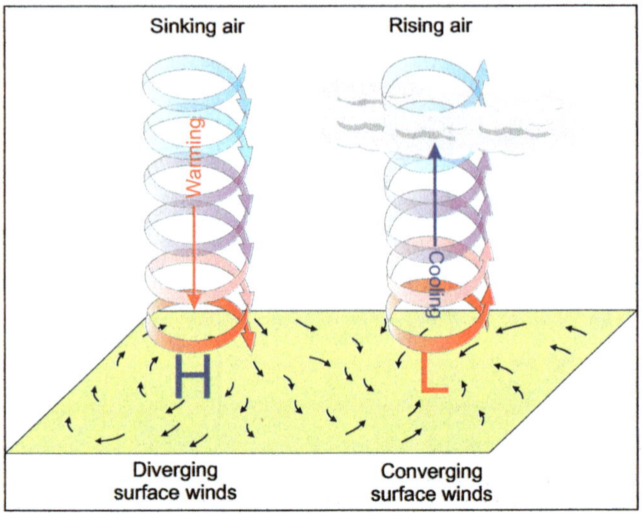

Figure 3.37: A cross-sectional view of high and low pressure systems.

Understanding the circulation around surface pressure systems allows one to see how effective they are in reducing temperature contrasts on a global scale. The high and low pressure systems transport warm air toward the poles and cold air toward the equator, often resulting in stormy weather where the unlike air masses meet along frontal boundaries. Figure 3.38A shows an idealized cross section of the Northern Hemisphere, with high-pressure at the center of cold and warm air masses. Temperatures are evenly distributed from north to south on this greatly simplified diagram. Eventually the air will spread out from both high-pressure centers and move toward the mid-latitudes. Clashes between polar and tropical air masses are quite common across much of the United States since

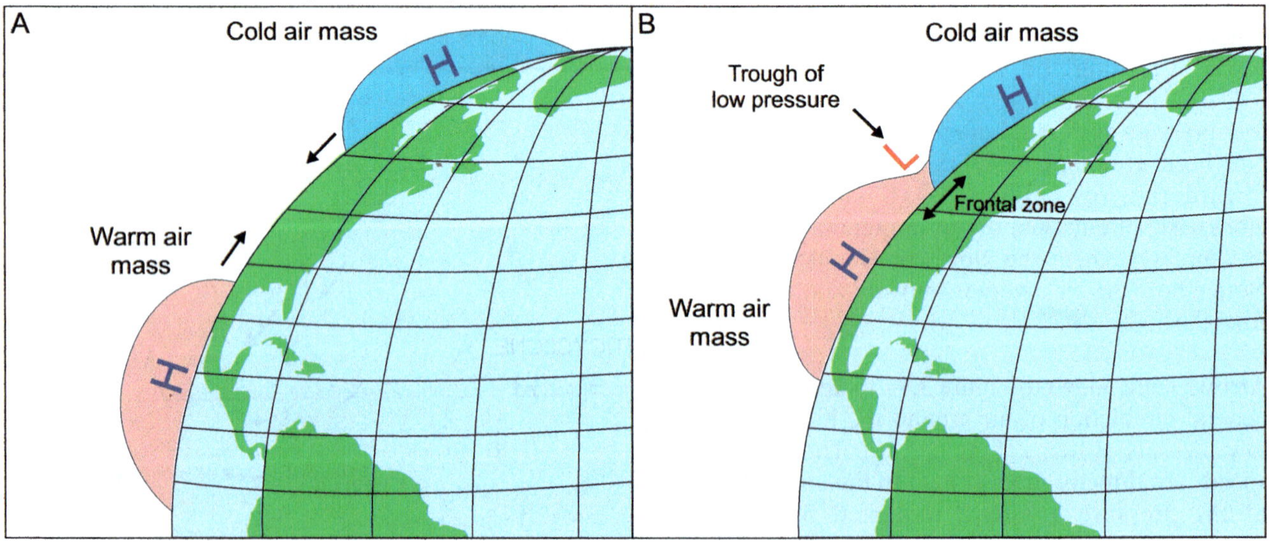

Figure 3.38: The formation of fronts. A. Cold air masses form over the polar regions and move towards the equator, while warm air masses form in the tropics and move towards the poles. An area of high pressure forms in the center of each air mass. B. When two air masses collide, a frontal zone forms and a trough of low pressure develops between the two ridges of high pressure.

there are no topographic barriers that prevent the cold air masses in Canada from moving south. Figure 3.38B shows a narrow zone of sharply contrasting temperatures, a frontal zone, where the unlike air masses meet over the mid-latitudes.

Frontal systems most often form along the periphery of air masses since their cores are marked by high-pressure. This means that fronts typically form in a zone of relatively low pressure, a trough, at the outer edges of the cold and warm air masses (Figure 3.39). Frontal zones are favorable areas for low-pressure centers to form and intensify because of the relatively low pressure and large temperature contrasts. Low-pressure centers can sometimes develop into very powerful and destructive storm systems.

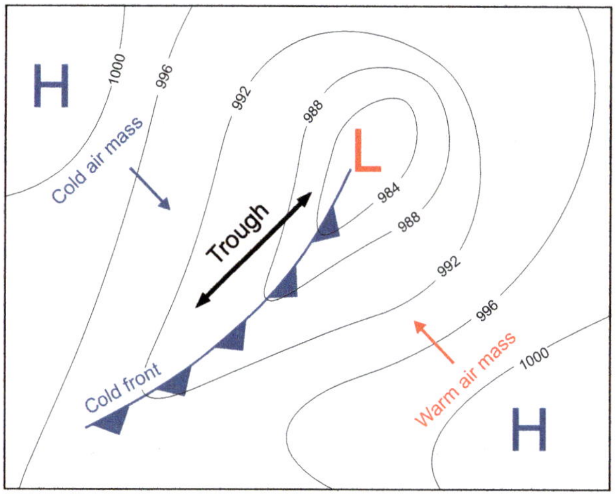

Figure 3.39: Figure 3.38 as viewed from above on an isobaric map. These situations are ideal for developing storms.

Section 3-8: Atmospheric Moisture

Water is distributed across the Earth and lower atmosphere in various forms. It is the only substance in the atmosphere that can exist as a solid, liquid or gas. Water can change phases continuously as it cycles between the Earth and atmosphere. Water vapor may return to a liquid state when the atmosphere contains sufficient moisture, producing clouds, fog, haze and a variety of precipitation types. As discussed above, fronts and low-pressure systems are associated with clouds and precipitation because they promote rising air. Atmospheric moisture is extremely important although it only accounts for only 0.001% of the Earth's water supply.

Evaporation is the process where liquid water is transformed to the gaseous state. When liquid water evaporates the water molecules break their bonds and enter the atmosphere as a gas. Water can easily evaporate as the weak hydrogen bonds continually form and break apart. Objects denser than water can be supported on its surface due to hydrogen bonding, this can be observed by floating a paper clip on the surface of a glass of water. Evaporation rates are controlled by mostly by temperature, the additional kinetic energy at higher temperatures allows the water molecules to oscillate faster and break their bonds more effectively. Evaporation rates increase as water temperatures increase, this can be observed by watching how quickly a puddle of water will evaporate on a hot summer day. What causes the water molecules to evaporate? The water molecules can break their bonds and enter the atmosphere as a gas with a sufficient amount of kinetic energy. Think of the water molecules in a puddle of water as dancers on a crowded dance floor, it is inevitable that they will bump into each other. Each time a dancer is bumped off the dance floor at a hip-hop club can be equated to the evaporation of a water molecule (Figure 3.40A). As the water molecules oscillate faster with increasing temperature, the rates of evaporation increase. This is similar to dancing in a mosh pit where more frequent collisions will cause more dancers to be bumped off the dance floor (Figure 3.40B). Coastal locations often experience distinct seasonal changes in humidity that are directly related to the surrounding water temperatures. Evaporation is a cooling process since it takes heat from the atmosphere to evaporate liquid water. This cooling process can be experienced by jumping out of a pool on a windy day or feeling a

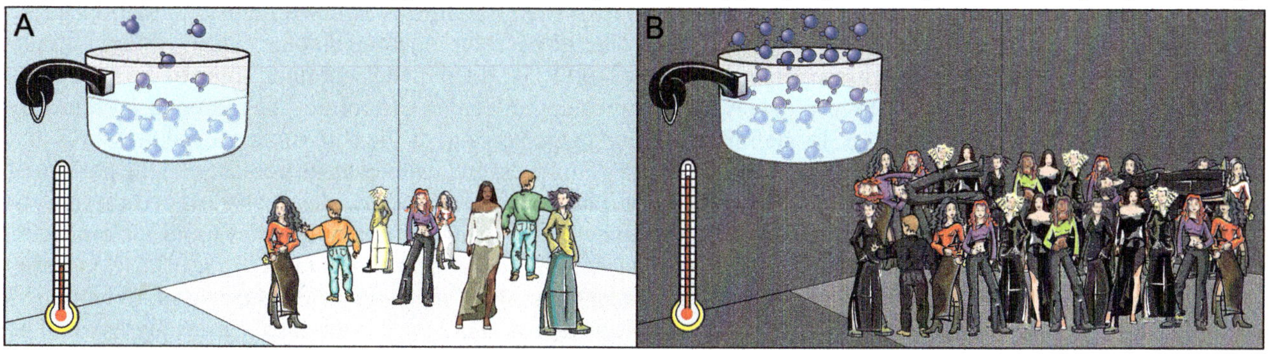

Figure 3.40: Rates of evaporation can be likened to dancing. A. People dancing to hip-hop at a club are like water molecules at lower temperatures: collisions don't occur often or with much force. B. People dancing in a mosh pit at a heavy metal concert are like water molecules at higher temperatures: Collisions are frequent and highly energetic.

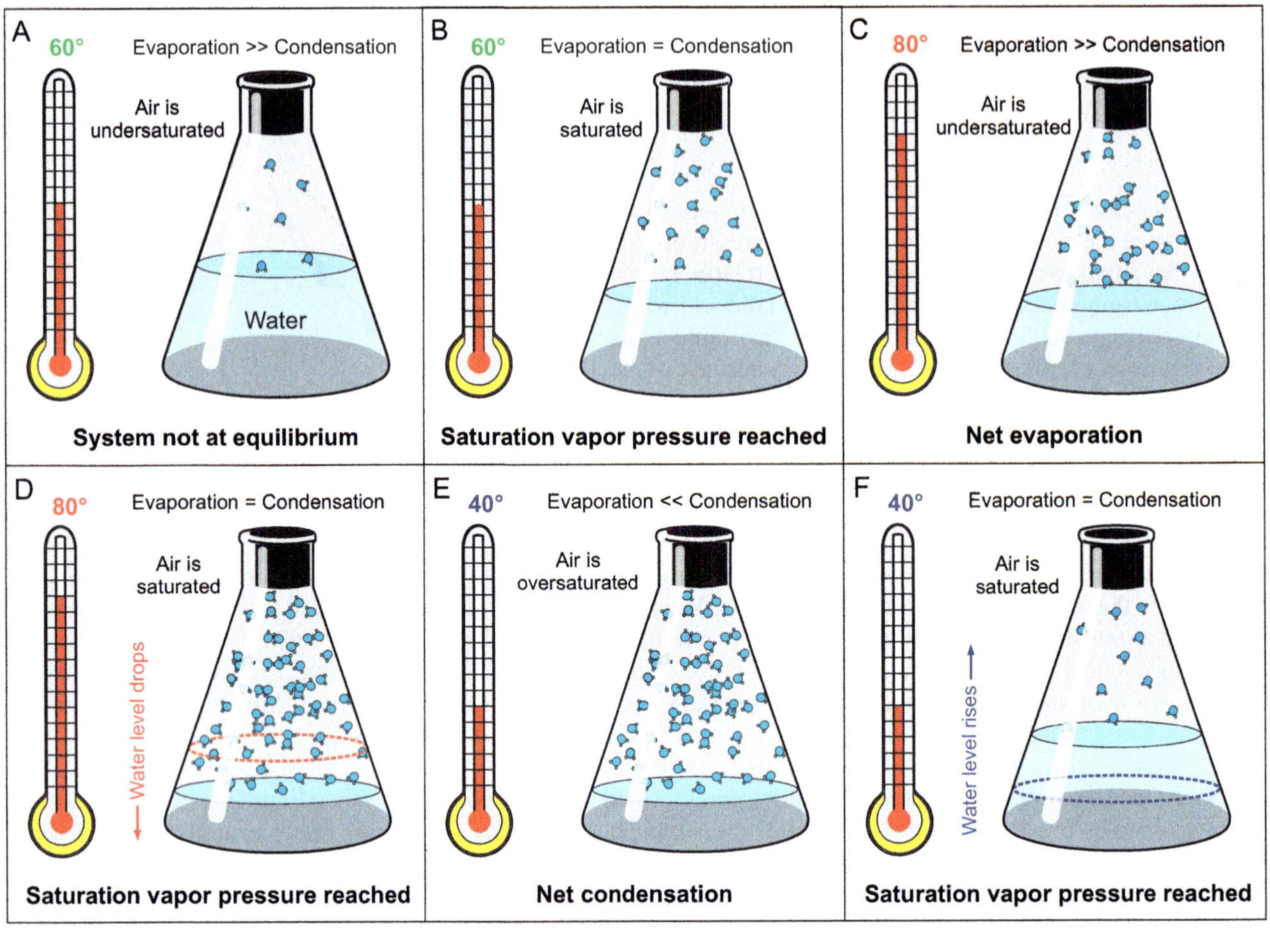

Figure 3.41: Vapor pressure demonstrated. A. Water is added to a flask. Initially, the air is unsaturated with respect to water vapor so rates of evaporation exceed rates of condensation. B. The saturation vapor pressure is reached and rates of evaporation equal rates of condensation. C. If the temperature is raised in the flask, net evaporation will occur. D. The system eventually reaches a new, higher saturation vapor pressure. Notice the decrease in the overall volume of liquid water as the liquid water is converted to vapor. E. If the temperature is lowered in the flask, net condensation will occur. F. The system eventually reaches a new, lower saturation vapor pressure. Note the increase in the overall volume of liquid water as the vapor is converted to liquid water.

rush of cool air immediately before it rains. Surface temperatures are significantly influenced by evaporation rates. Cities in warm climates often install fountains in the downtown areas to help reduce surface temperatures through evaporational cooling. This technique, utilizing evaporation, can greatly improve the comfort levels in areas that surround the fountain.

Condensation is the process where water vapor becomes a liquid, it is considered the opposite of evaporation. Evaporation and condensation should not be considered separate processes, they often take place at the same time. Consider a closed container like the one shown in Figure 3.41. Although this is not a realistic situation in the atmosphere, it does an effective job illustrating one phase change of water. Assume that the closed system initially contains no water vapor. As water at a temperature of 15.5°C (60°F) is added to the container, some of the molecules break their bonds and evaporate (Figure 3.41A). As the water molecules evaporate, they exert a pressure by bouncing off the walls of the container. By definition, atmospheric pressure is the sum of the partial pressures that are exerted by the mixture of gas molecules and water vapor. This pressure, called **vapor pressure**, is the portion of total atmospheric pressure due to water vapor in the air. The evaporation rate will initially exceed the rate of condensation as the air is still unsaturated. This results in an increase in vapor pressure as net evaporation proceeds. An equilibrium will eventually be reached between the evaporation and condensation rates (Figure 3.41B). At this point, when the rate of evaporation equal the rate of condensation, the air is said to be **saturated**. When air is saturated, the pressure exerted by the water vapor molecules is called the **saturation vapor pressure**. The equilibrium state is disrupted when the temperature of the water increases. The water molecules have more kinetic energy at a higher temperature, which makes it more likely for them to break their bonds and evaporate. In Figure 3.41C, the evaporation rate increases when the water temperature is raised to 26.5°C (80°F). The vapor pressure is increased in response to the additional water vapor supplied by evaporation. A new equilibrium state is eventually established between evaporation and condensation rates, but this time the equilibrium state exists at a higher saturation vapor pressure (Figure 3.41D). Figure 3.42 graphically displays how the saturation vapor pressure is a function of temperature. For any given temperature, **net evaporation** proceeds when the rate of evaporation exceed the rate of condensation. Net evaporation occurs anytime the vapor pressure is below the saturation vapor pressure curve. The atmosphere's capacity to contain more moisture at higher temperatures before reaching saturation is obvious from examining Figure 3.42. For this reason, warm and humid locations generally have significantly more annual rainfall compared to cold and dry locations. This also helps explain why thunderstorms can produce tremendous amounts of rainfall on a warm, humid day. For any given temperature, **net condensation** occurs when the rate of condensation exceed the rate of evaporation and the equilibrium state exists at a lower saturation vapor pressure (Figures 3.41E-F). Net condensation occurs anytime the vapor pressure is above the saturation vapor pressure curve, resulting in the formation of a cloud (Figure 3.42).

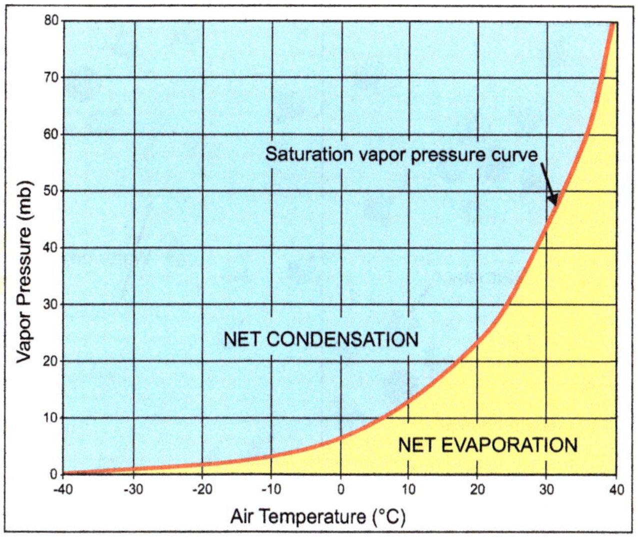

Figure 3.42: Vapor pressure as a function of temperature. The resulting profile is the saturation vapor pressure curve. Net condensation occurs above the equilibrium line, while net evaporation occurs below the line.

The evaporation of water from a water surface is partially dependent on water temperature. This means that the air contains a relatively high amount of water vapor for locations over and near relatively warm bodies of water. For example, locations surrounding warm bodies of water typically have a higher humidity as compared to regions that are surrounded by cooler bodies of water. This effect is particularly notable in the United States when comparing the average humidity between the Pacific Northwest and regions that surround the Gulf of Mexico.

Net Condensation and Cloud Formation

Clouds and fog form when condensation or deposition occurs in the atmosphere. A **cloud** is a visible aggregate of very small water droplets and/or ice crystals in the atmosphere. Clouds can become visible when water vapor condenses into a liquid onto minute airborne particles called **condensation nuclei**. When atmospheric temperatures are well below freezing clouds become visible through deposition as the water vapor changes to ice without first becoming liquid. In this situation the water vapor "deposits" onto **ice nuclei**, which are small airborne particles that serve as nuclei for the formation of ice crystals (Figure 3.43). Roughly half of the Earth is covered by a variety of cloud types at any given time. All forms of precipitation are produced by clouds but clearly not all clouds produce precipitation. Net condensation is required for a cloud to form, this means that the rate of condensation must exceed the rate of evaporation.

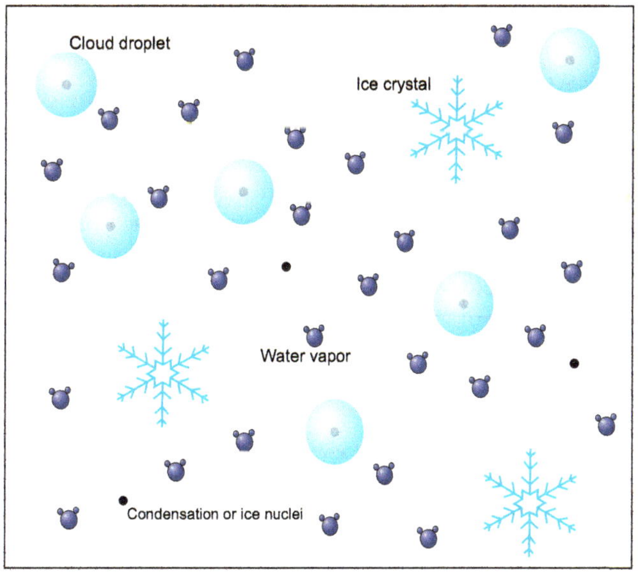

Figure 3.43: The "inside" of an average cold (below freezing) cloud. Water can exist in all three states in a cloud: Cloud droplet (liquid), water vapor (gas), and ice crystals (solid). Condensation or ice nuclei are critical for the formation of a cloud. These nuclei can be any airborne particle which encourages water to change states (i.e., soot, dust, pollen, volcanic ash).

The most common way for net condensation and cloud formation to occur in the atmosphere is to cool the air. Water vapor molecules slow down and crowd together around condensation nuclei as temperatures decrease

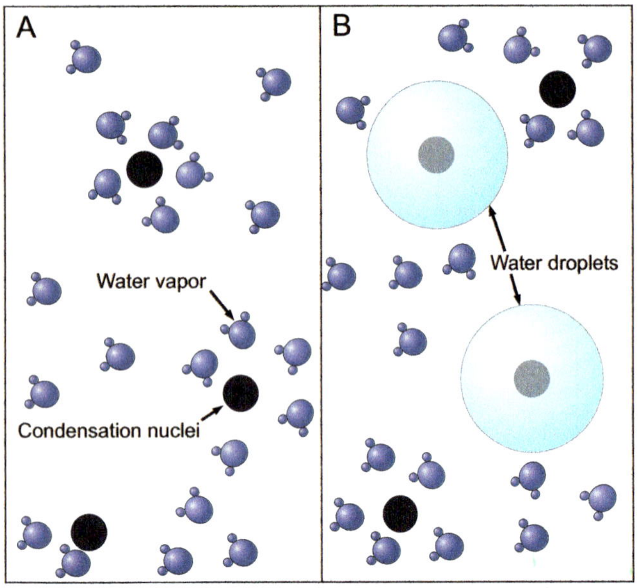

Figure 3.44: Condensation in the atmosphere. A. As the air cools, water vapor molecules "huddle" around condensation nuclei. B. Eventually, the molecules join together and condense as a water droplet.

(Figure 3.44A). Net condensation can proceed once there are a sufficient number of water vapor molecules available to form a water droplet (Figure 3.44B).

The easiest way to cool the air is to lift it off the ground, this is often accomplished by frontal systems and local topography. The air must be cooled considerably to reach net condensation and form clouds when water vapor concentrations are low. If clouds are present they are typically found at high altitudes as a result of the significant lifting and cooling that is necessary. When concentrations of water vapor are higher, a smaller degree of cooling is required for net condensation, resulting in low-altitude clouds that are located relatively close to the Earth's surface.

Section 3-9: How Precipitation Forms

Net condensation alone is not enough to cause cloud droplets to grow into raindrops or snowflakes, this process by itself is too slow to produce precipitation. There are too many (billions) of tiny cloud droplets that compete for a limited amount of available water. The transition from a fair weather cloud to a cloud that produces precipitation is not as simple as cloud droplets increasing in size through net condensation of water vapor until they grow large enough. The volume of a typical cloud droplet must grow by approximately one million times to form a single raindrop. As the cloud droplets grow bigger they become more efficient in removing water vapor in the surrounding cloudy air. The small water droplets that make up the cloud are initially too small to fall to the ground as rain or snow because their size makes them very buoyant. The normal turbulence of the atmosphere keeps them suspended. Even without turbulence, it would take at least 24 hours for the cloud particle to reach the ground. The majority of these slowly descending cloud drops easily evaporate in the drier air below the cloud. However, clouds can develop and begin to produce rain in less than one hour under favorable atmospheric conditions. This means that there must be other processes that allow cloud droplets to grow large enough to fall as precipitation. There are two important processes that are responsible for the production of precipitation: the Bergeron process and the collision-coalescence process.

Figure 3.45: Even in the summer, raindrops from storms like this start their life as snowflakes. Photo credit: B. Zaprowski.

The Bergeron Process

Most precipitation in the middle and high latitudes originates through the Bergeron process. Much of the rain that falls through the Bergeron process begins as snowflakes because the clouds at these latitudes typically extend high enough upward to have temperatures well below freezing. Even rainfall from a summertime thunderstorm over the middle and high latitudes almost always begins as snow near the top of the troposphere (Figure 3.45). The Bergeron process occurs in cold clouds, clouds that contain ice crystals. The process depends on the coexistence of water vapor, liquid water and ice crystals. Understanding that cloud droplets do not freeze as expected at 0°C (32°F) helps explain why almost all precipitation that falls over the middle and high latitudes begins as snow.

Water droplets that exist at temperature below freezing are

called **supercooled**. There are some water droplets suspended in the atmosphere can resist freezing down to temperatures as low as -40°C (-40°F). The supercooled cloud droplets can remain liquid at below freezing temperatures because the ice nuclei needed for ice crystal formation are rare at very low temperatures in the middle and upper troposphere. All water droplets freeze spontaneously at -40°C (-40°F), regardless if ice nuclei are present. These conditions allow water to exist in all three phases high in the troposphere. For a given temperature below freezing, it is more difficult for ice crystals to evaporate as compared to supercooled water. As a result, the saturation vapor pressure over ice is less than the saturation vapor pressure over water. Figure 3.46A demonstrates this principle using a partitioned box that contains water and ice at a temperature of -1°C (30°F). Water vapor molecules spread more uniformly though the box when the barrier is removed (Figure 3.46B). Net evaporation now proceeds over the water as there are not enough water vapor molecules needed for saturation. There are now more water vapor molecules over the ice than are needed for saturation, resulting in the net deposition of ice crystals.

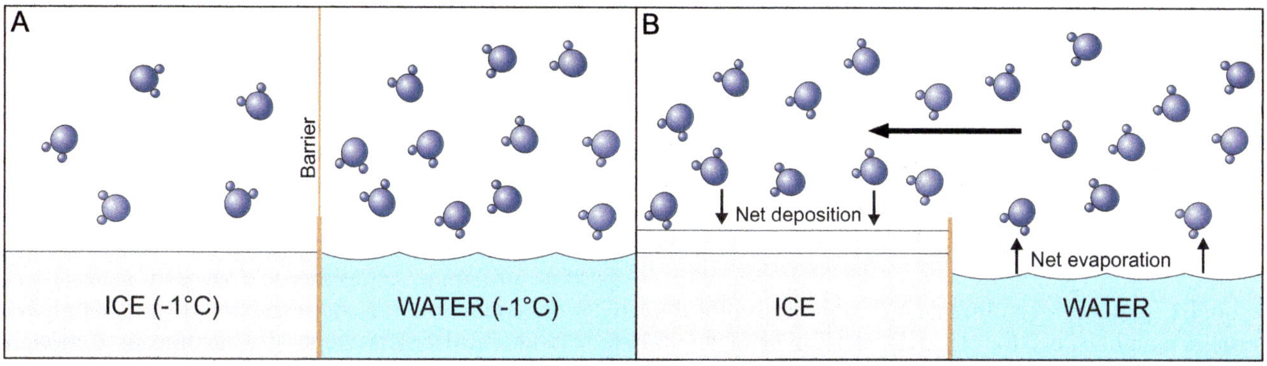

Figure 3.46: The Bergeron process explained using a box divided by a barrier. A. It is easier for water to evaporate than it is for ice to sublimate. As a result, more water vapor is found above the water than above the ice. B. When the barrier is removed, the water vapor spreads equally throughout the box. There is now more water vapor over the ice, exceeding the vapor pressure for ice. Thus, the deposition of ice from vapor occurs. Meanwhile, there is a lack of water vapor over the water, causing net evaporation to occur. In this way, water molecules go from the liquid state to the solid state via the vapor phase.

In the Bergeron process, ice crystals grow at the expense of supercooled water droplets. The Bergeron process is illustrated in Figure 3.47, the air is oversaturated for ice and undersaturated for liquid water. The ice crystals grow larger as they collide with other ice crystals and supercooled water droplets. The ice crystals eventually get heavy enough to fall out of the cloud as snowflakes. Ice crystals will continue to grow at the expense of supercooled water droplets until the supply of the cloud droplets is depleted. Snowflakes that form in the upper troposphere will usually melt into raindrops before they reach the ground unless the air temperature is below 4°C (39°F). Throughout the year, most of the precipitation that falls over the middle and high latitudes begins as snow. The temperature structure of the lower atmosphere determines if frozen or liquid precipitation will reach the ground. The Bergeron process can produce year-round precipitation in the middle and high latitudes if the upper portion of clouds are cold enough to promote the formation of ice crystals. Although the Bergeron process explains how most precipitation forms over the middle and high latitudes, it is not the only mechanism that causes cloud particles to grow large enough to precipitate.

Figure 3.47: The Bergeron process applied to clouds. A. The air is oversaturated with respect to ice, water vapor is deposited onto ice crystals, causing them to grow. B. Now the air is undersaturated with respect to liquid water. C. The cloud droplets evaporate to replace the water vapor lost to the ice crystal. The process continues until all of the cloud droplets are used up. At this point, snow crystal formation ceases.

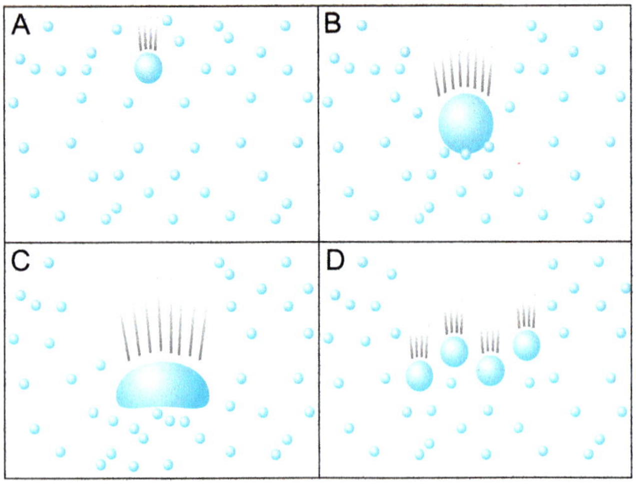

Figure 3.48: The collision-coalescence process explained. A. A large cloud droplet begins falling through a cloud. B. As the large cloud droplet collides with smaller droplets, it consumes them, growing in size. C. When the cloud droplet reaches a size of ~5 mm, it becomes unstable. D. The large droplet then breaks into smaller droplets, each of which then continues the process. Over time, this "domino effect" can produce heavy rains.

The Collision-Coalescence Process

Rainfall is produced by a different process when cloud temperatures are too warm for ice crystals to form. These conditions are especially common throughout the year in the tropics and during unusually hot days in the mid-latitudes. The **collision-coalescence** process explains how rainfall is produced in warm clouds, clouds that have temperatures above freezing. The collision-coalescence process requires that the cloud droplets have different sizes, which causes the fall speeds of the cloud droplets to vary. This means that the larger drops descend more rapidly than the smaller ones. As the larger cloud drops fall they overtake and collide with the smaller ones, merging with them. The merging of cloud droplets by collision is called coalescence. This causes the drops to grow in size and fall at a faster rate. Sometimes the raindrops become so large that they immediately break up into numerous droplets that are much smaller (Figure 3.48). The smaller raindrops act as new precipitation embryos that can grow large enough to break apart and create more raindrop embryos. This process can initiate a chain reaction if vertical uplift is strong enough to keep the cloud droplets suspended within the cloud. The droplets eventually become large enough to fall from the cloud and reach the Earth's surface as raindrops. The chance of a heavy burst of rain is increased when the raindrops rapidly grow and descend rapidly through this process.

Section 3-10: Measuring Atmospheric Humidity

Relative humidity is the ratio of the actual vapor pressure of the air to the saturation vapor pressure at a specific temperature. It is expressed as a percentage and is the most common way to describe atmospheric moisture even though it is often misunderstood. When the air is saturated the relative humidity is equal to 100%. Using relative humidity to describe the moisture content of the air can be confusing since it measures how close the air is to being saturated; it does not indicate the actual amount of water vapor in the air. Relative humidity is exactly that, a relative measure of humidity, it is relative because it depends on the temperature of the air. The relative humidity will increase or decrease depending on how the air temperature changes. Figure 3.49 graphically depicts the inverse relationship between relative humidity and air temperature over a 24 hour period. Notice that the relative humidity is the highest in the early morning when the air is closer to saturation at low temperatures. As the air temperature rises throughout the day the relative humidity drops due to an increase in the saturation vapor pressure. The relative humidity increases during the evening as the air becomes closer to saturation at a lower temperature. In many places the total water vapor content of the air changes only slightly throughout the day. This means that air temperature is the primary controller of relative humidity. This makes it hard to make direct comparisons of humidity between different locations. For example, the relative humidity can be identical between two locations even though the actual water vapor content of the air may be markedly different. Many people assume that the relative humidity approaches 100% on very hot, humid days. The relative humidity will never get close to saturation under these conditions since the saturation vapor pressure is so large. Relative humidity rarely exceeds 50% when air temperatures exceed 38°C (100°F). Although relative

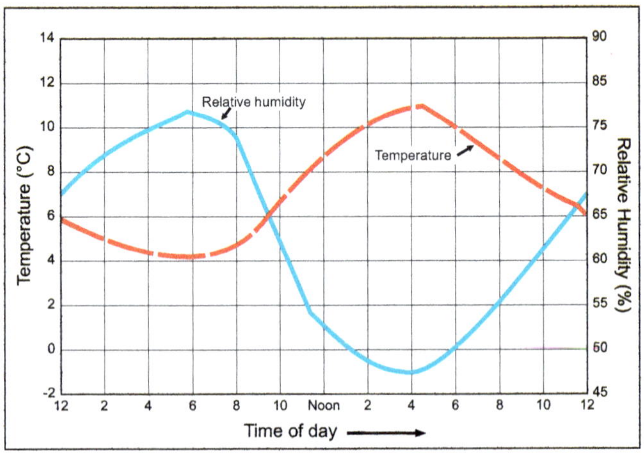

Figure 3.49: A plot of relative humidity and air temperature for an average day. Notice how there is an inverse relationship between the two variables. When the temperature goes up, relative humidity goes down. Thus, the highest values of relative humidity occur at night.

humidity cannot be used to make direct comparisons in humidity between locations, it is useful to determine how close the air is to saturation. Weather forecasters use relative humidity to determine the likelihood of fog formation, clouds and precipitation.

The **dew point** is another common way to express humidity, it measures the actual amount of water vapor in the atmosphere. The dew point is the temperature at which air must be cooled, at a constant pressure, to reach saturation. The dew point varies as the moisture content of the air changes, higher dew points indicate a greater concentration of atmospheric water vapor. The relative humidity is high when there is a small difference between the air temperature and the dew point. The temperature of a surface must be equal to the dew point for water vapor to condense as dew. Dew often forms on grass on clear and calm mornings as the air in contact with the ground is saturated and cooled down to the dew point. In this case the temperature at which dew formed on the grass would equal the dew point. The **frost point** is used to measure the moisture content of the air when air temperatures are at or below freezing. The frost point is the temperature to which air must be cooled at a constant pressure to reach saturation with respect to ice. There are several different types of frost that form when conditions are below freezing. The specific type of frost that forms is dependent on the atmospheric conditions and the temperature of the object that the frost forms on (Figure 3.50).

Figure 3.50: Frost has formed on one of the author's cars. Photo credit: B. Zaprowski.

When the air temperature is 26.5°C (80°F) and the dew point is 4°C (40°F), the humidity is low and the air must be significantly cooled to reach saturation. If the dew point increases to 21°C (70°F) with no change in air temperature, the humidity is high and not much cooling is needed to reach saturation. Ideal weather conditions can be expressed in terms of more than just air temperature. The dew point can also be used as an index related to human comfort. Most people are comfortable when the dew point is below 15.5°C (60°F). When the dew point exceeds 20°C (68°F) it will feel uncomfortably humid to most people. The danger of heat-related illness is elevated when the dew point exceeds 24°C (75°F), the high humidity will feel extremely oppressive to everyone. Dew points are typically between 21-24°C (70-75°F) for the Gulf Coast of the United States during July. The air during this time can contain roughly twice the amount of atmospheric moisture of inland locations with a dew point around 15°C (59°F). High humidity also directly relates to athletic performance. For example, the finishing times for marathons increase when runners are exposed to high levels of humidity.

The dew point is the measure of atmospheric moisture that is used on daily weather maps since it is easy to measure and is directly related to the amount of water vapor in the air. The dew point is always equal to or less than the air temperature, there will never be a situation where the dew point will exceed the air temperature. If no frontal systems are expected to pass over a region the dew point temperature can be used as an aid to predicting minimum temperatures since the temperature can never fall below the dew point. The dew point is also useful for predicting the likelihood of cloud formation, the probability for cloud formation will increase as the difference between the air temperature and the dew point temperature decreases. Clouds can generally be expected when the dew point temperature is within a few degrees of the air temperature. The height of the cloud base can also be estimated if the dew point temperature is known. Weather forecasters use the dew point temperature to identify areas that are at risk for severe weather. The dew point is also useful to predict the formation of dew, frost and fog.

The moisture content of the atmosphere can also be described in terms of its drying potential. The **vapor pressure gradient** is the difference in vapor pressure between the actual vapor pressure and the saturation vapor pressure at the same temperature. The size of the vapor pressure gradient indicates the atmosphere's evaporation potential. The vapor pressure gradient is large when there is a small amount of water vapor in the air. Net evaporation is high under these conditions since the actual vapor pressure is low compared to the saturation vapor pressure (Figure 3.51A). High vapor pressure gradients promote evaporation of moisture from your skin, which can quickly lead to dehydration. This explains why it is so important to stay hydrated in arid climates, you should continuously drink

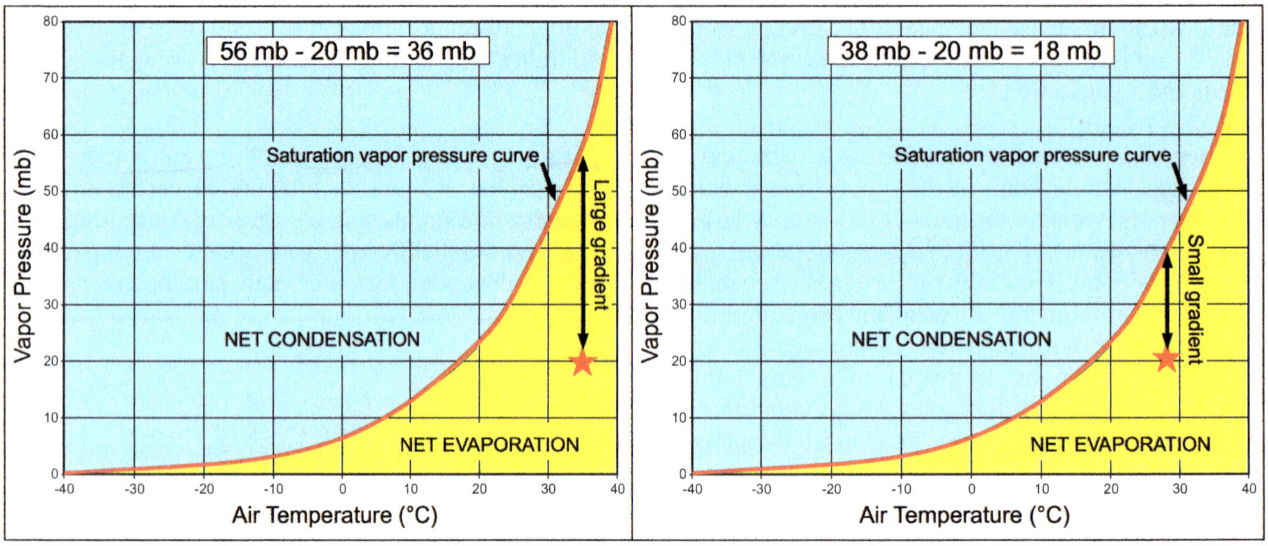

Figure 3.51: Two examples of the vapor pressure gradient. A. The vapor pressure difference is large so net evaporation would be large. B. The vapor pressure difference is small so net evaporation would be small.

fluids even if you don't notice perspiration on your skin. Vapor pressure gradients are small when the air is humid, resulting in low net evaporation (Figure 3.51B). Notice how most people tend to feel uncomfortable on a humid summer day. This is because high humidity reduces the overall evaporation of sweat from your skin. Evaporation from the skin is so minimal that that the body cannot cool itself even by sweating when the air is humid and at temperatures above the body temperature. Since evaporation is a cooling process, your level of discomfort is decreased when your sweat evaporates from your skin. Evaporational cooling is greatly reduced whenever the air is humid, leading to a higher degree of discomfort as compared to being exposed to similar temperatures with lower humidity. This explains the common saying often heard during the summer, "It's not the heat, it's the humidity." The best drying potential of the air can be related to drying clothes on a clothesline. Evaporation potential is high when conditions are windy with sunny skies and low humidity. Clothes will dry very quickly under these conditions compared to a cloudy day with high humidity and light winds.

Section 3-11: How to Change the Humidity of the Air

The various ways to change the humidity of the air can be discussed once that the basic ways to measure atmospheric humidity are understood. Evaporating water into the air is the most basic way to change atmospheric humidity. For example, evaporation of water is high when a cold, dry air mass from Canada moves over the relatively warm water of the Great Lakes. The heat and moisture is transferred from the water surface to the overlying atmosphere, providing the necessary conditions for "**lake-effect snow**" (Figure 3.52). The rate and intensity of lake-effect snow is determined by the difference between the lake's water temperature and the air temperature. The difference in temperature between the lake and the air must be at least 10°C (18°F) for lake-effect snow to develop. Larger temperature differences allow for more heat and moisture to be transferred to the atmosphere from the lake surface, providing more fuel for lake-effect snow. Lake-effect snow is also enhanced when the air passing over the lake surface is forced to rise due to topography, the additional uplift helps to cool and saturate the air.

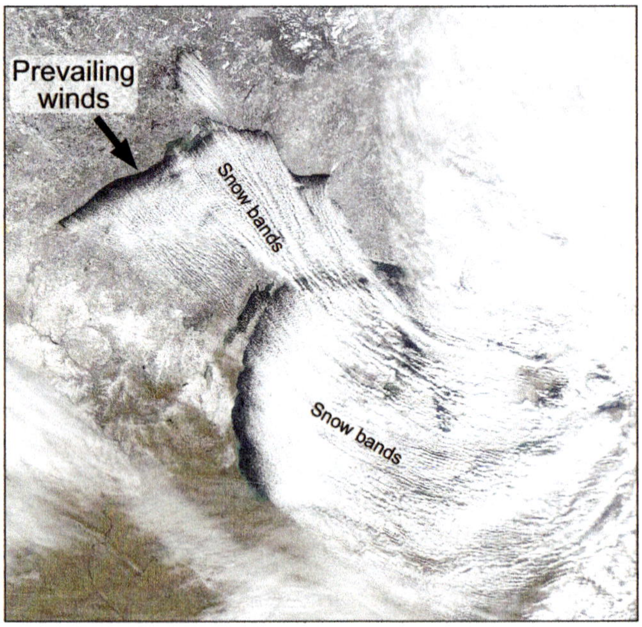

Figure 3.52: Lake effect snow. This picture from November 30, 2004 shows how cold winds blowing over relatively warm lake waters results in the formation of clouds and heavy snows. Credit: SeaWiFS Project, NASA.

Lake-effect snow tends to form east of the Great Lakes in narrow bands that are generally within about 50 miles downwind of the lakes. The lake-effect snow bands typically form parallel to the prevailing winds, making wind direction a critical element when forecasting which regions will have the highest snowfall totals. Large variability in lake-effect snowfall is common due to the narrow bands of snow that form. These bands may only be a few kilometers (miles) wide, making it very difficult to accurately forecast (Figure 3.53). It is not uncommon for one location to receive more than two feet of snow while little if any snowfall occurs for areas only 30 km (20 mi.) away. The Great Lakes' lake-effect snow season generally begins in November and lasts until the lakes become ice covered in February.

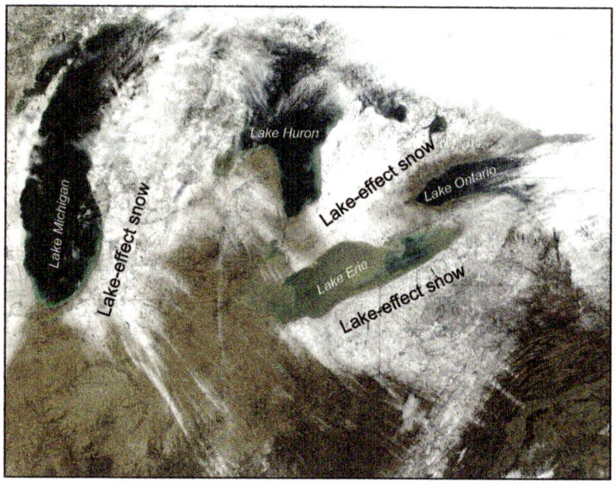

Figure 3.53: After a couple days of heavy snow, distinct bands of lake-effect snow have accumulate to the south and east of the Great Lakes. Photo credit: MODIS Rapid Response Team at NASA GSFC.

The dew point can also change when cold air is mixed with warm, humid air. Small eddies occur in the daytime when surface air is warmed by the Earth's surface and then rises. When eddies form a few thousand meters above the surface they mix drier air downward. This effect is most noticeable during the summer on a hot, humid and sunny day. As eddies are generated by solar heating, drier air above the Earth's surface is transported downward and mixes with the humid air near the ground, resulting in a noticeable decrease in humidity. Humidity can also increase or decrease as a humid or dry air mass is advected into a region. The term **advection** refers to the horizontal transport of an atmospheric property, in this case it refers to humidity.

When air parcels rise, they expand and cool adiabatically (adiabatic processes were discussed in Chapter 1, more details to follow in Chapter 4). The rising parcel will eventually cool to its dew point temperature, resulting in net condensation. The atmospheric pressure of the air parcel is initially the same as the surrounding air (Figure 3.54A). The pressure of the rising air parcel is reduced as it is lifted from the ground. The air parcel expands in response to the reduction in pressure with increasing elevation (Figure 3.54B). The air molecules will lose some of their kinetic energy as the parcel expands. This causes the temperature inside the air parcel to decrease, the temperature of the parcel will continue to decrease through expansional cooling as long as the air parcel continues to rise and expand. If the rising parcel is lifted high enough it will eventually cool to the dew point, resulting in net condensation (Figure 3.54C). The altitude where the parcel becomes saturated and clouds form is called the **lifting condensation level** (LCL). The LCL marks the cloud base of convective clouds, and it is clearly visible under normal sky conditions (Figure 3.54D). The LCL determines how far the air parcel needs to be lifted to produce clouds, it will be closer to the ground when the air is relatively humid. This is because less cooling is needed for the rising air parcel to reach saturation.

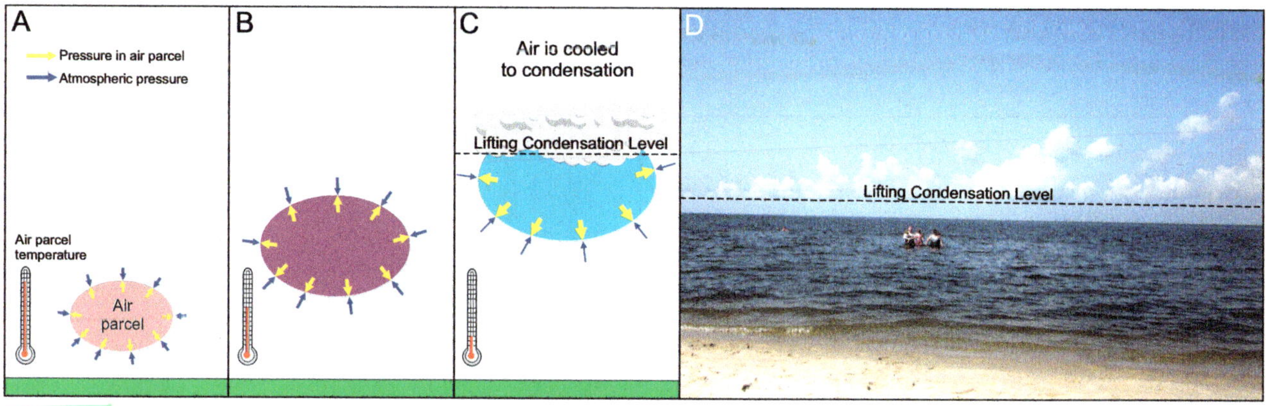

Figure 3.54: Air parcels cool as they rise. A. At ground level, the pressure in the air parcel is equal to atmospheric pressure. B. As the air parcel rises, the atmospheric pressure decreases, which in turn causes the air parcel to expand. As a result, the temperature of the air parcel goes down. C. With enough lifting and cooling, condensation may occur in the air parcel. D. An example of the lifting condensation level. The flat base of these cumulus clouds show the altitude where air masses are reaching the dew point. Photo credit: B. Zaprowski.

Lifting Processes

Rising air will cool to the dew point when lifted high enough, but what causes the air to rise? Air located near the ground has a tendency to remain in its original position, but when clouds form there is often a lifting mechanism that forces the air to initially rise. There are four principle mechanisms that cause the air to rise resulting in cloud formation.

As discussed earlier, convection is an important way to transfer energy away from the Earth's surface through the vertical movement of air, it is very important for cloud formation. Uneven surface heating can lead to the **convective lifting** of air parcels. On warm summer days rising air parcels are lifted convectively through uneven surface heating. This often results in rapidly developing afternoon thunderstorms that form in association with the rising convection currents. The rising convection currents are called **thermals**, birds of prey like hawks and eagles use these thermals to carry them to high altitudes when searching for food. Convective lifting occurs because the rising air is warmer and more buoyant than its surroundings, hot air balloons use this same principle to lift off the ground.

When unlike air masses meet a front is established at the transition zone between the two air masses. Clouds and precipitation are often caused by **frontal wedging** as the warm air is forced upwards over the cold air due to a difference in density. As the warmer air mass rises it will cool adiabatically, leading to clouds and possibly precipitation if the air mass contains enough moisture (Figure 3.55). Frontal lifting is responsible for a high percentage of precipitation across the mid-latitudes due to frequent collision between unlike air masses. Identifying the position and movement of fronts is extremely important for reliable weather forecasts since clouds and precipitation are associated with fronts.

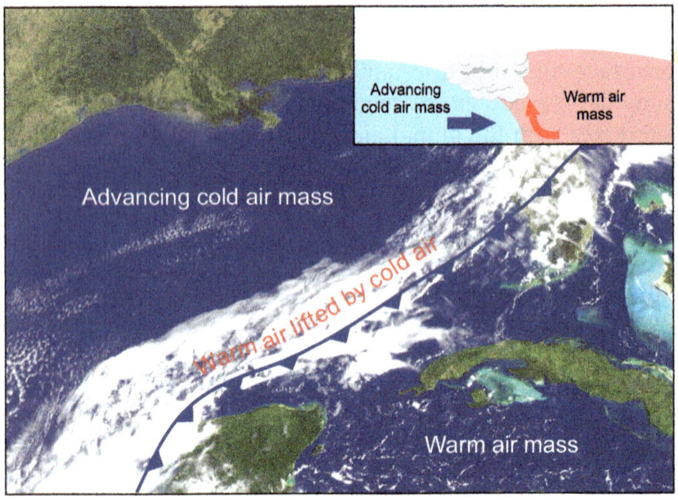

Figure 3.55: Frontal wedging, in this case, along a cold front.Credit: NASA-Goddard Space Flight Center, data from NOAA GOES.

Orographic lifting occurs when air is forced to rise due to topography. A mountain range acts as a natural barrier when it is oriented perpendicular to the prevailing wind direction. As air flows towards higher elevations, such as a mountain range, it will be forced to rise over the barrier. As the air ascends the mountain, it will cool and water vapor condense if the air is cooled to the dew point (Figure 3.56). Orographic lifting is the primary cause of precipitation in mountainous regions, it can produce a tremendous amount of variability in annual rainfall due to changes in elevation. Landscapes can look vastly different on opposing sides of a mountain range, one side of the mountain can be full of lush forest vegetation while the other side has the vegetation characteristics of a desert (Figure 3.56B). Some of the wettest places in the world are located on the **windward** sides of mountains, the side that is facing the prevailing wind direction. Most of the moisture will be lost before the rising

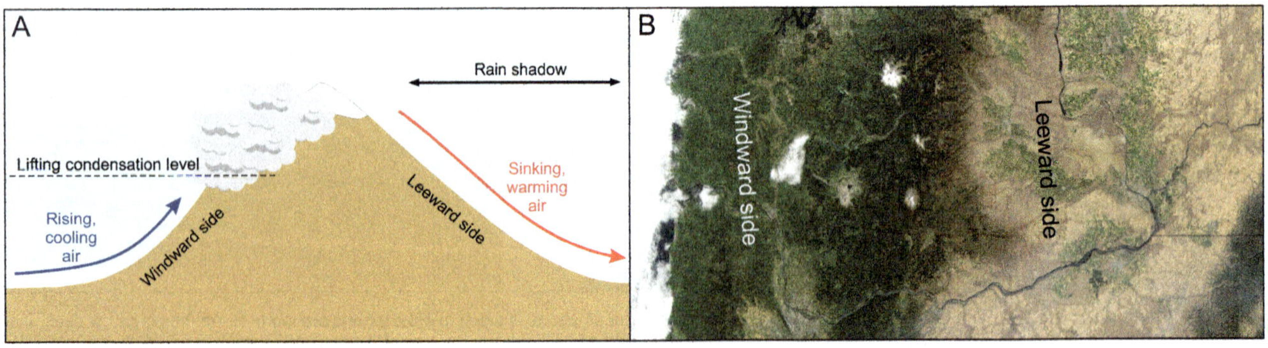

Figure 3.56: A. Orographic lifting explained. B. The Cascade Range in Washington and Oregon is an excellent example of this concept. The darker areas are temperate rainforests while the lighter areas are semi-arid grasslands. Photo credit: Jacques Descloitres, MODIS Land Rapid Response Team/NASA.

air reaches the opposite side of the mountain, known as the **leeward** side. The air will warm adiabatically as the air descends down the leeward side of the mountains. This will reduce the relative humidity of the air, resulting in clear skies and dry conditions. Mountain ranges often produce two distinct climate zones, a wet climate on the windward slopes and a dry climate on the leeward slopes.

A classic example of the orographic effect occurs along the West Coast of the United States. The development of clouds and resulting precipitation that extend from Northern California through the Pacific Northwest are mainly due to this process. Figure 3.57A is a topographic map of Northern California, while Figure 3.57B shows the average annual rainfall for the region. The prevailing winds over California are from the west, originating over the Pacific Ocean. The close relationship between average annual precipitation and elevation can be observed by comparing the two figures. The heaviest precipitation generally occurs at the highest elevations or where the elevation is rising. The Central Valley of California, where air descends off the mountains to the west, has little annual rainfall due to the effects of topography. The descending air warms up and dries out as it travels down the lee side of the mountain. Dry conditions often extend a great distance on the lee side of a mountain range, this zone is called a **rain shadow**. To the east of the Central Valley, precipitation increases again as the prevailing westerly winds encounter the Sierra Nevada Mountains. Many of the world's deserts are located on the leeward sides of mountains. Ironically some of these deserts are near an ocean but are cut off from the source of moisture by the mountains.

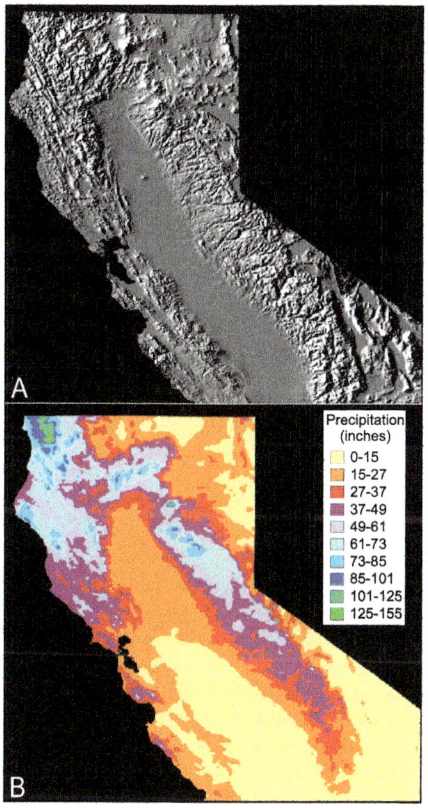

Figure 3.57: An example of the orographic effect. A. A shaded relief map of Northern California. Elevation data courtesy USGS. B. A map of mean annual precipitation for California. Note how the maximum precipitation generally corresponds with higher elevations. Precipitation data courtesy Oregon Climate Service.

Convergence is the least common of the four lifting mechanisms although it is significant in some locations. Air can't pile up at the surface when air flows in from different directions, it must go somewhere. When winds near the Earth's surface converge, upward vertical motion results; this leads to cooling and possibly cloud formation. When air moves over different surfaces the differences in friction affects the wind speed. Wind speed increases when air moves from a relatively rough surface to a relatively smooth surface. The opposite effect occurs when air moves from a smooth surface like the ocean to a rougher surface onshore. Florida is a great example of how convergence leads to cloud development (Figure 3.58). Air converges over the peninsula on warm afternoons due to intense solar heating. As the air converges over land, it is forced to rise and results in frequent afternoon thunderstorms throughout much of the year (Florida is one of the lightning capitals of the world!). The inward and upward motion of low-pressure systems is associated with convergence. There is a continuous band of cloud cover that exists in the tropical latitudes as a result of air converging from opposite hemispheres. The strength of convergence is related to the extent of vertical motions, clouds and precipitation are likely to form when convergence is strong. Convergence of air is not restricted to the surface, it occurs above the surface as well.

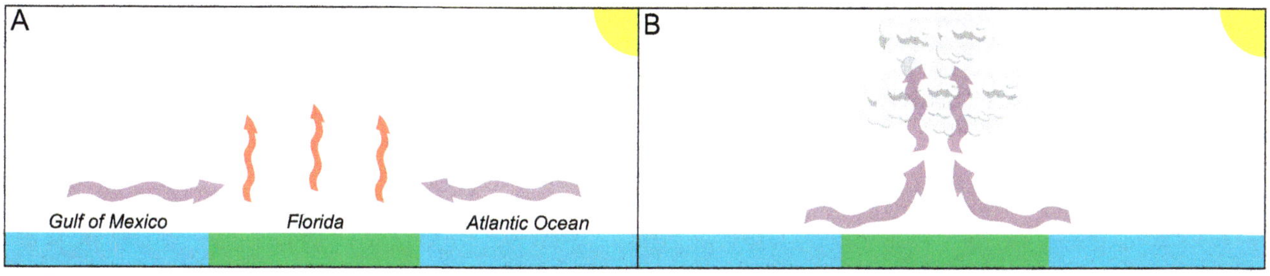

Figure 3.58: Convergence is a very common cause of lifting in Florida. A. The heating of the land leads to the formation of thermals which draw in humid air from over the ocean. B. The air converges over land and is forced to rise, resulting in clouds and precipitation.

NOAA

NOAA

Chapter 4 – Weather and Weather Forecasting

In the last chapter we looked at the structure and behavior of the atmosphere in some detail. It's one thing to know the parts of the atmosphere, how it behaves and why, but it is a completely different thing to actually be able to forecast the weather. In this chapter we will first look at global weather patterns, how they change throughout the year, and how they influence local weather patterns. Then we'll look at the role of upper-level winds and the jet stream in weather forecasting. Next we'll explore clouds and precipitation and the conditions that lead to the formation of each. Finally, we will look at one of the most dominant weather-producing phenomena in the middle latitudes, the mid-latitude cyclone.

Section 4-1: Global Atmospheric Circulation Patterns

Global atmospheric circulation is driven by the rotation of the Earth and the incoming energy from the sun. Meteorology, the science dealing with the atmosphere and its phenomena, is considered a youthful science compared to other scientific disciplines. Theories related to atmospheric behavior before the 1900s were based upon limited data. In 1735, George Hadley developed a simplistic air circulation scheme that was based on the known facts at the time (Figure 4.1). Although Hadley's theoretical model has been improved upon, he is credited for explaining the mechanism for convection in the atmosphere.

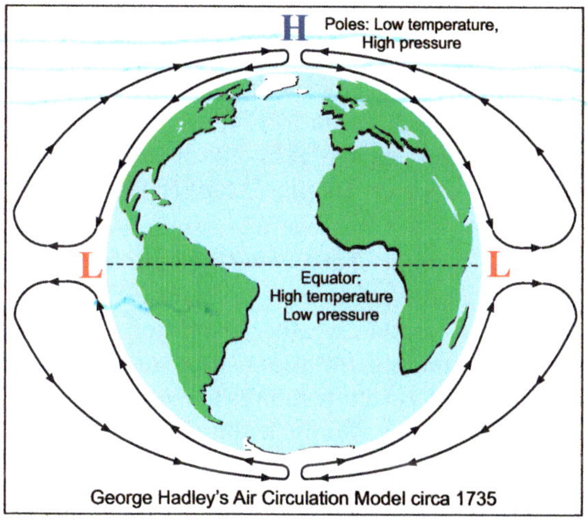

Figure 4.1: George Hadley's Air Circulation Model circa 1735. Although this model is not correct, Hadley's ideas were the foundation upon which future studies were based.

George Hadley was motivated to create his global model of atmospheric circulation because he was not satisfied with existing theories that explained the trade winds. He knew that temperature differences create pressure differences which drive winds. Latitudinal temperature bands provided a hypothesis about the distribution of pressure patterns and the resulting winds. Surplus energy and high temperatures near the equator should produce low pressures, causing the air to rise through convection. The rising air, being affected by gravity, is directed poleward in both hemispheres. Low temperatures and sinking air in the polar regions should produce high surface pressure. According to Hadley's model, surface winds should blow from poles to the equator, and upper-level winds should blow from the equator to the poles. This would create one complete convective cell in each hemisphere (Figure 4.1). Even though Hadley's model of global atmospheric circulation is simplistic and highly idealized, it was good for its time and it provides an explanation for the formation of the trade winds.

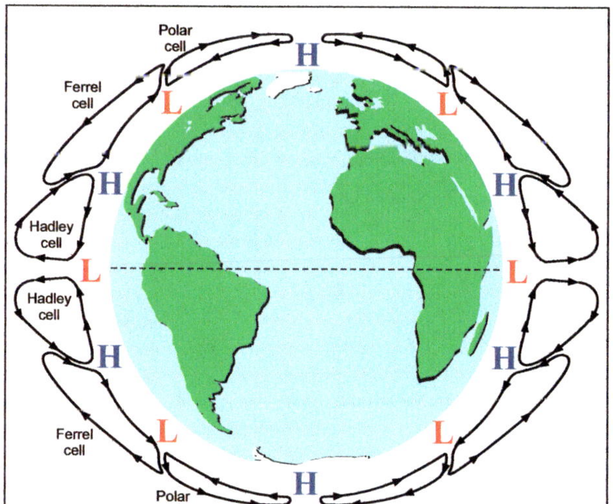

Figure 4.2: The general pattern of global atmospheric circulation. Each hemisphere has three circulation cells. *NOTE:* This diagram is only a generalization.

Unfortunately, Hadley's simple model does not fit reality. His concept of air circulation lacked consideration of factors other than the pressure gradient force that affect winds. First, his model ignores the influence of Coriolis force; a force that was not described until 100 years after his model was first devised. Second, rising air eventually sinks back towards the Earth's surface as it gets colder and more dense. This results in poleward-moving air being forced into a smaller space as density increases. Hadley's model also assumed that the sun remains in a fixed positon over the equator, creating equinox conditions at all times. In the 1920s a new conceptual model was devised with three cells in each hemisphere instead of a single cell (Figure 4.2). Meteorology textbooks today use a three cell model because it better represents the typical wind flow on a global scale. The tropical cells, named after George Hadley, are called the **Hadley cells**. The **Polar cells** are high latitude cells of circulation and the **Ferrel cells** are located in the mid-latitudes between the Polar cells and the Hadley cells. These three distinct convection cells in each hemisphere essentially divide the troposphere into regions of closed wind circulations.

Hadley correctly predicted that surplus energy in the equatorial region would result in continuously warm temperatures, low pressure and rising air. Winds are light near the equator due to a weak horizontal pressure gradient. Diverging air aloft helps to create a wide band of low atmospheric pressure at the surface in the tropics. This low-pressure region located between the subtropical high-pressure belts of the Northern and Southern Hemispheres is known as the **Equatorial Trough** or meteorological equator (Figure 4.3). In terms of wind characteristics, this region is sometimes recognized as the Doldrums or **Inter-Tropical Convergence Zone (ITCZ)**.

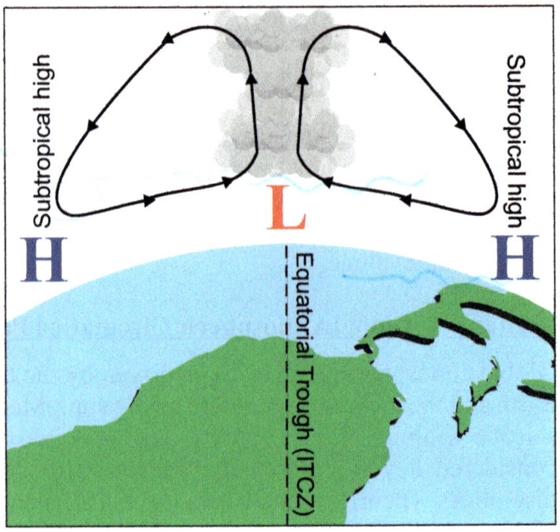

Figure 4.3: A closer look at the lower latitudes. Over the equatorial regions, warm, moist air rises producing low pressure systems (equatorial trough) and copious amounts of rain. Conversely, sinking air masses over 30° latitude produces a series of high pressure systems (subtropical highs) and dry, fair weather.

Tropical locations that receive high rainfall throughout the year are sometimes said to have a seasonless climate because they experience similar weather conditions day after day. Daily weather forecasts for these areas are very easy to predict. Expect afternoon temperatures to reach about 30°C (86°F) with a brief mid to late afternoon thunderstorm. The high humidity allows for the almost daily occurrence of thunderstorms and keeps the overnight lows around 21°C (70°F). The clouds release huge amounts of rainfall, well over 250 cm (100 in.) per year (~7 mm or ¼ in. per day) over a large region that straddles the equator (Figure 4.4). The humid atmosphere also releases massive amounts of latent heat through condensation, which helps fuel thunderstorms and provides energy to drive the Hadley cells (Figure 4.3). The rising air is directed poleward in both hemispheres when it reaches the tropopause. The stability of the tropopause acts like a "lid" for the troposphere and inhibits further upward motion. Because it is difficult for air to

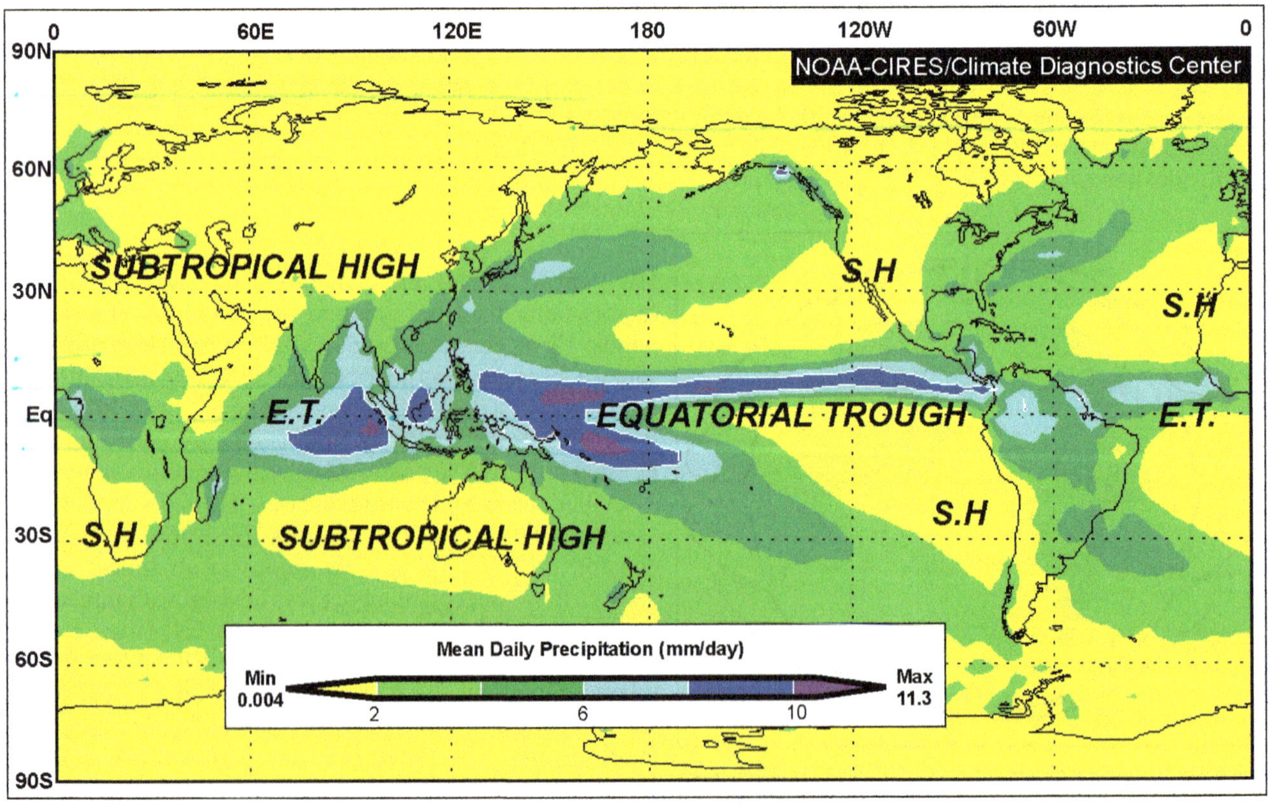

Figure 4.4: Long term mean daily precipitation (mm/day) for the entire earth. Data courtesy NOAA-CIRES/CDC.

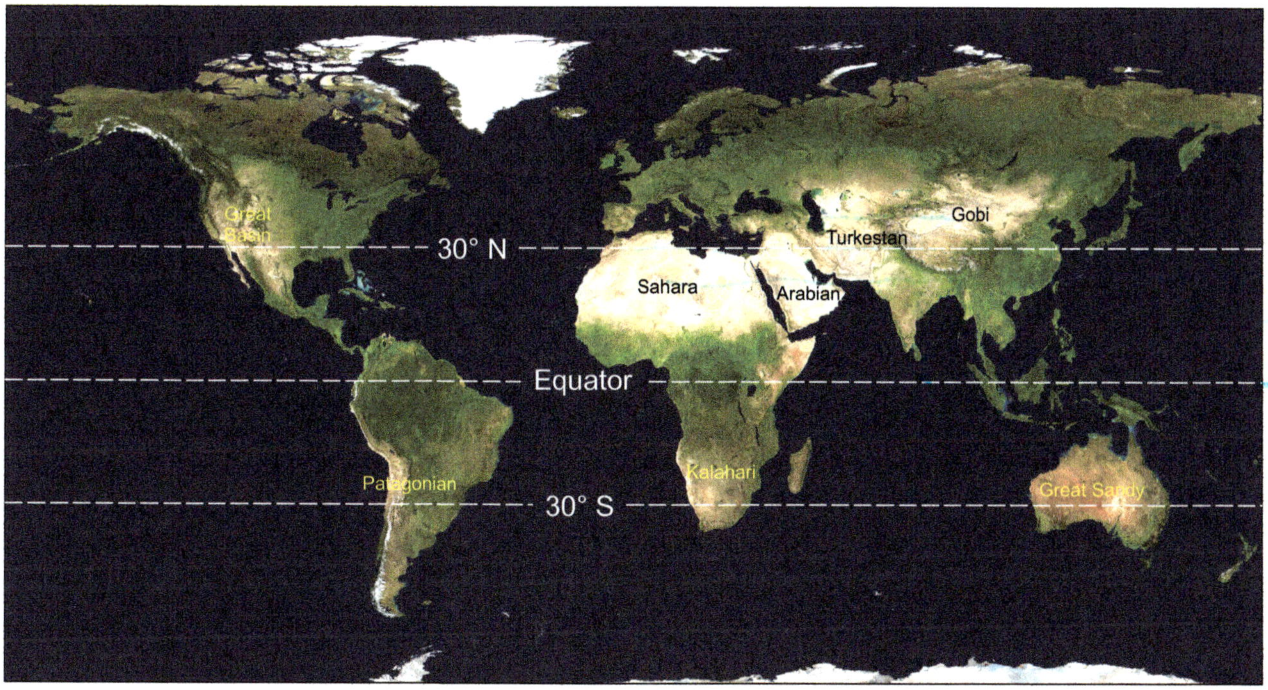

Figure 4.5: A map of the world's major deserts. Note how most of them are found near 30° latitude. Image credit: Reto Stöckli/NASA.

rise above the tropopause, clouds are typically confined below this stable layer. The poleward flow of air in the upper atmosphere is then deflected by the Coriolis force (Chapter 1) to create westerly upper-level winds in both hemispheres. The deflected air continues to cool as it moves poleward, causing it to sink around 30° on both sides of the equator. Surface pressure increases around these latitudes from the sinking air and produces bands of high pressure systems. These regions of high pressure are called the **Subtropical Highs** (Figure 4.3). The sinking air associated with these areas of high pressure forms warm, dry air masses and for this reason, clear skies, relatively high temperatures and very low precipitation are characteristic of these latitudes. Most of the major desert regions of the world are associated with the subtropical highs and are mostly located near 30° latitude (Figure 4.5). These latitudes are often referred to as the horse latitudes, a nautical term that refers to the subtropical latitude belt known for hot and dry weather with calm winds. According to legend, Spanish sailors who set sail from Spain to the New World had to throw their horses overboard when they encountered calm winds associated with the subtropical high to reduce weight and conserve water.

A latitudinal pressure gradient must exist for the convective cells to form the general pattern of global circulation. For example, a pressure gradient between the tropics and the mid-latitudes must exist for the Hadley cell to form. The pressure gradient is directed in both hemispheres from the subtropical high to the equatorial trough (Figure 4.6). The direction of the air is modified by Coriolis force and friction as it flows towards the equatorial trough to produce the most reliable winds on Earth. Air is directed to the right of its original path by Coriolis force in the Northern Hemisphere. Surface wind speeds are reduced by friction which reduces the magnitude of Coriolis force. The combined effect of Coriolis force and friction generates surface winds that travel from northeast to the southwest, creating the **Northeast Trade Winds** (Figure 4.6). All winds are named for the direction they blow from, the term "trade winds" was coined by sailors on trade ships centuries ago who used these reliable winds to cross the oceans frm East to West.

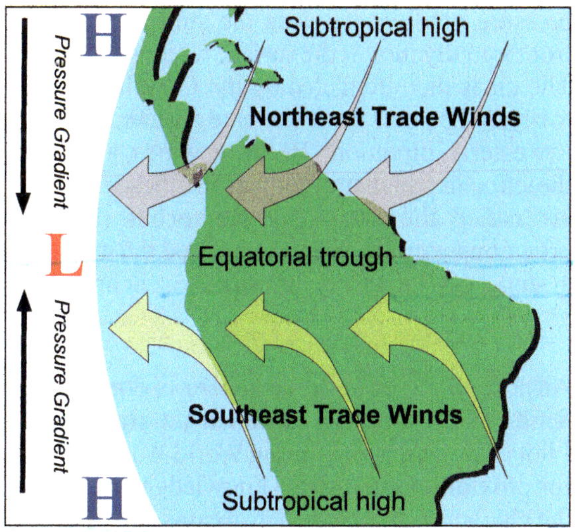

Figure 4.6: A look at why the Trade Winds form in the lower latitudes.

The predominant wind direction of Northeast Trade winds is from the northeast although they don't always blow from this direction. In the Southern Hemisphere air is directed to the left of its original path by Coriolis force to form the **Southeast Trade Winds**. The strength of the trade winds is mostly determined by the strength and location of the subtropical high pressure. The trade winds are stronger when the Subtropical Highs are stronger and closer to the equator, creating a stronger horizontal pressure gradient.

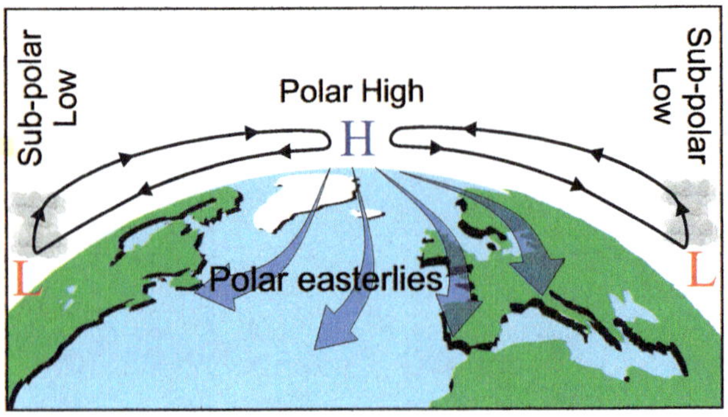

Figure 4.7: A closer look at the high latitudes. Cold, dense air masses sink over the poles, producing the polar highs. This cold air moves away from the poles towards the equator forming the polar easterlies.

High-pressure is always located over the polar latitudes due to the presence of cold, dry air. Air in the upper atmosphere converges as it moves from 60° latitude towards the Polar Regions. The cold, dense air sinks at the poles to form high-pressure systems referred to as the **Polar Highs** (Figure 4.7). The cold, dense surface air is deflected by Coriolis force as it moves towards the mid-latitudes in response to the pressure gradient force. The combined effect of the pressure gradient force, the Coriolis force and friction creates the **Polar Easterlies** (Figure 4.7)

The mid-latitudes (between 30° and 60°) are known for very active weather patterns due to their position between two unlike air masses in the tropical and polar latitudes. The polar easterlies clash with the relatively mild air at 60° latitude to create a band of stormy conditions. The warm air rises to create clouds and rainfall where the two air masses meet. The boundary between the Polar cell and the Ferrel cell in each hemisphere is known as the **Polar Front** (Figure 4.8). The position of the polar front migrates throughout the year as the positions of the subtropical high and polar high change. The polar front is located at higher latitudes during the summer and lower latitudes during the winter. The polar front is located in a region of relatively lower pressure called the **Subpolar Low**. This region exists because high pressure is found to the north and south of the polar fronts, resulting in lower pressure due to the convergence of surface air. The pressure gradient between the subtropical high and polar front initially moves the surface air towards higher latitudes. The air is then deflected by the Coriolis force as it moves towards the polar front to create a general airflow pattern in a westerly direction. Because winds are named for the direction in which they blow from, these mid-latitude winds are called the prevailing **Westerlies** (Figure 4.8). The prevailing westerlies are considered the most variable wind systems on Earth; they are strongest in areas where land is absent.

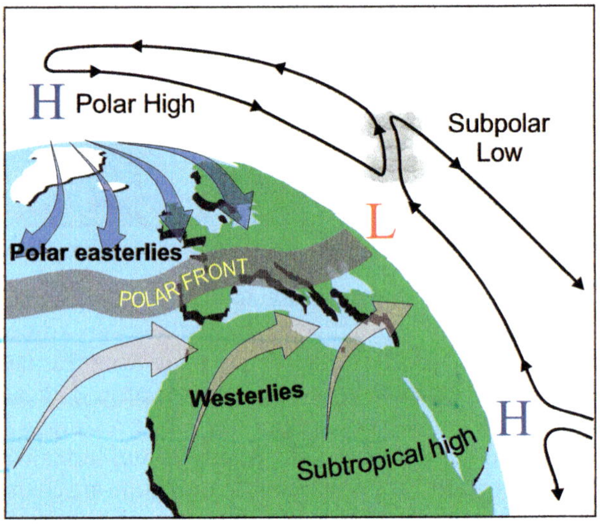

Figure 4.8: A look at the wind patterns in the middle latitudes. The polar front forms when cold air from the poles clashes with warm air coming from the tropics.

Airflow in the upper troposphere is dominated by narrow bands of fast flowing air called **jet streams** (Figure 4.9). Pilots flying missions during World War II are often credited for providing much of the knowledge we have today about the characteristics of the jet streams. Jet streams are caused

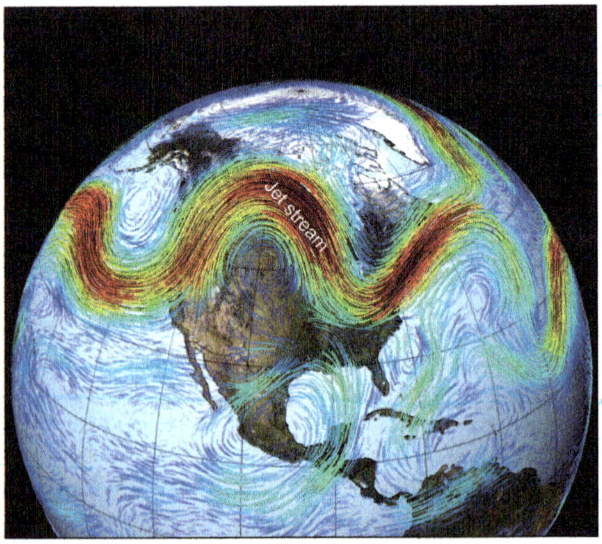

Figure 4.9: This visualization uses weather and climate observations from NASA's MERRA dataset to model 30 days of the jet stream's whirling journey over North America. Image credit: NASA.

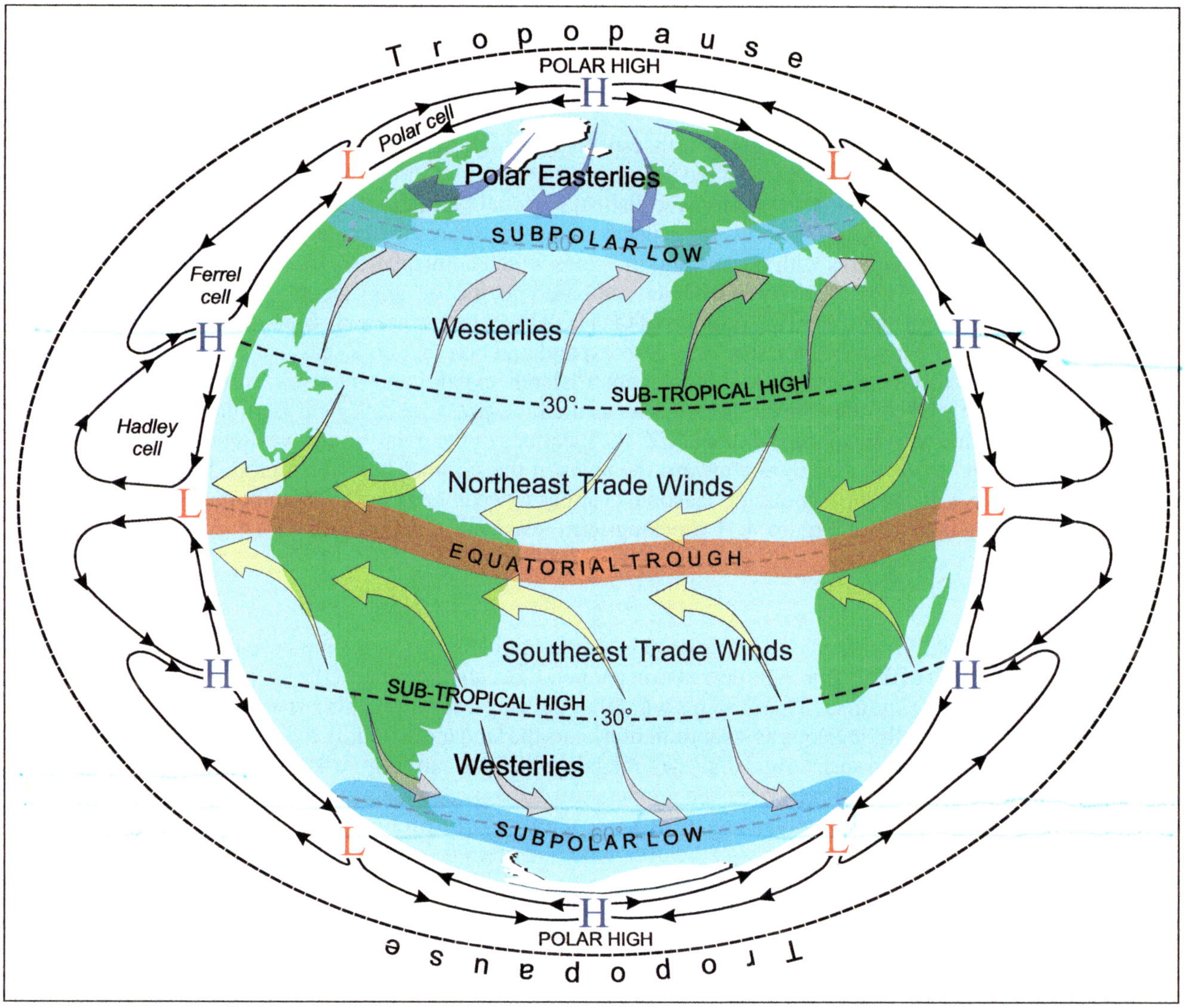

Figure 4.10: The general pattern of global atmospheric circulation and the resulting prevailing winds.

by a combination of a planet's rotation on its axis and atmospheric heating. There are typically two or three jet streams in each hemisphere that take a meandering path from west to cast. Jet streams play an important role in determining the weather because they separate colder air and warmer air. Meteorologists analyze the characteristics of the jet streams to note their influence on the steering of weather systems produce more accurate forecasts.

The general pattern of atmospheric circulation can be visualized as a global system once the components that comprise it are understood. Circulation around subtropical high-pressure systems creates the trade winds and westerlies in both hemispheres (Figure 4.10). Also notice how the equatorial trough is created by the convergence of the trade winds and how the polar fronts develop from the interaction between the prevailing westerlies and the polar easterlies. It is important to view the global pressure and wind pattern in Figure 4.10 as a long-term average and not as a predictive tool to forecast the weather over short time periods.

Local Variations in Atmospheric Circulation

The global model of atmospheric circulation is useful to explain the overall circulation of the atmosphere. This model effectively shows the long-term average of wind flow around latitudinal bands of high and low pressure. However, this model is of limited value explaining actual circulation patterns that exist at any given time and location. This model is only useful to illustrate the average conditions on a global scale. Keep in mind that the global model of atmospheric circulation is based primarily on latitudinal variations in solar radiation. and the resultant variations in atmospheric pressure. Actual conditions can vary considerably from the average due to local and regional differences in surface heating and seasonal changes in solar radiation (refer back to Chapter 3).

Seasonal Variations

Seasonal variations in solar intensity and day length, especially at high latitudes, produce large variations in surface heating throughout the year. The location of the pressure systems and wind belts migrate seasonally as the sun's declination changes. As a result some locations have distinct seasonal weather patterns that are associated with these variations. Figures 4.11A and 4.11B show average surface pressure patterns for January and July. The location of the equatorial trough (ITCZ) migrates with the sun and indicates which hemisphere is having its summer season. The ITCZ also moves further poleward over land as compared to the oceans due to greater variation of land temperatures. The seasonal changes in the position of the ITCZ are what are responsible for the wet and dry seasons in the tropics. The semi-permanent pressure cells also expand and contract on a seasonal basis. For example, the low-pressure cells located near Iceland and the Aleutian Islands expand and intensify during the winter. These pressure cells weaken and contract during the summer as the high-pressure cells near Bermuda and the Hawaiian Islands expand and intensity. In addition to the ITCZ, the locations of the polar fronts, westerlies, subtropical high-pressure systems, and trade winds migrate toward the poles in the summer hemisphere too. The high altitude jet streams also migrate with the sun, producing seasonal changes in weather as well. It is important to remember that the pressure systems shown on Figure 4.11 are long-term seasonal averages and actual conditions may vary significantly.

Daily Variations

When spending a day at the beach, a noticeable drop in temperature may occur during the early afternoon as a cool breeze begins to blow off of the water. This decrease in temperature can ruin a day at the beach in the spring but bring welcome relief during the summer. The onshore wind that develops during the day is known as a **sea breeze**. In coastal areas, circulation cells develops as solar radiation heats the land more rapidly than the water. Lower pressure develops over the warmer land surface and draws air inland from the ocean (Figure 4.12A). Sea breezes occur most often in the spring and summer months when the temperature difference between the land and water is the greatest. The strength of the sea breeze is directly proportional to the temperature difference between the land and sea. This is why sea breezes are typically the strongest in the afternoon and often form a line of clouds a few miles inland. At night, especially when conditions are clear and calm, temperature differences between the water and land can produce an offshore wind called a **land breeze**. Lower pressure develops over the warmer water as the land surface cools more quickly and draws air from the land out to sea (Figure 4.12B). It is not uncommon to see a line of clouds off the coast at night when the land breeze is particularly strong. Land breezes can occur at any time of the year but are the strongest during the fall and winter when water temperatures are still fairly warm and nights are cool.

In mountainous areas, another type of local wind develops through a process similar to sea and land breezes. During the day, a **valley breeze** develops as the sun heats up the mountaintops more than the valleys due to a more direct sun angle and thinner air. Areas of low pressure develop over the mountain peaks and cause air to rush up the mountain

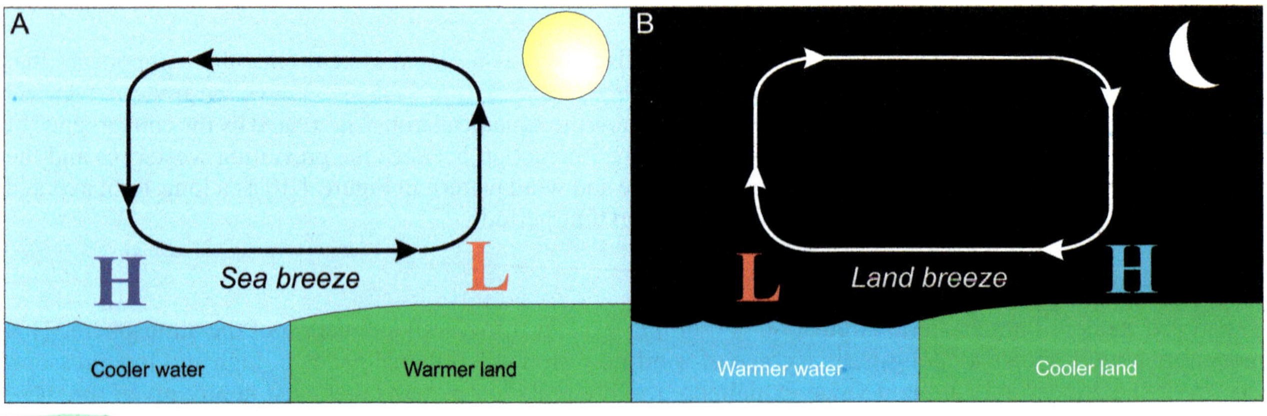

Figure 4.12: The development of land and sea breezes. A. Sea breezes develop during the afternoon. The wind blows from the sea towards the land as air moves from higher pressure over the relatively cooler water to lower pressure over the relatively warmer land. B. Land breezes develop late at night. The wind blows from land towards sea as air moves from higher pressure over the relatively cooler land to lower pressure over the relatively warmer water.

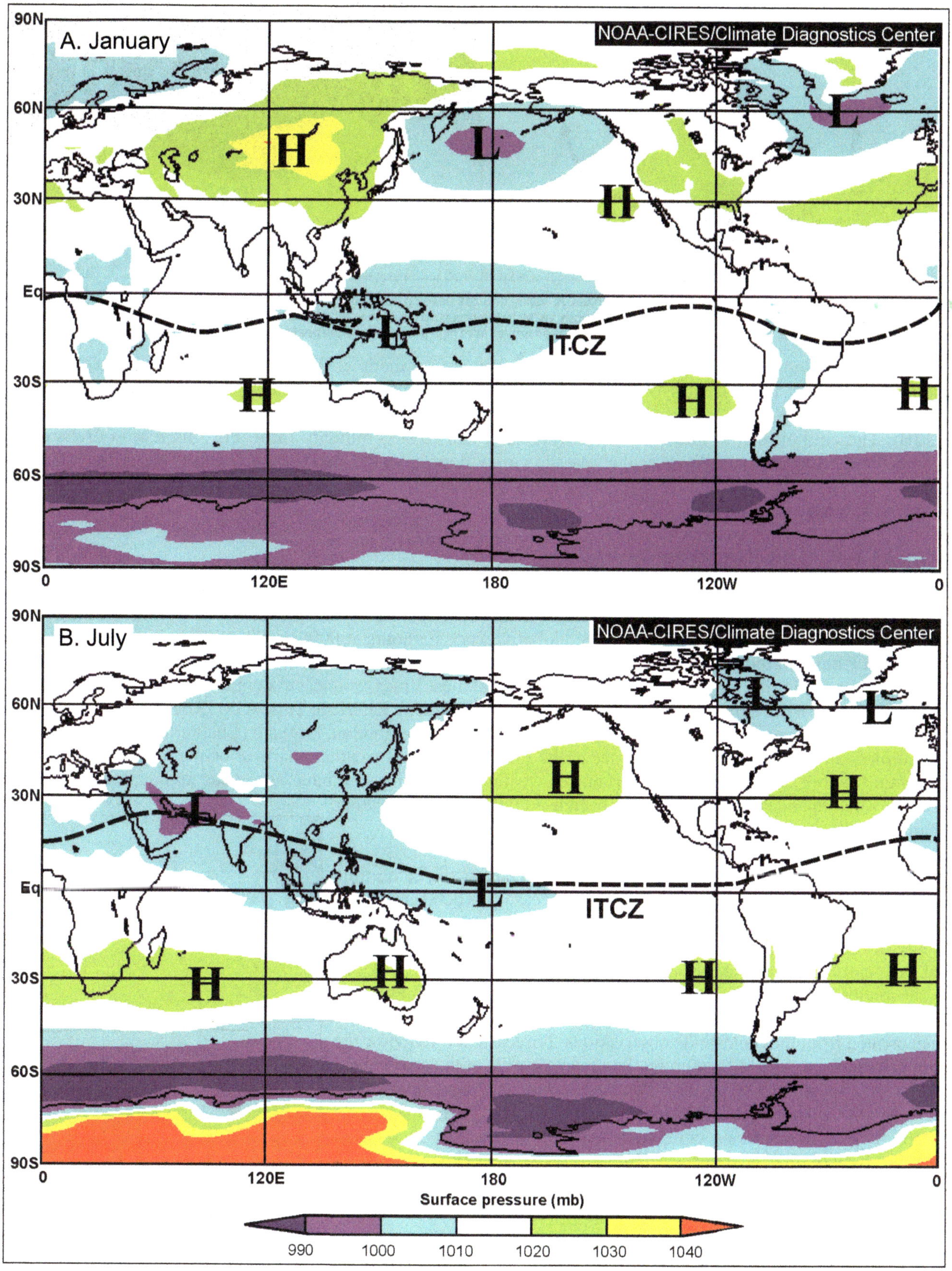

Figure 4.11: Long term average surface pressure from 1948-2005. A. January. B. July. Data courtesy NOAA-CIRES/Climate Diagnostics Center.

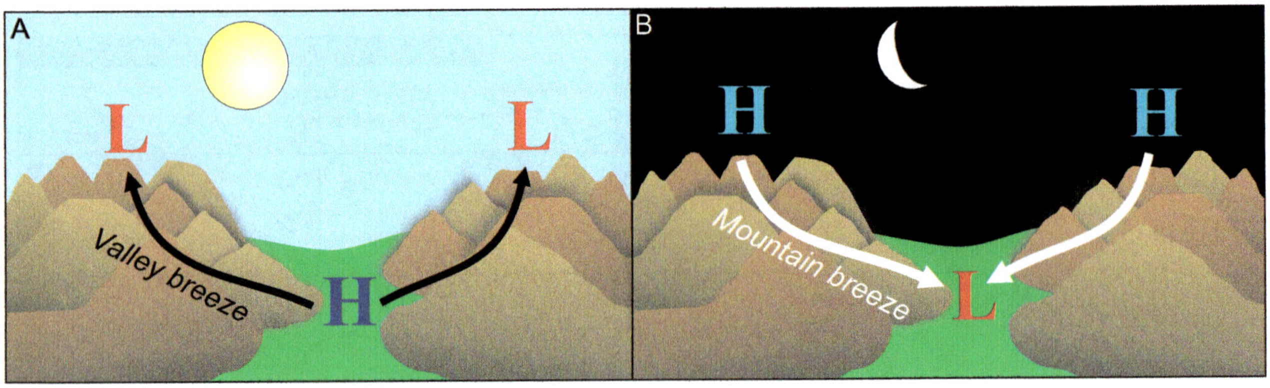

Figure 4.13: The development of valley and mountain breezes. A. Valley breezes develop during the afternoon. The wind blows up the valleys towards the peaks as the air moves from high pressure in the relatively cooler valley to low pressure over the relatively warmer peaks. B. Mountain breezes develop just before dawn. The wind blows from the peaks towards the valleys as air moves from high pressure over the peaks towards low pressure in the valley.

from the valley floor (Figure 4.13A).These rising currents can sometimes trigger sudden thunderstorms over the mountains. At night, the process is reversed as the mountaintops lose heat more quickly and grow colder. An area of lower pressure forms in the valleys and draws the cold, dense air downslope, producing a **mountain breeze** (Figure 4.13B). The cold air settles into the valley, where fog and frost commonly form. For this reason orchards in mountainous areas are typically placed above the valley floor along the intermediate slopes where they are less likely to freeze. Although mountain breezes are usually light, they can become a violent, gusty wind when the air is funneled through a narrow valley.

Section 4-2: Patterns of Upper-level Pressure and Winds

To describe weather patterns spatially, observations of basic weather elements need to be taken simultaneously over a large region. In order to do this nations have universally agreed to take weather observations at the same time four times each day. These observations are taken globally every six hours at 0000, 0600, 1200 and 1800 hours using **Universal Coordinated Time (UTC)**, the international standard for measurements. These four hours are called **synoptic hours**. Plotting the weather variables recorded at these times produces synoptic weather charts that depict the current state of the atmosphere. UTC time is also called Greenwich Mean Time (GMT) because it refers to the mean solar time measured at the Royal Observatory in Greenwich, London, England (Chapter 2). Ensuring that all meteorological measurements are recorded at the same time allows weather charts to be made on a global scale. Synoptic weather charts are very important for weather forecasting and have numerous applications. In recent decades satellite and radar imagery have been used with synoptic weather charts to understand the current state of the atmosphere.

Figure 4.14: Mandy Fetterman launches a 100-gram weather balloon and radiosonde as part of her duties to gather and record weather information aboard the USS Preble. Photo credit: U.S. Navy/Ramon Preciado.

Global upper-level observations are taken twice a day at the same time (0000 and 1200 UTC). These observations are necessary to detect and chart upper-level atmospheric features that are critical to weather forecasting. Observations are gathered from a radiosonde, which is an instrument package carried into the atmosphere by a weather balloon (Figure 4.14). Radiosondes collect measurements of temperature, pressure, humidity and wind up to heights of 15-20 kilometers (9-12 miles). Upper-air maps are then created from the radiosonde observations. Forecasters often view the data in a graphical format to interpret the vertical profile of the atmosphere at a specific location. The National Weather Service launches radiosondes from 92 stations across North America and the Pacific Islands. This is a small number of locations compared to the thousands of land-based weather stations that take surface observations several times each day. Radiosonde data are a critical component of numerical weather prediction because the data are used by computer models to forecast the future state of weather.

It is the upper-troposphere that dictates the weather conditions at the surface, not the other way around. Accurately forecasting the weather for an extended time period would be very difficult if one only had surface observations and surface maps to analyze.

Fortunately, consistent measurements of upper-air conditions have been taken since 1945, leading to significant advancements in weather forecasting. Many surface weather features are directly influenced by **upper-level troughs**, which are elongated areas of low-pressure in the upper atmosphere. Upper-level troughs produce divergence of air at high altitudes, leading to lower pressures at the surface (Figure 4.15). The convergence of surface air leads to the upward (vertical) motion necessary to form clouds and precipitation. Stormy weather and cold temperatures are often found at the surface under upper-level troughs. The formation and movement of surface low pressure areas and the locations of clouds and precipitation can be predicted through a careful analysis of upper-level conditions.

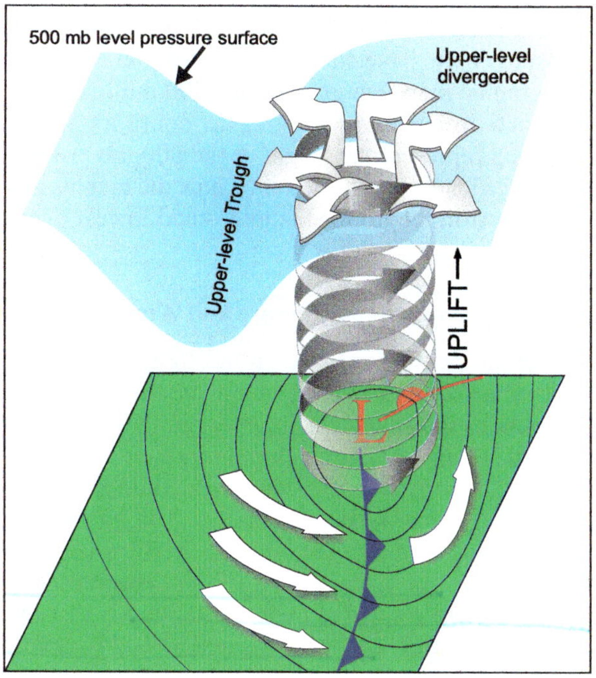

Figure 4.15: Upper-air troughs lead to upper-level divergence which in turn causes the formation of surface convergence, ultimately resulting in the formation of low pressure systems at the earth's surface.

Constant Pressure Surfaces

Upper-air maps display atmospheric pressure differently than sea-level pressure maps. While surface weather charts depict the pressure pattern at sea level, upper-air maps depict a pattern that indicates how heights vary along a **constant pressure surface**. A constant pressure surface is a surface where the pressure, not altitude, is constant. Meteorologists have drawn weather maps of upper-air conditions using constant pressure surfaces for the last 70 years. Maps of upper-air observations are called constant-pressure maps because all of the weather variables on a particular map are measured at the same pressure level above each station. These maps are drawn for several pressure surfaces to visualize pressure patterns and other atmospheric features at different altitudes. Constant pressure charts are useful to pilots because they provide a clearer picture of how the atmosphere behaves at different altitudes and pressures. It is very important to realize that the height patterns shown on upper-air maps give you the same information about changes in air pressure as isobars on a map of sea-level pressure, just at different altitudes. Constant pressure surfaces vary in height due to differences in air density, which is a direct function of air temperature. A constant pressure chart can be related to a poorly made bed with some folds and wrinkles (Figure 4.16A). Figure 4.16B shows a hypothetical constant pressure surface with these "folds and wrinkles" in constant pressure. Mapping these features on a constant pressure surface is necessary to identify upper-air features that influence weather conditions at the surface. Although radiosondes take observations at more than a dozen constant pressure levels, we will focus on only three of them. Constant pressure charts at 850 mb, 500 mb and 300 mb

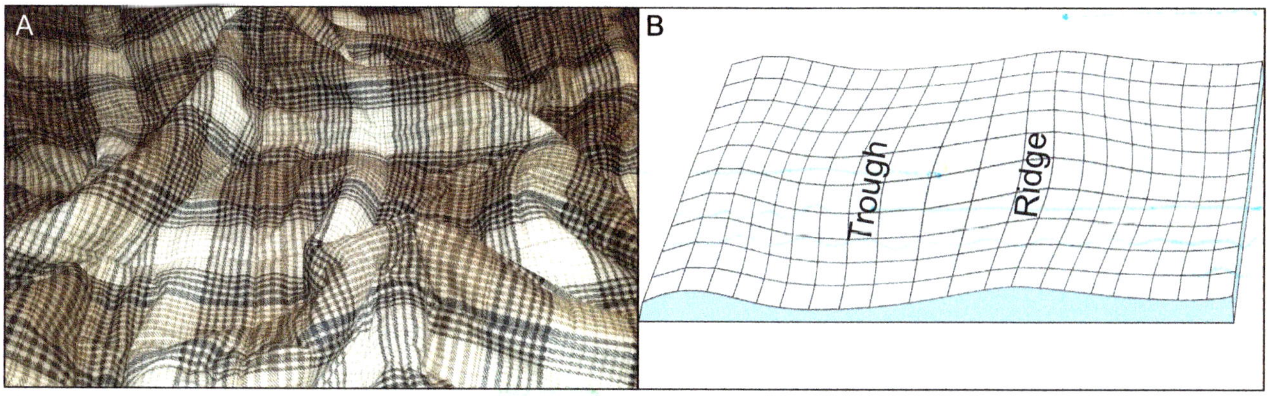

Figure 4.16: Visualizing constant pressure surfaces. A. A poorly made bed is much like a pressure surface with its wrinkles and folds. B. A computer generated 3-D model of a hypothetical pressure surface. The "folds" are called ridges and the "wrinkles" are called troughs.

represent the lower, middle and upper troposphere. Constant pressure charts have contours of equal elevation just like a topographic map. Figure 4.17 is a 500 mb map over North America in January 2004. The contours of constant elevation represent the height of the 500 mb pressure level above sea level in meters. Notice how the heights at 500 mb are highest in the tropical latitudes and lowest over central Canada.

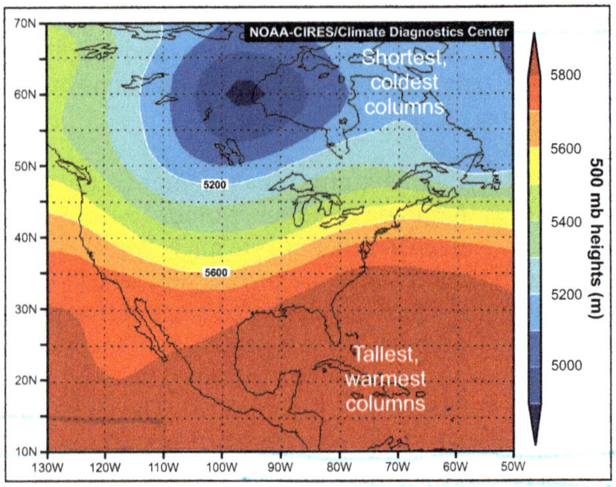

Figure 4.17: A 500 mb constant pressure chart from January 5, 2004. Data courtesy NOAA-CIRES/CDC.

Upper-air Patterns of Pressure and Wind

Important features of a constant pressure map can be interpreted once their basic characteristics are understood. Figure 4.18 shows three columns of air that extend from sea level to 300 mb. The left column represents the tropics; the middle column represents the mid-latitudes and the right column signifies the polar latitudes. The air columns vary significantly in height due to differences in their temperature. The air over the tropics is, on average, warmer and less dense compared to higher latitudes. You can see in Figure 4.18 that pressure decreases more rapidly with increasing altitude in cold air. Heights can vary significantly along a constant pressure surface when temperature contrasts are large between the tropical and polar latitudes. For example, the height of the 500 mb pressure level during the winter can vary by more than 1,200 meters. Due to the relationship between temperature and density, it makes sense that the highest heights on a constant pressure map are found where the air temperatures are the warmest (Figure 4.17). Constant pressure charts can be used to monitor fluctuations in upper-air height patterns as air masses move from their source regions to produce large changes in the weather.

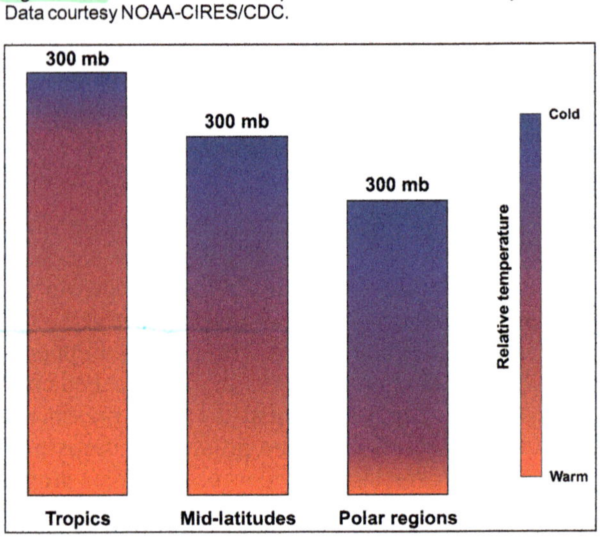

Figure 4.18: Schematic diagram showing the relative heights and temperature gradients of air columns over the tropics, the mid-latitudes, and the polar regions. The base of each column is at 1000 mb.

Figure 4.19 shows the 500 mb height pattern over North America in January 2006. Notice the **upper-level trough** over the Eastern United States. Troughs are identified on a constant pressure chart as regions where the height contours curve towards the equator. An **upper-level ridge** is present over the Western United States where the height contours curve to the north. The lowest and highest points on a constant pressure chart map relate to centers of low and high pressure for a specific pressure level. Airflow patterns in the upper-levels of the troposphere can be determined through a careful analysis of the height contours. Similar to isobars on a map of sea-level pressure, the wind speed can be inferred by the spacing of the height contours. Large height gradients relate to large pressure gradients and therefore fast wind speeds. The spacing of the height lines on a constant pressure map provides a good indication of the relative wind speeds. Figure 4.20 shows the relationship between wind speed and height gradients at 300 mb. The height lines are oriented in a southwest to northeast direction over the Atlantic Ocean. Notice how winds blow nearly parallel to the orientation of

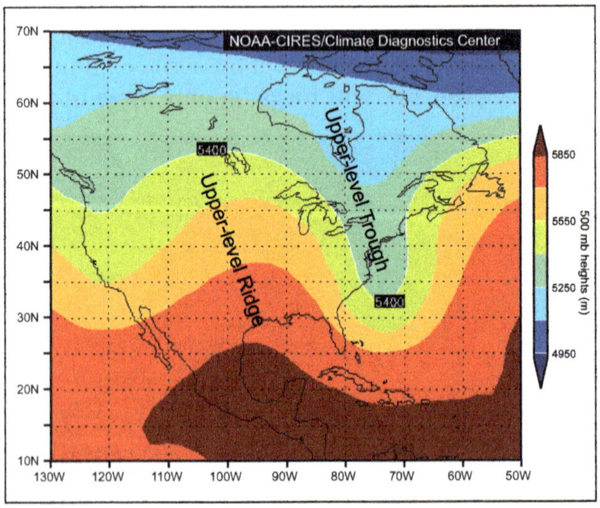

Figure 4.19: A 500 mb constant pressure map from January 15, 2006 showing an upper level ridge and trough. Data courtesy NOAA-CIRES/CDC.

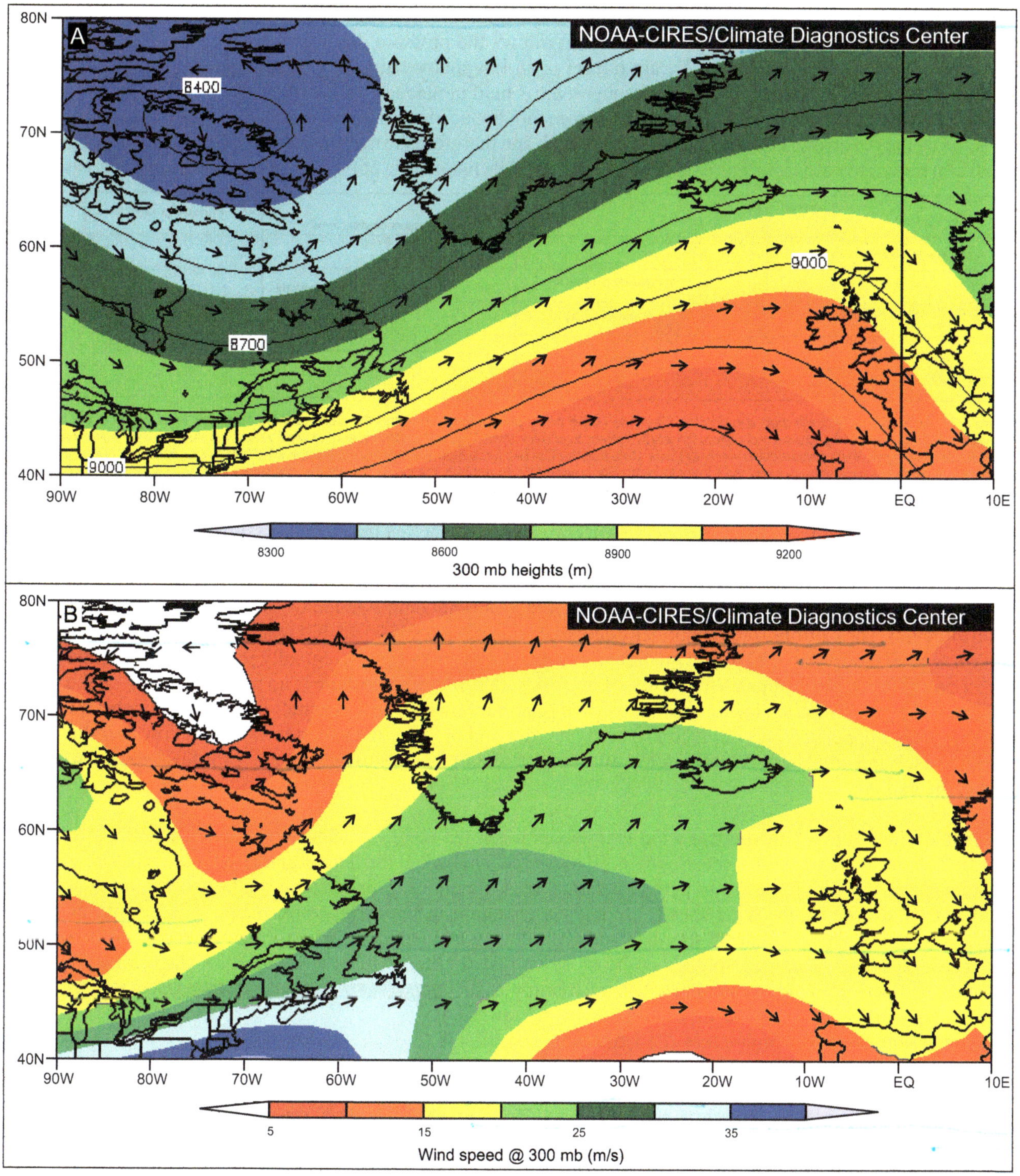

Figure 4.20: A look at upper level wind patterns over the North Atlantic Ocean for the months of January and February. A. A 300 mb constant pressure map with average heights with wind vector data. Note how the wind direction mimics the height lines. B. Average 300 mb wind speeds over the North Atlantic Oceanwith wind vector data. By comparing this diagram to the diagram above you can see that wind speeds are generally strongest where the height lines are spaced close together. Data courtesy NOAA-CIRES/Climate Diagnostics Center.

the height lines, resulting in a wind direction that is generally from the southwest. Also notice how the fastest wind speeds are found where the height gradient is the strongest. The closely spaced height lines typically extend from the East Coast of the United States to the British Isles during the winter months. Commercial pilots try to exploit these fast upper-level wind speeds over the North Atlantic region in order to fly more efficiently and conserve fuel.

Air flow patterns in the upper-troposphere are smoother and more wavelike than surface winds. This is primarily due to the absence of friction at high altitudes that allows the pressure gradient force and the Coriolis force to become equal but opposing forces. An **air parcel** is an imaginary volume of air that is often used to track air movement. Assume a parcel of upper-tropospheric air is held in place in the Northern Hemisphere. Figure 4.21A shows a parcel of air (the black circle) restrained from moving northward in response to the pressure gradient force, depicted by the red arrow. The velocity of the air parcel, and therefore the magnitude of the Coriolis force, is initially small as it moves towards lower pressure (Figure 4.21B). The magnitude of the Coriolis force steadily increases and

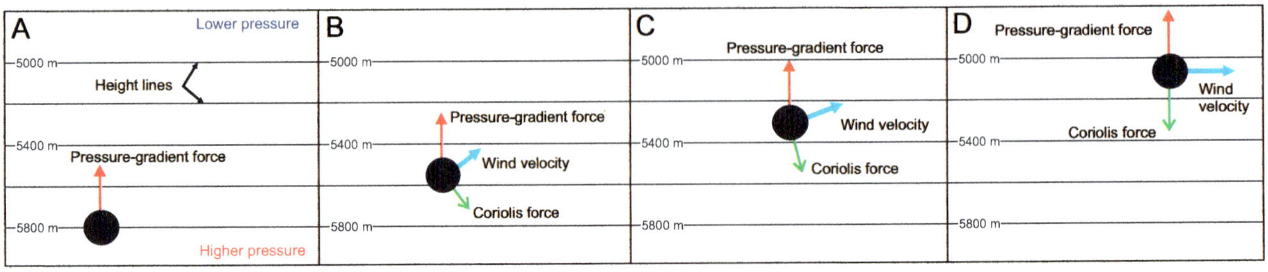

Figure 4.21: An idealized 500 mb map showing the influence of the pressure-gradient force and the Coriolis force on upper level winds in the Northern Hemisphere. The length of the arrows represents the relative strength of each force involved and the resulting wind velocity. A. An air parcel initially at rest. B. An air parcel moving northwards towards lower pressures. Note that the Coriolis force acts at 90° to the right of the wind velocity. C. As the air parcel continues to accelerate, both the wind velocity and the Coriolis force have increased. D. Eventually the Coriolis force and the pressure gradient force reach a balance and the wind reaches a constant velocity. The resulting wind, which blows 90° to the right of the pressure gradient, is called the geostrophic wind.

its direction shifts as the air parcel accelerates in response to the height gradient (pressure gradient) force (Figure 4.21C). Notice how the magnitude and direction of Coriolis force change as the parcel moves at different speeds. The idea that the air parcel eventually moves parallel to the height lines rather than directly toward the area of lowest height (pressure) may surprise you. This shows the important influence that the Coriolis force has on determining wind direction. The Coriolis force continually adjusts so that it is always 90° to the right of the path of the air parcel in the Northern Hemisphere. The height gradient force and the Coriolis force eventually reach a balance, influencing the air parcel in equal but opposite directions (Figure 4.21D). At this point there is no net force acting on the parcel and, according to Sir Isaac Newton, the parcel will continue to move in a constant direction at a constant speed. These equally opposing forces produce a westerly wind (wind blowing from west to east) that flows parallel to the height lines. Wind will always blow parallel to the height lines in the upper-troposphere regardless of the strength of the height gradient.

A **geostrophic wind** is a theoretical wind that occurs when the height gradient force and the Coriolis force are exactly balanced. By definition geostrophy is an idealized state when the height gradient force (pressure gradient force) and the Coriolis force are in perfect balance. It is important to understand that a geostrophic wind is an idealized wind because the wind speeds are always changing in the upper-troposphere. The magnitude of the Coriolis force changes as the wind speed changes, producing an imbalance between the two forces. Although the upper-tropospheric winds can sometimes be close to geostrophic, winds at the Earth's surface are far from geostrophic. Surface winds cross the isobars at roughly a 30° angle relative to the pressure gradient as it moves inward toward lower pressure. Surface winds cross the isobars at a smaller angle over smoother surfaces like water or ice due to the reduced influence of friction. Winds become closer to geostrophic with increasing altitude as the influence of friction is reduced. Winds well above the Earth's surface, in the middle and upper troposphere, tend to be nearly geostrophic and blow almost parallel to the height lines (Figure 4.22).

Figure 4.22: The linear clouds in this image reflect the presence of geostrophic winds. Photo credit: NASA.

The Jet Stream

Figure 4.23 shows two columns of air extending from the Earth's surface to the 300 mb pressure surface. Note how the temperature differences create height differences in the air columns between the Tropics and the Polar Regions. Because it helps explain why winds speeds in the upper-troposphere vary with the seasons, height differences across a constant pressure surface are related to differences in temperature. Assume that the height of the 500 mb pressure surface in Figure 4.23 is 5,160 meters in the cold air column. At this altitude the pressure in the warm air column is around 550 mb, which creates a pressure gradient between the two air columns at an altitude of 5,160 meters. Clearly the pressure decreases more rapidly with height in the cold air column due to increased density. The pressure gradient causes the winds to blow at both the Earth's surface and in the upper atmosphere. Large height gradients on constant pressure surfaces correspond to fast wind speeds. Large differences in temperature therefore create large height gradients and strong winds.

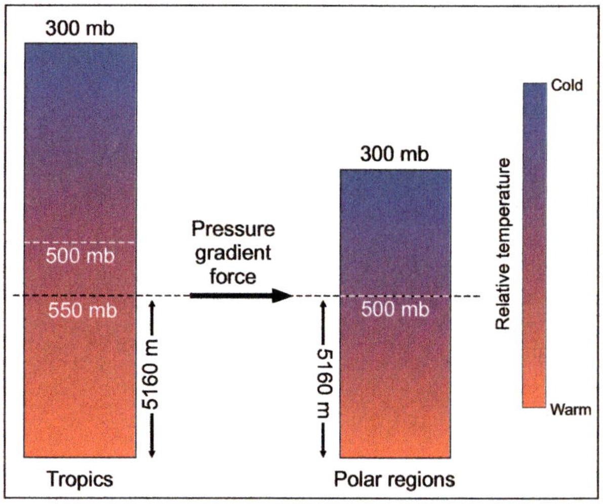

Figure 4.23: Horizontal temperature differences lead to differences in 500 mb heights. At the same altitude, the warm column of air has a higher pressure than the cold column. This in turn produces a pressure gradient force and upper level winds result.

The polar front jet stream is often thought of as a ribbon of air that flows from west to east in the upper troposphere. The jet stream typically takes a meandering path around the hemisphere that is broken up with segments of accelerated wind speeds (Figure 4.24). The position of the jet stream changes seasonally as it migrates with the sun. The jet stream moves towards the equator during the fall and winter months and moves towards the poles during the spring and summer months. Wind speeds in the upper-troposphere are strongly influenced by temperature differences at the Earth's surface. For example, wind speeds are generally much faster in the winter months due to the large temperature differences that exist. Temperature differences can exceed 55°C (100°F) between southern Florida and the northern United States in the winter months. The large temperature differences create a large pressure gradient in the upper-troposphere, resulting in extremely fast wind speeds. Wind speeds can sometimes exceed 320 kilometers per hour (~200 miles per hour) in the winter. Winds are generally much weaker in the upper-troposphere during the summer months due to a smaller range in surface temperatures.

Figure 4.25A shows the 250 mb heights over North America on a November day. Notice the spacing of the height

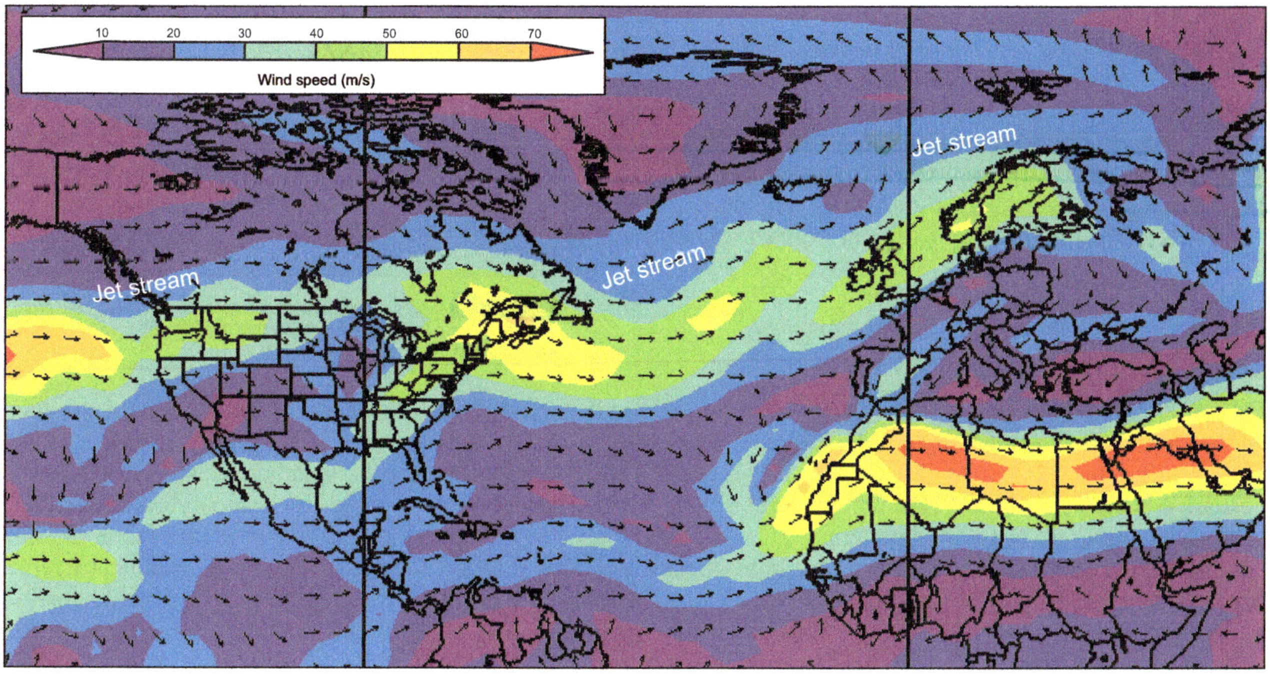

Figure 4.24: A 250 mb level winds speed map with winds vector data showing the discontinuous nature of the jet stream. Data courtesy NOAA-CIRES/Climate Diagnostics Center.

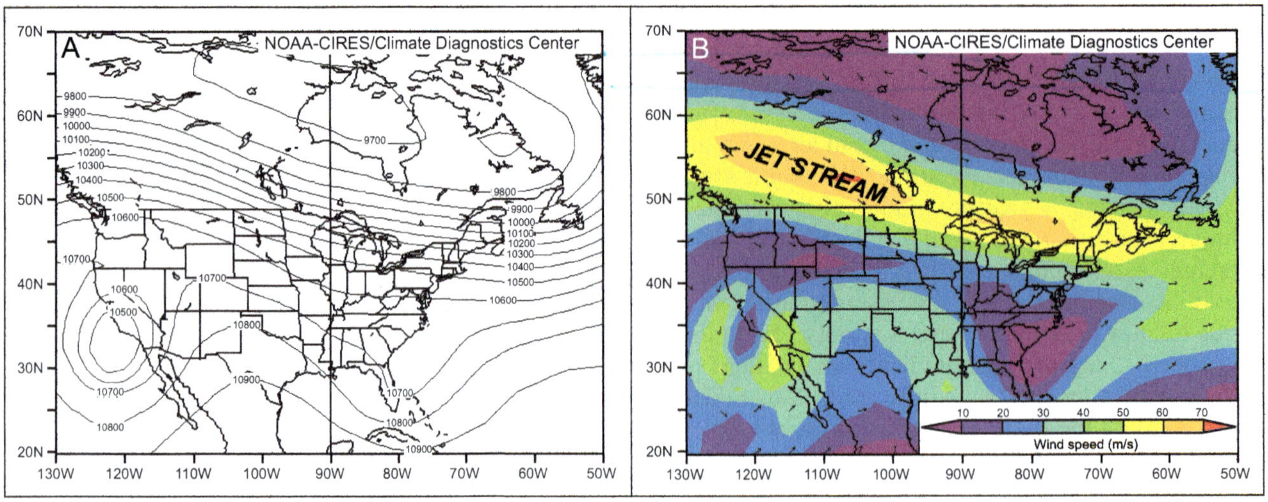

Figure 4.25: A. 250 mb constant pressure map from November 7, 2004 shows that the strongest winds are found where the height gradients are greatest. B. A wind velocity map at the 250 mb level for the same date shows a distinct area of high speed winds called the jet stream. Data courtesy NOAA-CIRES/CDC.

lines and regions that have large and small gradients. Figure 4.25B shows the 250 mb winds over the same time period. The fastest 250 mb winds correspond to the largest height gradient that extends southeastward from British Columbia, Canada to northern New England. The 250 mb winds are weakest from the Ohio Valley eastward to the Atlantic Ocean, where the height gradient is smallest. The strongest winds in the troposphere are typically found in a narrow zone where unlike air masses clash just below the tropopause. This fast flowing air current found around 10 kilometers (6.2 miles) above the surface of the Earth is the polar front jet stream. It is easy to identify the polar front jet stream on Figure 4.25A by locating the band of closely spaced height lines within the prevailing westerlies that typically cover the mid-latitudes. The polar front jet stream in the Northern Hemisphere is most commonly found between the latitudes of 30°N and 60°N. The width of a jet stream is typically a few hundred kilometers wide and is based on the width of the height gradient around 250 mb. The vertical thickness of the jet stream also varies by a few kilometers.

Strong cold fronts at the surface are always associated with the polar front jet stream due to the direct relationship between height gradients and temperature gradients. Figure 4.26 shows this relationship through a cross-sectional profile of a cold front. Notice the steep slope of the cold front that extends from the surface. The **frontal zone** represents the leading edge of a wedge of cold air that extends from the surface to just below the tropopause. The largest height gradient at 300 mb exists on the warm side of the frontal zone. The polar front jet stream, marked by the "X" in Figure 4.26, is located at around 300 mb on the warm side of the frontal zone.

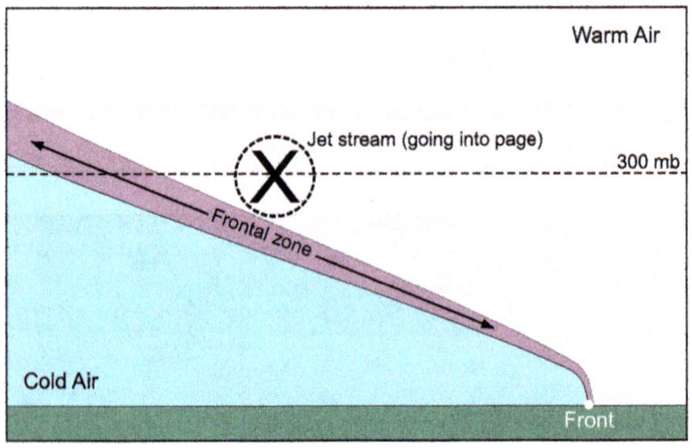

Figure 4.26: A cross-sectional profile across a surface cold front. The jet stream is usually found around the 300 mb level just to the warm side of the frontal zone.

Using Circulation Patterns as a Forecasting Tool

Pattern recognition is the ability of a forecaster to recognize weather patterns which produce certain weather conditions at the Earth's surface. Meteorologists use pattern recognition to identify circulation patterns along constant pressure surfaces, a critical skill that is needed to accurately predict the weather. Weather patterns on current weather charts that resemble previous weather patterns should produce the same type of weather. Identifying upper-air circulation patterns can sometimes be used to predict the general weather for several days in advance. This technique can also be used to help forecast a variety of severe weather events. For example, extreme temperatures can be predicted using previous weather patterns as a guide. A forecaster can recognize historical patterns on upper-level charts that are associated with significant cold air outbreaks to help predict them in the future. If the evolving weather pattern is similar to past cold air outbreaks, a more accurate forecast of expected weather conditions can be made.

Weather patterns continually change as the circulation pattern of the westerlies and the configuration of the jet stream change. When the wind flow in the upper atmosphere is predominately from west to east across most the United States the wind pattern is termed **zonal**. When the flow is zonal the upper level winds are parallel or nearly parallel to lines of latitude with no dips in the jet stream (Figure 4.27A). Temperature contrasts from north to south are usually small which prevents intense low-pressure systems from developing. The lack of strong ridges and troughs in zonal flow creates a calmer weather pattern and blocks the southerly flow of humid air from the Gulf of Mexico. As a result, the zonal pattern is generally associated with dry and sunny conditions in the Eastern United States. Zonal flow across the United States often produces warmer than normal winter temperatures as the east to west flow prevents cold Canadian air masses from moving south. Winds speeds, especially during the winter, tend to be accelerated in this flow pattern, often resulting in rapid changes in the weather. Forecasters must consider the speed of the upper-level winds when predicting the arrival of clouds and precipitation.

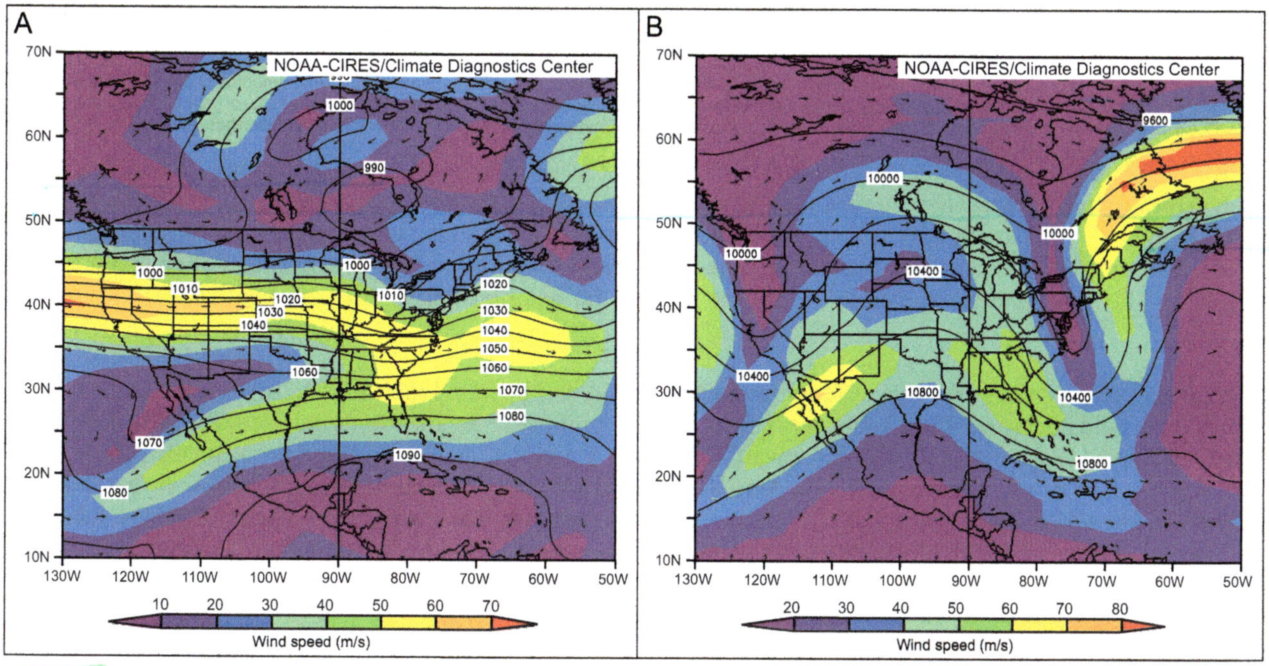

Figure 4.27: A. 250 mb vector wind speed map with height lines showing zonal flow. Since the jet stream flows west to east, very little mixing of air masses occurs, leading to generally fair weather. B. 250 mb vector wind speed map with geopotential height lines showing meridional flow. The jet stream flows north and south, forming ridges and troughs, increasing the potential for air masses to mix and produce storm systems. Data courtesy NOAA-CIRES/Climate Diagnostics Center.

The wind pattern is termed **meridional** when the wind flow in the upper atmosphere has a significant north to south component and crosses lines of latitude at a sharp angle (Figure 4.27B). The jet stream is highly amplified with strong troughs and ridges when the flow is meridional. This pattern allows unlike air masses to clash to create unseasonable temperatures and stormy weather. Forecasting becomes more challenging when the westerlies and jet stream have a meridional pattern. Meridional flow patterns have the potential to produce a variety of extreme weather, including record snowstorms and tornado outbreaks. Meteorologists use pattern recognition to forecast these severe weather events and provide timely warnings to the public.

Section 4-3: Stability, Clouds and Fog

Although we have focused primarily on the horizontal characteristics of wind and atmospheric circulation, vertical motions of the atmosphere are also very important in influencing the weather. Since rising air cools as it moves away from the Earth's surface, upward motions play a key role in the production of clouds and precipitation. Downward motions often dissipate clouds and contribute to fair weather. Meteorologists isolate an air parcel to track these vertical motions. The air parcel is assumed to be an isolated volume of air that doesn't mix with the air around it. Because air moves from higher to lower pressure according to the pressure gradient force, many assume the air would always rise upwards quickly. What prevents the air from rushing upwards which would cause us to suffocate? Although atmospheric pressure decreases much faster vertically than it does horizontally, gravity helps to balance

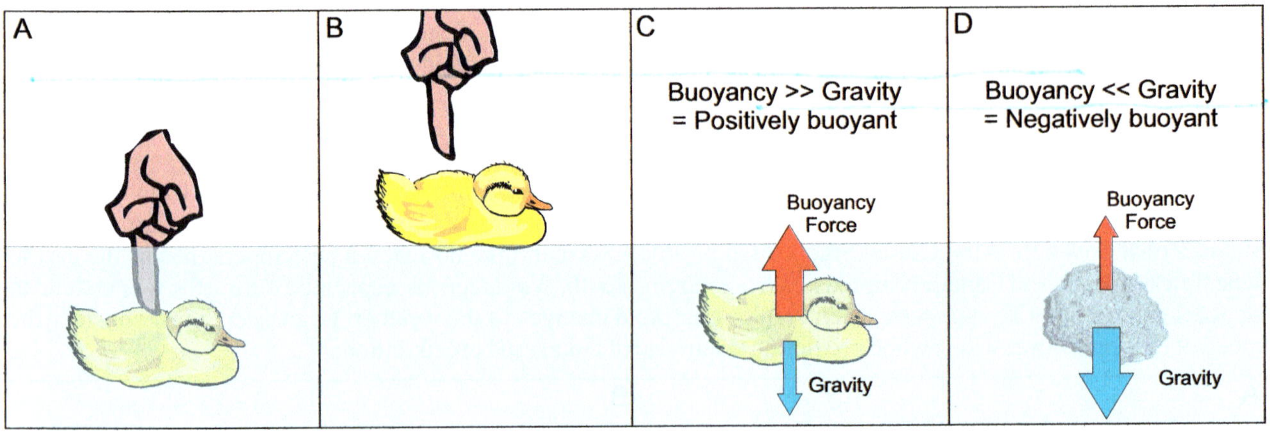

Figure 4.28: Buoyancy explained. A. It takes some force to hold a rubber duck under water. B. When released, the duck immediately bobs to the surface. C. The duck, when held under water, is positively buoyant because the buoyancy force pushing up on the duck far exceeds the force of gravity pulling down on the duck. D. A rock is negatively buoyant because the force of gravity far exceeds the buoyancy force. The rock will sink immediately when released.

out the vertical pressure gradient and limits the extent of vertical motions. The atmosphere is said to be in **hydrostatic equilibrium** when complete balance exists between the force of gravity and the vertical pressure gradient force. Vertical motions on a small scale, such as an individual thunderstorm, can be fast and reach speeds of over 100 kilometers (62 miles) per hour. On a large scale vertical motions are much slower and often go unnoticed by pilots. The atmosphere is generally in or near hydrostatic equilibrium, resulting in relatively small vertical velocities.

The concept of buoyancy as it relates to atmospheric stability is illustrated in Figure 4.28. Figure 4.28A shows a rubber duck held under water in a container. The rubber duck immediately rises to the surface when the downward force is removed (Figure 4.28B). The net upward force, buoyancy, is now the only force acting on the rubber duck. The rubber duck is now considered to be **positively buoyant** as the upward force of buoyancy exceeds the downward force of gravity (Figure 4.28C). An example of a **negatively buoyant** object is shown in Figure 4.28D. In this case the rock will sink to the bottom of the container since the downward force of gravity exceeds the upward buoyancy force.

The buoyancy force of an air parcel is related to the difference in density between the air parcel and the outside environment. The buoyancy of an air parcel, and how high it will rise or sink, is based on the temperature difference between the air parcel and its immediate environment. Temperature differences of only a few degrees between the air parcel and surrounding air are enough to cause rising and sinking motions in the atmosphere. The air parcel will remain positively buoyant and continue to rise as long as it is less dense than the surrounding environment. A positively buoyant air parcel ascends in a similar fashion as a hot air balloon (Figure 4.29). The hot air balloon becomes more positively buoyant and accelerates upward as long as the air inside the balloon is warmer and less dense than the surrounding air. If the balloon begins to sink, the air inside is heated so the balloon remains positively buoyant.

Figure 4.29: These hot air balloons demonstrate how temperature and buoyancy are related. Image credit: Jay Levine/NASA.

The buoyancy, and therefore stability, of an air parcel is identified by differences in temperature. Atmospheric stability gives us an indication of how easily an air parcel is lifted. Stability is important to determine the occurrence and strength of convection, which can help a forecaster identify the potential locations of afternoon thunderstorms. Stability is also important to determine the vertical mixing of the air, which directly influences concentrations of air pollution. Let's assume an air parcel is initially in hydrostatic equilibrium and is then nudged upward. The parcel is positively buoyant if it continues to move away from its initial position. In this situation the parcel is considered to be in an **unstable equilibrium** (top panel, Figure 4.30). The air is said to be **unstable** under these conditions and will continue to accelerate upward. Ideal conditions for an unstable atmosphere are warm, moist air at low levels with cool, dry air above it. An air parcel is said to be in a **stable equilibrium** if it is negatively buoyant and returns to its original position after it is nudged upward (middle panel, Figure 4.30). The air is said to be **stable** under these conditions, which are common during clear and calm nights. Stable atmospheric conditions are often associated with poor air quality because it causes pollutants to become trapped near the ground. An air parcel in hydrostatic equilibrium is said to be in **neutral equilibrium** if it stops rising a short distance after being nudged upward (bottom panel, Figure 4.30).

It is important to understand that there are a few important assumptions that must be considered when using an air parcel to demonstrate atmospheric stability. The air parcel is considered to be sealed from the outside air with no mixing or heat transfer between the parcel and its immediate surroundings. The air parcel can't be moving vertically when testing for stability; the test must be applied to the air parcel before it begins to rise. It is also assumed that the air parcel is flexible and adjusts to outside pressure changes. Understanding the basics of atmospheric stability and instability has important applications in weather forecasting.

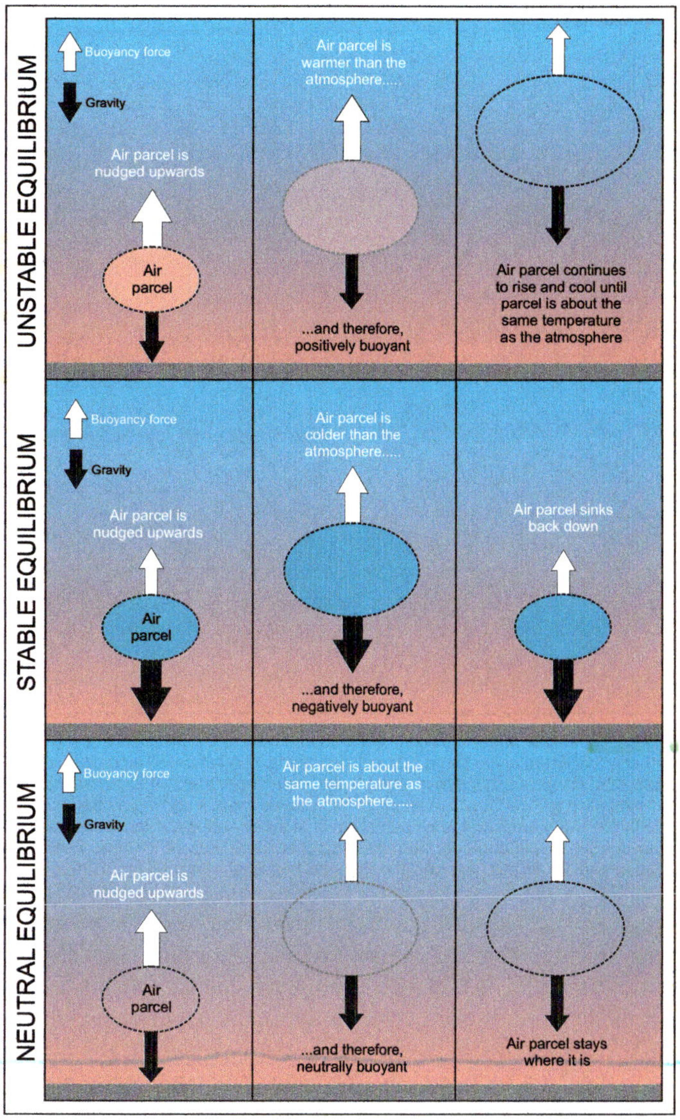

Figure 4.30: The relationship between the buoyancy of an air mass and the stability of an air mass. Positively buoyant air masses are unstable while negatively buoyant air masses are stable. Some air masses can also be neutrally buoyant.

Adiabatic Lapse Rates

The stability of an air parcel is identified by comparing the temperature of an air parcel to its outside environment at a specific altitude. It is assumed that the air parcel remains unsaturated as it rises through the lower troposphere since most cloud bases are generally at an altitude of at least one kilometer (~.6 mi.). An unsaturated air parcel cools at a rate of 10°C/km (5.5°F/1,000 feet) as it expands and rises through the troposphere (Figure 4.31). This rate of cooling is called the **dry adiabatic lapse rate** (DALR), referring to the rate of temperature decrease with height. The dry lapse rate does not imply that the rising air parcel is dry, only that it is not saturated. During an adiabatic process, no heat is exchanged between the air parcel and its surrounding environment. This means that the air parcel's temperature only changes as it expands or compresses. Expansional cooling and compressional heating of unsaturated air are considered adiabatic processes. The DALR allows us to predict the temperature change of an unsaturated air parcel as it moves vertically through the atmosphere. As an unsaturated parcel sinks it warms at a rate of 10°C/km (5.5°F/1,000 feet) due to compressional heating from increased pressure (Figure 4.31A).

If an air parcel rises high enough and expansion lowers its temperature to the dew point or frost point, the ascending air will no longer cool at the dry adiabatic lapse rate. Remember that the altitude at which saturation occurs is known as the **lifting condensation level (LCL)**. As the saturated parcel rises beyond the LCL, expansion continues to lower its temperature, but the cooling is partially offset by the release of latent heat due to condensation. Rising

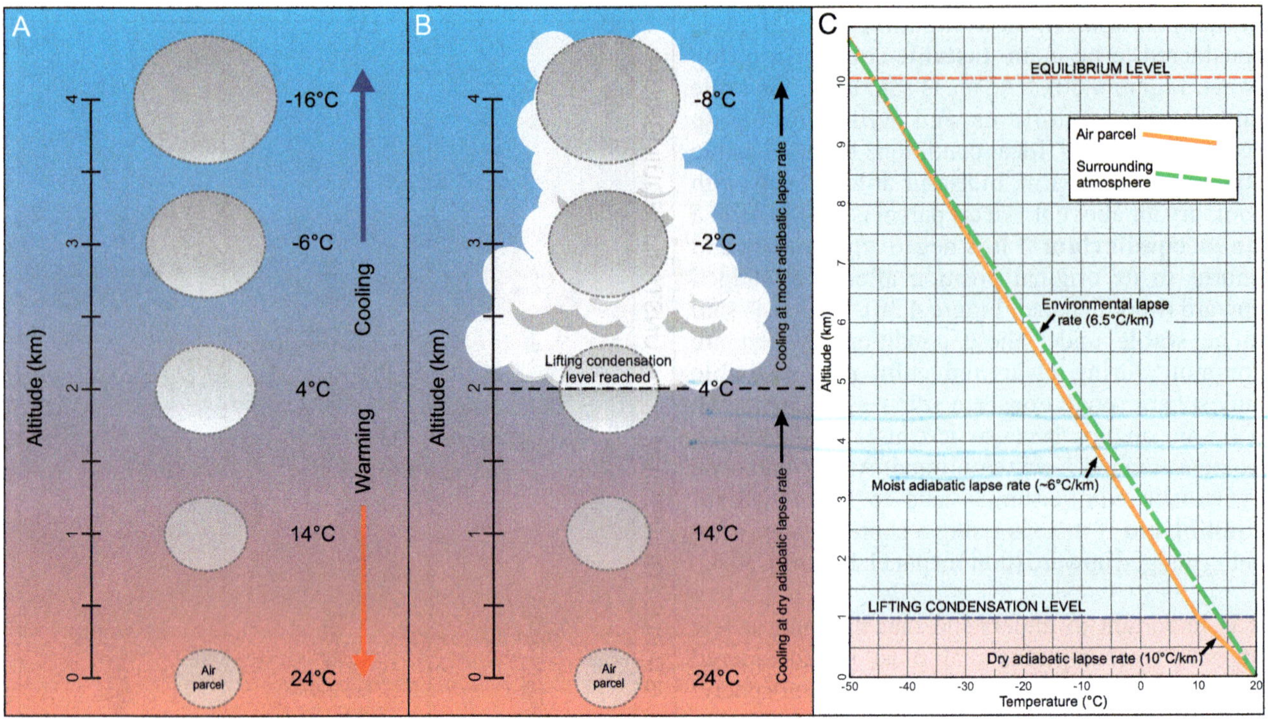

Figure 4.31: Lapse rates visualized. A. An unsaturated air parcel cools at a rate of 10°C per kilometer, also known as the dry adiabatic lapse rate. A sinking air parcel would warm at the same rate as it descends. B. A saturated air parcel cools at an average rate of 6°C per kilometer, also known as the moist adiabatic lapse rate. However, this rate is variable and can range from 5-9°C per kilometer. This rate of cooling occurs whenever an air parcel rises above its lifting condensation level (LCL). C. A graphical representation of a cooling air parcel. The atmosphere around the rising air parcel cools at the environmental lapse rate of 6.5°C/km. Initially, the air parcel cools at the dry adiabatic lapse rate of 10°C/km. When the air parcel becomes saturated at the lifting condensation level, the air parcel cools at the moist adiabatic lapse rate of 6°C/km. Eventually, the air parcel will cool to the same temperature as the surrounding atmosphere. When this happens, the air parcel has reached it equilibrium level.

saturated air parcels cool at the **saturated adiabatic lapse rate** (SALR) (Figure 4.31B). Although the DALR is fixed in value, the SALR varies with temperature. The release of latent heat slows the rate of cooling from 10°C per kilometer (5.5°F/1,000 feet) to about 6°C per kilometer (3.0°F/1,000 feet). The moist adiabatic lapse rate and the dry adiabatic lapse rate are very similar in the upper troposphere due to a limited amount of moisture. As an air parcel rises vertically from the Earth's surface it cools at the dry adiabatic lapse rate until it cools to saturation. Cooling continues at a reduced rate, the moist adiabatic lapse rate, above this altitude. The rising air parcel has a greater chance of staying positively buoyant when it cools at the moist adiabatic lapse rate. As we will see, this is why high levels of atmospheric moisture in the lower troposphere are so important to the development of thunderstorms. The parcel will continue to rise until the temperature of the air parcel is equal to the temperature of the surrounding environment, known as the **equilibrium level** (Figure 4.31C). In reality, due to momentum the air parcel will rise slightly above the equilibrium level.

Now that we have an understanding of the different rates of cooling that occur as an air parcel is lifted vertically we can predict its temperature as it moves from the surface through the troposphere. Figure 4.31C plots a rising air parcel with an initial temperature of 20°C (68°F) and a 10°C (50°F) dewpoint. It is assumed that the air parcel has the same temperature as the surrounding environment before it is lifted from the surface. The parcel will cool at the dry adiabatic lapse rate of 10°C per kilometer while the surrounding atmosphere cools at the environmental lapse rate (ELR) of 6.5°C per kilometer. The ELR can vary based on location, time and altitude; it is not a fixed rate. The rising parcel continues to cool at the dry adiabatic lapse rate until it reaches the LCL. The rising air parcel is initially stable as it cools at a faster rate than its surroundings. The air parcel will now cool at the reduced moist adiabatic lapse rate due the release of latent heat. The air parcel will continue to cool at this rate until it reaches the equilibrium level. The altitude of the equilibrium level is based on the temperature and moisture characteristics of the rising air parcel. You can see from Figure 4.31B how that the amount of moisture determines the cooling rate, and therefore stability, of a rising air parcel. Unless the air is extremely dry, most air parcels need to rise only a few thousand feet before they become saturated. Humid air parcels are much more likely to become unstable since they only need to rise a short distance before they become saturated, produce clouds and cool at a reduced rate. You can often get a general idea of the amount of humidity at a given location by noticing the elevation of the cloud bases. Humid air masses reach the LCL at a much lower altitude compared to a relatively dry one.

Cloud Classification and Stability

Even a casual sky observer notices that clouds come in a variety of forms. Cloud observations were relied on to predict the weather before the invention of modern weather instruments. Clouds give us a clue about what is going on in the atmosphere and can sometimes indicate how the weather might change. Each type of cloud forms in a different way, and each one brings different kinds of weather. Luke Howard, an English naturalist, is credited with devising the first classification of cloud types in 1803. Clouds are generally first classified by their appearance/form. The name of the cloud types come from Latin words that describe their appearance. Howard's classification system divided clouds into four basic categories based on their appearance. **Cirrus** clouds are thin, wispy clouds made up solely of ice crystals. Cirrus means "curl" in Latin, in reference to a lock of hair. **Stratus** clouds spread laterally to form layers; stratus means "spread over a large area" in Latin. **Cumulus** clouds are vertically developing clouds that form exclusively from convection. Cumulus means "heap" in Latin, which refers to the towering appearance of these clouds. Cumulus clouds are an example of **cumuliform** clouds, which are heaped or puffy in appearance. Clouds that produce precipitation are called **nimbus**, not surprisingly the word nimbus in Latin means "rain-bearing".

Clouds are also classified into four basic types on the basis of their altitude: high clouds, middle clouds, low clouds and clouds exhibiting vertical development. It is difficult to simply look up at the sky and determine a cloud's altitude. Other clues are needed to distinguish between high and middle altitude clouds. High, middle and low clouds form through the gentle uplift of air over large regions. These **stratiform clouds** have limited vertical development but spread laterally over large areas. In contrast, clouds that exhibit **vertical development** generally cover smaller areas and are associated with a much stronger uplift of air. The appearance/form of the cloud is combined with its altitude to classify clouds into ten specific cloud types.

High Clouds

The base of a high cloud is at an altitude above 6,000 m (~20,000 ft.). The prefix "cirro-" is included in the names of the three different types of high clouds. Temperatures are so cold at these altitudes that the clouds are composed entirely of ice crystals that originate from the freezing of supercooled water droplets. **Cirrus** (Ci) clouds are almost transparent due to their low temperatures and low ice content, their thickness is usually only about 1.5 km (~1 mi.). These clouds come in a variety of shapes and sizes; their ice composition gives them a thin, wispy appearance (Figure 4.32A). Ice crystals and snowflakes can slowly descend from cirrus clouds to produce **fall streaks**. Upper-level wind characteristics determine the size and shape of the streaks. Although cirrus clouds are associated with fair weather, they can indicate a change in the weather within the next 24 hours. Cirrus clouds are often the first sign of an approaching warm front; you can determine which direction the front is moving by noting the movement of the clouds.

Like cirrus clouds, **cirrostratus** clouds are nearly transparent and composed of ice crystals (Figure 4.32B). Cirrostratus clouds are darker than cirrus clouds and often appear as a hazy, sheet-like veil over the sky. These clouds are thin enough for the sunlight or moonlight to pass through them. Sometimes their presence is only

Figure 4.32: The basic types of high altitude clouds. A. Cirrus. B. Cirrostratus. C. Cirrocumulus. Photo credits: B. Zaprowski.

identified by an observed halo around the sun or moon that results from the refraction of light by the cloud's ice crystals. Cirrostratus clouds can be a reliable sign for deteriorating weather and typically form when a broad layer of air is gradually lifted over a large area. These clouds tend to thicken as a warm front approaches and indicate a chance of rain within the next 12-24 hours, especially if lower clouds are also present and the halo around the sun or moon becomes less visible. **Cirrocumulus** clouds are layered with small, rounded puffs that usually appear in long rows high in the sky (Figure 4.32C). These clouds are most common in the winter and often exhibit a wavelike pattern than look like ripples on the surface of a lake. When these clouds cover a large portion of the sky, the small, white puffs can create a "mackerel sky", a popular term for a cloud pattern that looks like the scales of a fish. Cirrocumulus clouds are usually white, but sometimes appear gray; they rarely cover the entire sky. These clouds often dissipate rapidly and are usually an indication of fair, but cold weather. However, they can be a precursor to precipitation and stormy weather when cirrus clouds give way to an increasing amount of cirrocumulus clouds.

Middle Clouds

Clouds with the prefix "alto-" are middle level clouds that are located between 2,000 and 6,000 m (~6,500 to 20,000 ft.). These clouds may be composed of liquid water droplets, ice crystals, or a mixture of both depending on their temperature. Temperatures of middle clouds generally range between 0 and -25°C (32 and -13°F).

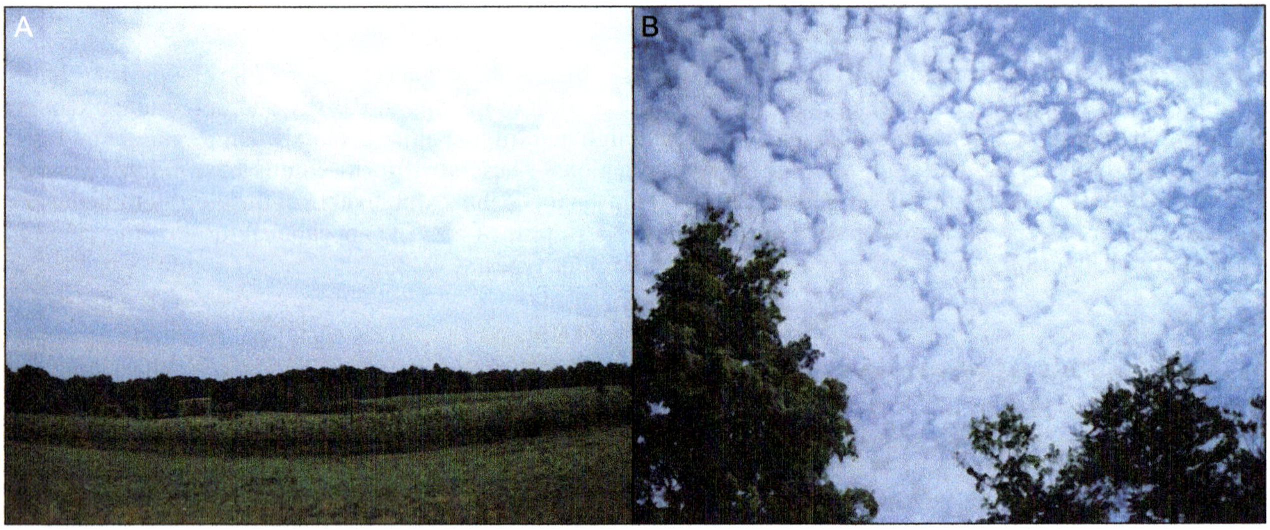

Figure 4.33: Examples of middle altitude clouds. A. Altostratus. B. Altocumulus. Photo credits: B. Zaprowski.

Altostratus clouds usually appear as featureless sheets with a uniform gray or bluish color that cover the entire sky (Figure 4.33A). These clouds are commonly seen in association with other cloud formations. Altostratus clouds can be thought of as a cirrostratus cloud that is found at a lower altitude. These clouds are generally thicker and more extensive than cirrostratus clouds since they are composed primarily of liquid water. Altostratus clouds are usually thick enough to scatter enough incoming sunlight back to space to make the sun appear only as a bright spot through the clouds. Altostratus clouds are commonly associated with warm fronts and are created by the lifting of a warm, stable air mass. As the warm front approaches, the clouds thicken into a dark gray layer. Altostratus clouds can indicate that a storm with continuous rain or snow might be approaching.

Altocumulus clouds are layered clouds that consist of roll-like patches or rounded masses that are arranged in a wavelike pattern (Figure 4.33B). These grayish-white clouds often form long parallel bands with some parts darker than the other. Altocumulus clouds sometimes look like floating cotton balls with a background of blue skies. They usually form in groups and may appear in several distinct layers. These clouds have a more distinct outline than cirrocumulus clouds because they consist primarily of liquid droplets. Altocumulus clouds with vertical development can indicate the presence of instability, especially in the morning. Afternoon thunderstorms commonly follow the appearance of these clouds on a warm, humid morning. Altocumulus clouds often indicate an approaching cold front since they can form from the lifting of air ahead of a cold front. Precipitation can be expected within 12 to 24 hours if the cloud coverage is increasing and thickening.

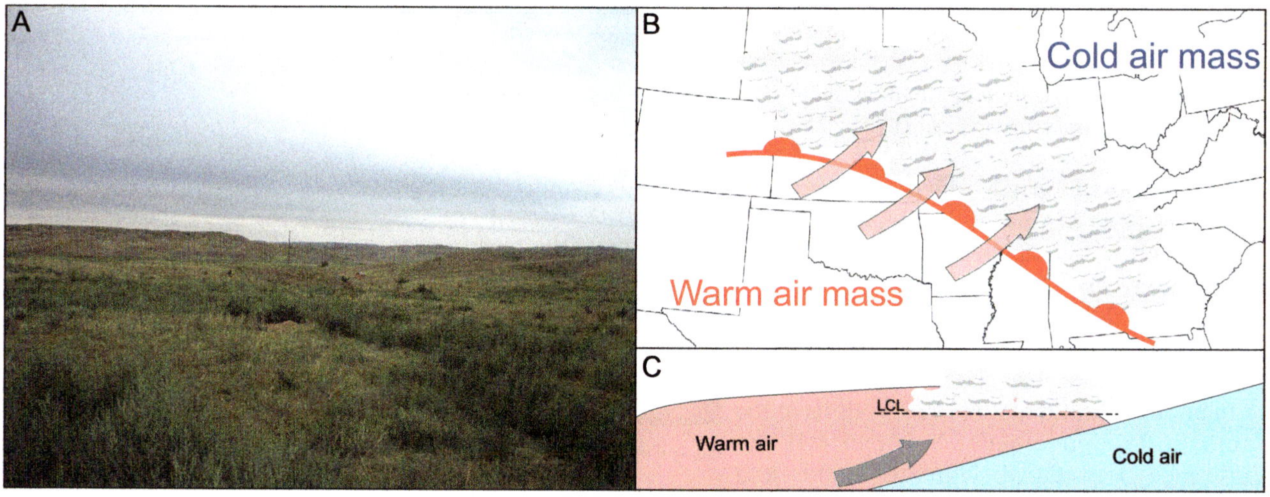

Figure 4.34: A. An example of stratus clouds. These clouds are very layered compared to cumulus clouds. Photo credit: B. Zaprowski. B. Stratiform clouds normally form to the north of a warm front. C. As the warm air runs into the colder air ahead of it, the air is forced to rise and stratiform cloud formation usually results.

Low Clouds

Low clouds have bases below 2,000 m (~6,500 ft.) and consist mostly of water droplets because they form at temperatures above -5°C (23°F). These clouds are not given a prefix, although, depending on their characteristics, their names can include the term "strato-" or "cumulo-". The two main types of low clouds include stratus, which develop horizontally, and cumulus, which develop vertically.

Stratus clouds are uniform gray clouds that can cover most or all of the sky (Figure 4.34A). Stratus clouds are wider than most clouds and form when extensive areas of stable air are lifted. This cloud type can sometimes look like a layer of fog that never reaches the ground. Stratus clouds generally form north of a warm front in a process known as **overrunning** (Figure 4.34B). When preparing a weather forecast, forecasters always look for situations when overrunning might produce continuous and uniform precipitation. The clouds are formed as the warm air moves northward and glides over the cold air mass. The cold, dense air at the surface acts like a lifting mechanism as the warm air runs up and over the cold air (Figure 4.34C). The lifted air stops rising and spreads out laterally when it becomes too negatively buoyant due to the steady cooling. As a result, stratus clouds that form north of a warm front are generally shallow and layered. Convective clouds only form north of a warm front when the air is extremely warm, moist and unstable. Low stratus clouds can sometimes persist for a few days and cause cold and gloomy conditions to linger. Precipitation rarely falls from stratus clouds since upward vertical motions are very weak, but light mist and drizzle can sometimes fall when they are present.

Low, layered clouds that produce precipitation are called **nimbostratus**; these clouds are associated with continuous rain or snow (Figure 4.35). Nimbostratus clouds resemble stratus clouds except they are darker gray and thicker with a more ragged cloud base. The cloud base of a nimbostratus cloud can be difficult to identify due to the falling precipitation and fog found beneath them. In contrast to a mid-level altostratus cloud, the sun and moon are not visible through nimbostratus clouds. Not surprising from their name, nimbostratus clouds generally produce continuous light to moderate precipitation over a relatively large area. Sometimes the precipitation will evaporate before it reaches the ground, producing virga. Nimbostratus clouds usually form in a stable atmosphere where warm, moist air is overrunning colder air at the surface. Visibility is often poor when these clouds are present; they often

Figure 4.35: An example of nimbostratus clouds. Photo credit: B. Zaprowski.

cover the entire sky. Nimbostratus clouds can produce precipitation that can last for several hours or more, they can indicate the passage of a cold front when they are breaking up.

Figure 4.36: An example of stratocumulus clouds. Photo credit: B. Zaprowski.

Stratocumulus clouds are low, layered clouds that show signs of vertical development. These dark to light gray clouds are formed by weak, shallow convection and appear as large, irregular rounded masses or rolls with breaks of clear sky in between (Figure 4.36). Stratocumulus clouds are darker in color than altocumulus clouds due to their additional mass but they both have the same fluffy appearance. You can tell the difference between altocumulus clouds and stratocumulus clouds by their size, altocumulus clouds are about the size of your thumbnail at arm's length while stratocumulus clouds are about the size of your fist. These clouds usually have a well-defined cloud base but exhibit limited vertical development due to the dry, stable air above them. Stratocumulus clouds normally dissipate by the late afternoon or early evening, improving weather can be expected if patches of blue sky are expanding. Precipitation rarely falls from stratocumulus clouds, only light rain or snow is produced when it does occur.

Clouds Exhibiting Vertical Development

Vertically developed clouds are all closely related and are associated with unstable air. These clouds have low cloud bases and significant vertical development due to their rapid vertical velocities. Compared to stratiform clouds, cumuliform clouds have much higher liquid water contents and possess a more cellular structure. Cumulus clouds can grow to be much taller than they are wide. There are two types of clouds that are classified by their vertical development.

Cumulus clouds are white or light gray clouds that resemble floating cotton balls on a fair day (Figure 4.37). These clouds are puffy and textured with rounded tops and a flat base. Cumulus clouds have sharply defined edges and their cloud tops can have the appearance of a head of cauliflower. Cumulus clouds most often form on clear days due to uneven surface heating, causing air to be lifted convectively. The development of cumulus clouds often follows the daily cycle of surface heating. When cumulus clouds are present in the morning hours, solar heating usually produces a noticeable increase in afternoon cumulus cloud coverage. Cumulus clouds indicate fair weather when they are widely separated with limited vertical development. Small cumulus clouds that form on sunny days are often called "fair-weather" clouds since they rarely produce precipitation. The cumulus clouds indicate rising air while the cloud-free areas indicate sinking air. Fair-weather cumulus clouds only last for a short time and when they produce rainfall it usually only lasts for a few minutes. Once cumulus clouds form, atmospheric stability determines the degree of vertical development. Although cumulus clouds are often synonymous with fair weather, they are

Figure 4.37: An example of cumulus clouds. Photo credit: B. Zaprowski.

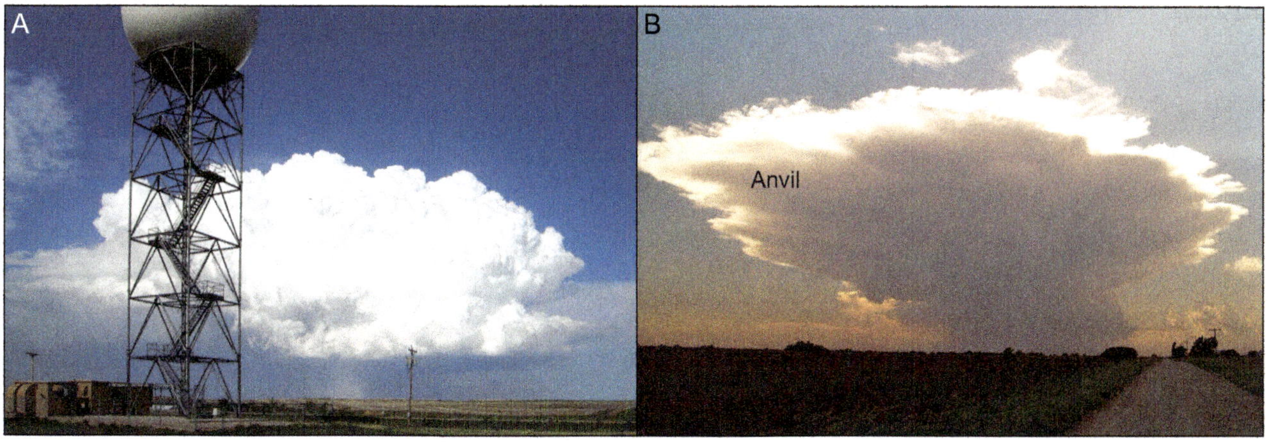

Figure 4.38: Examples of cumulonimbus clouds. A. A cumulonimbus cloud drops rain and hail near the New Underwood, SD radar station. Photo credit: Eric Kurth/ NOAA. B. A cumulonimbus cloud with a well-developed anvil looms over the Southern Great Plains. Photo credit: Brookhaven National Lab.

capable of producing heavy rainfall when they grow taller and become more organized

Cumulonimbus clouds are dense, towering clouds with a dark cloud base. These clouds are much larger and more vertically developed than cumulus clouds (Figure 4.38). Even small cumulonimbus clouds appear large compared to other cloud types. Under favorable atmospheric conditions, fair weather cumulus clouds can quickly develop into cumulonimbus clouds. These clouds are the most violent of all clouds; they produce the most intense thunderstorms and are responsible for most forms of severe weather. Fueled by intense convective updrafts, the tops of cumulonimbus clouds can easily reach 12,000 m (39,000 ft.) or higher. In warm, humid and unstable air, they have cloud bases just a few hundred meters above the surface with towers that can occasionally extend into the lower stratosphere. Lower portions of cumulonimbus clouds consist of liquid water drops while the higher elevations are composed of ice crystals. Their cloud tops flatten out in the shape of an **anvil**, resembling the tool used by blacksmiths (Figure 4.38B). The anvil is produced by fast upper-level winds that spread out the ice crystals, this feature usually points in the direction the storm is moving. This cloud type tends to develop during the afternoon and early evening when the effects of solar heating are the strongest. They can signal the approach of severe weather, including strong winds, heavy rain, hail and tornadoes.

Fog: A Cloud On The Ground

Fog is simply a cloud whose base is at or near ground level (Figure 4.39). The difference between fog and a cloud is how they form. Fog can form either by the cooling of near-surface air down to its dew point or by the addition of moisture to the air. Although fog is often not considered a newsworthy event, its impacts can be costly and even

Figure 4.39: Some examples of fog. A. A layer of radiation or ground fog. Notice how flat the top of this fog layer is. Photo credit: B. Zaprowski. B. Ground fog fills in valleys in Argentina. Credit: NASA Earth Observatory images by Jesse Allen and Robert Simmon, using USGS Landsat data.

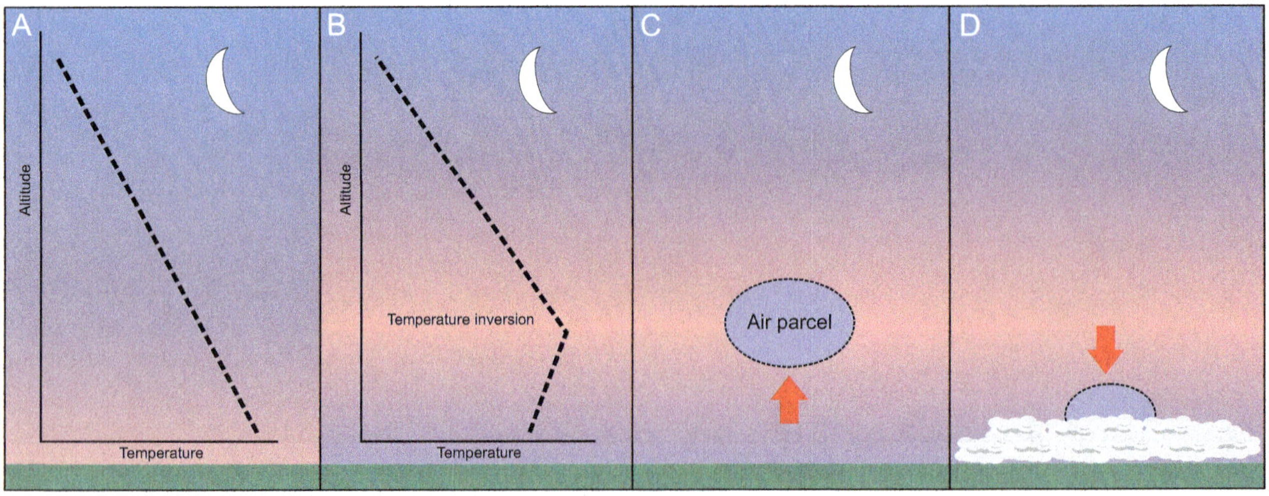

Figure 4.40: Nocturnal inversions explained. A. Initially, the atmosphere is warmest at the earth's surface and decreases in temperature as you move up in altitude. B. Radiational cooling at the earth's surface causes a layer of cold air to form. This produces a situation where the atmosphere above the earth's surface is warmer than the air at the earth's surface. C. An air parcel nudged upwards from the earth's surface will always be negatively buoyant as it rises into warmer air. D. As a result, the air parcel will sink back down to the earth's surface.

deadly. Various forms of transportation can be significantly impacted by fog. Several types of fog are classified according to their formation.

Radiation fog (sometimes called ground fog) is most common on clear nights with light winds. It usually appears stratified with the top layer appearing fairly flat (Figure 4.39A). It forms due to radiational cooling that cools a layer of air in contact with the ground to its dewpoint. **Valley fog** can form in hilly terrain as the cold, dense radiation fog drains into the adjacent valleys (Figure 4.39B). Radiation fog most commonly forms when the air near the ground is humid and the air above it is relatively dry. The high humidity at low levels is usually due to evaporation of water from a moist surface. As a result, radiation fog is most common in areas where the soil has been recently saturated by heavy rainfall or rapid snowmelt. On a clear night with light winds the ground can cool rapidly by a net loss of radiation. These conditions allow a **temperature inversion** to form near the surface (Figure 4.40). If a parcel of air is displaced it will sink back to its original position because it is colder than the environment and therefore stable. Stable layers of air near the ground tend to remain in their original position (Figure 4.40D). Radiation fog is likely to form in this situation if the air near the Earth's surface has enough moisture. Light winds are ideal for radiation fog to form; calm conditions favor the formation of dew or frost. Light winds gently mix the air and transfer heat, allowing the layer of air to cool to saturation, resulting in condensation. Radiation fog will either dissipate or not form if the winds are too strong. The winds cause the humid air near the ground to mix with the drier air above, reducing its humidity. Most radiation fogs persist for only a few hours after sunrise as the surface air is warmed and the fog droplets evaporate from the bottom up. This makes the fog appear to lift although it doesn't actually move.

Advection fog forms when relatively warm, humid air moves over a colder surface such as cold water or snow. This

Figure 4.41: Advection fog is best explained by looking at the famous San Francisco fog banks. A. Warm, moist Pacific air flows over a cold ocean current just offshore from San Francisco. The air is cooled as it passes over, leading to condensation and the formation of a fog bank. B. The Golden Gate Bridge wrapped in fog. Photo credit: Matt Macintosh/NOAA.

may allow the lowest layers of the air mass to be cooled to saturation. The term advection refers to the horizontal transport of air. These fogs can cover an extensive region if the air is cooled enough and winds are strong. A classic example of advection fog is around San Francisco's Golden Gate Bridge (Figure 4.41). San Francisco's famous fog forms as onshore winds advect warm, moist air from the Pacific over the cold California ocean current (Figure 4.41A). During the summer months it is not uncommon for the advection fog to roll in and fully engulf the San Francisco Bay and other West Coast locations (Figure 4.41B). Winds associated with advection fogs are stronger than winds associated with radiation fogs. There needs to be enough turbulence to cool a thick layer of air and transport the fog vertically. Advection fogs are much thicker and persistent compared to radiation fogs. This fog type is especially common across the Central and Eastern United States during the winter and spring if there is a snow cover.

Figure 4.42: An example of upslope fog from the Olympic Peninsula, WA. Photo credit: B. Zaprowski.

Fog can develop on hillsides and mountain slopes as air is forced to rise and cool. **Upslope fog** forms due to the gentle uplift of air along a sloping surface. The air expands, cools and eventually reaches saturation as it moves upward over a gently sloping surface. In the United States, upslope fog is common in the coastal mountain ranges of California and the eastern slopes of the Rocky Mountains (Figure 4.42). Upslope fog can be extensive when humid air is forced to rise up a steep mountain slope.

Figure 4.43: Steam fog dancing on the Potomac River. Photo credit: B. Zaprowski

There are two types of fog that result from the addition of moisture to the air. The air immediately above a water body can become saturated when enough moisture is evaporated from the water. A vertical vapor pressure gradient develops over the water and causes the humid air to rise. The rising water vapor molecules condense into tiny droplets when they meet the cool air above. This type of fog is called **steam fog** because the rising air looks like a cup of coffee with steam rising above it. Steam fog commonly forms over lakes and rivers on clear and cold mornings when the waters are still relatively warm (Figure 4.43). These conditions are most common during the fall as cold air frequently settles over a relatively warm body of water. It can also form on a cold day over a heated outdoor swimming pool or when the sun comes out over a wet highway after a recent rainfall. Steam fog is often shallow because the rising air streamers evaporate as they encountering drier air a short distance above the water surface.

Figure 4.44: An example of frontal fog. Photo credit: B. Zaprowski.

Frontal fog is a type of fog that is associated with frontal zones, it is typically is associated with a warm front. It forms near a front when light precipitation, falling from relatively warm air above the front, evaporates into cooler air close to the Earth's surface and saturates it. Fog forms as evaporational cooling and the addition of moisture saturates the air close to the ground. The frontal fog will extend through the saturated layer of air that exists from the ground to the clouds in the warmer air above. Although this type of fog primarily results from the mixing of unlike air masses along a frontal boundary, it can also form through the rapid cooling of air over a moist ground. A snow covered ground is an ideal condition for frontal fog to form (Figure 4.44). As the falling raindrops melt the snow, heat is extracted from the environment and cools the air.

Section 4-4: Different Forms of Precipitation

Precipitation is water in solid or liquid form that falls from the atmosphere and reaches the ground. Although everyone is familiar with rain there are many different forms of precipitation due to the variability of atmospheric conditions that exist on a global scale. Precipitation is observed and classified according to its form, type, intensity and character. All forms of precipitation have one thing in common; they are all produced by clouds. **Rain** is defined as liquid water droplets that are between 5.0 and 0.5 mm in diameter. **Drizzle** is composed of smaller liquid water droplets from 0.2 to 0.5 mm in diameter, while **mist** has water droplets that are only 0.05 to 0.005 mm in diameter. Tropical latitudes receive only precipitation in the form of rainfall due to continuously warm atmospheric conditions. Precipitation can occur in a solid or liquid form in the mid-latitudes, where temperatures at cloud heights can be cold enough to support ice crystal formation. Depending on the atmosphere's temperature profile, snow, sleet, freezing rain, or rain will fall. Continuous below-freezing temperatures in the high latitudes result in some locations only receiving precipitation in the form of snow. Although rain and snow are the most familiar types of precipitation, several other forms are possible.

Hail: Frozen Precipitation In The Summer

Hail is a type of frozen precipitation in the form of rounded or jagged lumps of ice that are at least five millimeters in diameter. Hailstones, single units of hail, come in a variety of sizes and shapes; they are always produced by thunderstorms. Large hailstones are often characterized by nearly concentric layers of ice that resemble an onion. Hail develops as strong convection currents within intense thunderstorms carry ice pellets above the freezing level. Hailstones begin as a small snow pellet, called **graupel**, covered with a thin layer of ice. The graupel particles have the consistency of a soft snowball and form as supercooled water droplets collide and freeze on impact. If these particles are held above the freezing level they will grow by forming new layers of ice to form the core of the hailstone. Eventually the balls of ice descend through the cloud as they become too large and heavy to be supported by the thunderstorm's updraft. They can also be tossed into the clear air ahead of the thunderstorm and fall towards the ground if they encounter the downdraft. Upon reaching the base of the cloud updrafts carry some of the hailstones back to the top of the storm, where they grow by collecting more supercooled water droplets and forming an additional layer of ice. This process can repeat itself as the hailstone moves above and below the freezing level. Stronger updrafts typically cause the hailstone to become very large by repeating this cycle many times. As a hailstone grows, it develops an onion-like pattern of ice that alternates between clear and opaque states (Figure 4.45). Opaque layers of ice form high in the clouds as rapid freezing traps the air bubbles in the ice. Layers of clear ice are produced in the lower and warmer regions of the cloud where the supercooled water droplets freeze more slowly, allowing the air bubbles time to escape.

Figure 4.45: This close-up view of a hailstone shows its clear and milky growth layers. Photo credit: NOAA Photo Library

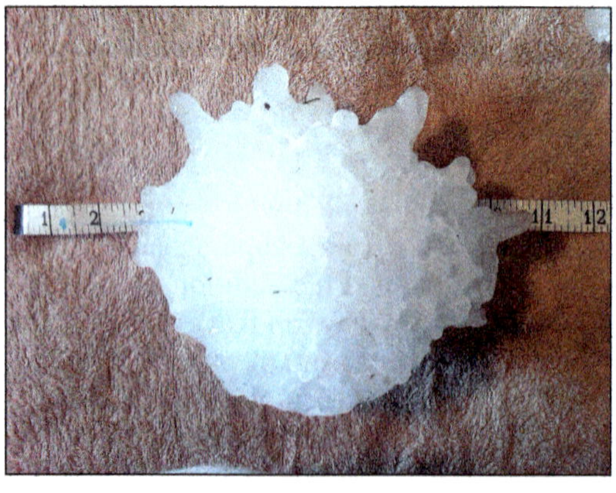

Figure 4.46: This record-setting hailstone fell in Vivian, South Dakota in 2010. Photo credit: NOAA.

The size of a hailstone is directly related to the strength of the thunderstorm's updraft. The intensity of the updraft determines how long the hailstone can remain above the freezing level and continue to collect supercooled water droplets. According to the National Weather Service, updraft speeds of at least 60 kilometers (37 miles) per hour are needed to produce dime-sized hail. Hailstones often cover the ground in long, narrow streaks called hailstreaks. Several hailstreaks may be produced from a large powerful thunderstorm; they parallel the path of the storm and indicate its direction of movement. Hailstreaks are typically about 2 kilometers (1.2 miles) wide and 10 kilometers (6.2 miles) long. Fortunately, most hailstones are harmless with diameters less than one centimeter. Sometimes severe thunderstorms may produce destructive hailstones that are the size of an apple or even larger (Figure 4.46).

Figure 4.47: Examples of hail damage. A. This cornfield was decimated by a hailstorm. Photo credit: © miropink/Shutterstock. B. Hail damage to a vehicle from baseball sized hail in Lubbock County, TX. Photo credit: NOAA.

Baseball sized hailstones can reach speeds of over 140 kilometers per hour (90 miles per hour) as they fall to the ground. The largest hailstone ever recorded fell on July 23, 2010 in Vivian, South Dakota. The hailstone was 20 centimeters (7.9 inches) in diameter and weighed over 0.87 kilograms (1.9 pounds). Many of the largest hailstones are a composite of multiple hailstones that are frozen together.

The devastating effects of hailstones are especially familiar to farmers, whose entire field of crops can be wiped out in minutes (Figure 4.47A). Crop losses from hail have been estimated to be more than one billion dollars annually in the United States alone. Because hail is generally a spring and summer phenomenon, it often occurs too late in the growing season to allow farmers to replant. Large hailstones can be large enough to smash windows and dent vehicles, causing extensive damage to buildings and cars (Figure 4.47B). In extreme cases snowplows have been needed to clear the hail from the roadways. The costliest hailstorm in United State history occurred on April 10, 2001 in St. Louis, Missouri. Nearly every home and business in northern St. Louis County suffered hail damage. Up to baseball sized hail was reported from eastern Kansas to southwestern Illinois, the total cost of the damages reached more than two billion dollars.

Winter Precipitation: Snow, Sleet And Freezing Rain

Snow is a type of frozen precipitation that is composed of white or translucent ice crystals that are commonly stuck together into snowflakes (Figure 4.48). Each snowflake is a delicately complex arrangement of ice crystals. No two snowflakes are alike although at the microscopic level they are all six-sided. Snow forms when the atmospheric temperatures are at or below freezing with a certain amount of moisture. For a snowflake to make it to the ground it must remain frozen, once it melts it can never turn back into a snowflake again. Under certain conditions snow can still reach the ground when the surface temperature is above freezing. The melting snowflakes create evaporational cooling as they encounter warmer air; this cools the air that surrounds the falling snowflakes and delays their melting. The temperature and moisture content of the air determine the size, shape and concentration of snowflakes.

Figure 4.48: The wide variety of snow crystals as seen under a scanning electron microscope. Credit: USDA.

Figure 4.49: A. Snow squalls make travel very dangerous. Photo credit: B. Zaprowski. B. U.S. Air Force Senior Master Sgt. Michael Sanders leads the pack despite snow flurries during the 20-kilometer upright bike portion of the Warrior Games in May 2010. Photo credit: Christopher Griffin/U.S. Air Force.

Snowflakes may be made up of ice crystals that are in the form of plates, needles, stars, columns or other complex structures. The "powdery" light, fluffy snow that skiers prefer forms when the air is very cold and contains little moisture. The ice crystals combine into wet, sticky snowflakes or giant puffs when temperatures are warmer than -5°C (23°F). These conditions produce snow with a higher moisture content that is easy to form into snowballs but hard to shovel. The appearance of especially large snowflakes often indicates that the snow is about to change to rain. The general shape of snowflakes can be predicted from the temperature and moisture characteristics of the air. Snowfall can sometimes occur in short, heavy bursts or accumulate slowly over time. Several terms are used to describe the rate and duration of falling snow. Snowfall intensity is determined by the accumulation over a given period of time. Snowfall rates are generally classified as light, moderate or heavy depending on visibility. A **snow squall** is a brief, intense snow shower that is accompanied by strong, gusty winds (Figure 4.50A). Snow squalls can produce significant accumulations and are commonly associated with lake-effect snow in the Great Lakes region. A **snow shower** is a short duration of light to moderate snowfall, and there is usually some spotty accumulation, typically around one inch, because they tend to come from cumuliform clouds. Snow showers are characterized by a sudden beginning and ending which creates dangerous driving conditions even though they are not long lasting. **Snow flurries** refer to light and brief snowfall with no accumulations or only a light dusting of snow expected (Figure 4.50B). They tend to come from stratiform clouds and usually only produce a small number of random snowflakes in the sky. Snow flurries are difficult to predict since they are generally not visible on radar and usually affect a small area.

Although astronomical winter arrives around December 21 in the Northern Hemisphere, meteorological winter runs from December through February, the three coldest months. Sometimes winter weather is unseasonable and doesn't follow the meteorological calendar. A few of the highest impact snowstorms to ever affect the Northeast United States, including the Blizzard of 1888 and the 1993 Storm of the Century, occurred in March.

Figure 4.50: A snow plow from the 436th Civil Engineer Squadron clears snow at Dover Air Force Base, DE. Wind-driven snow fell at a rate of one inch per hour at times, making visibility limited to a half-mile. Photo credit: U.S. Air Force/Roland Balik.

Forecasters sometimes refer to **whiteout** conditions when visibility is so severely reduced by snow that the horizon is not visible and causes people to become disoriented (Figure 4.50). Winter storm warnings are issued when heavy accumulations of snow are expected. Human behavior can be greatly affected by the anticipation of these winter storms, causing grocery stores to sell out of essential goods. The public often braces for the worst when forecasters predict a big snowstorm. Issuing a forecast for a major winter storm has serious economic effects and greatly affects local businesses. Often times the public forgets the correct snowfall predictions and only remembers the snow forecasts that were missed. Snow forecasts are continuing to improve but remain one of the most difficult challenges for meteorologists. Extremely small differences in near freezing temperatures make huge differences in forecasting accumulations for the various forms of liquid and frozen precipitation.

A **blizzard** is a dangerous winter storm characterized by low visibilities through a combination of blowing snow and strong winds. This term is commonly used for any heavy snowstorm that is accompanied by strong winds.The National Weather Service defines a blizzard by a specific set of weather conditions. For a winter storm to be classified as a blizzard it must meet the following criteria: at least three consecutive hours of falling or blowing snow with wind speeds of 56 kilometers (35 miles) per hour or greater that reduce visibilities to less than 0.4 kilometers (0.25 miles). A blizzard warning is issued by the National Weather Service when these conditions are expected. While heavy snowfalls and severe cold often accompany blizzards, they are not required. Temperature, geographical location and snowfall amounts are not factors used to define a blizzard. Snow already on the ground can be picked up by strong winds to create a ground blizzard.

Blizzards form when freezing cold air collides with warm, moist air (Figure 4.51). This collision of warm and cold air produces heavy snow and powerful winds. Blizzards most often develop on the northwest side of an intense low-pressure system. The strong pressure gradient in this location can sometimes produce hurricane force winds that blow any falling or existing snow, creating poor visibilities and the potential for large snowdrifts. Blizzards are notorious for trapping climbers and people in cars; there is little chance for survival if you are caught out in the open. Wind and snow are not the most dangerous part of a blizzard; cold temperatures are the biggest killer from these storms. Blizzards also can cause a variety of other problems including power outages and frozen pipes. Stay in your car if you are ever stranded in open country during a blizzard, many have died of exposure after getting out of their car trying to seek help.

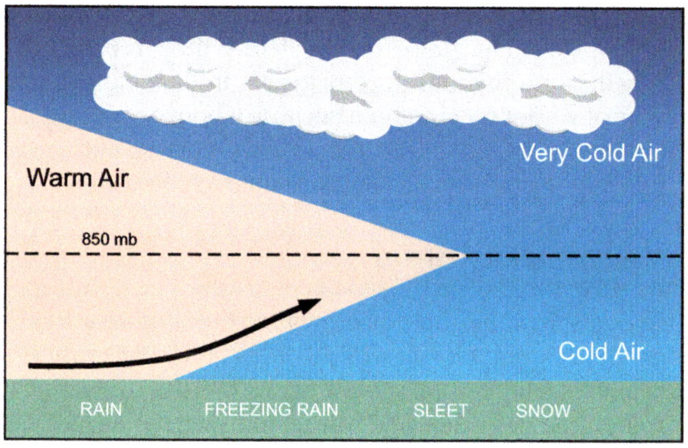

Figure 4.51: In the winter, overrunning can produce a wide range of wintery precipitation.

The easiest way for snowfall to occur is to have below freezing temperatures through the entire troposphere. Sometimes a thin layer of relatively mild air above the ground separates two layers of subfreezing air. This winter situation commonly occurs north of a warm front as the warm air advances northward and gradually rises over the cold, dense air ahead of an approaching low-pressure system (Figure 4.51). The snowflakes are transformed into raindrops as they melt and fall through the relatively warm air above the ground; the above freezing temperatures are generally found near the 850 mb pressure level. The raindrops have a chance to freeze when they fall through the subfreezing air near the surface. Depending on the depth and temperature of the cold air near the ground, the raindrops will reach the surface in the form of ice pellets or supercooled water that freezes on contact with surfaces below freezing. These types of winter precipitation require the presence of a temperature inversion since the colder air is near the surface.

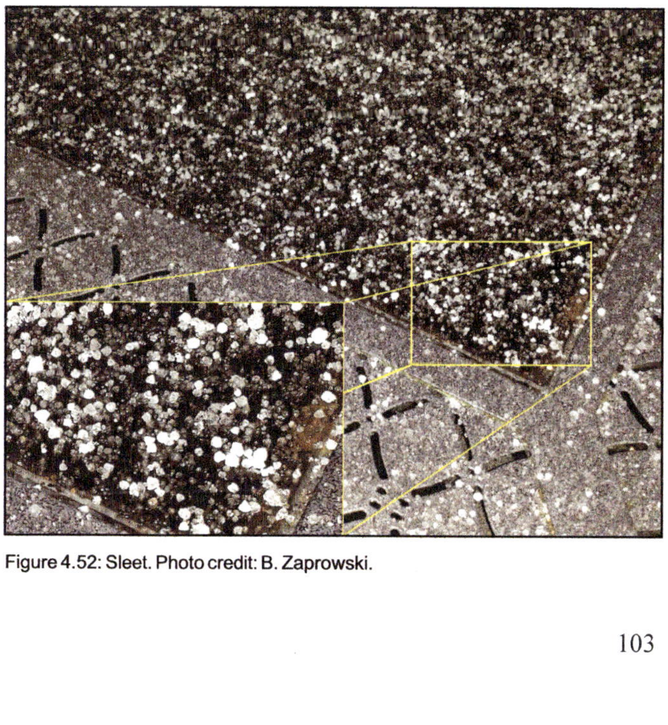

Figure 4.52: Sleet. Photo credit: B. Zaprowski.

Sleet is a type of winter precipitation that forms when raindrops freeze in the air while falling to the surface. Sleet first begins as falling ice crystals or snowflakes that melt into raindrops as they fall through warmer air (Figure 4.52). Sufficient cooling is needed for the

Figure 4.53: A. Freezing rain coating tree branches. Photo credit: ©Adam Radosavljevic/Shutterstock. B. Freezing rain brought down trees in Paris, TX. Photo credit: NOAA.

raindrops to freeze, they don't freeze instantaneously. If the surface layer of freezing air is deep enough the falling raindrops will have time to form sleet as they freeze into ice pellets prior to striking the ground (Figure 4.51). These ice pellets are five millimeters or less in diameter, smaller than hailstones. Sleet is easy to distinguish from other forms of winter precipitation because it bounces on impact with the ground or other objects without sticking. Sleet is difficult to forecast because it requires very specific atmospheric conditions to form. Accumulations of sleet can cause very hazardous driving and walking conditions.

Freezing rain is a type of precipitation that forms as snowflakes fall through a layer of warm air deep enough for them to completely melt and become rain. The raindrops pass through a thin layer of cold air just above the surface that cools the raindrops to below freezing (Figure 4.51). The raindrops do not freeze; they become supercooled and then freeze upon impact with surfaces below freezing (Figure 4.53). In extreme cases freezing rain can occur with surface air temperatures as low as -12°C (10°F). Freezing rain is more common than sleet, it is usually found in a thin band on the cold side of a warm front, where surface temperatures are at or just below freezing. Forecasting the occurrence of freezing rain is very challenging since the band of freezing rain is so narrow and dependent on the exact position of the frontal boundary. The freezing rain band is usually very thin and is rarely more than 160 kilometers (100 miles) wide.

Section 4-5: Mid-latitude Cyclones

A **mid-latitude** cyclone is a large low-pressure system that forms in the mid-latitudes. Steered by the westerly winds aloft, they generally move in an eastward direction. They are the principal weather maker for North America and are characterized by a distinct low-pressure center and well-developed fronts. The average life cycle of a mid-latitude cyclone is about one week, although their lifetime is largely determined by atmospheric conditions and topography. Most fully developed mid-latitude cyclones have a distinct cold front and a warm front ahead of it that extends from their low-pressure center. Mid-latitude cyclones are sometimes called the largest storm system on Earth; their typical size ranges from several 100's to over 3,000 km (~500 to 2,000 mi.) in diameter (Figure 4.54). They rotate counterclockwise, or cyclonically (in the

Figure 4.54: A very large, well-developed mid-latitude cyclone. Image credit: NOAA.

Northern Hemisphere!), and are associated with convergence, which leads to rising air, clouds and precipitation. These cyclones are effective at reducing temperature contrasts by moving warmer air to higher latitudes and colder air toward the equator.

Meteorologists have known since the 1800s that these cyclones form along the polar front and dominate the weather for the mid-latitudes. It was not until after World War I (1914-1918) that a model was developed to explain the life cycle of a mid-latitude cyclone. The polar front theory was developed by Vilhelm Bjerknes and colleagues to describe the formation, development and dissipation of mid-latitude cyclones. This model was initially created primarily from surface observations and later modified as upper-level data became available. Mid-latitude cyclones develop in this model as warm tropical air and cold polar air interact along the polar front. The polar front is easily identified as a nearly continuous band around the hemisphere on upper-air charts. According to the polar front theory, mid-latitude cyclones progress through a predictable life cycle that lasts over the course of a few days to over a week depending on atmospheric conditions. This model is very useful to visualize the evolution of a mid-latitude cyclone and forecast daily changes in the weather.

Mid-latitude cyclones form and intensify if conditions are favorable in the upper-troposphere. Any attempt to predict the evolution of a mid-latitude cyclone and the weather they produce must take the upper-tropospheric conditions into account. This is why broadcast meteorologists often illustrate the jet stream when describing their movement and evolution. The general conditions needed for the development of a mid-latitude cyclone in the U.S. are shown in Figure 4.55A. Low-level convergence is not enough to cause a low-pressure system to form; an upper-level trough approaching from the west is needed. The formation of a mid-latitude cyclone is often preceded by increasing divergence aloft. If the upper levels aren't favorable for cyclone development, the cyclone won't grow as the convergence of air into the surface low will cause it to decay. The upper-level trough provides the necessary

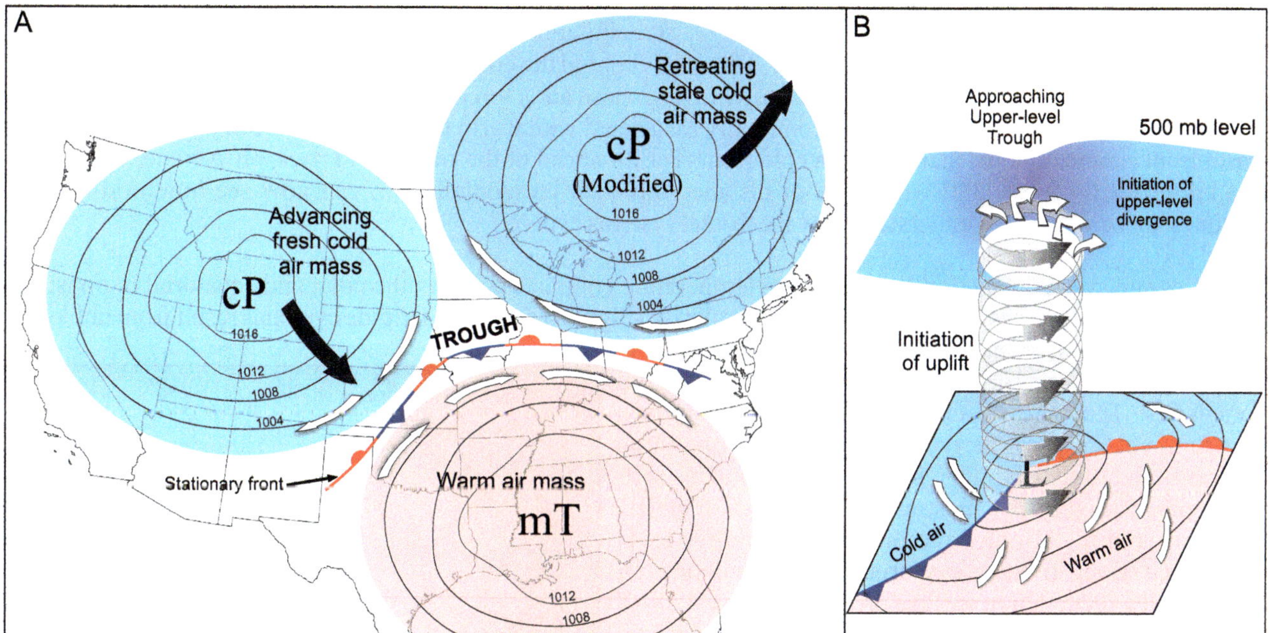

Figure 4.55: Idealized diagram for the formation of a mid-latitude cyclone. A. A stationary front develops in the trough between air masses. B. With the approach of a upper-level trough, upper-level divergence causes the formation of a low pressure center and the initiation of counter-clockwise air movement around the low.

divergence needed for the low-pressure system to develop (Figure 4.55B). **Cyclogenesis**, the development or strengthening of a cyclone, usually takes place along the polar front directly under an area of strong divergence aloft. Initially, the polar front separates the boundary between cold, dry (cP) air to the north and the warm, moist (mT) air to the south. Air pressure is reduced at the surface if divergence aloft exceeds the convergence of surface air. A mid-latitude cyclone can form once a horizontal pressure gradient is developed at the surface.

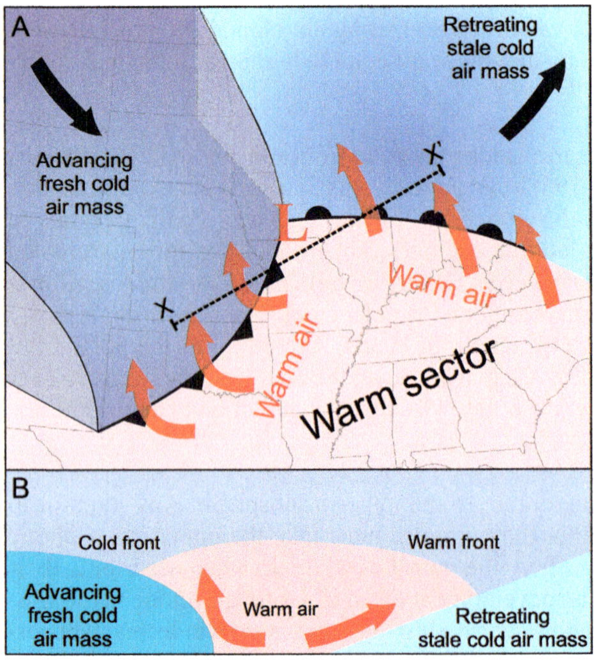

Figure 4.56: A. The structure of a young mid-latitude cyclone. As air circulates around the low, warm air rides up and over the retreating cold air mass while the advancing cold air mass pushes its way in. B. Cross-section profile along line X-X'.

A minor "kink" develops along the stationary front as cyclogenesis begins. The cyclonic circulation transports warm, moist air northward ahead of the low-pressure center and drags colder, drier air southward behind it. Figure 4.56A shows a typical mid-latitude cyclone in the Northern Hemisphere, with its associated cold and warm fronts. The fronts are shown as surfaces that extend vertically from the ground in order to visualize them as three dimensional features. It is easy to see how fronts can promote rising air, clouds and precipitation when they are drawn this way. Cold, relatively dry air advances north and west of the surface cold front, while the cold, dry air retreats north and east of the warm front. In both cases the warm air moves up and over the cold, dry air due to differences in density. The denser cold air is always wedged under the less dense warm air. A relatively mild region of air with similar temperature and moisture characteristics, called the **warm sector**, develops between the cold front and the warm front. Figure 4.56B shows a vertical cross section of both fronts between points X and X' that are depicted in Figure 4.56A. The cold air masses resemble wedges of cheese at the surface with warm air above them. The air behind the cold front thickens, on average, by roughly one kilometer for every 100 kilometers of distance traveled back into the cold air. Precipitation intensity is usually the strongest along a cold front as the warm air is more forcefully lifted. The winds behind the cold front advect increasingly colder air into a region as the cold front advances. The northward moving warm air is forced to glide up over the cold, dense air near the surface. Less intense precipitation is commonly associated with warm fronts as the slope of the air ahead of the warm front is more gradual than behind a cold front. The air ahead of the warm front rises, on average, by roughly one kilometer for every 300 kilometers of distance travelled horizontally. We can now see how mid-latitude cyclones intensify once we understand how the frontal systems evolve.

Low-pressure systems function to help reduce temperature contrasts between the polar and equatorial regions; the cloud pattern makes it easy to see the cyclonic circulation (Figure 4.57). Mid-latitude cyclones result from unlike air masses colliding over the mid-latitudes, resulting in large temperature contrasts focused along narrow frontal boundaries. They are largely responsible for their own development as they derive their energy from these contrasts and intensify; this process is known as self-development. Figure 4.58 illustrates the evolution of the life cycle of a mid-latitude cyclone. Initial upper-level divergence along a stationary front generates rising air motion and a reduction of surface pressure (Figure 4.58A). Cyclonic circulation becomes established around the low-pressure center as the approaching upper-level trough becomes positioned directly above the low-pressure center (Figure 4.58B). Distinct cold and warm fronts are now established as the cyclonic circulation advects warm air northward east of the low and cold air southward to the west of the low. The 500 mb heights fall as the fronts advance, causing the upper-level trough to become more cyclonically curved (Figure 4.58C). The upper-level pattern strengthens as divergence aloft exceeds surface convergence in a process known as deepening. Strong upper-level divergence

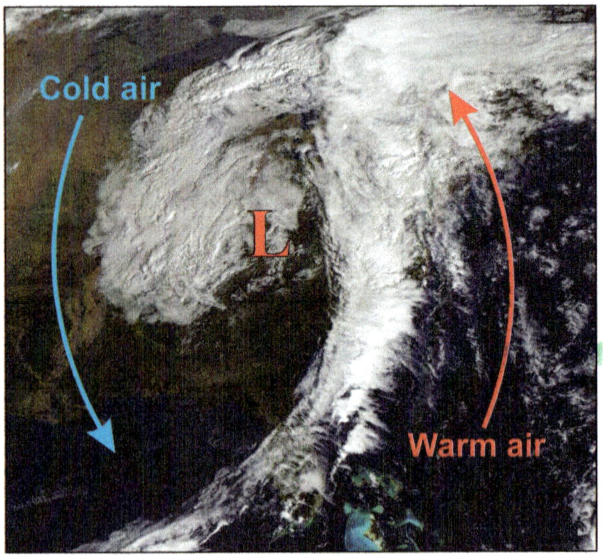

Figure 4.57: Mid-latitude cyclones help move heat poleward. Credit: NASA.

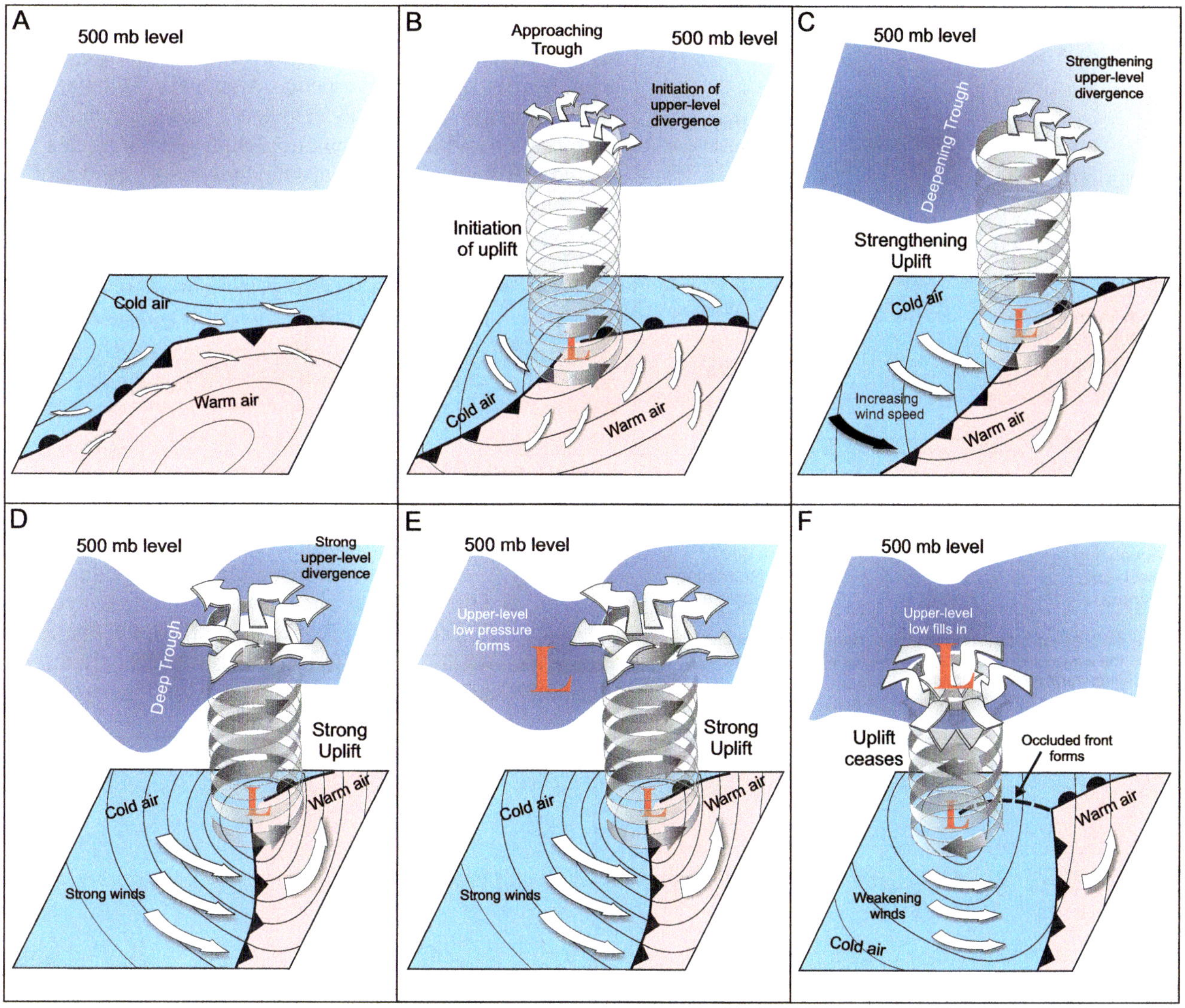

Figure 4.58: The evolution of a mid-latitude cyclone. A. Stationary front forms between warm and cold air masses. B. An approaching upper-level trough provides the upper-level divergence needed to trigger the formation of a surface low. C. A deepening trough causes a strengthening of the upper-level divergence and a resulting strengthening of the surface low. D. A now deep trough has caused a strong surface low to form with associated strong winds and uplift. E. An upper-level low forms. This marks the beginning of the end for the cyclone. F. As the surface low moves under the upper level low, forming an occluded front, it becomes surrounded by cold air. Meanwhile, the upper level low fills in, causing uplift to cease. As a result, the cyclone will slowly fall apart.

enhances uplift which reduces the surface pressure and strengthens the winds (Figure 4.58D).

Fully developed mid-latitude cyclones appear as distinctive comma clouds on satellite images. The head of the comma extends from the low-pressure center to the northwest while the tail of the comma represents the cold front. They can also be identified on a surface map by observing the cyclonic winds that flow around the low-pressure center. Distinct changes in all surface observations should be noted by a forecaster to identify the location of the frontal boundaries. The increased divergence aloft increases wind speeds and enhances upward motion, which leads to greater cooling. This cooling eventually marks the beginning of the end for the mid-latitude cyclone as the upper-level trough strengthens into an upper-level low (Figure 4.58E). As the cold front advances and the surface low migrates west towards the colder air, an **occluded front** forms as the cold front overtakes a portion of the warm front. Cold fronts usually travel faster than warm fronts, causing the air in the warm sector to be lifted into the upper atmosphere. The surface low moves directly under the upper-level low and loses direct contact with the warm sector when this happens (Figure 4.58F). At the surface, the warm sector is still present but is smaller and detached from the cyclone.

A cross section of an occluded front is shown in Figure 4.59. The occluded front forms as the surface low moves

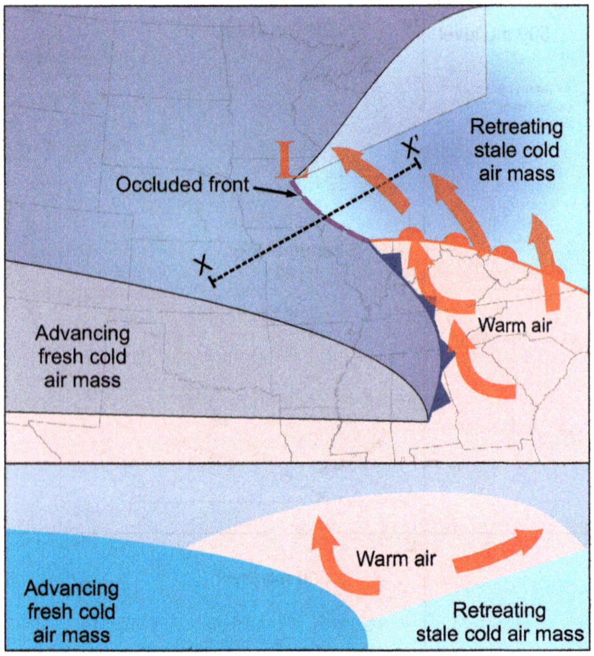

Figure 4.59: An occluded front. As the surface low moves into the cold sector, the advancing cold air mass collides with the retreating cold air mass, lifting warm air off the ground.

back into the cold air in response to the enhanced upper-level divergence. The surface low will weaken as it loses the support of upper-level divergence. Air converges toward the upper-level low once the system is occluded, which causes it to weaken and decay. The circulation usually moves the occluded front cyclonically around the low-pressure center. The mid-latitude cyclone has consumed all of its available energy at this point. The horizontal temperature differences between the unlike air masses are eliminated when all of the warm air has been displaced aloft. Although occlusion marks the end of the mid-latitude's life cycle, they can sometimes intensify as occlusion begins. Significant rainfall and snowfall are still possible during the dying stages of a mid-latitude cyclone.

Evolution of a Mid-latitude Cyclone

The life cycle of a mid-latitude cyclone evolves gradually over time, the changes from one stage to another are not easy to identify. Cloud patterns can be used as a guide of the expected weather changes ahead as a mid-latitude cyclone approaches from the west (Figure 4.60). High cirrus clouds and falling pressure are the first signs of an approaching mid-latitude cyclone; they are associated with the cyclone's

Figure 4.60: Cloud patterns expected with a mid-latitude cyclone. A. The approach of a mid-latitude cyclone begins with cirrus clouds. B. The cirrus clouds thicken into cirrostratus clouds. C. The cirrostratus clouds thicken, forming altostratus clouds. D. Thick nimbostratus clouds form as the low continues to develop. Photo credits: B. Zaprowski.

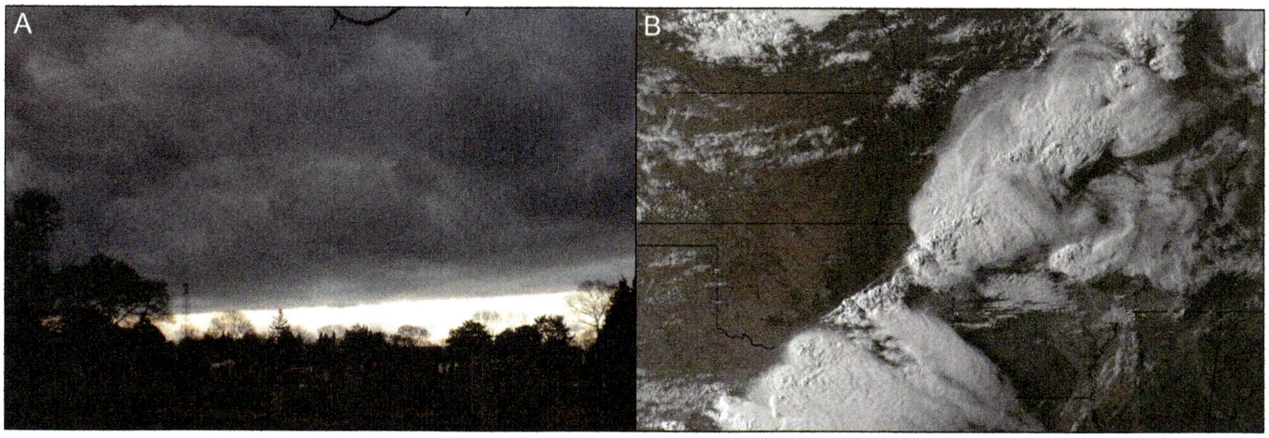

Figure 4.61: Weather patterns associated with a mid-latitude cyclone. A. The approach of a cold front from ground level. Photo credit: B. Zaprowski. B. This satellite image shows a line of storms forming along a cold front. Photo credit: NOAA.

warm front (Figure 4.60A). The cirrus clouds lower and thicken into cirrostratus clouds as the warm front advances (Figure 4.60B). The layer of cirrostratus clouds that covers the sky indicates the arrival of rain within the next 12-24 hours. Eventually the sun becomes barely visible as cirrostratus clouds give way to a layered deck of altostratus clouds (Figure 4.60C). As the front gets closer the sky darkens and precipitation starts to fall from thick nimbostratus clouds (Figure 4.60D).

Precipitation is less intense along and ahead of the warm front, varying from light to moderate showers. The arrival and passage of the warm front brings an increase in the rate of precipitation with a noticeable rise in temperature and a distinct wind shift. The strong southerly winds just before the cold front passes usually produce the warmest temperatures of the day. Conditions quickly deteriorate as the southerly winds and generally clear skies are replaced by a line of cumulonimbus clouds capable of producing severe weather as the cold front approaches (Figure 4.61). The passage of a cold front is easily identified by a shift in wind direction. Winds typically blow from the northwest behind the cold front, producing an obvious decrease in temperature, as the drier and cold air is advected to the southeast. Rising pressure behind the cold front indicates improving conditions. As high-pressure moves in behind the cold front the skies often clear with a noticeable decrease in humidity (Figure 4.62). The passage of the cold front usually brings a few days of fair weather with clear skies and light winds. Locations affected by the occluded front, on the northern end of the cyclone, generally experience continual cloud cover and a gradual drop in pressure as the cyclone approaches. Overcast skies keep temperatures cool during the passage of the storm, with snow or ice possible in the winter months. The slow movement of the occluded front causes these conditions to persist for an extended period.

Figure 4.62: Fair conditions following the passage of a cold front. Photo credits: B. Zaprowski.

Conveyor Belt Model

The conveyor belt model describes both the horizontal and vertical air motions of a mid-latitude cyclone. It is one of the best ways to visualize the three-dimensional flow around a mature mid-latitude cyclone and explain the distribution of precipitation. This model consists of three interacting airstreams, called conveyor belts, that typically develop as the cyclone matures. The conveyor belts are named for the type of air they transport. Once the low matures, warm air from the warm sector travels northward in a channel called a **warm conveyor belt** (Figure 4.63A). The warm conveyor belt flows near the Earth's surface along the warm side of the cold front. It represents

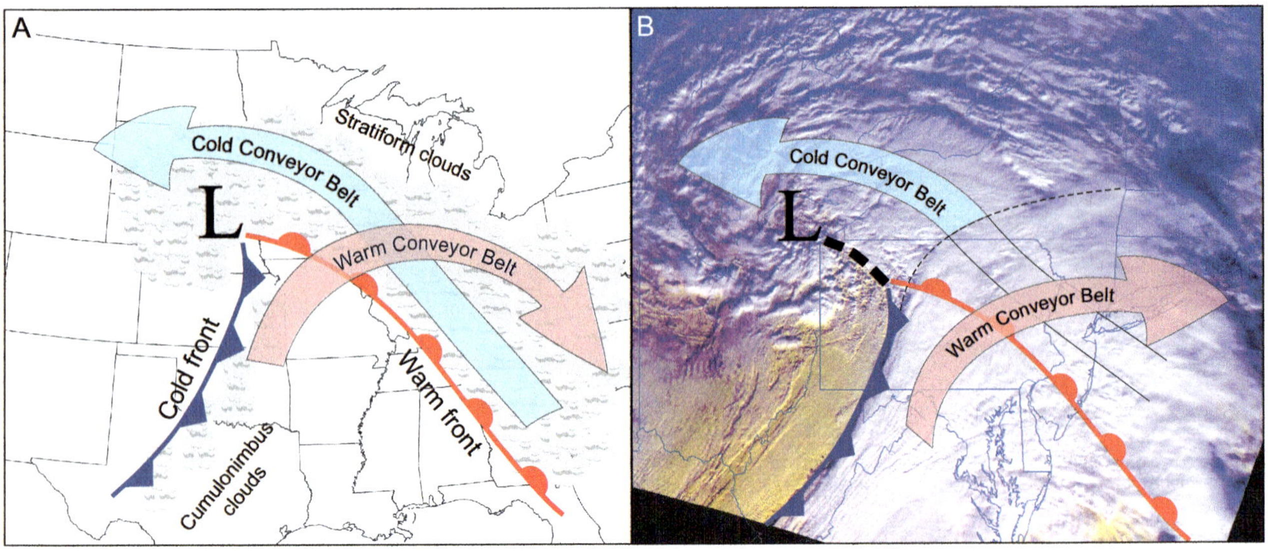

Figure 4.63: The Conveyor Belt Model. A. The parts of a mid-latitude cyclone that produce weather. As warm air rides up and over cold air along the warm front, it is turned eastward producing a wide swath of stratiform clouds and steady precipitation. Meanwhile, a cold conveyor belt forms just north of the warm front. This cold air wraps around the low and produces low-level clouds and moderate to heavy precipitation. Finally, cumulonimbus clouds and heavy precipitation forms as warm air is rapidly lifted upwards along the cold front. B. A real-life example of a mid-latitude cyclone. Image credit: NOAA/OSEI.

the main source of warm, moist air that fuels the cyclone. The amount of precipitation produced by a mid-latitude cyclone can be dependent on the extent of moisture contained within the warm conveyor belt. This warm and humid air, located south and east of the low-pressure center, produces clouds and precipitation as it overruns the low-level cold air. This airstream typically turns eastward as it ascends over the warm front and follows the upper-level westerly winds. The airflow of the warm conveyor belt is identified on satellite imagery by a swirling band of high clouds that extend eastward (Figure 4.63B). Note the streamers of cirrus clouds on the western and northern edge of the warm conveyor belt. Forecasters classify the cyclone as mature when this distinctive cloud structure is visible.

Figure 4.64: This mid-latitude cyclone displays a well-developed dry slot. Image credit: EO/NASA.

As the mid-latitude cyclone matures, the region west of the low-pressure center experiences downward motion as dry air near the tropopause sinks and moves towards the low. This airstream, called the **dry conveyor belt**, splits into two branches as it approaches from the west. One branch of the dry conveyor belt descends and turns southward to produce the characteristic clear skies behind the cold front. The other branch moves toward the low-pressure center and wraps cyclonically around it, forming what meteorologists call a dry slot. The dry slot helps to separate the head and tail of the comma cloud; it appears as a clear wedge on satellite imagery (Figure 4.64).

Isobars encircle the low-pressure center as the mid-latitude cyclone matures and intensifies. As this occurs, a conveyor belt of cold, initially dry, low-level air begins to circulate westward north of the warm front (Figure 4.63A). This **cold conveyor belt** originates north and east of the low-pressure center and flows underneath the warm conveyor belt. This airstream becomes saturated and splits into two branches as it slowly rises and gains moisture by the evaporation of falling precipitation associated with the warm front. The first branch rises abruptly as it approaches and wraps cyclonically around the northwest side of the low-pressure center, producing the distinctive head of the comma cloud.

Notice the swirl of clouds that flow underneath the warm conveyor belt and extend westward (Figure 4.63B).The cold conveyor belt often contributes to a band of heavy "wrap-around precipitation" in this region, the heaviest snowfalls are typically found on the northwest side of a mid-latitude cyclone. The path of winter storms are often depicted by a swath of heavy snow that forms on the northwest side of the low as it moves eastward (Figure 4.65). The second branch of the cold conveyor belt turns anticyclonically as it rises in response to the westerly flow aloft.

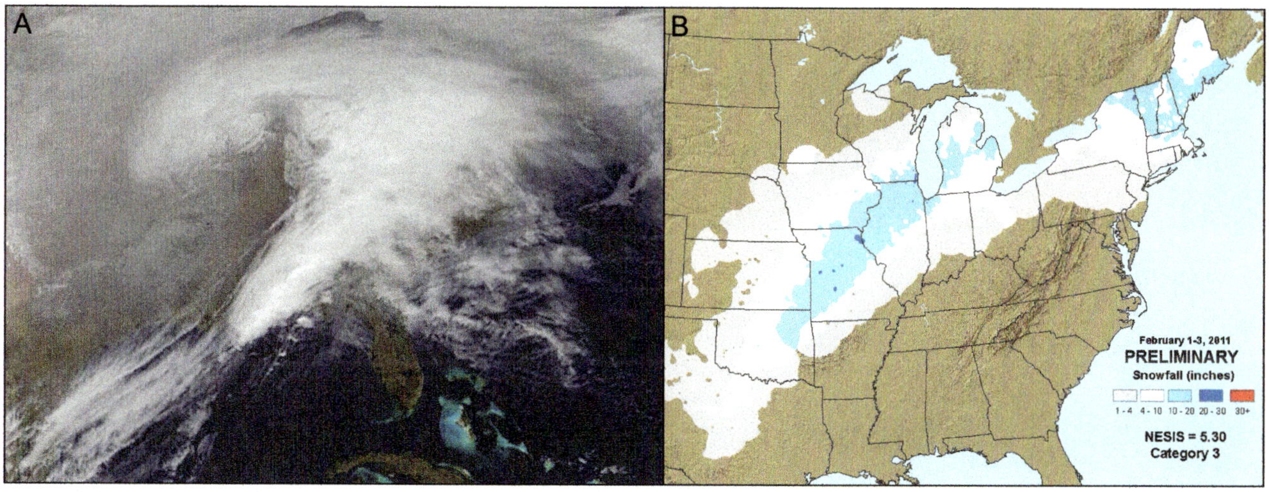

Figure 4.65: A. The mid-latitude cyclone responsible for the Groundhog Day Blizzard of 2011. Image credit: EO/NASA. B. The linear stripe of snowfall shows the path of the mid-latitude cyclone as it tracked across the U.S. Image credit: NOAA.

Andrea Booher/FEMA

Andrea Booher/FEMA

NOAA

NOAA/NSSL

NOAA/NWS

NASA

Chapter 5 – Severe and Hazardous Weather

In the last chapter, we explored weather and weather forecasting and discovered that weather has the ability to cause great damage and put human lives at risk (e.g. hail, freezing rain and blizzards). In this chapter we will look at the three most hazardous forms of weather: Thunderstorms, tornadoes and hurricanes.

Section 5-1: Thunderstorms

Almost everyone has experienced the weather associated with a thunderstorm. A darkening sky followed by strong wind gusts, torrential rain, lightning and thunder are common in some areas for much of the year. A **thunderstorm** is a storm produced by a cumulonimbus cloud that always produces lightning and thunder (Figure 5.1). They can bring welcome relief on a hot, humid afternoon or provide a farmer with much needed rainfall. However, thunderstorms can also be destructive, violent and even deadly. In the United States alone, thunderstorms produce about 25 million cloud-to-ground lightning strikes each year. Lightning is only one of the dangers associated with a thunderstorm; they can also create damaging hail, floods, wind gusts and tornadoes. Although there are many dangers associated with thunderstorms, many parts of the world depend on them for agricultural production and water resources. At any given time there are roughly 2,000 thunderstorms in progress across the Earth, with most occurring in tropical and subtropical latitudes.

Figure 5.1: A thunderstorm with lightning hovers over Bucharest, Romania. Photo credit: Cătălin Fătu.

A typical thunderstorm has an average diameter that ranges between 16 and 24 km (10 and 15 mi.). Although individual thunderstorms are generally small, they can develop into clusters and cover large areas (Figure 5.2). Thunderstorms are the product of convection; they are a working system of intense updrafts and downdrafts. The rapidly rising air, the updraft, is made visible as the cumulonimbus cloud develops vertically. Thunderstorms need both instability and a lifting mechanism to develop. They form when warm, humid air is lifted in an unstable environment; there are a variety of ways that the air can be lifted vertically. Diverging upper-level winds are also favorable for thunderstorm development since they enhance upward motion. If an air parcel is displaced upward from its equilibrium state, it will continue to rise if it remains warmer and less dense than its surroundings. A rising air parcel is much more likely to rise and gain momentum if the air in the mid-troposphere is cold. If the air parcel rises above its LCL, the release of latent heat reduces the parcel's rate of cooling and makes it more likely to remain positively buoyant. These conditions help explain why thunderstorms most commonly form when the lower atmosphere is warm and humid with cold air aloft. Thunderstorms can occur throughout the year at any time of day but are most likely to form during the afternoon or early evening hours in the spring and summer.

Figure 5.2: A cluster of thunderstorms form over Bangladesh. Photo credit: Image Science and Analysis Laboratory, NASA-JSC.

Thunderstorms are classified by the lifting mechanisms that

create them. Fortunately most thunderstorms are short-lived and dissipate shortly after forming without producing severe weather. Occasionally some thunderstorms can persist for several hours and become severe enough to uproot trees and damage buildings. Although lightning can be deadly, it is not used to define a severe thunderstorm since all thunderstorms are associated with lightning and thunder.

For a thunderstorm to be officially classified as severe, it must meet at least one of the following criteria:

1. Winds gusts at the surface of 50 knots (58 mi/hr) or greater
2. Hail with a diameter of at least 2.5 cm (1 inch, about the size of a quarter)
3. A tornado

Air Mass Thunderstorms

An **air mass thunderstorm** is a type of thunderstorm that forms in an unstable air mass that is warm and moist. These thunderstorms are the most common type and usually only last for about one hour. Air mass thunderstorms occur within an individual air mass far from frontal boundaries, they rarely produce strong winds or large hail. Although impacts of air mass thunderstorms are generally minimal, their greatest hazard is flooding when the storms move slowly or remain stationary. Uneven surface heating and weak low level convergence are the primary lifting mechanisms that generate air mass thunderstorms. Both of these lifting mechanisms are relatively weak and contribute to their brief life cycle. Because these storms are the product of uneven heating of the Earth's surface they tend to form as scattered, isolated cells.

Figure 5.3: A typical air-mass thunderstorm. Photo credit: Image Science and Analysis Laboratory, NASA-JSC.

Air mass thunderstorms typically develop in an environment with weak winds in the middle and upper troposphere; this is why this type of thunderstorm has a characteristic updraft tower that does not tilt. The weak winds prevent the updraft and downdraft from becoming separated. As a result, the storm quickly dissipates as the downdraft cuts off the storm's moisture supply. They commonly form during the spring and summer in the United States within maritime tropical (mT) air masses that move north from the Gulf of Mexico. These storms usually develop quickly on a warm, humid afternoon with abundant sunshine. Positively buoyant air parcels rise vertically after being heated by the Earth's surface to form a towering cumulonimbus cloud (Figure 5.3). The unstable air parcels that rise after being heated can be compared to the convection currents that form in boiling a pot of soup (refer back to Figure 1.19). Under the right atmospheric conditions cumulus clouds can sometimes develop into air mass thunderstorms. Air mass thunderstorms normally consist of one or more convection cells that each progress through a distinct three stage life cycle.

Cumulus Stage

The first stage of an air mass thunderstorm begins with the development of a cumulus tower as warm, unstable air begins to rise. A single column of rising air surges upward to form the updraft as the positively buoyant air parcel rises through the atmosphere (Figure 5.4A). Precipitation does not occur during the cumulus stage because it is dominated by updrafts that are strong enough to keep water droplets and ice crystals suspended. The parcel's rate of cooling is reduced as latent heat is released after it rises above its LCL, increasing the chance that the parcel will continue to rise. The tops of the cumulus clouds can reach altitudes of 8,000 to 10,000 m (26,000 to 33,000 ft.) in a little as 15 minutes. The edges of a cumulus cloud mix with cooler, drier air as it builds vertically above the LCL. This process is called **entrainment** and causes some of the water droplets to evaporate, making the edges of the cumulus cloud sharp and distinct. This evaporational cooling helps to reduce the buoyancy of the cloud and limits its vertical development. The cumulus cloud typically loses all of its positive buoyancy and stops growing vertically by

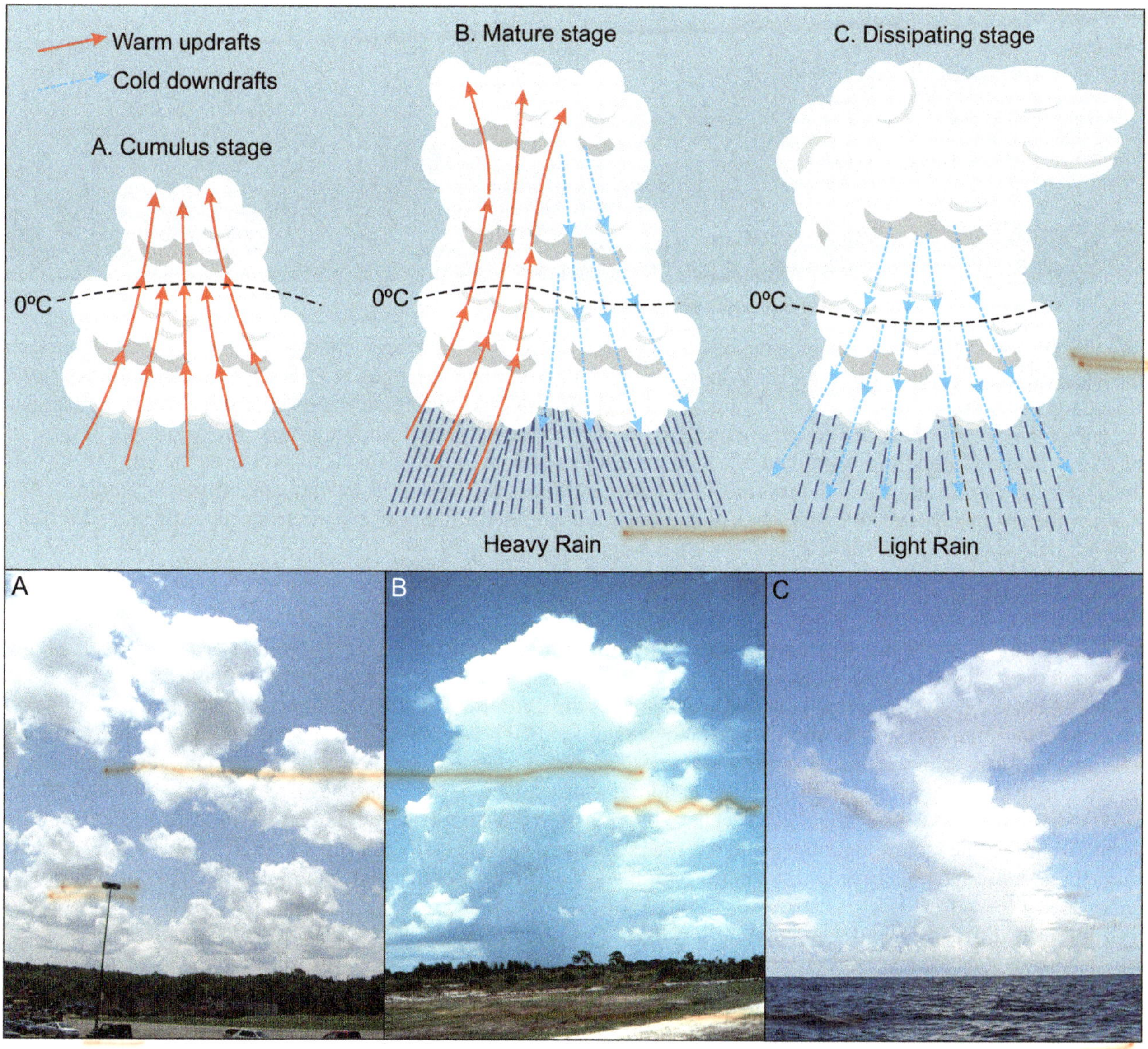

Figure 5.4: The stages of development for an air-mass thunderstorm. A. Cumulus stage: Warm updrafts cause the development of cumulonimbus clouds from cumulus clouds. Photo credit: B. Zaprowski. B. Mature stage: The warm updrafts are now balanced by cold downdrafts as heavy rain develops within the cumulonimbus clouds. Photo credit: NOAA. C. Dissipating stage: The cold downdrafts have killed off the warm updrafts. Without a source of warmth and moisture, the storm falls apart. Photo credit: Dr. Dwayne Meadows/NOAA.

the time the cloud builds to about one and a half times its diameter at the LCL. As a part of the cloud rises above the freezing level ice crystals begin to form and grow through the Bergeron process. Eventually the accumulation of precipitation becomes too great for the updraft to keep it in suspension.

Mature Stage

The mature stage of an air mass thunderstorm begins when the first raindrop from the cumulonimbus cloud reaches the Earth's surface (Figure 5.5). Rain, ice pellets and snow fall through the cloud as the weight of the water droplets and ice crystals become too heavy to be supported by the updraft. The falling precipitation particles drag air toward the surface to create a sinking column of air, the **downdraft**. The downdraft spreads laterally when it reaches the surface, and can be felt before the first raindrops reach the ground. This cool rush of air is created as falling precipitation evaporates before reaching the surface. The evaporational cooling enhances the downdraft and now the updraft and downdraft exist side by side (Figure 5.4B). The mature phase of an air mass thunderstorm is the most intense; this is when the cumulonimbus cloud gets larger and when rainfall, lightning and thunder are the most

Figure 5.5: A well-developed mature air-mass thunderstorm. Photo credit: B. Zaprowski.

intense. During the mature phase, the top of the cloud can reach the tropopause and often takes on an anvil shape as the ice crystals are spread horizontally by upper-level winds (refer back to Figure 4.38B for a good example of this). Cloud tops may grow vertically to an altitude of 14 km (~46,000 ft.) or more during its maximum intensity towards the end of the mature phase. Although precipitation is most intense during the mature stage, the beginning of the end of an air mass thunderstorm is marked once the raindrops reach the Earth's surface. Over time the downdraft will dominate the thunderstorm as the rainfall intensifies and cuts off the supply of moisture from the updraft. The mature phase typically lasts for only 15 to 30 minutes as entrainment enhances evaporational cooling and allows the downdraft to overwhelm the updraft.

Dissipating Stage

The dissipating stage begins when downdrafts form throughout the cloud; the storm collapses on itself during this stage and marks the end of thunderstorm activity (Figure 5.4C). The moisture supply of the thunderstorm, provided by the updraft, is now eliminated as sinking air replaces the updraft. The relative humidity drops and precipitation diminishes as compressional heating warms the subsiding air. The cumulonimbus cloud now gradually begins to evaporate as the cooling effect of entrainment and falling precipitation in the downdraft dominate this stage. Cloud droplets in the lower portion of cloud quickly evaporate as the thunderstorm dissipates, leaving only the remnants of the cirrus anvil behind. Only about 20 percent of the moisture that condenses within an air mass thunderstorm falls as precipitation, the remaining moisture is returned to the atmosphere through evaporation. The final stage of an air mass thunderstorm typically only lasts about 15 minutes, with the entire life cycle complete in roughly one hour.

Synoptically-Forced Thunderstorms

Some thunderstorms form through a combination of latent heat release and patterns of organized wind flow. Thunderstorms that form in response to circulation patterns associated with low-pressure centers, fronts and jet streams are called **synoptically-forced thunderstorms.** These thunderstorms can persist for much longer than air mass thunderstorms and can become severe thunderstorms under the right atmospheric conditions (Figure 5.6). These storms can remain organized for several hours or more because the updraft and downdraft of the thunderstorm are separated by **vertical wind shear**, this prevents the storm from collapsing on itself (Figure 5.7A). Vertical wind shear is a change in wind direction or wind speed (or both) between two altitudes. A strong jet stream is often a key ingredient for the development of these organized thunderstorms since it can be very effective at increasing wind shear. Vertical wind shear allows the updrafts and downdrafts to reinforce each other, allowing the thunderstorm to intensify. The updraft becomes tilted when strong vertical wind shear exists; this prevents precipitation from falling into the updraft, allowing it to continue to fuel the storm (Figure

Figure 5.6: A line of synoptically-forced thunderstorms developing over Florida. Image credit: Image Science and Analysis Laboratory, NASA-JSC.

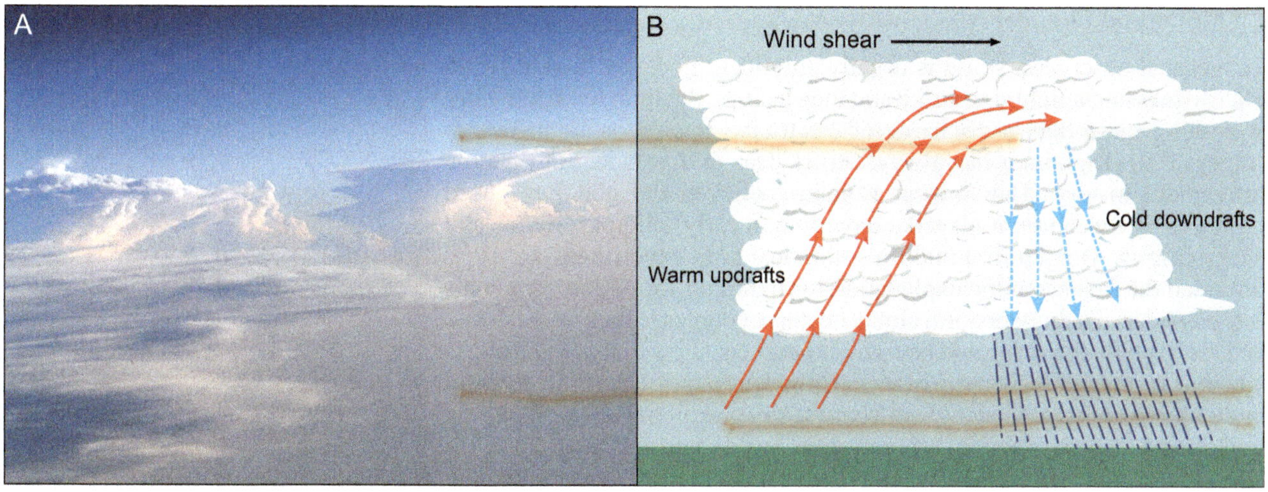

Figure 5.7: A. This image shows thunderstorms encountering strong vertical wind shear. Photo credit: Image Science and Analysis Laboratory, NASA-Johnson Space Center. B. Schematic diagram of a synoptically-forced thunderstorm.

5.7B). Strong vertical wind shear increases the lifetime of these thunderstorms, which increases the likelihood that they become classified as a severe thunderstorm. Usually only a small portion, roughly 10 percent, of thunderstorms that occur in the United States reach the status of a severe thunderstorm. The majority of thunderstorms that do reach severe status develop under a favorable set of atmospheric conditions. Thunderstorms that organize in response to synoptic-scale forcing usually need four key ingredients. Enhancing any of these ingredients increases their chances to become severe:

1. Warm, moist air in the lower troposphere
2. Cool, dry air in the upper troposphere
3. Upper-level divergence above 500 mb (~ 5,500 m or 18,000 ft.)
4. A synoptic-scale disturbance, between ~1,000 – 2,500 km (~ 620 – 1,550 mi.) in size

Synoptically-forced thunderstorms develop regularly in warm, moist unstable air in advance of eastward-moving cold fronts. Figure 5.8 shows a classic pattern for the development of severe thunderstorms. This pattern should be easily recognized by a weather forecaster as one with the potential to produce large hail, strong winds, and tornadoes. Occasionally, several states can be affected by large tornado outbreaks that occur with this type of weather pattern.

Supercell thunderstorms are the most intense synoptically-forced thunderstorm, they form in an environment with high instability and strong vertical wind shear. These large rotating thunderstorms can persist for several hours and travel hundreds of miles. They are responsible for producing the strongest tornadoes, large hail and damaging wind gusts. Supercell thunderstorms are extremely dangerous, but timely warnings are issued once the storm has been properly identified. Groups of synoptically forced thunderstorms can organize into a linear band of thunderstorms called a **squall line** or organize as a large circular cluster of interacting thunderstorms called a **mesoscale convective system (MCS)**. It is not uncommon for a MCS to be 1,000 times larger than a typical air mass thunderstorm.

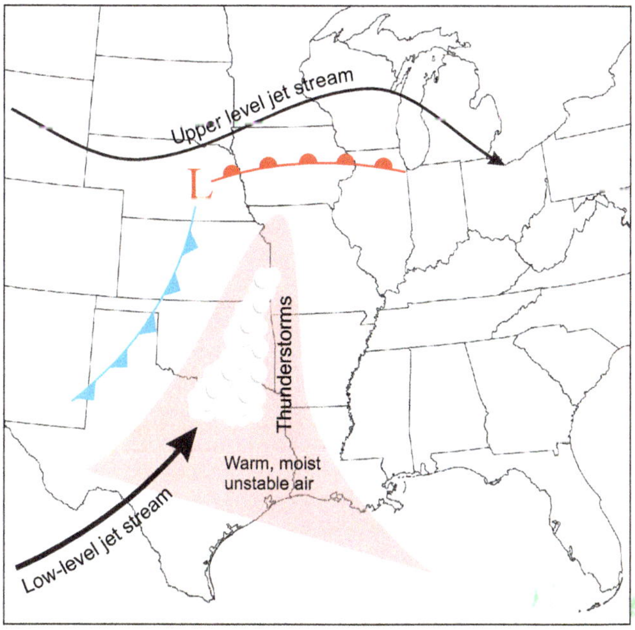

Figure 5.8: The atmospheric conditions necessary for the formation of synoptically-forced thunderstorms.

Lightning and Thunder

Figure 5.9: Lightning! Photo credit: NOAA.

To a casual observer a storm is usually only identified as a thunderstorm after witnessing a lightning strike and then hearing the thunder. **Lightning** is nothing more than a discharge of electricity, a giant spark in the air (Figure 5.9). Lightning has always struck fear and wonder into people throughout history. The Ancient Romans believed the god Zeus used lightning as a weapon; it played a large role in early religions, myths and superstitions. Benjamin Franklin's famous kite experiment in 1752 provided the first scientific explanation for lightning formation. When his kite passed near a thunderstorm cloud, negative charges leaked onto the kite and raced down the string and key. His experiment showed that lightning is simply a large, visible electric discharge, similar to static electricity that can build up on clothing when pulled from the dryer. You have experienced a buildup of excess electrons if you have ever shocked yourself on a doorknob in the winter when walking across a carpeted floor. Although lightning continues to fascinate people with its beauty, it is one of nature's deadliest phenomena, only flooding and tornadoes kill more people on average in the United States than lightning. Lightning can also cause costly disruptions to electrical services and is the most common cause of forest fires.

For lightning to occur, separate regions within a cumulonimbus cloud must contain opposite electrical charges, this is referred to as charge separation (Figure 5.10). The exact cause of charge separation within cumulonimbus clouds is not fully understood but has been extensively researched in recent years. The charge separation in clouds must be related to intense updrafts that separate the frozen precipitation particles. The creation of lightning must also be related to the formation of ice crystals since lightning is only observed in clouds that extend well beyond the freezing level. Precipitation and convection theories both attempt to explain the distribution of charges within clouds. These theories are related to a separation of charges within the cloud by the movement of precipitation particles through updrafts and downdrafts.

In clear air, the Earth's surface is negatively charged while the upper troposphere is positively charged. This distribution of unlike charges changes when a cumulonimbus cloud forms. Positively charged lightweight ice

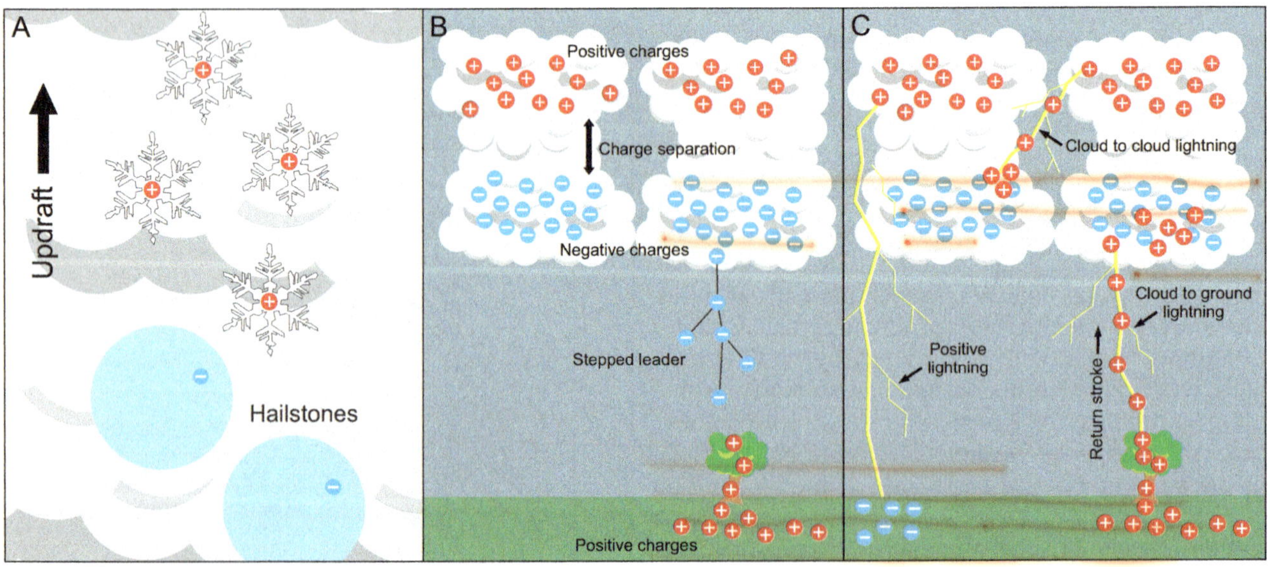

Figure 5.10: Diagram showing how lightning forms. A. Strong updrafts carry positively charged ice crystals to the top of the cloud, while negatively charged hailstones sink towards the bottom of the cloud. This leads to charge separation. B. As charge separation builds up within the cloud, a segment of negative charges called a stepped leader flows towards the ground. Meanwhile, positive charges build up on the ground beneath the cloud. C. When the stepped leader meets positive charges, the connection is complete and lightning results. In some cases, positive lightning can originate from the top of the cloud.

crystals are carried into the upper portion of the cloud while the larger negatively charged hailstones fall toward the bottom of the cloud (Figure 5.10A). The hailstones acquire a negative charge when supercooled water droplets freeze on its surface and positive ions are incorporated into the ice crystals. The ice crystals that form then break off, carrying the positive ions with them, and leaving a net-negative charge on the hailstone. The surface below the cloud acquires a positive charge as negatively charged particles are repelled beneath the cloud base. Lightning is now possible with charge separation between the upper and lower portion of the cloud.

Figure 5.11: Cloud to cloud lightning. Photo credit: NOAA

Lightning is possible between any two oppositely charged regions within and around the cumulonimbus cloud if the charge separation is large enough. Since the atmosphere behaves as an insulator, the charge differences must be large enough to overcome the atmosphere's natural resistance to the flow of electricity. There are two principle types of lightning: cloud-to-cloud lightning and cloud-to-ground lightning. Cloud-to-cloud lightning is the most common type and accounts for about 80 percent of all lightning (Figure 5.11). It is often called sheet lightning because the clouds obscure the view of the lightning flash and cause the bright white clouds to be illuminated. Cloud-to-ground lightning, the most damaging and dangerous type, accounts for the remaining 20 percent of all lightning strokes (Figure 5.9). An average thunderstorm produces roughly two cloud-to-ground lightning strikes per minute.

High speed photography has shown that a cloud-to-ground lightning strike consists of a regular sequence of events. The lightning stroke begins when streams of electrons flow from the cloud base toward the ground. This flow of electrons occurs in a series of short bursts with each burst about 50 meters (165 feet) long and only a few centimeters wide. This **stepped leader** is usually invisible to the human eye and is what gives a lightning stroke a forked appearance (Figure 5.10B). As the stepped leader approaches the ground it attracts positive charges from objects at the surface, especially tall objects like trees, power lines and the masts of ships. The stepped leader is met by a positively charged return stroke when a branch of the stepped leader gets within about 100 meters (328 feet) of the ground. Once a connection is made between the stepped leader and a positively charged object, a pathway is made for the flow of electrons from the ground upward (Figure 5.10C). The intensely illuminated discharge that flows upward from the Earth's surface to the cloud base is called the **return stroke** (Figure 5.12). The return stroke takes a path of least resistance to allow the flow of electrons between the cloud and the ground. This conducting channel follows the path of the stepped leader and is generally only about the thickness of a pencil. Even though the return stroke travels upward from the ground we see it as one continuous flash of light because it happens so quickly. The first stroke is usually followed by a few others along the same ionized channel to neutralize the charges. The lightning strokes after the initial return stroke are initiated by a series of **dart leaders** to give the lightning bolts a variety of shapes. Typically a flash of lightning consists of at least two dart leaders, with each followed by a return stroke. Some lightning flashes are comprised of a large number of individual strokes. The rapid succession of the lightning strikes occurs in a fraction of a second and is what sometimes makes lightning appear to flicker.

Figure 5.12: The return stroke of a lightning bolt. Photo credit: NOAA.

On rare occasions, a cloud-to-ground lightning stroke originates near the top of the cloud (Figure 5.10C). Although these **positive lightning strokes** account for less than ten percent of all cloud-to-ground lightning strikes they are especially dangerous. These lightning strikes are often called "bolts from the blue" since they can strike in areas with

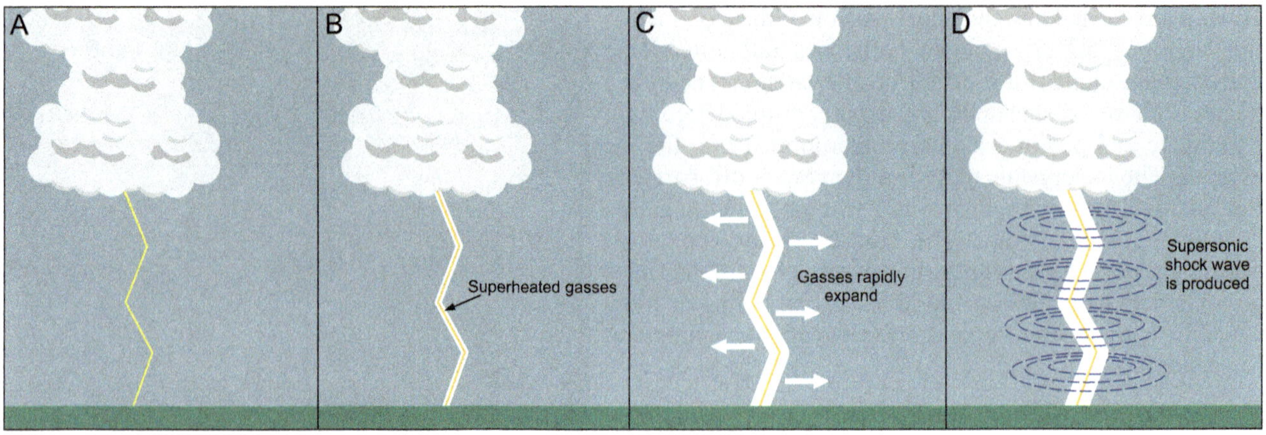

Figure 5.13: How thunder forms. A. The return stroke forms. B. The atmospheric gasses near the return stroke are instantaneously heated to 30,000°C (50,000°F). C. The superheated gasses rapidly expand outwards faster than the speed of sound. D. The expanding gasses form a supersonic shock wave which we hear as thunder when it reaches our ear.

clear skies several miles away from the thunderstorm. These powerful lightning strikes are thought to be the primary cause of forest fires and are known to damage power lines.

A lightning stroke heats the surrounding air to extreme temperatures, it can be heated to 30,000°C (~50,000°F) in less than one second. This rapid increase in temperature, about five times hotter than the surface of the sun, causes the air to expand in all directions and produces a supersonic shock wave we call **thunder** (Figure 5.13). There is always a delay between seeing the lightning and hearing the resulting thunder due to the difference in the speed of light compared to the speed of sound. We see a lightning bolt immediately since it travels at the speed of light, about one million times faster than the speed of sound. The sound waves reach your ears a short time later, travelling at roughly 1,200 kilometers per hour (750 miles per hour). The distance to a lighting strike can be approximated by counting the number of seconds between seeing the flash of lightning and hearing the resulting thunder. As a rule, the lightning strike is one mile (1.6 kilometers) away for every five second delay between the lightning and thunder. Lightning is considered close enough to be a hazard if the delay is 30 seconds or less. For example, you would be three miles (4.8 kilometers) away from the lightning strike if you see lightning and hear the thunder 15 seconds later. Using this technique, you can determine whether the thunderstorm is moving towards or away from you.

The old saying that lightning can be generated on a hot, humid night without a thunderstorm is a myth. Although sometimes lightning can be seen without hearing the thunder, all lighting strikes produce thunder. The sound of thunder usually can't be heard beyond 20 kilometers (12 miles) because the sound waves are distorted by the atmosphere. Local topography also influences how far the sound of the thunder travels; areas with trees or buildings can help absorb the sound waves and dissipate it before reaching your ears. This type of lightning is called "heat lightning", although it is no different than a lightning strike at closer range. The term heat lightning is most likely used because the effect most often occurs on warm, humid nights during the summer.

If a lightning strike hits nearby, the thunder will sound like a loud crack or bang. The thunder will sound like a rumble that slowly fades away if the lightning strike is a few miles away. The delay between observing lighting and hearing the thunder helps explain the rumbling sound of thunder. Figure 5.14 shows a hypothetical lightning bolt from Point 1 to Point 2. Assume the lightning strike is 2,000 meters long and strikes the ground 2,000 meters from your location. Thunder originating at Point 1 will arrive at your ears first; you will hear the clap of the thunder six seconds after you observe the flash. It takes time for the thunder from the farthest part to reach your ears since lightning

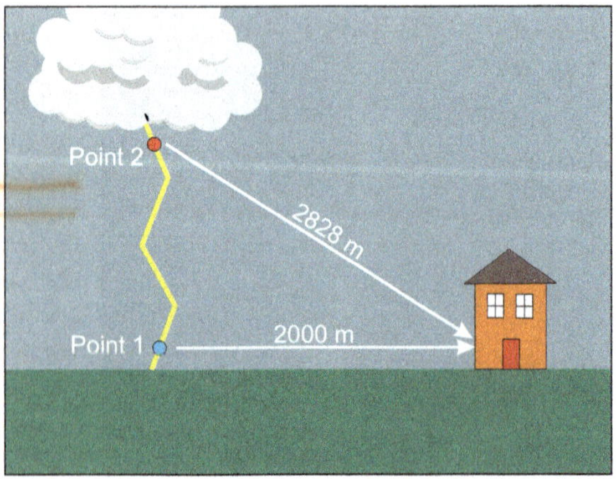

Figure 5.14: Diagram showing why thunder "rumbles". Since the lightning at point 1 is closer to your ear than point 2, the sound from point 1 reaches your ear first, while the sound from point 2 reaches your ear a couple seconds later.

bolts extend from the ground to the cloud base. A continuous rumble of thunder will now be heard for roughly two seconds from thunder that is produced in segments of the lightning strike that are farther from you. The end of the rumble is thunder from the farthest part of the bolt. Due to the zig-zag shape of lightning bolts the rumble of thunder often gets louder and softer as the sound waves travel from different angles.

Figure 5.15: A lightning strike split this tree into pieces. Photo credit: NOAA.

Lightning is an underrated killer, killing dozens and injuring hundreds more across the United States every year. Approximately 10 percent of lightning strike victims die instantly. The summer months are the most likely time to be hit by a lightning strike; this is when the occurrence of thunderstorms and people engaged in outdoor activities are the highest. You should always take cover when you hear thunder, especially when it is loud and sounds like a bang. Fortunately, lightning fatalities have declined in recent decades due to better infrastructure and improvements to farming equipment. Lightning fatalities have also steadily deceased in part to a larger percentage of the population that now resides in urban areas that provide more protection from lightning. Many victims are struck in open places such as agricultural fields and golf courses. You should seek shelter immediately if caught in a thunderstorm, making sure to avoid elevated places or being near tall isolated objects like trees (Figure 5.15). It is recommended to stay in a shelter until you have not heard thunder for 30 minutes. If caught in an open area, crouch down as low as possible and minimize your surface area in contact with the ground by staying on the balls of your feet. Lightning may be about to strike if your hair starts to tingle. You should also stay off electronic devices and avoid taking a shower or bath during a thunderstorm. Remember this important message: when thunder roars, go indoors!

Section 5-2: Tornadoes

With winds that can destroy everything in its path, a **tornado** is the most violent weather system on Earth. Fortunately, a very small percentage of all thunderstorms produce tornadoes. On average, over 1,200 tornadoes are reported in the United States each year. Major tornado outbreaks in the past have caused billions of dollars in damage and claimed hundreds of lives. The April 25-28, 2011 tornado outbreak is the largest ever recorded, affecting a large area of the Southern and Eastern United States. In total, 376 tornadoes were confirmed by the National Weather Service (NWS) across 21 states, with Alabama and Mississippi being the hardest hit. There were 62 confirmed tornadoes across the state of Alabama alone on April 27. There were 321 people killed over this four day period, with 240 of these fatalities occurring in Alabama. This tornado outbreak is one of the costliest natural disasters in United States history. It doesn't take an outbreak of numerous tornadoes to have devastating impacts; this was the case for the EF5 tornado that struck Joplin, Missouri on May 22, 2011 (Figure 5.16). As a result of one ill-placed tornado, over 160 people were killed and more than 1,100 were injured, the deadliest tornado in more than

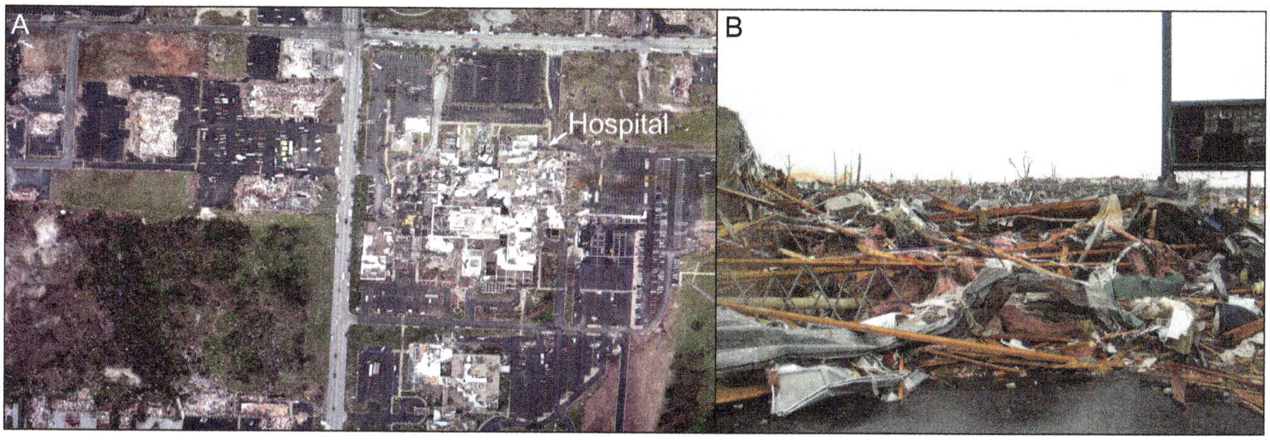

Figure 5.16: The aftermath of the Joplin, MO tornado of 2011. A. Aerial view. Even the largest, strongest buildings, like this hospital, were heavily damaged while smaller structures were completely leveled. B. Ground-level view of the damage. Photo credits: NOAA.

60 years. With roughly 7,000 homes destroyed, damage estimates from the 2011 Joplin tornado reached almost 3 billion dollars. Although decades of research have solved some of the mysteries surrounding the formation of tornadoes, there are still many things meteorologists don't fully understand. No amount of research will ever result in tornadoes becoming totally predictable due to their small size and erratic behavior. Once a tornado touches down it can destroy anything in its path, powerful tornadoes can demolish a house in less than 10 seconds.

Figure 5.17: A large tornado touches down. Photo credit: NOAA.

Tornado Characteristics

A tornado is defined as a rapidly swirling column of air that is in contact with the ground (Figure 5.17). The characteristic **funnel cloud** is often a precursor to a tornado; it appears as a rotating pendant that lowers beneath the base of a cumulonimbus cloud. Tornadoes are generally associated with powerful rotating thunderstorms called **supercells.** Supercell thunderstorms have a diameter of about 30 km (19 mi.), much larger than a typical thunderstorm. Tornado wind speeds are strong enough to level well-constructed houses and sweep a train off their tracks. Nearly all tornado fatalities are due to flying debris, large and small objects are tossed at great speeds by the fierce winds. Remember that the effect of friction is reduced with increasing altitude above the Earth's surface; people often risk greater danger by seeking shelter under a highway overpass because they subject themselves to the strongest tornadic winds found just above the surface.

Wind speeds in tornadoes were once thought to be over 500 mi/hr, possibly even faster than the speed of sound. Fortunately, most tornadoes have wind speeds less than 177 km (110 mi.) per hour. Recent measurements using mobile Doppler radar suggest that the highest wind speeds from the most powerful tornadoes are in the 450-483 km (280-300 mi.) per hour range. The fastest wind speeds ever recorded in a tornado were measured from a massive tornado that struck Oklahoma City, Oklahoma on May 3, 1999. Meteorologists measured wind speeds of 512 km (318 mi.) per hour using mobile Doppler radar; these are the highest wind speeds ever measured near the Earth's surface.

Although often funnel shaped, tornadoes come in a wide variety of shapes and sizes. It is difficult to generalize the common shapes of a tornado since they occur so frequently, they can even form without a funnel cloud. Tornadoes can have a characteristic funnel shape or resemble a wedge that is wider than it is tall. They can also appear as a symmetrical vertical column or have a long and slender rope-like appearance. When moving over grassy fields, where there is little dirt or debris to be picked up, tornadoes can sometimes appear almost invisible. Regardless of their shape all tornadoes are very dangerous.

Figure 5.18: Satellite photo showing the path of a tornado through La Plata, MD. Image courtesy Lawrence Ong, EO-1 Mission Science Office, NASA-GSFC.

The average tornado usually lasts only two to three minutes with a diameter of about 100 m (330 ft.), roughly the length of a football field. However, in extreme cases a tornado can remain on the ground for over two hours and produce a damage path over 2 km (1.3 mi.) wide that extends for more than 160 km (100 mi.). The damage path of a tornado is similar to a fingerprint; it leaves evidence of where the tornado struck (Figure 5.18). A tornado 4.2 km (2.6 mi.) wide struck El Reno, Oklahoma on May 31, 2013, the widest tornado path in recorded history. Tornadoes often appear motionless as they travel across a flat landscape with a distant horizon in the background; this is clearly not the case. Tornadoes and their parent thunderstorms generally move from southwest to northeast with an average forward speed of around 48 kilometers (30 miles) per hour, although forward speeds of over 97 kilometers (60 miles) per hour have been reported. Tornadoes sometimes take wildly erratic paths in the shape of a circle or figure eight.

Timing and Location of Tornadoes

Tornadoes primarily form in the mid-latitudes and the subtropics. No country in the world has nearly as many tornadoes as the United States. Globally, at least 75% of all tornadoes occur in the United States. This high concentration of tornadoes is a result of ideal topography and favorable atmospheric conditions across a large portion of the country. Relatively flat terrain combined with the absence of an east-west mountain range in the north allows unlike air masses to frequently clash across the Central and Eastern United States in the spring and early summer. The Rocky Mountains steer continental polar (cP) air masses into the Central United States, where they commonly mix with maritime tropical (mT) air masses to provide ideal conditions for the formation of thunderstorms. In addition, a dry continental tropical (cT) air mass from the Mexican plateau and the Southwestern United States moves into the Central United States. These three contrasting air masses interact in a concentrated zone to produce **tornado alley**, a region that produces the highest frequency of tornadoes in the United States (Figure 5.19). Roughly 95% of all tornadoes in the United States occur east of the Rocky Mountains.

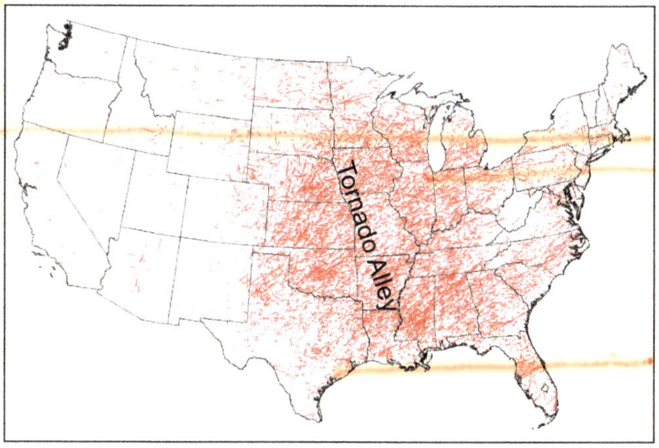

Figure 5.19: A map of tornado tracks from 1950-2014. The Midwest, with the most tracks, is also known as Tornado Alley. Image credit: NOAA.

The conditions necessary to form tornadic supercells across the United States generally start to come together over the Gulf States in late February or early March as intense mid-latitude cyclones clash with warm, humid tropical air from the Gulf of Mexico. Tornadoes gradually spread northward from late March through May as maritime tropical (mT) air reaches higher latitudes. Tornadoes don't generally form in the Central Plains or the Northeastern United States until moisture from the Gulf of Mexico is accessible in late May. The poleward migration of the jet stream from late winter through the summer is reflected by the northward spread of tornadoes during this time. The jet stream helps provide the upper-level conditions needed to spawn tornadic thunderstorms. The northward spread of tornadoes continues from the Dakotas into southern Canada throughout the summer.

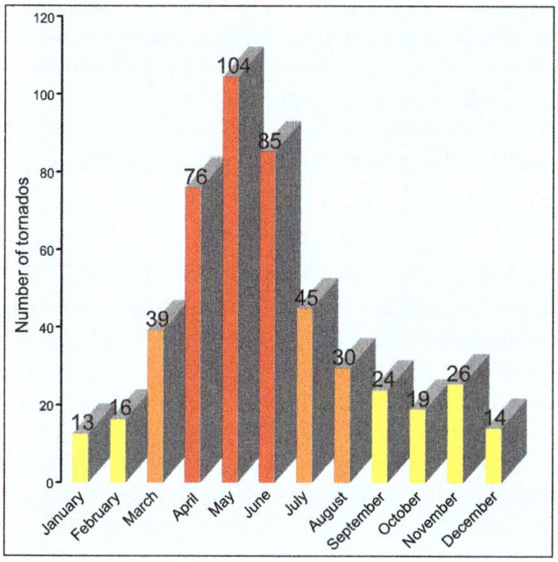

Figure 5.20: Average tornado occurrence by month for the United States from 1950-2004. Data from the National Weather Service - Storm Prediction Center/NOAA.

Figure 5.20 shows the monthly tornado frequency across the United States. May is typically the most active month, with the months April through June being the most active period of the year. This three month period is the time when the temperature contrasts between polar and tropical air masses are still large and there is a high sun angle to help produce unstable conditions, resulting in more intense thunderstorms. There were 1,817 reported tornadoes across the United States in 2004, the most to occur in a single year. The record for most tornadoes in a month (since modern tornado records began in 1950) was set in April 2011, with 875 confirmed tornadoes. April 27, 2011 set the record for most tornadoes to occur in a single day, when 226 tornadoes were reported. Tornadoes most often form during the afternoon and early evening hours, generally coinciding with the warmest time of the day. The peak time for tornado occurrence is between 4 pm and 6 pm local time. Only about 20% of all tornadoes form between midnight and noon.

Scale	3 Second Gust (mph)	Damage class
EF0	65-85	Light
EF1	86-109	Moderate
EF2	110-137	Considerable
EF3	138-167	Severe
EF4	168-199	Devastating
EF5	Over 200	Incredible

Table 5-1: The Enhanced Fujita Scale of tornado wind speed and damage.

Tornado Winds Speeds and Damage

Video footage, Doppler radar and other scientific techniques are used to estimate the wind speeds of a tornado. Videography uses frame-by-frame analysis to identify the distance that flying debris travels over a specific time period, allowing the wind speeds to be estimated. Wind speeds can also be estimated by observing the damage pattern caused by the tornado. In 1971, Dr. Ted Fujita developed a scale to rank tornado intensity based on estimated the wind speeds. The **Fujita Scale (F-Scale)** estimates the wind speeds based on the degree of damage caused by the winds. Although subjective, this scale allows trained meteorologists to scientifically estimate the wind speeds of a tornado after visually inspecting the damage. Wind engineers and meteorologists trained in damage surveys often arrive at different F-Scale rankings for the same tornado with the final ranking determined after a consensus is reached. To account for these differences, the Fujita Scale was updated and replaced with the **Enhanced Fujita Scale (EF-Scale)** in February 2007 to more accurately match the estimated wind speeds to the severity of tornado damage. Tornadoes will continue to be ranked using estimated wind speeds based on damage until there is a safe way to measure tornado winds speeds near the ground.

The Enhanced Fujita Scale has six categories of increasing intensity that range from EF0 to EF5 (Table 5-1). The lowest ranking on the scale is EF0, which applies to the weakest tornadoes with winds strong enough to rip branches off trees and hurl lightweight debris. Winds in an EF1 tornado are strong enough to overturn mobile homes and cause roof damage to buildings. Fortunately, over 60% of all tornadoes in the United States are ranked as an EF0 or EF1 and fall into the weak category. Winds in an EF2 tornado can uproot trees while an EF3 tornado is capable of ripping the roof off of a well-constructed house. Tornadoes ranked as EF2 and EF3 are classified as strong, they account for almost 35% of all United States tornadoes. EF4 and EF5 tornadoes are considered violent; they are capable of producing damage beyond comprehension. These tornadoes can hurl cars like toys and level well-constructed houses. These violent tornadoes often leave nothing but piles of rubble that have been pushed away from where houses originally stood. Less than 3% of all tornadoes in the United States are ranked as an EF4 or EF5, although they account for nearly 70% of the fatalities. Wind speeds in an EF5 tornado can exceed 482 kilometers (300 miles) per hour, the damage they produce is considered beyond comprehension. Less than 1% of tornadoes reach EF5 status and it is not uncommon for there to be several years between them.

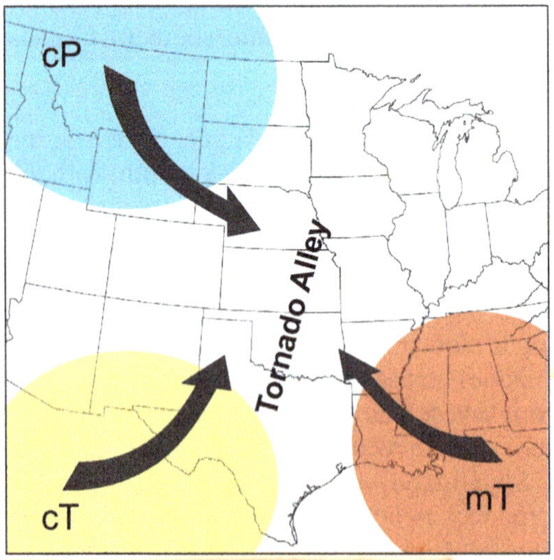

Figure 5.21: In the spring and early summer continental polar (cT), maritime tropical (mT) and continental tropical (cT) air masses meet over the Midwestern United States to form a region of severe weather called Tornado Alley.

Atmospheric Conditions Necessary for Tornado Formation

Tornadoes can develop from severe thunderstorms in a variety of situations. The Central United States is one of the few places in the world where topography and favorable weather conditions are ideal for the development of tornadic supercells. Three unlike air masses commonly interact in tornado alley across the Midwestern United States (Figure 5.21). East of the Rockies, low-pressure systems and a strong jet stream create unstable conditions by drawing maritime tropical (mT) and continental polar (cP) air masses together. A warm, dry continental tropical (cT) air mass from the Mexican plateau and the Southwestern United States allows the formation of supercell thunderstorms. This air mass, located at an altitude of about 3 km (10,000 ft.), acts like a lid on the atmosphere. The dry, stable cT air mass prevents unstable air parcels from rising, allowing the storms to gain strength (Figure 5.22A). Continued

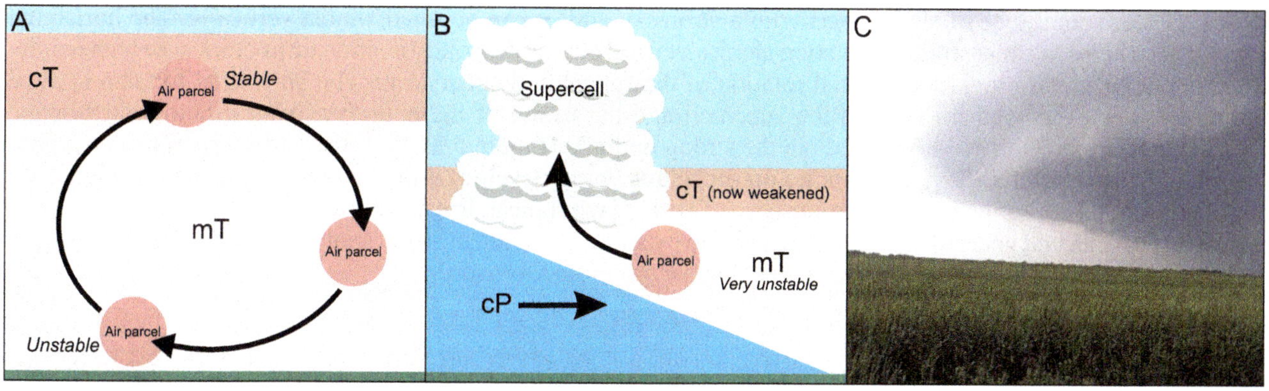

Figure 5.22: Diagram showing the conditions necessary for supercell formation. A. A continental tropical (cT) air mass aloft acts like a lid, trapping warm, moist maritime tropical (mT) air masses near the Earth's surface. As warm, moist unstable air parcels rise, they encounter the drier, cooler cT air mass and become stable. The air parcels then sink back towards the earth to repeat the cycle. B. As the day continues, the "lid" is weakened while the mT airmass near the Earth's surface has become very unstable. With the advance of a continental polar (cP) air mass (or some other lifting mechanism), the mT air mass breaks through the lid and cumulonimbus clouds explode upwards, leading to the formation of a supercell and severe weather. C. A supercell forms in Oklahoma. Photo credit: H. Hyre Hansen.

surface heating of the mT air creates more instability and slowly erodes the cap. The humid, unstable air parcels can rapidly rise through the troposphere as the cap weakens and a cP air mass advances from the west (Figure 5.22B). It is not a coincidence that the most likely time for a tornado to develop across tornado alley occurs during the mid to late afternoon when the lid is typically broken. During this time giant cumulonimbus clouds can rapidly develop into powerful supercell thunderstorms (Figure 5.22C).

The formation of a supercell by no means guarantees that a tornado will form; only about 30% of supercells produce a tornado. Tornadoes usually need strong speed vertical wind shear to form, that is, slower winds near the surface and faster winds aloft. Speed vertical wind shear can produce a horizontal tube of air, the first step in the formation of a tornado. A horizontal tube of air can form as wind speeds rapidly increase with height in the lower troposphere (Figure 5.23A). You can simulate this rolling motion by placing a pencil across the palm of your left hand. Use your right hand to roll the pencil toward the fingers of your left hand. This rolling motion is similar to speed vertical wind shear; your right hand represents faster winds aloft while your left hand represents the slower near surface winds. For a supercell to become tornadic, speed vertical wind shear values of at least 64 km (40 mi.) per hour are generally needed in the lowest 6 km (3.7 mi.) of the troposphere.

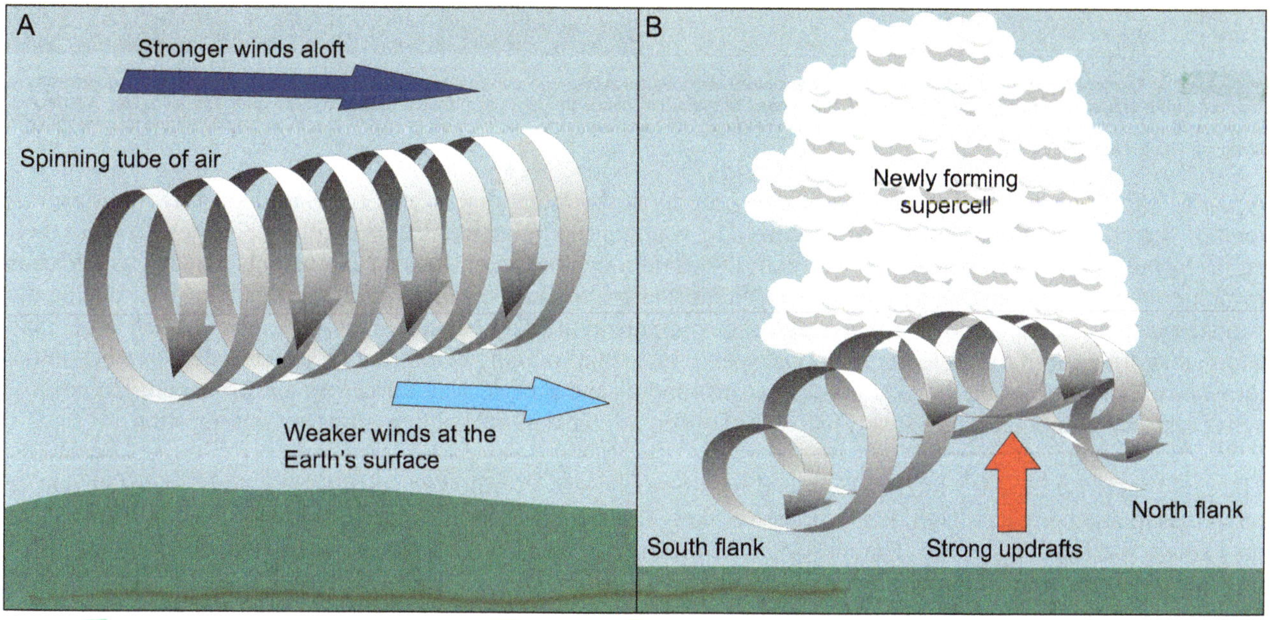

Figure 5.23: Diagram showing how a mesocyclone forms. A. Strong speed vertical wind shear can produce a horizontal spinning "tube" of air. The formation of this tube of air is the first step in forming a tornado. B. In the second step, the updraft of a thunderstorm can then "bend" the tube of air upwards, forming two vertical columns of rotating air. The column on the north flank usually falls apart, but the column on the south flank develops into a mesocyclone (Figure 5.24A).

When the horizontal tube of air interacts with a strong updraft it can be tilted almost vertically. The horizontal rotation also helps to concentrate the rotation along a vertical axis as the wind direction turns clockwise with height. The directional shear provides the initial rotation of the supercell thunderstorm. The horizontal rotation is tilted upright between the southern edge of the supercell and the center of the updraft as warm, humid air flowing northward from the Gulf of Mexico rises in the strong updrafts (Figure 5.24A). If the updraft rises fast enough it begins to rotate counterclockwise, forming a **mesocyclone**. A mesocyclone is the rotating updraft within a supercell thunderstorm; its formation requires the presence of vertical wind shear. It is usually about 5 to 10 kilometers (3 to 6 miles) in diameter and generally precedes the tornado by about 30 minutes. Storm spotters are trained to identify the formation of mesocyclones since roughly half of them evolve into a tornado.

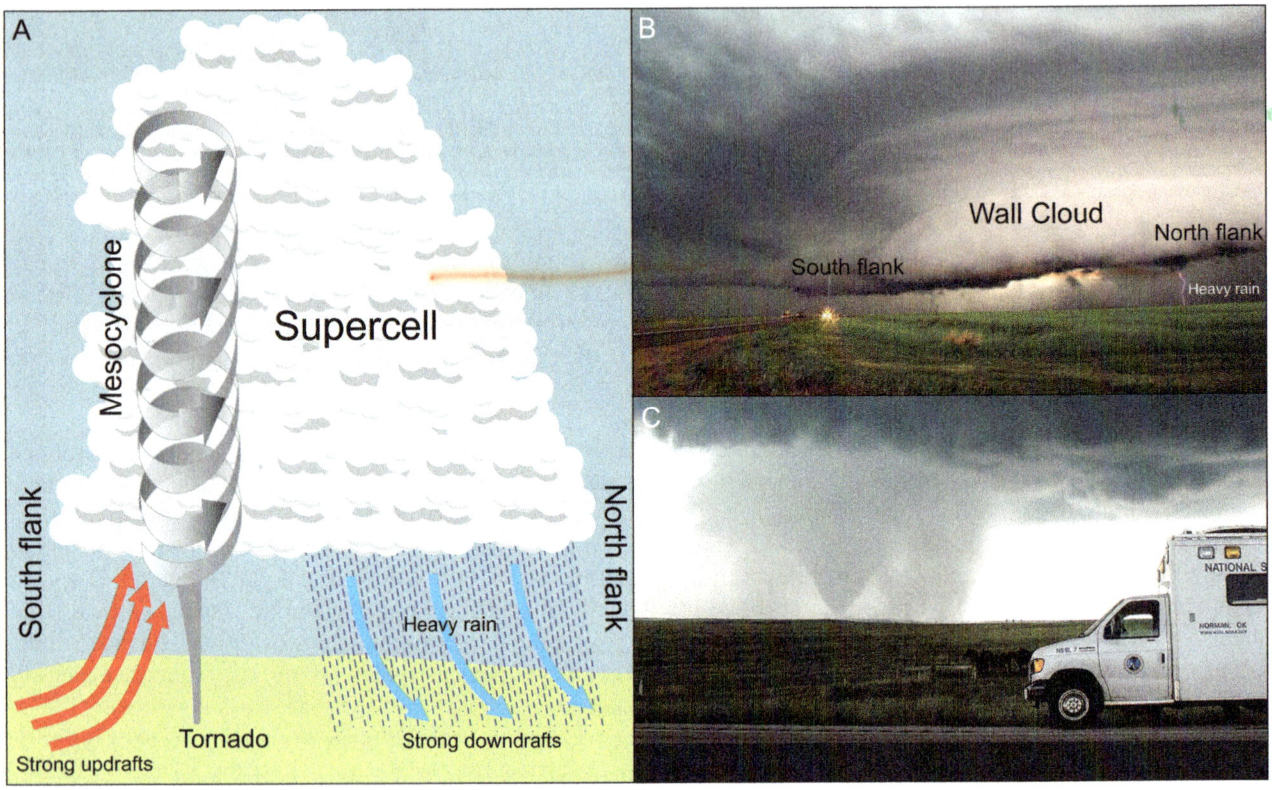

Figure 5.24. A. The final stages of tornado development. Cyclonically spinning air at the edge of the mesocyclone is drawn downwards by strong downdrafts at the front of the supercell, while the strong updrafts at the back of the storm cause the air to spin faster. As the column of air rotates faster, it stretches downwards, forming a tornado. B. A large supercell moves over the Nebraska countryside. Photo credit: NOAA. C. A rain-wrapped tornado touches down in Wyoming. Photo credit: NOAA.

According to the conservation of angular momentum, as the radial distance of rotation decreases, the speed of rotation increases. This concept can be visualized by watching spinning ice skaters accelerate as they decrease their radius by pulling their arms closer to their bodies. As a mesocyclone intensifies, its diameter is reduced which leads to an increase in rotational speed. The mesocyclone is also stretched vertically as it contracts inward, causing the winds to accelerate into a vortex. At this stage of development, the diameter of the mesocyclone is much larger than a tornado with a significantly lower rotational speed. How this spinning column of air is vertically stretched into a tornado remains one of the biggest mysteries surrounding the precise way tornadoes are produced. Decades of active research combined with a steady increase of observational data are helping to answer this question.

The rapid increase in video footage provided by storm chasers has allowed scientists to study the life cycle of tornadoes in much greater detail. It appears most likely that the mesocyclone first develops about 5 to 6 km (3 to 4 mi.) above the Earth's surface. The mesocyclone's diameter decreases as it stretches downward, leading to an increase in wind speeds (Figure 5.24). As the narrowing vortex of air stretches below the cloud base a **wall cloud** begins to form (Figure 5.24B). The wall cloud develops as low-level moist air condenses as it is wrapped around the rotating updraft. When tornadoes develop, they most commonly emerge from the wall cloud. Wall clouds with noticeable rotation and upward motion often develop before the touchdown of a violent tornado, usually preceding

Figure 5.25: This image of a truck wrapped around a pole shows the power a tornado's winds can pack. Photo credit: NOAA-NSSL.

the tornado by about 10 to 20 minutes. A narrow, rapidly rotating funnel cloud then emerges from the base of a wall cloud just prior to the formation of a tornado. The funnel cloud will be officially classified as a tornado if it makes contact with the ground.

Tornado Safety

It is very important to pay attention to radio and television when severe weather is threatening. Tornadoes are hazardous to people and property, they typically kill dozens of people and injure hundreds more each year in the United States. Wind speeds are strong enough to destroy buildings and blow down trees; flying debris causes many of the injuries and fatalities caused by a tornado (Figure 5.25). In the United States, tornado safety is both the responsibility of the individual and the National Weather Service (NWS). The most important responsibility of the NWS is to issue severe weather advisories, they are critical to the protection of life and property. Weather radios from the National Oceanic and Atmospheric Administration (NOAA) should be the public's first line of defense; the NWS broadcasts tornado warnings directly to them. The NWS encourages families, schools and businesses to have a safety plan in place when a tornado threatens. People should move indoors and seek shelter in a basement if possible. Go to an interior room or hallway on the lowest floor if a basement is not available to put as many walls between you and the tornado as possible. A bathroom provides extra protection due to the walls being supported by water lines. If possible, cover yourself with some padding to protect yourself from flying debris. Mobile homes should be evacuated since they offer little protection from tornadoes. If caught outdoors, lay down in the lowest spot available away from potential flying objects, such as a ditch. If you are driving and can't avoid the path of a tornado, pull off the road and seek shelter, approximately 10 percent of all tornado fatalities are motorists in cars.

Due to their small size and short lifetime, tornadoes are extremely difficult to forecast, it is a miracle that we can predict the possibility of a tornado as well as we can. Fortunately the average warning time before a tornado actually forms has been extended due to the use of Doppler radar. Before Doppler radar units were installed across the United States, the warning time averaged only six minutes. Forecasters now have the opportunity to warn the public of an impending tornado with an average lead time of 13 minutes. However, as development continues to put more people in harm's way, targets for tornado strikes will only get bigger and more numerous.

Section 5-3: Tropical Storms and Hurricanes

Figure 5.26: Hurricane Katrina as it approached New Orleans. Photo credit: Jeff Schmaltz, MODIS Rapid Response Team, NASA/GSFC.

With sustained winds sometimes exceeding 300 kilometers (185 miles) per hour, **tropical cyclones** are the most destructive storms on Earth. A tropical cyclone is a generic name for a synoptic-scale cyclone that originates over tropical or subtropical waters with organized convection and an observable cyclonic circulation (Figure 5.26). The most intense tropical cyclones are called different names depending on where they form. They are called **hurricanes** in the Atlantic and eastern Pacific, **typhoons** in the western North Pacific and **cyclones** over the Indian Ocean and near Australia. The typical locations where tropical cyclones form and travel are shown in Figure 5.27. Notice the absence of storms

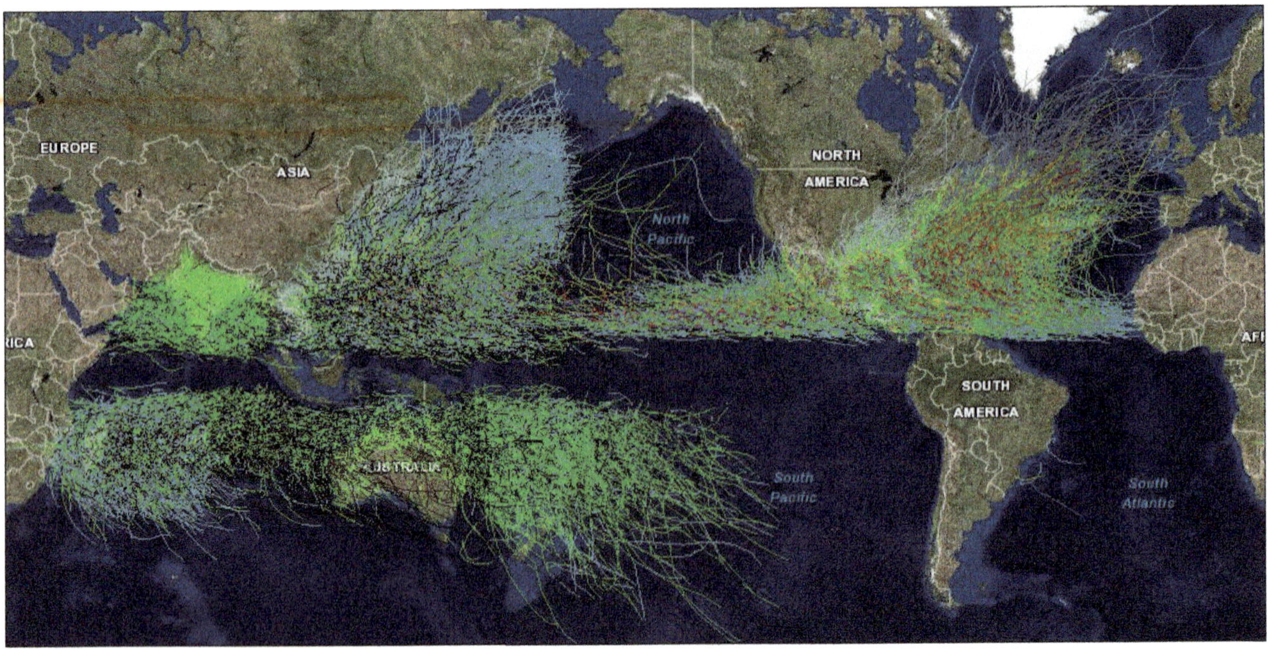

Figure 5.27: This map shows the tracks of all tropical cyclones which formed worldwide from 1842 to 2014. Each point shows the locations of each storm at six-hour intervals and uses a color scheme to show storm strength based on the Saffir-Simpson Hurricane Scale. Image created using the NOAA Historical Hurricanes website.

near the equator and over the South Atlantic Ocean. Although hurricanes attract the most media attention in the United States for obvious reasons, the western North Pacific is by far the most active region.

Tropical cyclones are intensively researched by scientists due to the tremendous damage and loss of life they cause in many parts of the world. The death toll and destruction of property can reach epic proportions when they pass over highly populated coastal areas that are near sea level. In addition to the threat of heavy rain, strong winds and tornadoes, these regions are especially vulnerable to an abnormal rise in sea level known as **storm surge**. As population and property values rise in coastal areas, economic losses from these destructive storms will continue to increase. To emphasize this point, four of the five costliest hurricanes in United States history have occurred since 2005. In August 2005, Hurricane Katrina produced the third most expensive natural disaster on record, resulting in over $100 billion in damage (Figure 5.28). It was also the deadliest hurricane to strike the United States since 1928. Although fatalities from hurricanes have been greatly reduced in the United States due to technological advances and improved warning and evacuation procedures, loss of life still continues to be a serious problem for many poorer island nations. Typhoon Haiyan cut a devastating path across the central Philippines in November 2013, resulting in at least 6,300 fatalities.

Figure 5.28: People sit on a roof waiting to be rescued after Hurricane Katrina. Photo credit: Jocelyn Augustino/FEMA.

A tropical cyclone evolves through a distinct life cycle on the way to becoming a hurricane. The birth of a hurricane begins with a **tropical disturbance**, a discrete system of apparently organized convection that maintains its identify for at least 24 hours. A **tropical depression** is an organized cluster of thunderstorms that has an observable circulation on satellite imagery with maximum sustained winds of 33 knots (38 mi/hr) or less (Figure 5.29A). When maximum sustained winds increase to between 34 knots (39 mi/hr) and 63 knots (73 mi/hr and the storm becomes more organized, the tropical cyclone is called a **tropical storm** and given a name (Figure 5.29B). A **hurricane** is a severe tropical storm with maximum sustained winds of at least 65 knots (74 mi/hr). The most characteristic feature of a hurricane is its eye in the center of the cyclonic circulation. The name of the tropical storm does not change if it

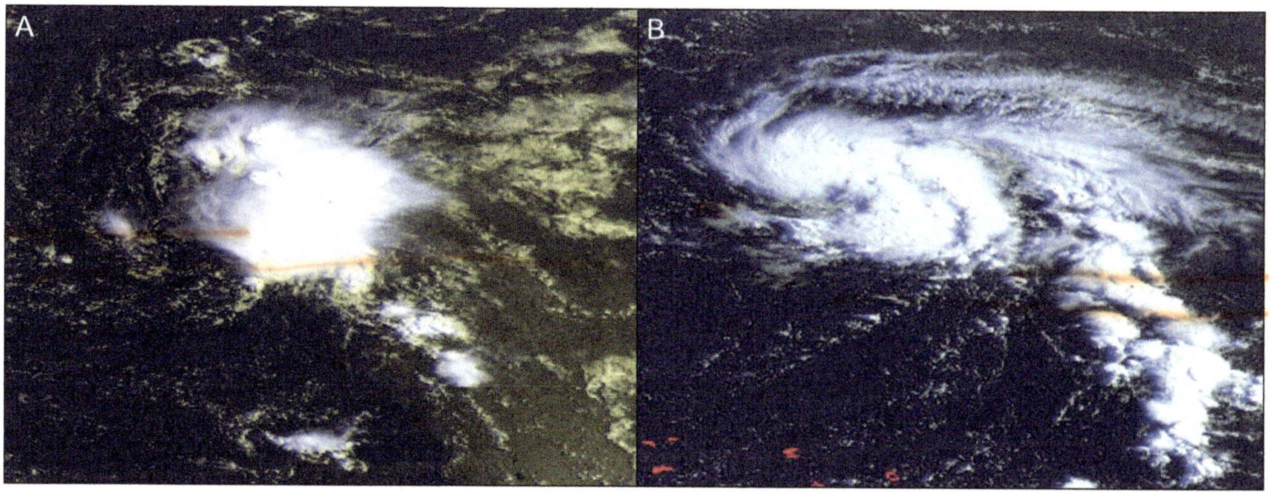

Figure 5.29: A. Photo of **tropical depression** Irene on August 9th, 2005. B. Photo of now upgraded **tropical storm** Irene taken on August 11, 2005. Note how the storm has become better organized in its circulation pattern. Photo credits: NOAA.

intensifies into a hurricane. The World Meteorological Organization (WMO) has created a rotating alphabetical list of male and female names for tropical storms and hurricanes. These names are used again every six years. Naming these storms makes it easier to remember them from a historical perspective and makes communication easier between the forecasters and the general public. The names of exceptionally devastating hurricanes are retired and replaced to avoid confusion in the future. For example, no one wants to hear of another Hurricane Andrew (1992) or Hurricane Sandy (2012) bearing down on the East Coast of the United States.

Globally, about 80 tropical storms develop each year, with about 50 to 70 percent of them reaching hurricane strength. Roughly two of every three tropical cyclones occur in the Northern Hemisphere due to a larger region of warm ocean temperatures compared to the Southern Hemisphere. Although the official hurricane season for the North Atlantic is from June 1 to November 30, the most active period is from August to October. Figure 5.30 shows the monthly frequency of Atlantic hurricanes from 1851 to 2014. The hurricane season clearly peaks in September. Tropical cyclones need a special set of conditions in order to form. It will become obvious why the late summer and early fall is the most active period for hurricanes once the conditions needed for development are understood.

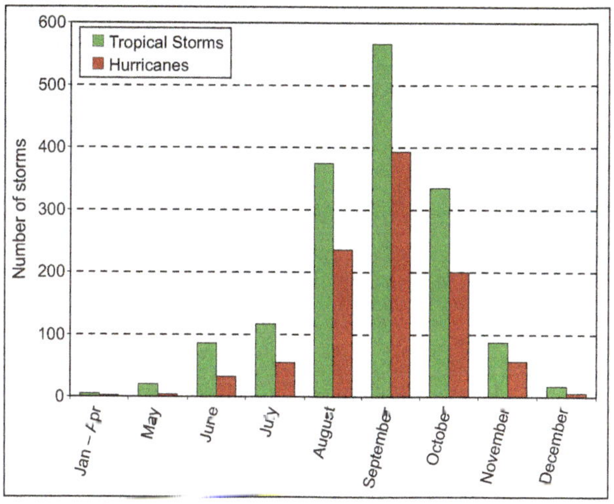

Figure 5.30: The total number of tropical storms and hurricanes by month for the Atlantic Basin from 1851-2014. Data from NWS/NOAA.

Conditions Necessary for Hurricane Formation

Tropical cyclones have organized convection around a center of circulation; they are sometimes referred to as a pinwheel of thunderstorms. There are a set of six specific conditions that must be in place before a tropical cyclone can form. These six environmental conditions can be related to the ingredients needed to make a recipe:

1) Warm sea-surface temperatures of at least 26.5°C (80°F) to a depth of about 46 m (150 ft.).

2) Conditional instability through a deep layer of the troposphere.

3) Relatively moist air in the middle troposphere.

4) Weak vertical wind shear between the surface and the upper troposphere.

5) A pre-existing disturbance with cyclonic circulation in the lower troposphere.

6) A formation location that is at least 5° latitude away from the equator.

All of these necessary environmental conditions are related to the key characteristic that all tropical cyclones exhibit: organized convection around a center of circulation that is observable on satellite imagery. Warm sea-surface temperatures also directly influence the degree of atmospheric instability and moisture.

Warm Sea-Surface Temperatures

Due to their enormous size, massive amounts of heat are needed to fuel hurricanes. Hurricanes are often thought of as giant heat engines, with their energy derived from heat and moisture supplied by the ocean. Latent heat is released through condensation of water vapor in thunderstorms as water molecules from the ocean surface evaporate and transfer their moisture to the atmosphere. Vast amounts of heat are released into the atmosphere as the ocean water evaporates, rises, cools and condenses. High evaporation rates are reliant on the presence of warm water, hurricanes only form over water that is warm enough to supply them with a sufficient amount of energy. Therefore, latent heat of condensation is considered the fuel source for a hurricane. Warm sea-surface temperatures (SSTs) ensure high evaporation rates and abundant moisture in the lower troposphere. Hurricanes only form where SSTs are 26.5°C (80°F) or higher, this explains why they most often develop in the late summer and early fall when the ocean waters are the warmest. The high SSTs ensure high dewpoints in the lower troposphere and help to destabilize it. In addition, stronger and more vertically-developed thunderstorms are able to form when they develop over a warm ocean surface. If atmospheric conditions are favorable, the intensity of a hurricane is often directly related to the SSTs it forms over.

The average SSTs over the North Atlantic throughout the official hurricane season are shown in Figure 5.31. Warm waters of 26.5°C (80°F) or higher extend from the west coast of northern Africa across the tropical Atlantic to the Caribbean Sea and the Gulf of Mexico. This region is referred to as "hurricane alley" since most North Atlantic

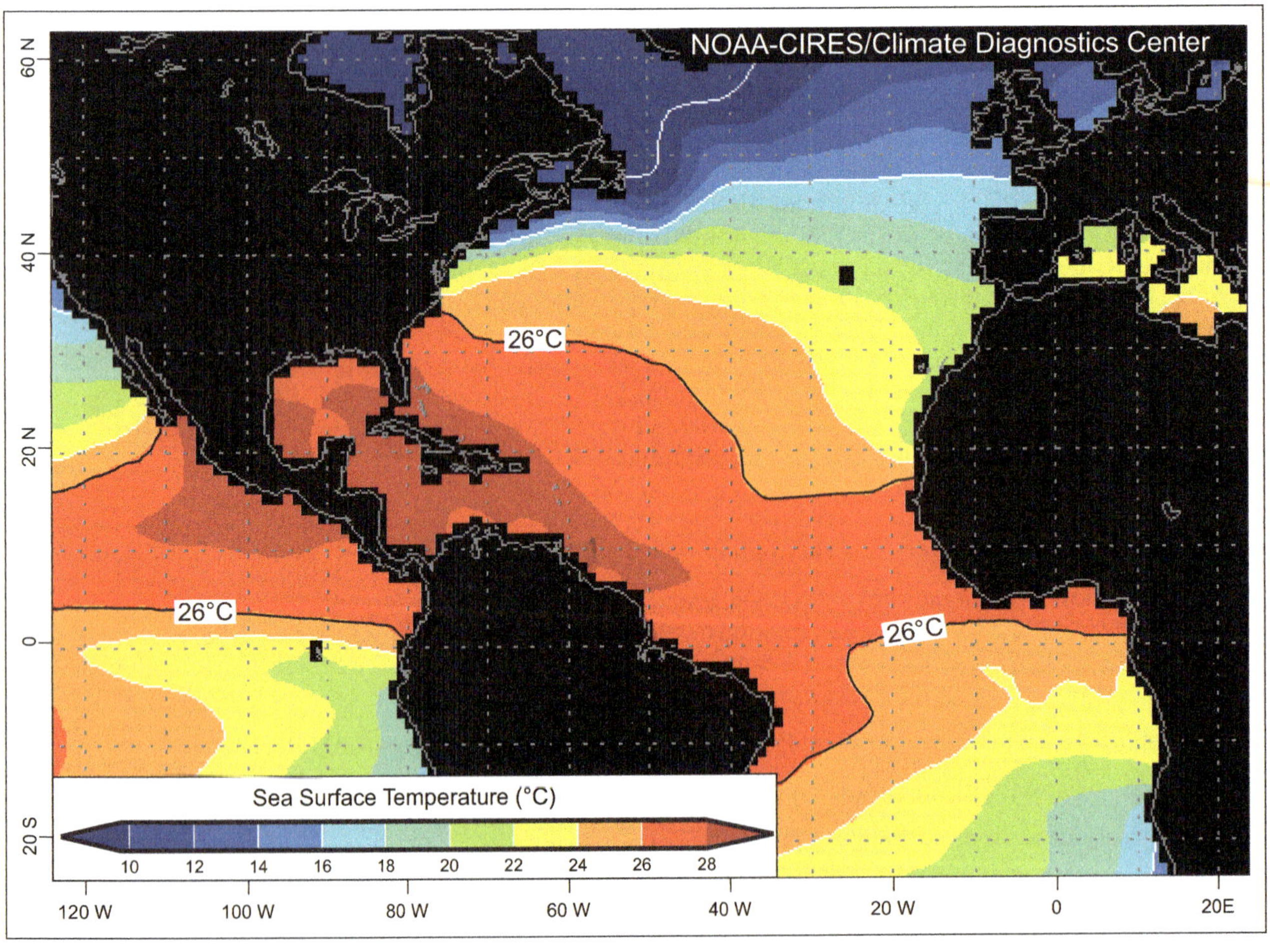

Figure 5.31: Average sea-surface temperature for the north Atlantic Ocean from June 1 to November 30. Areas where the water temperatures are above 26.5°C (80°F) are places where tropical storm development is likely to occur. Data courtesy of NOAA-CIRES/Climate Diagnostic Center.

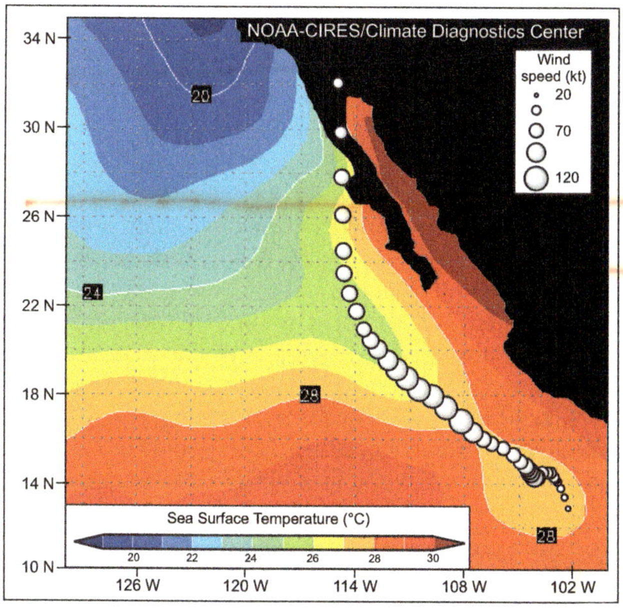

Figure 5.32: The track of Hurricane Nora, September 16-25, 1997, overlain on a map of the average sea surface temperatures for the week of September 16-23, 1997. Image credit: NOAA-CIRES/CDC. Hurricane track data courtesy NOAA/National Hurricane Center.

tropical storms and hurricanes form here (warm ocean temperatures must extend 46 m below the ocean surface for a hurricane to develop). If the uppermost layers of the water are not warm enough, mixing caused by the hurricane can bring colder water to the surface in a process called upwelling. When upwelling occurs the hurricane can quickly weaken as the cold water can't supply sufficient moisture and heat for the thunderstorms to maintain deep convection.

Hurricane Nora provides a great example of how upwelling can rapidly weaken a tropical cyclone. In 1997, Hurricane Nora moved along the southwest coast of Mexico from September 16-25 (Figure 5.32). Nora essentially stalled off the coast from the afternoon of September 18 to early on September 20. As a result, Nora's maximum sustained winds decreased from 90 knots (104 mi/hr) to 65 knots (75 mi/hr). As Nora's winds continued to stir the water, much like stirring some cream into a cup of coffee, cooler water upwelled from below to replace the warm surface water. In response to the upwelling, a pocket of anomalously cool water off the southwest coast of Mexico persisted for several days (Figure 5.32). As SSTs decreased by as much as 2°C (3.6°F) in this region, evaporation rates diminished and stabilized the atmosphere. The hurricane began to weaken and collapse as it lost its source of fuel.

Conditional Instability

We already understand the importance of low-level moisture in creating unstable atmospheric conditions. Unstable atmospheric conditions throughout the troposphere are essential to hurricane development. High dewpoints near the Earth's surface result in net condensation and the release of latent heat a short distance after the air parcel is lifted. The rising parcel remains positively buoyant as its rate of cooling is reduced, allowing it to rise through a deep layer of the troposphere. These conditions allow towering thunderstorms to form and organize around a center of circulation. High SSTs and conditional instability are not independent of each other; the warm ocean waters ensure that the air above it is rich in atmospheric moisture and conditionally unstable.

The cloud tops of hurricanes average about 15 km (~50,000 ft.), most of the water vapor has been depleted by the time the air parcels reach this altitude. Figure 5.33 shows that rising air parcels near the center of a developing tropical cyclone take one of two routes when they reach the tops of the thunderstorms. Most of the parcels move outward and away from the developing center, creating upper-level divergence that supports rising air motion and a reduction in surface pressure. The other parcels slowly descend in the center of the developing storm and warm adiabatically by compression. These sinking parcels heat up considerably because they contain so little moisture. As a result, the central column of a developing tropical cyclone has the highest average temperature and the lowest average pressure at the surface. Because the temperatures of tropical cyclones are warmer at their center than at their periphery they are called warm-core lows. Net divergence in the central air column occurs due to compressional heating

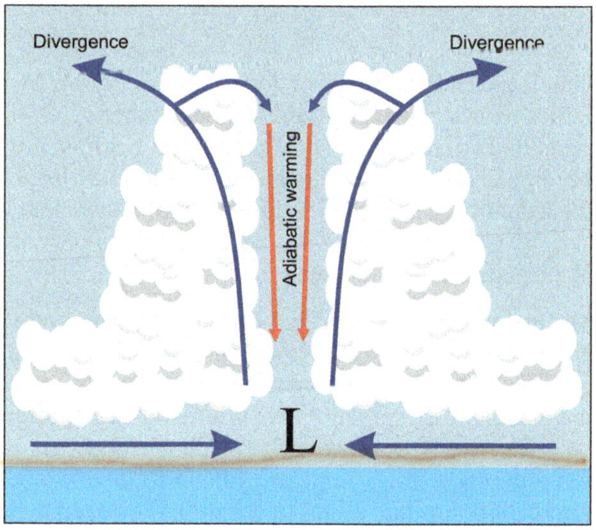

Figure 5.33: Diagram showing the general movement of air parcels in a hurricane. Once the air parcels reach the top of the storm, some of the air sinks back down in the eye, causing the air to warm adiabatically.

which also contributes to a reduction of surface pressure. There is an increase in compressional warming as the thunderstorms grow vertically and become more organized around a center of circulation. This leads to a reduction in surface pressure and therefore increased wind speeds which help the storm intensify.

Relatively Moist Air in the Middle Troposphere

Sufficient moisture in the mid-levels of the troposphere is another condition that is needed for a tropical cyclone to develop into a hurricane and maintain hurricane strength. In order for intense thunderstorms to remain organized around a center of circulation, relatively moist air in the middle troposphere is necessary. If dry air mixes with the thunderstorm's warm moist updraft, some of the water droplets evaporate through entrainment which reduces the parcel's buoyancy. As entrainment enhances evaporational cooling, latent heat supplied by condensation is reduced and the thunderstorm updrafts are weakened. Stronger thunderstorm downdrafts develop as dry air gets wrapped into the cyclone's circulation which can extinguish the formation of new thunderstorms near its core. If new convection stops forming near the cyclone's center, it will weaken rapidly as it becomes disorganized. Weather forecasters use satellite imagery to help predict the path and intensity of tropical cyclones. The amount of moisture in the middle troposphere has a large influence on creating a favorable or unfavorable environment for the tropical cyclone to intensify.

Figure 5.34: Hurricane Isabel as it consumes a parcel of dry air in the middle troposphere. A. This picture was taken on September 12, 2003 when Isabel was a strong hurricane. B. This was taken on September 16, 2003 when Isabel had weakened. Notice how the eye of the hurricane has fallen apart. Photo credits: NOAA/NHC.

Hurricane Isabel (2003) is a good example of how the entrainment of dry air can rapidly weaken a hurricane. Figure 5.34A shows Hurricane Isabel on September 12, 2003. At this time, Isabel was a major hurricane, with maximum sustained winds of 135 knots (155 miles per hour). In a little more than four days, Hurricane Isabel weakened and became disorganized as dry air from the middle troposphere was entrained into its circulation (Figure 5.34B). Over the Atlantic Ocean, dry air originating from the Sahara Desert often plays a role in suppressing hurricane activity.

Weak Vertical Wind Shear

Vertical wind shear, a change in wind speed and/or wind direction with height, is often the most critical factor that controls a tropical cyclone's formation and intensity. Tropical cyclones extend vertically through most of the troposphere; they are steered by the average wind characteristics through this layer. Hurricanes best form in an environment containing very little wind shear, allowing them to maintain a symmetrical structure. Wind shear weakens tropical cyclones by removing heat and moisture from their centers. An environment with high wind shear is unfavorable for a tropical cyclone to develop because the convection gets pushed away from the center of circulation. For hurricanes to develop, they need their energy concentrated around a clear center of circulation. As strong wind shear moves thunderstorms away from the center of circulation, the rotation weakens and the vortex near the center of the storm becomes tilted. This causes the storm to become disorganized and less efficient.

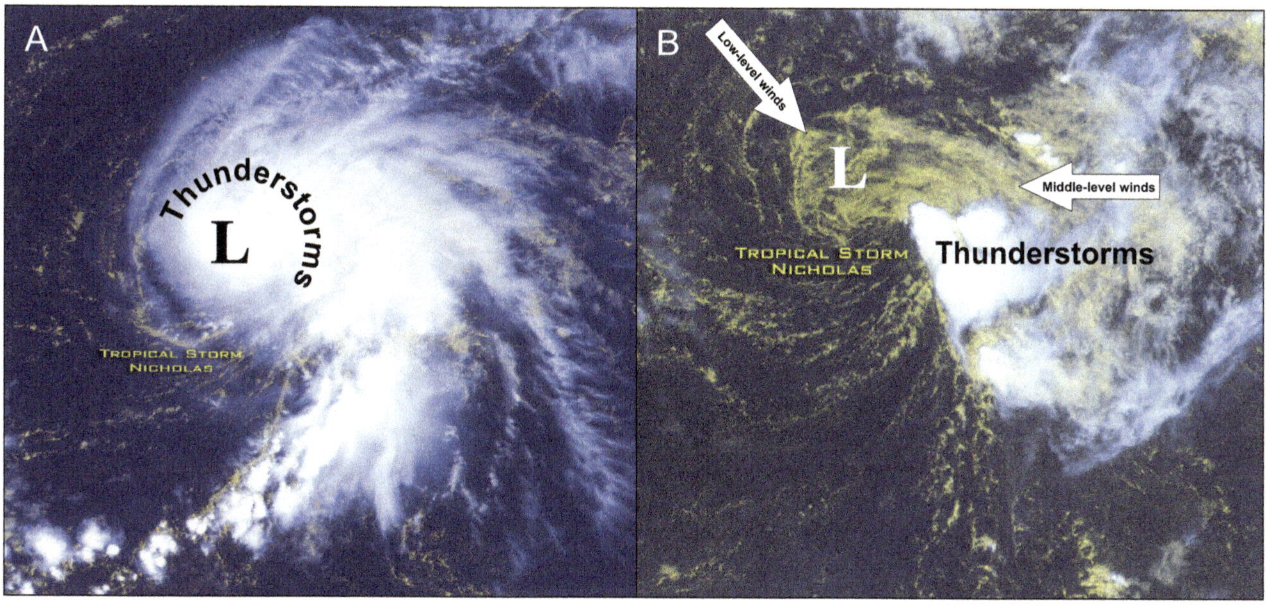

Figure 5.35: Tropical storm Nicholas as it succumbs to vertical wind shear. A. This picture was taken on October 17, 2003. B. This picture was taken on October 22, 2003. Strong low-level winds blew from the northwest while modest mid-level winds blew from the east. This separated the thunderstorms from the center of circulation (L), causing the storm to fall apart. Photos courtesy of NOAA/NHC.

Figure 5.35A shows a multi-channel satellite image of Tropical Storm Nicholas, located northeast of Guyana in South America on October 17, 2003. Nicholas weakened due to vertical wind shear as relatively weak easterly winds at 850 mb were accompanied by stronger northwesterly winds at 200 mb to create northwesterly wind shear. As a result, the National Hurricane Center (NHC) downgraded Nicholas to a tropical depression on October 23 as thunderstorms were separated from the center of circulation and the storm lost its constant supply of energy (Figure 5.35B). To calculate the magnitude of wind shear, simply subtract the wind speed at 850 mb from the wind speed at 200 mb. As a general rule, wind shear values must be below 20 knots (23 miles per hour) or less for the formation and development of tropical cyclones. The peak of the hurricane season, from August 15 to October 15, coincides with the time when wind shear values are the most favorable over the eastern and central tropical Atlantic.

A Pre-Existing Disturbance

A tropical depression doesn't always form even if the four conditions discussed above are favorable. Tropical cyclones can't be made spontaneously, a pre-existing low-level disturbance with sufficient spin and convergence of air is necessary. Under favorable conditions, a weakly organized system with cyclonic rotation and the convergence of warm, moist air can develop into a tropical cyclone. Forecasting a hurricane is challenging since only a small percentage of pre-existing disturbances evolve into a hurricane. Most of the pre-existing disturbances that develop into a hurricane in the North Atlantic Ocean originate as easterly waves. An **easterly wave** is a tropical disturbance that forms in the normal easterly trade wind patterns. These waves are generally complexes of thunderstorms that develop north of the equator over Africa and then move westward across the North Atlantic Ocean (Figure 5.36). Easterly waves can provide the low-level convergence and cyclonic spin needed to produce a tropical cyclone. The most intense hurricanes that threaten the United States usually develop from easterly waves.

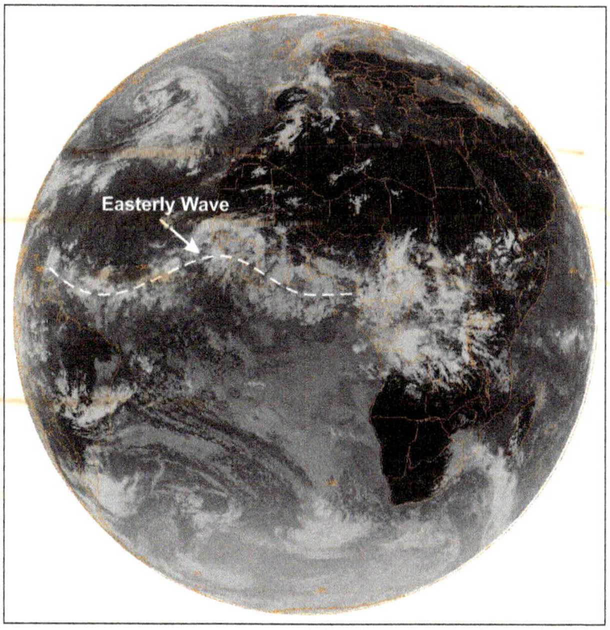

Figure 5.36: An easterly wave develops off the coast of Africa on September 23, 2003. This wave developed into a tropical storm as it moved across the Atlantic Ocean towards North America. Image courtesy of NOAA/NCDC.

If low-level convergence is strong enough the clusters of thunderstorms will continue to develop as they move westward. Easterly waves migrate off the coast of Africa every three to four days from June to October. They can easily be observed on satellite imagery. Since easterly waves typically move westward at about 15 to 35 km (10 to 20 mi.) per hour, it generally takes about seven to ten days for an easterly wave to migrate across the Atlantic Ocean. On average, approximately 60 easterly waves emerge from the West African coastline each year during this time (Figure 5.37). Every year approximately 12 named storms form on average during the Atlantic hurricane season. If only 10 percent of the average annual number of easterly waves develop into a named storm, they would account for half of the average number of named storms! Even more important, easterly waves initiate almost 85 percent of all major hurricanes, with maximum sustained winds exceeding 96 knots (111 mi/hr).

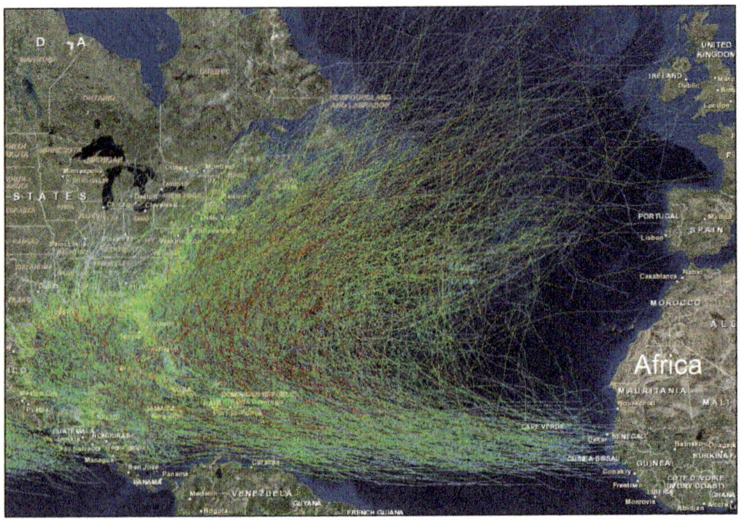

Figure 5.37: This map shows the tracks of all Atlantic hurricanes which formed between 1842 and 2014. Note how a large number of the tracks originate off the coast of northern Africa. The points show the location of each storm at six-hour intervals and uses a color scheme to show storm strength based on the Saffir-Simpson Hurricane Scale. Image created using the NOAA Historical Hurricanes website.

Formation Away From the Equator

Tropical cyclones are known for their distinctive circulation that is easily observable from above on a satellite animation. Tropical cyclone formation depends on the Coriolis force for its rotation. Since the magnitude of the Coriolis force at the equator is zero, tropical cyclones must form poleward of the equator to acquire enough spin for tropical cyclone development to be favorable. If the Coriolis force is too weak air will move almost directly towards the center of the low pressure, causing an increase in air pressure. This means that thunderstorms forming too close to the equator are not able to organize and concentrate around a distinct low-level circulation to become a tropical cyclone. At about 5° latitude, the magnitude of the Coriolis force is just enough for cyclonic circulation to be possible (Figures 5.27 and 5.37). For this reason tropical cyclones do not form within about 5° latitude of the equator, approximately 550 km (345 mi.).

The Life Cycle of a Tropical Cyclone

A cluster of thunderstorms must become organized for a tropical cyclone to develop. As latent heat is released from the thunderstorms the surface pressure drops to create a region of weak low pressure. As the pressure drops, warm, moist air converges inward toward the center of the thunderstorm cluster. Cyclonic circulation develops when the low forms far enough from the equator for the converging winds to be deflected by Coriolis force. As the pressure gradient increases, surface wind speeds increase and supply additional moisture to the newly-developing storm. As the air parcels spiral inward toward low pressure they rise and cool, which increases instability due to the release of latent heat.

As the surface pressure decreases, wind speeds and evaporation rates increase. This increases the number and intensity of thunderstorms, allowing them to organize around a center of circulation. Conditions are now favorable for the cluster of thunderstorms to transform into a giant heat engine, consuming vast amounts of latent heat released through condensation as fuel. As the thunderstorms intensify and organize around an area of low pressure, higher pressure forms at the top of the developing storm near the tropopause. This high pressure is caused by rising air that diverges away from the storm center. If there is no upper-level divergence to exhaust the storm, the convergence of low-level air would raise the surface pressure and prevent further storm development. The storm maintains itself by exhausting air outward, which supports low-level convergence and rising motion. If the central pressure continues to drop, the tropical disturbance will become a tropical depression, then a storm and eventually a hurricane. The maximum intensity of the hurricane is ultimately controlled by SSTs and other favorable atmospheric conditions.

Hurricanes can't last forever, at some point they begin to weaken and eventually dissipate. Their speed and path depends on complex interactions between the ocean and atmosphere. Most hurricanes last less than one week, and they diminish in intensity for three primary reasons. First, if the hurricane moves over colder water it loses its supply of warm, moist tropical air. As hurricanes move away from the tropics, they usually weaken considerably as they move over colder waters. Second, if the hurricane moves over land it loses its source of energy and moisture. In addition, the increased friction over land causes a rapid reduction in surface wind speeds when they make landfall. These weakening storms can still bring strong winds, intense rainfall and flooding across a large region. Lastly, wind shear can tear a hurricane apart (Figure 5.35). Tropical storms and hurricanes need light upper-level winds to develop vertically. Strong wind shear causes the storm to become disorganized by pushing convection away from the center of circulation. Strong wind shear causes the storm to dissipate because the storm's energy is not concentrated and is spread out over a larger area.

Features of a Hurricane

Hurricanes are typically about 480 km (300 mi.) wide although their size can vary considerably. The size of a hurricane doesn't always indicate its intensity; relatively small, compact hurricanes have had devastating impacts. Regardless of their size, the destructive winds and rain can impact a large area. Hurricane force winds can extend outward over 240 km (~150 mi.) from the eye of a large hurricane, with tropical storm force winds extending out 480 km (300 mi.). Hurricanes can sometimes remain almost stationary but typically move along their tracks at speeds of 16 to 32 km (10 to 20 mi.) per hour. However, in rare cases, they can travel at more than 56 km (35 mi.) per hour. Hurricane Wilma (2005) is the strongest hurricane on record for the Atlantic Basin. Wilma had a minimum central pressure of 882 mb and produced sustained winds of 280 km (157 mi.) per hour. In terms of wind speed, Hurricane Allen (1980) is the strongest Atlantic hurricane on record, with maximum sustained winds of 165 knots (190 mi/hr). Although the size, intensity, forward speed and structure of hurricanes can vary considerably, they all have three distinct features: An eye, and eyewall and rainbands.

The most distinguishing feature of a hurricane is the **eye**. The eye is the calm center of the hurricane, usually characterized by light winds and generally clear skies. Figure 5.38A is a satellite image of Hurricane Isabel on September 14, 2003. Notice how the eye of Isabel is not necessarily clear due to low-level clouds (Figure 5.38B). Sometimes gusty winds and clouds from the periphery of the eye mix inward. The average diameter of a hurricane's eye is 40 km (25 mi.); its size commonly fluctuates as the intensity of the hurricane changes. People sometimes wrongly assume the worst of the storm is over when the eye reaches them due to the clearing skies and weakening winds. Although winds in the eye are generally light, winds can be over 130 knots (150 mi/hr) in the most dangerous part of the storm a relatively short distance away, the eyewall.

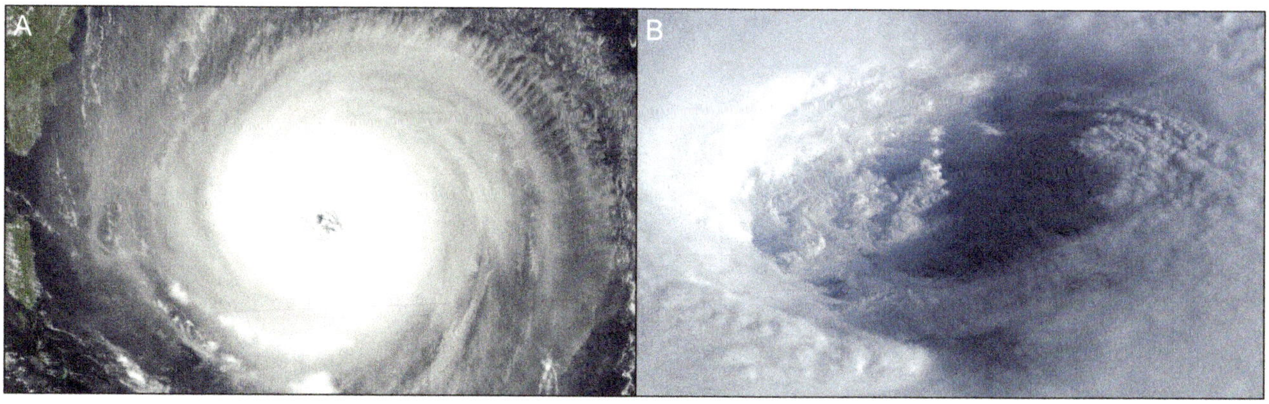

Figure 5.38: A. Hurricane Isabel with its well-developed eye. Image credit: Jacques Descloitres, MODIS Rapid Response Team, NASA/GSFC. B. The eye of Hurricane Isabel up close as seen from the International Space Station on September 13, 2003. Image credit: Image Science and Analysis Laboratory, NASA-Johnson Space Center.

The most dangerous part of a hurricane is located just outside of its relatively calm eye. The **eyewall** is an organized ring of intense thunderstorms surrounding the eye of a hurricane (Figure 5.39A). Airflow in the eyewall is upward and outward and extends from near the ocean surface to the cloud tops at about 15 km (~50,000 ft.). This is the location within a hurricane where the highest surface winds and heaviest rainfall are found. This part of a hurricane

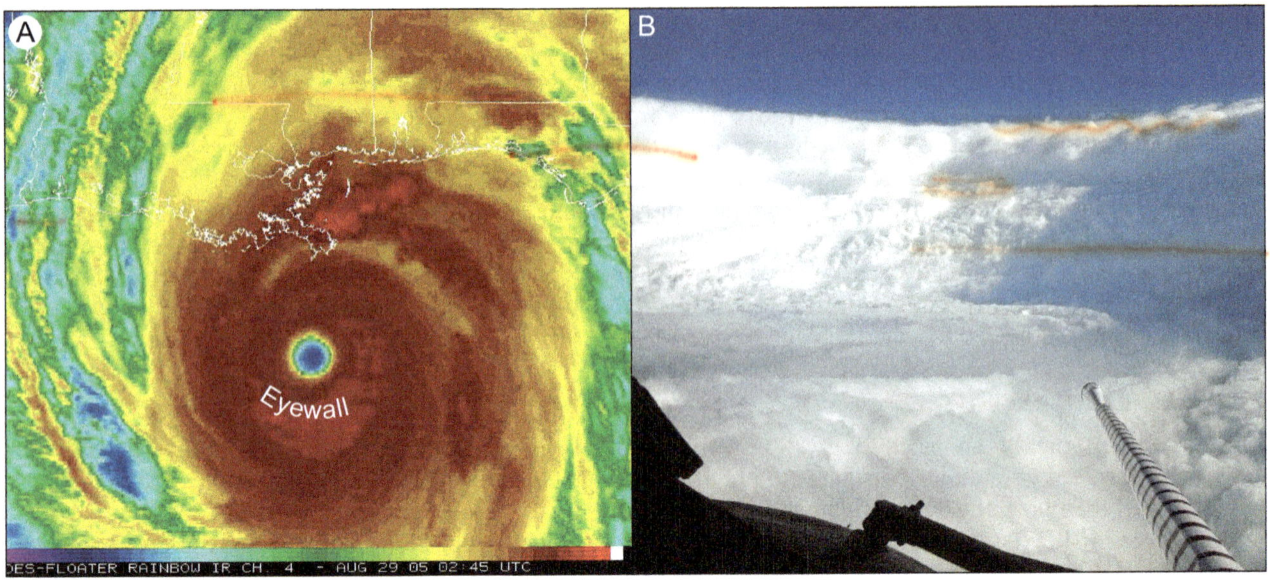

Figure 5.39: A. Infrared imagery of Hurricane Katrina as it approached the Louisiana coast. Note the heavy rain in the eyewall. Image credit: NOAA. B View of the eyewall of Hurricane Katrina taken on Aug. 28, 2005, as seen from a NOAA WP-3D Orion hurricane hunter aircraft before the storm made landfall on the United States Gulf Coast. Photo credit: NOAA.

is called the eye wall because it typically appears as a vertical wall of clouds that border the eye (Figure 5.39B). The most destructive part of a hurricane is in the eye wall on the side where the wind blows in the same direction as the storm's forward motion. For example, in a hurricane that is moving due north, the strongest winds would be found on the northeast side of the storm because the storm's counterclockwise motion also contributes to its swirling winds (in the Northern Hemisphere).

Outside the eyewall of a hurricane, **rainbands** curve cyclonically in toward the center of the storm to give the hurricane its pinwheel appearance. These bands of thunderstorms are typically 80 to 480 km (~50 to 300 mi.) long and range in width from a few kilometers to tens of kilometers. The rainbands appear to merge to form the eyewall as they spiral inward (Figure 5.40). Extreme rainfall and strong wind gusts are associated with the rainbands and sometimes tornadoes form when the rainbands come onshore. Wind speeds and rainfall from the rainbands decrease in intensity with increasing distance from the eyewall. There are often gaps between the rainbands with relatively

Figure 5.40: A. Tropical Cyclone Gafilo as it approached Madagascar. Note the distinct rain bands. The system soon became the most intense tropical cyclone in the South-West Indian Ocean on record. Photo credit: Jeff Schmaltz, MODIS Rapid Response Team, NASA/GSFC. B. This is a radar image of Hurricane Dennis, 2005, from NWS Eglin AFB NEXRAD radar. Note the distinct rainbands in yellow and orange. Data courtesy of NOAA.

calm winds and no rain. The outer rainbands of a hurricane can extend several hundred kilometers from the eye; they represent the outermost edge of the storm. The arrival of the outer rainbands is usually the first indication a hurricane will make landfall along a coastline, typically arriving about 24 hours prior to the eyewall coming onshore.

Hazards of Tropical Storms and Hurricanes

Statistically, the most likely areas of the United States to be impacted by hurricanes are the southern tip of Florida, the Outer Banks of North Carolina, and the Gulf Coast from Texas to Louisiana (Figure 5.37). Although many factors determine the degree of damage caused by a hurricane, the strength of the storm is unquestionably the most important factor. The **Saffir-Simpson Hurricane Wind Scale** rates the strength of a hurricane on a scale from one to five, with five being the strongest (Table 5-2). This five point scale is a way to relate the hurricane's maximum sustained wind speed to its damage potential. As soon as a tropical storm intensifies into a hurricane it is assigned a number. The hurricane is monitored throughout its lifetime and the category of the storm is adjusted as the intensity changes. The term major hurricane is used for hurricanes that are classified as a Category three or higher, with maximum sustained winds of at least 96 knots (111 miles per hour). Although all hurricanes are dangerous, major hurricanes have a high potential to cause significant loss of life and property damage. For example, a Category four or five hurricane typically produces about 16 times the amount of damage as compared to a Category two hurricane. The Saffir-Simpson Hurricane Wind Scale is useful to inform the public about the damage potential of hurricanes so proper precautions can be made. Hurricane evacuations must be taken seriously; evacuating the coastline well in advance of the storm's arrival is the most effective way to ensure the safety of coastal populations. The category of a hurricane does not always directly relate to the damage they produce, the damage is a function of several factors related to the storm itself and where they make landfall. Damage by hurricanes can be divided into three classes: (1) storm surge, (2) wind damage and (3) inland flooding.

Category	Central Pressure (inches)	Wind Speed (mph)	Damage (expected)
1	> 28.91	74-95	*Some damage*
2	28.50-28.91	96-110	*Extensive damage*
3	27.91-28.49	111-129	*Devastating damage*
4	27.17-27.90	130-156	*Catastrophic damage*
5	<27.17	>157	*Catastrophic damage*

Table 5-2: The Saffir-Simpson Hurricane Wind Scale.

Storm surge is an above normal, wind-driven rise in sea level that submerges the coastline as a hurricane begins to approach a coast (Figure 5.41). If the hurricane makes landfall during high tide, the flooding can be even more extensive. Storm surge is one of the most hazardous aspects of a landfalling hurricane; it is responsible for a large

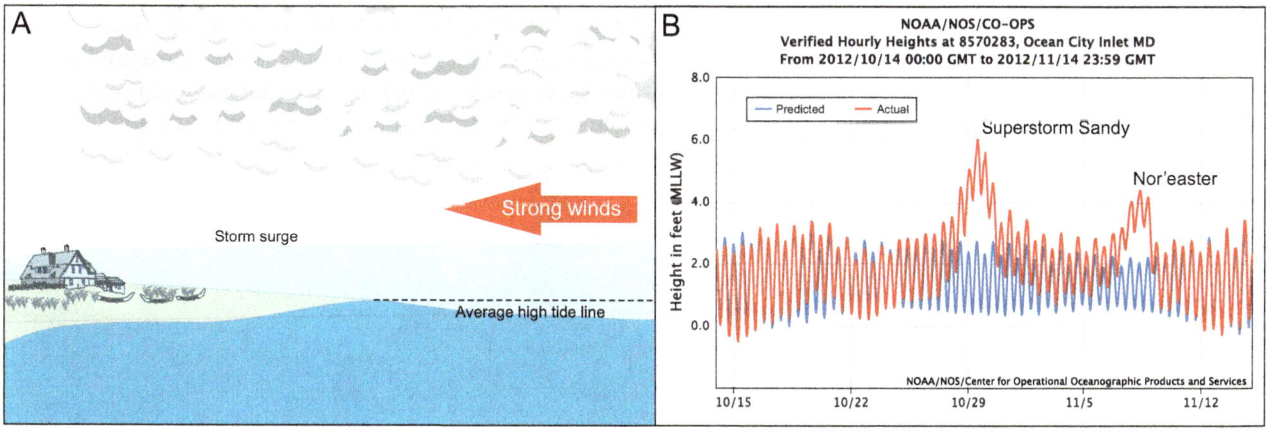

Figure 5.41: A. High winds cause water to pile up in shallow waters forming storm surge and flooding low-lying coastal areas. B. This tidal gage data from Ocean City, Maryland shows how much a storm surge event can elevate water levels above the normal predicted levels. Data courtesy of NOAA.

percentage of all fatalities related to hurricanes. Additionally, storm surge destroys property, erodes beaches and washes out road and rail transportation. A storm surge, usually lasting for several hours, is especially hazardous along low-lying coasts with no protective offshore barriers. Hurricane Isabel made landfall as a Category 2 storm on September 18, 2003 on the Outer Banks of North Carolina. To the north, Cape Hatteras was in the path of Isabel's strongest winds on the right side of the eye where the winds blow onshore. The onshore winds to the right of the eye

Figure 5.42: A. Hurricane Isabel caused extensive flooding in downtown Annapolis, Maryland. Photo credit: B. Zaprowski. B. Storm surge from Hurricane Isabel led to washovers of the Outer Banks in parts of North Carolina, cutting off some communities from the mainland. Photo credit: Mark Wolfe/FEMA.

produce a dome of water when a hurricane neared land, causing a rise in sea level. Maximum sustained winds reached 90 knots (104 mi/hr) and caused water to pile up along the coastline. As a result, the strong onshore winds and low atmospheric pressure caused the storm surge to inundate many coastal communities across a large area (Figure 5.42A). In extreme cases, historical storm surges of more than 7 m (23 ft.) have had disastrous impacts to coastal areas. Storm surge may also cause waves to completely wash over barrier islands, and sometimes form new inlets, cutting people off from the mainland (Figure 5.42B). It is also important to note that storm surge can also occur in conjunction with any type of strong coastal storm, like a nor'easter.

Wind damage is considered the most obvious type of damage caused by a hurricane (Figure 5.43). Strong winds can completely level buildings and cause flying debris to act like missiles. Tall buildings are especially vulnerable to wind damage since wind speeds increase with height. Although wind damage is usually not as costly as storm surge in areas where building codes are in place and enforced, it can impact a much larger area. Hurricane force winds can cause huge economic losses and impact areas more than 160 km (100 mi.) from the coastline. For example, the vast majority of damage associated with Hurricane Andrew (1992) was as a result of extremely high winds. Hurricanes frequently produce tornadoes once they make landfall, most commonly on the northeast side of the storm relative to the direction it is traveling. Hurricane Ivan (2004) holds the record for the most tornadoes produced by a hurricane. Ivan produced 117 tornadoes from September 15-17, with most occurring on September 17 (Figure 5.43B). Fortunately, the vast majority of tornadoes spawned by hurricanes are weak (EF0-EF1) and short-lived.

Extreme rainfall produced by hurricanes can be a major threat to locations hundreds of kilometers from where the storm made landfall. In many cases, inland flooding caused by heavy, prolonged rain events, not storm surge, is responsible for the most hurricane related fatalities. Hurricanes can still kill people and cause massive flooding long

Figure 5.43: Examples of wind damage caused by Hurricane Ivan. A. Uprooted trees and strong winds destroyed this home in Alabama during Hurricane Ivan. Photo credit: Jocelyn Augustino/FEMA. B. Damage from the Stokesdale Tornado during the passage of the remnants of Hurricane Ivan in North Carolina. Photo credit: NOAA/Guilford County Emergency Management, Stokesdale Fire Department, Doug Schneider, Darin Figurskey and Jeff Orrock.

after they have made landfall and weakened considerably. Over the last few decades, inland flooding has accounted for a significant number of deaths associated with tropical cyclones in the United States. The flooding potential of a hurricane is not directly related to its strength; some of the most severe inland flooding has been caused by weaker tropical storms. For example, Hurricane Irene (2011) produced significant inland flooding in New York State and elsewhere after being downgraded to a tropical storm (Figure 5.44). More than 15 centimeters (~6 inches) of rain fell across a large portion of the state in less than two days, resulting in one of the costliest floods in New York's history. Predicting where inland flooding will occur is related to several factors, making it especially challenging to forecasters. In addition to understanding the characteristics of the land they travel over, the specific track of the storm and where the heaviest rain will fall must be predicted accurately.

Figure 5.44: Flooding caused by Hurricane Irene. A. Flooded businesses in Bound Brook, New Jersey. Photo credit: Andrea Booher/FEMA. B. The lock 9 Mohawk river bridge near Rodderdam Junction, N.Y. was destroyed due to the heavy flooding caused from both Hurricane Irene and Tropical Storm Lee. Photo credit: Adam DuBrowa/FEMA.

NOAA
Lora Koenig/NASA

Chapter 6 - Climate and Climate Change

Section 6-1: A Description of Climate

In a very general sense, weather is simply the condition of the atmosphere at a particular place at a specific time. It is defined by various weather elements, commonly the ones discussed on a weather forecast by a broadcast meteorologist, such as temperature, pressure, humidity, wind and precipitation. **Climate** is only different from weather in terms of the time period involved; it describes the variation of weather for a given location from a long-term average of observed conditions. The same atmospheric variables are used whenever the weather or the climate is described for a particular region. The climate of a locality for a specific time period is usually defined by the average, or expected, weather conditions in addition to the extremes that have been recorded. This information can be used by the public as a general guide for various planning purposes. The famous quote "the climate is what you can expect; the weather is what you get" describes the difference between weather and climate well. The most likely or expected weather conditions are generally used when describing the climate; they are considered "normal" or "average". When the weather of a given day is described as "unseasonable" the daily atmospheric conditions are being compared to a long-term average (usually 30 years). In the United States, 30-year averages are calculated for pressure, temperature and precipitation. This technique allows the weather to be put in perspective by comparing it to what is "expected" for a given period of time.

There are many kinds of climate data needed for a range different purposes. One of the primary reasons detailed weather observations are recorded is to enable meteorologists to forecast the weather. In order to forecast the weather accurately the current state of the atmosphere must be understood, which makes the continuous recording of weather observations essential. Accurate weather forecasts are essential to aviation; pilots need information on winds, visibility and hazardous flying conditions. Transportation officials need to know if weather conditions require the roads to be treated while local residents want to know the possibility of thunderstorms producing severe weather. In addition to forecasting the weather on a daily basis, climate data are useful for planning in the longer term. For example, a vegetable grower wants to know the most likely time to operate within the growing season to avoid damaging frosts. Electric companies employ climatologists to predict the intensity and frequency of temperature extremes to ensure that enough energy is available for the upcoming season. As you can see, climate data are valuable and used in a variety of applications to help both with daily operations and long-term planning.

We have all experienced times when the weather changes suddenly. Weather consists of short-term variations of the atmosphere with seasonal characteristics. Variations in climate occur slowly and over much longer periods of time (Figure 6.1). Similar weather conditions can often persist for several days; it is common for the weather to become locked in a repeating pattern. For example, clear and sunny conditions or cloudy and rainy skies can last for several days in a row. In many areas the weather can remain hot and dry for several weeks or more during the summer months.

For planning purposes, it is very important to remember that the climate of an area consists of the average "expected" conditions in addition to daily and seasonal extremes in the observed atmospheric variables. A great example of this occurred in the Central United States during the 1920s and early 1930s. Rainfall during this time had been above average and as a result settlers from across the country moved into western Oklahoma, northern Texas, northeastern New Mexico and southwest Texas (Figure 6.2A). The abundant rain encouraged an increase in agricultural activities as farmers extensively plowed and planted crops across the region. When a drought struck from 1934 to 1937, the crops dried up and the loose soil turned to dust.

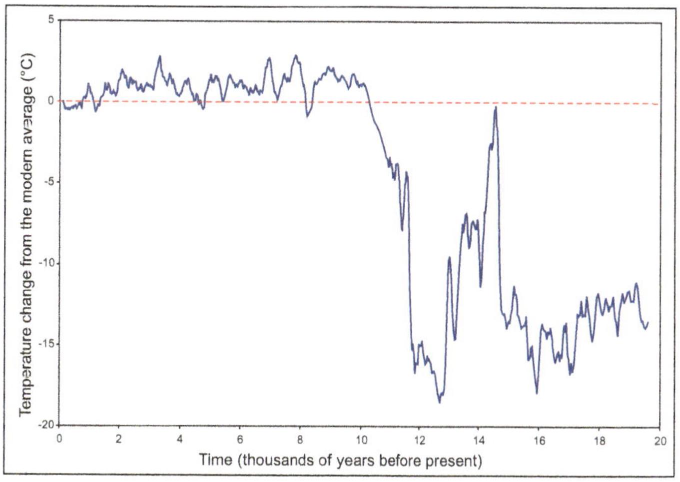

Figure 6.1: Northern Hemisphere temperature changes for the past 20,000 years as inferred from the GISP-2 project ice cores. Data from Alley, R.B., 2004. GISP2 Ice Core Temperature and Accumulation Data. IGBP PAGES/World Data Center for Paleoclimatology Data Contribution Series #2004-013. NOAA/NGDC Paleoclimatology Program, Boulder CO, USA.

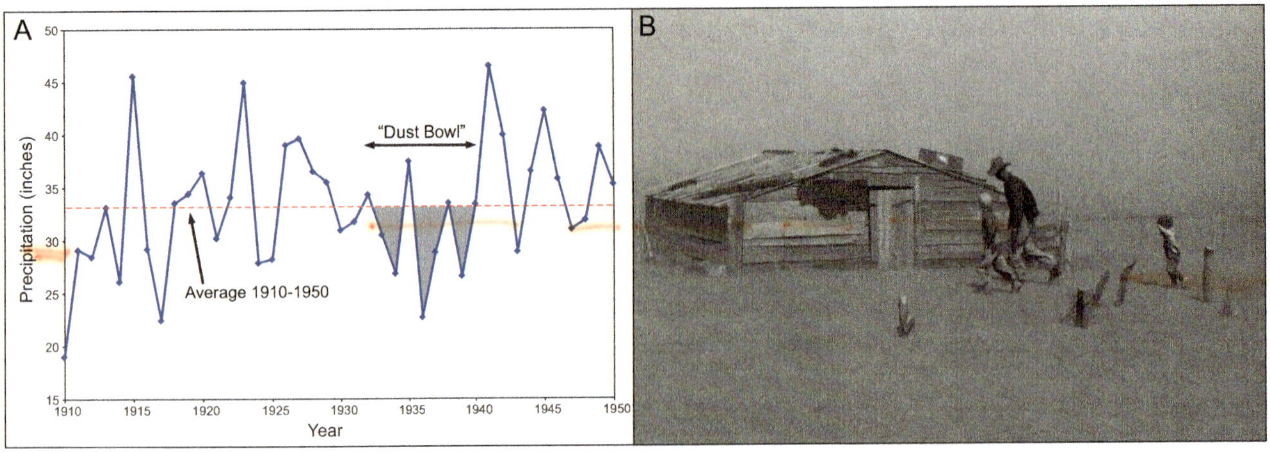

Figure 6.2: A. Average annual precipitation for Oklahoma for the years 1910-1950. Data courtesy NOAA/NCDC. B. Farmer and sons walking in the face of a dust storm. Cimarron County, Oklahoma, USA, 1936. Photo credit: Arthur Rothstein/Farm Security Administration/U.S. Library of Congress.

The dry conditions combined with poor farming practices led to the "dust bowl" of the 1930s and forced tens of thousands of families to abandon their farms (Figure 6.2B). The dust bowl is considered both a human-made and natural disaster. Scientists are trying to identify the causes of weather cycles like this that occur on a fairly regular basis. Short-term weather cycles must be considered when characterizing a region's climate.

Section 6-2: Global Patterns of Climate

Understanding how the Earth's climate system works is a challenging task since it is influenced by a variety of interacting factors. It has been assumed for a long time that the climate is largely controlled by geographic location. There are a number of factors that account for the global climate pattern. The general climatic controls, including latitude, continental position, proximity to water, and prevailing winds affect large areas in a similar way. Local factors such as elevation, topography and vegetation make the actual climate pattern very complex. The same factors that control the climate of a particular location also produce their day-to-day weather. The Earth's climates are so diverse that no two places experience the exact same climatic conditions. However, when viewed globally the climate does exhibit some regular patterns.

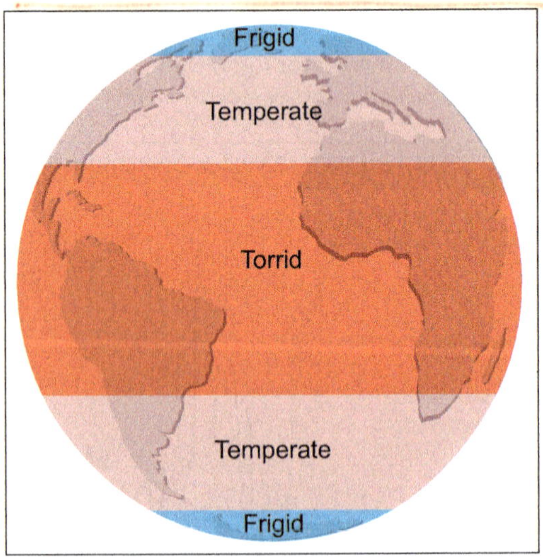

Figure 6.3: A generalized climate map of the world as envisioned by the Greeks.

Broad similarities among local climates allow climatic regions to be grouped into geographic regions. There have been numerous different climate classification schemes devised, one of the first was developed by the ancient Greeks around 350 BC. Some Greek scholars understood the size and shape of the Earth and recognized that temperatures were hotter to the south and cooler to the north. At this time little was known about the tropical or Polar Regions, the Greeks believed the area near the equator was too hot for habitation. They also suspected that regions near the poles were permanently frozen and too cold to inhabit. The Greek climate classification divided the Earth into three types of climatic zones based on latitude (Figure 6.3). The zone of the Earth's surface between the Tropics of Cancer and Capricorn was assumed to be a "winterless" climate and called the Torrid Zone. Regions poleward of the Arctic and Antarctic Circles were called Frigid Zones and considered a "summerless" climate. The Greeks believed the only areas suitable for habitation were the two Temperate Zones located between the Torrid and Frigid Zones. Based on their own experience the climates of the Temperate Zones were thought to have an "intermediate" climate. The Greek climate classification scheme is unrealistic and greatly simplified; some of the wettest and driest areas of the world are located in the same climate zone. We know now that tropical latitudes and areas north of the Arctic Circle are indeed habitable. A large percentage of the world's population and some of the world's largest cities are located in the Torrid and Frigid Zones. Despite its shortcomings, this scheme was used for more than 2,000 years and finally discarded in the 20th Century. Although modern climate classifications are much more complex, they still recognize comparable zones. Today the Torrid Zone is associated with tropical latitudes, the Temperate Zone is synonymous with the mid-latitudes and the Frigid Zone refers to polar latitudes.

The most successful and frequently used climate classification scheme was designed by a German botanist, Wladimir Köppen (1846-1940). Köppen recognized that vegetation has specific temperature and moisture requirements and therefore the native vegetation of an area can be used as a natural indicator of the regional climate. The **Köppen system** defines climate boundaries based on vegetation patterns and average monthly temperature and precipitation. Any location's climate can be classified using the Köppen system as long as average monthly temperature and precipitation data are available.

Each of the five main groups of world climates is classified using letters in the Köppen system (Figure 6.4). The Köppen system defines four of the major climate zones based on temperature and one based on precipitation. Increasing in latitude from the equator, the five main groups of world climates are: (A) tropical humid; (B) dry; (C) temperate mid-latitude; (D) cold mid-latitude; and (E) polar. These climate zones generally coincide with latitude, continental position and location with respect to topographic features. The A climates of the tropics are often considered winterless while moisture deficiency characterizes the B climates. The C and D climates are distinguished by the severity of the winter; the C climates typically have mild winters while the D climates' winters are severe. The E climates of the high latitudes are often considered summerless. Each of the major climate zones is further subdivided into subtypes using a series of second and third letters according to regional characteristics of temperature and precipitation.

Although the Köppen system of climate classification is useful, the climate zones and vegetation boundaries do not

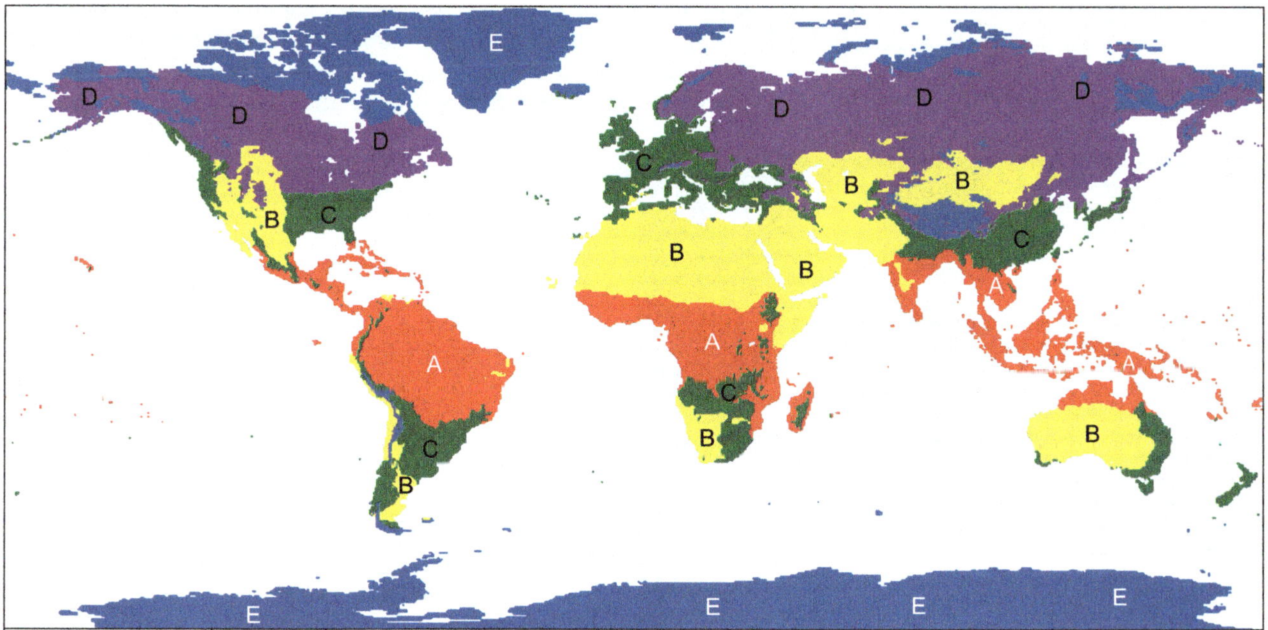

CLIMATE ZONE	CODE	GENERALIZED DESCRIPTION
TROPICAL	A	Coolest mean monthly temperature > 18°C
DRY	B	<10 inches of rain per year with hot days and a large daily temperature range
TEMPERATE	C	Warmest mean monthly temperature > 10°C and coldest mean monthly temperate > 0°C
COLD	D	Warmest mean monthly temperature > 10°C and coldest mean monthly temperate < 0°C
POLAR	E	Warmest mean monthly temperature < 10°C

Figure 6.4: A generalized climate map showing Köppen's five main climate zones. Base map image source: NOAA.

always correspond to each other. In addition, the sharp boundaries shown in Figure 6.4 are usually gradual transition zones. The Köppen system has been modified several times since it was introduced in 1918. It is still widely used today and is considered a useful tool to understand the basic climate types on a global scale.

Section 6-3: Paleoclimatology

Paleoclimatology is the study of past climates throughout geologic time, and the causes of their variations. Indirect evidence must be used to reconstruct past climates since the modern instrumental record of historical weather and climate conditions is limited to the past few centuries. Climate conditions before this time can be estimated using a variety of less precise natural environmental records. Unfortunately, these records can only provide a rough idea of what the past climate was like. Although we only have a general understanding of the Earth's past climate, it is amazing that we know as much as we do. Climate information becomes less reliable and more sporadic the farther back in time we go. Paleoclimatologists use a variety of **proxy records** to reconstruct past climate conditions. Proxy records are a substitute for instrumental data, they are indirect sources of climate information and must be used since temperature and precipitation data are not available for prehistoric times. Examples of proxy records include documentary evidence, ice cores, tree rings, lake sediments, corals and pollen records. These proxy records can reflect changes in the past climate in various ways. A great deal of information on Earth's past climates can be obtained once all the available proxy records are carefully examined and pieced together.

A proxy record can be a personal diary or harvest log that contains weather information during the growing season. Valuable information regarding climate extremes, such as catastrophic floods or temperature extremes is often preserved in this documentary evidence. Ice cores are valuable proxy records that are used to interpret the past climate for hundreds of thousands of years. Climatologists analyze the annual layers of glacial ice that form to track changes in air temperature and snowfall (Figure 6.5A). The layers of ice also contain dust, which can indicate past volcanic activity and wind patterns. Air bubbles trapped in the ice can be analyzed to record variations in atmospheric composition. Fluctuations in atmospheric concentrations of carbon dioxide and methane are related to temperature variations, allowing climatologists to determine how changes in climate may be related to changes in atmospheric composition (Figure 6.5B). Although Figure 6.5B doesn't show that past temperature changes are driven by fluctuations in carbon dioxide, it is obvious from the data that there is a strong relationship between the two.

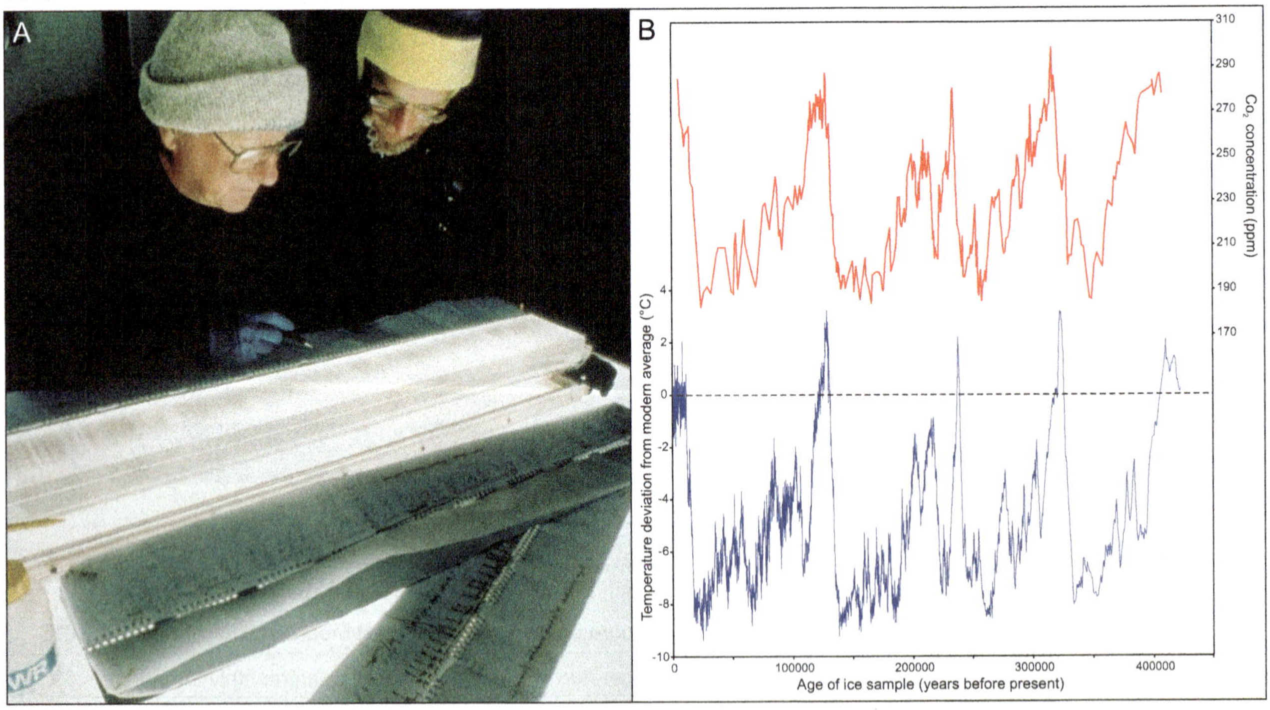

Figure 6.5: A. Scientists study the annual layers in an ice core from Greenland. Photo credit: NOAA. B. CO_2 and temperature data from the Vostok ice core from Antarctica. The data show a very strong relationship between temperature changes and changes in atmospheric CO_2. Temperature deviations are calculated from changes in deuterium concentrations in the ice. CO_2 concentrations were measured directly from entrapped air inclusions. Data courtesy NOAA/NGDC.

Information about past climates can also be determined through the analysis of the annual growth rings of certain trees, a technique called **dendrochronology**. Tree rings can be used to measure annual changes in climate over hundreds or thousands of years (Figure 6.6). Dendrochronologists study the changes in the thickness of tree rings to indicate the climatic conditions that occurred each year the tree was alive. Patterns in the width of the tree rings from hundreds of trees for a particular tree species are compared to the modern instrumental record. The tree ring growth patterns for some tree species are more sensitive to temperature variations while others are more sensitive to precipitation variability. The width of each tree ring depends on the temperature and/or moisture conditions for a given year. Wide rings indicate favorable growth conditions in terms of temperatures and/or moisture conditions during the growing season. Narrow rings represent stressful climatic conditions that result from excessive warmth and/or lack of moisture. Temperature and precipitation changes in the past can be estimated with a high degree of certainty due to the strong relationship between modern temperatures and tree ring growth patterns. Dendrochronology has been used to estimate annual average temperature and precipitation conditions in many regions of the world going back thousands of years. Detailed tree ring chronologies have been constructed at several locations by matching the patterns in tree rings from tree ring samples. Since the yearly patterns of tree rings are similar for a given region, a continuous record can be made by matching overlapping patterns of successively older pieces of wood (Figure 6.7). This technique, known as cross-dating, can be used to reconstruct the past climatic conditions over thousands of years.

Figure 6.6: A close look at this log illustrates the variation in ring width over the course of several decades. Photo credit: B. Zaprowski.

Using the data from a variety of proxy records, paleoclimatologists can create paleoclimate maps that represent the regional climate conditions for a specific time period. For example, the series of maps in Figure 6.8 shows changes in vegetation over the past 18,000 years based on pollen records obtained from a variety of sites around Eastern

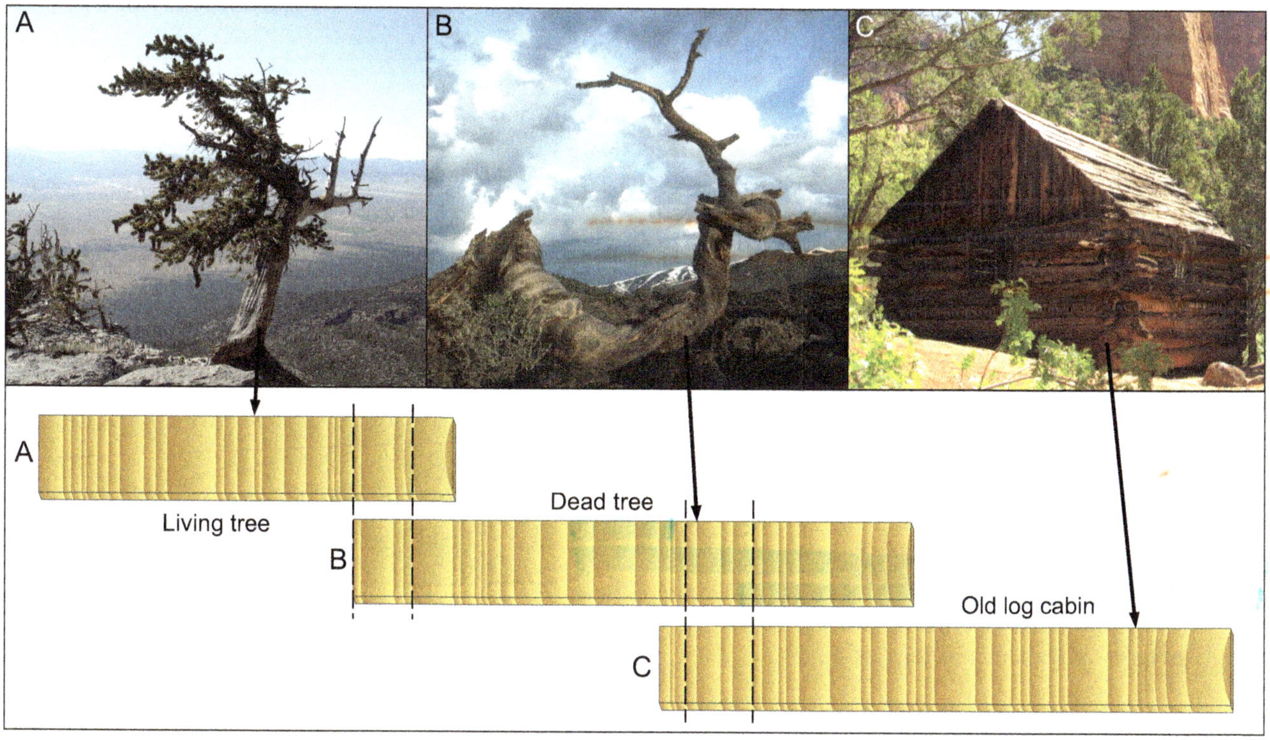

Figure 6.7: This illustration shows how tree rings from different pieces of wood can be matched using distinct ring patterns. This helps climatologists produce a tree ring chronology back through time. Photo credits: NPS.

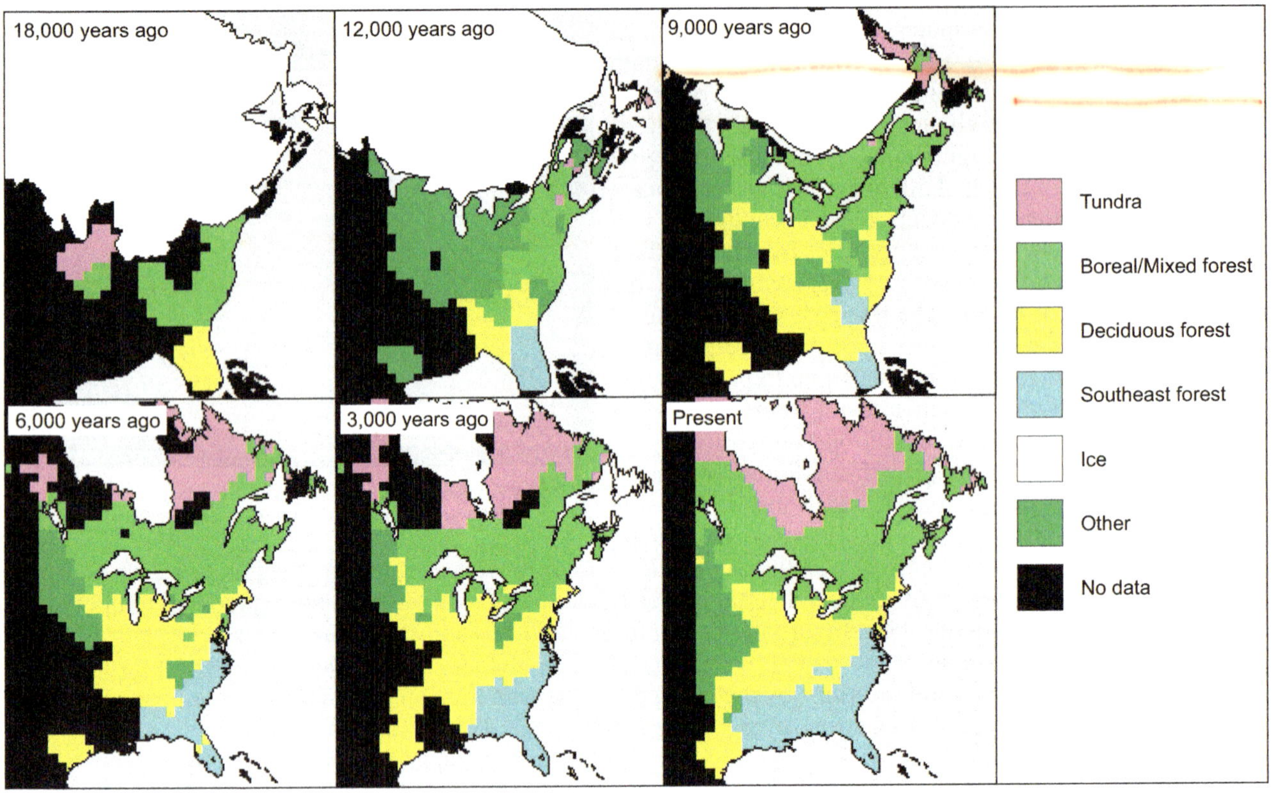

Figure 6.8: Paleovegetation maps for Eastern North America. These maps were constructed using 11,700 fossil pollen samples, and 1744 modern pollen samples. Image credit: NOAA/NCDC Paleoclimatology Program.

North America. Pollen records indicate the types of plants that were growing when the pollen was deposited. This allows scientists to estimate the climatic conditions and climate changes based on the types of past vegetation. Although this proxy record can't be used to estimate the specific temperatures for a region, it does provide a general idea of environmental changes over long periods of time.

Section 6-4: Natural Causes of Climate Change

The Earth's climate has varied significantly over a wide range of time scales; evidence of climate change in the geologic past is obvious. For example, tropical animal and plant fossils have been discovered in polar locations while evidence of humid climates has been preserved in desert soils. Evidence of glaciers in the United States as far south as the Ohio River and Long Island, New York indicate that the past climate was much different than today (Figure 6.9). There have been several theories proposed in an attempt to explain climate change. It is clear that the Earth's climate results from the combination of many factors. Due to the complex relationship between atmospheric variables it is not possible to fully explain all of the changes in the Earth's climate with a single theory. Climate variability is also influenced by interactions between the atmosphere, oceans, ice and the Earth's surface. Although there are numerous processes that influence the climate, they operate simultaneously and interact in various ways.

Figure 6.9 These glacial sediments in upstate New York were transported by glaciers and deposited here during the last Ice Age. Photo credit: B. Zaprowski.

Climate change has occurred in the past and will continue in the future. The rate and magnitude of climate change projected over the next 50 to 100 years is a highly debatable topic. Temperatures have been rising at many locations since the middle of the 20th century. In recent years there has been increased concern about the consequences of global climate change as the environmental and social impacts of a changing climate have become better understood. The changing climate impacts the environment and society in a broad variety of ways. For example, rising sea levels, temperature extremes and increased air pollution have the potential to significantly alter the environment in ways that can impact human society. Changes in water resources are currently a primary concern in many parts of the world.

Variations in Solar Output

It seems logical that variations in the sun's solar output (intensity) would be associated with climate variability. Chinese astronomers began observing sunspots more than 2,000 years ago by viewing a reflected image of the sun's surface off a still body of water. The variation of the sun's intensity has been measured since the 1600s by counting the number of sunspots on the surface of the sun. A **sunspot** is a large magnetic storm that appears as a relatively dark irregular shaped area on the surface of the sun. Sunspots appear dark because they are cooler than the much brighter material that surrounds them. They change position from day to day and are usually found in pairs or groups due to their magnetic nature. Figure 6.10 shows a plot of the number of sunspots observed each year since 1750. The number and intensity of sunspots on the sun's surface vary over a fairly regular 11 year cycle. During periods of maximum sunspot activity the sun's solar output can increase by 0.1 to 0.2 percent (Figure 6.11).

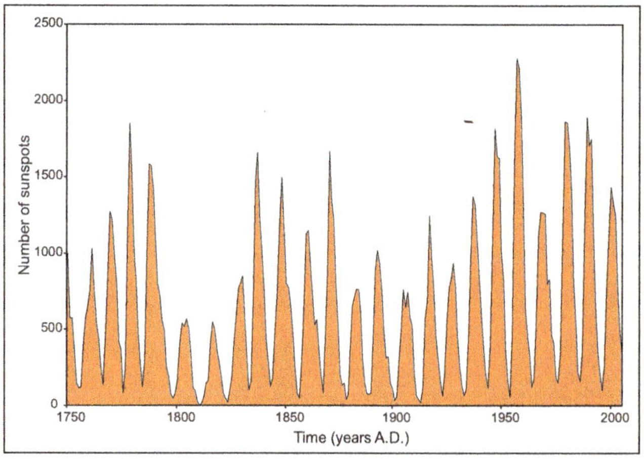

Figure 6.10: Variation in the number of sunspots per year from 1750-2005. Data courtesy of World Data Center for Solar-Terrestrial Physics and NOAA/NCDC Paleoclimatology Program.

Some scientists believe that variability in solar output due to sunspot activity is correlated with the Earth's past climatic changes. This belief is supported by the fact that few sunspots were observed from about 1645 to 1715, a period that corresponds to the peak of the Little Ice Age. This cold spell during a period of low sunspot activity is called the Maunder minimum. Periods of high sunspot activity in the past have also been associated with warmer times. Recurring droughts in the Great Plains and Western United States also seem to occur on an approximately 22 year cycle. However, there have been similar periods of sunspot activity that do not coincide with changes in climate. It has been argued that the increases in solar output during active sunspot cycles are not strong enough to significantly affect the Earth's climate. Even though solar output variability due to sunspot activity can be measured, no correlations between sunspot cycles and climate have been well-established. The influence sunspots have on climate will likely continue to be a debatable topic.

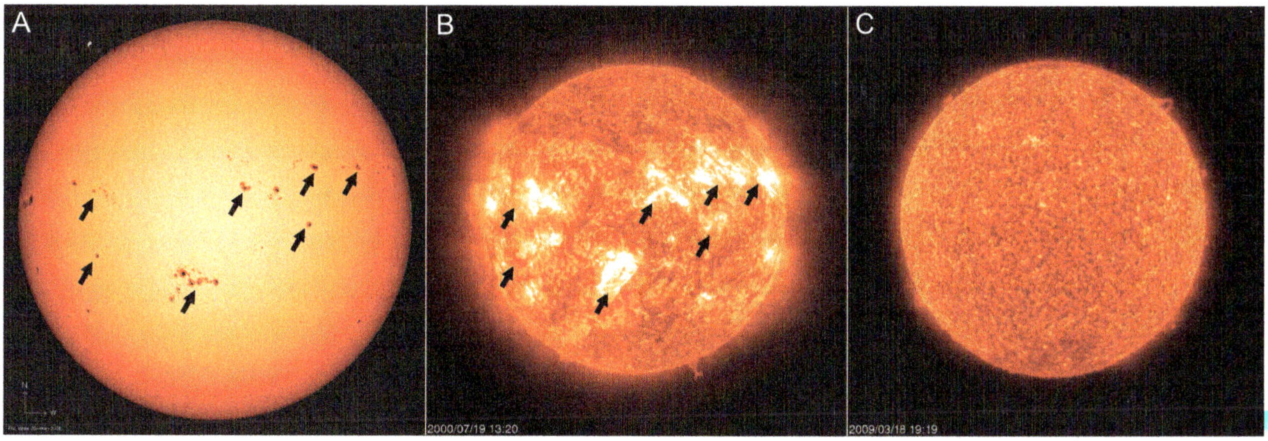

Figure 6.11: A. Sunspots during a year with high sunspot activity. B. An infrared image of the sunspots shown in A. C. An infrared image from a year with low sunspot activity. Notice how much less radiation is visible in C as compared to B. Image credits: SOHO/NASA/ESA. SOHO is a project of international cooperation between ESA and NASA.

Variations in the Earth's Orbit

Changes in the orientation of the Earth relative to the sun cause significant climate changes by affecting the Earth's heat budget. The term **insolation** is often used to describe the total amount of incoming solar radiation received at the Earth's surface. Variations in the Earth's orbit influence the climate by altering the latitudinal distribution of insolation, which causes changes in the severity and timing of the seasons. Regular shifts between glacial and interglacial cycles correspond to these orbital cycles over the past few million years. Many scientists believe that orbital variations are the trigger mechanism for the Earth's glacial and interglacial cycles. These orbital variations,

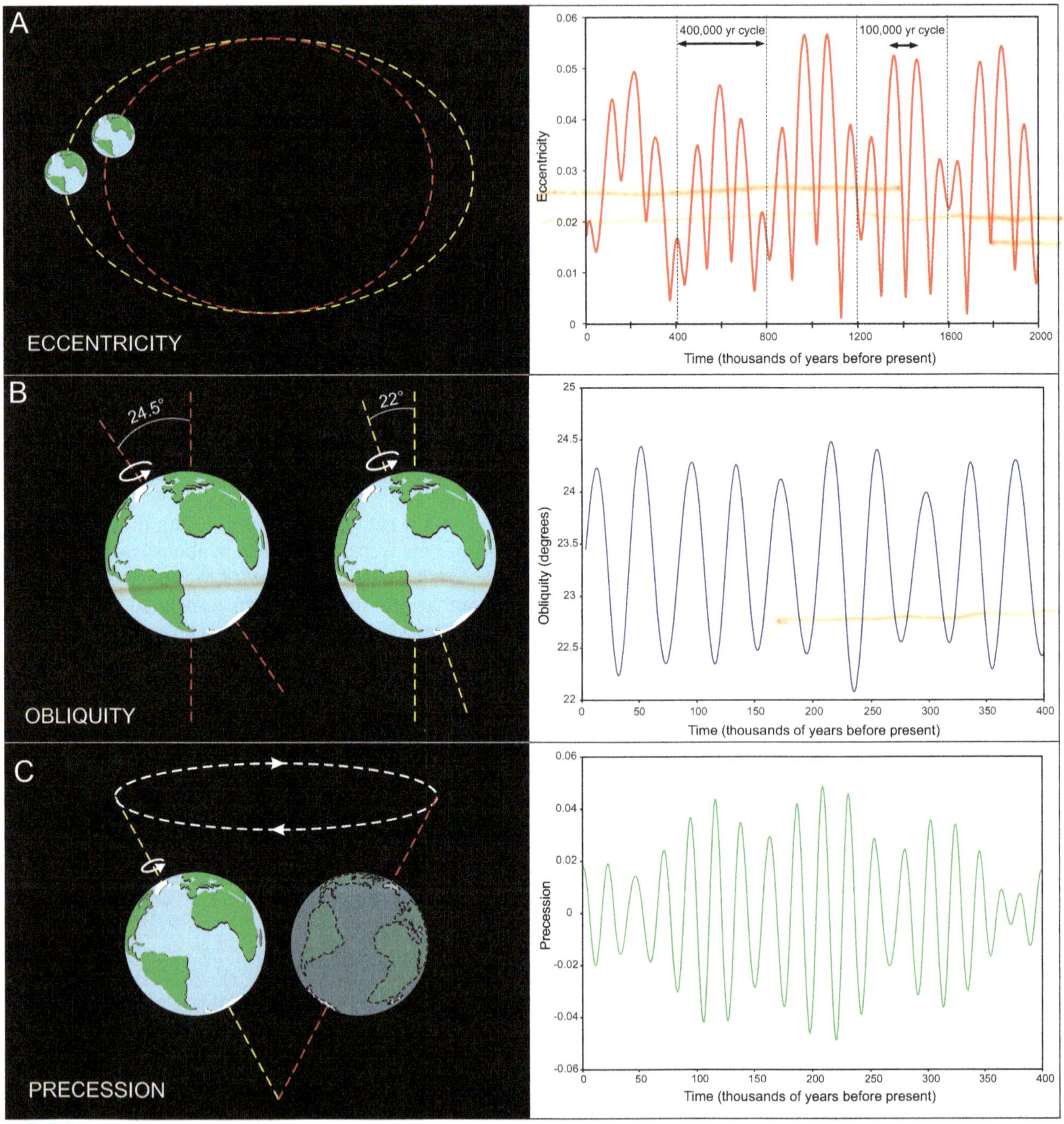

Figure 6.12: The Milankovitch Cycles visually and graphically depicted. A. Eccentricity describes the variation in the Earth's orbital path around the Sun. The path becomes more and less elliptical over time. There are two distinct cycles associated with eccentricity: A 100,000 year and a 400,000 year cycle. B. Obliquity is the change in the angle of the Earth's rotational axis relative to the plane of the ecliptic. Obliquity can be highly variable from one cycle to the next. C. Precession is the "wobble" in the Earth's axis. As you can see in the graph, the orientation of the Earth's rotational axis slowly changes over thousands of years. Data courtesy NOAA/NGDC Paleoclimatology Program, Boulder CO, USA.

called **Milankovitch cycles**, were studied in great detail in the 1920s and 1930s by the Serbian mathematician Milutin Milankovitch (Figure 6.12). There are three distinct Milankovitch cycles that account for the cyclical variations in the orbit of the Earth around the sun. These cycles influence climatic fluctuations over tens to hundreds of thousands of years.

Eccentricity describes the elliptical shape of the Earth's orbit around the sun. The Earth's path around the sun slowly changes from being elliptical to being nearly circular. The eccentricity of the Earth's orbit varies on cycles of 100,000 and 400,000 years (Figure 6.12A). Variation in solar radiation received at the top of the atmosphere is greater between the aphelion and perihelion when the Earth's orbit is more elliptical. The Earth receives approximately 20 to 30 percent more solar energy in January compared to July when the Earth's orbit is the most elliptical. A more elliptical orbit also contributes to the variation in the length of seasons by changing the length of time between equinoxes. We are currently in a period of low eccentricity, which results in a smaller difference in the receipt of solar radiation between July and January. Large variations in the shape of the Earth's orbit result in significant climate changes on a global scale.

Obliquity refers to the variation in the angle of the tilt of the Earth's axis. The axial tilt of the Earth varies cyclically between 22° and 24.5° from perpendicular on an approximate 41,000 year time scale (Figure 6.12B). The present tilt of the Earth is 23.5° from perpendicular, about halfway between extremes in obliquity. The severity of the seasons is influenced by obliquity since it determines how solar radiation is distributed throughout the year. A reduced axial tilt decreases the seasonal variation between winter and summer, resulting in milder winter and cooler summer temperatures. These conditions tend to promote the growth of glaciers and ice sheets in high latitudes due to more snow accumulation and less snow melt throughout the year. Conversely, polar latitudes have a greater amount of seasonality when the tilt is increased.

The third Milankovitch cycle, **precession**, is related to the slow wobble of the Earth's axis of rotation. The wobbling of the Earth's axis of rotation is similar to a spinning top that wobbles as it is starting to run down. The precession of the Earth shifts 180 degrees over a time scale of approximately 24,000 years (Figure 6.12C). The Earth's axis is currently pointing at Polaris (the North Star); it is slowly drifting and in about 12,000 years it point at the star Vega. When this happens the timing of the winter and summer seasons will be reversed. Additionally, the Northern Hemisphere will experience winter when the Earth is farthest from the sun and summer when it is the closest. This will make the seasons in the Northern Hemisphere more severe as a result of colder winters and warmer summers in the middle and high latitudes. Seasonal differences will be less severe in the Southern Hemisphere when the star Vega becomes the North Star.

The relative influence of the three Milankovitch cycles on the climate is not equal. It is also important to understand that the climate of the high latitudes is much more affected by these orbital cycles than tropical latitudes. Obliquity has the most significant impact on climate while eccentricity has the least. These cyclical variations do not change the amount of insolation for the Earth as a whole, but they do influence the latitudinal distribution of insolation throughout the year. Although a decrease in insolation is not the only trigger mechanism for glacial and interglacial cycles, it seems obvious that Ice Ages only occur when insolation at high latitudes is weak (Figure 6.13). The amount of insolation received in July at 65°N is generally considered the most reliable method to indicate the timing of the Milankovitch cycles, past Ice Ages correspond well to summer insolation at this latitude.

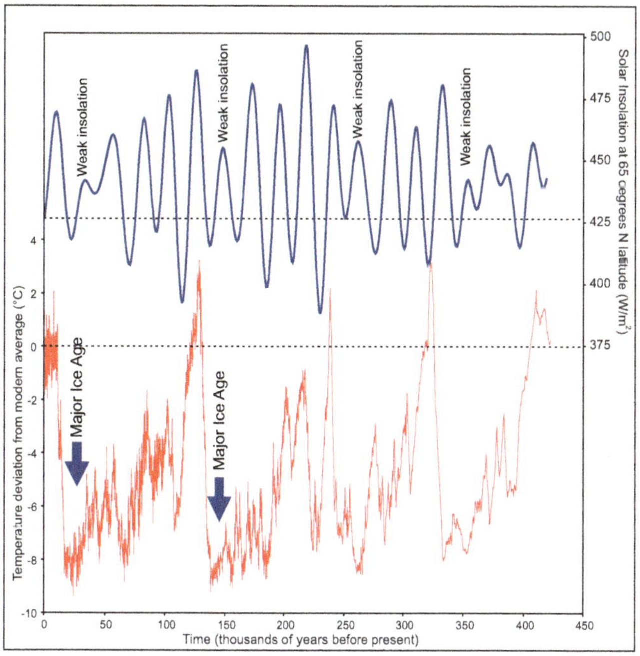

Figure 6.13: Variation in solar insolation at 65° N latitude for the past 400,000 years versus the temperature data from the Vostok ice core. Variations in insolation at the high latitudes is one of the key mechanisms for triggering ice ages. Data courtesy NOAA/NGDC Paleoclimatology Program.

Although Milankovitch cycles help explain climatic changes over long timescales, the exact connection between glacial and interglacial cycles and Earth-sun geometry is still not fully understood. For example, Milankovitch cycles can't be used to explain the extended periods of a less turbulent climate before the Ice Ages. The effects of precession should cause warmer temperatures in one hemisphere and cooler temperatures in the opposite hemisphere. However, the climate record indicates that climate changes have been similar between hemispheres, not opposite, during changes in the Earth's precession. Evidence also suggests that the positions of the continents are related to the Ice Ages, widespread glacial conditions can only exist when large portions of the continents are situated at higher latitudes. In addition to Milankovitch cycles, a host of other factors may be operating to influence the global temperature variations that control the growth and retreat of glaciers. Even though the Milankovitch cycles influence on climate fluctuations has been disputed in recent years, they do help explain the timing of long-term temperature changes.

Volcanic Activity

Figure 6.14: Mount Pinatubo ejected vast quantities of volcanic ash and sulfur dioxide when it erupted in 1991. Photo credit: USGS.

Explosive volcanic eruptions in the past have caused significant effects on the climate (Figure 6.14). Particulate matter and gases ejected into the stratosphere from an erupting volcano can reduce global temperatures by reflecting solar radiation (Figure 6.15). A notable cold spell linked to volcanic activity produced the "year without a summer" in 1816. This cool period followed the massive eruption of Mount Tambora in Indonesia the previous year. Frost occurred in every month of the year in New England with heavy snow reported across the Northeast United States in June. Many parts of Western Europe also reported unusually cold summer temperatures during this time. Unless the eruption is extremely large, fine dust and ash particles settle out of the stratosphere rather quickly and therefore only cause short-term effects on the climate. Large-scale climate impacts from volcanic eruptions depend primarily on the amount of sulfuric gases that are emitted. Sulfate

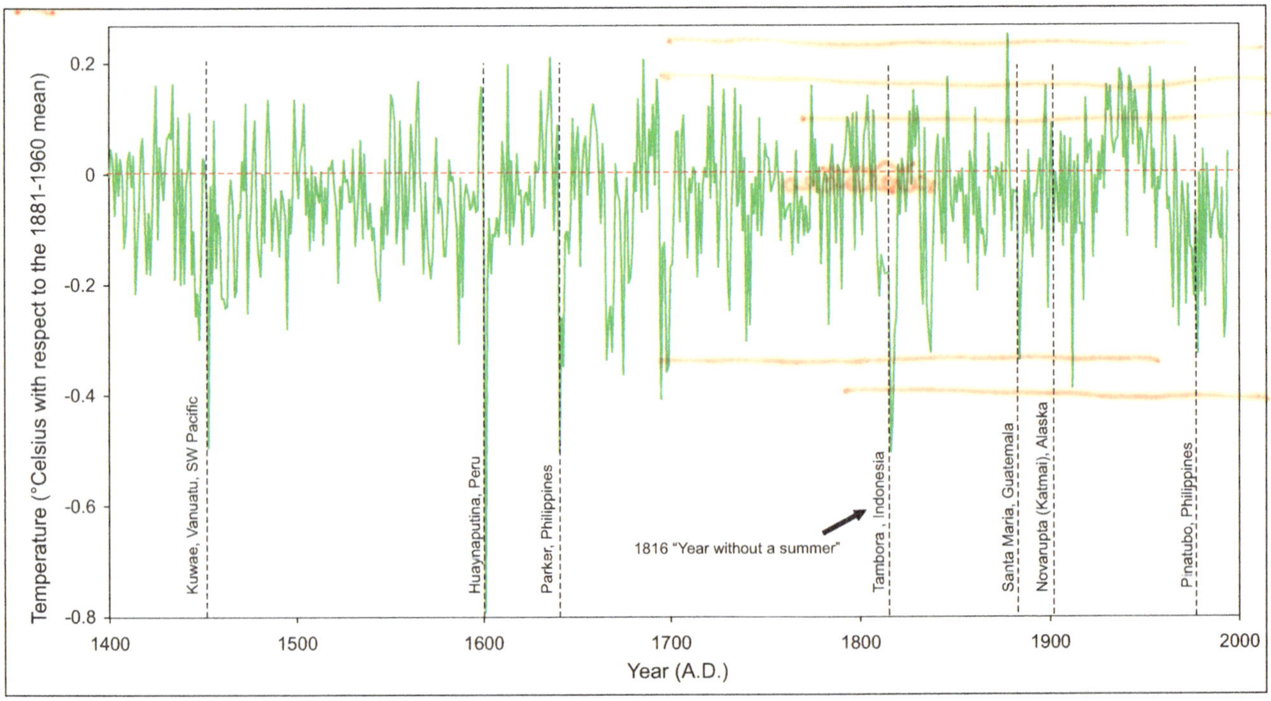

Figure 6.15: Northern Hemisphere growing season temperatures estimated from tree ring densities for 1400-1994 A.D. Some of the most explosive volcanos from the past 600 years are plotted to show how volcanic eruptions have had a significant impact on temperatures. Data from NOAA/NGDC Paleoclimatology Program.

particles are produced when sulfuric gases combine with moisture to produce sulfate aerosols. These aerosols are more effective than volcanic ash at reducing the amount of solar radiation available because they can remain suspended in the atmosphere for several years. The eruption of Mount Pinatubo in June of 1991 (Figure 6.14) ejected an estimated 30 million tons of material into the stratosphere, mostly in the form of sulfur dioxide (SO_2). Global temperatures were reduced the following year by 0.5°C (0.9°F) as the sulfate aerosols and volcanic ash increased the planetary albedo. Some have suggested that volcanism can also alter the atmospheric circulation from the addition of particulate matter and gases. Unusual weather events following a significant volcanic eruption may occur as a result of the disrupted circulation pattern. It is possible that the heavy rainfall that preceded the Great Flood of 1993 may have been in response to the 1991 Mount Pinatubo eruption. Meteorite impacts, which scatter debris from the explosion high into the atmosphere, also cause similar short-term climate impacts (Figure 6.16). In theory, a significant cooling event could occur if a high frequency of massive volcanic eruptions occurs over an extended time period. The addition of millions of tons of gases and volcanic ash to the stratosphere would greatly diminish the amount of solar radiation that would reach the surface, greatly reducing global temperatures.

Figure 6.16: The mile-wide Meteor Crater in Arizona formed in a fraction of a second, throwing huge amounts of debris into the atmosphere. For a sense of scale, if this hit Kansas City, MO, the blast radius would take out the entire city. Photo credit: NASA.

Other Natural Causes of Climate Change

In addition to the natural causes of climate change discussed above, there are other natural mechanisms that can produce gradual climate changes over extremely long time periods. Continental position related to plate tectonics has a large influence on the Earth's climate; continents cool more rapidly when positioned at higher latitudes. Ocean circulation is also affected by the position of the continents which influences the global distribution of heat transfer. Periods of mountain building have acted as mechanisms of global climate change throughout Earth's history. Mountains produce ideal conditions for glaciers to form by providing cooler locations for them to develop. Mountain building over large areas can decrease regional temperatures due to a natural decrease in temperature with increasing elevation. Large mountain ranges can modify atmospheric circulation and influence the geographical distribution of clouds and precipitation (Figure 6.17).

Figure 6.17: We know from paleoclimate records that the Gobi desert had a much wetter climate before the rise of the Himalayas. Image credit: NASA/EO.

Climate provides the heat and moisture needed for vegetation to thrive; it is the primary factor that controls the global distribution of natural vegetation patterns. Temperature and precipitation influence the type of trees and other vegetation found at a particular location. Climate induced vegetation change modifies the physical characteristics of the land surface. For example, it is evident that climate change over the last 18,000 years has resulted in continuous changes in natural vegetation across Eastern North America. The Earth's energy balance is affected when interactions between the atmosphere and the land surface are altered; changes in regional vegetation patterns influence the energy balance by modifying the albedo and water availability of the land surface. Natural vegetation patterns will continue to adjust and adapt to changing climatic conditions.

Earth's climate has changed on various time scales, including long before human activity could have played a role. There have been at least five major Ice Ages over the last two and a half billion years of Earth's history. These lengthy glacial periods have been separated by shorter and warmer interglacial periods. The modern climate is in a warmer interglacial period that began about 12,000 years ago. This is an unusual time in geologic history since it has been more common for large portions of the continents to be covered by huge ice sheets or glaciers. It is possible that the glacial and interglacial cycles will continue and the Earth will experience another period of extensive glaciation.

Section 6-5: Human Impacts on Global Climate

A reconstruction of the Earth's past climate using modern instrumental records and various proxy records indicates that the Earth's climate has varied naturally over a wide range of time scales. Changes in the Earth's climate before the Industrial Revolution in the late 18th Century were almost entirely due to natural causes that are not related to human activity. The natural factors that cause climate change do not match the observed trend in temperatures since 1900. Although these natural causes are still operating today, recent changes in climate since the Industrial Revolution cannot be explained solely by natural processes. For example, the burning of fossil fuels due to human activity can also cause significant climate changes (Figure 6.18). Human influences on the climate have increased substantially since the Industrial Revolution, especially through the addition of heat-trapping greenhouse gases to the atmosphere. Changes in land cover and land use also have significant impacts on the climate. The soil moisture characteristics and reflectivity of the Earth's surface are altered whenever the landscape is transformed. The significance of the Earth's ability to absorb or reflect insolation in producing climate change has been observed in the past through natural changes in reflectivity. Although natural changes in reflectivity have contributed to climate change in the past, humans are modifying the landscape at an alarming rate. Human activity also affects the atmosphere's ability to absorb or reflect solar radiation by contributing to the amount of aerosols in the atmosphere. Atmospheric aerosols are considered a significant factor related to climate change on a local or regional scale. They modify the climate by scattering and absorbing radiation as well as influencing the formation of clouds and precipitation. It is possible that the local and regional effects of aerosols on the climate may accrue to produce effects globally as the concentrations of atmospheric aerosols continue to increase. The various ways that humans influence global climate change needs to be fully understood in order to quantify the degree to which humans are influencing the climate. Human's impact on global climate change has been a focal point of climate research over the last 30 years due to its significance.

Figure 6.18: Smog and air pollution hover over Las Vegas, NV. Photo credit: B. Zaprowski.

Changes in Radiation-Absorbing Gases

The addition of greenhouse gases to the atmosphere from the burning of fossil fuels is the most significant human activity affecting global climate change. The addition of greenhouse gases can cause an increase in global surface temperatures by increasing the Earth's natural greenhouse effect. Humans have been adding huge quantities of radiation-absorbing gases to the atmosphere since the Industrial Revolution. Carbon dioxide represents the vast majority of greenhouse gas emissions from fossil fuel combustion, atmospheric levels of carbon dioxide have increased (from about 285 to over 400 ppm) since ~1850 (Figure 6.19). While the level of atmospheric carbon dioxide seems insignificant, it has a large influence on

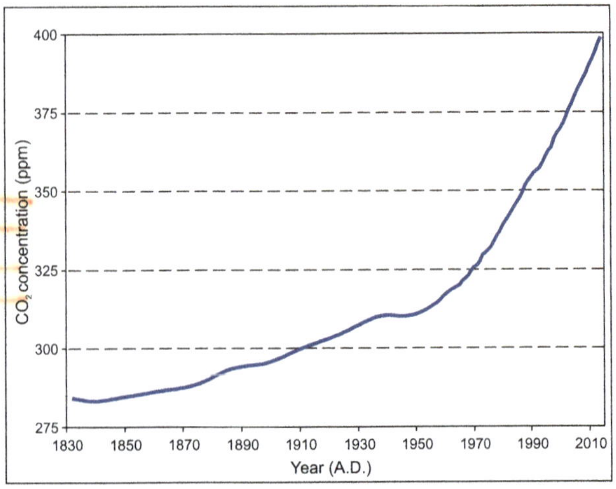

Figure 6.19: The change atmospheric CO_2 concentration between 1832-2014. The 1832-1958 data is from D.M. Etheridge, L.P. Steele, R.L. Langenfelds, R.J. Francey, J.-M. Barnola and V.I. Morgan, 1998. 1959-2014 data from NOAA/NCDC.

regulating the Earth's temperature. Although active volcanoes continue to emit carbon dioxide to the atmosphere, their contribution is minor compared to human emissions. Since fossil fuel emissions are the only significant source of atmospheric carbon dioxide that is primarily non-natural, the observed increases are often seen as a "human fingerprint" in producing climate change. A strong relationship has been established between trends of temperature and carbon dioxide in the recent geological past (Figure 6.5B). Atmospheric levels of carbon dioxide are currently about 40% higher than they were in 1750 and are projected to continue to increase. The increased use of alternative fuels and renewable energy sources is essential to reduce the concentration of carbon dioxide in the atmosphere.

Human activity has also increased the concentrations of atmospheric carbon dioxide through deforestation, the widespread clearing of forested landscapes (Figure 6.20). Forests act as a carbon sink by removing carbon dioxide from the atmosphere through photosynthesis. The destruction of forests also contributes to the buildup of

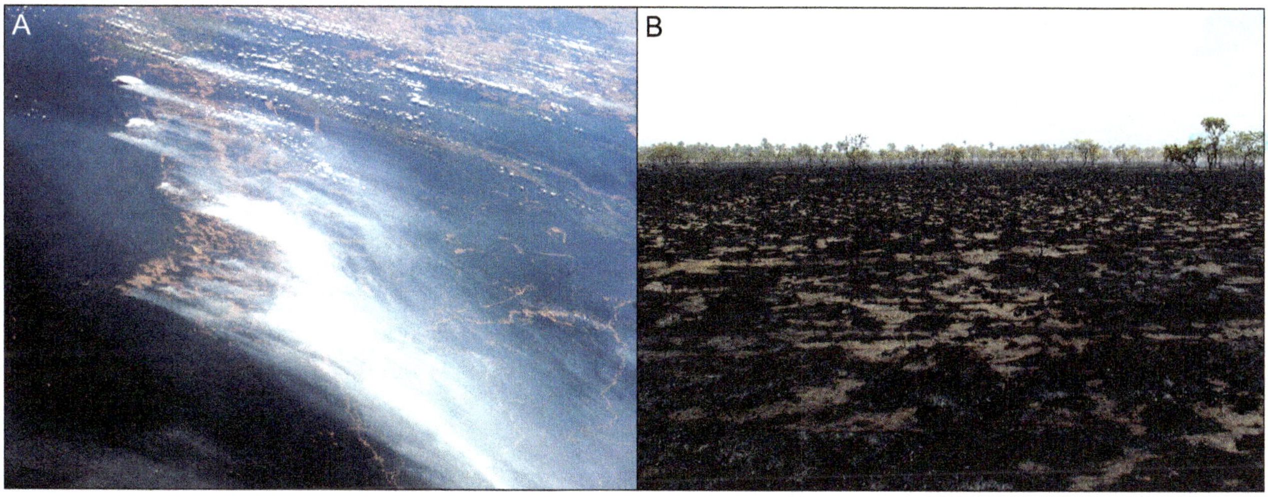

Figure 6.20: A. Deforestation in Amazonia. Image credit: NASA. B. The aftermath of a burning event in Amazonia. Photo credit: Doug Morton/NASA.

atmospheric carbon dioxide through the direct burning and decay of vegetation (Figure 6.20B). Widespread deforestation in the tropics is occurring as large tracts of land are converted for other uses. For example, the expansion of cattle ranching is one of the leading causes of deforestation in the Brazilian Amazon and parts of Central America. Tropical forests need to be protected in order to reduce the increasing trends of atmospheric carbon dioxide. It seems logical that increases in atmospheric carbon dioxide could influence the Earth's temperature in the lower atmosphere.

There are several other radiation-absorbing gases besides carbon dioxide that are added to the atmosphere through human activity. These gases are called trace gases due to their small concentration compared to carbon dioxide. These gases have the potential to significantly enhance the greenhouse effect even though they represent a small fraction of the total atmosphere by volume. Methane, the second most important greenhouse gas, is produced through both natural and human activities. The most significant source of methane comes naturally from the decomposition of organic matter in wetlands. The flooding of rice fields for rice cultivation (Figure 6.21) and the digestion processes of animals, especially livestock, emit methane to the atmosphere. Other sources of atmospheric methane include biomass burning, landfills and fossil fuel extraction and combustion. Methane was initially assumed to play only a minor role in the greenhouse effect due to its low abundance. However, methane is a stronger greenhouse gas than carbon dioxide because it absorbs terrestrial

Figure 6.21: Rice paddies like this one contribute methane to the atmosphere. Photo credit: Lisamarie Windham-Myers/USGS.

Pollutant	Major Sources	Potential Effect
Carbon dioxide	Burning coal, oil, gas, biomatter	Global warming, faster plant growth
Carbon monoxide	Burning coal, oil, gas, biomatter	Global warming, health risks
Chlorofluorocarbons (CFCs)	Refrigerants, polystyrene, solvents	Loss of stratospheric ozone, global warming
Carbon tetrachloride	Solvents	Loss of stratospheric ozone, health risks
Methane	Agriculture, anaerobic bacteria	Loss of stratospheric ozone, global warming
Methyl chloroform	Solvents	Loss of stratospheric ozone
Nitrous oxide	Agriculture	Loss of stratospheric ozone, global warming
Nitrogen dioxide	Burning coal, oil, gas, biomatter	Acid rain, global cooling
Sulphur dioxide	Burning coal, oil	Acid rain, global warming
Ozone (surface level)	Burning coal, oil, gas, biomatter	Global warming, health risks

Table 6-1: A list of some common anthropogenic emissions to the atmosphere, their potential sources, and the potential risks each emission incurs. Modified from Martin and Nunnally, 2003.

radiation much more effectively. Human activities have increased methane concentrations significantly since the Industrial Revolution. Methane and trace gases, although important, are thought to be much less significant in potentially affecting the climate when compared to carbon dioxide. Table 6-1 lists several of these gases, their sources, and their potential or suspected effects. The Earth's natural greenhouse effect will continue to be enhanced as human activities add greenhouse gases to the atmosphere. Computer models are predicting a continued increase in global average temperatures over the next several decades as atmospheric carbon dioxide levels are projected to rapidly increase.

Changes in Aerosol Concentrations

Atmospheric aerosols can alter the climate through their ability to absorb and reflect solar radiation. Although human activities add a wide variety of aerosols to the atmosphere, most aerosols are naturally produced. Aerosols that contain considerable amounts of black carbon warm the atmosphere by absorbing solar radiation. These soot particles are created and accumulate in the atmosphere through human activities such as industrial emissions, automobile exhaust and biomass burning. As mentioned earlier, volcanic eruptions are effective in reducing temperatures by reflecting incoming solar energy back to space. Fine light-colored particles, particularly sulphate aerosols, are effective in reflecting incoming solar energy back to space. Daily high temperatures can be reduced in areas where considerable sulphate concentrations result in a highly reflective layer of smog in the lower atmosphere (Figure 6.22). The daily temperature range is decreasing slightly in areas with high concentrations of sulphates. Aerosols can also increase cloud cover by acting as condensation nuclei, which provide opportunities for the condensation of water vapor. The increased cloudiness, especially high clouds, decreases insolation by intercepting and reflecting sunlight.

Figure 6.22: A layer of highly reflective smog covers Salt Lake City, UT. Photo credit: B. Reed.

Human activity is clearly increasing the amount of atmospheric aerosols, the concentration of aerosols is roughly three times higher in the Northern Hemisphere where about 90 percent of the human population lives. Increased aerosol concentrations are especially evident in active industrial regions like the Eastern United States, Europe, the Middle East and East Asia. It is important to understand that most of the direct effects of increased aerosol concentrations are local or regional. Most of the aerosols stay in close proximity to the source of emissions and remain in the lower atmosphere for a relatively short time. A decreasing trend in in insolation has been observed over the past few decades, especially in densely populated areas. It has been speculated that some of the late 20th Century warming due to the addition of radiation-absorbing gases may have been masked due to a net loss of solar radiation from the addition of atmospheric aerosols.

Land Use Changes

Land use and land cover changes cause significant climate impacts by affecting the Earth's radiation balance. Modifying the land surface also affects evapotranspiration rates by changing soil moisture characteristics. Humans have transformed the land surface in a variety of ways through processes such as deforestation, desertification, urbanization, irrigation, and damming rivers. These processes can have both warming and cooling effects on the climate by influencing the amount of insolation that is absorbed or reflected. Although human activities have been modifying the landscape for thousands of years, the current rates, scale and intensity of land use change are unprecedented. Land use and land cover changes affect the climate on scales ranging from the macroscale to the microscale. Deforestation leads to locally higher surface temperatures as the loss of vegetation reduces the amount of evapotranspiration from the surface (Figure 6.20). Continued deforestation is projected to lead to significant warming trends and shift thunderstorm activity in tropical and equatorial regions. Desertification, the expansion of deserts into semiarid regions through human activity, can influence climate change on a regional scale. The overgrazing of cattle and overuse of water through ranching activities has resulted in widespread desertification in many semiarid regions of the world (Figure 6.23A). Desertification results in increased temperatures and reduced water availability in regions where large areas of land are being converted to a desert landscape. In places like China,

Figure 6.23: A. A Food Security Assessment in Somalia found severe impacts on livestock due to drought conditions. Photo credit: USGS. B. A dust storm covers China. From A.D. 300 to 1949, northwestern China experienced an average of one dust storm every 31 years. After 1990, the rate increased to a dust storm almost every year. In 2005, scientists warned (in the journal Nature) that China's dust storm season was getting worse for a variety of reasons, including desertification and human-exacerbated drought. Image credit: NASA.

desertification and overgrazing have led to a sharp increase in dust storm activity (Figure 6.23B). As discussed in Chapter 3, urbanization and the urban heat island effect can cause cities to be significantly warmer than the surrounding rural areas, especially at night. The intensity and spatial scale of the urban heat island is influenced by several factors, the highest temperatures are typically located within the urban core. Land use and land cover changes play a significant role in the observed warming trend in global average surface temperatures. All of these processes need to be included in climate models in order to more accurately project long-term global climate patterns.

Section 6-5: Why Do We Care? Why Climate Change Matters

Climate change is currently a focal point across the scientific community and is of great interest to the general public as well. Reconstructions of past climate conditions indicate that the climate has varied significantly over a wide range of time scales. This suggests that the future climate conditions are more likely to differ from the present than to stay the same. Climate change projections are not expected to be uniform across the Earth. For example, some locations are expected to become warmer with increased precipitation while cooler and drier conditions are projected for other regions. Climatologists use climate reconstructions from warmer periods in the past to predict how local climate patterns could possibly change in the future if global temperatures continue to rise. For example, Figure 6.24 shows a reconstruction of climate conditions across Europe 6,000 years ago when global average temperatures were ~1°C warmer than present (Figure 6.1). Even through the average global temperature was warmer at this time some areas' winter temperatures were actually colder than present. Decreases and increases in precipitation were also observed during this warmer period. Global climate change does not always produce a similar change in local climate patterns, climatologists must consider several local factors when trying to determine how an area's climate may change in the future. Despite our best efforts to understand future climate change, the potential for "climate surprises" exists. We might experience abrupt, unexpected shifts in the Earth's climate system due to its complexity.

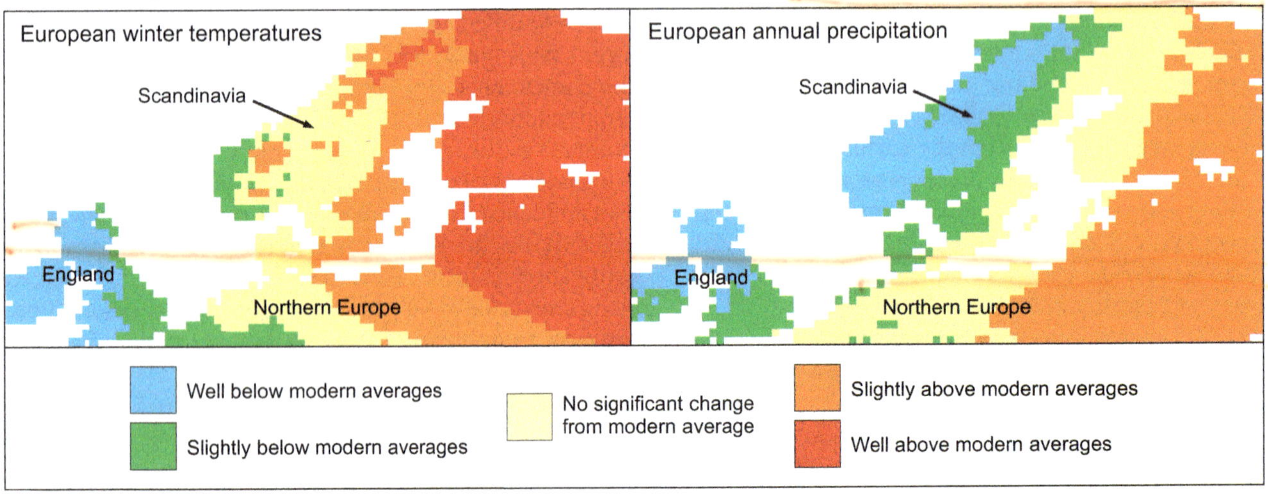

Figure 6.24: A look at temperature and precipitation values 6,000 years ago compared to the present. As can be seen from the data, western Europe winters were cooler and precipitation was lower while eastern European winters were warmer and precipitation was higher. If the Earth's conditions were to change due to global warming, not all locations on Earth would see an increase in temperatures or a decrease in precipitation. Rather, patterns of temperature and precipitation would change, like in the figures above. Data from Cheddadi, R., et al, 1997, European Climate 6K Reconstruction. IGBP PAGES/World Data Center-A for Paleoclimatology Data Contribution Series # 97-006. NOAA/NGDC Paleoclimatology Program.

Are the observed temperature changes since the late 1800s due to human activities or part of a natural cycle? Few climate scientists today question if there has been a warming trend over the last 120 years. However, the debate over the cause and the significance of the increased temperatures still persists. Increasing greenhouse gas concentrations are widely considered the primary cause for the recent warming trend. Evidence from proxy records and instrumental data indicates a strong relationship between trends of temperature and carbon dioxide in the recent geological past. An enhanced greenhouse effect combined with the addition of aerosols and land use changes all point towards a recent warming trend that is closely tied to human activities. These concerns have led to an increased effort to phase out fossil fuels and shift towards renewable sources of energy. The consequences of climate change have the potential to directly affect human civilization and the environment.

There are some climate scientists who believe that the recent warming trend is primarily due to natural variability in the climate and debate its significance. They contend that there is too much uncertainty in our current understanding of the numerous interacting processes that influence the climate. Many of these processes operate simultaneously on different time scales. For example, increased temperatures should result in increased cloud cover, which should offset some of the projected warming trends by reducing the amount of insolation that reaches the Earth's surface. Some climate scientists note that the natural changes in solar output may cause more significant climate change than all of the human impacts on the global climate combined. Issues related to the reliability of climate data and the methods of constructing long-term temperature records have some climate scientists questioning whether the abrupt warming trend is realistic. There are concerns that the quality and selection of climate data used for global temperature reconstructions may have amplified the observed warming trend. There are some climate scientists who believe that higher temperatures are actually beneficial and do not view the current warming trend as a serious problem. The history of the Earth's climate change indicates that civilization began during a warm period. Human health, food production and wildlife have been observed to flourish during warmer times. Agriculture may benefit

Figure 6.25: A. Rising sea levels put low-lying coastal areas at risk for more frequent and more severe storm surge flood events. Photo credit: Charles Powell/ FEMA. B. This image shows the projected changes to the Blackwater National Refuge in Maryland by 2100 at current rates of sea-level rise. Image credits: USGS.

from warmer temperatures and a greater concentration of atmospheric carbon dioxide. The projected temperature increase may enhance recreation and tourism and cause a reduction in traffic accidents. Society may benefit more from an increase in global temperatures over the next 100 years as compared to a decrease.

Due to its complexity there is no simple answer to explain why the climate varies. The spatial and temporal scale must be considered when answering any question about future climate change. There are currently too many unresolved questions to determine the relative contributions of human activities and natural causes of the recent warming trend in global temperatures. Continued research is necessary to help provide answers to these questions which can minimize the potential environmental and human health impacts related to the projected increases in global temperatures. Even though there is uncertainty over the cause and potential impacts of the recent increase in global temperatures, reducing greenhouse gas emissions and pollutants has positive benefits. The potential exists for global climate change to be greater in magnitude and occur at a faster rate than during any time in the history of civilization. Considering that roughly 40 percent of the world's population lives within 100 km (62 mi.) of the coast, sea-level rise is considered to be the most serious consequence of continued rising temperatures (Figure 6.25). Some of the most significant potential consequences of climate change are listed in Table 6-2.

Potential Consequences of Global Climate Change
Sea-level rise
Increased number of heat waves
More intense precipitation events
Increased intensity of tropical cyclones
Decreased water supply
Human health impacts
Plant and animal extinction
Increased risk of wildfires
Disruption to agriculture
Coral bleaching
Ocean acidification

Table 6-2: A list of some potential consequences of global climate change.

Climate change has occurred in the past and will continue in the future. The future of the Earth's climate will remain a mystery until the accuracy of global climate models is greatly improved. Climate models currently predict broad regions of expected positive and negative changes in temperature and precipitation. These models are in need of considerable refinement, their accuracy can be improved with a denser weather observation network. A better understanding of the interaction of all the processes that influence climate variability also needs to be included in climate models. It is imperative that quality and unbiased climate data are used for reconstructions and projections of climate change, especially if they are used to drive changes in policy. The impacts of global climate change will not be uniform due to the magnitude of the changes and the ability of regions to cope with them. Despite not knowing the specifics of the Earth's future climate, climate projections can be utilized to develop strategies to help deal with the potential threats of global climate change.

NASA

NASA

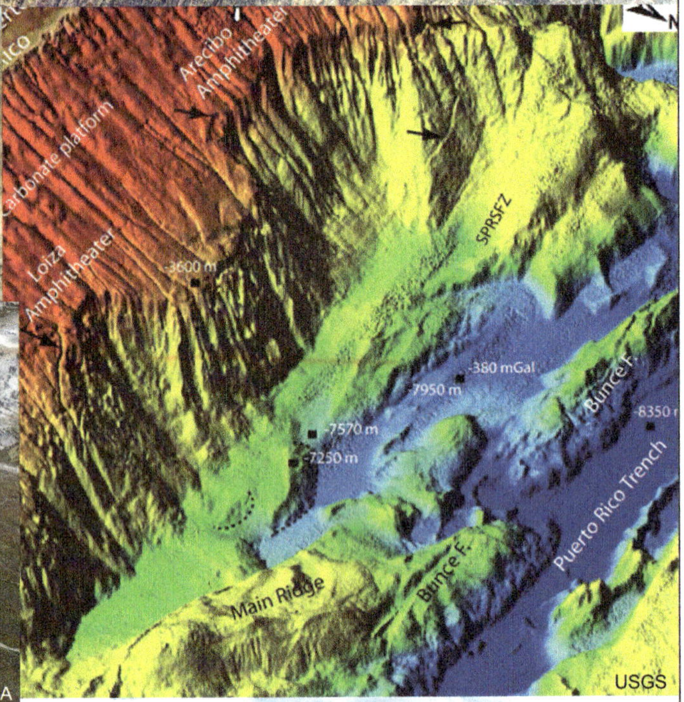
Arecibo
Amphitheater
Carbonate platform
Loiza
Amphitheater
-3600 m
SPRSFZ
-380 mGal
-7950 m
-7570 m
-7250 m
Bunce F.
Puerto Rico Trench
Main Ridge
Bunce F.
USGS

Jesse Allen/NASA

Chapter 7 - Plate Tectonics

Figure 7.1: The Crab Nebula is a remnant of a supernova and is a good analog for the formation of the solar system. Image credit: NASA, ESA, J. Hester and A. Loll (Arizona State University).

At the largest scale, the physical character of the Earth's surface is a reflection of the whole Earth process called plate tectonics. The existence of both the land masses and the ocean basins are a direct result of plate tectonics. The location of mountain chains and the general topography of the land surface is controlled by plate tectonics. Most of the features found on the seafloor develop and evolve over time as a direct result of tectonic processes. Hazardous events like earthquakes and volcanism are also controlled by plate tectonics (more on that in Chapters 8 and 9). In this chapter we will explore how the Earth formed, how this eventually led to the physical processes we call plate tectonics and how plate tectonics shapes our world.

Section 7-1: The Earth Forms

The Earth and the other planets in the solar system were formed by ~4.6 billion years ago. Although there are no known rocks on the Earth from this period of time, various meteorites that have fallen to earth have been dated and no material older than 4.6 billion years old has yet been discovered. From this observation, scientists have inferred the age of the Solar System and the Earth. We will discuss how scientists date rocky material further in Chapter 11.

The leading hypothesis for the formation of the solar system is the nebula hypothesis, which was first proposed in the 1700's. The nebula hypothesis states that approximately 5 billion years ago the solar system began as a large mass of dust and gas, perhaps occupying billions of cubic meters of space. This large cloud of material is believed to have been composed of the remains of a massive, short-lived star that died and scattered its debris across space during a huge, cataclysmic explosion called a supernova (Figure 7.1). The principle element in these clouds would have been hydrogen, but carbon, oxygen, silicon, magnesium, aluminum and iron, the common building blocks of rocks, were also abundant.

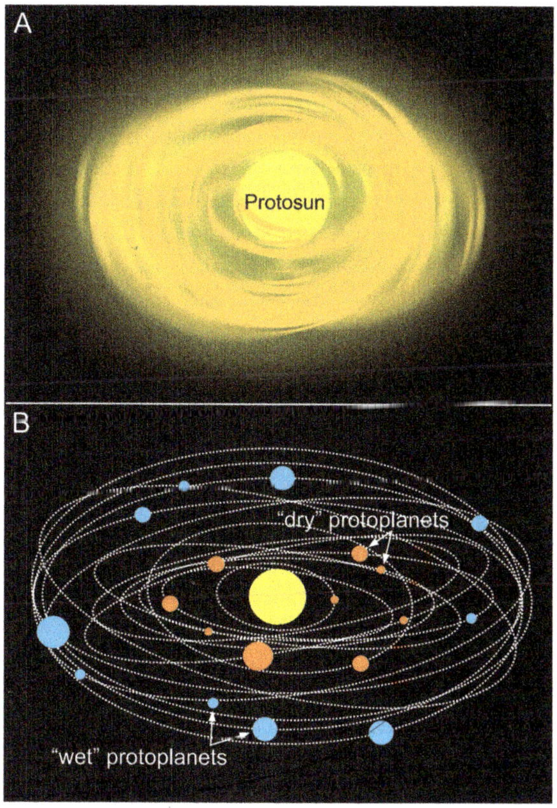

Figure 7.2: The formation of the early solar system. A. When the cloud of dust and gas from the novae began to coalesce and rotate, a large disk formed with a newly-forming protosun in the middle. B. Eventually "wet" and "dry" protoplanets developed, many of which coalesced to form the eight main planets.

This massive cloud of debris began to rotate and contract into a disk as gravitational forces pulled the material towards its gravitational center. Approximately 90% of the matter accumulated in the central part of the cloud to form a protosun (Figure 7.2). Outside of the core, clumps of material began to form protoplanets. At first, there were probably millions of smaller protoplanets orbiting the sun. Those protoplanets which were closer to the sun were hot, dry and rocky while those further from the sun were cold, wet and icy.

As the protoplanets began to move around the protosun and gravitationally interact with each other, the larger protoplanets accreted (added on) more mass by colliding with and "consuming" the smaller protoplanets. Over time, the largest of these protoplanets grew into the eight main planets. The terrestrial planets, being close to the sun, lost most of their gases as the solar wind drove them off. The outer planets, being farther from the sun, managed to hold on to their extensive gaseous atmospheres. It is believed that the planets formed in about 10 million years or so. The remnant material left over from the accretion of the planets formed the abundant numbers of asteroids, meteoroids and comets found throughout the solar system (more on this in Chapter 16). This is not the end of Earth's story, however, as the Earth had not yet reached its current size. Prior to 4.5 billion years ago, Earth was smaller than it is now, and it probably lacked any significant water. But things were about to take a radical turn for the better.

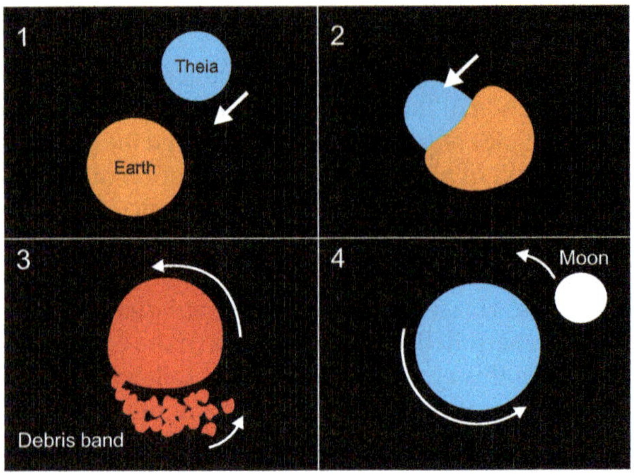

Figure 7.3: The giant impact hypothesis states that Earth was struck by Theia, throwing a debris band into space. This band then coalesced into the moon.

Scientists believed that an icy, Mars-sized protoplanet named **Theia** was in orbit around the sun, near the Earth's orbital plane. The **giant impact hypothesis** proposes that at around 4.5 billion years ago, Theia and Earth collided and this event made a lasting impression on the Earth. Because the Earth was the larger of the two planets, most of Theia's iron core was incorporated into the Earth. Some of the lighter materials were thrown into space where the debris quickly coalesced into the Moon (Figure 7.3).

There are several lines of evidence to support the giant impact hypothesis. First, the oldest Moon rocks date to ~4.5 billion years old. These are the rocks which were brought back to Earth by astronauts during the Apollo missions. Second, the isotopic signatures of the rocks on the Earth and rocks of the Moon are virtually identical. It is virtually impossible to imagine a scenario where the moon formed separately from the Earth and took up orbit around the Earth at a later time. Third, the moon lacks a significant iron core. This also suggests a lunar origin which must be intertwined with Earth's history.

It is now believed that the additional mass the Earth gained from its collision with Theia is what allowed the Earth to retain its atmosphere, one of the key components of life. Had the earth never collided with Theia, it may not have had the gravitational force necessary to keep the solar wind from driving away our (relatively speaking) thin and tenuous atmosphere. Many scientists believe that Theia may have been the source of most of the Earth's water, although comets and asteroids also contributed water during subsequent impact events. The earliest evidence of liquid water on the Earth's surface is from ~4.4 billion years old minerals in Australia, minerals than can only form in the presence of liquid water. The first clear evidence of the existence of widespread oceans is from rocks that date to ~3.8 billion years old. So although the early Earth would have been very hot and molten right after the Theia impact, it appears the Earth cooled quickly and water vapor began to condense into liquid water.

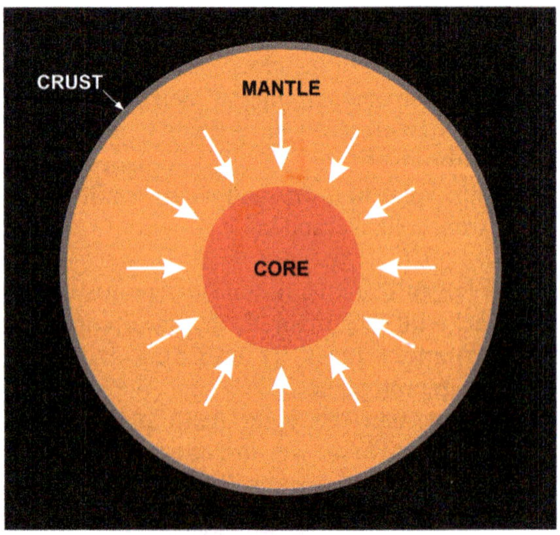

Figure 7.4: The differentiation of the early Earth into layers.

Section 7-2: The Earth's Interior

This was not the last of the collisions: Intense meteoric bombardment continued until ~4 billion years ago, adding more mass to the Earth. Although there is no visible evidence of this on the Earth today, the Moon's heavily cratered surface records this active period of time. This continuous bombardment, along with the heat produced by radioactive decay, would have heated the Earth's interior, leading to the **differentiation** of the Earth into layers based on density (Figure 7.4). The densest materials (iron and nickel) sank to the center of the Earth, forming the Earth's core. This core, being partially molten, led to the formation of the Earth's magnetic field. Our magnetic field protects us from most of the sun's harmful radiation, a key step towards the appearance of life on Earth. The presence of a molten core is critical in regulating the Earth's temperature. If the Earth's interior was cold, the temperature at the Earth's surface would likely be too cold for life to exist. Partial melting

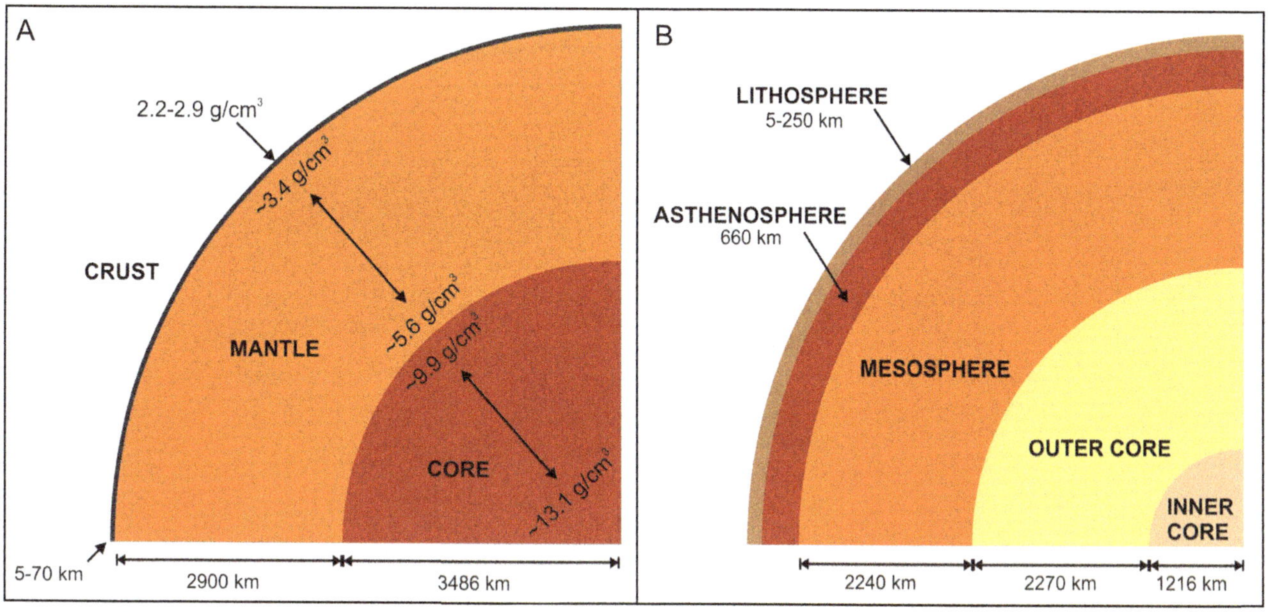

Figure 7.5: A. The interior structure of the Earth based on chemistry (density). Our knowledge of the Earth's interior is based mostly on seismic evidence. B. The interior structure of the Earth based on rheology. The physical nature of the Earth's layers helps to explain plate tectonics.

of the Earth's interior also led to the release of gases, particularly water vapor, carbon dioxide, carbon monoxide, hydrogen sulfide, and hydrogen chloride. However, free oxygen did not begin to accumulate in the atmosphere until ~2.5 billion years ago as a result of photosynthesis (More on that in Chapter 12).

Initially, the Earth's exterior would have been a thin, unstable layer of iron-magnesium rich rocks but over time, volcanism produced the first continental rocks. The earliest known continental rocks are ~4.04 billion years old while the earliest evidence of a land surface is from a fossilized soil layer found in Australia that dates to ~3.46 billion years ago. The Earth is still an active planet. The high temperatures and pressures in the Earth's interior keeps the interior mobile. This mobility has consequences for humans as you will see later in the chapter. There is no single model of the Earth's interior that can explain all of the phenomena associated with Earth's interior. Instead, two models are used to describe the structure of the earth's interior: a **chemical** model and a **rheological** model.

The Chemical Model

The chemical model emphasizes the structural properties of the Earth's interior in terms of rock type, which is controlled by chemical variations within the crust (Figure 7.5A). In this sense, *rocks are stratified based on density* which, in turn, is controlled by mineral composition. The **core** is composed chiefly of iron and nickel. Above the core is a thick layer of rocks called the **mantle**. The mantle is composed of rocks dominated by iron-rich minerals, making the mantle less dense than the core, but more dense than the thin crust above. During differentiation, the lowest density minerals worked their way towards that surface, cooled and formed a thin brittle **crust** composed predominantly of silica-rich minerals. It should be pointed out that the composition of the crust itself is highly variable from place to place: Continental crust (found under the continents) is composed of silica-rich (silicate) minerals while oceanic crust (found under the ocean basins) has significant amounts of iron-rich silicate minerals, making it slightly more dense than continental crust. This model of the Earth's interior, although useful in explaining why distinct continents and ocean basins exist, is not sufficient in explaining how plate tectonics operates. To explain phenomena like volcanos and earthquakes, we need to look at the **rheology** of the Earth's interior.

The Rheological Model

Rheology is simply the study of the deformation and flow of material, in this case, rocks. The rheological model divides the Earth's interior up based on *the physical properties of the rocks*. For example, the core can be divided into an inner core and an outer core (Figure 7.5B). The **inner core** is a solid ball of iron and nickel while the **outer core** is liquid layer of iron and nickel. The inner core, being in the center of the Earth, is under the greatest pressure so

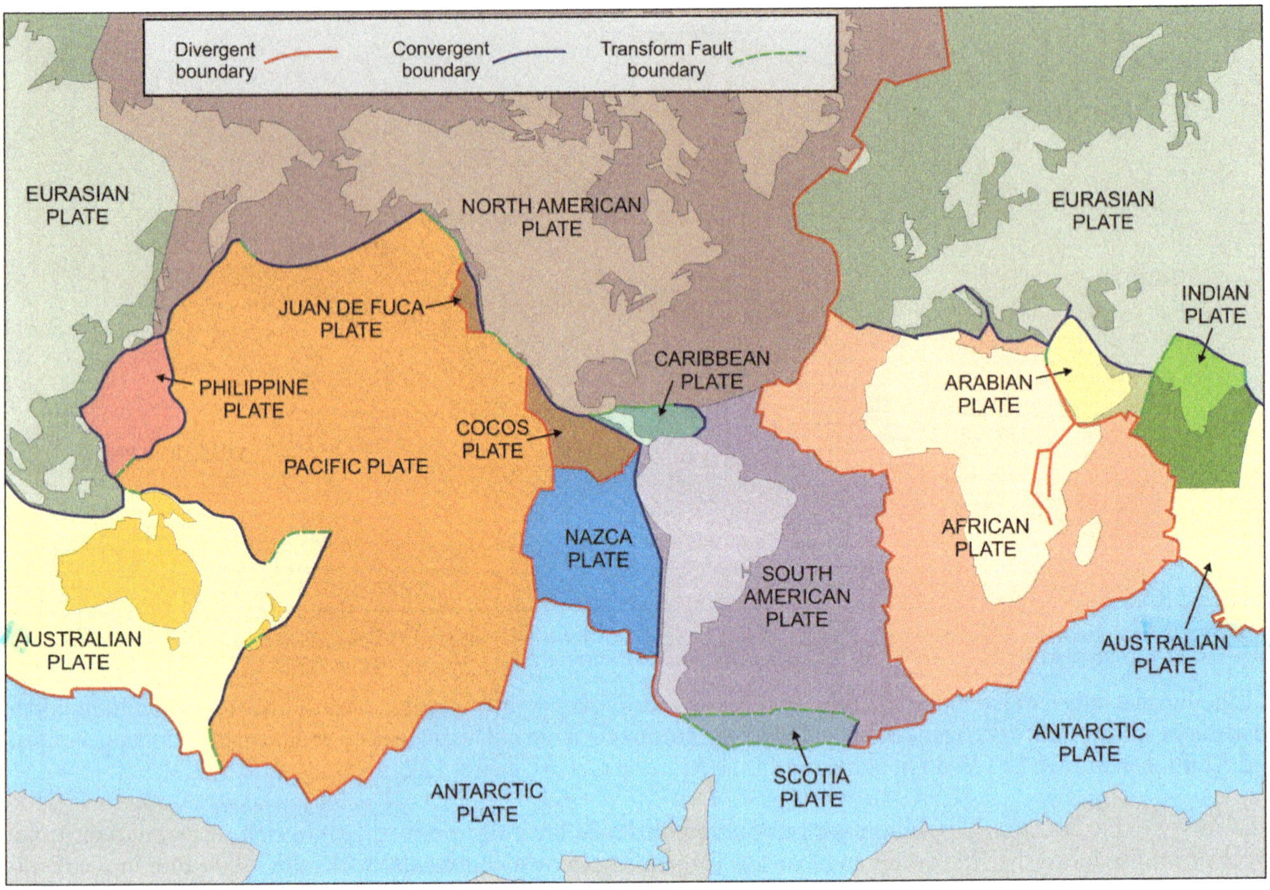

Figure 7.6: A map of the world's tectonic plates. Many plates contain both continental crust and oceanic crust.

molecules are forced into a solid state, while the outer core is under less pressure so the iron and nickel behave like a viscous fluid. The mantle is subdivided into three parts based on rheology: the mesosphere, the asthenosphere and the basal part of the lithosphere. The **mesosphere**, is a "solid" layer of rocks which are able to slowly move and flow over time, much like glaciers made of solid ice can flow. This happens because the rocks in the mesosphere are very hot. The **asthenosphere** is a "plastic" layer of rocks. A plastic solid can be deformed easily while still retaining its solid state, much like a ball of clay when you squeeze it.

The **lithosphere** has two distinct layers based on composition and rheology. The upper part of the lithosphere is composed of low-density rocks, silica-rich minerals (the crust of the chemical model) while the lower part of the lithosphere has a higher concentration of iron-rich minerals (the uppermost part of the mantle). Rocks in the lower lithosphere are just hot enough to be folded and deformed without breaking, while the rocks in the upper part of the lithosphere are cold and brittle, and not able to flow. When rocks in the upper lithosphere are subjected to stresses, they tend to break or rupture forming faults (and sometimes earthquakes).

Because the lithosphere is brittle, it has been broken into large pieces called **plates** (Figure 7.6). The uppermost layer of the plates can have one of two different types of crust: **oceanic crust** or **continental crust** (Figure 7.7). Oceanic crust is enriched in iron-based minerals and has a higher density, while continental crust is enriched in silica and aluminum-based minerals and has a lower density. The thickness of continental crust is highly variable: it ranges from 16 km (~10 mi) thick under the thinnest areas up to 70 km (~43.5 mi) thick under the world's mountain ranges with an average thickness of about 40 km (~25 mi). In contrast, the ocean crust is a relatively consistent 5 km (~3 mi) thick. Although most plates have both oceanic and continental crust in them, some plates, like the Pacific Plate or the Philippines Plate, are composed almost exclusively of oceanic crust, while other plates, like the Arabian Plate, are composed chiefly of continental crust.

These plates are essentially "floating" on the plastic asthenosphere beneath them, like a ship floats on the water. The difference in density between oceanic crust and continental crust explains why there are ocean basins and continents (Figure 7.7). Although oceanic plates are thinner, their dense nature causes them to "float" lower on the asthenosphere, while the thicker, less-dense continental plates "float" higher on the asthenosphere. This "floating" action also allows the Earth's plates to actively move around and interact with each other, causing changes in the location, size and shape of the continents and ocean basins over time. This motion is driven by the physical process called plate tectonics which will be explained further in the next section.

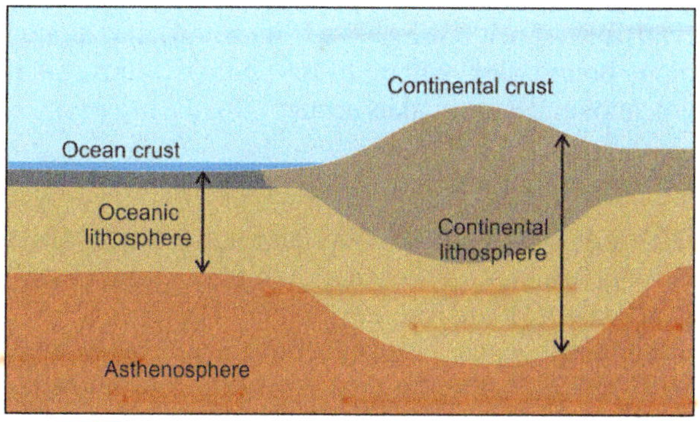

Figure 7.7: A comparison of ocean crust and continental crust. Continental crust is thicker and less dense, so it "floats" higher on the asthenosphere relative to the thinner, denser oceanic lithosphere.

Section 7-3: Plate Tectonic Processes

Plate tectonics is a comprehensive scientific theory which explains large-scale geologic phenomena such as the volcanism, earthquakes and formation of mountains. Plate tectonic theory explains how ocean basins slowly grow and shrink in size over millions of years, and how this leads to the breaking apart and recombining of the continents (Figure 7.8). Plate tectonics is also the driving force behind the formation of rocks and the rock cycle. Plate tectonic processes generate magma for igneous rocks, they lift up the land surface, accelerate rates of rock erosion, drive the production of sedimentary rocks, and produce the conditions necessary for the formation of metamorphic rocks.

The plates continually move and change shape over millions of years, mostly through the creation and destruction of oceanic lithosphere. The motion of the plates is driven by convection in the lower mantle (Figure 7.9). By conventional standards, convection occurs due to the circulation of a fluid. In this case, the mantle, although solid, can still move and act like a fluid, transporting heat produced in the Earth's interior towards the surface, where it radiates into space. Some of this heat is left over from the creation of the Earth but most of it is generated by radioactive decay. To understand how the plates (and the continents which ride along on them) can move and change over time, we have to discuss the nature of the three types of plate boundaries: **Divergent**, **convergent**, and

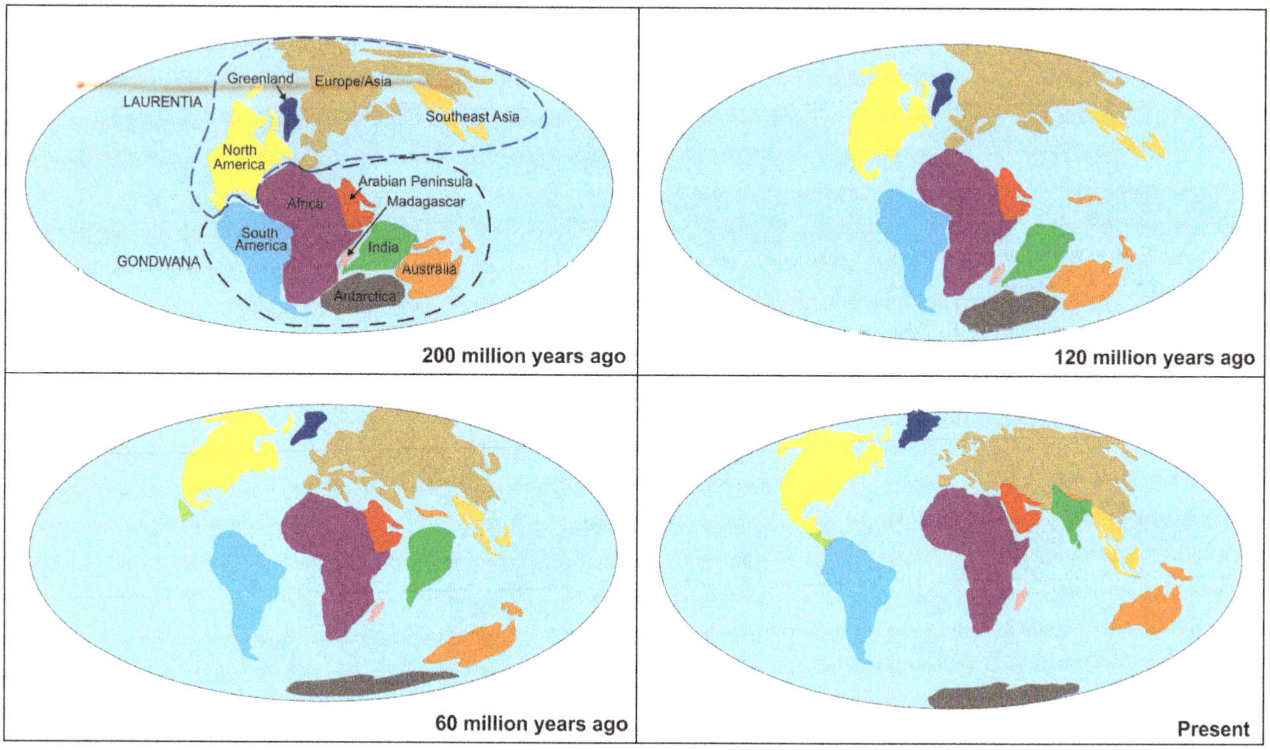

Figure 7.8: The movement of the continents over the past 200 million years. The movement of the continents occurs because of the creation and destruction of ocean crust along plate boundaries. In this case, the Atlantic Ocean is growing at the expense of the shrinking Pacific Ocean.

transform-fault boundaries (Figure 7.6). It is along the plate boundaries where most geologic hazards like volcanos and earthquakes occur.

Divergent boundaries

Divergent boundaries are places where plates are moving away from each other in a process called **seafloor spreading** (Figure 7.10, inset). More importantly, when the plates move apart, magma wells up from the asthenosphere and new crustal material (rocks) is added to the edge of each plate. Over time, the each plate move away from the divergent boundary in a conveyor-belt like fashion as the plate continues to grow (Figure 7.9). This is also why the Mid-Atlantic Ridge runs right down the center of the Atlantic Ocean (Figure 7.10). Also, the age of the ocean crust on each side boundary gets progressively older.

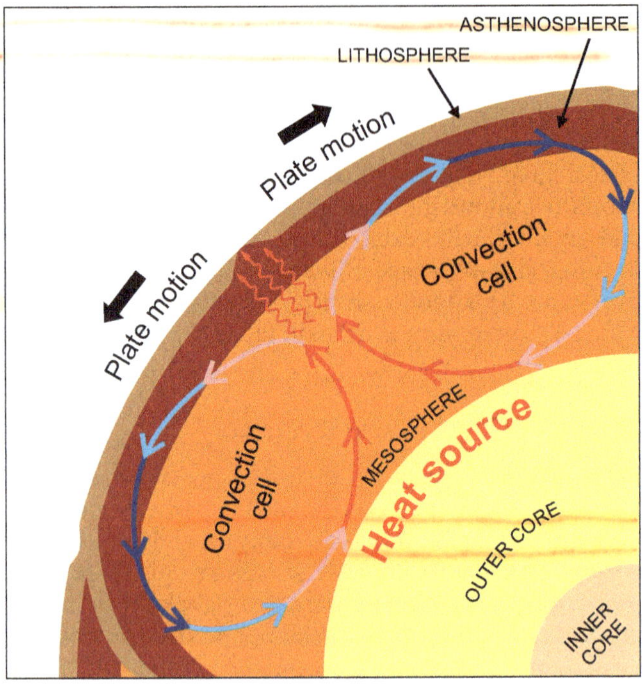

Figure 7.9: Convection cells form in the mesosphere as heat from the core escapes to the Earth's surface. The convection cells subsequently move the overlying plates around.

Given that 71% of the Earth is covered in water, it should be no surprise that most divergent boundaries are found at the bottom of the world's oceans. These divergent boundaries are basically long chains of underwater volcanic mountains called **mid-ocean ridges** (Figure 7.10). These undersea features have an average relief equal to any of the highest mountain peaks on land. It is important to note that **only oceanic crust** forms at a divergent boundary. Rates of seafloor spreading vary from ridge to ridge. For example, the East Pacific Rise is spreading at a relatively fast rate of 7 cm/yr, while the Mid-Altantic Ridge is creeping along at a relatively slow rate of 2 cm/yr. Although the overall rate of seafloor spreading may seem too small to matter (at least in the perspective of human life spans) the continuous production of ocean crust has led to the growth of ocean basins which are thousands of kilometers wide. If you refer back to Figure 7.8, you can see how this process led to the birth and growth of the Atlantic Ocean over the past 200 million years. Of course while the Atlantic Ocean has grown, the

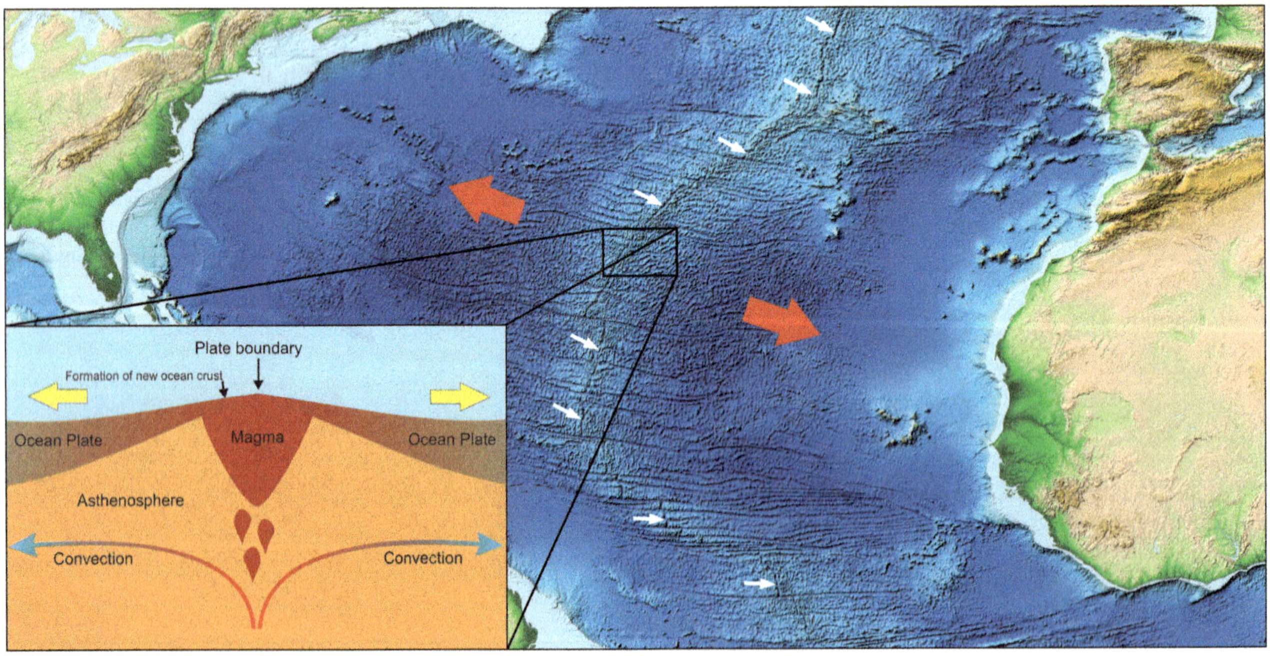

Figure 7.10: The Mid-Atlantic Ridge splits the Atlantic Ocean in half. Seafloor spreading occurs along the ridge, forming new oceanic crust as the plates move apart (as shown by the red arrows). Image credit: NGDC/NOAA.

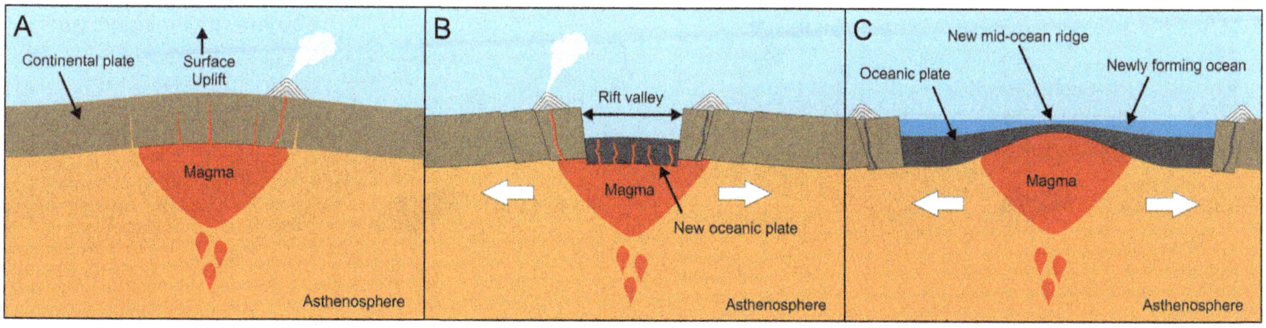

Figure 7.11: Continental rifting explained. A. At first, magma accumulating under a continental plate causes it to warp upwards. B. Eventually, the plate is split apart, and new ocean crust begins to form. C. As the plates continue to move apart, a new ocean forms.

Pacific Ocean has shrunk through the destruction of ocean crust, a topic we will discuss further in the next section.

Some divergent boundaries cut through the continents as part of a process called **continental rifting**. When the Earth's convection patterns shift, a new divergent boundary may form under a continent. So continental rifting is really just an early stage of divergent plate boundary formation (Figure 7.11). First, magma begins pushing up on the continental crust, thinning it. As convection begins to pull the continental crust apart, parts of the crust begin to break and collapse downwards, forming a rift valley. Volcanic activity is also common in active rift zones. As the valleys continue to deepen and widen, they eventually become flooded by the oceans, forming an inland sea. At this point, continental rifting ceases and sea-floor spreading begins, ushering in the birth of a new ocean basin.

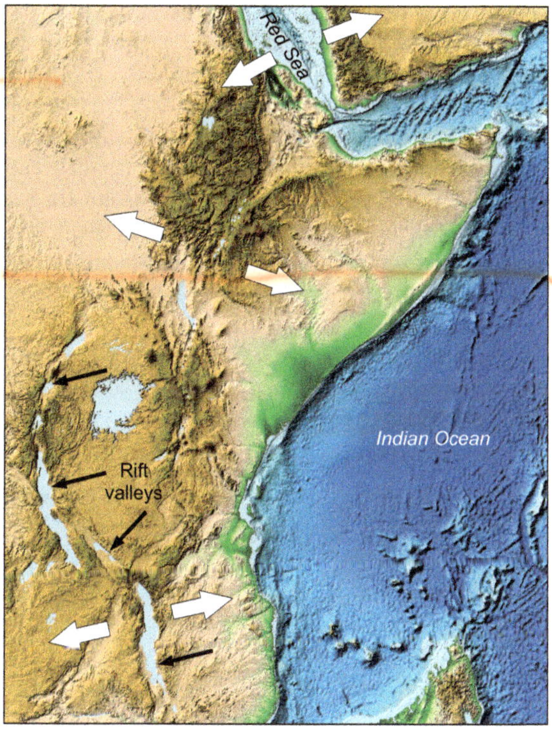

Figure 7.12: The East African Rift Zone is a good example of a continent in the early stages of rifting. Likewise, the Red Sea is a good example of a newly forming sea produced by rifting. Image credit: NGDC/NOAA.

In the modern world, the East African Rift Zone is an excellent example of this rifting process in its early stages (Figure 7.12). The eastern portion of the African continent is beginning to rift apart and move to the east. If you trace the Indian Ocean ridge up towards the Arabian Peninsula, you will see it goes right down the center of the Gulf of Aden and then takes a turn up the middle of the Red Sea. The Red Sea is growing through sea-floor spreading, pushing the Arabian Peninsula towards the northeast and forming a large mountain range. Meanwhile, you can also trace another divergent boundary down from the Red Sea across the Afar Triangle and south through a series of rift valleys, connecting back to the Indian Ocean near Mozambique. Many of these rift valleys are at elevations lower than sea level and they commonly contain deep lakes. Eventually, over millions of years, these lakes will all link together to form one long sea and eastern Africa will become an island.

Convergent Boundaries

Since new crust is constantly being created at the divergent boundaries, it only seems reasonable to expect the destruction of crust to occur somewhere else (if that wasn't the case, the Earth would just keep growing in size!). **Convergent boundaries** are places where two plates are moving towards each other and usually one of the plates is being forced down under the other in a process called **subduction** (Figure 7.13). Subduction occurs when a plate with oceanic crust sinks down into the mesosphere. As it sinks, the plate carries seafloor sediments and water with it. The water causes partial melting of asthenosphere above it sending plumes of magma towards the surface. This magma rises because it is less dense than the rocks around it. The rising magma will ultimately lead to volcanism and the formation of mountain or island chains. Once the plate has traveled deep into the mesosphere, it melts completely and is recycled as part of the rock cycle. It is important to

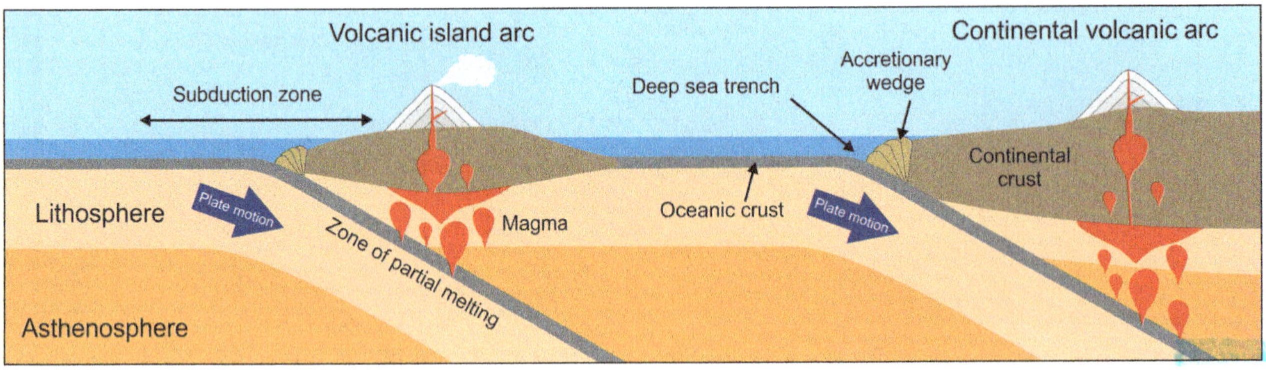

Figure 7.13: The two primary types of convergent boundaries: Ocean-ocean (left) and ocean-continental (right). Ocean-ocean type boundaries produce a volcanic island arc, while ocean-continental boundaries produce a continental volcanic arc.

note here that continental crust cannot be subducted; it is too buoyant. Two other features associated with subduction are **accretionary wedges** and **deep sea trenches** (Figure 7.13). Accretionary wedges form when sediments scraped off of the subducted plate are "plastered" onto the edge of the adjoining plate. Deep sea trenches are long, linear features that form parallel to the plate boundary; these trenches are the deepest parts of the oceans and they mark the place where subduction is occurring (Figure 7.14).

There are two main types of convergent boundaries: **Ocean-continental** and **ocean-ocean** (Figure 7.13). At an ocean-continental convergent boundary, an "oceanic" plate sinks down beneath a "continental" plate. As plumes of magma rise from the subducting plate, the magma works its way through the plate to feed a chain of volcanos at the surface. These volcanos are collectively known as a **continental volcanic arc**. The Cascade Range, which extends from northern California to Washington, is a great example of this. The Cascade Range is a string of volcanos which parallel the Pacific Coast where the Juan de Fuce plate is subducting under the North American plate (Figure 7.14A). A similar scenario occurs at an ocean-ocean convergent boundary. Here, one ocean plate subducts under another ocean plate forming a **volcanic island arc**. Many of the chains of islands in the Western Pacific Ocean are good examples of this (Figure 7.14B).

So what happens when a subducting plate carrying continental lithosphere reaches a subduction zone? Since continental crust is too buoyant to be subducted, the whole subduction process slowly grinds to a halt as the two continental plates go through a massive collision. The Himalayan Mountains in Asia are still tectonically active from a massive collision that started over 40 million years ago (Figure 7.15). Prior to colliding with Asia, India and

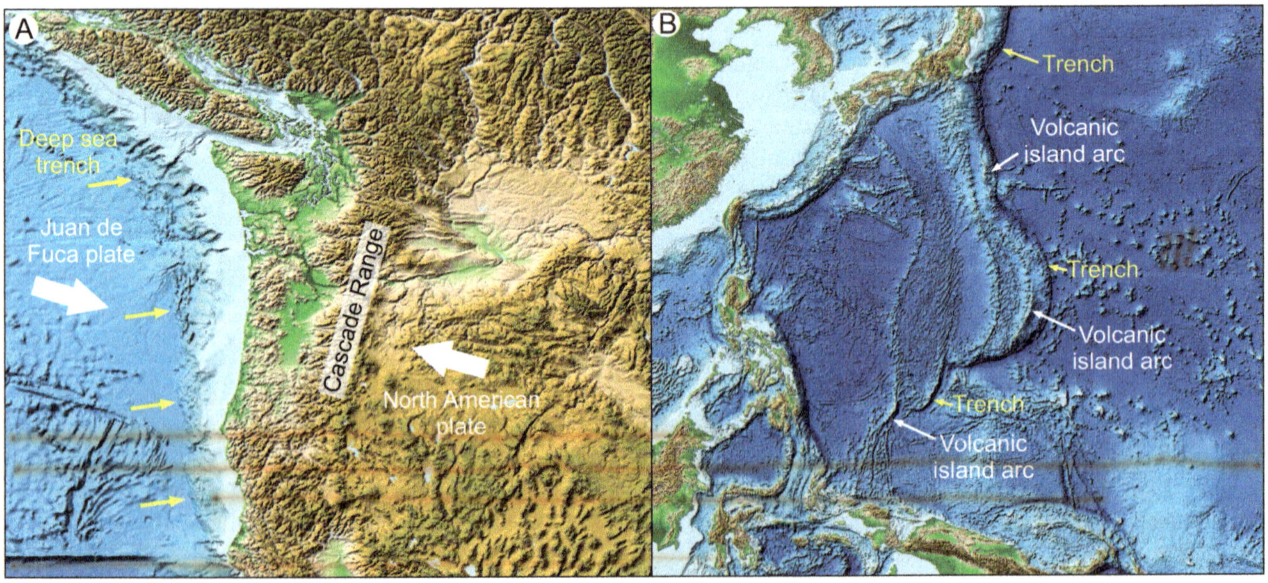

Figure 7.14: A. The Cascade Range of Washington, Oregon and northern California is a good example of a continental volcanic arc. B. The western Pacific Ocean has many deep sea trenches and volcanic island arcs. Image credits: NGDC/NOAA.

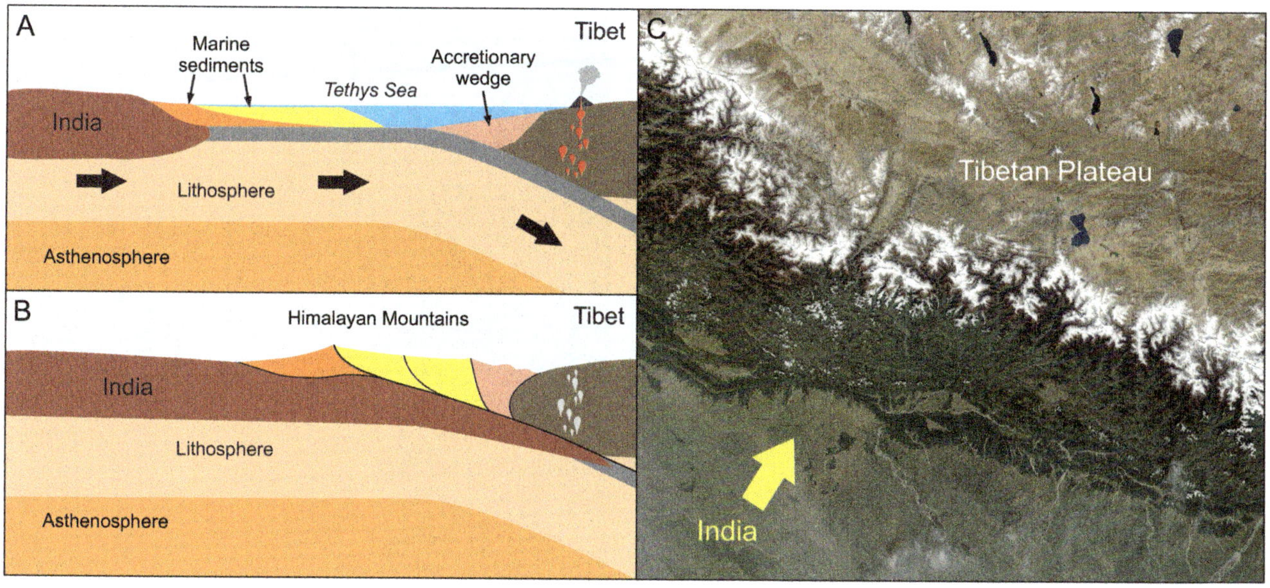

Figure 7.15: The Himalaya Mountains are the classic example of mountain building caused by continental collisions, in this case, between the Indian Plate and the Eurasian Plate. A. Prior to collision, the Indian Plate was subducting under the Eurasian Plate. B. India, being too buoyant, collided with Asia, bringing subduction to a halt. C. The Himalayas as viewed from Landsat-7. You can see how the southern edge of the Eurasian Plate has been "crumpled up" by its collision with India. Image credit: Jacques Descloitres, MODIS Rapid Response Team, NASA/GSFC.

Asia were separated by an ocean, the oceanic portion of the Indian Plate was subducting under the Eurasian Plate and modern day Tibet was the shoreline (Figure 7.15A). When all of the oceanic crust between Asia and India was subducted, the inevitable collision ensued. Rocks from the Indian Plate and the former seafloor sediments in between were subsequently pushed up and onto the edge of the Indian Plate while the rocks along the edge of the Eurasian plate were severely deformed (Figure 7.15B). The modern Himalayas are the result of that collision (Figure 7.15C).

Not all collisions occur on such a large scale. Smaller plate features, like islands and microcontinents, also cannot be subducted. However, they are also not massive enough to bring the subduction process to a complete stop either. Instead, small islands can be "scraped" off the subducting plate and "plastered" onto the edge of the overriding plate. Over time, this process adds new land to a continent. Large portions of western North America, including most of Alaska, are "exotic" landmasses that have been "plastered" to the edge of the North American continent over the last 350 million years (Figure 7.16).

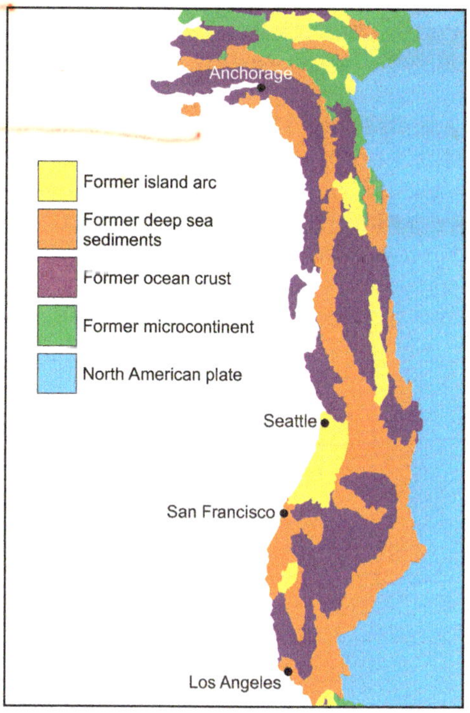

Figure 7.16: This illustration shows how the west coast of North America is a collage of old pieces of land, seafloor sediments and ocean crust plastered onto the edge of the continent during millions of years of subduction.

Transform Fault Boundaries

The final type of plate boundary is a **transform fault boundary**. Transform fault boundaries are places where two plates are moving past each other laterally. This lateral motion produces earthquakes, but volcanism is pretty rare. An example of a transform fault boundary is the San Andreas fault system (Figure 7.17A) which lies along the boundary between the Pacific Plate and the North American Plate. Part of California rests on the Pacific Plate, which is moving to the northwest relative to the North American Plate (Figure 7.17B). The slow but continuous motion of the two plates along this boundary produces frequent, and many times severe, earthquakes. The tectonic forces are also ripping southwestern California apart, leaving a large visible scar on the landscape (Figure 7.17C). Other well-known continental transform faults can be found in New Zealand, the Middle East, and British Columbia (Figure 7.6).

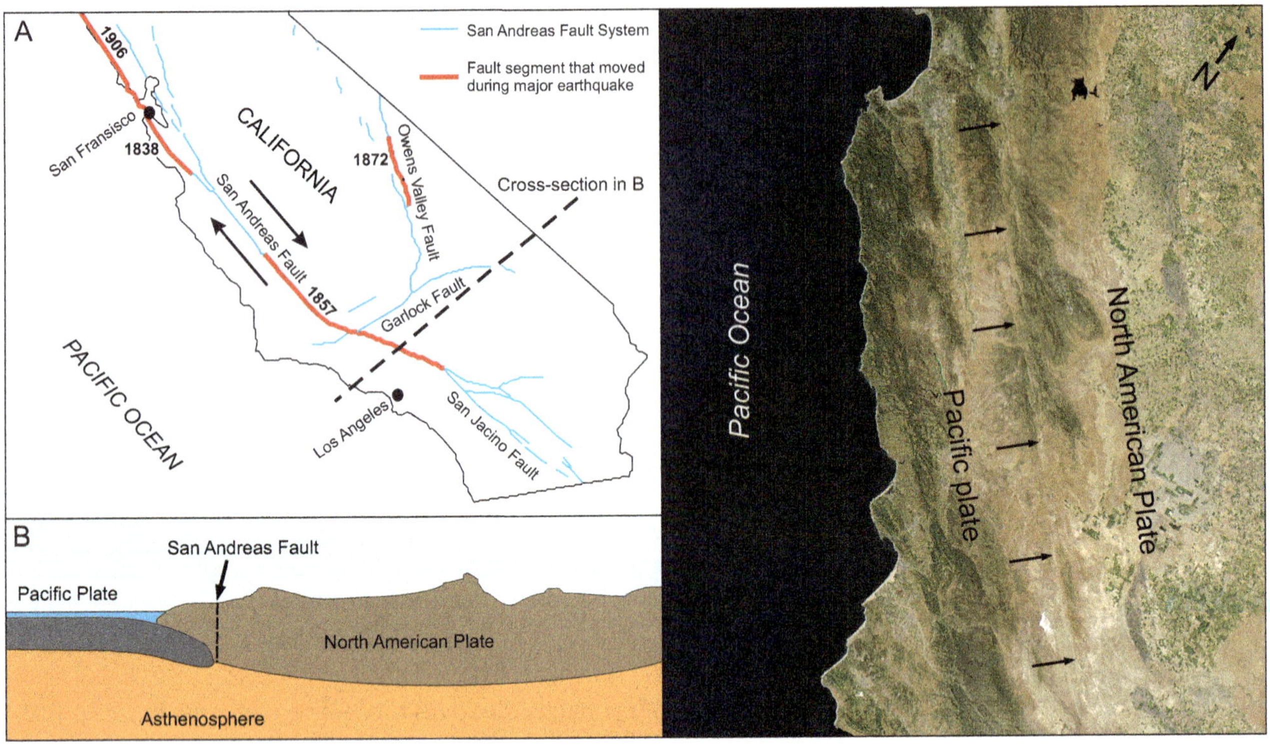

Figure 7.17: A. The San Andreas Fault system. B. A cross-sectional profile through the San Andreas Fault. C. The San Andreas Fault leaves a very noticeable fault scarp across the landscape. Photo credit: Jacques Descloitres, MODIS Rapid Response Team, NASA/GSFC.

Hot Spots

Why do volcanos occur on Hawaii? Hawaii is not anywhere near a plate boundary, so subduction isn't the answer. Hawaii sits over a **hot spot**, a place where a deep mantle plume has burned through the crust (Figure 7.18A). **Deep mantle plumes** are "blobs" of hot, low density material that rise up to the lithosphere and "burn through" the plate, leading to volcanism at the surface. Deep mantle plumes are thought to form at the core-mantle boundary. It is the constant supply of magma from a deep mantle plume which formed the Hawaiian Islands (Figure 7.18B). Hot spots don't move but instead, tectonic plates move over them and a string of volcanos result over time (Figure 7.19A). Most of these hot spots are found in the world's ocean basins and strings of islands and seamounts form as a consequence of plate motions. **Seamounts** are commonly (but not always) extinct volcanos that have sunk down below the ocean surface. Many of these seamounts have a flat top, the result of wave erosion (Figure 7.19B). Upon closer inspection of the Pacific Ocean basin, you can see that the Hawaiian Islands are just a few of the dozens of islands and seamounts that the Hawaiian hot spot has created over millions of years (Figure 7.19C). In fact the newest member of this chain, Lo'ihi, is starting to grow off the southeastern coast of the main island (Figure 7.18B).

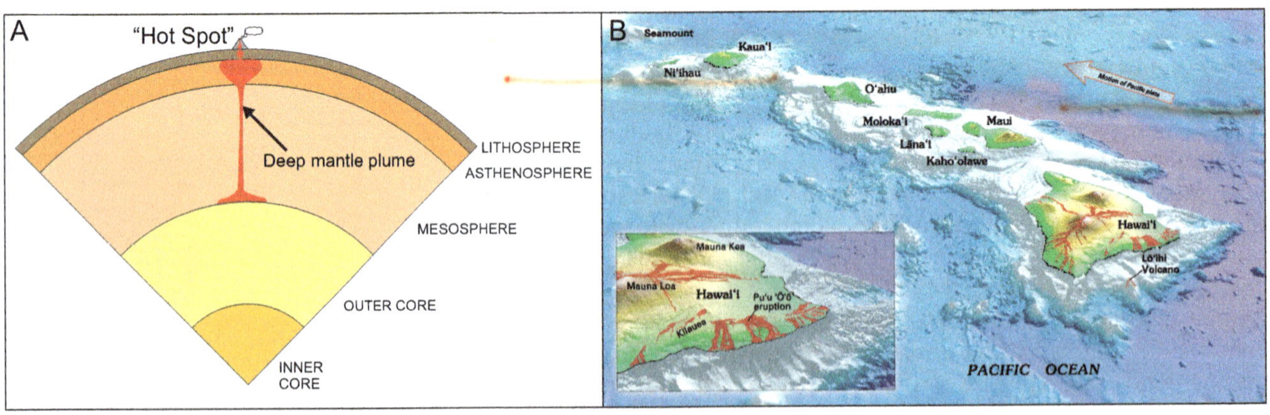

Figure 7.18: A. Diagram demonstrating deep mantle plumes and the resulting "hot spot" that forms. B This image shows how the Hawaiian Islands are really just the top of an immense pile of lava flows which have been fed by the Hawaiian hot spot. Image credit: Joel E. Robinson, USGS.

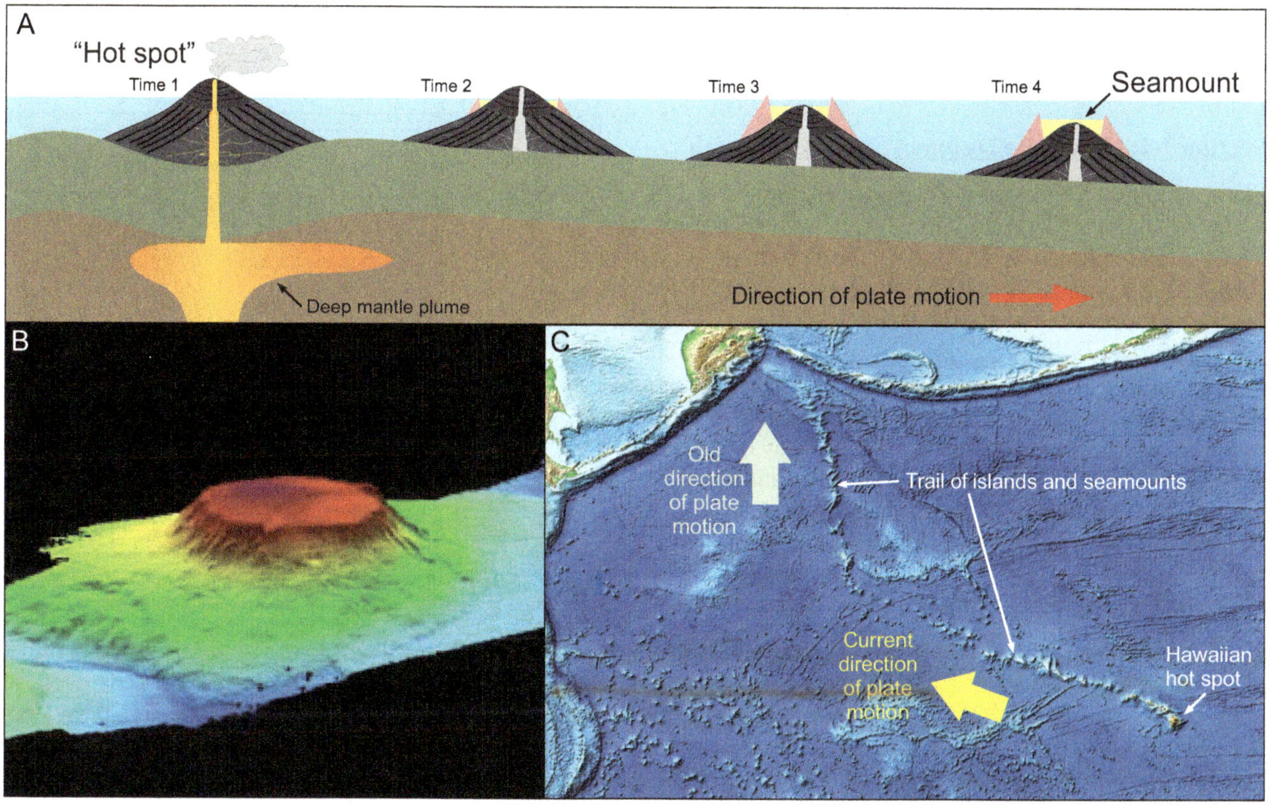

Figure 7.19: A. When a plate moves over a hot spot, a continuous stream of magma builds a volcano which can build up to the ocean surface, forming an island. Once the island moves off the hot spot, the crust underneath begins to sink, and the island sinks down with it, eventually forming a seamount. B. A 3-D depiction of Bear Seamount which is off the coast of Massachusetts. Image courtesy of NOAA. C. This image of the Pacific Ocean basin shows the trail of islands and seamounts created as the Pacific Plate moved over the Hawaiian hot spot. Image credit: NGDC/NOAA.

Other hot spot volcanic islands include the Canary Islands, the Azores, the Galapagos Islands, and Easter Island. Most hot spots are found in the ocean basins, but there are a couple places where hot spots occur on land. One of the most notable is the Yellowstone hot spot, which current sits under Yellowstone National Park and feeds the Yellowstone Supervolcano (Figure 7.20A). We will talk more about the interesting and unique features related to this hot spot when we discuss volcanism in Chapter 8. Although the hot spot is currently under Yellowstone, over the

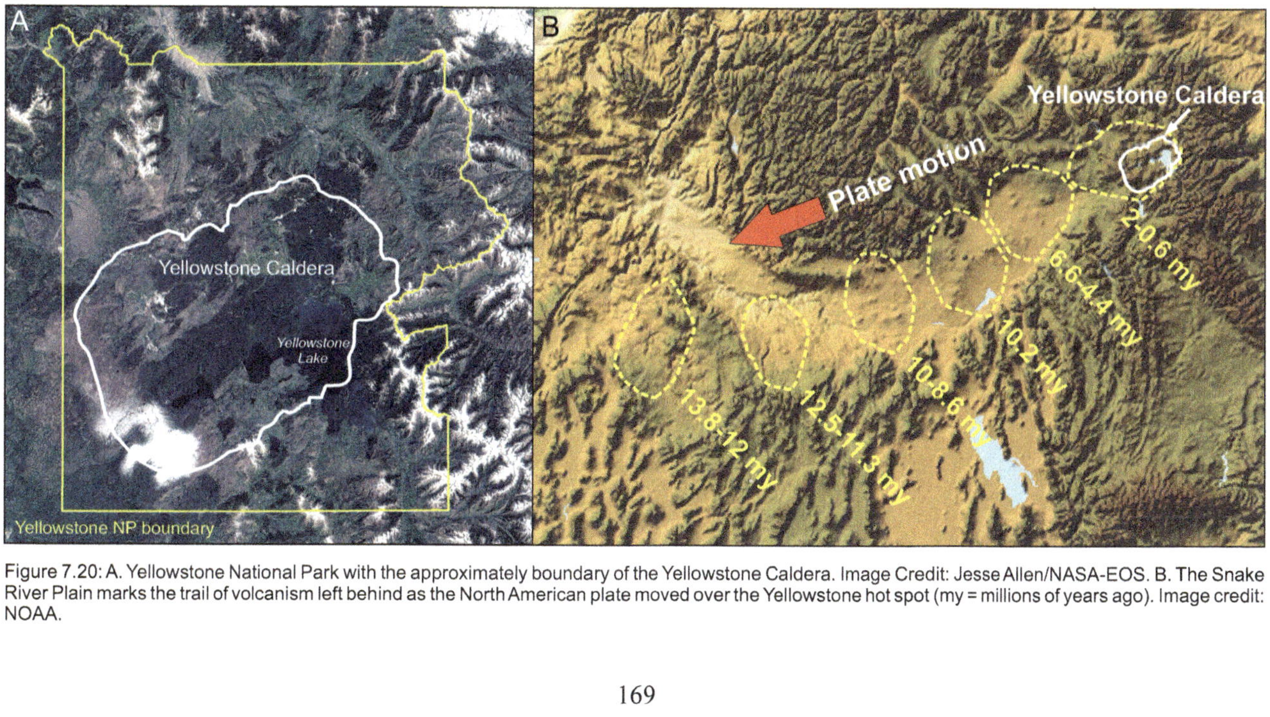

Figure 7.20: A. Yellowstone National Park with the approximately boundary of the Yellowstone Caldera. Image Credit: Jesse Allen/NASA-EOS. B. The Snake River Plain marks the trail of volcanism left behind as the North American plate moved over the Yellowstone hot spot (my = millions of years ago). Image credit: NOAA.

past 16 million years, the Yellowstone hot spot has left behind a trail of volcanic features from Nevada to Wyoming as the North American plate moved over it (Figure 7.20B). The Snake River Plain follows in the wake of the hot spot's trail.

Section 7-4: How Plate Tectonics Shapes the Earth

Plate tectonics is the dominant force shaping the Earth's surface. Whether you are looking at the topography of the Earth's land masses or looking at the bathymetry of the ocean basins, much of what you see is dictated by plate tectonic processes in one way or another (Figure 7.21). In this section, we will investigate how plate tectonics dictates the topography of the land surfaces and the ocean basins and the coasts that fringe them. First we will look at how tectonic processes shape the ocean basins and how they influence the transition zone between the oceans and the continents. Then we will investigate how plate tectonics shapes the land through different types of mountain building processes.

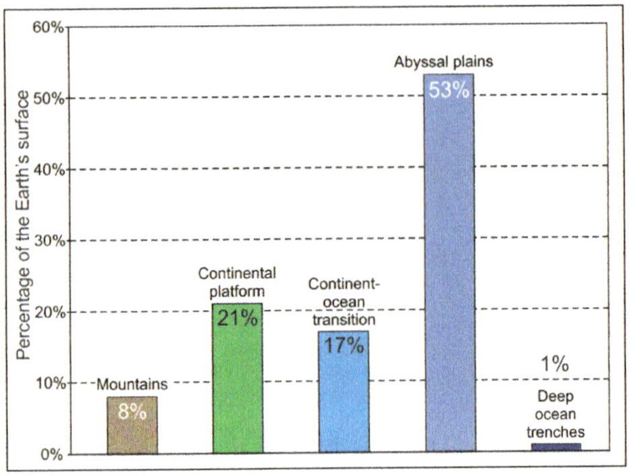

Figure 7.21: The different surface features of the Earth. In sum, 29% of the Earth is covered by land and 71% is covered by the oceans.

Tectonic Features of the Ocean Basins

Over half of the Earth's surface is covered by flat, featureless plains of sediment called the **abyssal plains** (Figure 7.22). The abyssal plains are underlain by iron-rich oceanic crust. Although the abyssal plains aren't the deepest parts of the oceans, they are what we call "the bottom of the ocean" with depths upwards of 6 km (3.6 mi.) on average. Down at the bottom, it is a cold, dark world which has barely been explored. Each new manned expedition to the abyssal plains brings new surprises. The abyssal plains are broken up by many of the tectonic features we discussed in the previous sections: Mid-ocean ridges, transform faults, deep sea trenches and seamounts. In comparison, the deep ocean trenches, which are found cutting though the abyssal plains, cover only a measly 1% of the Earth's surface (Figure 7.21).

Although the continent-ocean transition zone only covers 17% of the Earth's surface, it plays a large role in our lives as it encompasses the coastal zones where some of the largest populations on Earth reside. The continent-ocean transition zone has three parts: The **continental shelf**, the **continental slope** and the **continental rise** (although the

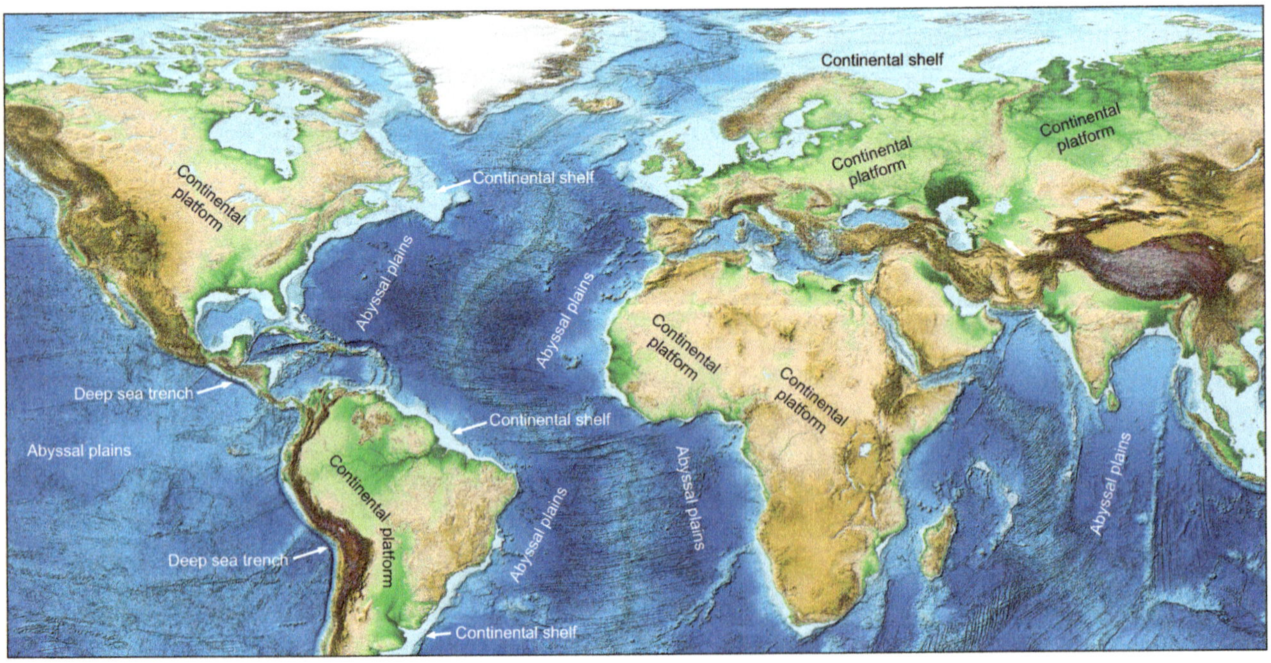

Figure 7.22: A map highlighting the main surface features of the Earth. Image credit: NCGC/NOAA.

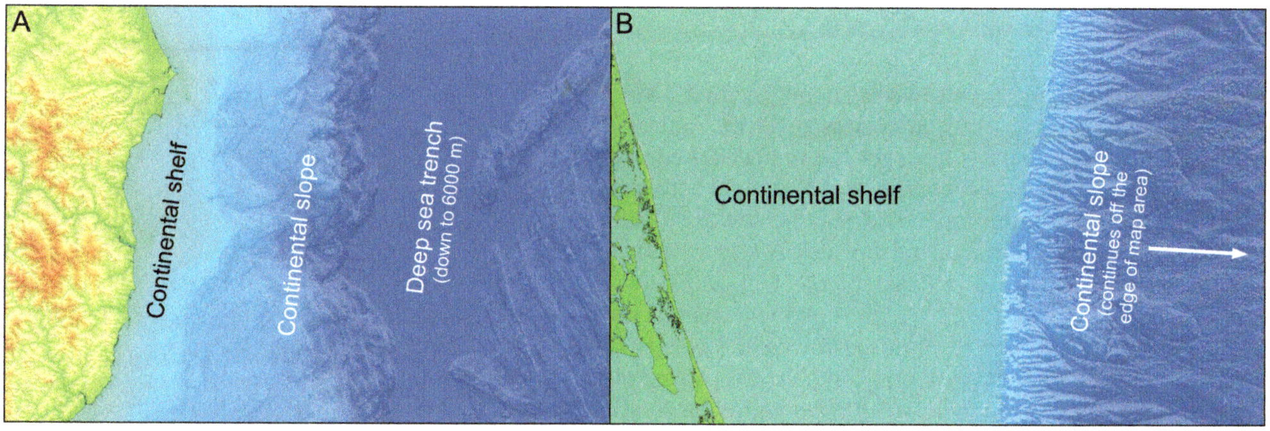

Figure 7.23: A. The transition zone along a tectonically-active part of the West Coast. B. The transition zone along tectonically-quiet East Coast near the Outer Banks of North Carolina. Image credits: NGDC/NOAA. *Note: Both of these images are at the same scale.*

continental rise may be missing along some coasts). All of the continents have some portion of them flooded by the oceans. In fact, during much of the Earth's history the continents have been covered by huge, shallow continental seas. In contrast, the modern oceans are at a fairly low level and only the edges of the continents are submerged. These areas are known as the continental shelves (Figure 7.22). The ocean reaches a maximum depth of about 100 m (~325 ft) on the continental shelves, which is relatively shallow compared to the abyssal plains. The width of the continental shelves is highly variable, even along the same continent. For example, the east coast of North America has a wide continental shelf while the west coast has a relatively narrow continental shelf (Figure 7.23). This is due primarily to plate tectonics. The continental shelf is commonly incised by **deep-sea canyons** (Figure 7.24). These canyons are commonly found offshore from large rivers like the Susquehanna River, the Hudson River and the Delaware River, and they are thought to have formed during the ice ages when global sea-levels were much lower

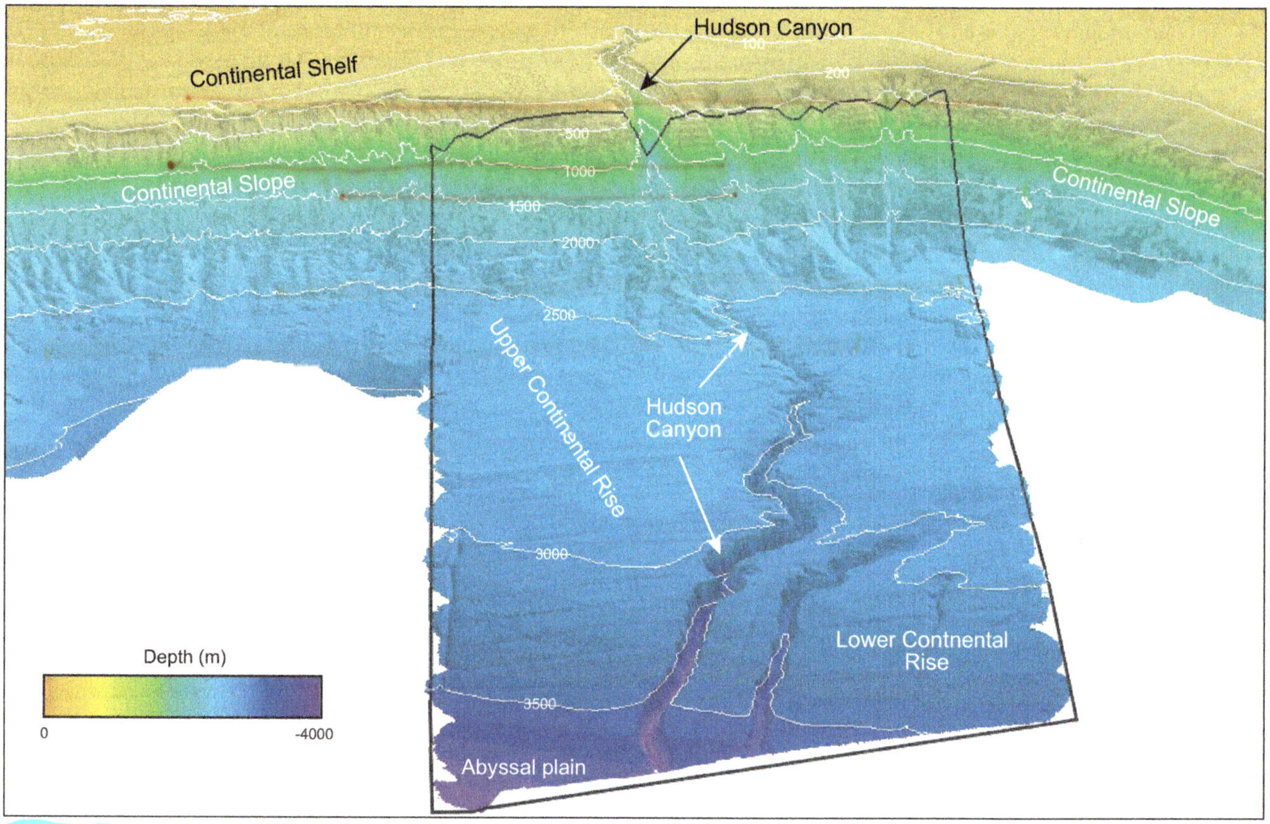

Figure 7.24: Perspective view, looking toward the northwest, of the continental slope and rise incised by the Hudson Canyon. The black line outlines the limits of the survey carried out by the NOAA Ship Ronald H. Brow. Image credit: Peter Dartnell, William R. Normark, Neal W. Driscoll, Jeffrey M. Babcock, James V. Gardner, Rikk G. Kvitek, and Pat J. Iampietro/USGS.

than they are today.

The transition zone between the continental shelf and the abyssal plains is known as the **continental slope** (Figures 7.23 and 7.24). The continental slope is relatively steep (as compared to the continental shelf) because the transition zone between the higher "floating" continental crust and the lower "floating" oceanic crust is fairly abrupt in most places. Along tectonically-active coasts, like the West coast of the United States, the transition zone between the edge of the continental shelf and the adjacent deep-sea trench is quite narrow and steep (Figure 7.23A), while along passive margins, like the East Coast of the United States, the transition between the continental shelf and the abyssal plain is a wider, more gradual slope (Figure 7.23B).

Passive coasts also differ from active coasts because they have a large, gently sloping wedge of sediment at the base of the continental slope called the **continental rise** (Figure 7.24). Over time, undersea "landslides" called **turbidity currents** have transported fine-grained sediments from the continental shelf to the base of the continental slope, forming **deep sea fans** (Figure 7.25). Many of these turbidity currents occur in deep-sea canyons. Along older coasts, overlapping fans have built up over hundreds of millions of years to produce the continental rise as we see it today.

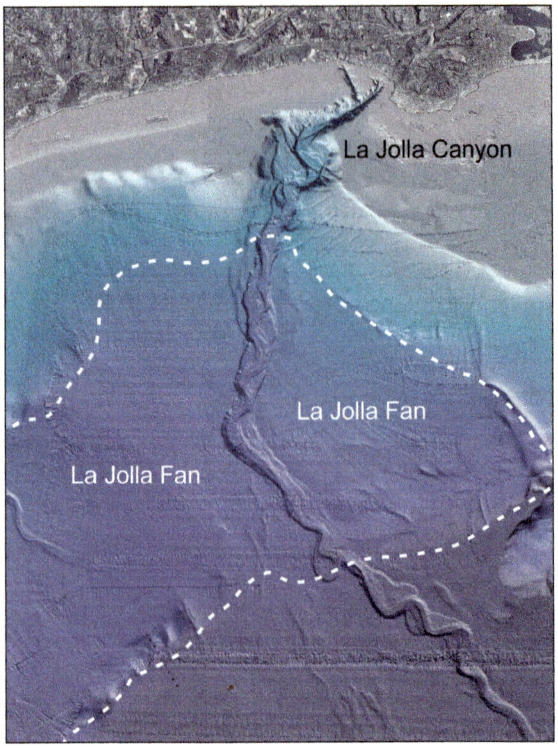

Figure 7.25: The La Jolla Canyon off the southern coast of California has started to develop a sizable deep-sea fan. Image credit: USGS.

As you can image, there is a lot of variation from coast to coast. In general, most continental margins can be placed into one of two groups: **passive continental margins** and **active continental margins**. Along a passive continental margin (Figure 7.26A) there is a rather gradual transition from continental crust to oceanic crust. Because this coastline is not near a plate boundary, tectonic processes are minor and the coastline is fairly stable. This stability allows sediments to gradually accumulate into thick layers, forming a broad, flat continental shelf. This gradual transition simply reflects the gradual transition from continental crust to ocean crust. Along the east coast of the United States, for example, the continental slope is underlain by faulted blocks of continental crust which transition into oceanic crust (Figure 7.26A). This transition zone formed when North America rifted apart from Africa over 200 million years ago. After rifting occurred, the continent-ocean boundary was buried under millions and millions of years worth of sediments which have developed into the wide, flat continental shelf as we see it today with its wide, sandy beaches. In many cases, a broad, flat sandy plain called a **coastal plain** will develop in these settings (Figure 7.26B). The sediments that underlie the coastal plain are also part of this massive wedge of sediments which were deposited when sea level was higher than it is today. Most passive margins begin their lives as a tectonically-active divergent boundary (e.g., the Red Sea) but once a mid-ocean ridge begins to develop and oceanic crust forms,

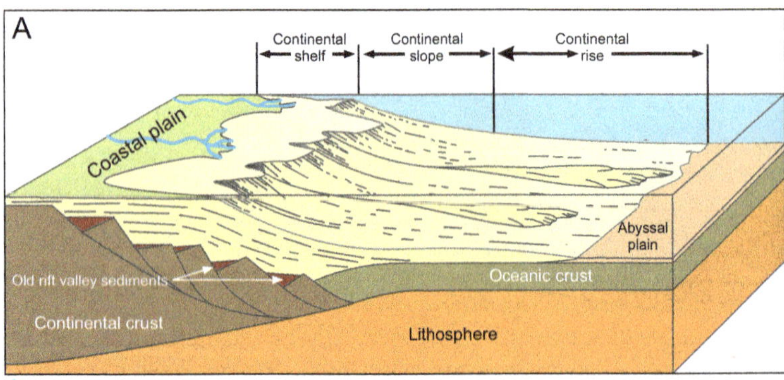

Figure 7.26: A. An idealized cross-sectional profile showing the nature of a passive continental margin like the East Coast of the United States. B. This view of Chincoteague National Wildlife Refuge in Virginia is typical of a passive continental margin. Photo credit: U.S. Fish and Wildlife Service Northeast Region.

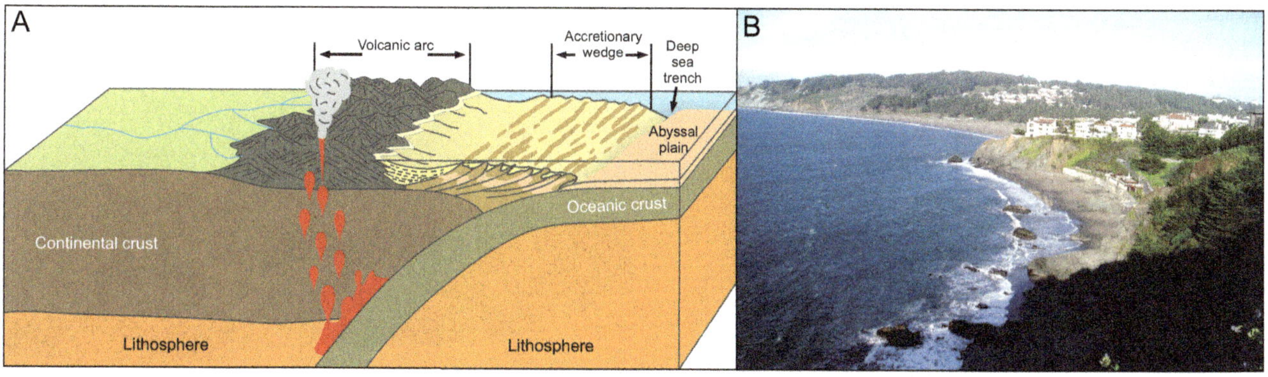

Figure 7.27: A. An idealized cross-sectional profile showing the nature of an active continental margin like the northwestern coast of the United States. B. This stretch of the California coast near San Francisco is typical of an active continental margin. Photo credit: B. Zaprowski.

these coastal margins become tectonically inactive.

In contrast, an **active continental margin** is constantly being modified by tectonics. Active coastlines form along plate boundaries, usually subduction zones (Figure 7.27A). When oceanic lithosphere is subducted, abyssal plain sediments are scraped off the down-going plate and accreted to the edge of the overlying plate in the form of an accretionary wedge. As the wedge of sediment grows, coastal mountain ranges rise up, lifting the coastline up with them. This constant lifting limits the amount of sediment which can accumulate along the edge of the continent and thus leads to a narrow continental shelf. Beaches along these types of coasts tend to be poorly-developed and may even be gravelly (Figure 7.27B).

Tectonic Features of the Land Surface

About two-thirds of the Earth's land surface is fairly flat with low relief and at a relatively low altitude above sea level. These **continental platforms** (Figure 7.22) are found in the areas of the continents where the bedrock has not experienced any type of tectonic deformation in a long time. This is primarily due to their distal proximity to any type of plate tectonic boundary or hot spot. The Great Plains region of the United States and Canada, Siberia, the Eurasian Steppe, and the African Savannahs are all good examples of continental platforms (Figure 7.28). Here the dominant process shaping the Earth's surface is erosion, a topic we'll discuss further in Chapter 13.

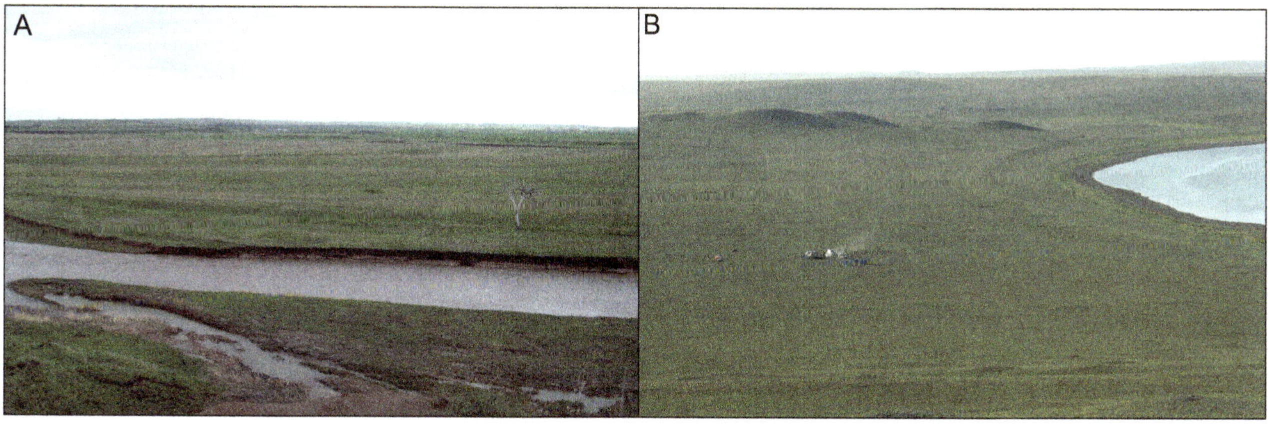

Figure 7.28: A. The Great Plains of the United States in South Dakota, shows the nature of the continental platforms. Photo credit: B. Zaprowski. B. Grassland steppe of eastern Mongolia is also a good example of a continental platform. Photo credit: N. Batbayar, Mongolia Wildlife Science and Conservation Center

The remaining parts of the Earth's land surface have high relief and are generally mountainous. Some areas have high relief landscapes because of extensive downcutting by rivers or glaciers, but most of the world's major mountains chains are (or were) produced by volcanism or some type of tectonically-driven mountain building process along a plate boundary. Note that high relief is not always caused by rocks being lifted upwards (like in the case of the Himalaya Mountains). Relief can also be produced by rocks sinking downwards, like in the case of the Dead Sea in the Middle East. The Dead Sea sits in a transform-fault zone which extends northward from the Red Sea. Slowly the land along the fault zone is sinking downwards, producing a rugged terrain (Figure 7.29A). As a

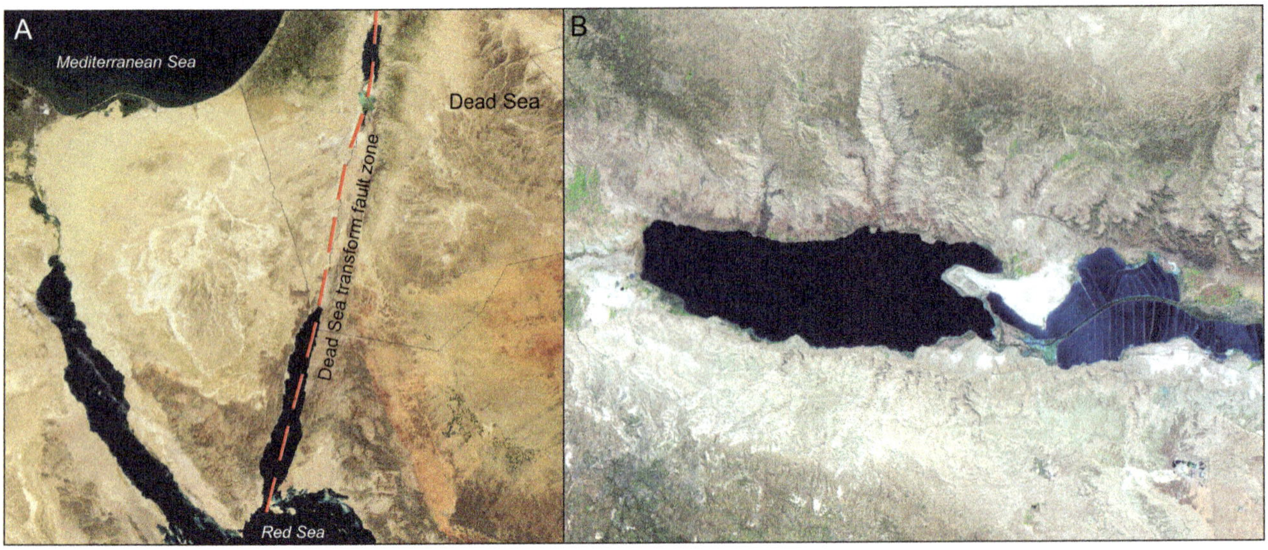

Figure 7.29: The rugged terrain of this part of the Middle East is mostly due to the crust sinking downwards along the Dead Sea transform fault zone. Photo credit: Jacques Descloitres, MODIS Rapid Response Team, NASA/GSFC. B. The Dead Sea is the lowest elevation on Earth at 429 m (1,407 ft.) below sea level. Photo credit: Jesse Allen/USGS.

result, the Dead Sea is the lowest elevation on Earth at 429 m (1,407 ft.) below sea level (Figure 7.29B).

In this final section of the chapter we are going to briefly explore the basic types of mountain building. Here we will put the different types of active mountain building processes into three groups: **Extensional mountain building, compressional mountain building**, and **upwarping** (Figure 7.30). Extensional mountain building occurs in places where the crust is being pulled apart, like in the case of the Dead Sea or the rift basins of Africa (Figure 7.13). Compressional mountain building occurs when the crust is being compressed, usually in a collisional setting like the Himalayas (Figure 7.16) or the Appalachian Mountains. Upwarping is a little different in that it can occur away from a plate boundary and it is commonly associated with magma pushing up from below, like along the edge of a subduction zone or from a deep mantle plume. The Yellowstone hot spot is a good example of this (Figure 7.20B).

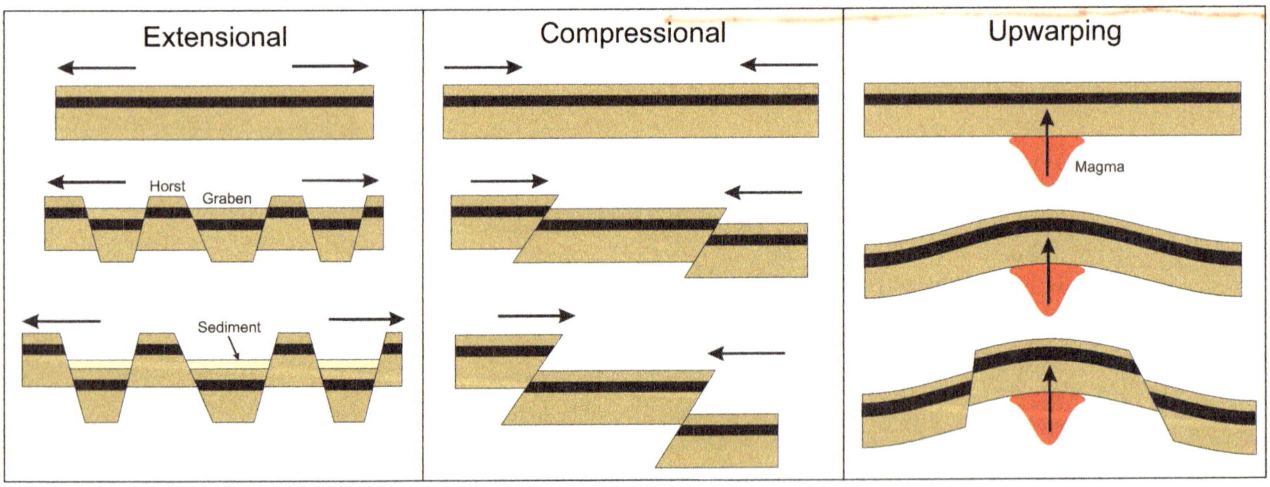

Figure 7.30: The three types of mountain building, from left to right: Extensional, compressional and upwarping.

Extensional Mountain Building

Extensional mountain building is accomplished mostly through faulting. As the crust is pulled apart, it begins to thin and rupture. The crust breaks into pieces along these faults; Some of the blocks (horsts) move upwards and form ranges, while other blocks (grabens) sink downwards and form basins (Figure 7.30). As extension continues, the floors of the basins begin to fill with sediment eroded off of the adjacent ranges. In some cases, the basin drops down so much that a lake forms. The lakes in the East African Rift Zone are a good example of this (Figure 7.12). In arid

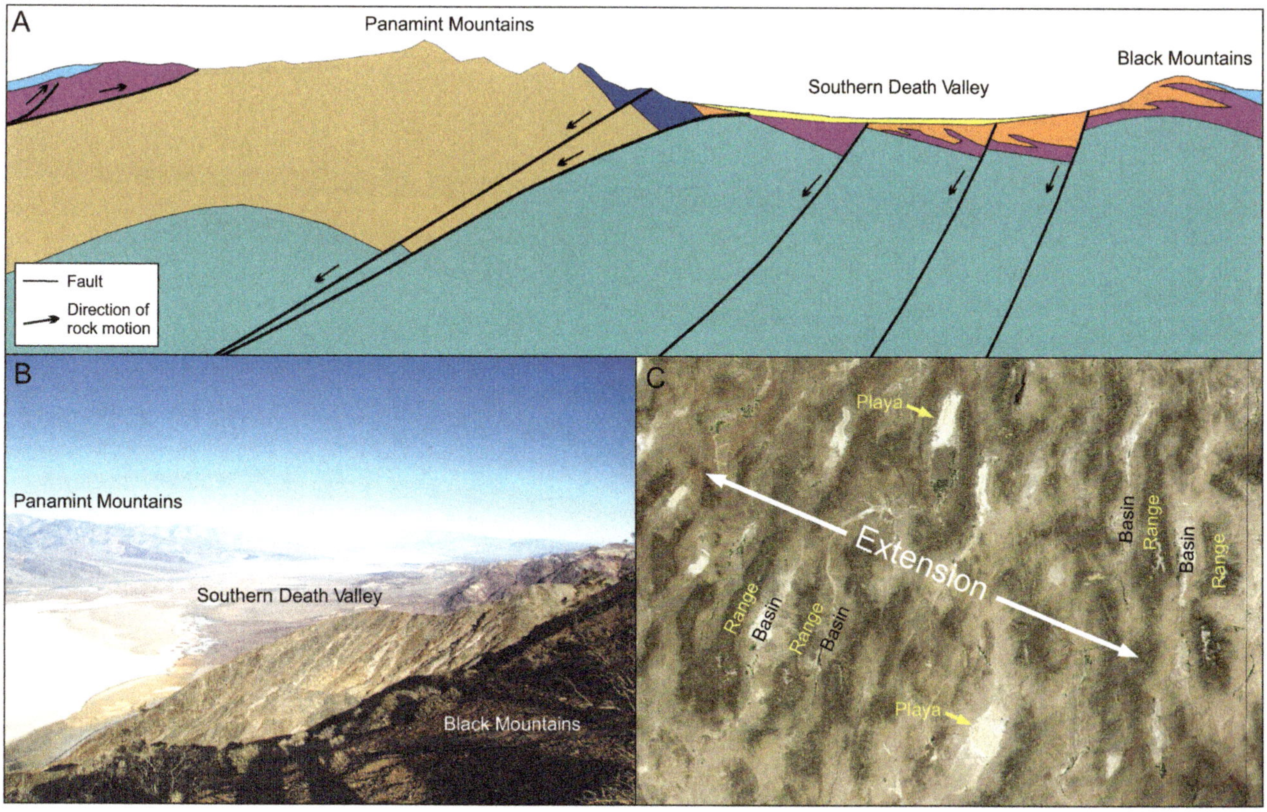

Figure 7.31: A. Generalized cross-sectional profile through southern Death Valley. As the crust is pulled apart and extended, the valley is dropping down. B. The southern end of Death Valley as seen from on top of the Black Mountains. Here, the floor of the valley is actively sinking down along faults which run along both sides of the valley. Photo credit: B. Zaprowski. C. This satellite image of central Nevada displays a prime example of a landscape undergoing extensional mountain building with its prominent ranges and down-dropping basins. Photo credits: Jacques Descloitres/MODIS Land Rapid Response Team at NASA.

climates, flat shallow lakes called **playas** may form in the basin. Playas commonly dry up, leaving behind flat deposits of clay or even salt deposits (Figure 7.31C). Some of the flattest places on Earth are the surfaces of dried up playas (Figure 7.32). Death Valley, CA is great example of how this process occurs (Figure 7.31A). As tensional forces have pulled the Panamint Mountains and the Black Mountains away from each other, the area in between (Death Valley) has sunk down, filling it with thousands of feet of salt and sediment deposits (Figure 7.31B). This is also why Death Valley is the lowest elevation in the Western Hemisphere. In these cases, high relief is produced because the valleys or basins are sinking downwards, leaving high-standing ranges in between. Death Valley sits at the edge of a tectonically-active part of North America called the Basin and Range province. High levels of heat flow under the continent is extending the crust, causing it to thin. As a result. dozens of ranges separated by deep, flat basins have formed, especially in places like Nevada (Figure 7.31C). When the extension occurs over a wide area, it produces a landscape that has sometimes been described as "An army of caterpillars." The East African Rift Valley is also a good example of extensional mountain building, but in this case the ranges run in long parallel lines and many have deep lake-filled basins in between.

Figure 7.32: Bonnie Claire Playa in Nevada. Notice how flat the surface of the playa is. Photo credit: B. Zaprowski.

Compressional Mountain Building

Compressional mountain building occurs through a combination of faulting and folding. When rocks are more brittle (like in the shallow crust), faulting is

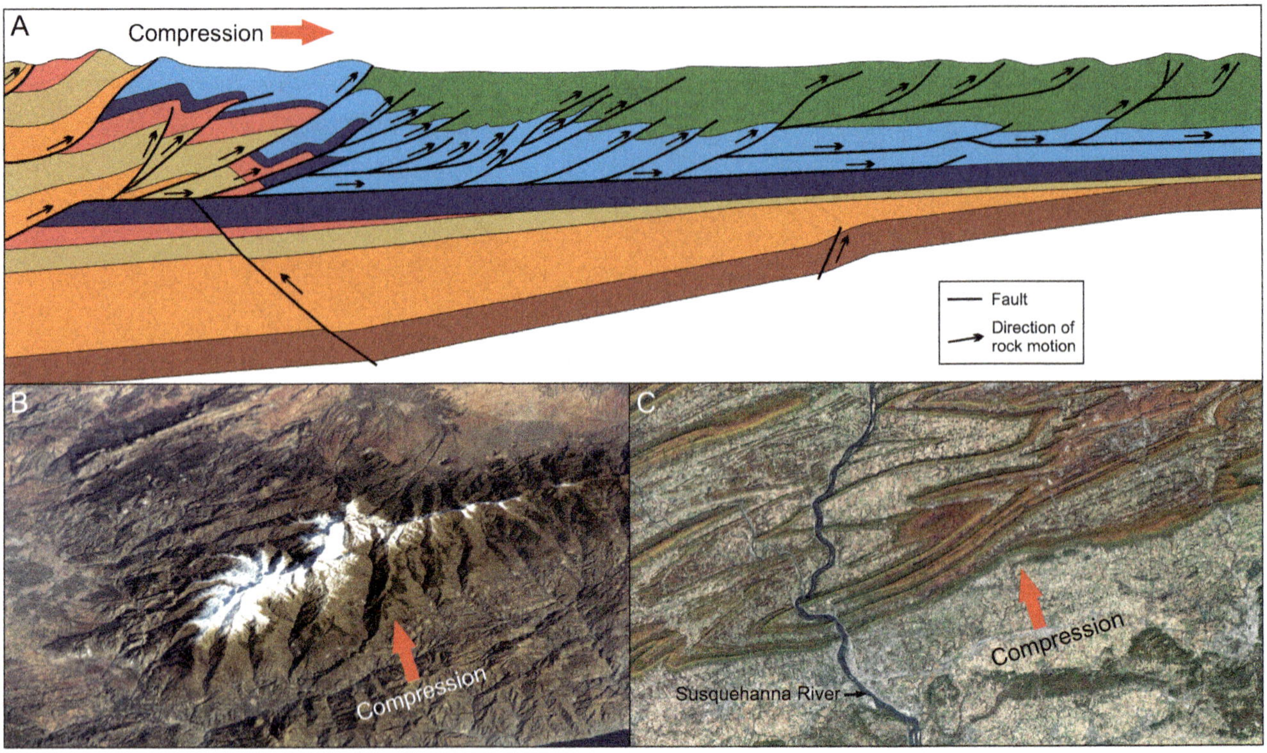

Figure 7.33: A. Generalized cross-sectional profile through the foothills of the Canadian Rocky Mountains. This profile typifies what happens during a compressional mountain building event with the rocks being both faulted and folded in multiple places. B. The Sierra Nevada in Spain is a young mountain range that formed when Europe and Africa collided. Photo credit: Image Science and Analysis Laboratory, NASA-JSC. C. The Appalachian Mountains in Pennsylvania are mere remnants of the mighty mountain range that once stood here. They formed when Africa and North America collided. Photo credits: Robert Simmon/NASA.

common. When the rocks are ductile, folding is more common. In contrast to extensional mountain building, high relief is created as layers of rocks are sheared off and stacked on top of each other. Note on the left side of Figure 7.33A how the same group of rocks units are repeatedly stacked onto one another. As the stack of rocks grows higher, so does the elevation of the mountain range. It should also be pointed out that the rocks in the core of the collision are likely going to be highly deformed and metamorphosed, a topic we'll explore further in Chapter 10. Generally speaking, the more energetic the collision, the larger the mountain range can grow.

Examples of compressional mountain building are numerous: the Appalachians, the Alps, and the Himalayas (Figure 7.15) to name a famous few. Some once mighty mountain ranges, like the Appalachians, have been actively eroding for hundreds of millions of years, while others, like the Alps, are still actively rising. This produces a noticeable difference in relief between "young" mountain ranges (Figure 7.33A) and "old" mountain ranges (Figure 7.33B).

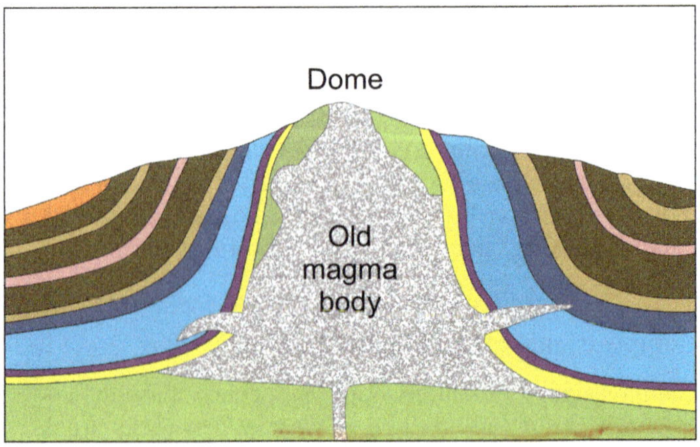

Figure 7.34: A generalized cross-section through Antoine Butte in Montana. Although small in scale, it exemplifies upwarping and the formation of a dome.

Upwarping

Upwarping involves the pushing up of the Earth's crust from below, usually by magma, causing the rocks to be folded upwards, forming a dome (Figure 7.34). On a larger scale, continental volcanic arcs also form high relief mountain chains, and in those cases, the relief is due in part to the emplacement and growth of magma chambers beneath the mountain range (e.g., the Cascade Range in Figure 7.14A). Sometimes, like in the case of the Black Hills of South Dakota, the lifting occurs over a relatively small area, producing a mountain range

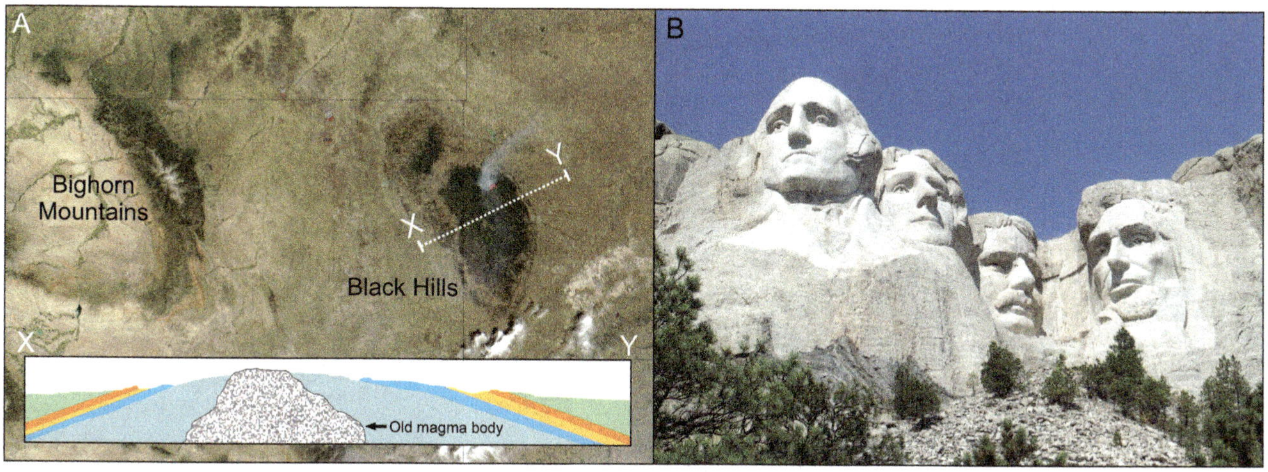

Figure 7.35: A. Satellite image of western South Dakota and eastern Wyoming showing the Black Hills and Bighorn Mountains. The inset contains a generalized cross-section through the Black Hills. This upwarp occurred approximately 60 million years ago. Photo credit: Jacques Descloitres/MODIS Land Rapid Response Team at NASA. B. Mount Rushmore is carved into the same rocks that formed the Black Hills. Photo credit: NPS.

and exposing deep crustal rocks at the surface (Figure 7.35A). In fact, the famous Mount Rushmore is carved into the rocks at the core of the Black Hills uplift (Figure 7.35B)! Note in figure 7.35 the Bighorn Mountains; they are also the result of deep crustal uplift. Uplift can also occur over a wide area, forming high plateaus. We already discussed one example, the Yellowstone Plateau, which is being lifted by the Yellowstone hot spot. The Colorado Plateau is also a good example of this (Figure 7.36A). In the case of the Colorado Plateau, warm buoyant rocks from the mesosphere have welled up under that region, much like a hot air balloon rises. The buoyant rocks have vertically lifted approximately 337,000 km^2 (130,000 mi^2) of land to an elevation of several thousand meters (Figure 7.36B). This uplift occurred in conjunction with volcanism around the edge of the plateau. Subsequent erosion by the Colorado River over millions of years is what eventually led to the formation of the Grand Canyon as we know it today (Figure 7.36C).

Figure 7.36: A. The Colorado Plateau. Image credit: NOAA. B. An overlook of the Colorado River near Hite, Utah. You can see the profile of the Colorado Plateau in the distance. Photo credit: B. Zaprowski. C. The Grand Canyon with the profile of the Colorado Plateau in the distance.

USGS

Leslie Gordon/USGS

Tim Orr/USGS

Alicia Burtner/USGS

USGS

USGS

Chapter 8 - Volcanism

When most people think of volcanism, they think of fountains of lava shooting into the air like in Hawaii or volcanic mountains like Mt. St. Helens (Figure 8.1). But volcanos and volcanic eruptions are only part of a larger process called **volcanism**. Most volcanism occurs deep in the ground, never even reaching the Earth's surface. Based on this simple distinction (above ground / below ground) we can divide volcanism into two basic types: **Intrusive volcanism**, which occurs beneath the Earth's surface, and **extrusive volcanism**, which occurs on the Earth's surface. Although we can't see intrusive volcanism in action, we can see some of the rocks that form as a result of intrusive volcanism. Let's explore.

Figure 8.1: Mount St. Helens in Washington. Photo Credit: B. Zaprowski.

Section 8-1: Intrusive Volcanism

Intrusive volcanism involves the emplacement of large bodies of magma in the Earth's crust. The source of this molten rock can vary from place to place. Along convergent boundaries the partial melting of subducting crust produces bodies of magma which rise up through the Earth's crust and feed volcanic arcs. Along divergent boundaries, magma rises up from the mesosphere to form new ocean crust. Deep mantle plumes carry magma from deeper down in the mesosphere to produce hot spot volcanism. In all of these cases, only a relatively small percentage of the magma actually erupts onto the Earth's surface. Most of the magma remains below the surface in the form of a **magma chamber** (Figure 8.2A). Magma chambers are the sources of magma that feed volcanism occurring at the surface. Over time the source of magma disappears and magma chambers cool and crystallize into igneous rocks. At this point, we refer to the old magma chamber as a **batholith** (Figure 8.2B). We cannot see the magma cooling and hardening into rock at this point, but over time erosion strips away the rock above the batholith, exposing it at the Earth's surface. The Sierra Nevada mountain range in California is the remnant of a long string of batholiths that used to feed a continental volcanic arc along the Pacific Coast. When the volcanism stopped around 60 million years ago, the magma chambers cooled and erosion (especially through the action of glaciers) removed several kilometers of overlying rock, exposing the ancient batholiths below (Figure 8.3A). Many of the world's granite quarries are found in batholiths, like the ones found in New England and North Carolina.

As the magma forces its way into the Earth's crust, the magma can radiate up and out from the magma chamber in the form of "sheets" and "pipes" of magma. Some of this magma may reach the surface and cause volcanic eruptions, but most of the magma cools and hardens below the Earth's surface (Figure 8.2B). These cooled features, when

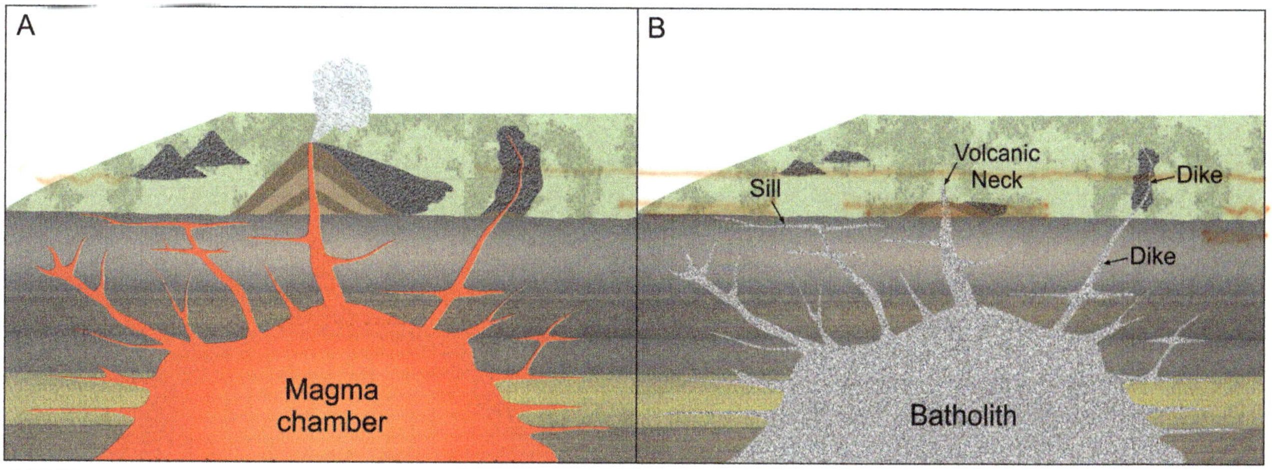

Figure 8.2: A. The intrusion of a magma chamber into the crust. B. When cooled, the old intrusion produces a variety of features that may be exposed on the Earth's surface as erosion removes rock from above.

Figure 8.3: Common intrusive volcanic landforms. A. El Capitan is a granite monolith that rises about 900 m (3,000 ft.) from the valley floor and is the batholith of an old volcanic arc. Photo credit: Alex Demas/USGS. B. This distinctive dark-colored sill in Antarctica is a remnant of a massive volcanic plumbing system that produced major eruptions about 180 million years ago. Photo credit: Robert Simmon/NASA/EO. C. A dike near Torrey, Utah. Photo credit: B. Zaprowski. D. Devil's Tower in Wyoming, is probably one of the most famous examples of a volcanic neck. Photo credit: B. Zaprowski.

exposed by erosion, may form distinct landforms. Horizontal sheets of cooled volcanic rock are called **sills** while the vertical sheets of cooled volcanic rock are called **dikes** (Figures 8.3B-C). The cores of old volcanos, when exposed by erosion, may also form a distinct landform called a **volcanic neck**. A very famous volcanic neck is Devil's Tower in Wyoming (Figure 8.3D). It was featured prominently in the movie *Close Encounters of the Third Kind*.

Section 8-2: Extrusive Volcanism

When people hear the term volcanism, they usually visualize extrusive volcanism. In extrusive volcanism, magma erupts onto the Earth's surface in a variety of ways. Once magma reaches the Earth's surface it is known as **lava**. Before discussing the hazards associated with the different types of extrusive volcanism, it is important to first discuss why some volcanos are explosive (like Mt. St. Helens) while others produce massive amounts of lava but don't really explode (like Kilauea in Hawaii). The differences in explosiveness can be explained by looking at differences in magma composition (Table 8-1). In particular, there are two properties of the magma which explain why certain types of eruptions occur at certain locations: The silica content and the gas content.

Silica content is important because it controls the **viscosity** of the magma. By definition, viscosity is the resistance to flow. Magma with a lower silica content has a low resistance to flow, while magma with a higher silica content resists flowing. To help illustrate this, consider three common foods: spaghetti sauce, salad dressing and maple syrup (Figure 8.4). The thick spaghetti sauce hardly flows anywhere when the plate is tipped, demonstrating high viscosity (that is, a high resistance to flow). More importantly, if a substance has a high resistance to flow, then it will take considerably more force to set it in motion. So it should be no surprise that explosive volcanos have magmas with high viscosity. It allows the kinetic energy in the

Silica Content	Viscosity	Gas Content	Explosiveness
50%	Low	1-2%	Low
60%	Moderate	3-4%	Moderate
70%	High	4-6%	High

Table 8-1: This table shows how a magma's silica content, viscosity and gas content all contribute to volcano explosiveness.

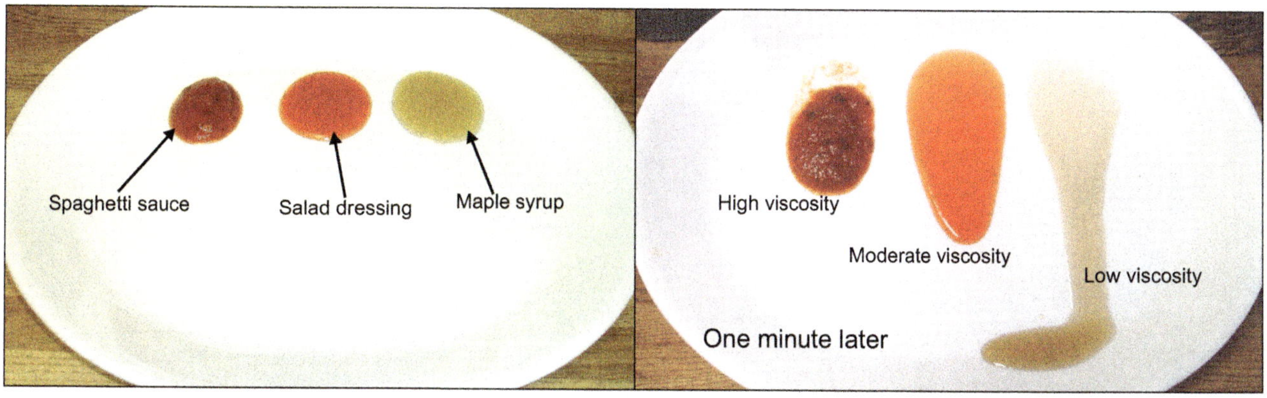

Figure 8.4: Spaghetti sauce, salad dressing and maple syrup can be used to demonstrate different levels of viscosity (the resistance to flow). The platter was held at an angle for 1 minute. Photo credit: B. Zaprowski.

volcano to build up to high levels before the eruption occurs. In contrast, substances with a low viscosity, like the maple syrup, flow very easily. Volcanos with low viscosity magma do not tend to be explosive because it takes considerably less energy to overcome the resistence to flow and very little kinetic energy builds up.

Figure 8.5: The gas bubbles in soda are a good analogy for the gas dissolved in magma. Photo credit: B. Zaprowski.

But that is only part of the story. Magma can also contain significant quatities of dissolved gas, especially when the magma is under high pressure, deep in the asthenosphere. For comparison, carbonated beverages, like soda (Figure 8.5), have about 0.5% gas dissolved in it while magma may contain 1-6%. As the magma works its way towards the surface, the gasses begin to expand and pressure builds up (kind of like when you open a bottle of soda and bubbles start to appear). It should be no surprise then that explosive volcanoes also tend to have large amounts of dissolved gas in their magma. An explosive volcano is like a bottle of champagne. At first, the cork has enough resistance to withstand the pressure of the gas inside the champagne bottle. If you start shaking the bottle, however, the pressure builds as the gasses begin to bubble up and expand. Eventually, the force exerted by the expanding gasses overcomes the cork's resistance to move and "POP" you have an "eruption," and a violent one at that. The cork flies out at a dangerously high speed and the champagne sprays everywhere.

The process is not too different with an explosive type volcano. As the pressure (kinetic energy) builds inside the volcano, the whole mountain can swell. Eventually the pressure exerted by the expanding gasses and the continuous supply of magma from the magma chamber overcomes the confining pressure above and an eruption occurs. The magnitude of this eruption is proportional to the amount of kinetic energy that builds up beforehand. The violent ejection of magma, volcanic ash and other rocky material (called pyroclastic debris) is common during the initial stages of the eruption, and the top of the mountain may be literally blown off (Figure 8.6).

Figure 8.6: The 1980 Mt. St. Helens eruption is a good example of an explosive type volcanic eruption. Photo credit: Donald A. Swanson/USGS.

What about the non-explosive (Hawaiian) type eruptions? The case is the opposite for them. Typically, the magma associated with Hawaiian type eruptions has low viscosity and low amounts of dissolved gas in it, so the less viscous magma spills out onto the surface under low pressure to form lava flows like those in Figure 8.7A. Think of a champagne bottle with a greased cork in it (low viscosity magma). It would take very little shaking to make the cork come out of the bottle, and it would do so with very little force because there wasn't

Figure 8.7: A. This photo from Kīlauea in Hawaii is a good example of a non-explosive type eruption with its lava fountains and rivers. Photo credit: HVO/USGS. B. This photo from Hawaii shows just some the large areas of land covered by the lava flows from the very active Kilauea Volcano. Photo credit: Tim Orr/USGS.

any significant build-up of kinetic energy beforehand. The champagne will now just flow out of the bottle. Moderately explosive volcanos fall somewhere between the two extremes with moderate viscosity and moderate amounts of dissolved gas.

There is one more piece of this story we must explain. Typically, certain types of volcanism occur in specific tectonic settings. Volcanic eruptions along divergent boundaries and above many deep mantle plumes tend to be the non-explosive type. In general, the magmas that feed this type of volcanism have low silica content and low gas content. Eruptions with massive amounts of lava spilling out onto the land surface are quite common in these tectonic settings. For example, Kilauea, in Hawaii, has been erupting continuously since 1983, producing an estimated 2.7 km^3 (0.6 $mi.^3$) of lava covering 117 km^2 (45 $mi.^2$). Figure 8.7B shows just a mere fraction of these lava flows. On the other side of the coin, explosive type volcanos are more common along convergent boundaries. When an oceanic plate is subducted, some of the water-saturated sediments are carried down with the plate. The water-saturated sediments cause partial melting of the asthenosphere above and add dissolved gasses (including water vapor) to the newly-generated magma. The magma then works its way towards the surface to feed the volcanic arc above, as in the case of Mount St. Helens. Rifting events tend to produce less-explosive type volcanic eruptions, typically somewhere between the two extremes.

Section 8-3: Volcanic Hazards

There are hazards associated with volcanism of any kind. The simple release of a gas cloud from a fracture in the side of a volcano can be fatal for thousands of people. Some volcanos can display a mixture of different hazards depending on the geographic setting. For example, a volcano with a snowy peak is much more likely to produce a volcanic mud flow (a lahar) than a volcano without one (Figure 8.8). Although most volcanic hazards are quite destructive in terms of property, deaths can be minimized by evacuating people to a safe distance from the eruption. Sometimes it can be difficult to convince people to stay away when they have been living near a volcano which hasn't erupted in thousands of years. Unfortunately millions of people currently live near active volcanos. For those people, there are a number of volcanic hazards which pose a threat to their property and their lives.

Figure 8.8: Snowy volcanoes like these can produce volcanic mudflows without even erupting. Photo credit: USGS.

Lava Flows

Although lava flows are somewhat unpredictable, people can generally get out of harm's way with adequate warning. On the other hand, there isn't much

Figure 8.9: Lava flow hazards. A. A portion of charred, lava covered forest along the east rift zone of Kilauea. Photo credit: USGS. B. Close up of a lava flow from the Kilauea eruption of 2014 burning a shed near Pahoa, Hawaii. Photo credit: Kelly Hudson/FEMA.

people can do to stop the lava flows from running over their property. Many attempts have been made to divert or stop lava flows in the past, but only a handful have ever succeeded. Lava flows, though quite destructive, are easy enough to mitigate in terms of human lives, but property damage is almost inevitable (Figure 8.9).

The nature of the lava flow reflects its viscosity. Some lava flows are like spaghetti sauce, while others are more like maple syrup (Figure 8.4). Low viscosity lavas, which flow like maple syrup, are known as **pahoehoe** (Figure 8.10A). These lavas flow smoothly and easily, sometimes for long distances, before finally slowing down and solidifying. The resulting cooled lava produces a smooth dense rock (Figure 8.10B). Higher viscosity lavas, called **aa**, flow much slower than pahoehoe. They behave more like the spaghetti sauce. Because aa flows move slowly, they tend to form a rocky crust on their surface as they flow. This crust then gets broken up and churned around as the lava slowly flows along (Figure 8.10C). This produces a "chunky" lava flow that can build up to considerable

Figure 8.10: A. This photo from Kilauea shows the smooth nature of pahoehoe. B. Cooled pahoehoe lava is smooth or ropy in texture. C. This aa lava flow is slowly pushing its way through a forest in Hawaii. D. This photo shows the jagged nature of cooled aa flows. Photo credits: HVO/USGS.

heights. The resulting moving "wall" of lava and hot jagged pieces of rock consumes everything in its path. The cooled lava produces a sharp jagged rock with lots of voids and spaces known as **scoria** (Figure 8.10D). Scoria is commonly used in gas grills because of its excellent ability to transmit heat evenly.

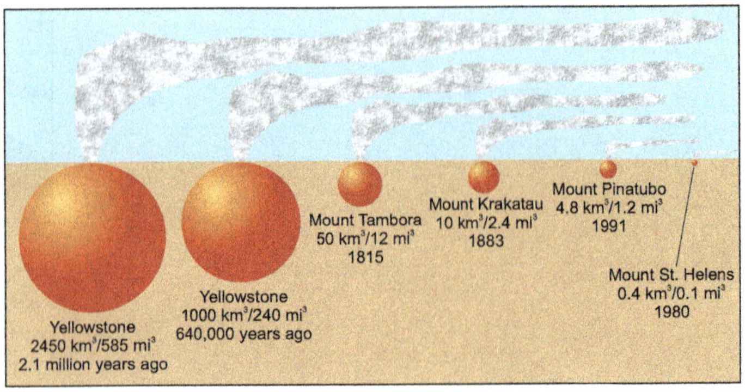

Figure 8.11: Volume of pyroclastic material erupted from different volcanos in terms of the original volume of magma released.

Pyroclastic Debris

Pyroclastic debris of some form is common in almost all types of eruptions. **Pyroclastic debris** is a catchall term which refers to the rocks and volcanic deposits which are ejected during a volcanic eruption. In the case of explosive volcanism, some of the debris is pieces of the volcano itself that are ripped up during the eruption. An enormous volume of pyroclastic material may be ejected during a major eruption, much of it in the form of volcanic ash (Figure 8.11). Mt. St. Helens produced about 0.4 km^3 (0.1 mi.^{3}) of pyroclastic debris when it erupted in 1980, which is actually not very much when compared to other eruptions like that of the Yellowstone Supervolcano 2 million years ago (2450 km^3/585 mi.^{3})!

Pyroclastic debris is grouped into three main categories based on size. The biggest pieces, called **bombs**, are larger than 64 mm (~2.5 in.) and don't land too far from the volcano (Figures 8.12A-B). Interestingly, many of these projectiles take on a streamlined form as air resistance shapes the hot, soft magma as it flies through the air. Pieces of pyroclastic debris which range in size from 64 mm to 2 mm are called **lapilli** or **cinders** (Figure 8.12C). Any piece of pyroclastic debris which is smaller than 2 mm in size is classified as **volcanic ash**, although most volcanic ash is microscopic in size (Figure 8.12D). Pyroclastic debris deposits are subdivided into two categories: **Pyroclastic falls** and **pyroclastic flows**.

Pyroclastic fall deposits are the result of magma which is flung through the air, forming a wide variety of different

Figure 8.12: Examples of pyroclastic debris. A. A very large bomb. Photo credit: B. Zaprowski. B. Smaller bombs. Photo credit: USGS. C. Lapilli. Photo credit: B. Zaprowski. D. Volcanic ash. Photo credit : D. Weiprecht/USGS; inset: A.M. Sarna-Wojcicki/USGS

sized-particles that rain down on the areas surrounding the volcano. In the case of less explosive eruptions, like in Hawaii, blobs of magma may fly into the air, forming bombs and cinders, but generally there is very little volcanic ash produced and the material doesn't travel far from its source (Figure 8.13). In the case of explosive eruptions, huge amounts of volcanic ash may be produced in addition to large pieces of the mountain which are thrown in the air during the eruption. In terms of hazards, the larger pieces of debris aren't likely to cause a high number of deaths if people are properly evacuated from the hazard zone.

Figure 8.13: Blobs of magma fountain upwards and cool to form piles of cinders around a fissure. Photo credit: USGS.

Volcanic ash is actually the most dangerous type of pyroclastic fall debris. Volcanic ash can be fatal when inhaled and wearing a mask at all times is the only way to prevent the tiny shards of volcanic glass from damaging your lungs (Figure 8.14A). In areas closer to the eruption the build-up of thick layers of volcanic ash may cause buildings to collapse (Figure 8.14B). The ash can also cause car and plane engines to seize up as the ash (which is

Figure 8.14: A.These two workers are wearing masks to prevent accidental inhalation while they clean ash up in their town. Photo credit: USGS. B. These buildings lie in ruins after collapsing from the weight of ash produced by the eruption of Mount Pinatubo in 1991. Photo credit: Willie Scott/USGS.

harder than metal) grinds up the metal in the pistons. As a result, volcanic eruptions, like the Iceland eruptions in 2010, can cause major disruptions in airline traffic. And unlike all the other volcanic hazards, volcanic ash can be spread over a wide area, impacting the lives of thousands or millions of people.

Pyroclastic flows are probably the most dangerous volcanic hazard of all and several famous volcanologists have been accidentally killed by pyroclastic flows while filming volcanic eruptions up close. Pyroclastic flows are high-speed clouds of hot gasses, rocks and volcanic ash which flow down the sides of a volcano (Figure 8.15). Pyroclastic flows can reach temperatures of 1,000 °C (1,800 °F) and will incinerate anything they come in contact with. Pyroclastic flows occur because many of the gasses ejected by a volcano during an eruption are heavier than air (i.e., carbon monoxide, sulfur dioxide). These gasses flow down the flanks of the volcano and sweep up rocks,

Figure 8.15: A pyroclastic flow descending the south flank of Merapi, May 14, 2006. Photo credit: USGS.

ash and whatever else stands in their way, forming a towering wall of debris which can level everything in its path. Pyroclastic flows can reach speeds of 700 km/hr (450 mi/hr) and their speed is a function the steepness of the slope, the overall density of the flow, and the rate at which the material erupts from the volcano. Pyroclastic flow deposits come in a range of sizes from 1 to 200 m (1 to 650 ft.) thick and hundreds to thousands of cubic meters in volume. Smaller pyroclastic flows may flow a few kilometers from the volcano, but some ancient pyroclastic flows are known to have traveled hundreds of kilometers and were more than a thousand cubic *kilometers* in volume! Predominantly, pyroclastic flows tend to flow in valleys. Because they move fast they are difficult or impossible to escape from. In 2010, the Mount Merapi volcano in Indonesia killed 341 people, destroyed more than 2,000 homes and forced the evacuation of nearly 350,000 people (Figure 8.16A). Much of the damage was due to pyroclastic flows and lahars (more on that below). But this wasn't just an isolated event. Mount Merapi is an active volcano and pyroclastic flows of various sizes occur every few years (Figure 8.16B). Luckily with proper warning most people escape unharmed. Other have not been so fortunate. In 1902, 30,000 people were killed by a pyroclastic flow which swept over the capital city of Martinique. More recently, in 2010, Soufrière Hills Volcano on the Caribbean island of Montserrat experienced a partial dome collapse. Pyroclastic flows destroyed many buildings in the village of Harris before flowing out to sea (Figure 8.16C). Montserrat has been experiencing frequent eruptions and pyroclastic flow events since 1995.

Lahars

Lahars can be just as dangerous as pyroclastic flows and perhaps even more so. **Lahars** are fast-flowing slurries of volcanically-derived rock, mud, ash and debris and they occur mostly in river valleys adjacent to the volcano (Figure 8.17A). Lahars commonly have the consistency of wet concrete and have the ability to carry pieces of rock

Figure 8.16: A. The aftermath of the Merapi volcano in 2010. This house in the village of Kopeng, 6 km (3.7 mi.) from summit of Mount Merapi, was destroyed by the explosive eruption. B. This house was buried by pyroclastic flows during the 2006 eruption of Mount Merapi. Photo credits: USGS. C. Pyroclastic flow depsoits (tan) produced by the active Soufrière Hills Volcano on the island of Montserrat. Image credit: Robert Simmon/NASA.

Figure 8.17: A. The leading edge of a lahar triggered by heavy rain crashes down the Jiangjia Gully in China. The flow front is about 5 m (16 ft.) tall. Photo credit: K. Scott/USGS. B. This lahar in South America traveled over 100 km (60 mi.), killed over 23,000 people and destroyed more than 5,000 homes. Photo credit: J. Marso/USGS. C. The town of Armero (to the upper left in image B) was hardest hit with over three quarters of the residents killed and the town totally destroyed. Photo credit: N. Banks/USGS. D. Lahars triggered by the 1980 eruption of Mt. St. Helens buried this house. Photo credit: Lyn Topinka/USGS.

that are more than 10 m (30 ft.) in diameter! Lahars travel at speeds of 30-65 km/hr (20-40 mi/hr) and can travel more than 100 km (60 mi.) from their source area. Lahars can range in size from centimeters deep and a few meters wide to 10's of meters deep and 100's of meters wide. They grow in size by incorporating rocks, trees, mud and water as they travel down the flanks of the volcano, and may grow to more than 10 times their original size before eventually slowing and coming to a stop. Lahars most commonly occur on or near stratovolcanos. Most stratovolcanos are tall and their peaks are high enough to be covered in snow for most or all of the year. Lahars are also associated with other types of volcanos in polar regions, like those found in Iceland.

Lahars are dangerous for several reasons. First, the muddy debris can knock down structures and bury people alive (Figure 8.17B-D). For example, a lahar triggered by the 1985 eruption of the Nevado del Ruiz volcano in Columbia killed 23,000 people, injured 5,000 people, and destroyed more than 5,000 homes. Particularly hard hit was the town of Armero where three-quarters of it residents were killed (Figures 8.17B-C). Second, lahars can occur well after the main eruption has occurred; sometimes days later. In the case of the Armero lahar, the lahar traveled over 100 km (60 mi.) and hit the town four hours after it left the summit. Third, lahars can occur without an actual eruption occurring. Heat from the magma just below the surface of a snow-capped volcano can cause the snow above to melt and trigger lahars, and in many cases there is no warning the lahar is about to occur! In addition, volcanic eruptions can produce natural dams by producing river-blocking landslides, pyroclastic flows and/or lava flows (Figure 8.18). These dams

Figure 8.18: Castle Lake formed when a lahar from Mount St. Helens' blocked this former tributary of the North Fork Toutle River. Photo credit: Robert L. Schuster/ USGS.

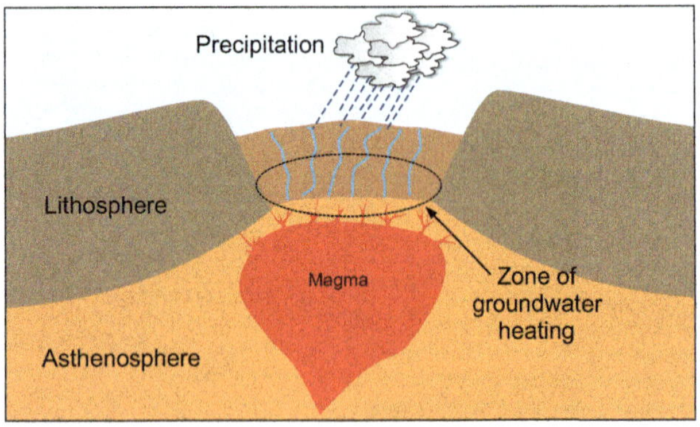

Figure 8.19: When bodies of magma get really close to the Earth's surface, groundwater can be heated to produce hydrothermal features like geysers and hot springs.

may cause rivers to back up for a period of time, flooding the land behind them. Even worse, the rocky debris may be weak and unstable, so dam failure is a real risk. This may lead to catastrophic flooding in the valley below. Heavy rain events during or after a volcanic eruption may lead to repeated and prolong lahar events as well. Thankfully in some areas the development of lahar detection equipment may be able to provide residents with ample warning time so they may evacuate the hazard zone.

Geothermal hazards

Geothermal hazards occur in places where the groundwater can interact with the magma just below the surface. As groundwater circulates through the bedrock, it is heated, sometimes converting it into superheated water and steam (Figure 8.19). This produces a wide variety of features, most of which are displayed at Yellowstone National Park which sits over the Yellowstone hot spot (Chapter 7). Hundreds of thousands of people visit the park each year to see these beautiful and fascinating displays which are produced by a magma chamber mere miles beneath their feet. Be warned however: these features, although beautiful, are still hazards and the danger of being seriously burned or even killed is very real.

Hot springs are places where this hot groundwater bubbles out at the Earth's surface. Because the water is hot, it has a lot of dissolved minerals in it, and the water in hot springs can appear as a beautiful array of different colors (Figure 8.20A). Many famous resorts and spas have been built near hot springs because the mineral-rich waters are thought to contain healing properties. When hydrothermally-heated water flows through limestone bedrock and then flows out onto the Earth's surface, calcium carbonate is deposited. These carbonate deposits produce unique features

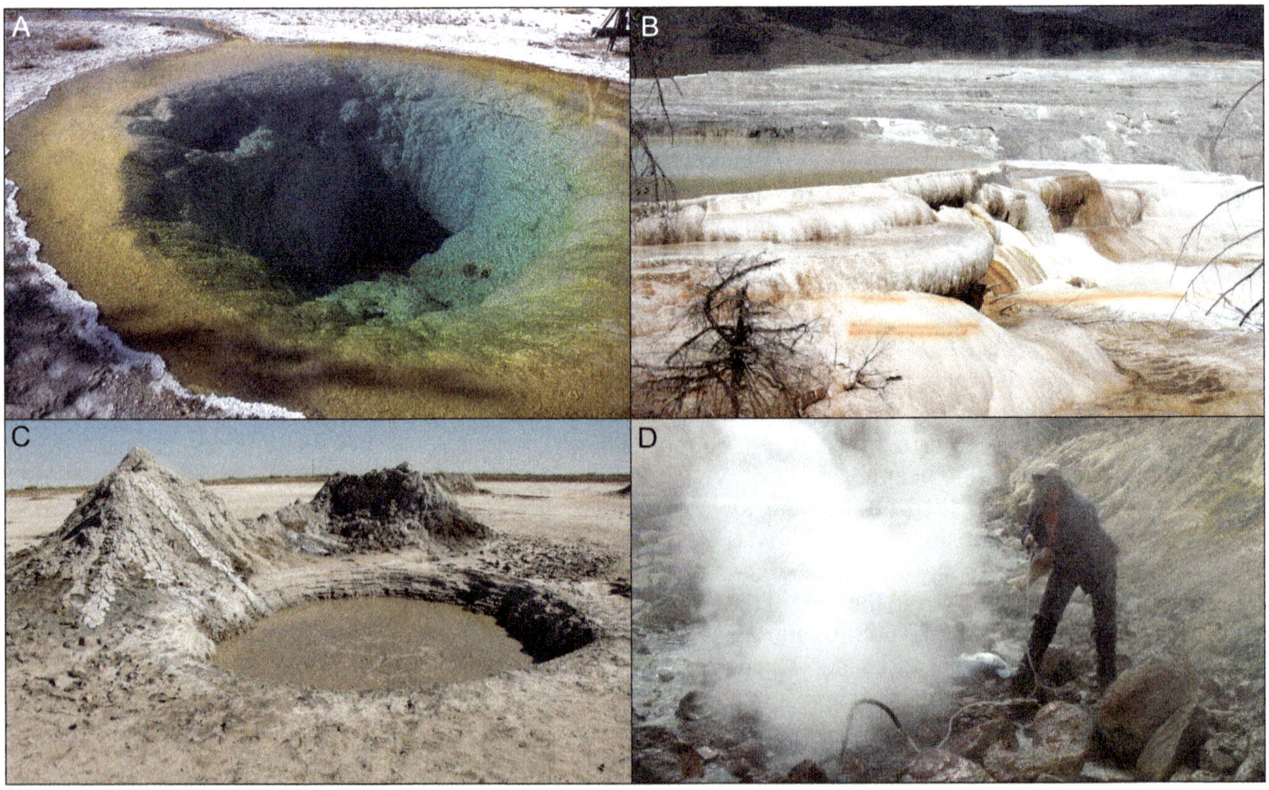

Figure 8.20: A collage of hydrothermal features. A. Morning Glory hot spring in Yellowstone National Park. Photo Credit: Jim Peaco/USGS. B. Travertine terraces in Yellowstone National Park Photo credit: Jim Peaco/USGS. C. A mudpot bubbles away near the Salton Sea, CA. Photo credit: Douglas Barnum/ USGS. D. A geologist prepares to sample the gasses emanating from this fumerole in Little Hot Springs Valley at Lassen Volcanic National Park. Photo credit: Fred Murphy/USGS.

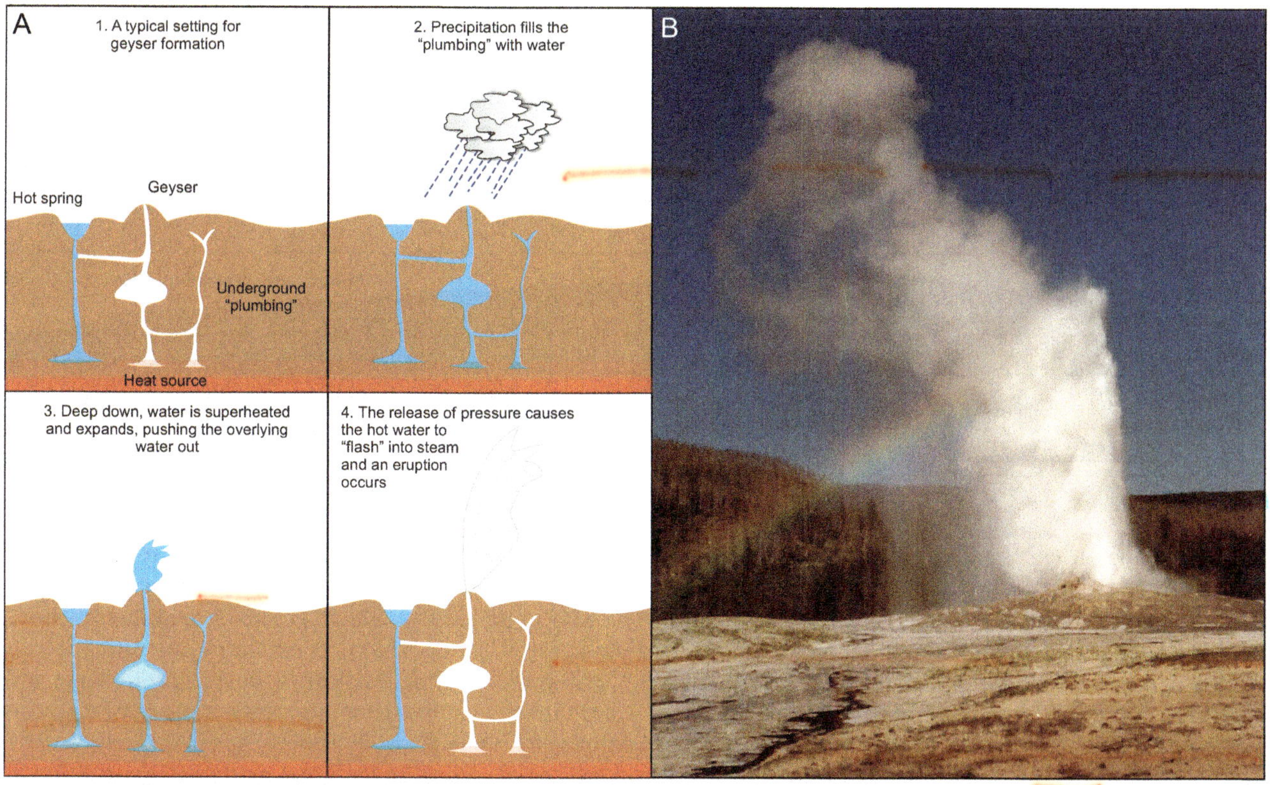

Figure 8.21: A. How geysers work. Geysers only form when the right type of "plumbing" (cracks in the rocks) exists in the subsurface. As the cracks fill with water, the water is heated to the boiling point. As the water expands and escapes, it flashes into steam, causing a violent eruption of water and steam. B. Old Faithful erupts with a rainbow. Photo credit: B. Zaprowski.

called **travertine terraces** (Figure 8.20B). If the hydrothermally heated water flows through clay, shale or volcanic ash, hot bubbling **mudpots** may form (Figure 8.20C). Sometimes only steam and noxious (potentially deadly) gasses hiss from openings in the ground, forming a feature called a **fumerole** (Figure 8.20D).

Geysers are probably the most famous of all hydrothermal features (Figure 8.21). To form a geyser, you need to have a geologic situation where the water stored in the subsurface is temporarily confined, usually in a complex series of connecting caves. The water stored in the lower chambers is confined by the weight of the cooler water above. Over time, the groundwater is heated, causing water temperatures to rise, and more importantly, causing pressure to build. Eventually, the pressure generated by the superheated groundwater below overcomes the weight of the overlying water holding it back, pushing it up. When this happens, the high-pressure, high temperature water at the bottom "flashes" into steam. The hot water and steam erupt through a hole in the ground to produce the geyser. The most famous geyser of all, Old Faithful in Yellowstone National Park erupts on a regular basis but most geysers have very irregular eruption schedules; years may pass between eruptions in some cases (Figure 8.21B).

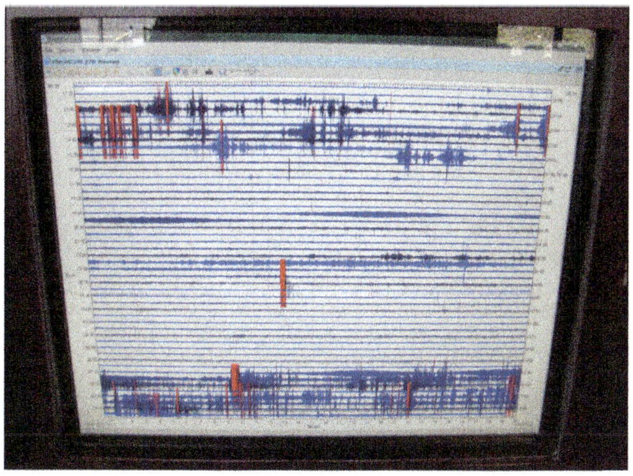

Figure 8.22: This computer display shows the small tremors being produced by the magma activity inside Mount St. Helens. Photo credit: B. Zaprowski.

Earthquakes

Earthquakes and tremors are also quite common in areas experiencing some form of volcanism (Figure 8.22). Much of the motion and slippage of the rocks is due to the movement of the magma inside the volcano. As the magma chamber grows, the volcano swells outwards, causing rocks to break, producing earthquakes. Swarms of earthquakes tend to occur just prior to an eruption, so volcanologists constant monitor earthquake activity for signs of an eruption. These earthquakes can trigger landslides and lahars, damaging nearby properties.

Section 8-4: Volcanic Structures

When many people think about what a volcano is shaped like, they probably have a picture in their heads of a peaked, cone-shaped mountain peak like they commonly see in the movies. But in fact there is a range of different sizes and shapes that volcanos come in. We can group all volcanos into four basic types of volcanic structures: Shield volcanoes, stratovolcanos, cinder cones, and lava plateaus. Each type of volcanic structure reflects the type of extrusive volcanism that forms them. Figure 8.23 shows the relative size and shape of the first three types of volcanos.

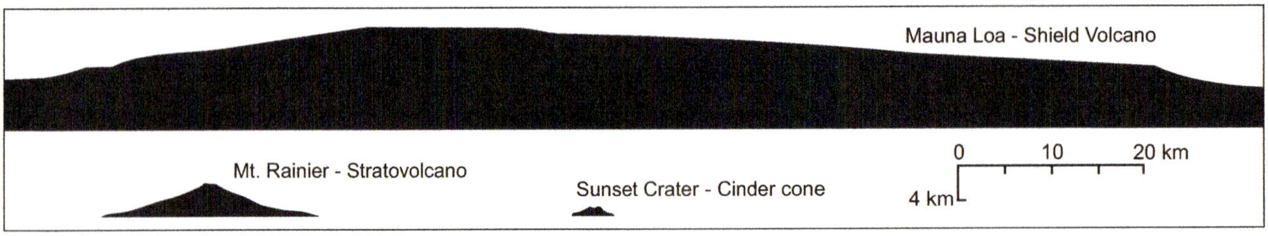

Figure 8.23: This diagram shows the relative size differences between the three of the main types of volcanos (lava plateaus are too big to be shown).

Mauna Loa is an excellent example of a shield volcano. In fact, all of the Hawaiian islands are former shield volcanoes associated with the Hawaiian Hot Spot (Figure 7.18B). **Shield volcanos** are the largest type of volcanic structure, and they are very broad, much like a warrior's shield (Figure 8.24A). Shield volcanos are typically non-explosive in nature and don't commonly produce much volcanic ash. They do however erupt large quantities of lava (Think of the lava flows produced by Kilauea in Figures 8.7, 8.9 and 8.10). Shield volcanos, besides being large, are also quite complex. Lava can erupt from multiple places simultaneously. Over time, successive layers of lava can cause shield volcanos to grow to very large sizes. The lava commonly pours out of large openings in the flanks of the volcano called **craters** (Figure 8.24B) or it may erupt out of large cracks forming **fissure eruptions** (Figure 8.24C). The lava also flow just beneath the surface via **lava tubes**, sometimes for a distance of several kilometers (Figure 8.24D).

Mount St. Helens and its sister volcanos of the Cascade Range are excellent examples of **stratovolcanos**. Stratovolcanos are generally the explosive type of volcanos and are commonly associated with convergent plate boundaries. They are built from alternating layers of pyroclastic debris and lava flows. This gives stratovolcanos

Figure 8.24: A. Mauna Loa Volcano in Hawaii is one of the largest shield volcanos in the world. Photo credit: K. Baccala. B. Lava lake in Halema'uma'u Crater on Kilauea. Photo credit: HVO/USGS. C. Lava pours from a fissure. Photo credit: HVO/USGS. D. A geologist takes a sample of lava flowing out of a lava tube. Photo credit: HVO/USGS.

that "peaky" shape that most people associate with a volcano (Figure 8.25, also see Figures 8.1 and 8.8). Stratovolcanos, being explosive in nature, are very dangerous and produce almost all of the different types of volcanic hazards discussed previously in Section 8.3.

Figure 8.25: Mount Rainier in Washington is a classic example of a stratovolcano. Photo credit: D. Harris.

Cinder cones fall somewhere in between shield volcanos and stratovolcanos in terms of their explosiveness and the volcanic hazards they produce. Cinder cones are composed almost exclusively of pyroclastic debris (Figure 8.26). As a result, they don't grow particularly big, relative to the other primary types. They can produce lava flows, but these are much smaller in scale and shorter-lived compared to the vigorous Hawaiian-type lava flows. Cinder cones can form in a variety of tectonic settings, including on the flanks of shield volcanos and in areas where extensional mountain building is occurring.

Figure 8.26: A. An active cinder cone in Hawaii. Photo credit: J.D. Briggs/USGS. B. A variety of different sized cinder cones in Craters of the Moon National Park. Photo credit: D. Harris.

Lava plateaus are different than the three types of volcanism explained above. Lava plateaus are the result of flood volcanism. In flood volcanism, huge volumes of very low viscosity lava pours out from large cracks or fissures in the ground, producing layer upon layer of relatively flat volcanic rock. Because the lava has such low viscosity, it can flow very long distances. As a result, these lava flows don't produce a single "volcano." Instead, successive eruptions of lava build up layers of rock that can cover hundreds of thousands of square miles and pile up thousands of meters thick! Large portions of Oregon, Washington and Idaho are covered by lava plateaus which formed between 14-6 million years ago (Figure 8.27). Other famous lava plateaus include the Deccan Traps in India and the flood basalts of Siberia.

Figure 8.27: The Columbia Plateau in central Washington. If you look at the right side of the image, you can see seven of the many lava flows which built up the Columbia Plateau over millions of years. Photo credit: B. Zaprowski.

Walter D. Mooney/USGS
USGS
USGS
Dan Ponti/USGS
Katherine Mueller/NCDC/NOAA

Chapter 9 - Earthquakes

Earthquakes are nothing more than the release of potential energy stored in rocks along a fault. A **fault** is a break through a group of rocks caused by shifts in the Earth's crust (this makes them distinct from cracks and fractures caused by weathering). For example, in Figure 9.1 the red-colored rocks are separated from the tan-colored rocks by a fault, and the red-colored rocks have been moved upwards. Although a new fault may form in conjunction with an earthquake event, most earthquakes occur along pre-existing faults. Earthquakes occur all the time in the Earth's crust, but most earthquakes are not large enough to be noticeable. Large, damaging earthquakes occur much less frequently and require the build-up much larger of amounts of potential energy. Since the energy needed to generate large earthquakes comes from plate tectonics, most large earthquakes occur along plate boundaries (Figure 9.2). And, as discussed in Chapter 8, they occur in areas which are volcanically active because migrating plumes of magma deform and rupture the rocks they are pushing through.

Figure 9.1: This fault is very obvious because the rocks on each side of the fault are such different colors. Photo credit: B. Zaprowski.

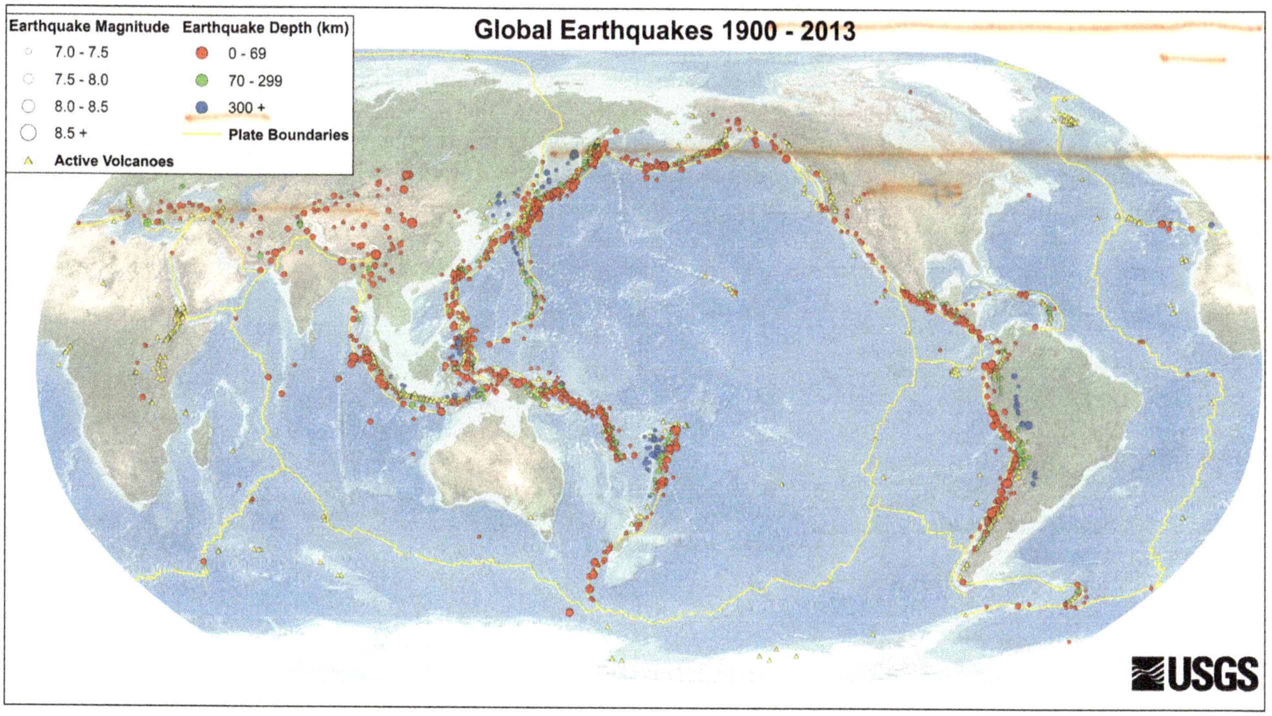

Figure 9.2: A global seismicity map of earthquakes with a magnitude 7.0 or higher from 1900-2013. Image credit: U.S. Geological Survey.

On the other hand, small earthquakes occur just about anywhere, even in the middle of a tectonic plate. These earthquakes are commonly produced by the slow and gentle vertical (up and down) motions of different parts of a tectonic plate. Since the plates are "floating" on the asthenosphere, they can rise and sink when mass is removed from or added to them by weathering, erosion and deposition (Figure 9.3). As mass shifts around on the tectonic plates, stresses build-up and small earthquakes occur, commonly along older pre-existing faults. In this context, small earthquakes are much like the groans and squeaks you hear in an old sailboat. As the boat rocks in the waves, differential stresses cause the wood to squeak and groan. This is why the 2011 earthquake outside of Washington, D.C. occurred. In this chapter, we will review the basic properties of earthquakes, how we measure them, and the various hazards they can cause.

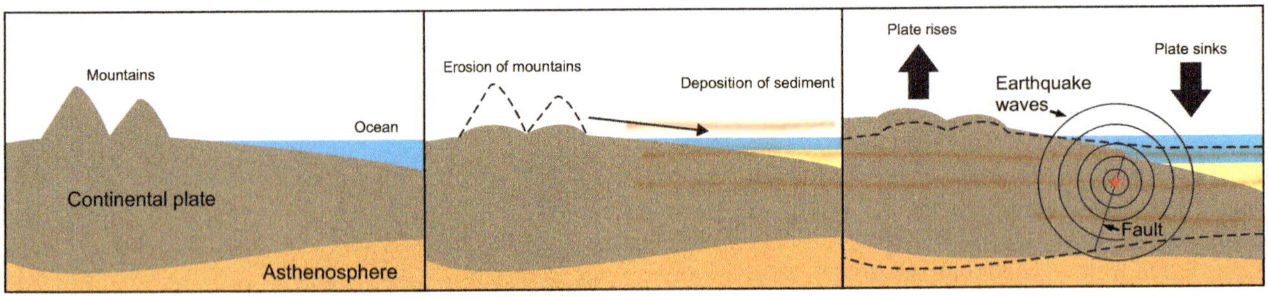

Figure 9.3: Many small earthquakes are caused by slow vertical motions of the plates. Over time, as the land surface is eroded down, the plate rises up (or rebounds), just like a cargo ship rises up out of the water after its cargo has been removed. Meanwhile, the weight from the sediments deposited at the edge of a continental plate pushes the crust down. This slow tectonic movement produces lots of small earthquakes over time (red star).

Section 9-1: Anatomy of an Earthquake

When tectonic stresses are applied to a group of rocks, the rocks will either break and fracture or bend and be deformed in order to accommodate the forces acting on them. Cold "brittle" rocks tend to break along **faults** when put under stress, while warm "plastic" rocks, like those found deeper in the lithosphere, tend to deform into **folds**. Structural geologists study this deformation because it directly reflects the type of stress that was exerted on a group of rocks and from what direction the stress came. This allows geologists to infer where earthquakes may have occurred in the past and where they might occur again in the future.

The place on a fault where an earthquake occurs (that is, the place where energy is released from the rocks) is called the **focus** (Figure 9.4A). The energy released by an earthquake moves away from the focus in all directions, kind of like a ripple forms on a pond after you throw a rock in it. This energy can move through the Earth in several different wave forms. However, when earthquakes are reported in the news, the location of the **epicenter** is given. The epicenter is the place on the Earth's surface *directly above the focus*. This is the place on the Earth's surface where the earthquake's energy will be strongest (and the damage greatest) but it is **not** where the earthquake actually occurs. If motion along the fault causes the rupture of rocks to occur at the Earth's surface, a **fault scarp** is produced (Figure 9.4B).

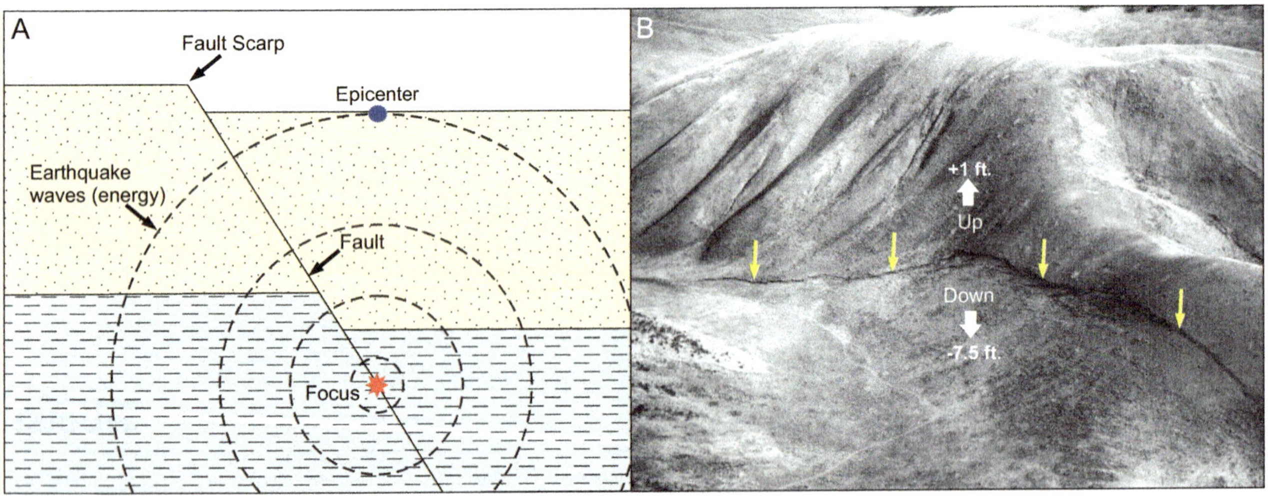

Figure 9.4: A. The anatomy of an earthquake. The energy release occurs at the focus while the epicenter is the location on the Earth's surface directly above the focus. If the fault breaks the Earth's surface, a fault scarp is created. B. This large fault scarp (yellow arrows), formed during the 1983 Idaho earthquake, cutting across the foothills of the Lost River Range, and lifting the range 1 foot. Photo credit: R.E. Wallace/USGS Photo Library.

All faults can be placed into one of three groups based on the relative motion of the rocks along that fault: **Dip-slip**, **strike-slip** and **oblique** faults (Figures 9.5A-C). Dip-slip faults involve the vertical (up/down) motion of rocks along a fault, while strike-slip faults involve horizontal motion (shear) along the fault. Oblique faults combine both types of motion and most faults are oblique to some degree. These different faults reflect the different ways rocks respond when stresses are applied to them. Dip-slip faults generally result from the compression or extension of a group of rocks (Figure 9.5D), while strike-slip faults generally result from the shearing of a group of rocks (Figure

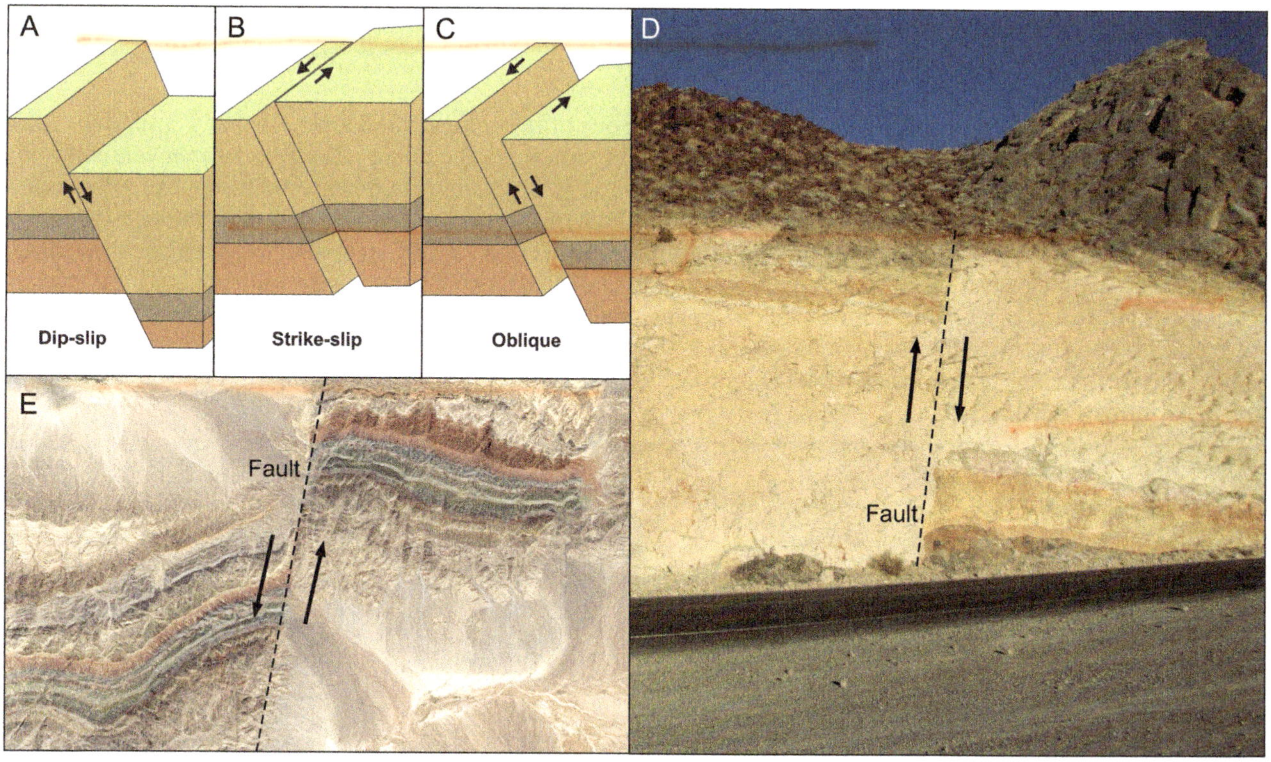

Figure 9.5: The three main groups of faults. A. Dip-slip. B. Strike-slip. C. Oblique. D. An example of a dip-slip fault showing the vertical offset of rock layers. Photo credit: B. Zaprowski. E. An aerial view of a strike-slip fault displaying the horizontal offset of the colorful rock layers. Photo credit: Robert Simmon and Jesse Allen/NASA.

9.5E).

In contrast to faults, **folds** occur when stress is applied a group of rocks which are easily deformable or ductile, and they bend rather than break. There are two basic kinds of folds: Anticlines and synclines (Figure 9.6). **Anticlines** form when rocks are warped upwards, while **synclines** form when rocks are warped downwards. Anticlines and synclines rarely occur alone but rather they tend to occur as a series of folds, much like a throw rug on a wooden floor

Figure 9.6: A. Folding in rocks: Anticlines and synclines. B small anticline near Flintstone, MD. Photo credit: Thomas Cawthern. C. A syncline exposed in a road cut through Siedling Hill, MD. Photo credit: B. Zaprowski.

can get bunched up if you slip on it. Anticlines and synclines primarily occur in conjunction with compression, but minor folding can occur with extension and shearing too.

When tectonic forces begin to build up in the Earth's crust, one of two things will happen. In many cases, the rocks along the fault are able to slide past each other, immediately releasing the energy before the strain can build up. Since fault creep occurs slowly and continuously without any large strain build-up, the rocks don't necessary rupture. Instead, they can slowly be deformed over time without producing an obvious fault scarp. In these cases, we say the rocks on each side of the fault are undergoing **fault creep** (Figure 9.7A). In this scenario, large earthquakes are not likely. In other cases, the rocks on each side of the fault are not able to slide past each other. The frictional resistance of the rocks to movement allows much larger amounts of energy to be stored within the rocks. Eventually, the strain exceeds the breaking strength of the rocks, the rocks catastrophically fail, and the rocks along the fault move, usually in a rapid and violent manner (Figure 9.7B). This rapid release of energy is what causes the shaking you feel in an earthquake.

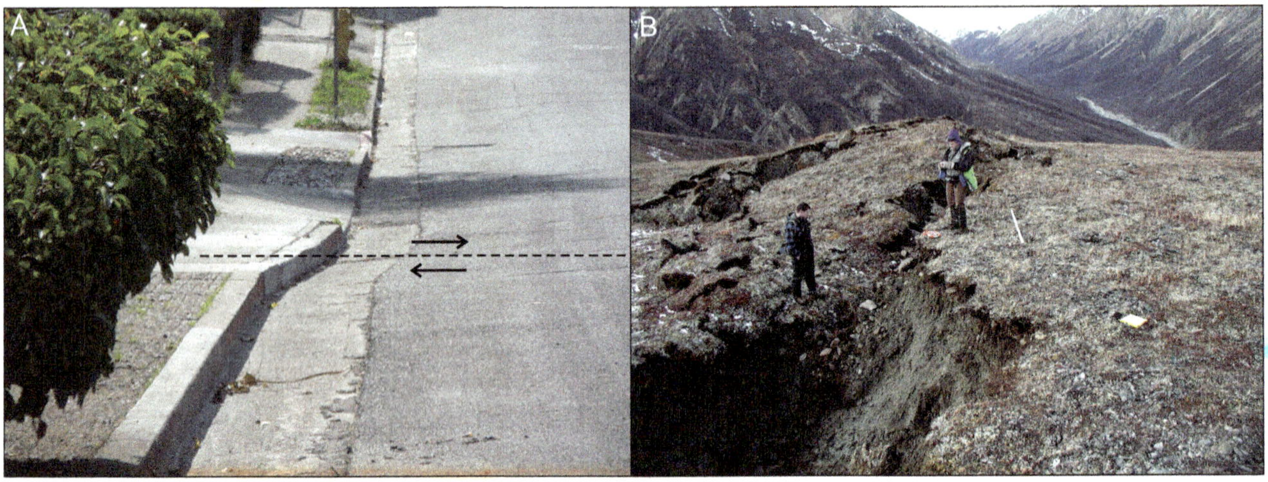

Figure 9.7: A. An example of a fault displaying creep (the arrows show the general direction of motion along the fault). The road and curb are able to slowly bend without breaking, so no obvious fault scarp has formed. Photo credit: USGS. B. The fault scarp in this photo was taken on a survey along the Denali fault in Alaska after a 7.9 magnitude earthquake on November 3, 2002. Photo credit: Peter Haeussler/USGS.

Earthquake waves are measured using a device called a **seismometer** or **seismograph**. A recording sensor measures the energy moving through the Earth as it passes by (Figure 9.8A) and transmits this data back to the seismometer, which records the data. Old-style seismographs used to plot this data with a pen and rotating drum (Figure 9.8B), but modern seismometers use computers to store and plot the data (Figure 8.22). A **seismogram** is the plot of the data recorded by the seismograph (Figure 9.8C). It should be pointed out that a seismometer records *any* energy passing through the Earth, even if it is not generated by an earthquake. For example, seismometers located near a coast may record the energy of the breaking waves.

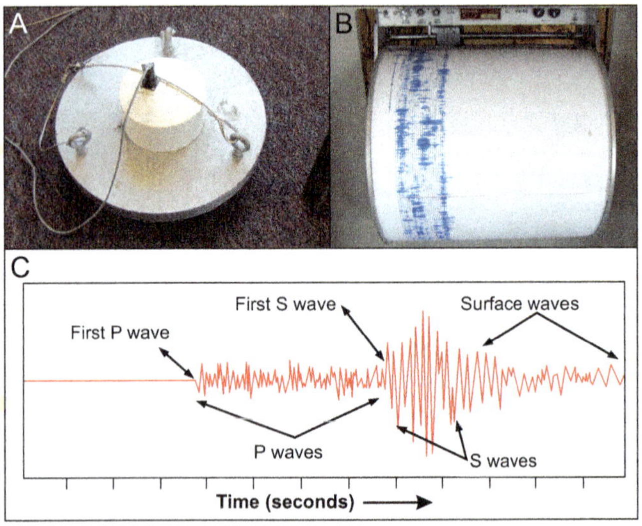

Figure 9.8: A. A modern seismograph sensor B. An old-style paper seismograph. Modern seismologists use computers to display earthquake data. C. An example of a seismogram and the different types of earthquake waves it records. Photo credits: B. Zaprowski.

A seismogram provides a wealth of data about an earthquake. When earthquakes occur, they release their energy in a number of different wave forms. The two most important earthquake waves are **P-waves** (primary) and **S-waves** (secondary). P-waves are also known as compressional waves. In the case of a compressional wave, energy is transmitted through the compression and extension of the medium it travels through. In this way, a compressional wave is much like

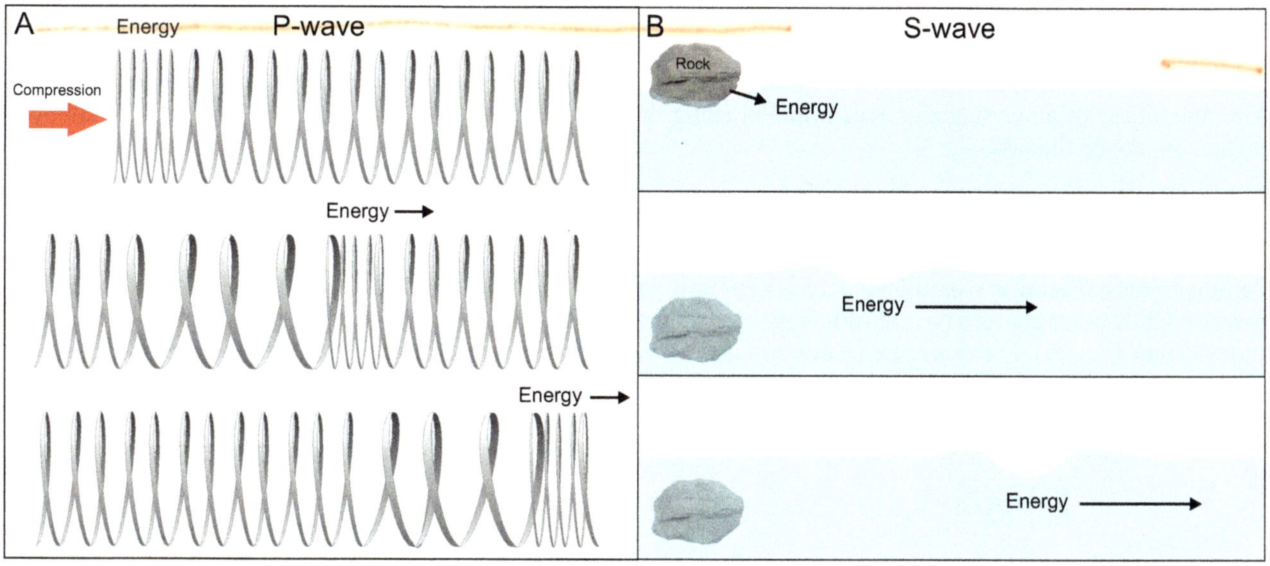

Figure 9.9: A. A demonstration of P-Waves in action. Energy is transmitted through compression and expansion of the medium the wave travels through, in this case, a spring. B. A demonstration of S-waves in action. A rock thrown in water imparts energy to the water. This energy moves out in all directions away from the place of impact, disturbing the water surface as it goes.

a slinky (Figure 9.9A). When a stretched-out slinky is "pushed" forward, the pulse of compression moves down the coil rapidly. S-waves are known also as shear waves. Shear waves cause the medium they are passing through to move perpendicularly relative to the direction of travel (Figure 9.6B). Ocean waves are S-waves: Water moves up and down as energy passes through. In the case of earthquakes, the rocks of the Earth move up and down in a vertical motion as the energy from the S-waves passes through them. Although other earthquake waves play a role in understanding the details behind earthquakes, P and S waves provide two of the most important pieces of information: The location of the earthquake and the magnitude of the earthquake.

P-waves travel faster than S-waves. Since the waves leave from the exact same place at the exact same time (from the focus) seismologists can use the difference in arrival times between P and S waves to calculate the distance to an earthquake (Figure 9.10). However, for any one location, this only tells us how far away an earthquake is from the seismograph, not what direction it came from. At least 3 seismograms are needed to triangulate the exact location of the earthquake. Figure 9.10 shows how this process works. The circle drawn around each seismograph location represents all of the possible places the earthquake could have occurred, based on the station data. The three circles overlap to identify the unique location of the epicenter. We can see that this particular earthquake occurred near Northridge, CA. Earthquake magnitude is most commonly measured using the Richter scale. The **Richter scale** uses the amount of vertical (shear) displacement produced by an earthquake as a measure of earthquake magnitude.

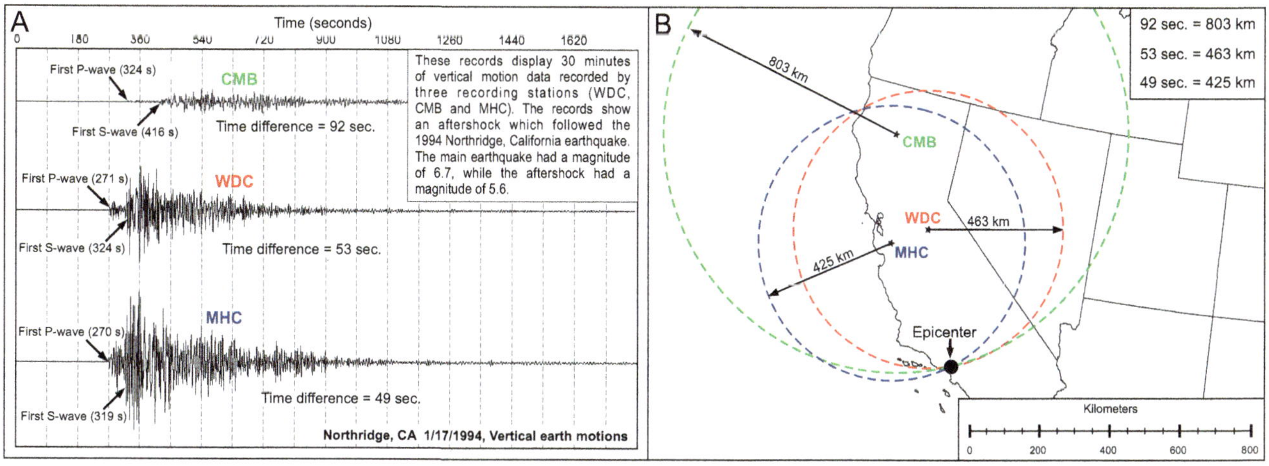

Figure 9.10: A. Three seismograph records from an earthquake event In southern California. B. This illustration shows how the P and S wave data from three seismograms can be used to pinpoint the epicenter of an earthquake.

The Richter scale is logarithmic, that is, an increase of one magnitude unit represents a ten-fold increase in amplitude. However, in terms of energy, each increase in magnitude results in about 31 times more energy being released! To put that in perspective, that is equivalent to the size difference between a pea and a large cantaloupe. The magnitude of an earthquake is determined using the amplitude of the S-waves on the seismograph and the distance to the earthquake.

Section 9-2: Earthquake Hazards

Earthquake hazards are highest in areas where active tectonism and volcanism is occurring (Figure 9.2). But as we demonstrated in Figure 9.3, earthquakes can occur most anywhere. Having said that, some places have an extremely low risk while other places have a surprisingly high risk, even if an earthquake has never occurred there in historical times (Figure 9.11A). For example, Seattle has never experienced a major earthquake, but paleoseismic evidence shows that they have occurred in that region and one will almost certainly happen again in the future (Figure 9.11B). Geologist will commonly dig trenches across fault zones in order to determine how often a fault is active.

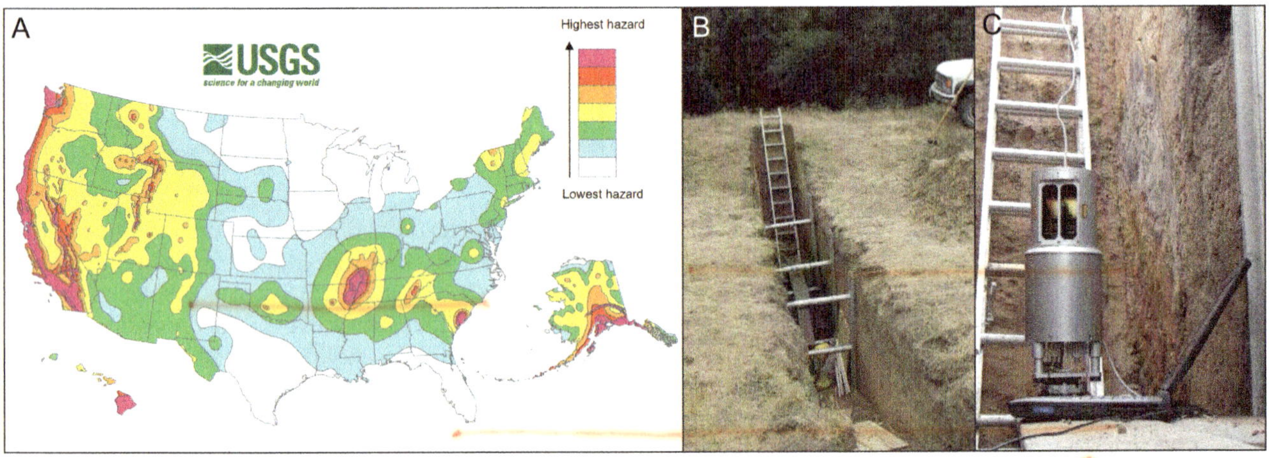

Figure 9.11: A. This figure depicts the various earthquake hazard zones in the U.S. Image credit: USGS. B. Geologists typically use trenches to gather paleoseismic data about a fault. C. Down in the trench. Photo credits: USGS.

The most common hazard associated with earthquakes is the *shaking of the ground.* The rapid motion causes objects on the Earth's surface to be violently shaken for a time period of seconds to minutes. This causes rigid objects (like buildings and bridges) to rupture and perhaps even collapse (Figure 9.12). In populated areas, fires can be a risk if water, gas and electric lines are broken during the earthquake. Large areas of San Francisco burned down after the 1906 earthquake because there was no water available to put the fires out. The amount of shaking produced by an earthquake is also a function of geology. Objects sitting on top of loose sediments like mud or sand will experience a much greater amount of shaking than objects sitting on solid bedrock (Figure 9.13A). Thus, the damage produced by

Figure 9.12: Examples of the damage caused by shaking during the 1989 Loma Prieta Earthquake in Northridge, CA. A. A townhouse has fallen over. Photo credit: D. Perkins/USGS. B. Collapsed Interstate 880 in Oakland, CA. Photo Credit: E.V. Leyendecker/USGS.

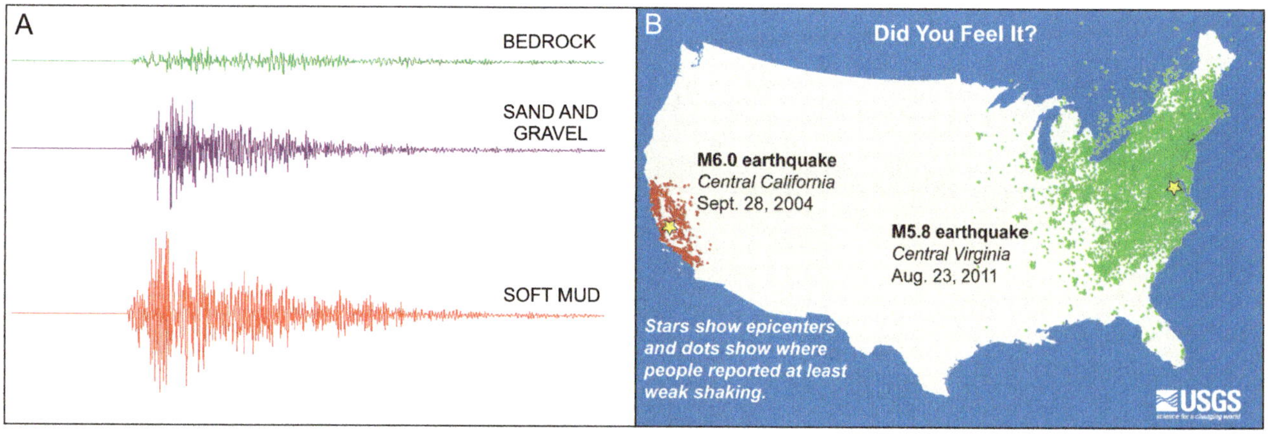

Figure 9.13: A. This diagram illustrates how bedrock affects the amount of shaking felt at any given location. As you can see on the seismograph records, more shaking occurs on loose sediments than on solid bedrock. B. A comparison of the reported shaking associated with the 2011 Virginia earthquake and an equivalent earthquake in California. Image credit: USGS.

an earthquake due to shaking can be greatly minimized if the structure is built on solid rock with firm connections to the foundation. Unfortunately, many of the world's largest cities are built on soft marine sediments. Geology also effects the size of the area which will experience shaking. For example, the 2011 Virginia earthquake was felt over a much wider area when compared to a similarly-sized earthquake in southern California (Figure 9.13B). This happened because of differences in crustal rock properties. Rocks west of the Rockies don't transmit seismic energy as easily or as far as rocks east of the Rocky Mountains.

Mass wasting events such as rock slides, earthflows and avalanches are also common in mountainous areas during an earthquake (Figure 9.14A). In areas underlain by water-saturated sediments, a phenomenon known as **liquefaction** may occur. When the shaking occurs, the sediments begin to flow in a fluid-like manner, much like

Figure 9.14: A. This landslide occurred during the 2001 El Salvador quake. Photo credit: USGS. B. Liquefaction during the Northridge, CA Earthquake caused silt and sand to erupt from the ground. Photo credit: USGS. C. Collapse of Fourth Avenue near C Street in Anchorage, AK as a result of liquefaction during the Good Friday Earthquake, March 27, 1964. Photo credit: USGS.

Figure 9.15: A time series of photographs showing what a tsunami is like. A. First, the water recedes well below normal low-tide level. B. The tsunami wave approaches as a bulge of water. C. The tsunami wave engulfs the building. D. The wave retreats. Several more waves followed this one. Photo credits: NOAA/NGDC, Chris Chapman, Schlumberger Cambridge Research, Cambridge, U.K.

when you tap your feet in saturated sand at the beach. This may cause sediment to erupt at the surface in the form of sand boils (Figure 9.14B) and/or it may lead to ground subsidence. For example, the liquefaction and failure of the ground beneath Anchorage, Alaska was a primary cause of widespread destruction after the 1964 Good Friday Earthquake. Whole city blocks subsided during this earthquake (Figure 9.14C).

Tsunamis are a hazard along most any coast, but the risk is much higher along tectonically-active coasts like those which fringe the Pacific Ocean. **Tsunamis** are a series of large waves which form when a large volume of water is displaced. The displacement forms large waves which radiate out in all directions. Tsunamis have very long wavelengths (kilometers in length!) and very high speeds (up to 900 km/hr or 560 mi/hr) and travel long distances; but in the open ocean, their wave height is only a few meters. As such, they are not observable by ships or from the air. The true nature of the monster wave only appears when the wave reaches shallow water (Figure 9.15). First, water will begin to recede from the shoreline. Then, a large bulge of water surges up onto land. Finally, the water will run out of energy and recede, carrying debris with it. Unfortunately, tsunamis typically occur as a series of waves, and the later waves carry the debris from the earlier waves. This can lead to further damage to structures.

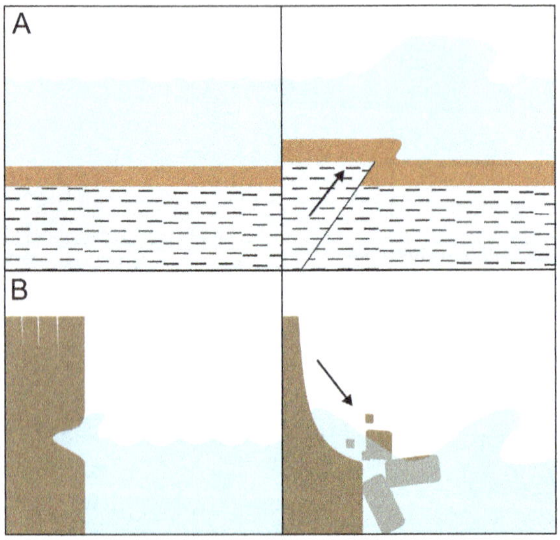

Figure 9.16: Two ways a tsunami can form. A. Fault rupturing: A fault which breaks through the sea floor will push a bulge of water upwards, forming a wave. B. Landslides: Landslides that occur along a seacoast or even underwater can produce tsunamis. These landslides can displace a large amount of water, producing a wave.

Tsunamis form in two ways: earthquakes and landslides. When an earthquake occurs, motion occurs along a fault plane. If this motion ruptures through the sea floor, a large mass of rock may be thrust upwards, which, in turn, displaces an equivalent volume of seawater (Figure 9.16A). A large amount of energy is transferred to the water and a wave forms. In the case of the

Figure 9.17: A. This picture shows the Yala Safari Resort after the 2004 tsunami hit Sri Lanka. Photo credit: USGS. B. An aerial view of damage to Sukuiso, Japan, a week after a 9.0 magnitude earthquake and subsequent tsunami devastated the area in 2011. Photo credit: NOAA/NGDC, Dylan McCord/U.S. Navy. C. Tsunami damage along the waterfront at Kodiak, Alaska after the 1964 Good Friday Earthquake caused an underwater landslide. Photo credit: USGS Photo Library. D. A mega-tsunami in Lituya Bay, Alaska washed over this spur of land and removed trees that were 520 m (1,720 ft.) above sea level! Photo credit: J.D. Miller/USGS.

largest fault-genetated tsunamis, the waves can be 10's of meters high. The 2004 tsunami off the coast of Sumatra formed in this way when a magnitude 9.0 earthquake occurred along the edge of the Burma Plate. For those areas in close proximity to the epicenter, the results were devastating (Figure 9.17A). The 2011 tsunami in Japan also formed in this manner (Figure 9.17B). Landslides and other forms of mass wasting can also produce tsunamis (Figure 9.16B). A good example of this is the tsunami that formed along the southwestern coast of Alaska shortly after the 1964 earthquake. The tsunami formed when a large bulge of clay liquified and flowed rapidly out into the ocean, forming a huge wave. The wave quickly hit the nearby port towns of Kodiak, Valdez and Seward (Figure 9.17C). Most of the fatalities related to the 1964 earthquake were caused by this tsunami. Even more ominous are **mega-tsunamis** which can form when very large landslides occur along a coast. Waves *hundreds* of feet high have been documented following coastal landslides, and the only limit to their size is the volume and mass of the rock involved. In 1958, a landslide in Lituya Bay, Alaska produced a wave which washed away trees up 520 m (1720 ft.) above sea level (Figure 9.17D)!

In many cases, tsunami fatalities (but not damage) can be minimized by evacuating people to higher ground. Most coastal communities along high-risk coasts have tsunami warning systems and evacuation routes for just this reason. One of the reasons the 2004 tsunami was so devastating was because there was no warning time. The epicenter of the magnitude 9.0 earthquake occurred at a convergent plate boundary between the Burma plate and the Indian plate, producing over 5 m (15 ft.) of vertical sea floor displacement. This caused a wave of 17 m (50 ft.) high to form and break at some locations mere minutes after the earthquake, washing whole villages away! The tsunami also caused significant damage along the coasts of India and Sri Lanka. Thankfully tsunamis of this size are fairly rare and do not commonly form near densely populated coasts, so there is sufficient time for warnings to be given.

Chapter 10 - Rocks and Minerals

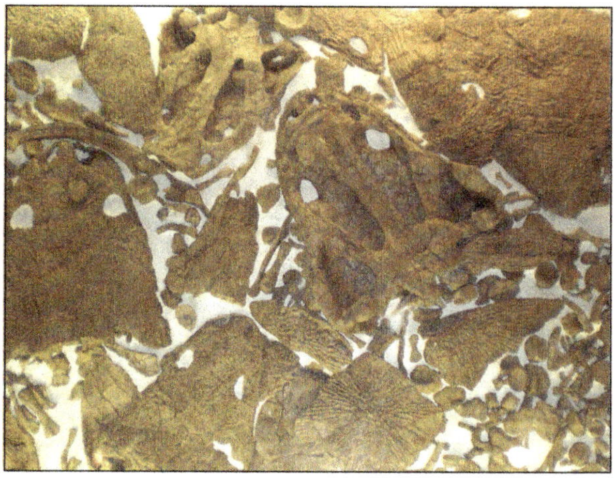

Figure 10.1: This mass of giant amphibian skulls on display at the Smithsonian Museum of Natural History is just another page in the long story of Earth's history. Photo credit: B. Zaprowski.

Rocks have been and continue to be an essential part of human activities. From the first stone tools to the majestic Egyptian pyramids to the Brooklyn Bridge, we have been tapping into the Earth's crust for a wide variety of resources for millennia. The stones, minerals, gems, and fossil fuels we extract from the Earth all play a significant role in modern society. So rocks have a great economic value, and geologists help to find new mineral resources in as yet unexplored parts of the world. But rocks also have another important role.

Everything we know about the Earth's past comes from information trapped and stored in sediments and rocks. Every rock potentially contains information about how, where and when it formed. If you understand how and where rocks are formed, you can interpret the history of a particular group of rocks. During its formation, a rock may also preserve other useful types of information about its surroundings such as what lived there and what the climate was like (Figure 10.1). Sometimes this information is obvious and easy to interpret, but many times geologists have to use multiple lines of evidence and/or circumstantial evidence to interpret the information contained within a rock. In this way, geologists are like archeologists, deciphering and putting together a giant book of Earth history, and the rocks of the Earth are the "pages."

So why do we care? Understanding how, where and when catastrophic events such as earthquakes, volcanos, and floods will occur in the future requires us to look at the past. Most geologic events, although devastating on human time scales, may only occur once every few hundred years. Geologists use the information in the rock record to determine where and how frequently geologic events have occurred in the past. They can then use this information to assess when and where catastrophic events might happen in the future, and potentially save lives in the process. In this chapter we first look at the big picture and discuss the rock cycle and the three main groups of rocks. Then we will delve a little deeper into minerals, the different types of rocks, and how geologists classify them. We'll also explore the economic significance of each rock group along the way.

Section 10-1: The Rock Cycle

The first step in deciphering the meaning of a rock is by understanding the rock's place in the **rock cycle**. The rock cycle is one of the most fundamental concepts in Earth Science. The rock cycle explains how rocks and sediments are slowly and continuously cycling through the geosphere. Regardless of how they form, all rocks have one thing in common: They are composed of **minerals** (Figure 10.2A). A mineral is a naturally occurring inorganic solid that possesses a definite chemical structure, giving it a unique set of properties. By definition a **rock** is a combination of

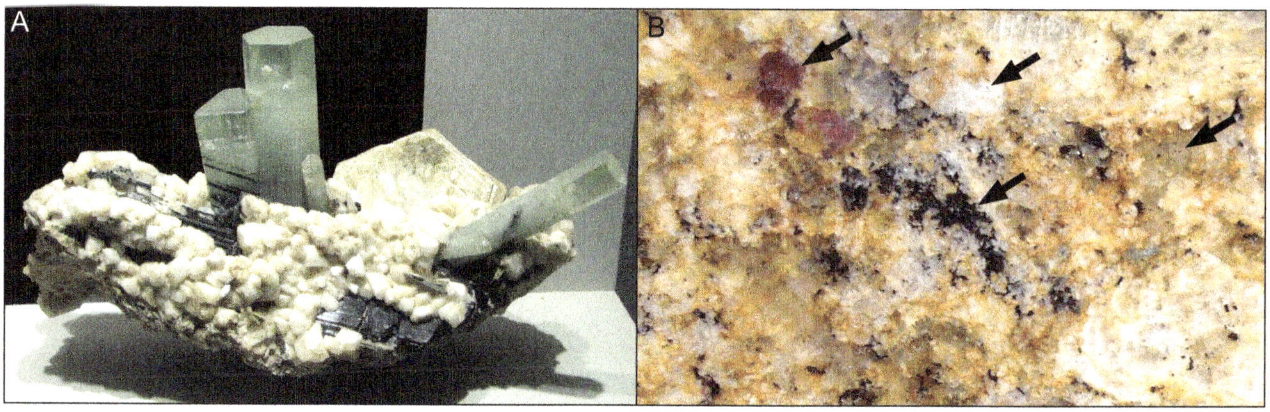

Figure 10.2: A. A large cluster of minerals from Brazil. B. Rocks are usually composed of several different kinds of minerals, like in this example. Photo credits: B. Zaprowski.

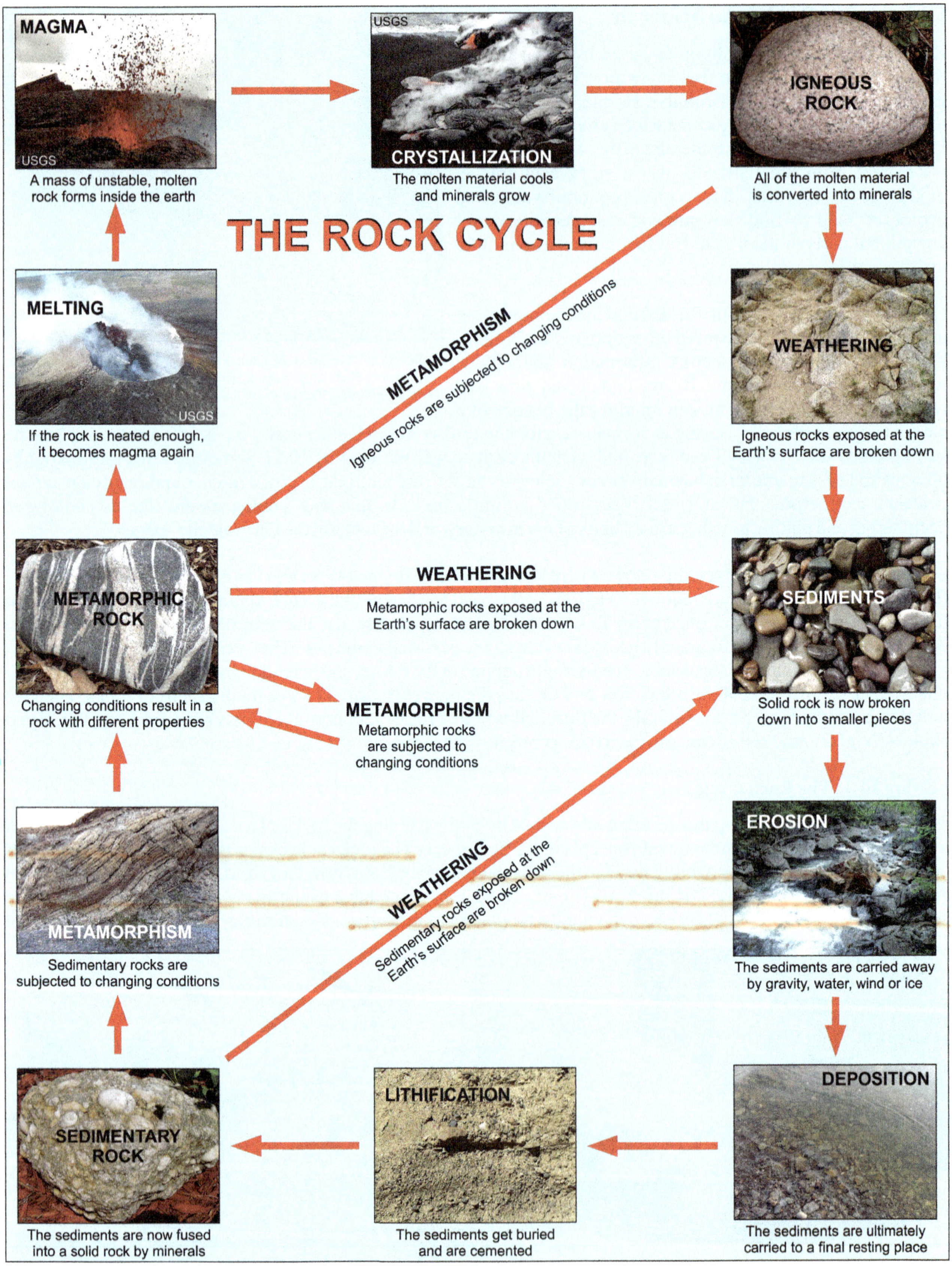

Figure 10.3: The rock cycle. All photos by B. Zaprowski except where noted.

one or more different minerals (Figure 10.2B). Rocks are then subdivided into one of three primary groups of rocks based on how the rock formed.

1. **Igneous Rocks** are rocks formed directly from molten rock and includes volcanically-derived debris, like ash.

2. **Sedimentary Rocks** are rocks formed from one or both of the two main by-products of rock weathering: Sediments and ions dissolved in water. Most sedimentary rocks form in aqueous environments.

3. **Metamorphic Rocks** are rocks which are altered in some way to form a new rock with different properties *without melting it.* These alterations can result from older rocks being subjected to an increase in pressure, an increase in temperature, and the exposure of the rock to chemically active fluids.

Far from static, rocks are constantly being formed, destroyed, and reformed as new rocks. Each of the three rock types, and the processes that form them, are a reflection of this never ending process we refer to as the **rock cycle** (Figure 10.3). Understanding how and why different types of rocks form starts with understanding the different parts of the rock cycle.

The most logical place to start explaining the rock cycle is with **magma** (Figure 10.3). Magma is molten rock that forms inside the Earth. Eventually this magma cools and solidifies, forming minerals. This mineral-forming process is called **crystallization** and the rocks that form as a result of the crystallization process are called **igneous rocks**. Most igneous rocks are **intrusive igneous** rocks and form deep in the Earth. Intrusive igneous rocks are only exposed at the Earth's surface in certain regions or in deep mines (Figure 10.4A). Closer to the Earth's surface, volcanism produces **extrusive volcanic** rocks. Extrusive igneous rocks are commonly created by lava flows and volcanic debris like the rocks pictured in Figure 10.4B.

Figure 10.4: A. Intrusive igneous rocks exposed in a granite quarry in Vermont. Note the blue porta-potty for scale. Photo credit: B. Zaprowski. B. Recently formed extrusive igneous rocks in Hawaii. Photo credit: USGS.

Although most people think of rocks as hard and indestructible, the vast majority of rocks when exposed at, or just below, the Earth's surface, are attacked by the agents of weathering: primarily water, but also ice, fire, wind, plants and animals. Slowly but steadily, weathering can cause even the biggest, hardest rocks to crumble away into smaller pieces of rock and minerals called **sediments**. There are four basic types of sediment: Gravel, sand, silt and clay (Figure 19.5). Sediments are chiefly classified based on the size of the particles; gravel (which includes cobbles and boulders) is the largest-sized type of sediment while clay is the smallest-sized type of sediment. To understand the nature of any sedimentary rock, we first have to look at how sediments form.

Physical weathering processes break rocks down into small fragments or individual mineral crystals but *do not chemically alter* the minerals. Physical weathering is most dominant in polar and arid climates, although it can also be vigorous in temperate climates during the winter. Physical weathering is basically the splitting or prying loose of

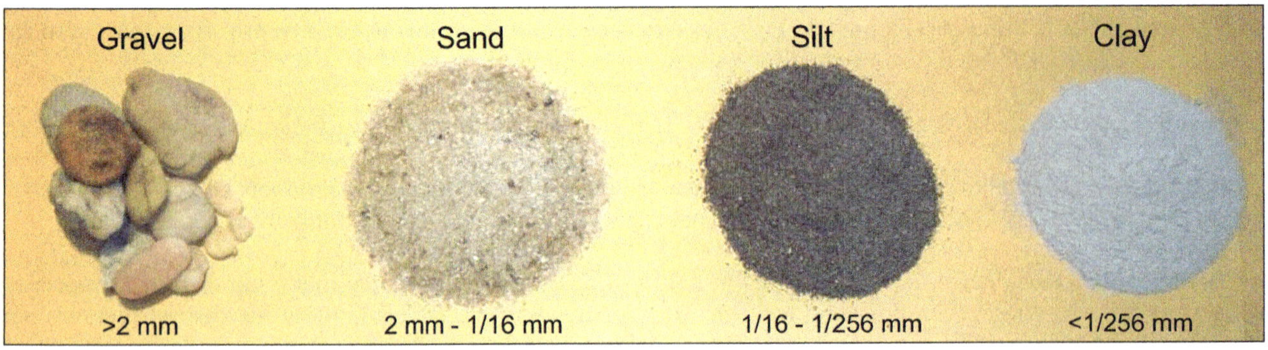

Figure 10.5: The four basic types of sediment. Photo credit: B. Zaprowski.

chips, flakes and chunks of rocks from a solid mass of rock. One of the most common agents of physical weathering is ice. When water fills a crack in a rock and freezes, the ice causes stress to build-up at the ice-rock interface. Over time, this stress weakens the rock, enlarging the crack through a process called **hydrofracturing**. (Figure 10.6A). Heat can also break rocks down. In arid regions, the surface of a rock can be heated up drastically during the day, causing it to expand. Meanwhile, the cooler interior of the rock hardly expands at all. This causes stress to build up at the intersection between the hot and cold parts of the rock and the rock cracks (Figure 10.6B). Rocks can also be worn away by the wind. In arid environments, especially those that lack a cover of vegetation, small pieces of sand can be blown around at high speeds. These sand grains chip away at any other rocks they hit. Some rocks can be polished into intricate yet smooth rocks called **ventifacts** (Figure 10.6C).

Figure 10.6: Examples of physical weathering. A. A hydrofractured rock. B. These layers of rock are "flaking" off this boulder because of repeated heating and cooling. C. A wind-polished ventifact in Death Valley, CA. Photo credits: B. Zaprowski.

Chemical weathering involves the chemical alteration of individual minerals within the rock, usually through the action of water (even in small quantities). Because of the importance of water, chemical weathering occurs primarily in temperate and tropical climates. Many minerals are unstable at the Earth's surface and break down quickly when they come in contact with water, especially warm, flowing water (Figure 10.7). The minerals break down into various dissolved ions like sodium, calcium, potassium, and carbonate (CO_2^{-2}) among others. Some minerals decompose completely into dissolved ions, but most minerals go through chemical restructuring and recrystallize as stable minerals such as clays or iron oxide (rust). Clay particles are the most abundant size of sediment on Earth because a large number of the minerals found in rocks are altered into clay when they come in contact

Figure 10.7: Chemical weathering has led to the formation of this cave passage in Puerto Rico. Photo credit: B. Zaprowski.

with water. It is important to note here that the term clay refers to both a specific group of minerals and a sediment size class.

Once broken down, the rocky debris can be easily carried away from its place of origin by a process called **erosion**. Erosion is the transport of sediments by gravity (landslides), water (streams and rivers), wind (sand storms), or ice (glaciers). As the sediments are transported along, they are physically and/or chemically reduced in size. Virtually all sediments are transported to some location where they are deposited. **Deposition** occurs when sediments in motion come to a rest, forming layers of sediment. Most sediment is deposited at the bottom of some large body of water, commonly called a **basin**. The Gulf of Mexico is a good example of a basin (Figure 10.8). Over 9,000 m (30,000 ft!) of sediments have accumulated on the bottom of the Gulf of Mexico over hundreds of millions of years, and most of the sediment was delivered there by the Mississippi River. These sediments also contain large quantities of organic material, making this region one of the world's richest sources of oil and natural gas. Once sediments are deposited in a basin, they are unlikely to be disturbed any further for a long period of time (unless of course plate tectonics is interfering, then all bets are off).

Figure 10.8: The Mississippi River delivers tons of sediment to the Gulf of Mexico each year. Photo credit: NASA.

In most cases, the layers of sediment keep piling up, getting thicker and thicker. The older layers at the bottom get compacted and the water they contain is slowly squeezed out. As this water moves through the sediments, minerals dissolved in the water solidify and fill in the spaces (called pores) between the sediment particles "cementing" the sediments together (Figure 10.9). Eventually the pores are completely filled in and the loose pile of sediment has now become a more substantial solid rock. This process, called **lithification**, is very slow; it takes thousands to millions of years for the cementation process to be completed. Sedimentary rocks composed of sediments cemented together in this manner are known as **clastic sedimentary rocks**. Over 75% of all sedimentary rocks are clastic.

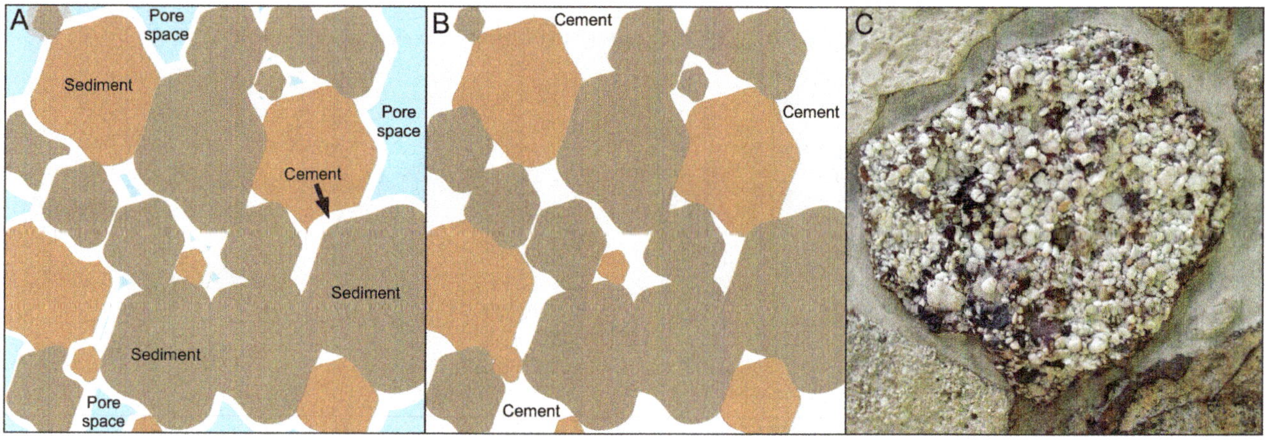

Figure 10.9: The lithification process up close. A. Because sediments are irregular in shape, pore spaces exist between them. As water circulates in the pores between the grains of sediment, a mineral cement is deposited. B. Eventually the pore spaces are all filled in with cement and a sedimentary rock is born. C. This piece of conglomerate is a good example of a clastic sedimentary rock. Photo credit: B. Zaprowski.

However, another group of sedimentary rocks called **chemical sedimentary rocks** can form in much different ways. For example, some of the dissolved ions produced by chemical rock weathering are also washed into the basin. Many of these dissolved ions are spontaneously transformed into new minerals through the chemical process of **precipitation**. Precipitation is the formation of solid crystals from a solution of dissolved ions. The thick layers of salt on the floor of Death Valley, CA are a good example of this (Figure 10.10A). When water from the adjoining mountains washes into Death Valley, it carries minute quantities of salt. When the water evaporates away in the hot sun, the salt is left behind to form crystals. Biological processes can also produce chemical sedimentary rocks. All marine organisms with hard parts, like a shell, extract the elements they need for its construction directly from

Figure 10.10: A. Salt crystals forming on the floor of Death Valley, CA. B. This fossiliferous limestone from the Niagara Gorge, NY is full of fossil crinoid stems (also known as sea lilies). Photo credits: B. Zaprowski.

seawater. Eventually, the creatures die and the shelly debris accumulates in thick layers at the bottom of the sea where it may eventually be lithified into a fossiliferous limestone (Figure 10.10B).

Some rocks can subsequently be buried deeply in the crust where they may be exposed to higher temperatures, higher pressures and/or chemically active fluids. Initially, the increase in temperature and/or pressure may result in the re-orientation of the minerals in a rock, but ultimately the agents of metamorphism can completely change the composition of the minerals in the rock, creating a new rock in the process (Figure 10.11). This process is called **metamorphism** and it results in the formation of metamorphic rocks. It is important to note that metamorphism changes the nature of rocks ***without melting them***. Metamorphic rocks can go through the process of metamorphism several different times as temperatures and pressures change over time. Ultimately, if a rock is subjected to high enough temperatures and pressures, it will undergo melting and becomes magma once again, thus completing the

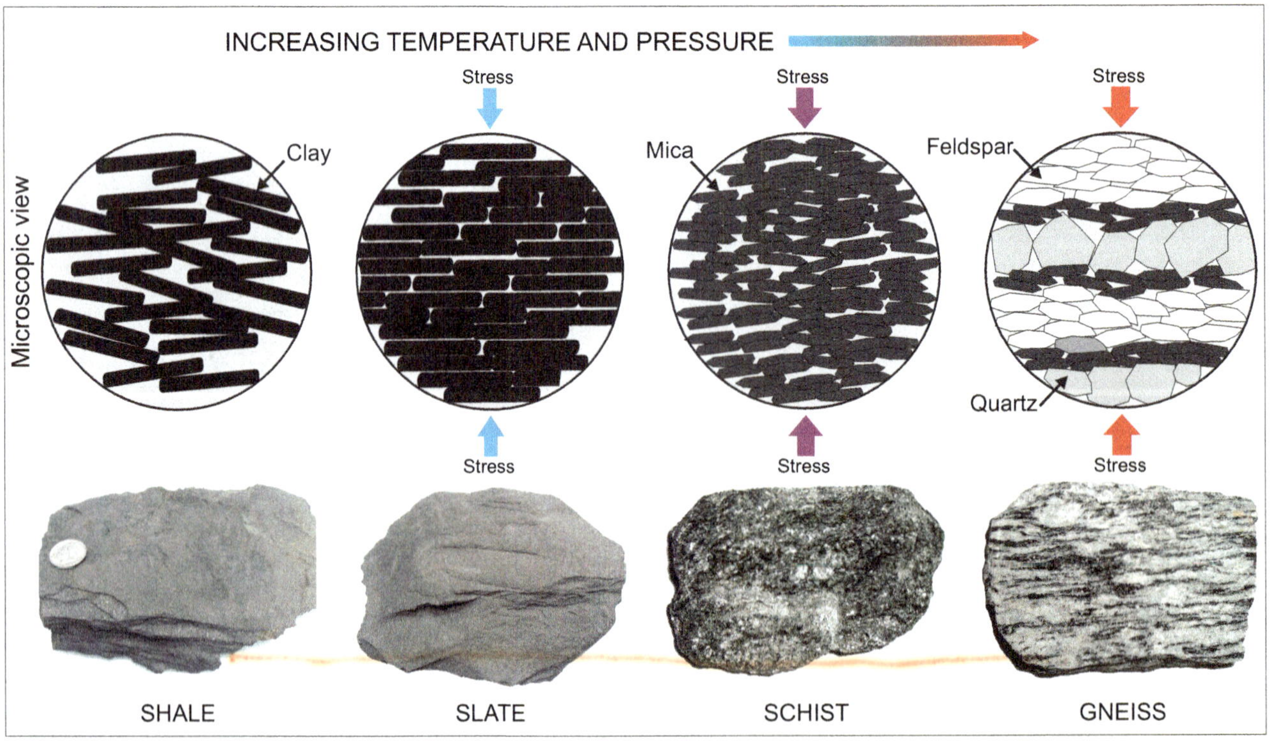

Figure 10.11: This diagram shows how a rock can be metamorphosed into a variety of different rocks with different minerals due to increases in temperature and pressure. Photo credits: B. Zaprowski.

rock cycle. It is important to note that the rock cycle is highly generalized. In the real world, some rocks never complete the cycle but instead are recycled repeatedly through a series of loops within the rock cycle without ever reaching the melting stage (or even the metamorphism stage). For example, sedimentary rocks can be recycled through the weathering-erosion-deposition-lithification process numerous times as plate tectonics lifts the Earth's crust upwards (Figure 10.3). Now that we've taken a step and looked at the big picture, we can now focus more on the basic properties of rocks and the minerals they are made of.

Section 10-2: Minerals

In the last section we briefly defined a mineral as a naturally occurring inorganic solid that possesses a definite chemical structure, giving it a unique set of properties. Let's dissect that definition a little further by using Figure 10.12 as an example.

A mineral must be naturally occurring. Minerals can be created synthetically but, strictly speaking, they are not considered true minerals. For example, synthetic diamonds are not considered a true mineral, even though their structure is identical to real diamonds, because they were not formed through natural processes. It is possible to create a synthetic amethyst, but not a large massive one like in Figure 10.12A.

A mineral must be inorganic. By definition, a substance which contains both carbon (C) and hydrogen (H) is considered to be organic, while an inorganic substance does not have both carbon and hydrogen. The amethyst in Figure 10.12 is made of just two elements: silicon and oxygen, so it clearly meets this standard. Of course the definition of inorganic mineral can get a little fuzzy. For example, fossil fuels (coal and hydrocarbons) are still considered mineral resources even though they are organically-derived substances.

A mineral is a solid. This is true in almost every case, although there are a couple of rare exceptions like elemental mercury, which can naturally occur as a liquid.

A mineral possesses a definite chemical structure. Each type of mineral has a unique arrangement of atoms and molecules in a crystalline lattice (Figure 10.12B). As a mineral grows in size, this unique pattern repeats itself. There are many different ways that elements can be arranged to form a mineral. In the case of Figure 10.12B, the minerals grow into 6-sided prisms ending in 6-sided pyramids. As the mass of amethyst grew, each of the individual crystals retained that unique form by simply repeating the 6-sided molecular pattern in Figure 10.12.

A mineral has a unique set of properties. This is because the unique chemical structure dictates the unique physical properties of a mineral. There are over 4,000 known minerals in the world, and each one has its own unique set of properties. Learning what the key properties of a mineral are and how a mineral's structure dictates these properties is one of the most important tools in geology. *Minerals reflect the conditions under which they form, giving the geologist more information about the Earth's history.*

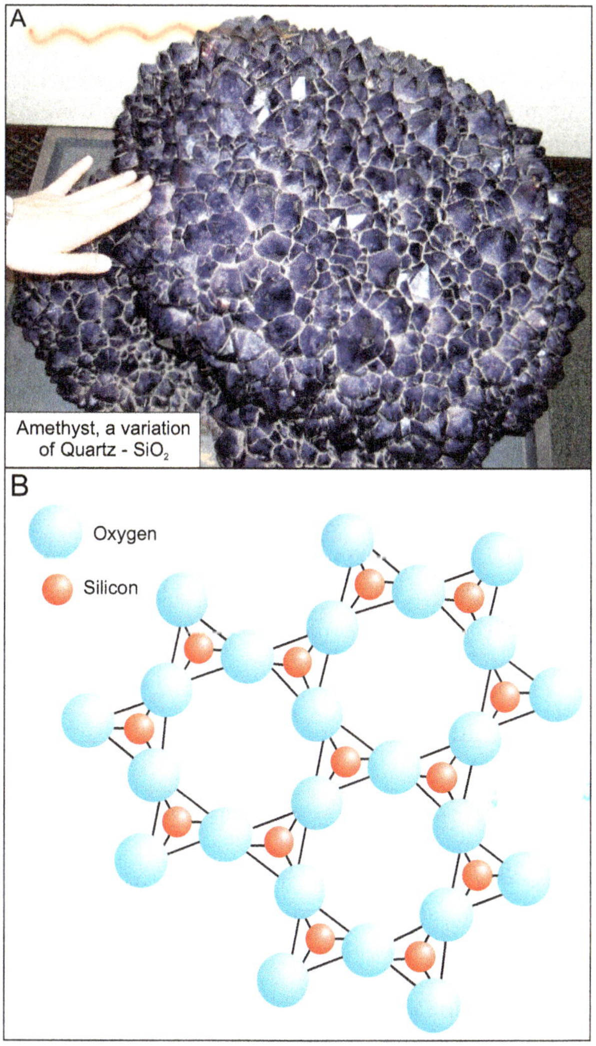

Figure 10.12: A. A large mass of amethyst in the Smithsonian Museum, Washington D.C. Photo credit: C. Zaprowski. B. The crystalline structure of amethyst.

Minerals form in a wide variety of environments. Many

minerals begin their life in magma and develop through the process of crystallization. Other minerals, like those found in the chemical sedimentary rocks in Figure 10.10, develop through the processes of precipitation. Minerals also commonly form during rock weathering. For example, the mineral potassium feldspar ($KAlSi_3O_8$) is not stable at the Earth's surface, but the mineral kaolinite ($Al_2Si_2O_5(OH_4)$) is. So when orthoclase is exposed to water and carbon dioxide at the Earth's surface (chemical weathering!), its atoms spontaneously rearrange themselves into a more stable mineral arrangement, forming a new mineral, kaolinite (Figure 10.13).

Figure 10.13: The weathering of potassium feldspar produces the mineral kaolinite. Photo credits: B. Zaprowski.

Of course stability is not the only controlling factor. You can only make minerals if the elements needed are available. So if you have magma with silica, oxygen, iron, aluminum, magnesium and potassium in it, you can't make the mineral calcium plagioclase ($CaAlSi_3O_8$) because there isn't any calcium available to form it. Because the availability of elements is a controlling factor in mineral development, there is a fairly strong relationship between the most common minerals in the Earth's crust and the most common elements in the Earth's crust. Table 10-1 shows the eight most common elements in the Earth's crust by weight and the eight most common minerals. These minerals are collectively known as the **rock-forming minerals** and almost any "average" rock you pick up will contain at least one of these eight minerals. You can see in Table 10-1 that these eight minerals are composed exclusively of the eight most common elements in the Earth's crust (by weight) and the most abundant element in the universe, hydrogen.

Element	Abundance by weight (%)	Mineral	Composition
Oxygen (O)	46.6	Orthoclase	$KAlSi_3O_8$
Silicon (Si)	27.7	Plagioclase	$(Ca,Na)AlSi_3O_8$
Aluminum (Al)	8.1	Quartz	SiO_2
Iron (Fe)	5.0	Olivine	$(Mg,Fe)_2SiO_4$
Calcium (Ca)	3.6	Pyroxene	$(Mg,Fe)SiO_3$
Sodium (Na)	2.8	Amphibole	$Ca_2(Fe,Mg)_5Si_8O_{22}(OH)_2$
Potassium (K)	2.6	Biotite	$K(Mg,Fe)AlSi_3O_{10}(OH)_2$
Magnesium (Mg)	2.1	Muscovite	$KAl_2(AlSi_3O_{10})(OH)_2$
All others	1.7		

Table 10-1: On the left, the 8 most abundant elements in the Earth's crust by weight %. On the right, the eight common rock-forming minerals and their chemical compositions. Notice how all of these minerals have oxygen and silicon in them. They are the by far the two most abundant elements in the crust.

Common Mineral Properties

The physical and chemical properties of a mineral are what allow you to identify it. Some of the more common properties used to identify minerals are explained below:

Crystal form (also known as **crystal habit**): Each mineral has a unique shape that it takes on when it grows. Some crystals are long and slender, some have complex geometric shapes and some have some very unique shapes (Figure 10.14). The unique shape that forms as a mineral grows is a direct reflection of its crystalline lattice. For example, you are shaped like a human because of the arrangement of the bones in your body (your "crystalline lattice"). It is

Figure 10.14: Some examples of different crystal forms (habits) from the Smithsonian. Photo credits: B. Zaprowski.

important to note that some minerals do not have a well-defined crystal habit, while other minerals may display more than one type of crystal habit.

Color: Although color is the easiest mineral property to recognize and identify, it is also the least useful property for identifying most minerals. Given that there are thousands of minerals and only a few basic colors, only a handful of minerals have a color which is truly unique. In addition, many minerals display a range of colors. A mineral's color is dictated by how light reflects off of the mineral's surface. Some wavelengths of the electromagnetic spectrum are absorbed by the crystalline lattice while other wavelengths are reflected back to your eye. The amethyst in Figure 10.12 is colored purple because the crystalline lattice absorbs all of the colors in the spectrum except purple.

Luster: Luster is defined as the appearance or quality of light reflected from a mineral. Like color, this mineral property is dictated by how light interacts with the crystalline lattice. So a mineral which is *transparent* has a lattice which lets light pass through unobstructed while a mineral which is *translucent* lets some light through, making it appear "glassy". A mineral which is **opaque** has a crystalline structure which causes light to bounce off of its surface. (Figure 10.15). An important type of mineral luster which is useful for identifying minerals is a **metallic** luster. A metallic luster can be shiny or dull, but in either case, the mineral has a metallic appearance. Rusty looking minerals are also considered a form of metallic luster (Figure 10.15). In many mineral guides, luster is one of the first properties used for mineral identification.

Figure 10.15: Some examples of mineral luster. Photo credits: B. Zaprowski.

Streak: If a mineral has a metallic luster, then the mineral property called "streak" also becomes important. Streak is simply the color of the mineral in its powdered form. In most cases, the mineral is dragged across a white ceramic plate (called a streak plate) to form a streak (Figure 10.16). The color of this streak can then help in identifying a mineral. Sometimes, the streak color and the mineral color are the same, but sometimes streak color is different than the mineral color. For example, pyrite (fool's gold) is a shiny, brassy-colored mineral, but it produces a greenish colored streak .

Cleavage: Cleavage is a term which refers to the planes of weakness in a mineral. If a mineral is broken, it will break along these planes of weakness. For example, halite (salt) is a cubic mineral with 3 distinct planes of weakness (Figure 10.17). If you hit a chunk of salt with a hammer, it will break into smaller pieces, and each of the pieces will have the same form as the original piece of salt. Minerals can have 1, 2, 3, 4 or more planes of cleavage, that is, planes of weakness (Figure 10.18). Some minerals have no cleavage at all; these minerals are said to have **fracture,** that is, the complete absence of any planes of weakness. More importantly, since a plane of weakness is a 2-dimensional plane, every mineral will have *two* distinct surfaces defined by *one* plane. To use an analogy, we can compare a cleavage plane with a piece of paper. A piece of paper defines one plane in space but has two sides to it. A mineral with one cleavage plane will have two matching (paired) crystal faces. *Not every flat surface on a mineral is a cleavage plane!* The key here is to have a *pair of flat, parallel sides*, that is, they are both parallel the same cleavage direction. In addition, not all cleavage planes intersect each other at a perfect 90° angle (Figure 10.18). The angle between the intersection of 2 cleavage planes can also help you identify a mineral. It should be noted that not all

Figure 10.16: Some examples of streak. Photo credit: B. Zaprowski.

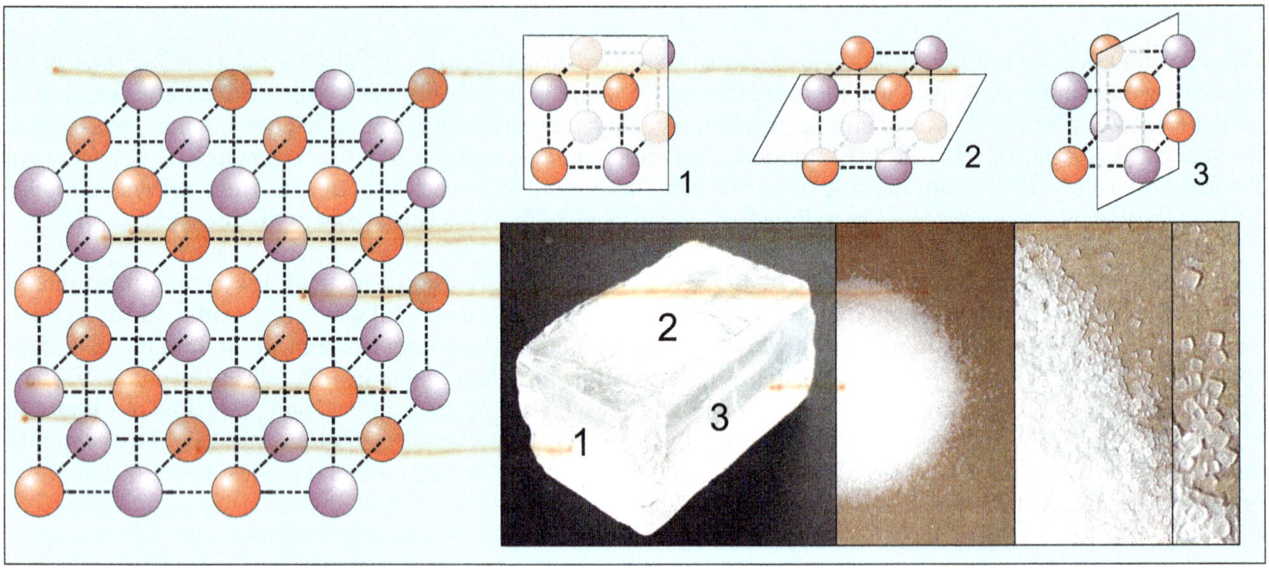

Figure 10.17: Halite (common salt) can be used to demonstrate the concept of cleavage. Halite has three directions of cleavage, that is, three planes of weakness in the crystalline structure. If you break a halite crystal up into little pieces, like table salt, each little piece still has the same cubic form. Photo credit: B. Zaprowski.

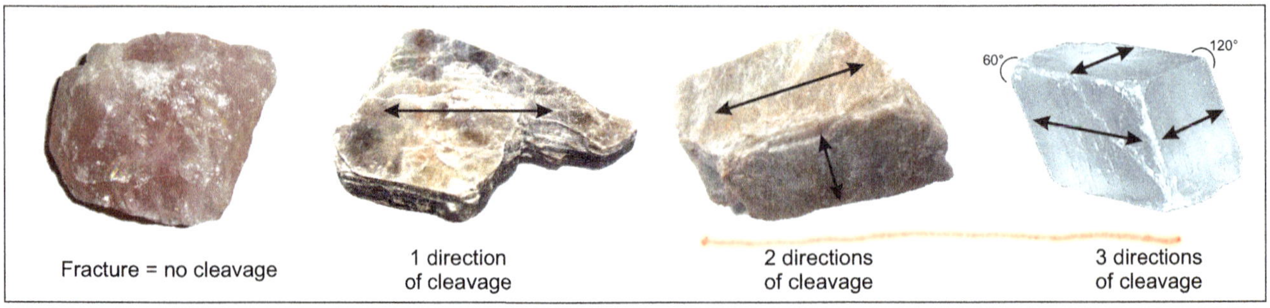

Figure 10.18: Examples of minerals with 0, 1, 2, and 3 directions of cleavage. Minerals without cleavage are said to have fracture. Note that each **direction** of cleavage is two-dimensional plane of weakness, and each plane of weakness is defined by two "flat" parallel sides. Photo credits: B. Zaprowski.

cleavage planes will be perfectly smooth. Some minerals have excellent cleavage and some have poor cleavage. *Regardless of how "smooth" the sides are, the emphasis with cleavage is that the mineral breaks the same way every time.*

Figure 10.19: A Mohs hardness kit with reference minerals. Diamonds are not usually included with these kits! Photo credit: B. Zaprowski.

Hardness: Minerals display a wide range of hardness from diamonds (the hardest) to talc (the softest), a mineral so soft it can be used to make baby powder. Strictly speaking, mineral hardness is a measure of the mineral's resistance to scratching. In fact, the test for hardness involves scratching the surface of an unidentified mineral sample with a reference mineral. A mineral's hardness is a direct function of the bonding strength within the mineral's crystalline structure. Hardness is a difficult property to quantify, but geologists use a relative hardness scale called the *Mohs scale*. The Mohs scale is used to determine the hardness of a mineral by comparing it to one of 10 standard minerals (Figure 10.19). The Mohs scale ranges from 1 to 10, where the standard mineral for 1 is talc and the standard mineral for 10 is diamond. All of the other minerals in the world fall between these two extremes. Note that the Mohs scale is an *ordinal scale*, that is the reference minerals are simply ranked in order. For example, diamond (10) is four times harder than corundum (9) but corundum is only twice as hard as topaz (8).

Specific Gravity: Specific gravity is the formal term for a mineral's density. Thus, if you measure the volume of your mineral and divide it by the mass of your mineral, you can calculate your mineral's specific gravity. In general, minerals with high specific gravity (like fool's gold) are rich in "heavy" elements (like iron or lead) while minerals with low specific gravity, like quartz, tend to lack "heavy" elements.

Mineral Classification

All minerals can be grouped into two broad categories: silicates and non-silicates. 75% of minerals fall into the **silicates** category, and indeed, most of the rocks you will encounter in your life are composed of silicates. All of the rock forming minerals are silicates too. Although the non-silicates make up a smaller proportion of mineral types, some non-silicates like the minerals calcite and halite (salt) are abundant and economically important.

All **silicate** minerals have one commonality: They have **silica tetrahedra** in their crystalline lattices (Figure 10.20).

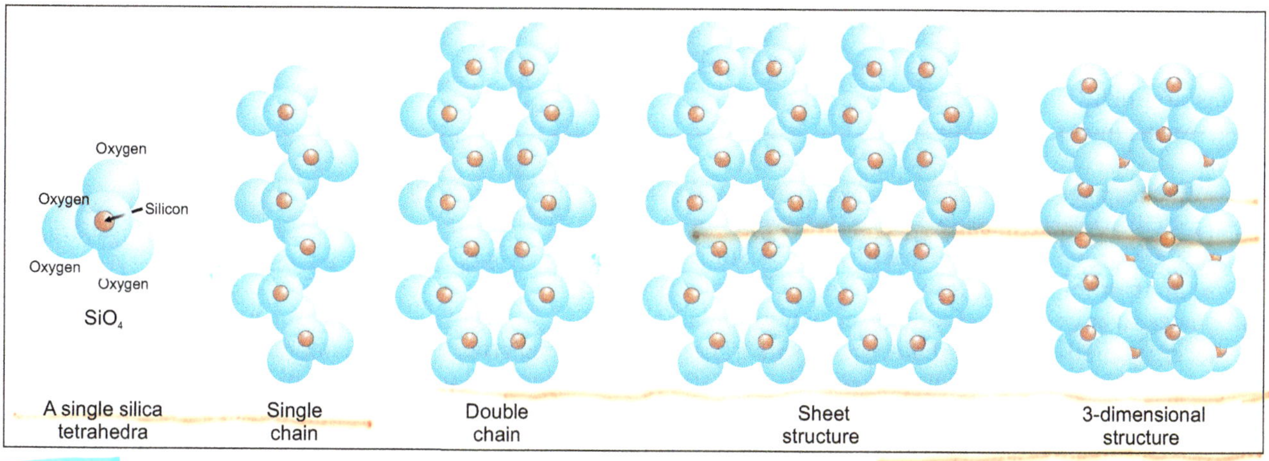

Figure 10.20: The five basic groups of silicate minerals. Each group is based on the arrangement of the silica tetrahedra. The arrangement of the tetrahedra strongly control many of the minerals' properties.

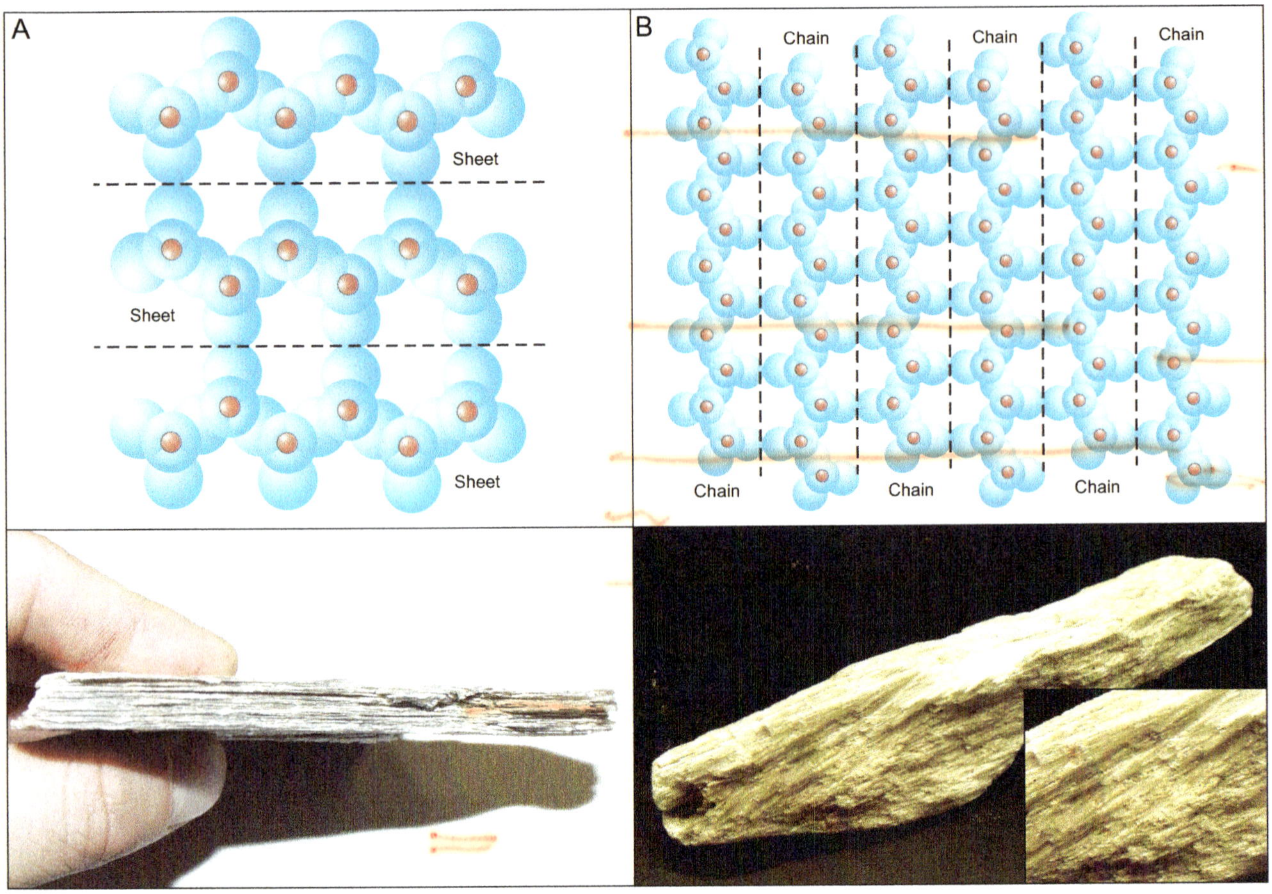

Figure 10.21: A. An example of a sheet silicate, muscovite, one of the basic rock-forming minerals. Muscovite looks like a stack of paper when viewed from the side. B. The mineral tremolite is an excellent example of a chain silicate. Notice how the chain-like pattern is repeated in the crystalline structure in the form of long, slender crystals. Photo credits: B. Zaprowski.

The silicate tetrahedra is a molecule comprised of four oxygen atoms and one silicon atom, arranged in a tetrahedral arrangement. If you combine these silica tetrahedra with other elements, you can build many different types of silicate minerals, including all of the rock-forming minerals. All silicates fall into one of five broad categories based on the arrangement of the silica tetrahedra in their crystalline lattices (Figure 10.20). For example, the mineral muscovite is a type of mineral called a sheet silicate because the silica tetrahedra are bonded in flat, two dimensional sheets. These sheets then are stacked on top of each other, just like a stack of paper, to form the mineral (Figure 10.21A). The mineral tremolite is a single chain silicate because the silica tetrahedra are arranged in chains. These chains are then bonded to each other, forming masses of long, slender minerals (Figure 10.21B).

Group	Anion	Mineral	Composition
Carbonates	CO_3^{2-}	Calcite	$CaCO_3$
Sulfides	S^{2-}	Pyrite	FeS
Oxides	O^{2-}	Hematite	Fe_2O_3
Halides	F^- or Cl^-	Halite	NaCl

Table 10-2: Some common non-silicate groups.

Non-silicates have no commonality except for the fact that they *lack* silica tetrahedra in their crystalline structure. Non-silicates are grouped together based on the nature of the *anion* in their chemical formula. So for instance, the carbonates all have CO_3^{2-} as the anion in their formula(Table 10-2). Many of the minerals we have highlighted in this section are non-silicates (e.g., halite and calcite). There is one special group worth mentioning, the **native elements**. Native elements are elements that occur naturally in their pure elemental form. For example gold, silver, and copper (Figure 10.22) are all metals which can be found in their pure elemental state (that is they are *NOT* bonded into molecules with other types of atoms). Other examples of native elements are charcoal, graphite and diamond - three different elemental forms of carbon!

Figure 10.22: An example of three common native elements: gold, silver and copper. Photo credits: B. Zaprowski.

Section 10-3: Igneous Rocks

The process by which molten rock cools to form solid rock is called **crystallization** (as discussed in Section 10-1). When the magma (or lava) is still very hot, the individual atoms have too much kinetic energy to form permanent bonds. To use an analogy, think about how hard it is to rake leaves on a windy day. Each time you get a few leaves gathered into a pile, the wind blows them away again - and so it is with the atoms and molecules within hot molten rock. Eventually, as the magma cools, permanent bonds do begin to form and crystallization begins. Once the individual crystals form, they grow by adding more and more atoms and molecules to their faces in a systematic fashion (Figure 10.23). Crystal growth continues until one of three things happens: 1. The mineral runs into its neighbors - there is literally no more room for continued crystal growth; 2. One of the elements needed to form the mineral runs out - at that point, a new mineral will begin crystallizing; 3. The rock is cooled extremely rapidly, stopping the crystal-forming process.

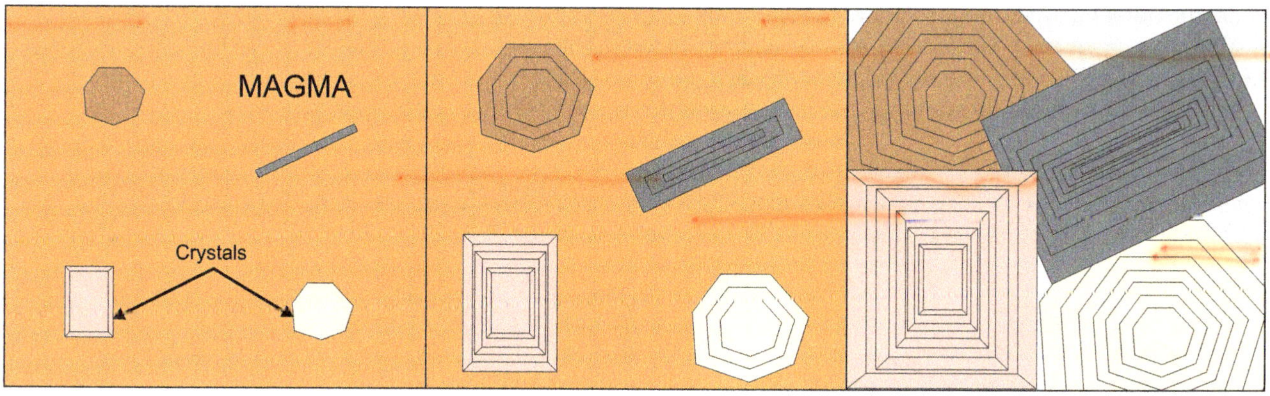

Figure 10.23: This series of illustrations shows how the crystallization process works. Once small crystals form in a body of cooling magma, the crystals will grow by adding successive layers to their crystalline lattices.

Igneous Rock Properties

Igneous rocks are classified using two basic properties: **Texture** and **composition.** Texture refers to the size and arrangement of the minerals in an igneous rock whereas composition refers to the types of minerals the rock is composed of. By identifying the texture and composition of an igneous rock, you can correctly identify it by name.

Igneous Rock Textures

The size of the crystals in an igneous rock is directly proportional to the rate at which the magma or lava cooled. If the igneous rock cooled relatively quickly (hours to weeks), the crystals will be very small, so small in fact, that a

Figure 10.24: A-C. Some examples of igneous rocks with an aphanitic texture. D-F. Some examples of igneous rocks with a phaneritic texture. Photo credit: B. Zaprowski.

majority (but not necessarily all of them) *will not be visible* to the naked eye. These rocks are said to have an **aphanitic** texture (Figure 10.24 A-C). Aphanitic rocks form during extrusive volcanism, that is, they form from rapidly cooling lava. Heat quickly dissipates from the cooling lava and so crystals don't have much time to grow.

If, on the other hand, the magma cooled slowly (hundreds to millions of years), the crystals can grow large and are visible to the naked eye. These rocks are said to have a **phaneritic** texture (Figure 10.24 D-F). Most phaneritic rocks form in association with intrusive volcanism. Magma below the Earth's surface cools very slowly as it takes time for heat to dissipate into the surrounding rocks. This gives the crystals plenty of time to grow to larger sizes.

Some rocks can cool very, very slowly, resulting in the growth of extremely large (sometimes meters long!) crystals. These rocks are said to have a **pegmatitic** texture (Figure 10.25). Pegmatitic rocks are the source of many gemstones such as rubies, emeralds, sapphires and garnets (but not diamonds). It is only in the center of a slow-

Figure 10.25: These igneous rocks display a pegmatitic texture, that is, the crystals are very large. Note the dime for scale. Photo credits: B. Zaprowski.

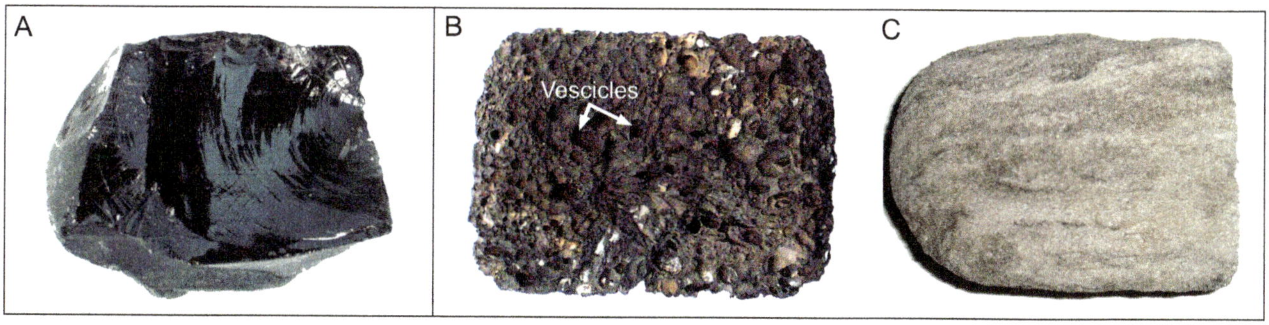

Figure 10.26: More examples of igneous rock textures. A. This igneous rock, called obsidian, displays a glassy texture. Rocks with a glassy texture have a completely random or *amorphous* molecular structure. B-C. These igneous rocks display a vesicular texture. Photo credits: B. Zaprowski.

cooling magma chamber, in the last little bit of cooling magma, that crystals can grow big enough to be of gem quality and size. In contrast, other igneous rocks can cool very quickly when lava flows into water (a process also known as quenching). The lava cools so quickly that minerals have no time grow at all, forming an amorphous solid, that is, a mineral-like solid without a crystalline lattice. This gives the rock a **glassy** texture (Figure 10.26A). These types of igneous rocks are also commonly refered to as volcanic glass. Magmas with high gas content that cool quickly form igneous rocks with a **vesicular** texture, that is, the rocks have lots of voids or *vesicles* in them (Figures 10.26B-C). One type of vesicular rock, pumice (Figure 10.26C), has so many vesicles that it can float on water!

Some rocks can have a more complex, two-stage cooling history. At first, the magma cools slowly in the Earth and some larger crystals begin to form. Then the magma erupts onto the Earth's surface and finishes cooling quickly, forming smaller crystals around the distinctly larger crystals. A rock which has two distinctly different-sized groups of crystals is said to have a **porphyritic** texture (Figure 10.27A).

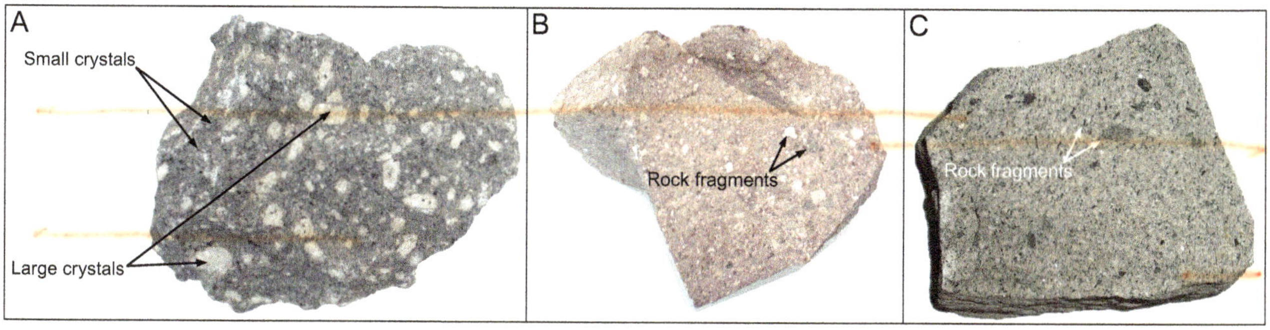

Figure 10.27: Examples of igneous rock textures. A. An example of a rock with a porphyritic texture, that is, it has two distinctly different-sized groups of crystals. B-C. These igneous rocks display a fragmental texture. These rock formed when hot volcanic ash and volcanic debris fused together. Photo credits: B. Zaprowski.

And finally, there are some igneous rock that do not form directly from cooling magma or lava through the processes of crystallization. When volcanos erupt, huge volumes of volcanic ash and rock fragments may be thrown into the air. As this volcanic debris rains down and accumulates on the Earth's surface, the residual heat can fuse the debris into a hard rock called a **welded tuff.** Rocks formed this way are classified as **fragmental** because these igneous rocks commonly have fragments of rock in them (Figure 10.27 B-C).

Igneous Rock Composition

So why does the composition of igneous rocks vary? The composition of an igneous rock typically reflects the tectonic setting in which it formed. For example, magma generated deep in the earth, like those feeding deep mantle plumes, is generally richer in iron and less rich in silica, so iron-bearing minerals are more common. In contrast, magmas that form in subduction zones will have more silica and less iron in them, so silica-rich minerals are more likely to dominate.

The formation of a specific type mineral within a rock is also a function of stability. A given type of mineral forms because the environmentalconditions (i.e., temperature andpressure) are just right for the atoms to bond in a specific crystallinearrangement. For example, let's say there is a large body of magma which contains silicon, oxygen, iron,

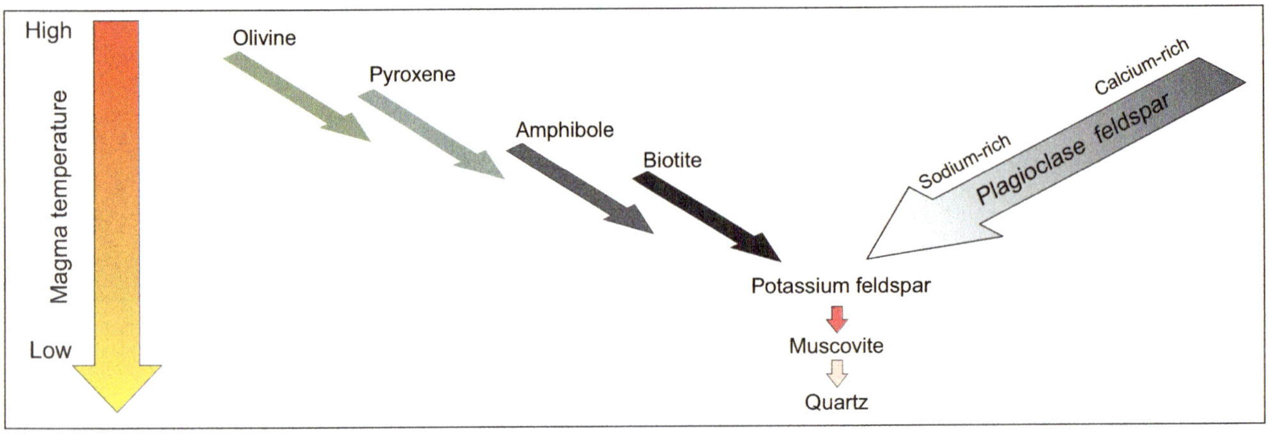

Figure 10.28: Bowen's reaction series. In any cooling magma, the types of minerals that form are directly related to the temperature of the cooling magma. Thus, at high temperatures we expect olivine and calcium-rich plagioclase to form, while at lower temperatures, we expect potassium feldspar, muscovite and quartz to form.

aluminum, magnesium and potassium. At high temperatures, the most stable arrangement of atoms is to form the mineral olivine $(Mg, Fe)_2SiO_4$. However, at lower temperatures the most stable arrangement of atoms is to form the mineral biotite $K(Mg,Fe)_3AlSi_3O_{10}(OH)_2$. The formation of the most common rock-forming minerals as a function of temperature is displayed in Figure 10.28. Figure 10.28 isknown as **Bowen's reaction series** and geologists use it to determine what the temperature of the magma was when minerals in an igneous rock were crystallizing. Even within the same magma chamber, the composition of the rocks at the edge of the chamber that cool first may be different than the rocks in the middle of the chamber that cool last, kind of like an onion has layers.

The identification of an igneous rock's composition can be a difficult concept to grasp for the amateur Earth Scientist. However, on a first order basis, composition can be approximated based on the color of the minerals in an igneous rock. By and large, most igneous rocks are composed of the eight basic rock-forming minerals and perhaps a few other rarer minerals. There are **four** basic types of igneous rocks based on composition, although it should be noted that ultramafic rocks are very rare (Figure 10.29).

Mafic rocks are composed chiefly of pyroxene, calcium-rich plagioclase, some amphibole and/or sodium-rich plagioclase. As a result, mafic rocks tend to be composed almost exclusively of dark green, dark gray and black minerals, *although some light gray colored sodium-rich plagioclase crystals may also be included.*

Figure 10.29: A summary of the different types of igneous rock composition. Photo credits: B. Zaprowski.

Intermediate rocks are composed primarily of sodium-rich plagioclase, amphibole and biotite mica with some pyroxene. As a result, intermediate rocks are typically composed of half black or dark green minerals and half white or light gray minerals.

Felsic rocks are composed chiefly of potassium feldspar (orthoclase), muscovite mica and quartz with minor amounts of biotite mica and/or amphibole. Thus, most felsic rocks are composed of light colored minerals: whites, light grays, reds and pinks, *although small amounts of black-colored minerals are almost always present.*

Ultramafic rocks, which are very rare, are composed dominantly of olivine with some calcium-rich plagioclase and pyroxene. Ultramafic rocks tend to be olive green in color.

Common Types Of Igneous Rocks

Figure 10.30 shows the most common types of igneous rocks. As you can see, texture and composition are the rock properties used in this classification scheme. All igneous rocks can be grouped as either intrusive or extrusive in nature. Rocks with phaneritic or pegmatitic textures are **intrusive igneous rocks**, while rocks with aphanitic, glassy, vesicular or fragmental textures are **extrusive igneous rocks**. Porphyritic rocks are the oddballs, as they are both intrusive and extrusive in nature. Some rock names, like obsidian, pumice and scoria have no specific composition associated with them. Fragmental rocks, as mentioned at the beginning of the section, are commonly referred to as welded tuffs or volcanic breccia.

Figure 10:30: Diagram showing the most common igneous rock types.

Some final notes about identifying and describing igneous rocks. Porphyritic rocks are named and classified by identifying the texture and composition of the smaller of the two crystal sizes. So, for example, the smaller minerals on the rock in Figure 10.27A have an aphanitic texture and the composition is intermediate so we would classify that rock as a porphyritic andesite.

Section 10-4: Sedimentary Rocks

Although sedimentary rocks make up only 4% of the Earth's crust by volume, 75% of all rocks exposed at the Earth's surface are sedimentary. Thus, 3 out of every 4 rocks you will look at in your lifetime are likely to be sedimentary in origin. In fact, the number one mineral resource in the United States is not gold or iron but sand and gravel. Compared to igneous rocks, sedimentary rocks display a huge range of variation, as a consequence the numerous geographic and tectonic settings in which they form and the physical and chemical processes operating in those settings. There really is not any general "blanket" statement that can adequately describe the origin of *all* sedimentary rocks. The origin of many sedimentary rocks has to be described individually on a case-by-case basis. However, there is one commonality all sedimentary rocks share: they are all composed of the products of rock weathering: Sediments and precipitated ions are fused together by lithification. As we discussed in Section 10-1, there are two principle types of sedimentary rocks: *Clastic sedimentary rocks* and *chemical sedimentary rocks*. Clastic sedimentary rocks are formed when sediments are cemented together into solid rocks. Clastic rocks are the predominant type of sedimentary rock by volume. (Figure 10. 31). Chemical sedimentary rocks are a catchall group of sedimentary rocks with a wide variety of origins, but all of them go through a (sometimes complex) series of chemical alterations during their genesis, many times (but not always) forming directly out of a hydrothermal solution.

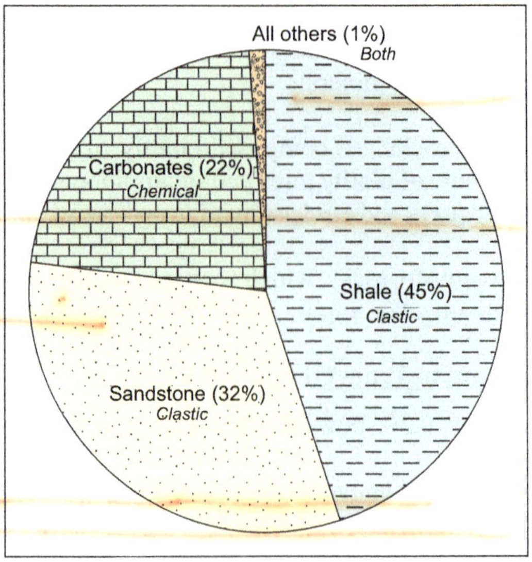

Figure 10.31: Diagram showing the relative abundance of different sedimentary rocks in the Earth's crust.

Clastic Sedimentary Rocks

Clastic sedimentary rocks are classified based on the texture (size and shape) and composition of the sediment they contain. For instance, a sedimentary rock composed of sand sized sediment is called sandstone. However, there are many different types of sandstone based on the composition of the sand (Figure 10.32). Quartzose sandstone is made up almost exclusively of tiny pieces of quartz, a very durable mineral, while greywacke is composed of a wide variety of different sand-sized sediments "floating" in a matrix of silt and clay (Figure 10.32B). The term arkose is commonly used to describe a special group of clastic rocks which are distinctly terrestrial in origin (most

Figure 10.32: An example of the range of sandstone compositions and the settings they form in. A. Quartzose sandstones form along beaches like this one in Maryland. B. Greywackes form along tectonically-active coasts like this one in Washington. C. Arkosic sandstones form in desert streams like this one in the Arches National Park in Utah. Photo credits: B. Zaprowski.

sedimentary rocks are marine in origin). An arkosic sandstone has a significant amount of plagioclase and iron oxide in it, giving it a pink to brick-red color.

The composition of a sedimentary rock is important in that the differences in composition reflect the origins of the rock. Quartzose sandstone typically forms in coastal environments (like the East Coast of the United States) where the action of the waves grinds up all of the softer minerals and leaves behind the really hard, chemically-resistant minerals like quartz (predominantly), garnet, corundum and topaz (Figure 10.32A). In contrast, greywacke forms in deep ocean environments in association with undersea landslides along tectonically-active coastlines (Figure 10.32B). Arkosic sandstones are terrestrial sedimentary rocks that commonly form in association with extensional faulting and rift basins, especially in arid environments (Figure 10.32C).

Overall, the classification scheme for clastic sedimentary rocks is pretty straightforward (Figure 10.33). Sediment texture is the primary characteristic used for identifying clastic rocks at this level, but sediment shape plays a role too. Notice that there are two types of clastic rocks made of gravel-sized sediments: conglomerates and breccia. The difference between the two is the shape of the gravel. Conglomerates are made up of rounded pieces of gravel while breccia is made up of angular pieces of gravel. This distinction is related to their origins. Rounded gravel is commonly formed through the action of tumbling in water, either in a river or at the seashore, whereas angular gravel is commonly associated with physical weathering in cold and arid environments, especially in mountainous terrain. Coquina is a sedimentary rock made up almost exclusively of shell fragments. Coquinas form in shallow marine environments where the action of waves breaks up the shells into small pieces. Once the shell fragments are deposited they are cemented together, forming a rock.

Sediment type	Rock name	Sediment type	Rock name
Round gravel	Conglomerate	Sand	Sandstone
Angular gravel	Breccia	Silt	Siltstone
Shell fragments	Coquina	Clay	Shale

Figure 10.33: The basic types of clastic sedimentary rocks. Photo credits: B. Zaprowski.

Chemical Sedimentary Rocks

What do we mean by a chemical sedimentary rock? Unlike clastic sedimentary rocks which are composed of sediments cemented together, most chemical sedimentary rocks form in conjunction with one or more chemical processes, like precipitation. Because chemical sedimentary rocks form in a wide variety of environments and have a wide variety of compositions, there are no obvious classification schemes like as for clastic rocks. Nonetheless, we can roughly group different types of chemical sedimentary rocks together based on chemical process and rock composition. The classification of chemical sedimentary rocks is based on careful observations of the rock sample and identifying the key properties associated with each chemical sedimentary rock group. One common property that almost all chemical rocks share is that they are crystalline in nature, the exception here being coal and amber.

The **carbonates** are a very important group of sedimentary rocks, making up almost 22% of all sedimentary rocks (Figure 10.31). Sedimentary rocks in the carbonate group are overwhelmingly made up of the mineral calcite ($CaCO_3$ or calcium carbonate), although some carbonates have cations other than calcium. Limestones are the most dominant rocks in the carbonate group. Limestones can be further divided into two subgroups, *crystalline*

Figure 10.34: A. Crystalline limestones (inset) commonly form in warm, tropical lagoons like this one in the South Pacific Ocean. Photo credit: Image Science and Analysis Laboratory, NASA-Johnson Space Center. Inset photo credit: B. Zaprowski. B. Cave features like these in Crystal Cave, PA form when a type of carbonate called travertine is deposited by water dripping from the cave ceiling. Photo credit: B. Zaprowski.

limestones and *biogenic limestones*, and the real difference between the two types is one of texture and origin.

Many crystalline limestones form in warm, shallow tropical lagoons where tiny crystals of calcite precipitate out of seawater, forming a calcite "mud" (Figure 10.34A). Over time, this mud lithifies into a solid rock with a distinct crystalline texture to it, much like the sugary coating in a bowl of frosted flakes (inset). Crystalline limestones that form on land are known as travertine. When groundwater moves through limestone bedrock, it can dissolve the calcite crystals a limestone is made of. If the groundwater seeps onto the Earth's surface, the now dissolved calcite precipitates out to form layers of travertine. The stalagmites and stalactites found in many caves are also made of travertine (Figure 10.34B). We'll explore caves and cave features more in Chapter 13.

Fossiliferous or biogenic limestones form directly from organic debris such as pieces of shell or corals, usually in warm tropical marine waters. Biogenic limestones start out as coquinas. Over time, the spaces between the fossils fills in with precipitated calcite crystals. As the rock continues to lithify, the fossils themselves begin to dissolve away and the resulting empty voids fill in with even more calcite. The resulting rock is called a fossiliferous limestone, meaning it contains abundant fossils (Figure 10.35A). Not all fossiliferous limestones have visible fossils in them. Chalk is a carbonate made up of millions of tiny microscopic shells and other skeletal debris (Figure 10.35B). This debris is produced by planktonic (i.e., they float near the surface) single-celled marine plants and animals. When they die, their tiny shells sink to the bottom of the ocean and pile up into thick layers. Over millions of years, the shells are slowly fused into layers of chalk.

The **Evaporites** are another important group of chemical sedimentary rocks. As the name would seem to imply,

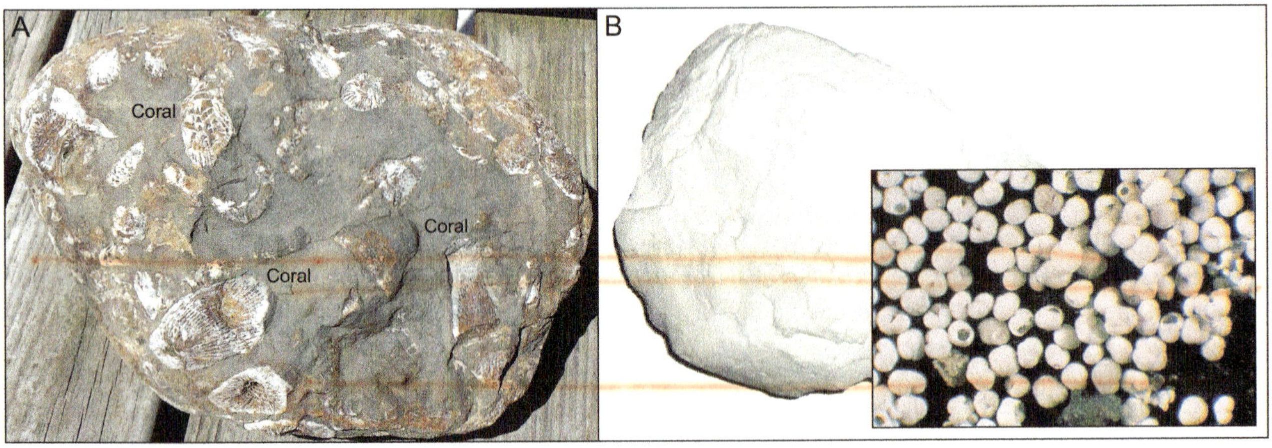

Figure 10.35: A. A fossiliferous limestone with pieces of coral in it. Photo credit: B. Zaprowski. B. Under a microscope, you can see chalk is just a mass of shells. Photo credit (chalk): B. Zaprowski. Photo credit (inset): Anne Jennings, University of Colorado-Boulder/NOAA.

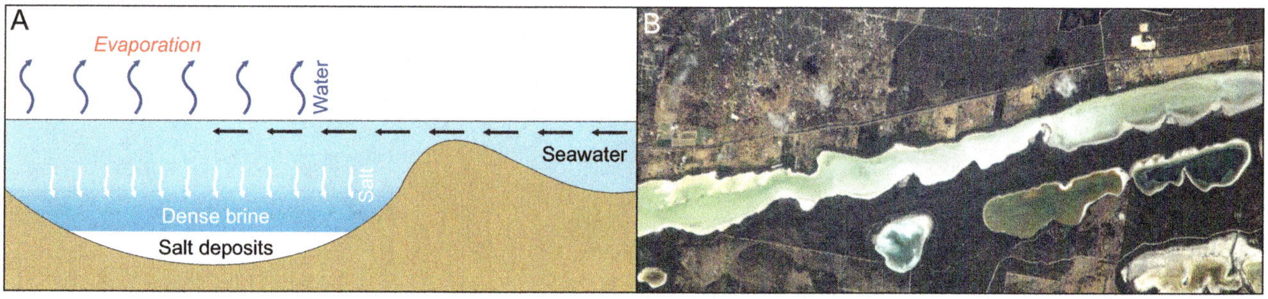

Figure 10.36: A. A common geologic setting for evaporite formation is in restricted coastal embayments. Water enters the bay through a narrow inlet but can only leave through evaporation. The evaporation leads to the formation of dense brines and eventually triggers the precipitation of salt crystals. B. These hypersaline lagoons along the coast of Australia are fringed with salt deposits. Photo credits: Image Science and Analysis Laboratory, NASA-Johnson Space Center.

evaporites form through the process of evaporation. Any time water is evaporating, it leaves behind the ions that were dissolved in it. Over time, in the right geologic setting, the concentration of salts in the water becomes more and more concentrated, eventually forming dense brines. When the brines become supersaturated, crystals of salt and other minerals begin to precipitate out of solution. These minerals then form layers at the bottom of the basin and eventually lithify into solid rock. One location where evaporite formation is common is along arid coastal lagoons (Figure 10.36). The key here is that the lagoon must be partially connected to the ocean such that water can slosh into the lagoons during storms or high tide, but the water can only leave via evaporation (Figure 10.36). Over time the water in the lagoon becomes hypersaline and the precipitation of evaporite minerals results. Evaporite deposits also commonly form in mountainous desert regions, like Death Valley (Figures 10.10A and 10.37). Although most evaporite minerals are rare, two common types of evaporites are economically important: Rock salt (halite) and gypsum, a common building material used for making drywall and plaster.

Figure 10.37: A close-up of evaporite deposits on the floor of Death Valley, CA. Photo credit: B. Zaprowski.

The **silicates**, a small but important group of chemical sedimentary rocks, are composed of amorphous quartz. The mineral quartz can exist in many forms, including a special microcrystalline form called chert or flint (Figure 10.38A). The texture of these rocks is very smooth, kind of like the pieces of frosted glass you might pick up at a beach, and they are very hard. Flint and chert have been prized by human civilizations all around the world for hundreds of thousands of years because they could shape and chisel these rocks into sharp-edged tools like axes and spear points (Figure 10.38B). Chert and flint are commonly found as nodules in limestone but they may also be found in bedded layers. Some silicates can be biological in nature (some marine organisms like diatoms, radiolaria and sponges have hard parts made of silica) but most bedded layers of chert are associated with volcanic ash deposits. Over time, chemical processes convert the fossils or volcanic ash into silica gels which are then lithfied into hard silicate rocks.

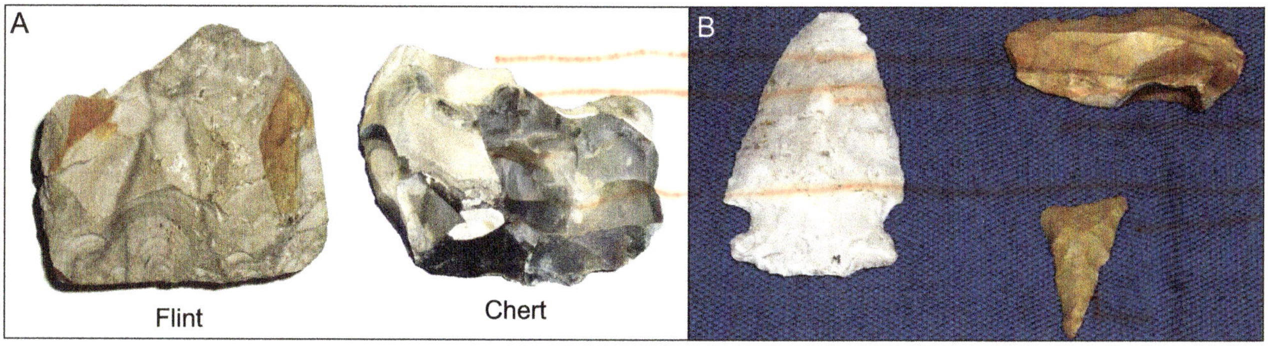

Figure 10.38: A. Examples of flint and chert, two common silicates. B. Silicates have been used to make stone tools like these for tens of thousands of years. Photo credit: B. Zaprowski. Tools courtesy of the Nabb Research Center at Salisbury University.

Figure 10.39: Biogenic sedimentary rocks. A. Some examples of amber. Amber can preserve even the most delicate and fragile insects. Photo credits: B. Zaprowski and USGS (inset). B. A seam of coal in southwestern Pennsylvania. Photo credit: A. Zaprowski.

The last group of chemical sedimentary rocks is **biogenic** in origin, that is, they form chiefly from organic material through a slow chemical alteration process. The two chief examples of this are amber and coal. Amber is simply tree sap or resin. If the resin is buried in water for some time and tumbled around for a little bit, it slowly hardens into a rock-like object we call amber (Figure 10.39). It doesn't have true mineral structure in the classical sense; it is really more like "fossilized tree sap" than a kind of rock. Amber is one of the prized sources of fossil insects. Most sedimentary environments are too rough to preserve something as delicate as an insect, but the gentle entombing sap encases the insect, preserving it.

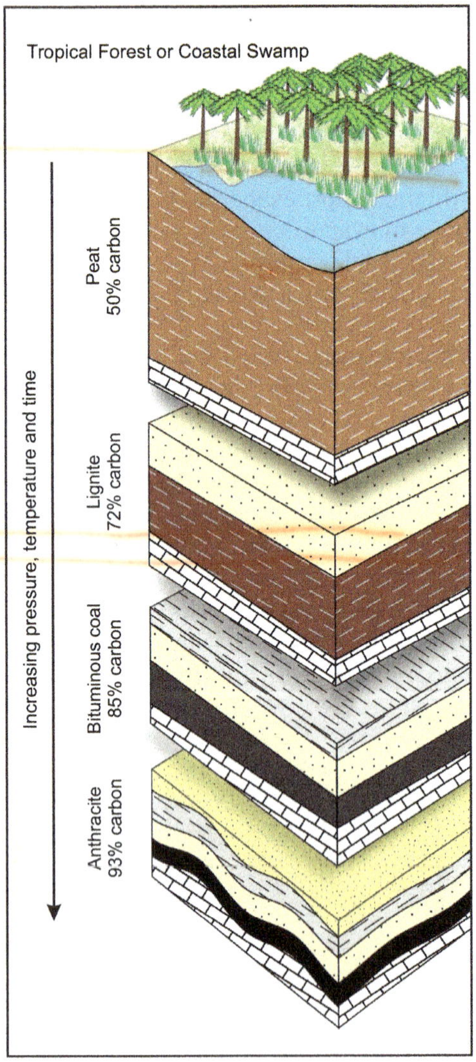

Figure 10.40: The different stages of coal formation.

Coal is simply metamorphosed plant material. In fact, coal offers us a nice segue into metamorphism and metamorphic processes. There are three key elements to coal formation. First, you need to have LOTS of plant material. For example, the coal seam in Figure 10.39B would have required a pile of organic material about 8 meters (25 feet) thick to form! Huge tropical rainforests like in the Amazon River Basin are a good example of places where biomass can accumulate in the massive quantities needed for coal formation. Second, you need to bury the plant material rapidly. Plant material which is exposed to oxygen is quickly decomposed. The best way to seal the plant material off from the air is to bury it. Again, the Amazon River Basin is a good example of this. The Andes Mountains to the west have been rising and eroding for millions of years. The mighty Amazon River carries large quantities of sediment through the vast Amazon River lowlands, flooding millions of acres of forest floor annually. These floods can quickly deposit a thick layer of sediment on top of the organic litter which has piled up on the forest floor, preserving it. Third, you need lots of time for pressure and heat to chemically alter the plant material. It takes millions of years for coal to form because the metamorphism of plant material into solid coal is a slow process (Figure 10.40).

When the plant material is first buried, it forms a loose but organic-rich soil called **peat**. As the plant debris is buried deeper and deeper under the weight of the sediments above, it is slowly compressed, water is squeezed out, and the carbon is concentrated into a dense layer of

Figure 10.41: The basic types of coal: A. Lignite or brown coal. B. Bituminous coal. C. Anthracite coal. Photo credits: B. Zaprowski.

concentrated organic substance called **lignite** or brown coal (Figure 10.41A). Even deeper burial causes the plant material to break down completely. When the mass has a carbon content of at least 85%, it is called **bituminous coal** (Figure 10.41B). In the right tectonic setting, bituminous coal may be further metamorphosed into a very hard, dense form of coal called **anthracite** (Figure 10.41C).

Section 10-5: Metamorphic Rocks

By definition, metamorphism means "to change form" and metamorphic rocks do just that: they change their form, either physically and/or chemically *without melting*. Metamorphism can form in a wide variety of environments, and under a range of different conditions, but the bottom line is that metamorphic rocks change their structure and composition to *become more stable*. Any crystalline rock possesses the most stable arrangement of ions possible for the geologic environment it was subjected to during its formation. For example, think back to Bowen's Reaction Series in Section 10-3. Certain minerals form at certain temperatures. Thus, if you take a pre-existing rock and subject it to a higher temperature for a period of time, its minerals will slowly and spontaneously change to more stable minerals which reflect the new environmental conditions the rock was subjected to (in this case, warmer temperatures).

So where do we find metamorphism occurring? Pretty much anywhere active tectonism is occurring. For example, a subduction zone is an environment where high pressures can build up and form metamorphic rocks. As sediments are scraped up off the subducting plate, they pile up to form an accretionary wedge (Figure 10.42). As the wedge grows, the pressure on the sediments increases, and they lithify into sedimentary rocks. As the pressure continues, those rocks are metamorphosed into a type of rock with high pressure-low temperature minerals called a blueschist. Because there is no real source of heat nearby (like magma), the temperatures in the rocks never get particularly hot. The most widespread type of metamorphism occurs in association with mountain building. During mountain building, rocks are subjected to both high temperatures and high pressures, forming a wide variety of different metamorphic rocks with high pressure and high temperature minerals in them. Because it generally occurs over

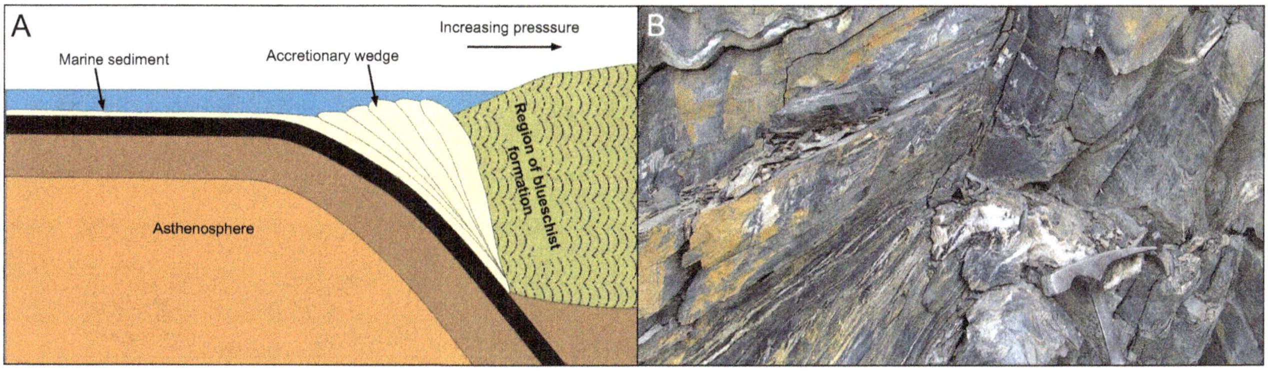

Figure 10.42: A. A subduction zone is the ideal location for forming metamorphic rocks with high pressure, low temperature minerals. B. Blueschist-type metamorphic rocks exposed in an outcrop in Vermont. Photo credit: J. Kim/Vermont Geological Survey.

Figure 10.43: A. This photo of a new road cut in Vermont is a good example of metamorphic rocks produced by regional metamorphism. Photo credit: Tom Eliassen/Vermont Geological Survey. B. A close-up image of the metamorphic rocks. Photo credit: J. Kim/Vermont Geological Survey.

wide areas, this type of metamorphism is called regional metamorphism. Many of the rocks in New England are this type. The rocks are the remnants of deep mountain roots that formed during a collision between North America and Europe over 400 million years ago. These rocks have been exposed only after hundreds of millions of years of erosion removed the kilometers of rock that used to overlie them (Figure 10.43).

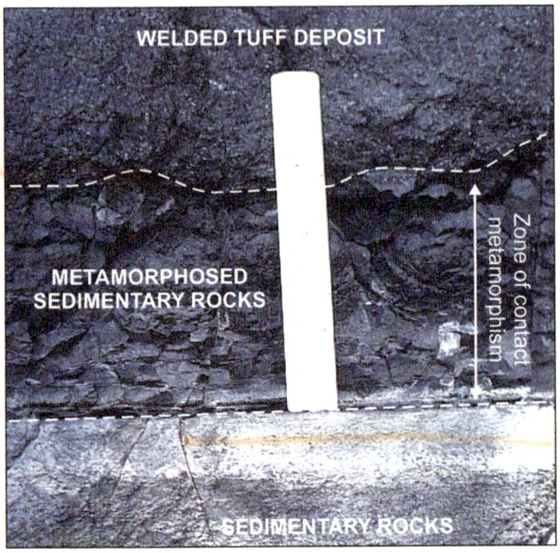

Figure 10.44: This picture from Yellowstone National Park shows a narrow zone of contact metamorphism between a welded tuff and some marine sedimentary rocks. Photo credit: Robert L. Christiansen/USGS.

Volcanism also a common cause of metamorphic rock formation. Any time you have magma intruding into the crust or lava flowing on the Earth's surface, layers of metamorphic rock can form. This contact metamorphism causes the rocks near the magma to be "cooked" - the rocks are heated up, but not hot enough to melt (Figure 10.44). At the same time, the pressures may not increase significantly during the intrusion so these metamorphic rocks tend to have high temperature-low pressure minerals in them. On the surface, lava flows and volcanic ash deposits can "cook" the land surface they are deposited on, producing a thin layer of metamorphic rocks just below them. The movement of chemically-active fluids through these rocks and the deposition of minerals is an important part of this type of metamorphism.

Causes of Metamorphism

Changing temperature, as shown in Figure 10.44, is one of three ways you can metamorphose rocks. **Increasing pressure** can also cause a rock to change its structure and/or composition. When increasing amounts of pressure are applied to the "parent" rock, the minerals in the rock will re-orient themselves into a denser, more stable form. And finally, some metamorphic rocks result from the circulation of **chemically active fluids** through rocks (Figure

Figure 10.45: A. The Homestake Gold Mine in Deadwood, SD formed when the light-colored igneous intrusions penetrated through the local dark-colored metamorphic bedrock, depositing gold. B. Gold veins cutting through quartz. C. A geode filled with quartz. Photo credits: B. Zaprowski.

10.45). Fluids play a role in almost all kinds of metamorphism, but they can also act alone to alter rocks. For example, many deposits of gold, silver and copper form this way (Figure 10.45). When a magma chamber forms just below the Earth's surface (and perhaps feeds a volcano above), fluids from the magma can circulate into the rocks surrounding the magma, altering the bedrock and depositing minerals. "Veins" of gold form in cracks which were filled in over time with gold deposits as hot, mineral-rich fluids moved through them (Figure 10.45B). *Geodes* form when hot fluids fill-in rock cavities with an array of different minerals such as quartz, calcite and other low-temperature minerals (Figure 10.45C).

There are different degrees of metamorphism - some rocks are only changed slightly (**low-grade metamorphism**) while other rocks experience a radical change in mineral composition (**high-grade metamorphism**) because of changes in their environment. In low-grade metamorphism, the minerals in the original parent rock are rearranged into a different, more stable form *without being chemically altered*. For example, slate is metamorphosed shale. Both are made of clay minerals, but the clay minerals in the slate are arranged into a denser, more compact arrangement than in the shale (Figure 10.46A). Coarse-grained crystalline rocks (like igneous rocks) also experience a rearrangement of their minerals. In these cases, banded metamorphic rocks called gneiss may form

Figure 10.46: Examples of low-grade metamorphism. A. Slate. B. Gneiss. Photo credits: B. Zaprowski.

(Figure 10.46B). In both of these examples, the minerals realigned themselves *perpendicular* to the stress. This is of importance to geologists because the orientation of the minerals can be used to determine the direction tectonic forces were coming from during a particular event. In high-grade metamorphism, the minerals in the parent rock have been chemically altered into new minerals. The old minerals are able to slowly metamorphose into new minerals through a rearrangement of the ions in the crystalline lattice. When the new minerals form and grow, they grow *perpendicular* to the stress. So just like the low-grade metamorphic rocks above, the orientation of the minerals says something about the direction of stress. Schist is a rock that forms in this way (Figure 10.47A). Schist may for through the alteration of several different types of parent rocks, but most commonly schist forms from the metamorphism of rocks with clay minerals in them, like shale and slate. The clay minerals are chemically altered

Figure 10.47: Examples of high-grade metamorphism. A. Schist. Photo credit: B. Zaprowski. B. Migmatite. Photo credit: USGS.

into a group of sheet silicates called mica. When mica minerals grow, they grow in flat, thin layers, perpendicular to the stress. Some gneisses are also high-grade metamorphic rocks. These gneisses form when coarse crystals like feldspar and quartz grow in planes which are perpendicular to the direction of stress. Some gneisses become a highly-metamorphosed rock called a *migmatite*. Migmatites are banded metamorphic rocks, but these rocks display distinct patterns of "swirled" mineral bands in them (Figure 10.47B).

Types of Metamorphic Rocks

Metamorphic rocks are the most difficult for a novice Earth Scientist to identify and classify because they are highly variable in form and appearance. The reality is that each type of metamorphic rock has its own unique set of characteristics that has to be learned individually. Luckily there are only a handful of common metamorphic rocks, many of which are economically important because of their hardness and beauty. The simplest way to group them is by their texture, that is, the size and orientation of the minerals in them. The three main metamorphic textures are: **Banded**, **foliated** and **non-foliated**. Let's take a brief look at these basic metamorphic rock textures.

Banded metamorphic rocks, like gneiss, have thick "rods" or layers of minerals in them, usually in the form of coarse minerals (Figure 10.48A). Feldspar, mica and quartz are common minerals in banded metamorphic rocks. A foliated metamorphic rock exhibits layers of thin, flat minerals in a preferred orientation. This preferred layering within a metamorphic rock is called **foliation** (Figures 10.48B-D). Foliated rocks, like schist and phyllite, usually contain mica and other sheet-like minerals which give them a unique layered look, like a stack of paper.

Non-foliated metamorphic rocks have no preferred mineral orientation. A good example of this is marble (Figure 10.48E). Marble is a metamorphosed limestone. The calcium carbonate in the limestone is re-crystallized into a

Figure 10.48: Examples of metamorphic rock textures. A. Banded. B-D. Foliated. E-F. Non-foliated. Photo credits: B. Zaprowski.

Group	Rock name
Foliated	Slate
	Phyllite
	Schist
Banded	Gneiss
	Migmatite
Non-foliated	Quartzite
	Anthracite
	Marble

Figure 10.49: Common metamorphic rocks. Photo credits: B. Zaprowski.

denser, harder more compact form of calcite. Some marble can even preserve fossils - decorative marble, like the type used in banks, commonly has shells and coral pieces in it. Quartzite (Figure 10.48F) and anthracite (Figure 10.41C) are also common non-foliated metamorphic rocks.

Once you have determined what texture group your rock belongs in, the luster of the rock can help you determine which type of metamorphic rock you have (Figure 10.49). Among the foliated rocks, slate has a fairly dull luster (Figure 10.48B), phyllite has a glossy-like sheen (Figure 10.48C), and schists tend to be coarse-grained and have a shiny luster (Figure 10.48D). Among the non-foliated rocks, marbles tend to have a shiny, crystalline luster and distinct, flat crystal faces may be seen (Figure10.48E), whereas quartzites have more of a microcrystalline luster (Figure 10.48F). Gneisses, migmatites and anthracite are distinctly recognizable for their unique textures.

Although it may seem like a rock which has been so significantly altered would not be of much use to geologists seeking to decipher Earth's history, metamorphic rocks can prove quite a bit of information about the past. All metamorphic rocks have a **parent rock**, that is, the original rock they were metamorphosed from. If you know what the parent rock was, you can infer what the geologic environment was *before* metamorphism occurred. For example, quartzite is metamorphosed sandstone (Figure 10.50). Since we know the rock used to be sandstone, we can infer its *original* environment of formation might have been a beach (coastlines commonly get caught in the "squeeze" during continental collisions). Figure 10.50 shows the parent rocks of many common metamorphic rocks. As you can see, some metamorphic rocks have more than one parent, and, some parent rocks can form several different types of metamorphic rocks. Metamorphic rocks are complicated - in many cases they can have a complex history of heating and cooling, making them a challenge to interpret and date. Geologists who specialize in metamorphic rocks commonly use high-powered microscopes and electron microprobes to study the nature of individual minerals within the rock.

Now that we have thoroughly reviewed rocks and the rock cycle, the next chapter can delve a little deeper into the interpretation of rocks and the information they contain.

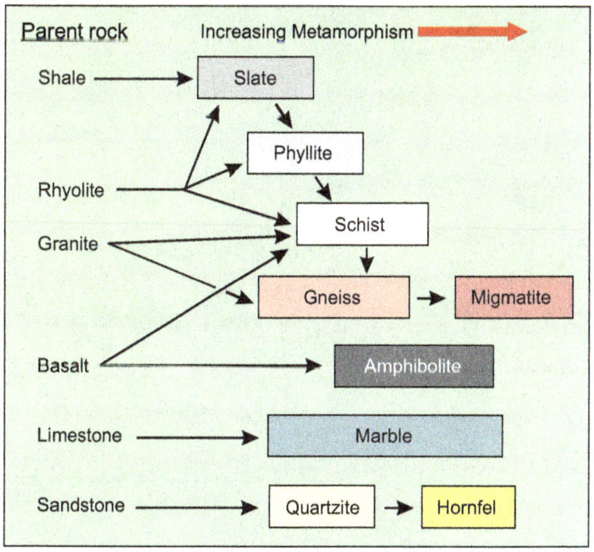

Figure 10.50: Parent rocks for common metamorphic rocks. Some metamorphic rocks can have more than one parent.

Chapter 11 - The Principles of Historical Geology

Recall that at the beginning of Chapter 10 we said: "Everything we know about the Earth's past comes from information stored in rocks. Every rock potentially contains information about how, where and when it formed. If you understand how and where rocks are formed, you can interpret the history of a particular group of rocks.....In this way, geologists are like publishers putting together a giant book of Earth history, and the rocks of the Earth are the pages." Because geologists know from observations where different types of rocks form and the environmental conditions necessary for those rocks to form, they can interpret the rock record. For example, consider the ripple-marked sandstone in Figure 11.1A. Sandstones can form in a wide variety of environments, but in this case, we can conclude that when this sandstone formed, this part of Utah was flooded by a shallow sea. How do we know this? First, the composition and color of the sandstone is indicative of sediments deposited in a marine setting. Second, the ripple marks preserved in this sandstone are only formed in shallow waters by the sloshing action of waves approaching shore (Figure 11.1B). Using this type of logical reasoning, geologists can infer a lot about what the environment was like at a given place and time in Earth's history just from studying rocks. Geologists use the term **historical geology** to describe the sub-discipline of geology that deals with interpreting the Earth's history.

Figure 11.1: Using the present to interpret the past. A. Sand ripples preserved in a sandstone in Utah. B. Sand ripples exposed during low tide at the mouth of a creek in Puerto Rico. Because of their similarity, we can infer the rock in A formed in a similar setting. Photo credits: B. Zaprowski.

The real challenge for geologists is putting the "pages" in the correct order and figuring out where to put new pages when they are discovered. What makes this task particularly challenging is that an untold number of rocks have been destroyed by the rock cycle and the information they contained about a certain place and time on Earth is lost forever. Other rocks we can't seem to fit anywhere in the book - random pieces of the past without "page numbers" if you will. At the same time, there are many, many more rocks yet to be discovered and interpreted. For a geologist, the excitement comes from knowing the next rock discovered could be another critical page needed to fill in the details about the Earth's past. It is important to note that the further back in time we go, the less detail we know about Earth's history because there are fewer and fewer rocks of the right age to look at (thanks to plate tectonics and the rock cycle). Nonetheless, geologists have gathered enough pages to piece together a fairly clear picture of how the Earth and its biosphere have changed over the past 4.55 billion years. We'll explore that topic further in Chapter 12. For now, let's look a little closer at how geologists interpret and date rocks in order to weave that information into a coherent picture of past events.

Section 11-1: The Table of Contents for Earth History - The Geologic Time Scale

Now that you understand a little bit about how the "pages" of Earth's history are made, we need to discuss how geologists determine the "age" of a rock. First, let's be clear here, geologists can determine the age of a rock with a relatively reasonable amount of accuracy but there are rarely any absolutes in geology. Even the best laboratory techniques have some statistical range of error. Nonetheless, geologists have been able to date rocks with enough precision to put the "pages" of Earth's history in order. Some techniques involve common sense and simple logic; others can involve high-tech physical and chemical analyses by specially-trained professionals. In general, the techniques geologists use to put the pages of Earth's history in order can be broken down into two categories: **Absolute dating** and **relative dating**.

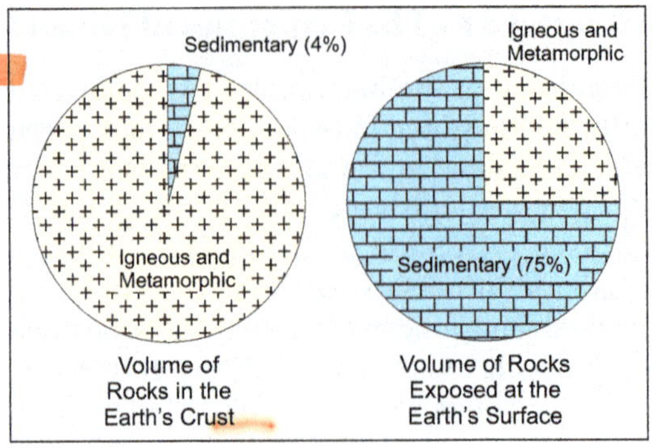

Figure 11.2: The distribution of sedimentary rocks in the Earth's crust.

Absolute dating techniques allow a geologist to calculate how many years have elapsed since a given rock formed. If you can determine the absolute age of a rock, it is very easy to determine where that rock fits in Earth's history. We'll discuss exactly how the process works in Section 11-3. Absolute dating techniques were first developed in the middle of the 20th century; prior to that, no one really had any idea how old the Earth was (although there were a lot of clever attempts to estimate it). Early geologists could only put the information contained within rocks in a sequential order; that is, they knew which rocks were older and which ones were younger. That information is still very useful and geologists were able to use it to infer what happened in the past. The only information early geologists lacked was knowing exactly when these events occurred.

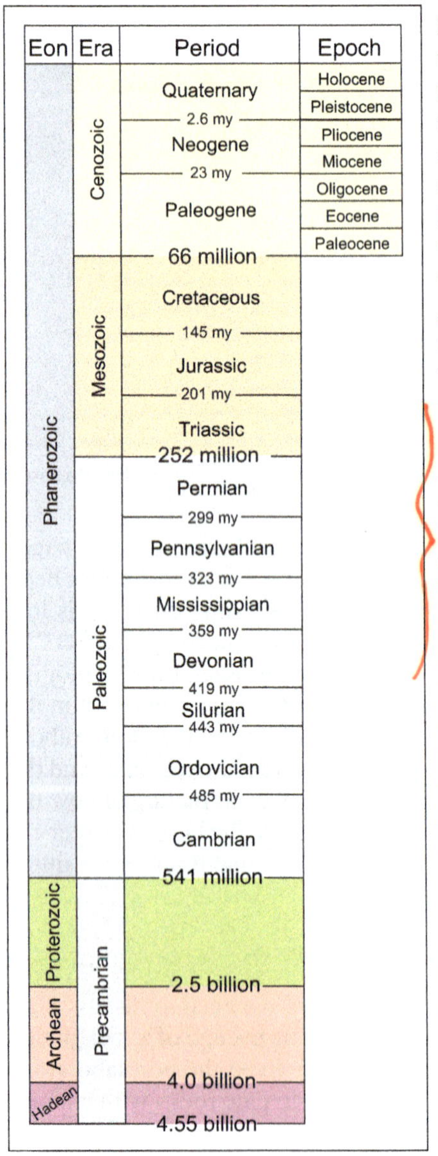

Figure 11.3: The Geologic Time Scale.

For example, imagine a pile of leaves on the forest floor. Since you weren't there to watch them fall, you don't know when each leaf fell, but you can tell the order in which they fell, since the leaves that fell last would be on top of the leaves which fell first. Geologists use that same kind of basic logic approach with rocks when they use the principles of relative dating to develop a time-line for the sequence of geologic events that have led to the Earth as we know it today. Even if early geologists had absolute dating techniques available to them, those techniques alone wouldn't have helped them put all the events in correct order because *the age of most sedimentary rocks cannot be determined using absolute dating techniques* (more on this in Section 11-3). This is unfortunate because 75% of the rocks exposed at the Earth's surface are sedimentary and they contain pretty much all of the fossils (Figure 11.2).

Although you may think the lack of absolute dating would be a major obstacle to deciphering Earth's history, relative dating is a powerful tool all by itself. Even without any absolute dates, early geologists were still able to piece together the **Geologic Time Scale** (Figure 11.3). You can think of the geologic time scale as the "table of contents" for the book of Earth's history. It's the way geologists describe the relative age of rocks. To use an analogy, human history is commonly divided into ages like the "Stone Age" or the "Bronze Age" based on changes in technology. A piece of pottery discovered in an archeological site might subsequently be described as "Bronze-aged" because it was known to have been made during that age's time range based on studies at other locations. Even though you may never know *exactly* when this piece of pottery was made, you know from its composition, shape and location *approximately* when it was made.

Likewise, the rock record has been broken into periods of time, like the Jurassic period, and any rocks which formed during that time period would be described as Jurassic-aged. In this case, early geologists quickly realized that certain fossil assemblages were only found in rocks of a given relative age. This happens because over time plants and animals have evolved and subsequently gone extinct, many times in mass extinction events (more on this in Chapter 12). Geologists used the sudden appearance and disappearance of major fossil groups in the rock record to break the Earth's history into divisions of time. For example, the Mesozoic Era is commonly referred to as "The Age of the Dinosaurs" because during that period of time, and only during that period of time, do we find dinosaur fossils (Figure 11.4).

As the years progressed, geologists began to refine and subdivide the eras and periods of Earth's history using relative dating and the fossil assemblages found in rocks. Later on in the 20th century, when geologists developed absolute dating techniques, they were able to finally assign ages to the different time divisions on the geologic time scale as we know it today (Figure 11.3). So we know now that the Mesozoic Era, as defined by the fossil record, spans the period of Earth's history from 252 to 66 million years ago. We'll discuss fossils and fossil formation in more detail in Chapter 12, but for now let's delve into the details behind relative and absolute dating.

Figure 11.4: Dinosaur fossils are only found in the Mesozoic Era. Photo credit: B. Zaprowski.

Section 11-2: Putting the Pages of Earth's History in Order - Relative Dating

The principles of relative dating are based on logic, the laws of physics and direct observations of processes occurring in the modern world. For example, we can directly observe water flowing downhill and transporting sediment (Figure 11.5). We know from the laws of physics that the water flows downhill under the influence of gravity. We also know that water in motion has kinetic energy and that kinetic energy has the ability to move sediment. Simple logic dictates that it is reasonable to assume that the laws of physics have not changed during the entirety of Earth's history, that water has always flowed downhill, and that water has always had the ability to erode bedrock and transport sediment. Geologists commonly use the term **uniformitarianism** to describe this logical thought process. By observing geologic processes carefully as they occur in the present, we can determine what may have happened the past. Of course, there are plenty of geologic events we haven't directly observed in the whole of human history (i.e., a major meteorite impact or the eruption of a supervolcano), but that's why interpreting the rock record is so important. It helps us understand the range of possibilities for the future.

Figure 11.5: Uniformitarianism states that streams like this one have always been flowing downhill. Photo credit: B. Zaprowski.

Some of the most basic principles of relative dating are based on the process of sediment deposition. As we discussed in Chapter 10, sediments are carried by water, wind and/or ice to some location and deposited. If the deposition continues and the sediments continue to thicken, the older layers of sediment will become buried by progressively younger layers of sediment (Figure 11.6A). *As long as the layers are not moved from this original position*, the oldest layer will always be at the bottom and the youngest layer will always be on top. This is known as the **principle of superposition**. Although this may seem like a trivial observation, it has important implications for the fossils contained within the rocks. If you walk up to a tall, continuous exposure of undisturbed sedimentary rocks, you will know with absolute confidence that the fossils at the bottom are older than the fossils at the top (Figure 11.6B).

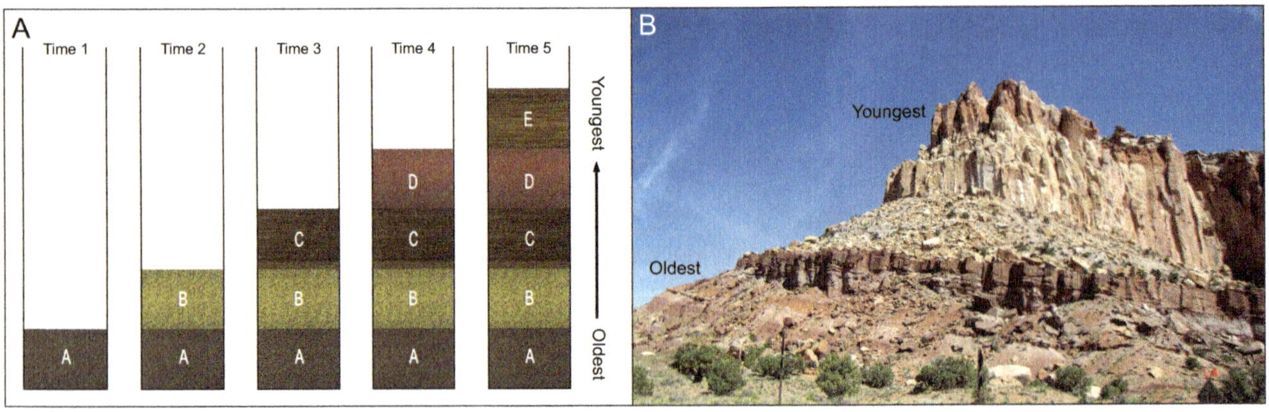

Figure 11.6: A. The principle of superposition explained schematically. As sediments are deposited in a basin, the sediment deposited first (A) will always be the oldest, while the sediment deposited last (E) will always be the youngest. B. Since this exposure of rock has not been moved from its original position, we can infer that the sedimentary rocks on the bottom of the outcrop are older than the ones on top. Photo credit: B. Zaprowski.

The key here is to be sure that the rocks haven't been disturbed and moved from their original orientation. Geologists know from observations that when sediments are deposited in water, gravity causes the sediment to accumulate in relatively horizontal layers. The **principle of original horizontality** basically states that sedimentary rocks will initially form in relatively horizontal layers. If you find layers of sedimentary rock that are tilted, folded or faulted, you can assume they have been moved from their original position (Figure 11.7). Some sedimentary rocks can even be turned upside-down! In these cases, all bets are off and the oldest rock may not be on the bottom. That's not to say that you can't figure out the relative age of the rocks, it just means you have to study the rocks in more detail to find other clues as to which rocks are older.

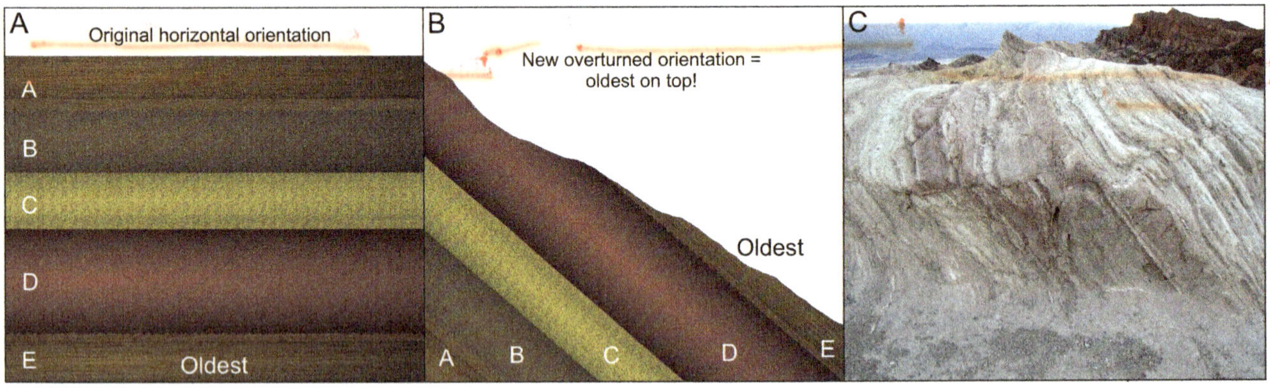

Figure 11.7: A-B. The principle of original horizontality explained. When tectonic forces push rocks out of their original, horizontal position, the principle of superposition *might not* apply. C. These sedimentary rocks have clearly been moved from their original horizontal position so it is impossible to tell which rocks are younger and which are older without a more thorough examination of the individual layers. Photo credit: B. Zaprowski.

Another very helpful guiding principle is the **principle of cross-cutting relationships**. Any time one geologic feature "cuts" through another geologic feature, the "cutting" feature will always be younger than the feature it is cutting through. A simple example of this would be a fault. A fault can only occur once the rocks it is cutting through have formed - it would be impossible for the fault to form before the rocks have! In this case, the fault must be younger than the rocks it cuts. One very common "cutting" phenomenon is erosion. For example, when rivers erode the landscape, they "cut" down through the underlying bedrock and leave behind an erosional surface (Figure 11.8A). Most rivers are constantly shifting their position laterally too, and sometimes they may abandon a channel altogether and take up a new path (more on this in Chapter 13). These abandoned channels will eventually fill in with more sediment, burying the erosional surface. In this scenario, the channel deposits *must* be younger than the rocks they are in contact with. Another common "cutting" phenomenon is volcanism (Figure 11.8B). When magma intrudes into another rock and cools, the resulting igneous rock *must* be younger than the rocks around it.

There are other principles worth noting. For example, if you walked up to an exposure of sedimentary rocks that looked undisturbed, it would be safe to say that the rocks on top are the youngest and the rocks at the bottom are the

Figure 11.8: A. This outcrop in Pennsylvania displays the principle of cross-cutting relationships. In this instance, a river channel cut through the older rocks and then the channel was filled in with rocks. B. The principle of cross-cutting relationships for igneous rocks. In this example from Titus Canyon, CA, light-colored igneous rock cuts through dark-colored sedimentary rock. Photo credits: B. Zaprowski.

oldest. However, this doesn't automatically mean the rocks represent a *continuous* stretch of time. There may be long breaks or pauses in the deposition process. So although the rocks themselves are physically unbroken, there may be **unconformities** or breaks in time within that sequence of rocks. Commonly erosional surfaces are indicators of a missing piece of time. For example, in Figure 11.8A, we know that there must have been some period of time that elapsed between when the river first started cutting downwards through the older rocks and when the channel was subsequently filled in with the younger rocks. Thus, the old erosional surface in Figure 11.8A is also an unconformity. Sometimes these periods of erosion may span just a few thousand years; sometimes they may last for millions, or even billions of years. As a result, geologists have to be careful to recognize when pieces of time are missing from a section of rocks.

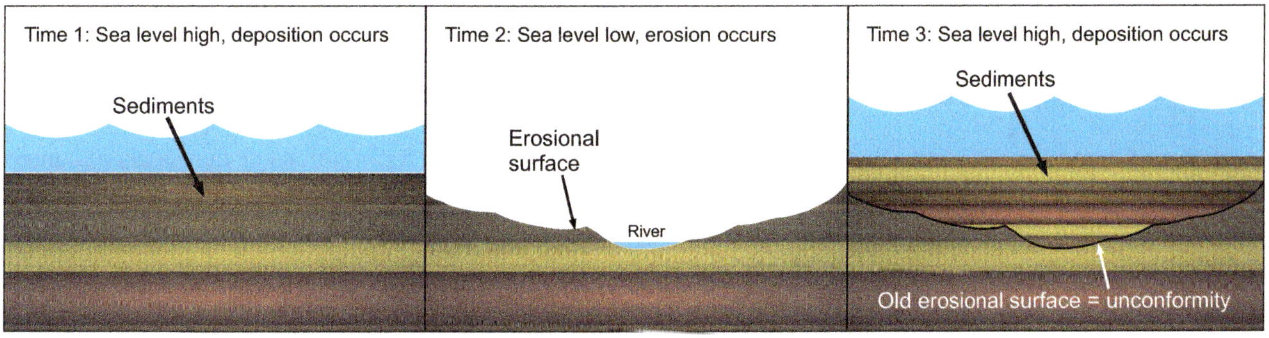

Figure 11.9: A schematic diagram showing how the unconformity in Figure 11.8A formed. First, sediments were deposited and lithified. Then sea level dropped, exposing the sedimentary rocks to weathering and erosioin. Finally, sea level rose high enough to bury the erosional surface with more sediments which were subsequently lithified.

Many unconformities are the result of changing sea-level. Throughout geologic time, the world's sea-level has fluctuated between 100 m lower to 100 m (350 ft.) higher than it is today, causing the oceans to repeatedly flood back and forth across the continents. During times of higher sea-level, the continents are flooded and sediments are deposited. During times of lower sea level, rivers cut down through those sedimentary deposits. In many places, multiple unconformities have formed within the thick sequences of sedimentary rocks as global sea-level rose and fell. This is how the unconformity in Figure 11.8A formed (Figure 11.9). Plate tectonics can also cause unconformities. Mountain building and volcanism can lift rocks upwards, deforming them and exposing them to erosion. The rocks shown in Figure 11.10 are a good example. During a period of high sea-level sediments were deposited on the flooded, low-lying parts of the continent. Then, those sedimentary rocks were deformed and lifted above sea-level where erosion could attack them, forming an erosional surface. After millions of years, the now eroded rocks were buried under a second set of younger sedimentary rocks. In the final stage, the rocks were lifted up once again by a second mountain building event, forming the modern erosional surface.

Figure 11.10: An example of an unconformity caused by mountain building. This outcrop near Catskill, NY displays a clear example of an old erosional surface where the younger rocks cut across the older rocks. The older rocks have also been subjected to the effects of two separate mountain building events. Photo credit: B. Zaprowski.

Up until now we have been focusing mostly on sedimentary rocks because they contain most of the information about what has been happening on the Earth's surface and what lived there. Igneous and metamorphic rocks (or "crystalline" rocks) also have a place in the pages of Earth's history. We know that most (but not all) igneous and metamorphic rocks form deep in the Earth. It takes a long time for rocks that formed several kilometers down in the crust to be exposed at the Earth's surface by weathering and erosion. Conversely, sedimentary rocks form at the Earth's surface in completely different types of geologic environments. If you see crystalline rocks and sedimentary rocks directly touching each other, you have found a type of unconformity called a **nonconformity**.

A nonconformity is a surface where igneous or metamorphic rocks are in direct contact with sedimentary rocks. For example, in Figure 11.8B the contact between the darker sedimentary rocks and the lighter igneous rocks which intruded into them is one type of nonconformity. The importance of the nonconformity is that in this situation we know that the two rocks could *not* have been formed at the same time so there *must* be a gap in time between the formation of the two rocks in question. We also know the sedimentary rocks must be *older* that the igneous rocks that cut through them. Lava flows represent a similar type of non-conformity, but in this case, the lava flow must be younger than the rocks it flowed over (Figure 11.11A). Another common type of nonconformity occurs when older igneous or metamorphic rocks are exposed by erosion and then subsequently buried by sedimentary rocks (Figure 11.11B). Even though a nonconformity represents a gap in time, we can still use the crystalline rocks to help us determine the general age range of sedimentary rocks. The sedimentary rocks in Figures 11.8B and 11.11 can't be dated using any absolute dating techniques, but we can at least narrow down their age using the age of the crystalline rocks they are in contact with. More on this in the next section.

Another consideration is one of scale. Sedimentary rock layers can extend for hundreds or thousands of square

Figure 11.11: Examples of nonconformities. The arrows point to the nonconformity. Both of these photos were taken in the Grand Canyon. A. The dark-colored rocks are from lava flows which flowed over the reddish-colored sedimentary rocks. In this example, the sedimentary rocks must be older than the lava flow. B. The reddish-colored sedimentary rocks were deposited on the dark-colored metamorphic rocks after they were exposed by erosion. In this case, the sedimentary rocks are younger than the metamorphic rocks. Photo credits: D. Harris.

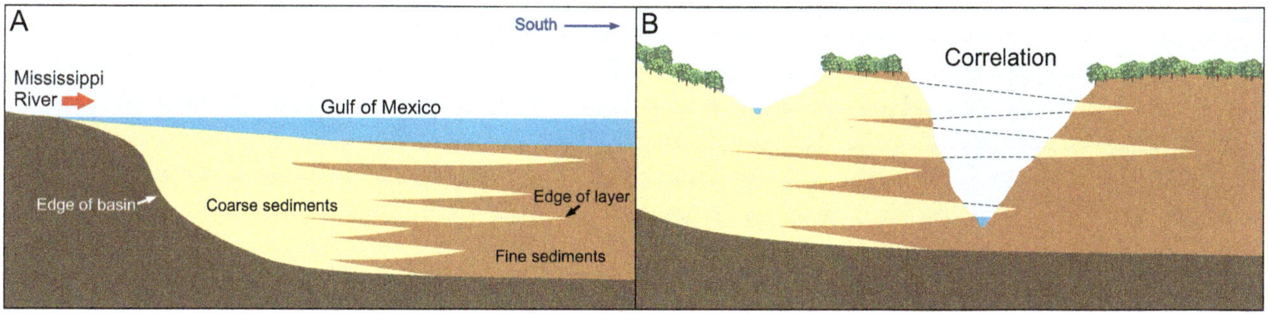

Figure 11.12: The principle of lateral continuity and correlation explained. A. Layers of coarse-grained sediment accumulates at the mouth of the Mississippi River but thin out with distance. The sediments also pile up along the edge of the basin. B. Hypothetical diagram showing what would happen if those sediments were exposed to erosion. Using the principle of lateral continuity, we can confidently correlate the rocks exposed along each side of the river valley.

miles, but in most cases, only small exposures of these rocks may occur at the Earth's surface (like those seen along a highway). So how can geologists be confident that two similar-looking sedimentary rocks exposed at two different locations are in fact part of the same rock layer? Again we turn our attention to the process of deposition. **The principle of lateral continuity** states that a layer of sediment being deposited in a basin will extend continuously outwards in all directions until it reaches a barrier or thins out completely. A good example of this is the Mississippi River and the Gulf of Mexico (Figure 10.8). When the sediment-laden water from the Mississippi reaches the Gulf, sediments are deposited in layers (Figure 11.12). The coarse-grained sediments are deposited closest to the river mouth while the finer-grained sediments drift south into the Gulf before being deposited (Figure 11.12A). The further you travel south from the river mouth, the thinner the layer of coarse sediments from the Mississippi River will be. Eventually if you travel far enough from the river mouth the layer of course sediments will disappear completely. If you reverse direction and follow a layer of sediments northward, you eventually run into land (the edge of the basin), so the layer of coarse sediments must end there as well. Now imagine a time millions of years in the future when those sediments are now exposed as land (Figure 11.12B). If a river has now cut down through those sediments, we can reasonably infer that the rocks on each side of the valley were once connected as part of larger, more continuous layers of sediment.

The principle of lateral continuity allows geologists to reasonably infer that two rocks exposed in two different places are part of one larger continuous layer of rock, even if most of the rock layer has been eroded away or is buried under soil and vegetation. For example, in Figure 11.13, it is reasonable to assume that the rocks exposed in A and B are the same rock layer because of the similarities in rock type, color, composition and distinct layering. Inferring the equivalent relationship of two rocks in time and/or space is known as **correlation**. Correlation is a powerful tool because it allows geologists to put individual exposures of rock into a regional context. For example, geologists have correlated the sequences of rocks at Grand Canyon, Bryce Canyon and Zion National Parks (Figure 11.14).

Figure 11.13: A demonstration of correlation. In this example, the rocks in A are exposed in Canyonlands National Park in east-central Utah while the rocks in B are exposed in Zion National Park in southwestern Utah. Because of similarities in color, composition and distinctive bedding pattern, we can infer that the rocks at these locations are both part of the same rock layer, so we can correlate them to each other. Photo credits: B. Zaprowski.

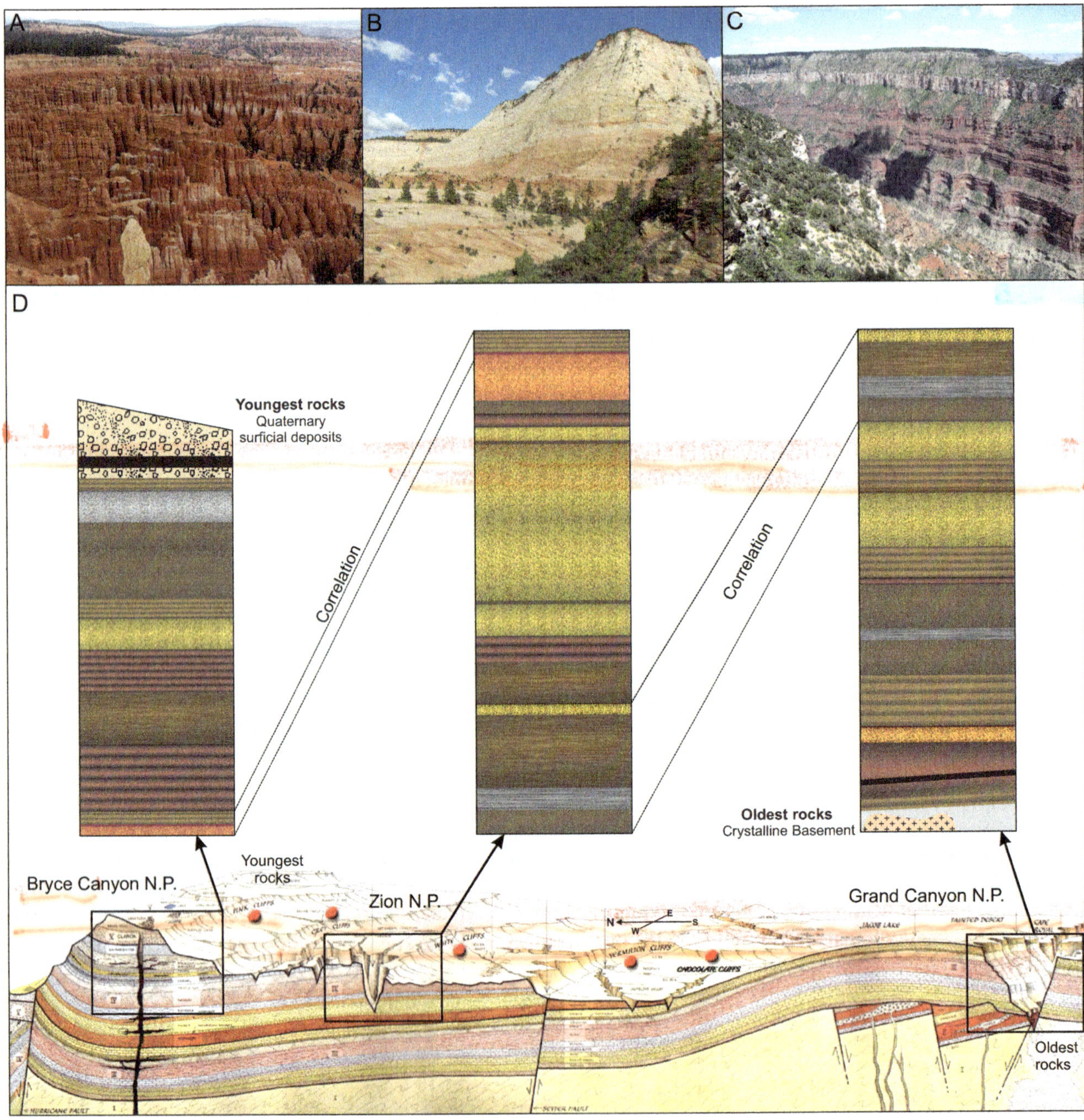

Figure 11.14: A real-world example of correlation. A. Bryce Canyon National Park. B. Zion National Park. C. Grand Canyon National Park. D. We can correlate the rocks exposed in three National Parks using common rock layers. Using correlation, the geologic cross-section of the area could be determined (lower image). Photo credits: B. Zaprowski. Cross-sectional diagram courtesy of the National Park Service.

They were able to do this by correlating the distinctive layers in the sedimentary rocks to create the diagram shown in Figure 11.14. There is no one place where all of these rocks are exposed. Instead, geologists visited all the locations where some of the rocks are exposed, described and mapped them in detail, and then used correlation to start making drawings about what the sequence of rock layers under the Earth's surface probably look like between the exposures. Well drilling records and/or seismic profiles also aid geologists in interpreting the structure of the earth's crust.

In the previous example, we correlated rocks based on their physical properties, but you can also correlate rocks based on their age. Return back to Figure 11.12A. At any one instance, both coarse-grained sediments and fine-grained sediments are being simultaneously deposited. Although the resulting sedimentary rocks may have different properties, they formed at the same time, and thus are the same age. This is where fossils play an important role in relative dating. Fossils form and become entombed in sediment at the time the rock is forming, so if you can identify the fossils and you know the age of the fossils, you know the age of the rock. This technique works because of the **principle of fossil succession**. The principle of fossil succession states that life on Earth has evolved over time in a known, orderly manner and we can use this knowledge to determine the relative age of a layer of sedimentary rocks. Each fossil species in the rock record has a unique **age range**, that is, the portion of Earth's history when it was known to have existed. Some species, like horseshoe crabs, have been around for hundreds of millions of years, so they aren't very useful for dating rocks, but other fossil groups only existed for a (relatively) short period of time. If these short-lived fossil species are also found in wide abundance over a large area, they are described as **index fossils**. Index fossils are extremely useful for determining the age of a rock unit. For example the trilobite fossil shown in Figure 11.15 is only known from rocks of Devonian Period. If you were to find this particular species of fossil trilobite in a rock, you would know the rock must be Devonian-aged.

Figure 11.15: This trilobite species is only found in rocks from the Devonian Period. Photo credit: B. Zaprowski.

If you don't have an index fossil, you can still possibly date a rock using the overlapping age ranges of several different fossils found together in the rock (Figure 11.16). This allows you to narrow down the age range of the rock. Thus, by identifying the fossils in a rock, you can get a pretty good idea of how old that rock is. So in Figure 11.16, we can determine the age of Rock A because it contains an index fossil. However, to determine the age of Rocks B and C we need to use the overlapping age ranges of all the fossils it contains. The larger the number of species you identify, the more accurate the age you can assign to that rock. The drawback of course is that not all rocks have fossils in them and the technique is not nearly as precise as absolute dating. Still, it worked pretty well for hundreds of years before modern absolute dating techniques were available and even today it can be a handy tool for geologists working in the field.

Section 11.3: Dating the Pages of Earth History - Absolute dating

So how do geologists determine the absolute age of a rock? There are a number of techniques used, but most are based on the principle of **radioactive decay**. Radioactive decay is the spontaneous breakdown of a radioactive

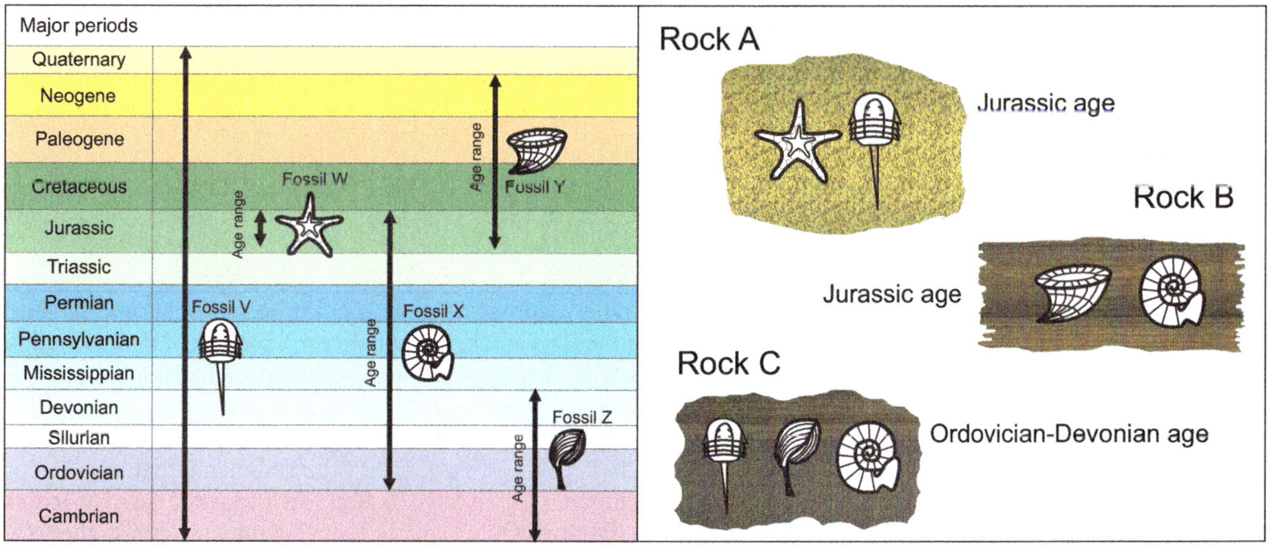

Figure 11.16: The principle of fossil succession and how it can be used to date rocks. On the left is an example of age ranges for a hypothetical group of fossils. Fossil V is of no use in dating rocks, it has too large of an age range. Fossil W is a good example of an **index fossil** because it has such a narrow age range. Rock A is easy to date as Jurassic age because it contains an index fossil of that period. We can use the overlapping age ranges to date Rock B and C. Rock B is also Jurassic-aged because fossils X and Y only overlap during the Jurassic. We can only narrow down the age of Rock C to a range of Ordovician- to Devonian-aged.

isotope (the parent) into a new element (the daughter) through the addition or release of protons, neutrons, and/or electrons. Energy release is also common during decay. For example, one of the radioactive isotopes of uranium (U-235) spontaneously decays into thorium (Th-233) as part of the decay process. It does this through the release of two protons and two neutrons plus some energy (Figure 11.17). Not only does the uranium spontaneously decay into thorium, it does so at a fixed rate called a **half-life**. One half-life is the time it takes for half of a mass of a radioactive isotope (the **parent**) to decay into one or more new elements (the **daughter product**). So for example, after one half-life, 50% of the parent is left, after two half-lives, 25% of the parent is left, after 3 half-lives, 12.5% of the parent is left and so on (Figure 11.18). We know from laboratory tests precisely what the half-life of every known radioactive isotope is.

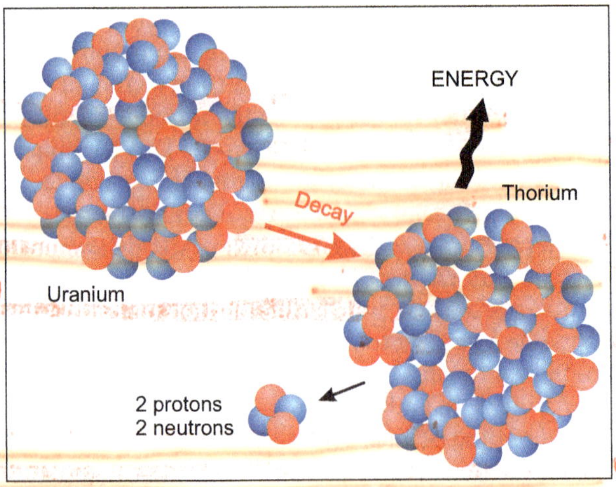

Figure 11.17: The decay of Uranium-235 into Thorium. Initially, the radioactive element U-235 has 92 protons in its nucleus but after radioactive decay, the atom is now the element Thorium-233 with only 90 protons in its nucleus.

Radiometric dating is the dating technique that takes advantage of the natural process of radioactive decay. Because each element decays at a fixed rate, a mineral which contains some of that radioactive isotope essentially has a built-in "clock". If you carefully measure the ratio of parent element to daughter element in a mineral sample, you can calculate the age of the rock based on the number of half lives that have elapsed. It is important to note that a rock has to contain the correct radioactive minerals in order to be datable, so not all crystalline rocks are datable (Table 11-1). So when does the "clock" start? The clock starts when a mineral crystallizes, either from magma or

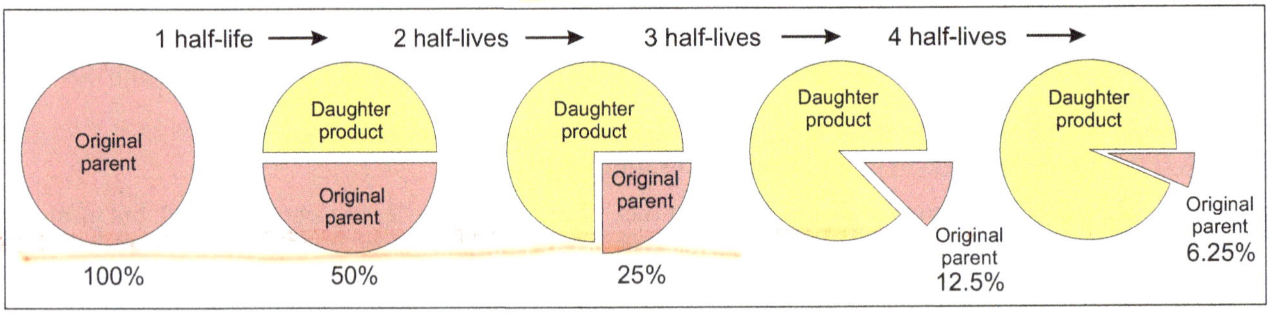

Figure 11.18: How a radioactive element decays. When the original parent isotope is solidified as part of a mineral crystal, it begins to decay into the daughter product. After one half-life has elapsed, 50% of the original parent is left. After two half-lives, 25% of the original parent is left (half of 50%). This process continues as the parent decays into the daughter product.

from solution. The only exception to this rule is ^{14}C-dating. This radioactive isotope of carbon can be used to date organic-based material (i.e., wood, bones, shells) which is younger than 50,000 years old. Although ^{14}C-dating is very useful in many applications from geology to archeology, it isn't very practical for dating most rocks because of its age restriction.

Radioactive parent	Daughter product	Half-life value
Uranium-238	Lead-206	4.5 billion years
Uranium-235	Lead-207	713 million years
Thorium-232	Lead-208	14.1 billion years
Rubidium-87	Strontium-87	47.0 billion years
Potassium-40	Argon-40	1.3 billion years
Carbon-14	Nitrogen-14	5730 years

Table 11-1: Some common radioactive elements used for absolute dating, their daughter products and half-life values.

So why can't geologists use the same technique to date sedimentary rocks? Sedimentary rocks contain minerals that have been recycled as sediments. If you attempted to date a sedimentary rock using absolute dating techniques, the age you would calculate would be incorrect (Figure 11.19). The minerals within the sediments that make up a sedimentary rock have to first form and then be exposed to weathering. Once weathered, the minerals must then be transported to a site of deposition and lithified into a solid rock. Thus, the absolute age tells us how old the original, recycled minerals are, but the absolute age tells us nothing about when the sedimentary rock itself formed. So

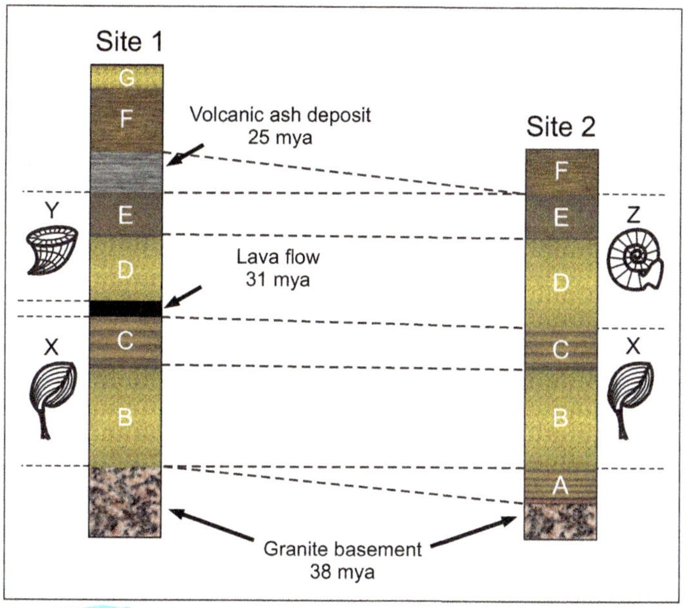

Figure 11.19: How to determine the age of a sedimentary rock using both radiometric dating and fossils. The sedimentary rocks are lablelled A-G.

determining the age of a sedimentary rock and the fossils it contains, requires a different approach.

Since geologists can date most crystalline rocks, they can use the principle of cross-cutting relationships in combination with absolute ages from crystalline rocks to date the age of a group of sedimentary rocks. For example, let's consider two hypothetical outcrops at two different sites which are separated by hundreds of miles (Figure 11.19). Site 1 happens to be located near an active volcanic arc. As a result, there are several instances where igneous rocks are in contact with sedimentary rocks. The sedimentary rocks sit on top of an old land surface which is underlain by granite. In this case, the granite is dated to be 38 million years old (or in geologist speak we would say **mya**). That rock is the oldest and all of the sedimentary rocks on top must be younger than 38 mya. But how much younger? At Site 1 there is a lava flow siting on top of rock layer "C" and it has an age of 31 mya. Therefore, we can conclude that rock layers "A", "B" and "C" are older than 31 mya but younger than 38 mya. At Site 1 there is also a volcanic ash deposit on top of layer "E". Since volcanic ash is technically an igneous rock, it can also be dated radiometrically, in this case to 25 mya. Now we know that rock layers "D" and "E" must be older than 25 mya and younger than 31 mya. Likewise, we know layers "F" and "G" are younger than 25 mya. We can also conclude that rock layers at Site 2 have the same ages too. You have just applied the principles of relative and absolute dating to determine the ages of a group of rocks.

What about the fossils contained within those rocks? In the case of Fossil X, we can conclude it has the same age range as the layers "B" and "C" it is found in (31-38 mya), even at Site 2. In fact anywhere we found sedimentary rocks with Fossil X in them, we could conclude the rocks had an age range of 31-38 mya. The same logic would apply to Fossil Y as well (25-31 mya). But what about Fossil Z? Since Fossil Z is only found in rock layers "D" and "E" then we can conclude it would also have the same age range. Now, you could find either fossil "Y" or fossil "Z" in a rock layer anywhere else and you would be confident the age range of the rocks is 25-31 mya.

That's basically how the geologic time scale was created. First, early geologists began to correlate rocks using the principles of relative dating and fossils, and from this data they developed the Geologic Time Scale. Later, when radiometric dating was discovered, geologists scoured the Earth in search of places where crystalline rocks are found in direct contact with sedimentary rocks (and they continue to do so today). Putting that information together, geologists began to put dates on the boundaries between the Eras, Periods and Epochs of the Geologic Time Scale. And they are not finished (Figure 11.20). The Geologic Time Scale is constantly being refined as new sites and better dates are discovered. What follows in Chapter 12 is the story those rocks and fossils tell. But keep in mind, as new rocks are discovered, the story may change - so stay tuned!

Figure 11.20: Geologists-in-training learning how to measure and describe rocks. Perhaps one of them will discover a new page in Earth's history? Photo credit: B. Zaprowski.

Chapter 12 - Fossils and the History of Life

Fossils are any record of past life on Earth, even if it is just a simple worm trail (Figure 12.1). The minute anything dies, it becomes a fossil. The vast majority of life on Earth does not get preserved in rocks - the Earth's life cycle usually decomposes organic material and recycles it back into the environment long before it can become fossilized. Nonetheless, in the right geologic environment under the right set of circumstances, plants and animals of the past have been preserved and entombed in rocks, leaving behind clues about the Earth's history for geologists to unravel. Let's explore how fossils are made before we delve into the fossil record itself.

Figure 12.1: Fossil trails in a sandstone. Photo credit: B. Zaprowski.

Section 12-1: Fossils

As noted above, the moment a living organism dies, it becomes a fossil. However, the odds of any one organism becoming preserved in the fossil record is extremely low. The potential for an organism to be fossilized requires a number of different independent variables to occur simultaneously. Some of these variables include:

1. Where does the organism live? Fossils are much more likely to be preserved if they live in an environment where sediments are being deposited, like a river, a large lake or the ocean. Large numbers of plant and animal species will never be known in the fossil record because they didn't live in an environment that readily preserves fossils.

2. The rate at which an organism is buried by sediments. The faster an organism is buried, the less likely scavengers and decomposers are to destroy it.

3. The presence or absence of oxygen. The bacteria that quickly break down organic material need oxygen. In anoxic environments, like very deep lakes, organic material is readily preserved as fossils.

4. The grain size of the sediment. Finer-grained sediments restrict the circulation of oxygen-rich waters, slowing the decay process down.

5. What is the morphology of the organism? Is it a rugged, solid species with hard parts (like an oyster) or is it a fragile, delicate one (like a butterfly)?

6. Chance. Only a minute fraction of all organisms that ever lived are preserved as fossils. In many cases, it is simply just a matter of being in the wrong place at the wrong time. For example, would you want to be an animal living on a side of a volcano when it erupts? Yet some unfortunate animals have met just such a fate (Figure 12.2).

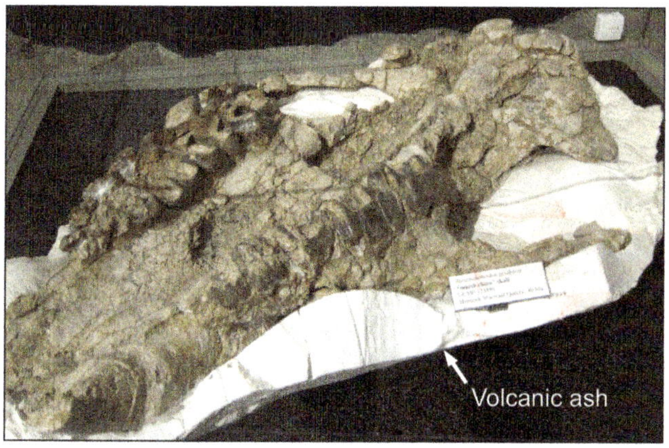

Figure 12.2: Volcanos near the John Day Fossil Beds National Monument emitted large volumes of ash and dust, much of which settled in the John Day basin. The rapid deposition of ash preserved the remains of plants and animals living in the region, like the skull of this "marsh rhino". Photo credit: NPS.

Thus, organisms with hard parts that live in or near an aqueous environment are much more likely to be preserved in the fossil record. Being aware of such shortcomings is important when interpreting the meaning of fossils. Of course the final step is that someone has to find the fossil, collect it, and incorporate it into the fossil record before weatherig and erosion destroy it.

Fossilization is really just a slow progressive change in the chemical and physical properties of the original

Figure 12.3: Sometimes melting permafrost deposits like these yield Ice Age fossils, like mammoths and mastodons. Photo credit: Benjamin Jones/USGS.

organic remains of a plant or animal. Sometimes these changes will eventually lead the remains to be completely destroyed, but in other cases, the changes lead to the preservation of the fossil by minerals (**mineralization**). Fossils begin as original material which has not been significantly altered physically. Soft tissues like skin, organs, and muscles will first begin to decompose unless they become mineralized first. That happens fairly rarely. In very arid environments, the soft tissues may become desiccated and preserved for thousands of years. In polar environments, plant and animal remains may become trapped within glacial ice or permafrost and frozen, which can also preserve the soft tissues for thousands of years (Figure 12.3). Finally, plants and animals which are deposits into very acidic water, like a swamp or peat bog, may also be preserved. In essence, the organic material becomes "pickled" and the soft tissues may be preserved. In each of these case, the soft tissues are preserved for a while, but it is not likely that they will become mineralized unless the remains become deeply buried and encased within sedimentary rocks.

The hard parts, being more durable, are more likely to become mineralized. Hard parts include wood, bone (phosphate), shells (calcium carbonate), and chitin (the exoskeleton of crustaceans and insects). Even materials like hair, fur and horns are fairly durable and may be preserved. In many cases, the original hard part may be preserved intact for millions of years with very little physical or chemical alteration occurring. Shells in particular are very durable because they are already composed of a mineral. Usually the first step in mineralizing hard parts is the process of **replacement** (Figures 12.4 A-B). In replacement, the original organic material is replaced by a mineral.

Figure 12.4: Examples of mineralized hard parts. A. Petrified wood is the classic example of replacement. B. A whale bone from Calvert Cliffs, MD that has been replaced by silica. Note the porous marrow inside the bone. C. A dinosaur bone which has been permineralized and replaced. Note how all the empty spaces in the marrow are filled in with minerals. Photo credits: B. Zaprowski.

For example, petrified wood forms when carbon is replaced by silica (Figure 12.4A). The key to identifying fossils which have undergone replacement is the internal structure of the fossil. If the organic material has been replaced, then the fossil will retain its internal structures (e.g., tree rings will be visible in petrified wood). As the mineralization process continues, any empty spaces or voids in the hard part (like the marrow inside of your bones) may begin to fill in with more minerals. This process is referred to as **permineralization**, and it commonly occurs in conjunction with replacement (Figure 12.4C). In some cases, the organic remains may be flattened out into a thin film of black or brown carbon and preserved through the process of **carbonization**. This leaves behind a fossil that can have exquisite detail, but it is a significantly flattened replica of the original organism (Figure 12.5).

Figure 12.5: This fern leaf fossil is an example of carbonization. Photo credit: B. Zaprowski.

Many fossils don't become mineralized. Instead, they are replicated by

Figure 12.6: Mold and cast demonstrated. First, the horseshoe crab was buried by sediment, forming the mold. Later, the mold was filled with sediment, producing a cast of the crab. On the right is the internal cast of a snail shell. Photo credits: B. Zaprowski.

sediments through a process called **mold and cast** (Figure 12.6). When an organism is buried, an imprint of the fossil will form a **mold** in the sediment surrounding it. This mold is a negative replica of the outsides of the original fossil remains. If the original organic material subsequently dissolves away (instead of being mineralized), the resulting space can be filled in with sediments, forming an **external cast** of the original organic remains. Fossils with hollow interiors (i.e., the inside of a snail shell) can also be filled in with sediment, forming an **internal cast**. Keep in mind that fossils preserved in this manner will have no internal structure.

Before we move on, it is important to note that the **fossil record is inherently biased!** There are approximately 250,000 different species known in the fossil record but this number only represents about 5-10% of all organisms that ever lived. To put it in perspective, insects make up about 75% of all organisms in the modern world, but they only make up 3.2% of the fossil record. As you may well imagine, insects don't make very good candidates for fossilization. Plants, vertebrate animals and worms are all very important in terms of evolutionary progress (especially for us, a land-dwelling species) but they do not make up any significant portion of the fossil record. Most of the fossil record is represented by just 9 major invertebrate phyla (Table 12-1). Nonetheless, geologists have been able to piece together a fairly cohesive picture of how life has changed over time. Let's see what the fossil record tells us about the past.

Phylum	Modern examples
"Protozoans"	Foraminifera, diatoms
Cnidaria	Corals, jellyfish
Porifera	Sponges
Bryozoa	"Moss animals"
Arthropoda	Insects, crustaceans
Echinodermata	Sea urchins, starfish
Brachiopoda	Fairly rare filter-feeders
Mollusca	Snails, clams, squid
Archaeocyatha	Extinct

Table 12-1: The nine most abundant phyla in the fossil record.

Section 12-2: A Brief History of Life on Earth

There are billions of fossils in the world's museums, schools and homes and many more are yet to be discovered in the thick layers of sedimentary rock draped over the Earth's continents. Early geologists quickly realized that certain fossils were only found in certain rocks. More importantly, using the principles of relative dating, they could determine which ones were older and which ones were younger. As more and more fossils were discovered, the pattern became clear: **Life on Earth has changed and evolved in a definite and determinable order** (Figure 12.7). In particular, geologists have noted that many life forms have evolved and/or gone extinct in response to the Earth's changing environmental conditions through time. Let's take a brief journey into the Earth's past to see what we mean. It would take a lot of pages to even highlight how all life on Earth has changed over time. In this chapter we're going to focus on one particular story, the story of all the twists and turns in the evolution of life that led to us. Our discussion will follow the red line on Figure 12.7, starting at the beginning with the prokaryotes.

The origins of life are a mystery, but it must have begun with very simple, single-celled **prokaryotes** (organisms without a nucleus) like bacteria (Figure 12.8A). To date the fossilized remains of this ancient ancestor have not been

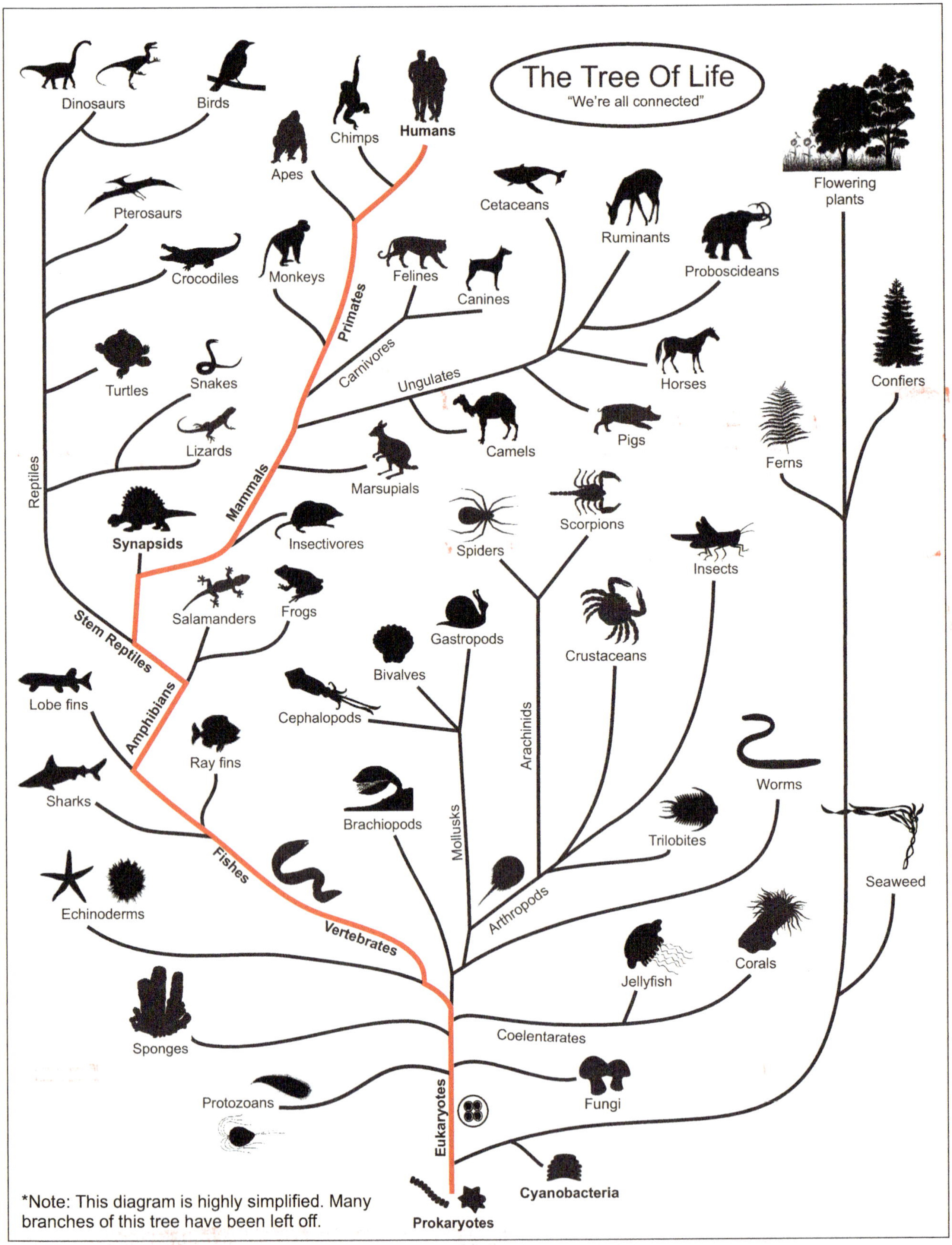

Figure 12.7: The tree of life. This highly generalized diagram highlights the genetic relationships between the major groups of organisms. In this chapter we will be focusing on the red line, the path that led to us.

Figure 12.8: Some examples of early fossils. A. The first forms of life were very simple prokaryotes like these. Photo credit: USGS. B. Fossil stromatolites (blue-green cyanobacteria). Stromatolites are the oldest known fossils. Photo Credit: B. Zaprowski.

conclusively found, and likely never will be. The earliest *fossil* evidence of life are fossilized blue-green cyanobacteria called **stromatolites** (Figure 12.8B) found in 3.5 billion year old deposits in both Africa and Australia (ironically, the only place in the modern world where you can observe living stromatolite colonies is in Australia). There is also trace chemical evidence of life from rocks as old as 3.8 billion years, but that data doesn't reveal much about what type of life it was. Regardless of when it started, life was very simple for the first 3 billion years of history. Though life was simple, the presence of living organism began to fundamentally alter the Earth's chemistry.

In particular, the cellular evolution of photosynthesis was a key event in Earth's history because photosynthetic organisms produce oxygen. Photosynthetically-produced oxygen slowly began to accumulate around 3.5 billion years ago, about the same time the stromatolites appear in the rock record. At first, this free oxygen was absorbed by "oxygen-loving" elements like iron which were abundant in seawater at the time. When iron and oxygen combine, they form iron oxides, and these iron oxides began to precipitate out of seawater to form massive **banded iron formations** (Figure 12.9). Banded iron formations are the primary source of iron ore worldwide. By around 2 billion years ago, the oxygen had saturated the oxygen "sinks" in the Earth's rocks and oceans, and began to accumulate in the atmosphere and oceans. We know this because around this time the formation of banded iron deposits came to a halt. The build-up of free oxygen led to several important changes that were critical to evolution of life as we know it today.

Figure 12.9: A piece of a banded iron formation. Photo credit: B. Zaprowski.

One consequence of the oxygenation of the atmosphere was the formation of the ozone layer. As we discussed in Chapter 3, the ozone layer absorbs harmful UV radiation and without it, complex forms of life could never have evolved (at least not on the Earth's surface). Second, as oxygen levels increased in the oceans, the oxygen reacted with other elements to form important compounds like nitrates. These nutrients are critical for the growth of more complex forms of life. Third, aerobic metabolism developed. Prior to the oxygenation of the Earth, the primitive life forms that did exist were chemotrophs that use elements like iron and sulfur as a source of energy. But aerobic respiration is more efficient and it can produce the energy needed to fuel larger and more complex forms of life. So once oxygen was available in abundance, new life forms evolved to take advantage of it. This led to the development of the **eukaryotic** cell (cells with a nucleus), which could use the oxygen in the atmosphere to produce energy. It is around 2 billion years ago that the first fossil evidence of eukaryotes appeared in the rock record, although some chemical evidence suggests they

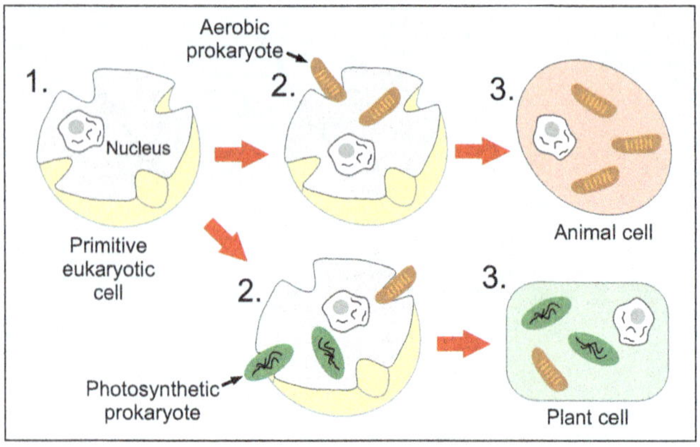

Figure 12.10: The endosymbiotic theory of the evolution of eukaryotic cells. The theory states that primitive eukaryotic cells engulfed prokaryotic cells and entered into a symbiotic relationship which led to animal and plant cells.

could have evolved as early as 2.7 billion years ago. Currently, the leading explanation for the appearance of the eukaryotes is the **endosymbiotic theory** of the evolution of eukaryotic cells. The theory states that the mitochondria and chloroplasts found in eukaryotic cells are former prokaryotic cells which were engulfed by a proto-eukaryotic cells, and formed a symbiotic relationship (Figure 12.10). The eukaryotic cells received a source of energy while the former prokaryotic cells received protection from both predators and oxygen, which is poisonous to most primitive prokaryotes. In the modern world, most of these types of primitive prokaryotes are only found in oxygen-poor environments (i.e, hot springs or deep sea thermal vents). This theory is supported by similarities in the physical structure of prokaryotes and the organelles in eukaryotic cells, DNA processing and sequencing evidence, and membrane structure. In addition, symbiosis has been observed in the modern world.

Even though these new eukaryotic cells were bigger and more complex, the eukaryotes did not expand and diversify until around 1.2-1.0 billion years ago. This expansion probably occurred because of the ever-increasing amounts of oxygen in the ocean, but the advent of sexual reproduction may have also spurred diversification. This diversification led to the first **metazoans** around 630 million years ago. Metazoans are multi-cellular animals that have multiple types of cells organized into specialized organs and tissues. The development of the metazoans followed the end of a series of three glacial episodes known as "**Snowball Earth**." Geological evidence from around the world suggests that during these snowball earth glaciations, the entire Earth was covered with a layer of ice, even in the tropics! Although there is no direct fossil evidence, many researchers have suggested that the appearance of metazoans was triggered by the favorable environmental conditions that followed the last snowball earth episode. Early metazoans include soft-bodied creatures like jellyfish, sponges and worms as well as some fossils of unknown affinity (Figure 12.11).

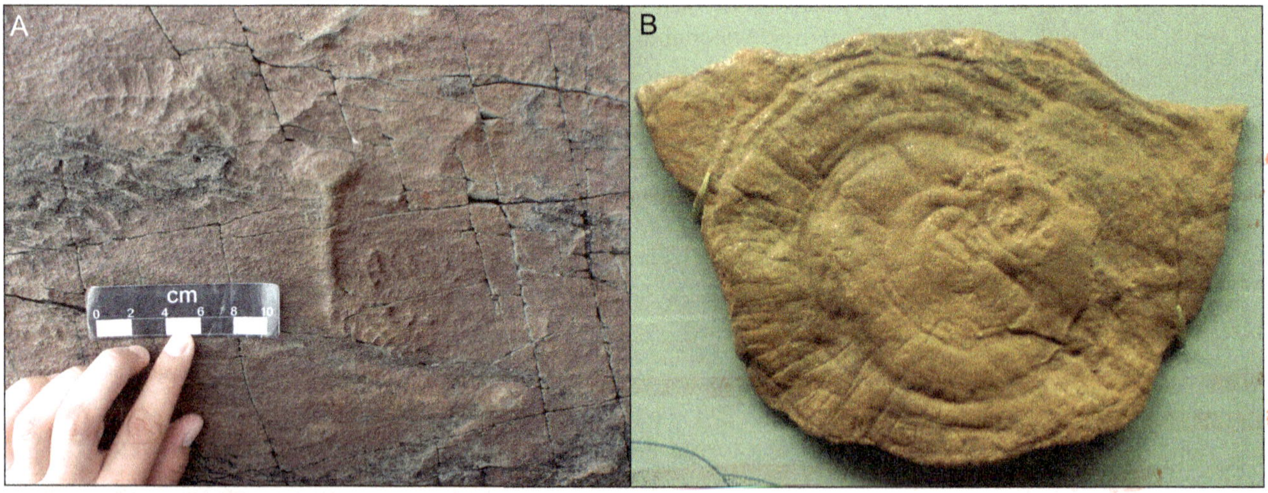

Figure 12.11: Examples of the earliest metazoans. A. This mysterious fossil from Newfoundland does not appear to fit into any modern animal group. Photo credit: Greg Burzynski. B. A 600 million year old fossil imprint from Australia. Photo credit: B. Zaprowski.

The first metazoans with hard parts appeared in the fossil record around 540 million years ago during an event known as the **Cambrian explosion**. Almost all of the modern animal phyla appear in the fossil record during the Cambrian explosion, including the first **chordates**, the group of animals which includes all vertebrates (Figure 12.12A). The chordates appear to be most closely related to the echinoderms, an early group that includes sea urchins, starfish, and sea lilies. Other common early fossils include the relatives of sponges, corals, clams, snails,

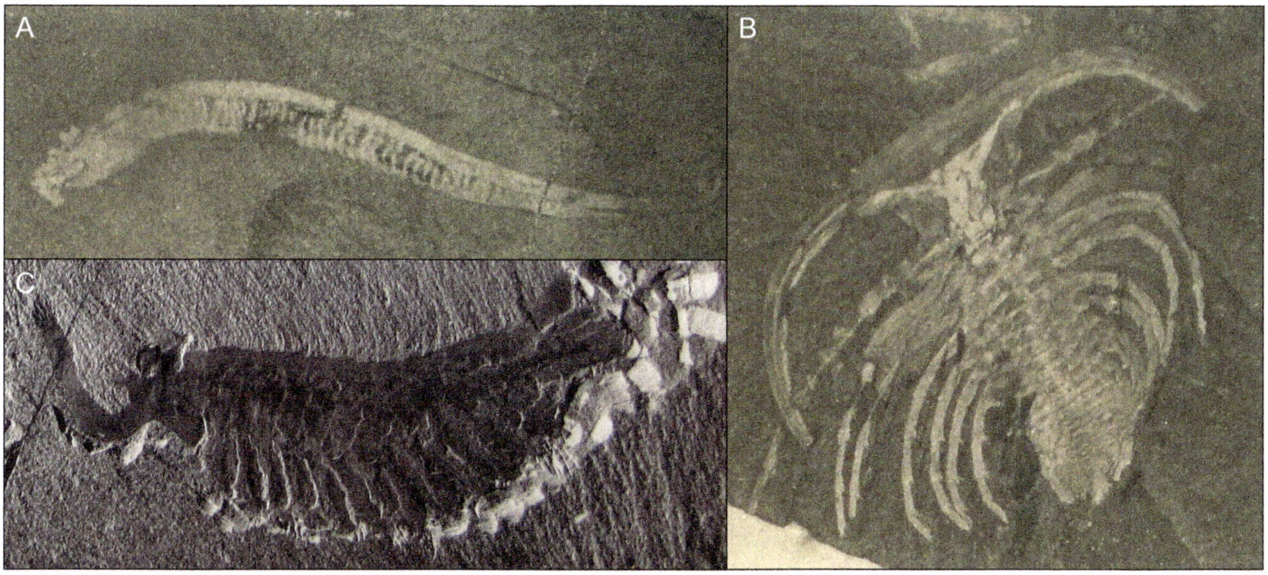

Figure 12.12: Cambrian fossils from the world famous Burgess Shale. A. An early chordate called *Pikaia*. B. *Marrella*, an arthropod also known as a "lace crab". C. *Opabinia* does not appear to have any relationship with modern faunas. Photo credits: B.Zaprowski.

cephalopods (squid) and arthropods (Figure 12.12B). This diversification continued as the metazoans adapted to new environments and new food sources. Some of the oddest forms of life known from the fossil record are also found in rocks of this age. It is not clear what modern animal group they belong to, if any (Figure 12.12C).

Figure 12.13: *Dunkleosteus* was a very large predatory fish covered in bony plates. Photo credit: B. Zaprowski.

The first bony fishes appear during the Cambrian but they really expanded in abundance during the middle of the Paleozoic. The earliest fishes were jawless fish, like lamprey eels, but by the early Silurian, jawed fished appear in the fossil record. Many of these early fished were covered with plates instead of fish scales (Figure 12.13). One group of jawed fish in particular took a significant "step" in the evolutionary process: the Sarcopterygians (Figure 12.14). Unlike, other fish groups, the Sarcopterygians had muscular, bony fins which allowed them to "walk" on land and they had a primitive set of lungs which allowed them to breathe air (they also had gills for in the water). The modern lungfish in Australia is a "living fossil" which has similar traits. This group led to the rise of the four-legged land-dwelling animals we call **tetrapods** in the Devonian.

The first tetrapods were **amphibians** and they were the dominant land animals throughout the Paleozoic (Figure 12.15). So what drew the amphibians from the water? Probably the appearance of new food sources such as plants,

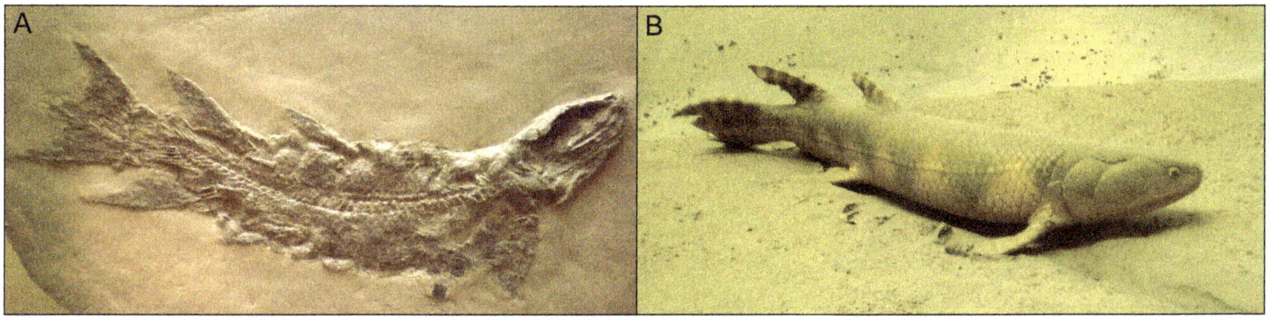

Figure 12.14: An example of a sarcopterygian. A. *Eusthenopteron*, a 360 million year old sarcopterygian from Quebec, Canada. B. A model of the same fish. Photo credits: B. Zaprowski.

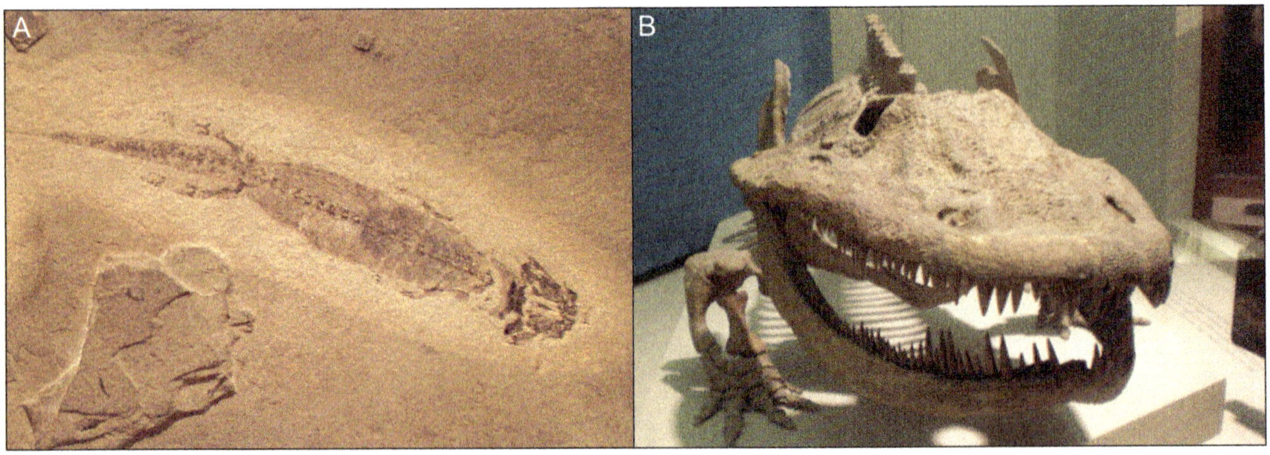

Figure 12.15: Early inhabitants of the land: A. A small, immature amphibian *Pelosaurus* from 270 million years ago in Germany. B. The large amphibian *Eryops* from 210 million years ago in Texas. Photo credits: B. Zaprowski.

spiders and insects. The first plants appeared on land in the late Silurian and by the Devonian period, the earth was covered in lush forests of moss, herbs, ferns, and fern trees. The first known animals which crawled on land, scorpions, also appeared in the Silurian. During the Devonian, spiders and other primitive insects appear in the rock record. Now bursting with life (and new food sources), the land was ready to host larger animals. Plus, there were always predators lurking in the water, and where better to escape to than land? However, the amphibians as a group are limited. Most have skin that needs to be kept moist, and they need water to reproduce, so they were only able to inhabit areas of land close to the water.

Figure 12.16: Fossilized amniotic eggs, a critical evolutionary step for land dwellers. Photo credit: B. Zaprowski.

A new group of animals, the amniotes, solved this dilemma by producing an **amniotic egg** (Figure 12.16). The amniotic egg is self contained and needs no water to develop so this huge evolutionary step allowed animals to leave the water and seek out other ecological niches in which to live. The first amniotes were **stem reptiles**, and they appear in the fossil record around 300 million years ago (Figure 12.17). Stem reptiles weren't quite like modern reptiles, but they gave rise modern reptiles like snakes and turtles, as well as crocodiles, dinosaurs, and birds. The stem reptiles also gave rise to a group of animals called the **synapsids**. Synapsids were not quite like a reptile and not quite like a mammal. They had characteristics of both. Like a reptile, they still laid eggs. But like a mammal, they could regulate their body temperature and likely had fur or hair of some sort. Some of the most famous early

Figure 12.17: A. Model of one of the earliest reptiles, *Hylonomus*. B. This *Cotylosaurus* is a stem reptile - not quite like modern reptiles as we know them today. Photo credits: B. Zaprowski.

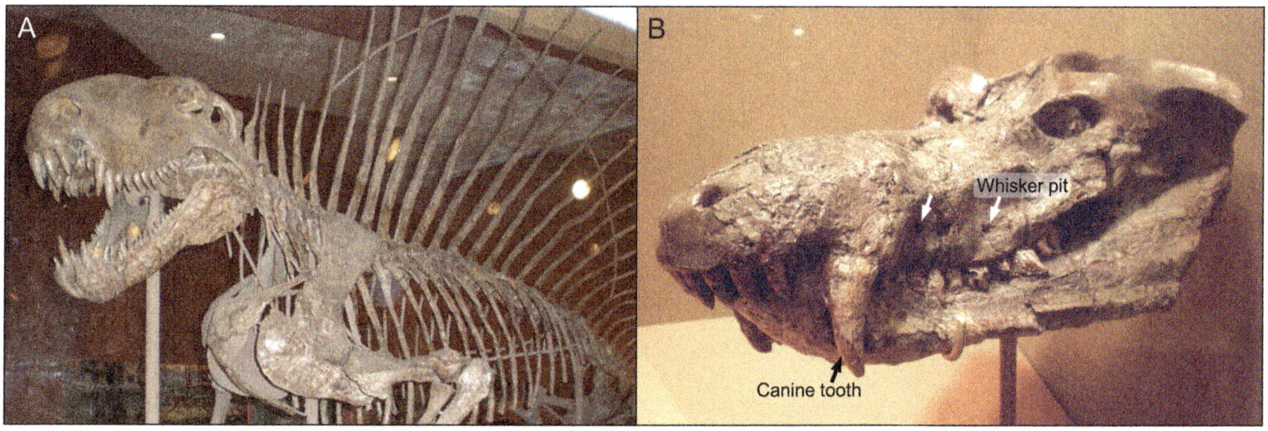

Figure 12.18: Examples of synapsids. A. *Dimetrodon*, a meat eating sail-back synapsid. B. This *Cynognathus* is a therapsid, the group of synapsids which is thought to have given rise to the mammals. Note the canine teeth and whisker pits. Photo credits: B. Zaprowski.

synapsids include the incorrectly-labeled "sailback" dinosaurs (they are not dinosaurs!) which were both plant-eaters and flesh-eaters (Figure 12.18A). One particularly important group of synapsids was the **therapsids**. **Therapsids** have hair and specialized teeth (like canine teeth) and they are thought to be the group which eventually evolved into the mammals during in the Mesozoic (Figure 12.18B). In the Paleozoic, the reptiles were not a very large group of animals but things were about to take a change for the better (for them, anyway). In the Mesozoic the reptiles rapidly evolved and colonized just about every ecological niche on the Earth. However, the rise of the reptiles came at the expense of the amphibians and other species of plants and animals during the **Late Permian Mass Extinction Event**.

Mass extinction events have occurred five times in the Earth's past: In the Late Ordovician, the Late Devonian, the Late Permian, the Late Triassic and Late Cretaceous. Mass extinction events are caused by world-wide changes in the environment. For example, the Late Ordovician Mass Extinction occurred in two phases. During the first phase, the rapid growth of ice sheets caused global cooling and a lowering of sea level. Plants and animals which could not adapt to the changing conditions went extinct. The second phase of the event occurred when the climate rapidly warmed back up and and the ice sheets melted. Organisms which had become adapted to the cooler climate could not survive the rapidly warming world and went extinct. Mass extinctions also pave the way for new groups of plants and animals to expand and diversify. The Late Permian Extinction Event *was the most catastrophic of all the extinction events*: 95% of all life on Earth went extinct. The exact cause of this extinction event is not known, but several different factors such as volcanism and the formation of the super-continent Pangea probably worked simultaneously to make living on the Earth extremely difficult during this time period. At the same time, the environmental factors that killed off most of the life on Earth also paved the way for the prolific expansion of the reptiles. The reptiles were so successful during the Mesozoic that it is known as the "Age of Reptiles."

Reptiles are divided into two major subgroups: The **lepidosaurs** (lizards and snakes) and the **archosaurs**, a subgroup which includes dinosaurs, flying reptiles, marine reptiles, crocodiles and modern birds. The first true dinosaurs were small and appeared in the Triassic (Figure 12.19). Throughout the Mesozoic Era, the dinosaurs grew larger in size and became more diversified. Some reptiles retained the four-legged stance of their stem reptile ancestors, while others developed bipedal locomotion. During the Jurassic, one group of dinosaurs, the **therapods** gave rise to the birds. Although the origin of birds has been a hotly debated subject for decades, it is clear from paleontological studies that birds have a very strong evolutionary connection with the dinosaurs. Recent fossil evidence even suggests that many dinosaurs were covered in feathers, an important step toward evolving into birds

Figure 12.19: *Heterodontosaurus tucki* is one of the oldest known dinosaurs. Photo credit: B. Zaprowski.

Figure 12.20: Examples of archosaurs. A. *Archaeopteryx lithographica* was a theropod with bird-like characteristics. B. The marine reptile *Mosasaurus*. C. The flying reptile *Pteradactyl*. D. *Stegosaurus* was well-equipped to fend off predators with his armor plates. E. *Triceratops* was also well-equipped to fend off predators. F. The fearsome *Tyrannosaurus rex* is one of the reasons why *Stegosaurus* and *Triceratops* needed armor. Photo credits: B. Zaprowski.

(Figure 12.20A). Feathers did not evolve specifically for the function of flight. Initially, feathers would have been used for camouflage, insulation and/or attracting a mate. It just so happens that feathers turned out to be useful for flying too.

By the late Mesozoic, huge dinosaurs such as *Allosaurus, Tyrannosaurus, Supersaurus* and *Stegosaurus* lumbered across the land while large *Pterodactyls* flew in the skies and the fearsome *Mosasaurus* swam in the oceans (Figures 12.20B-F). But their time was about up. A combination of catastrophic environmental changes led to the demise of the dinosaurs and many other plant and animal groups during the Late Cretaceous Mass Extinction. Although the famous Chicxulub Impact Crater in Mexico is commonly cited as "the" cause of the Cretaceous extinction, many geologists believe that extensive world-wide volcanism may have contributed to the end of the dinosaurs as well. Massive outpourings of volcanic ash and gasses would have had devastating impacts on the environment, and plants and animals not able to adapt to these frequent and rapidly changing environmental conditions would have gone extinct. By the time the Chicxulub meteorite hit the Earth, groups like the dinosaurs were probably already on the brink of extinction.

But one group's demise usually leads to the rise of another group, and the end of the Age of Reptiles heralded the beginning of the "Age of Mammals" in the Cenozoic. Although the reptiles were the dominant animal group in the Mesozoic, the mammals also appeared in the Triassic Period (Figure 12.21A). Throughout most of the Mesozoic, mammals were mostly small insectivores, but by the end of the Mesozoic, larger and more diverse mammals began

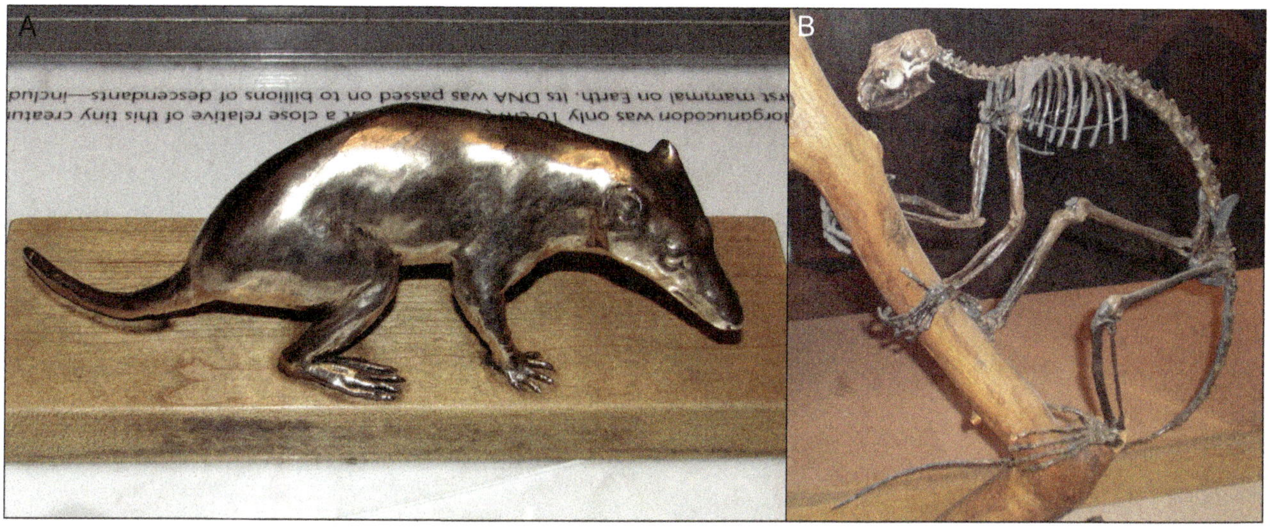

Figure 12.21: A. A model of the first known mammal from the fossil record, *Morganucodon*. B. The early primate *Smilodecton* from Wyoming. Photo credits: B. Zaprowski.

to evolve, including one important group, the primates (Figure 12.21B). After the Late Cretaceous Extinction, mammals quickly took advantage of the ecological niches that were vacated by the reptiles at the end of the Mesozoic. Many mammal groups were able to take advantage of a new food source that appeared in the late Mesozoic, **angiosperms** or flowering plants. Flowering plants (which include grasses and deciduous trees) are a great source of food in addition to the nectar, nuts and berries they produce. Animals such as squirrels evolved to eat the nuts, while herbivores such as horses, deer, antelope and elephants quickly evolved to take advantage of extensive grasslands spreading across the land (Figure 12.22). Simultaneously, new groups of carnivorous

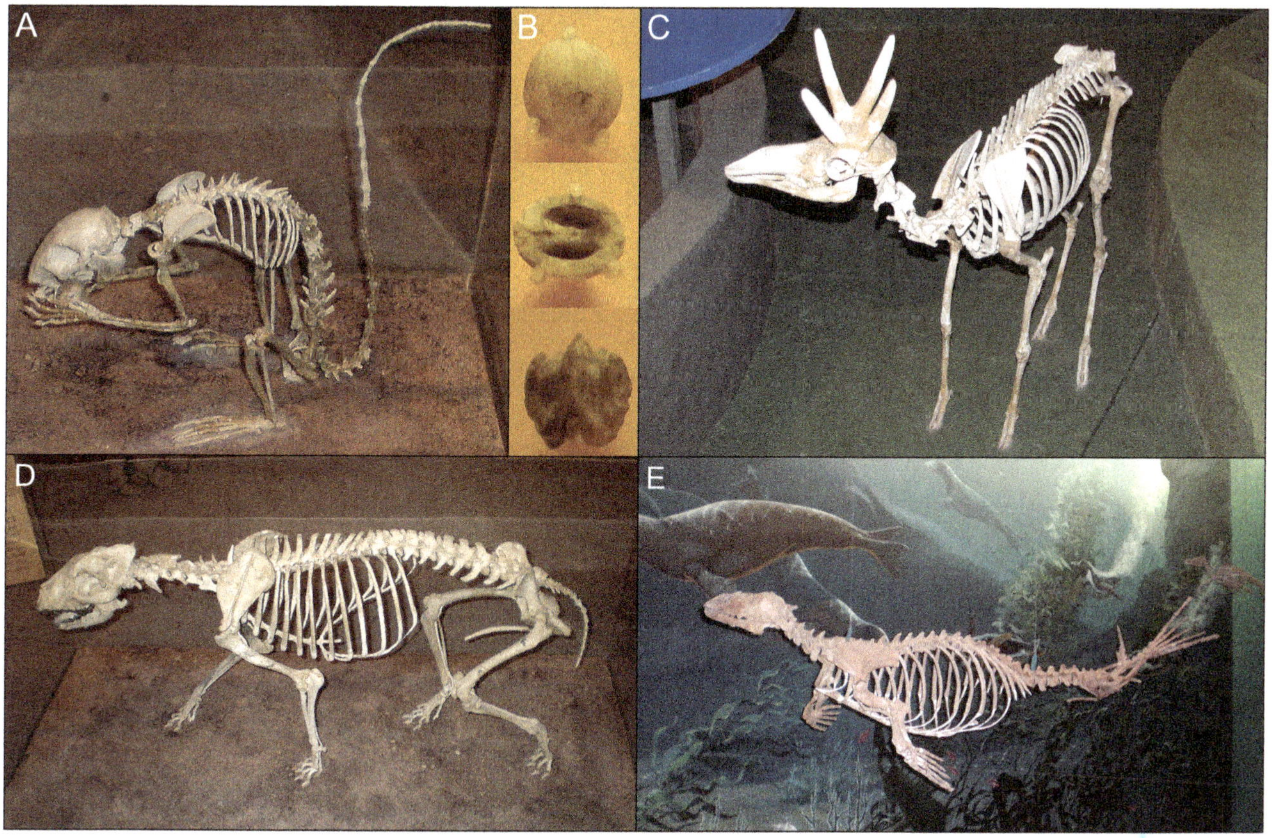

Figure 12.22: Some early mammals: A. One of the oldest fossil squirrels along with some fossilized walnuts (B). C. An extinct antelope. D. *Strobodon* was an early canine. E. A seal fossil. Seals are really just aquatic canines. Photo credits: B. Zaprowski.

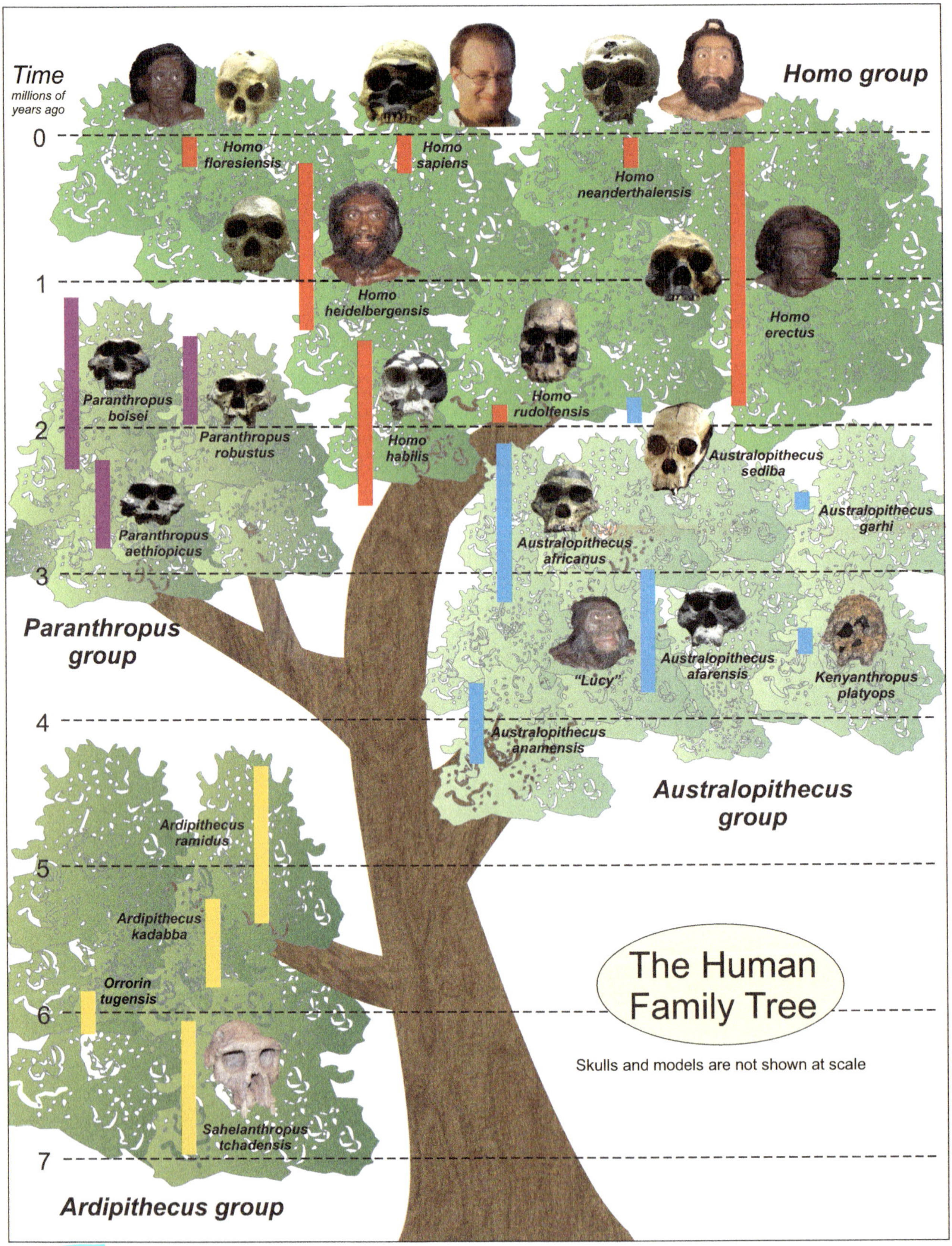

Figure 12.23: The human family tree. This highly generalized diagram highlights the genetic relationships between the major groups of hominids. Photo credits: B. Zaprowski.

mammals such as felines and canines evolved to catch those herbivores. Mammals also spread to marine realm in the form of seals, dolphins and whales.

Primates were no exception, and as they evolved to adapt to a life in the forest canopy, they began to develop many of the same traits we as humans have today. They developed stereoscopic vision which gave them good depth perception, they developed a hand with an opposable thumb, better hand-eye coordination, and they developed a smarter brain (Figure 12.21B). Given that early primates were well adapted for life in the trees, why did later primates like us eventually leave? Once again, it is a change in the environment that led to the rise of the **hominids**, a group of upright walking primates that evolved into modern humans, *Homo sapiens*. Around 7 million years ago the climate in Africa was changing the landscape from lush forests into vast grasslands. In the treeless world of central Africa, walking upright allowed the hominids to see predators, thus allowing them to escape to safety. It also freed their hands to perform other tasks (like using tools). This was the first "step" towards becoming human. Finally, we have reached the end of our journey as we have reached the base of the human family tree (Figure 12.23).

Figure 12.24: A model of *Australopithecus afarensis*, better known as the famous "Lucy". Photo credit: B. Zaprowski.

The human family tree has four major groups in it. The oldest group, the ***Ardipithecus* group**, is the first group of hominids which displayed one of the key hominid traits of walking upright, although they probably didn't do so all the time. *Ardipithecus ramidus* is considered to be the hominid which first displayed true bipedalism and thus it sits towards the top of the *Ardipithecus* branch. The next group of hominids, the ***Australopithecus* group**, were much better at walking upright but still spent time in the trees too (Figure 12.24). Although early hominids were probably smarter than most of their predators, the trees still offered the best protection. There is a third, lesser known group of hominds, the ***Paranthropus* group**, which also displayed many of the same hominid traits as the *Australopithecus* group, but they had large teeth and powerful jaws. These features likely reflect a diet of tough nuts and vegetation. One thing all three of these groups didn't appear to do was use tools. That second major "step" on the way to us was taken by the last group, the ***Homo* group**.

Figure 12.25: A. Paleolithic hand axes from Kenya, Africa. These date between 1.7 million and 500,000 years old. Photo credit: B. Zaprowski.

The oldest fossils of the genus *Homo* appears in Africa around 2.5 million years ago. And around the same time, stone tools begin to appear in the same layer of sediments as *Homo habilis* ("the tool-maker") fossils. Although the tools may appear fairly primitive (Figure 12.25), they require a level of cognitive ability likely beyond that of the earlier hominid groups. In fact, they display a level of cognitive ability beyond that of any other animal group on Earth. It marks the beginning of technology, a trait that only humans and their recent ancestors possessed. Tools gave early hominids the ability to manipulate their environment, like the ability to chop down tress to make a shelter. Tools also gave early hominids the ability to hunt and eat meat from large animals. By 1.8 million years ago, an advancement in stone tool technology appeared. It consisted of large cutting tools, primarily hand axes and cleavers. With both weapons to hunt with and tools to cut the meat up with, early hominids got a big boost in their food supply. It also gave them more free time to do other things with those newly-discovered tools. Around this time we also see a significant increase in brain

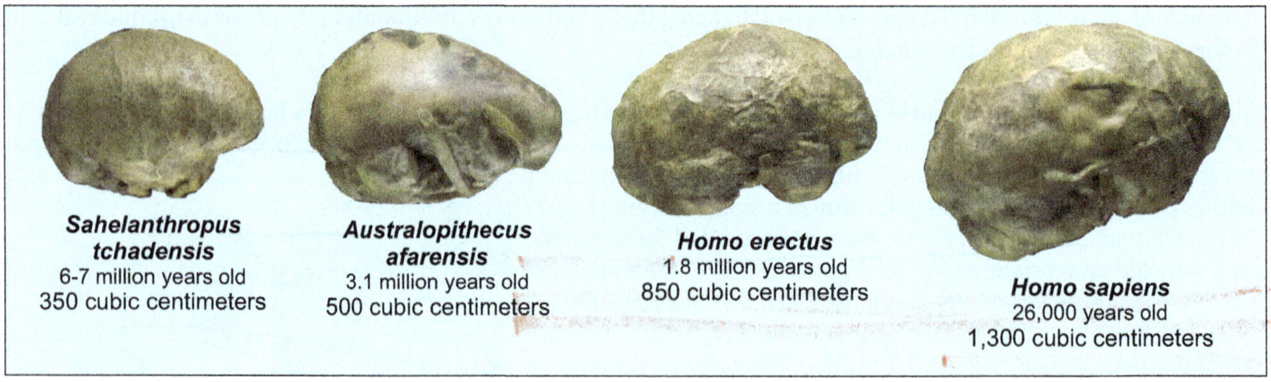

Figure 12.26: Brain endocasts show how brain size began increasing around 1.8 million years ago, leading us to the big brains we have today. Photo credits: B. Zaprowski.

size (Figure 12.26). The cause(s) of this size increase certainly must include having a better and more regular food supply, but that alone is not sufficient to explain the evolutionary changes that occurred in the *Homo* group at this time. Although the other cause(s) of these changes are still being debated, one "cooking hypothesis" suggests that the next major "step" played an important role: Controlling fire.

The species *Homo erectus* is thought to have diverged from earlier Homo populations roughly 1.8 million years ago, and then spread into Asia. Not only were *Homo erectus* accomplished toolmakers, but there is clear evidence the species was able to control and use fire. Most all of the animals in the animal kingdom fear fire, and likely early hominids groups did too. It is not clear what led *Homo erectus* to overcome their fear and start using naturally-caused lightning fires to their benefit, but once they did, it changed how hominids lived forever. Fire would have been useful for both protection from predators and as a source of heat. It also allows humans to work on activities after dark and live in harsh climates. And it may have also played a role in improving socialization and communications skills. Even today, people circle around fires to tell stories and cook food (Figure 12.27). The ability to cook food in particular may have spurred the evolutionary changes occurring at this time. Cooked food is easier to chew and digest, and the body can extract about 30% more energy from cooked food as opposed to raw food. This decreases the amount of food one has to consume. Thus, the "cooking hypothesis" states that we see the rapid increase in hominid brain size coincide with the appearance of fire and the subsequent cooking of food. In addition to a change in brain size, there is a decrease in tooth size, a change in tooth wear patterns, and a decrease in stomach size. These are all traits they archeologists would expect to see with a change to a cooked-food diet. Of course to prove this hypothesis true, archeologists will have to find a site with clear evidence of controlled fire and cooking of food. Currently, evidence for the use of controlled fire by *Homo erectus* in a hearth dates to about 400,000 years ago. There is also evidence of fire use from a wide variety of sites that date back as far as 1.7 million years ago, but most of that data is still controversial. For now, it is clear that fire and cooking food played an important role in the final steps to modern humans.

Figure 12.27: Fire and all of its benefits really helped hominids make the next major leap forward in both technology and brain size. S'mores came later. Photo credit: B. Zaprowski.

Homo erectus was the longest-lived and the most successful hominid species of all the members of the *Homo* group (They were around for almost 1.9 million years!) and their skeletons have been found at sites in Africa, China, Indonesia and the Middle East. But they were not the only hominids around. *Homo heidelbergensis* ("Heidelberg Man") is the species which is the most likely direct ancestor of *Homo sapiens* (Figure 12.28A). *Homo heidelbergensis* was the first hominid species to live in colder climates, and evidence for wood and rock structures has been found as well. They also performed ceremonial burials. Most *Homo heidelbergensis* fossils date between 700,000 and 200,000 years old, although there is some debate about whether or not this species may even date back to as far as 1.3 million years ago. Sometime between 200,000 and 500,000 years ago *Homo neanderthalensis, Homo*

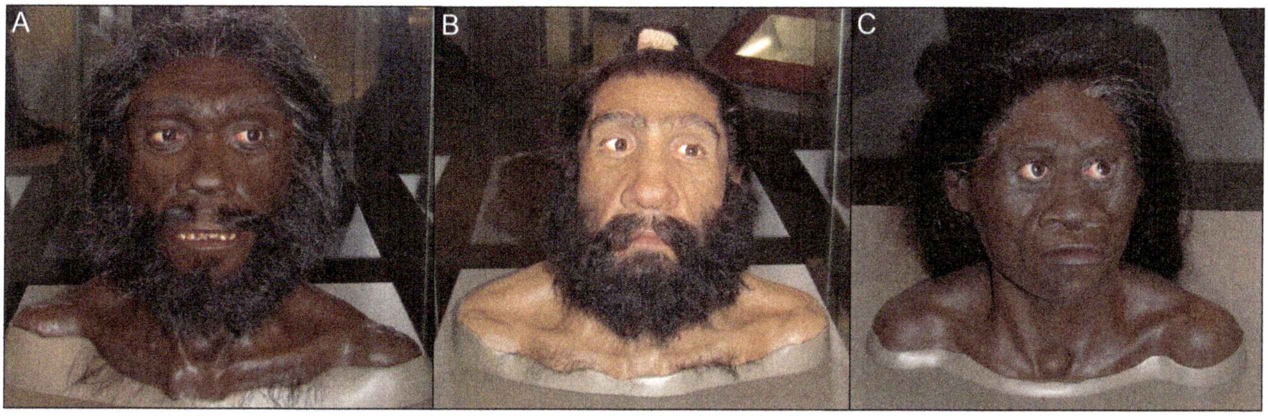

Figure 12.28: Meet some of your closest hominid relatives. A. *Homo heidelbergensis*. B. *Homo neanderthalensis* C. *Homo floresiensis*. Models were on display in the Hall of Human Origins exhibit at the Smithsonian Museum of Natural History in Washington, D.C. Photo credits: B. Zaprowski.

floresiensis and *Homo sapiens* diverged from *Homo heidelbergensis* (Figure 12.28). *Homo neanderthalensis* settled in Europe, *Homo sapiens* settled in Africa and *Homo floresiensis* fossils are only known from Indonesia.

Homo neanderthalensis first appeared in the fossil record about 400,000 years ago. *Homo neanderthalensis* was slightly larger than modern humans and even had a larger brain. They made and wore clothing, lived in shelters, made complex weapons, made ornamental objects and they performed rituals. They ate mostly meat and their bodies were well adapted to living in colder climates. This hominid species existed side by side and interbred with *Homo sapiens* for tens of thousands of years, but they eventually disappear from the fossil record around 40,000 years ago. *Homo floresiensis* ("Hobbit People") is a species of hominids that have only been found on the Island of Flores in Indonesia. The fossil samples that have been collected date from between 95,000 and 17,000 years old. Their small stature (average height was 3'6" tall) was probably the result of dwarfism, a common phenomenon that occurs to animals trapped on isolated islands. It is not clear when *Homo floresiensis* evolved or when they first reached Indonesia.

The oldest known *Homo sapiens* fossils that possess skeletal features typical of modern humans date to ~200,000 years ago. Modern *Homo sapiens* are thought to have evolved in Eastern Africa and then spread to Europe and Asia before finally making it to the Americas. DNA evidence now shows that *Homo sapiens* interbred with *Homo neanderthalensis* and other archaic hominids along the way and this may explain the wide variety of regional traits expressed among modern humans today. *Homo sapiens* is capable of making very complex weapons (Figure 12.29A) and applying technology in new and creative ways (i.e., the creation of fish hooks). We eat a diverse diet of meat, fish, vegetables, fruits and nuts, both cooked and raw. We are good communicators, expressive, creative, and have a distinct sense of spirituality (Figure 12.29). So why us? Why are we the only hominid species left? Our other hominid ancestors were certainly intelligent and able hunters too. For one thing, our highly varied diet gave us more food options, more energy and better cognitive skills as compared to *Homo neanderthalensis*, who had a diet almost exclusively of meat. Some archeologists have suggested that it was the development of complex social interactions and the sharing of knowledge that allowed *Homo sapiens* to out-compete the other contemporary hominid species. We were simply better fed, better organized and better equipped, and now we stand alone as the last hominid species on Earth. And hopefully you now appreciate the 3.5 billion-year journey of evolutionary twists and turns through time that led to you being you!

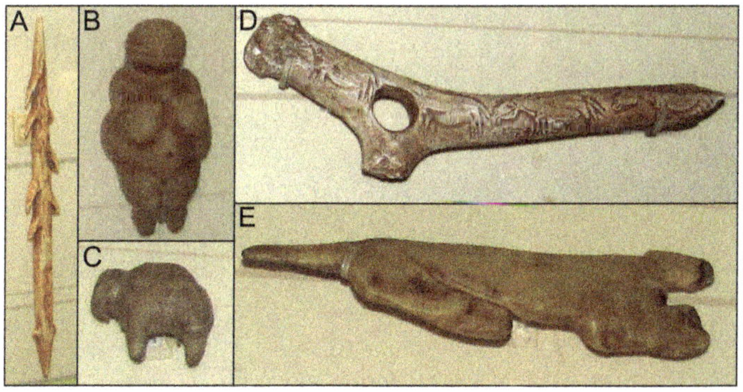

Figure 12.29: Artifacts of early *Homo sapiens*. A. Antler harpoon. B. Fired clay Venus figurine C. Fired clay bison. D. Carved antler baton. E. Carved ivory lion. Photo credits: B. Zaprowski.

Dean Gesch/USGS

Chapter 13 - Landscapes and Earth's Surface Processes

The landscapes we see around us are the result of many factors coming together to shape the rocks of the Earth's crust. In this chapter we will look at the major processes that work in combination to sculpt the Earth's surface, including the hazards these processes pose to humans.

Section 13-1: Introduction to Geomorphology

Geomorphology is the study of landforms, landscapes and the processes that shape them. Anytime you look at a landscape, you are seeing the combined result of five major factors working together: 1) Process, 2) Tectonics, 3) Rock type, 4) Climate, and 5) Time. In this section we will briefly review how these five factors dictate the nature of the Earth's surface at any one locale.

The primary processes which wear down the Earth's land surface over time are weathering and soil formation, gravity and mass wasting (i.e., landslides, rock falls, etc.), running water (streams and rivers), and ice (glaciers) among others. Landscapes undergoing one or more of these processes usually have distinct landforms unique to those processes. We'll discuss these processes in more detail in the following sections. Tectonically-active landscapes tend to look different than tectonically-inactive landscapes (For example, go back and compare Figure 7.28B with Figure 7.30B). As you saw in Chapters 7, 8, and 9, plate tectonic-driven mountain building, volcanism, earthquakes and their resulting landforms leave a distinct fingerprint on the landscapes in which they occur.

Rock type plays an important role in dictating how weathering will shape the landscape. The relative hardness of different rock types controls how fast rocks will break down when exposed to the agents of weathering. In this sense, "hard rocks" are both physically *and* chemically resistant to weathering. Because of their resistance to weathering, the hardest rocks produce topographic highlands while valleys form in areas underlain by softer rocks (Figure 13.1A). Rock type also dictates how steep a hillslope may be in a given area. Hard, weathering-resistant rocks tend to produce steep cliffs of exposed bedrock while softer rocks tend to produce gentle slopes covered in soil and vegetation (Figure 13.1B).

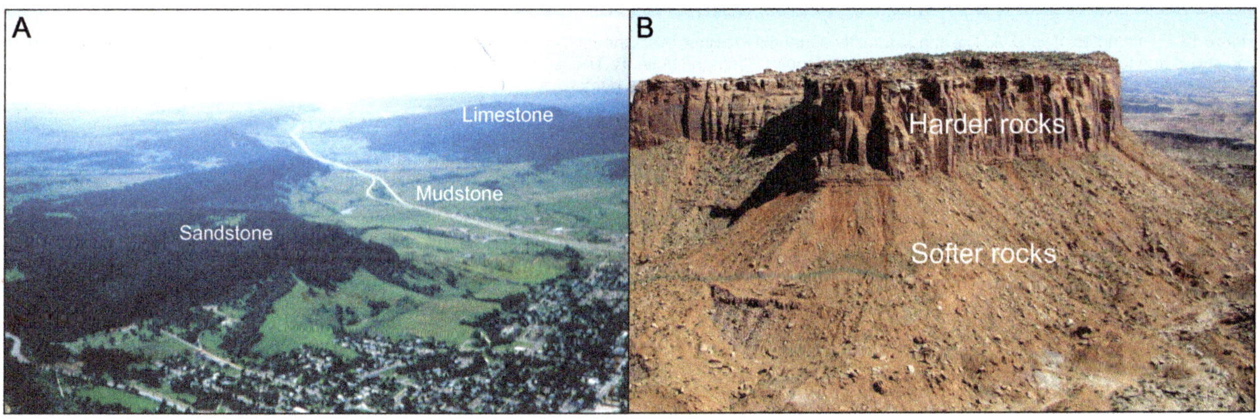

Figure 13.1: Examples of how rock type influences the formation of landscapes. A. Rock type dictates where uplands and valleys may form. In the Black Hills of South Dakota, the uplands are underlain by sandstone and limestone, while the valleys are underlain by soft mudstones. B. Rock type dictates how steep a hillslope may get. Harder rocks form steep cliffs, while softer rocks form gentler hillslopes covered with regolith. Photo credits: B. Zaprowski.

It should be no surprise that there are exceptions to the rule. For example, in warm, humid climates limestone bedrock can literally "dissolve" away to form a rolling landscape full of caves and sunken depressions called **karst topography** (Figure 13.2A). Distinct rounded "haystack" hills commonly form between the depressions (Figure 13.2B). Limestone is fairly easy to dissolve when it comes in contact with warm, acidic water, and most groundwater tends to be acidic. As groundwater moves through pores and fractures in limestone bedrock, it dissolves the limestone, producing larger and larger cavities in the rock. Over time, the cavities may grow large enough to form a system of caves and depressions in the land surface above called **sinkholes** (Figure 13.2C). In some cases, the roof of a cave may collapse, rapidly forming a sinkhole. Sometimes these sinkholes form under urban areas, destroying roads and houses (Figure 13.2D). Residents in central Florida in particular live in hazardous areas with a very high rates of sinkhole occurrence.

Figure 13.2: Examples of karst. A. The image from the Guangxi Province in southeast China displays a classic example of karst terrain. The "textured" pattern you see is due to limestone dissolution. Image credit: Robert Simmon/NASA. B. These "haystack hills" in Puerto Rico are common in well-developed karst landscapes, like the one in Figure A. Photo credit: B. Zaprowski. C. All of the numerous lakes that dot central Florida are due to karst. Many of them are sinkholes as well. Image credit: Jacques Descloitres, MODIS Land Rapid Response Team, NASA/GSFC. D. More than 110 sinkholes formed in the Dover, Florida area in January 2010, destroying homes, roads and sections of cultivated areas. Photo credit: Ann Tihansky/USGS.

Although the hazard posed by sinkholes in karst terrains is very real, the caves beneath their surface can be beautiful and fascinating places to visit. When the groundwater moves through the subsurface, it carries the dissolved limestone with it. If the water reaches the ceiling of a cave, it drips to the floor, leaving behind a thin layer of calcium carbonate attached to the ceiling. As more and more water drips from the ceiling, the deposits of calcium carbonate build up to form a **stalactite** (Figure 13.3). Not all of the calcium carbonate is left behind however. When the water drops hit the cave floor, more calcium carbonate is deposited, forming a **stalagmite**. Stalactites and stalagmites both grow very slowly, taking hundreds of years to form. Sometimes the stalactite and stalagmites grow together into complex columns or pillars.

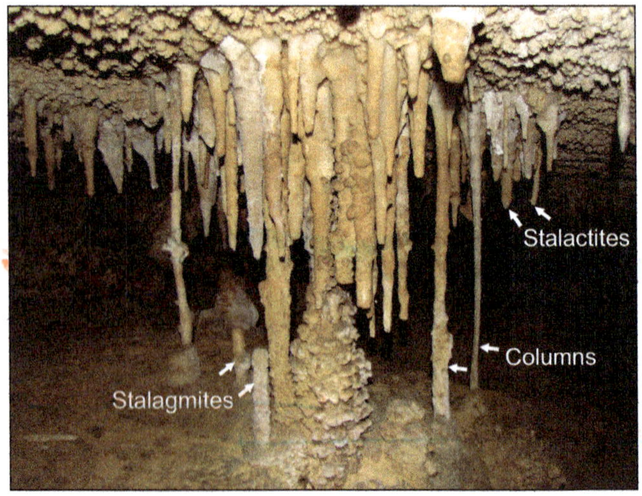

Figure 13.3: Examples of common limestone cave features. Photo credit: B. Zaprowski.

Climate plays an important role in several ways. First, the types of processes that may occur at a given location is strongly a function of climate. For example, glaciers will only form in cold climates, while large sand seas can only form in arid climates. Climate also controls the type of weathering that occurs. In humid climates, the chemical weathering process is extremely vigorous and the bedrock is quickly altered to clay and other residual sediments (Figure 13.4A). At the other end of the spectrum, physical weathering dominates in arid and

Figure 13.4: Comparing humid landscapes with arid landscapes. Both locations are underlain by the same type of bedrock. A. Humid landscapes tend to have thick soils and lots of vegetation due to intense chemical weathering. B. Arid landscapes tend to have thin, rocky soils and sparse vegetation, if any at all. Photo credtis: B. Zaprowski.

polar environments because the lack of liquid water greatly slows down chemical weathering processes. Coarse-grained sediments tend to be abundant in these settings while clays occur minimally (Figure 13.4B). Temperate climates fall in the middle of the spectrum: Physical weathering dominates in the winter and chemical weathering in the summer, so their landscapes are covered in a mixture of clays and coarser-grained sediments.

It is important to note that plants and animals also play a role in rock weathering. **Biological weathering** can be physical or chemical in nature, but it generally occurs in temperate to humid climates where plant and animal life is abundant. Animals like snails and sea urchins can break down rocks while they are grazing on the algae which grows on the rocks. Slowly, over time, they can wear deep depressions into a rock, weakening it and making it more susceptible to wave energy (Figure 13.5A). Plants play a huge role in breaking down minerals during the soil forming process. Even small, simple plants such as algae, lichens and moss can slowly eat away at a rock, weakening it. Plant roots chemically attack rocks by excreting chemicals that break the minerals down (Figure 13.5B). Once dissolved in solution, the roots can take up the ions and transport them to other parts of the plant's system. In this way, plants "digest" rocks as their roots penetrate the subsurface. This is how the tree in Figure 13.5B is able to grow on the side of the rock.

Finally, the age of the landscape dictates the nature of the landforms found there. Landscapes "age" over time, just

Figure 13.5: Examples of biological weathering. A. Even small animals like snails can weather a rock. As the snails feed, they slowly scrape the rocks down. B. Tree roots are a huge weathering force. This tree is even able to grow on the side of the outcrop! Photo credits: B. Zaprowski.

Figure 13.6: Models of landscape evolution for tectonically quiet areas. A landscape "begins" when rocks are exposed to erosion following a drop in sea level. As the landscape matures, the relief and hillslope angle decreases while the valley width increases. Keep in mind this model is highly idealized and doesn't account for variations in rock type.

like people do. Although simplistic, the model in Figure 13.6 helps to demonstrate what we mean. Generally speaking, a "landscape" is born when sea level falls and rocks are exposed to erosion. Initially, the landscape is dominated by broad, flat upland areas, steep hillslopes and narrow valleys as rivers cut downward. Over time, the rivers meander back and forth, widening the valleys and creating floodplains. The hillslope angle decreases and the flat uplands are replaced by more distinct ridges. Eventually the uplands are worn down to low hills with very gentle hillslopes and broad flat river valleys in between. We see the same type of pattern with mountains (Refer back to Figures 7.33B and 7.33C). "Young" recently active mountain ranges tend to be rocky, steep, high in elevation and lack significant soil and vegetation. The may have glaciers as well. In contrast, "old" inactive mountain ranges may still have steep slopes, but generally that have much less relief, and the slopes are covered with soil and vegetation with much less exposed bedrock visible. Keep in mind that this model is highly idealistic. It does not take into account rates of uplift, weathering, variations in rock type or glacial erosion. Now that we've taken a look at the big picture, let's look at some of the processes in more detail, starting with the weathering of rocks and the formation of soil.

Section 13-2: Soil and Soil Formation

A majority of the Earth's land surfaces are covered with a layer of weathered rock debris known as **regolith** (Figure 13.7). **Soil** is regolith that can support plant life; this ultimately makes soils the basis of the food chain on land. At the same time, soils are fragile and easy to destroy because the soil forming process is slow: It takes about 1,000 years for 1 foot of good, productive topsoil to develop. Thus, it is important to protect and care for soil just like any other natural resource because once soil has been eroded away, it is essentially gone forever. We'll discuss this more later in the section.

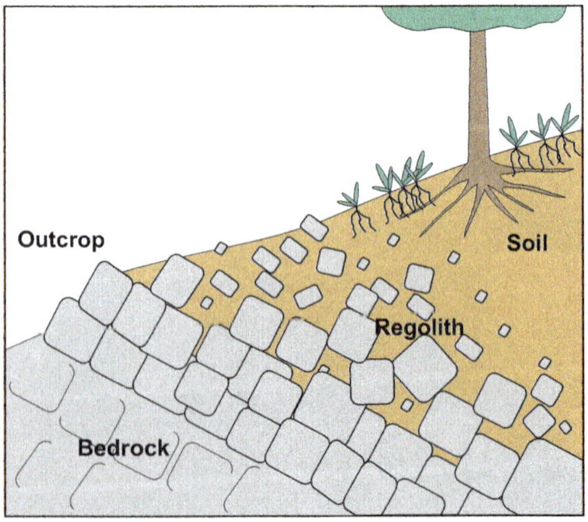

Figure 13.7: Bedrock exposed at the Earth's surface is attacked by the agents of weathering, producing regolith and soil. Any exposure of bedrock above the soil is called an *outcrop*.

Soils are described and classified based on the physical and chemical characteristics of the **soil profile** and the layers within it (Figure 13.8). The soil profile develops over time as weathering breaks down rocks and mixes the resulting mineral matter with organic debris called **humus**. The humus is dead plant and animal material which falls on the ground and breaks down through decomposition. The volume of an average topsoil is about 45% mineral matter and 5% humus by volume while the remaining 50% of a soil's volume is occupied by pore spaces filled with water or air (Figure 13.8). If it just rained, most of the spaces will be filled with water but if it hasn't rained in a while, then the spaces will be mostly filled with air.

The nature of the soil profile at any location is a unique product of 5 independent factors: Rock type, climate, land stability, biota, and time. Collectively, these factors control how thick a layer of soil may be, what the composition of that soil will be like and what types of layers will form in the soil profile.

Factors Which Control Soil Formation

It should be no surprise that soil composition is primarily a function of rock type. Quartz-rich rocks are likely to produce sandy soils while quartz-poor rocks are more likely to produce clay-rich soils, as many minerals readily decompose into clay minerals in the presence of water (i.e., Figure 10.13). Some of the rarer soil types are only found in association with specific types of bedrock. For example, *andisols* are soils that only form in association with volcanic deposits. The relationship between rock type and soil thickness is just as straightforward: Softer, chemically-susceptible types of bedrock (like shale) generally produce a thicker layer of soil than hard, chemically-resistant bedrock (like sandstone) given all else is equal (Figure 13.9).

Climate plays two important roles in soil formation. First, climate dictates how thick the soil profile will be. As you can see in Figure 13.10, soils are thickest in tropical locales, thinner in temperate climates and practically non-existant in cold and arid climates. This relationship reflects the importance of both liquid water and warm temperatures in the soil forming process. Climate also plays a role in soil composition by controlling what the products of rock weathering will be like. The same type of bedrock may produce different types of soil depending on the climatic conditions.

In warm, humid climates, rocks are quickly broken down and transformed into very thick, fine-grained, clayey soils called *oxisols*. The oxisols also have varying amounts of residual quartz in the form of silt or sand. Deep red-colored soils called *ultisols* are also common in tropic locales. The red color comes from iron oxides produced during intense chemical weathering (Figure 13.11A). The chemical weathering is so intense in some areas that weathering will remove

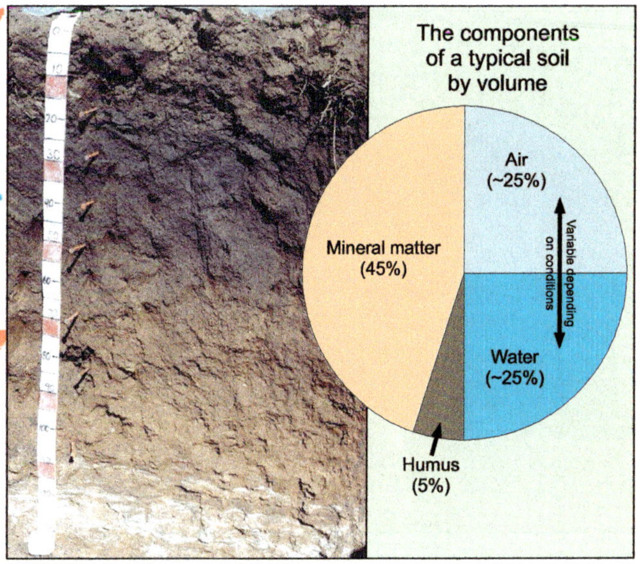

Figure 13.8: An example of a soil profile and the bulk composition of an average soil by volume. Photo credit: USDA/NRCS.

Figure 13.9: A contrast in soil thickness due to parent material. On the left, the rocky, silica-rich bedrock has produced a soil profile which is only about 30 cm thick. On the right, clay-rich bedrock has produced a soil profile which is about 150 cm thick. Photo credits: NCRS/USDA

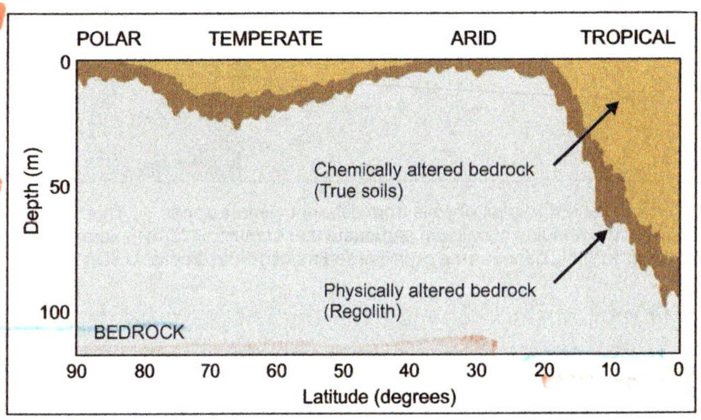

Figure 13.10: A schematic diagram of depth to bedrock versus latitude. Note how the thickness of chemically-altered bedrock (true soils) is practically negligible in arid and polar climates.

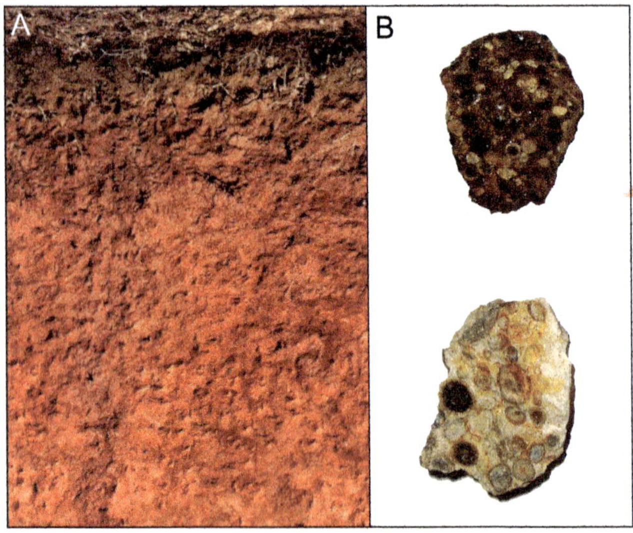

Figure 13.11: A. Ultisols are soils that form in warm, humid climates and they tend to be reddish-colored in nature. Photo credit: USDA/NRSC. B. Two examples of the mineral bauxite, an aluminum ore commonly found in tropical soils. Photo credit: B. Zaprowski.

aluminum ions from the topsoil and deposit "nodules" of aluminum further below in the soil profile, forming a layer of aluminum ore called bauxite (Figure 13.11B).

In temperate-humid climates, soils called *alfisols* tend to form. Alfisols have clay-rich layers in the subsurface, and this clay helps to retain water and nutrients, making them highly productive in terms of plant growth (Figure 13.12A). In contrast, temperate semi-arid grasslands are commonly underlain by soils called *mollisols* which have calcium carbonate in their subsurface layers (Figure 13.12B). The calcium carbonate helps to make these soils very fertile, and most of the world's major agricultural areas are underlain by these types of soils (i.e., the Great Plains of the U.S.).

In arid climates, the soil tends to be thin and have a sandy or rocky texture because the chemical weathering process occurs at a much slower rate. Without the water needed for chemical weathering to proceed, minerals are simply broken down into smaller and smaller sizes, *but very little clay is produced.* These soils, which are called *aridisols*, tend to accumulate salt, carbonate and gypsum in the subsurface while very little chemical alteration of the bedrock occurs. Because of this, aridisols can't support most forms of plant life (Figure 13.12C). In addition, the lack of plant life makes the soil very unstable and easy to erode, keeping soil thickness to a minimum.

Figure 13.12: Examples of soils from different climate zones. A. This soil profile with clay layers is typical for humid-temperate climates. B. This soil profile displays the nodules of calcium carbonate that commonly form in semi-arid climates. C. Aridisols are really just a form of regolith with salts and other mineral deposits forming between the grains of sediment. Photo credits: USDA/NRSC.

The relationship between soil thickness, soil development and the topography of the land surface is also fairly straightforward. Steeper slopes have thinner soils because gravity works quickly to remove loose rock debris from the hillslope (Figure 13.13A). Highly unstable areas like floodplains, sand dunes and coastlines are also places where soils tend to be thin and/or poorly developed. At these locales, the constant erosion and deposition of sediment doesn't allow the soil development process to proceed very far before the soil profile is disrupted or buried (Figure 13.13B). As a result, poorly developed soils called *entisols* tend to form in these areas (Figure 13.14B).

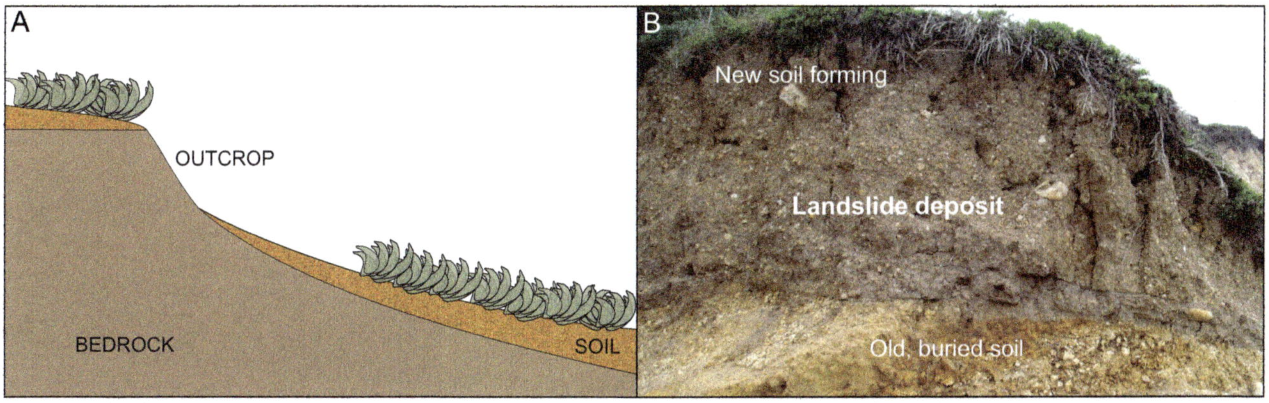

Figure 13.13: Examples of how land stability affects soil formation. A. Generalized illustration showing the relationship between slope steepness and soil thickness. B. This landslide deposit along the coast of California buried an older soil profile. Photo credit: B. Zaprowski.

We can't ignore the role of the biota (including bacteria and fungi) in the soil-forming process. In most soils, humus is only a small component of a soil's bulk by volume, but in terms of plant growth, humus plays a key role in soil formation. Humus is an important source of weak organic acids which aid in the breakdown of minerals into soluble ion forms which plant roots can absorb. Burrowing action by animals also helps in the soil forming process by mixing the mineral matter and humus together. In some locations, extremely humus-rich soils called *histosols* can form (the organic content can be as high as 80-90% by mass). Histosols are more commonly known as peat. Histosols commonly form in poorly drained areas where the soil is constantly saturated with water and levels of dissolved oxygen are low. Organic debris which falls into these wet areas decomposes at a very slow rate, producing thick piles of organic-rich sediment. Common peat-forming locales include coastal salt marshes, swamps and bogs.

And finally, some soils simply take time to develop. Some soils are "mature" and some soils are "immature". Mature soils have well-developed soil profiles with distinct layers while immature soils lack a soil profile or the soil profile is only weakly developed. For instance, oxisols and entisols are two types of soils which can form in humid climates. The differences between these two soil types are primarily due to a difference in their ages. Oxisols are mature soils that form in areas where the land surface has been stable for a long periods of time, allowing soil development to reach its maximum potential, whereas entisols are young, poorly developed soils (Figure 13.14).

Figure 13.14: Examples of soils of different ages. A. This oxisol soil profile with its bright-orange iron oxide layers is typical for very old soils. B. This entisol soil profile displays no evidence of soil development. Photo credits: USDA/NRCS.

Causes of excessive soil erosion

Even as soil is forming, surface erosion and mass wasting events are removing topsoil and carrying it away to some distant location. In natural settings, rates of soil production are equal to or greater than rates of soil erosion. Thus, new soil is constantly being produced even as the land surface erodes downwards over time. However, sometimes the balance is tipped and soil erosion rates exceed soil production rates. This most commonly happens any time vegetation is stripped from the land surface, because vegetation is very effectively at retaining soil. For example, rates of soil erosion can soar after a forest fire because there is no vegetation to hold the soil back (Figure 13.15A). Likewise, any time humans strip the land of vegetation and expose the soil, they increase rates of soil erosion, potentially to excessive levels. The two main human activities that lead to excessive soil erosion are agriculture and construction. Although the action of plowing a field is generally beneficial for growing plants, it also loosens the soil and makes it more susceptible to erosion, particularly during the winter when the soil lays bare and unprotected (Figure 13.15B). For this reason, farmers are now encouraged to grow "cover crops" in their fields during the off season. Likewise, the exposure of loose soil during construction projects (like building a road or a shopping plaza) can also lead to permanent soil loss from the construction site. Soil loss from these sites is typically minimized through the use of barriers like "silt fences."

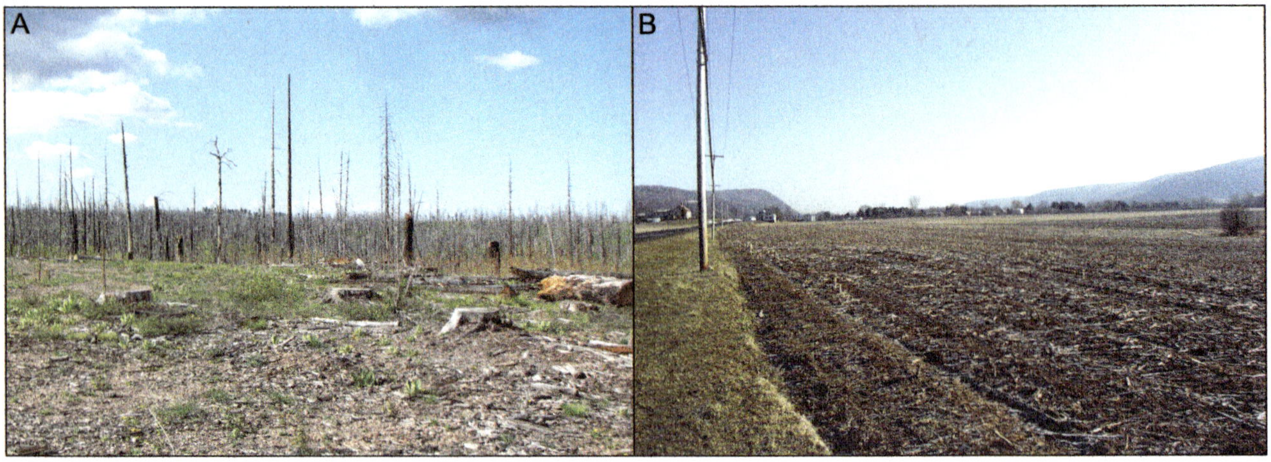

Figure 13.15: A. The vegetation is just now beginning to come back after a forest fire near the Grand Canyon. B. This barren field in upstate New York is susceptible to soil erosion during the winter months. Photo credits: B. Zaprowski.

Section 13-3: Mass Wasting

Rock weathering and soil formation occurs *in situ*, that is, in place. Although the action of wind, water and ice can erode soil and sediments directly from their place of origin, most weathered rock debris is first transported down hillslopes through a process called **mass wasting** (more commonly referred to as landslides). Mass wasting is the movement of weathered rock debris (regolith) down a slope under the influence of gravity. Once the regolith reaches the valley below, it may be removed by erosion. We will discuss erosive processes further in Sections 13-4 and 13-5.

Technically speaking, mass wasting events occur when gravity overcomes friction. When bedrock begins to weather and regolith begins to accumulate, friction is the only force that prevents rocks and rocky material from moving. When that friction is reduced, gravity is able to move the earthen material downhill. Common causes of mass wasting events are heavy rain and earthquakes. In the case of earthquakes, the shaking of the earth can reduce friction and cause rocky outcrops to fail and collapse (Figure 13.16). In the case of heavy rains, the water plays a dual role: not only does the water reduce friction, but the water adds weight to the earthen material, causing gravity to pull even harder on the materials.

Figure 13.16: The Hebgen Lake slide in Montana was triggered by an earthquake. Over 1.7 million cubic yards of rock slid down. Photo credit: B. Zaprowski.

When we classify the different types of mass wasting, three factors are considered: The type of movement, the rate of movement and the type of material in motion (Figure 13.17). Earthen materials basically move in two ways: through the process of **slip** and through the process of **flow** (Figure 13.18). Slip usually involves dry, cohesive earthen materials like rock or packed earth while flow generally involves water-saturated sediments. Slip is a form of movement where semi-cohesive "blocks" or "slabs" of earthen material slide down a surface (called a *slip plane*), much like a child goes down a slide at the park. In this case, the child is the block of material and the slide is the slip plane (Figure 13.18B). Flow involves the movement of earthen material in a churning, fluid-like motion. Instead of

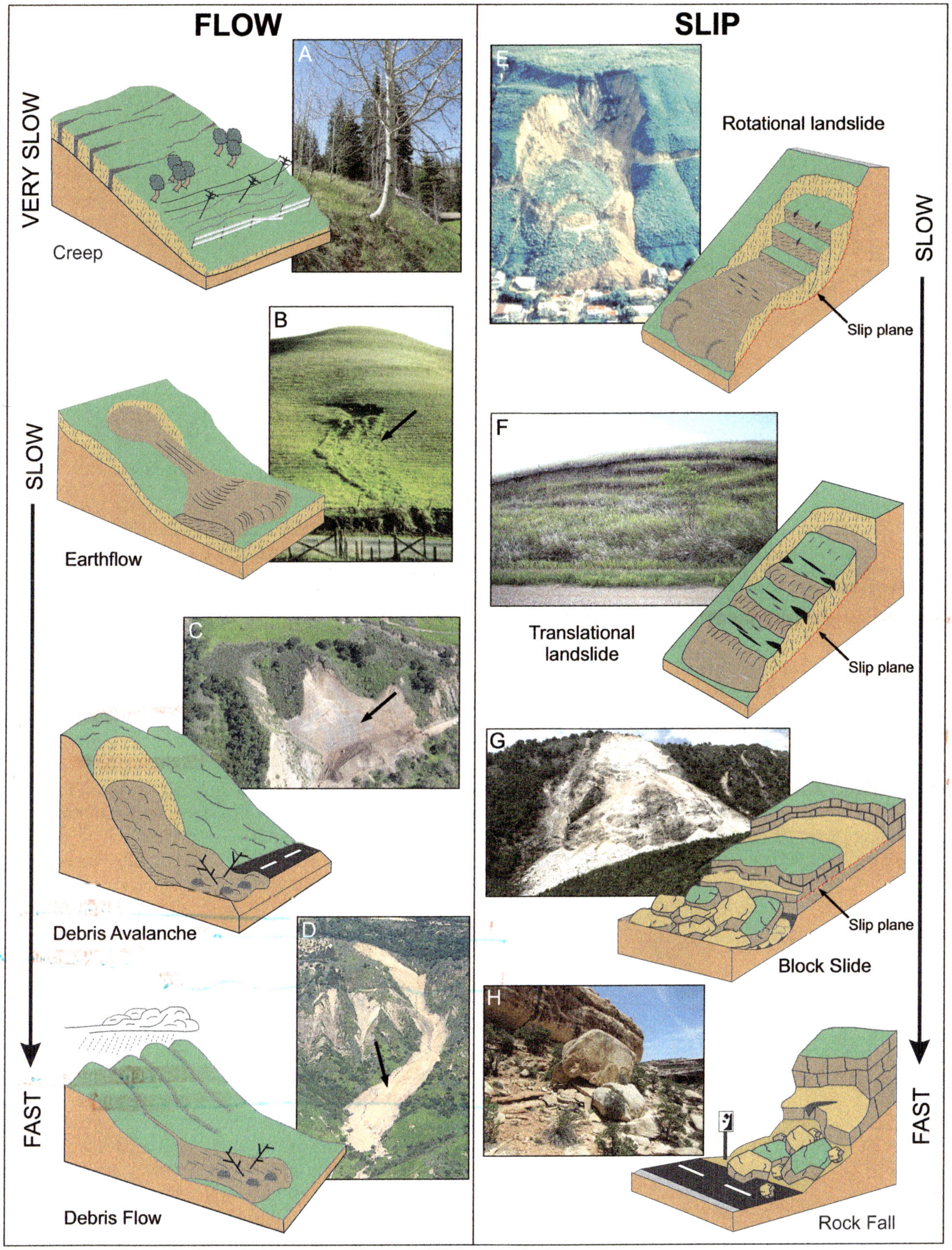

Figure 13.17: Common types of mass wasting. Based on the scheme devised by the USGS. Photo credits: (A, F, and H) - B. Zaprowski; (B) - B. Bradley/University of Colorado/NCDC/NOAA; (C and D) USGS; (E) - R.L. Schuster/USGS. (G) - Brian Collins/USGS.

Figure 3.18: A. An illustration demonstrating the difference between slip and flow. Slip involves sliding along a slip plane, while flow involves a fluid-like motion. B. A rock slide in Summit County, Utah. If you look near the arrow, you can see where large slabs of rock have begun to slide downhill. Photo credit: USGS. C. A debris flow caught in action on camera. You can see how it is moving in a very fluid-like manner. Photo credit: Jeffrey Coe/USGS.

a child going down a slide, think of wet concrete flowing down the slide instead (Figure 13.18C).

Once we consider the type of movement, we further classify mass wasting events based on the rate of motion. The rate of mass wasting is controlled by the steepness of the slope and by the water content of the material, which acts as a lubricant to reduce friction and increase movement. The slowest rates of mass wasting might be as low as centimeters per year, so slow that it is barely perceptible. It is like trying to watch your hair grow. On a daily basis you can't really observe it happening, but every few weeks you need to get a haircut. Other rates of mass wasting vary from meters per day to meters per second! The fastest form of mass wasting is a rock fall - you can't go any faster than falling straight down! Finally, certain types of mass wasting are specific to the material in motion. For example, an earthflow is made up exclusively of loose sediment while a block slide generally involves solid bedrock. Although there are many different ways to classify mass wasting events, the USGS model presented in Figure 3.17 will serve our purposes adequately. Following the discussions above, mass wasting events are first broken into two groups: those which display forms of flow and those which display forms of slip. From there, the types of mass wasting are arranged by rate with the slower forms on top and the faster forms on the bottom. Keep in mind that each type of mass wasting may display a range of rates depending on the setting.

Figure 3.19: Creep is causing this retaining wall to lean over. Photo credit: B. Zaprowski.

The slowest form of mass wasting is **creep**. Creep is the imperceptibly slow flow of material down a slope, but the evidence of movement may not appear for years. This makes creep dangerous. Over time, creep can lead to fences, telephone poles, walls and even gravestones toppling over (Figure 3.19). Creep can even move whole houses off their foundations. One good sign of creep is curved tree trunks (Figure 3.17A). If a slope is undergoing creep, then the trees on that slope will continuously adjust their growth pattern to offset it, resulting in curved trunks.

Earthflows are the slow movement of earth or mud downslope as a viscous fluid. They don't move particularly fast (you could outrun one) and they usually occur on over-saturated steep slopes underlain by weak rock types. Still, they can be quite large and unstoppable. In 1983, heavy rains triggered a large earthflow 1,000 feet wide, nearly 200 feet thick, and over one mile long near the town of Thistle, Utah (Figure 13.20). The earthflow reached a maximum speed of about 3.5 feet per hour. The earthflow dammed the river, cut off railroad service, flooded two major highways, and submerged the town of Thistle, which had to be abandoned. The earthflow continues to remain active though on a smaller scale. In contrast, the earthflow in Figure 13.17B is about as small as an earthflow will get.

Figure 13.20: This large earthflow, although slow moving, caused a lot of damage, especially by flooding the former town of Thistle, Utah. Photo credit: R.L. Schuster/USGS.

Debris avalanches and **debris flows** are very similar to each other as they are both fast-moving mixtures of bedrock, dirt, trees and whatever other debris gets swept up along the way (like houses and cars). Both are commonly triggered by heavy rainfall, and both occur on steep slopes. Both can be very damaging as they can occur without warning and quickly bury structures under several feet of earthen material. The main difference is the composition of the mixtures. Debris avalanches tend to have more rocky debris and not as much mud (Figure 13.21A) whereas debris flows (also called mudflows) have a high amount of mud which causes the material to flow in a very plastic manner (Figure 13.21B).

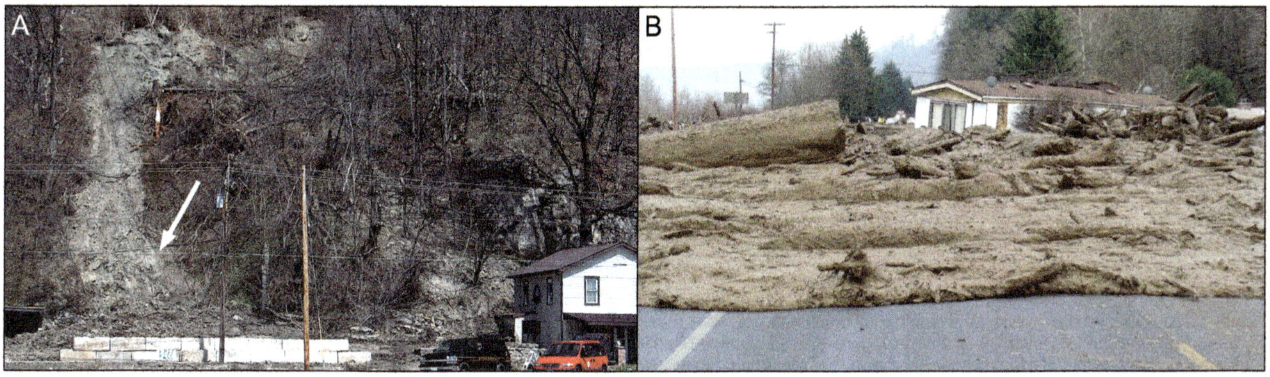

Figure 13.21. A. An example of a debris avalanche from Grafton, Illinois. Thankfully it missed the house to the right. Photo credit: Robert Bauer, Illinois State Geological Survey/USGS. B. Unfortunately, this house wasn't so lucky as a mudflow swept it into the road. Photo credit: Clatskanie, Oregon, PUD employee Kerry Kallunki/USGS.

In contrast to the forms of flow above, when the earthen material is more stiff, it tends shear off from the hillslope and slide down as a more or less cohesive block of material. **Rotational landslides and translational landslides** are very similar in this manner (Figure 13.22A-B). Both involve large slabs of earthen material which slide down along well defined slip planes. Both may spread out at the toe of the slide when the slide reaches the valley below. Both don't usually travel too far from their source. The main difference is that rotational slides have a curved slip plane and the resulting scarp is typically bowl-shaped, while translational slides tend to spread out along a horizontal plane. Although the slides move along as a cohesive unit, heavy rains are commonly the trigger for these types of landslides. They form in a range of sizes from meters to hundreds of meters in width.

Unlike the two forms of slip above, rock slides (or rock avalanches) are composed primarily of hard rocks (Figures 13.16 and 13.18 are both examples of rock slides). They may be triggered by heavy rain, but very commonly they are triggered by earthquakes. Pictures of rock slides may be misleading as they make it appear that the rocks move down the hill in the form of small boulders, but the rocks typically start out as large slabs and only break up into smaller pieces when they reach the valley below (Figure 13.22C). Many times they have enough momentum to slide across the valley floor and up the opposite hillslope. For example, the Hebgen Lake slide in Figure 13.16 was triggered by an earthquake in 1959. 40 million cubic yards rushed down into the Madison River Valley, slid across, and dammed up the Madison River before running several hundred feet up the opposite side of the valley (The photographer was

Figure 13.22: Types of slip. A. An example of a rotational landslide. Note the bowl-shaped scarp. Compare this photo to Figure 13.17E. Notice the similarity? This slide occurred in 2005 while the slide in Figure 13.17E occurred in 1995. As you can imagine, these types of landslides are common in Southern California. B. An example of a translational landslide. The 2014 Oso slide in Washington was exceptional because the slide spread out more than a kilometer, burying a subdivision and blocking the river. C. This rock slide blocked one of the roads leading to Yosemite National Park. D. Large boulders have fallen onto to this road in Washington State. Photo credits: (A-C) - Mark Reid/USGS, (D) Washington State DOT/USGS.

standing on some of the landslide deposits when he took the picture). Unfortunately, the rock slide occurred at night, and several campers were buried in their sleep. Rock falls are similar to rock slides, but rock falls involve pieces of rock breaking off from a cliff face and dropping more or less straight down (Figure 13.22D). Although the impact may not be a wide-reaching, falling rocks can be a real hazard for passing motorists and pedestrians who travel near rocky cliffs.

Although heavy rain and earthquakes are commonly the triggers for mass wasting events, it is the geology and climate of an area which dictates how susceptible an area is to different types of mass wasting. This includes not only rock type, but also hillslope steepness and soil thickness. It should be intuitive that the risk of mass wasting rises as hillslopes get steeper. So it is should be no surprise that areas which have a low relief have a low risk for mass wasting events and areas with high relief have a high risk for mass wasting events. Easily erodible types of bedrock like claystone, mudstone and shale also increase the likelihood of a mass wasting event occurring. In terms of climate, almost any type of mass wasting can occur anywhere, but in arid climates, the risk is much lower because water is usually the lubricant which allows most mass wasting events to occur, whereas humid climates are generally at higher risk for landslides due to high amounts of rainfall, and the presence of weathering-produced clays in the subsurface. For example, the risk of mass wasting is pretty low in the desert regions of the southwestern U.S., even though the relief is fairly high. Mass wasting also occurs more frequently along roadways which have been cut into a hillside (Figure 13.22C).

Section 13-4: Groundwater and the Hydrologic System

We don't usually think too much about what happens to precipitation when it falls to the Earth. Contrary to what you might think, most of the precipitation that falls onto the land surface soaks or **infiltrates** into the subsurface becoming **groundwater**. This groundwater then works its way through the cracks and spaces in the bedrock below until it eventually trickles back out of the ground and onto the Earth's surface, forming into stream channel. Thus, surface waters and groundwater are all part of the same hydrologic system. On a day-to-day basis, most of the water flowing in streams and rivers originates as groundwater. When rainfall is particularly heavy and/or the ground is saturated, water **runoff** from the adjoining hillslopes will quickly flow across the land surface and directly into nearby streams and rivers, sometimes leading to flash floods.

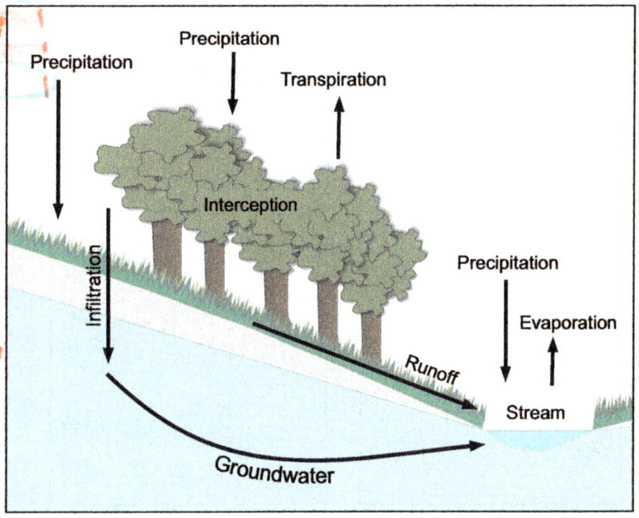

Figure 13.23: This diagram shows the various paths precipitation can take to reach a stream channel. Most of the water becomes part of the groundwater.

Although most precipitation becomes part of the groundwater system, there are a number of other paths water can take as well. When rain falls, some of the water is **intercepted** and absorbed by plants (Figure 13.23). In forested areas, up to 35% of the annual precipitation that falls is intercepted by trees. Some of this intercepted water is used by the plants and some of it returns back to the atmosphere through transpiration. Precipitation may also evaporate back into the air from bodies of water and from land surfaces (like from the soil). In colder climates, snow can accumulate to form glaciers and vast ice sheets, a topic we'll cover in Section 13-6. In this section, we'll look at groundwater in more detail and then in Section 13-5 we'll look at stream and river dynamics.

Groundwater is the single most important source of fresh water for human consumption and use. There is over 650 times as much water in the subsurface as there is in all the streams, rivers and lakes combined. To understand how we can extract water from the ground and how it feeds the network of streams and rivers running across the land surface, we have to first understand how water is stored deep in the ground. The groundwater beneath our feet resides in millions of tiny fractures and voids called **pore spaces** (Figure 13.24). Although the fractures and pore spaces are generally very small, they are capable of storing vast quantities of water. However, not all rocks are good at yielding water when you put a well into them. To understand why, we need to consider two terms: **Porosity** and **permeability**.

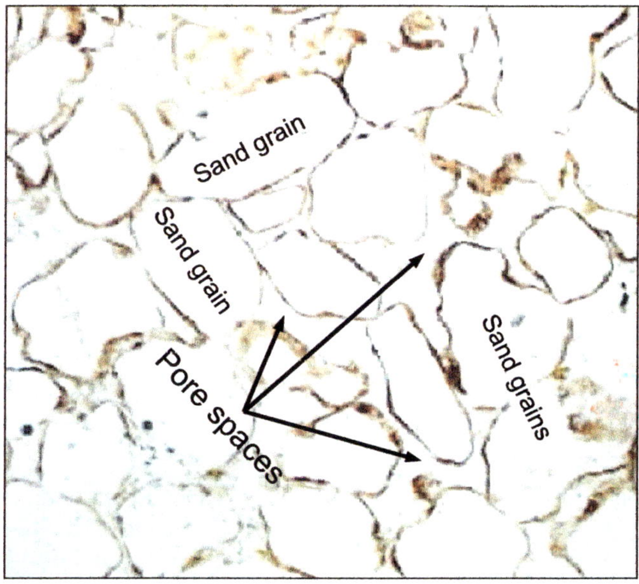

Figure 13.24: A microscopic thin section of a sandstone showing pore spaces. These spaces can collectively hold a lot of water! Photo credit: B. Zaprowski.

Porosity is the measure of how much empty space a rock contains relative to the total volume of the rock. The higher the porosity, the more voids there are which can potentially hold water (Figure 13.25). **Permeability** is a measure of how well connected the pores are. The higher the permeability, the easier it is for water to move through the rock. So, for example, a block of Swiss cheese has lots of large cavities in it (good porosity), but they aren't connected to each other (bad permeability). So although the cheese can hold a lot of water, it isn't going to be easy to get the water out of it. In contrast, consider a sponge. A sponge also has lots of voids in it, but the voids are well connected. Thus, water passes in and out of the sponge very easily. So some rocks are like

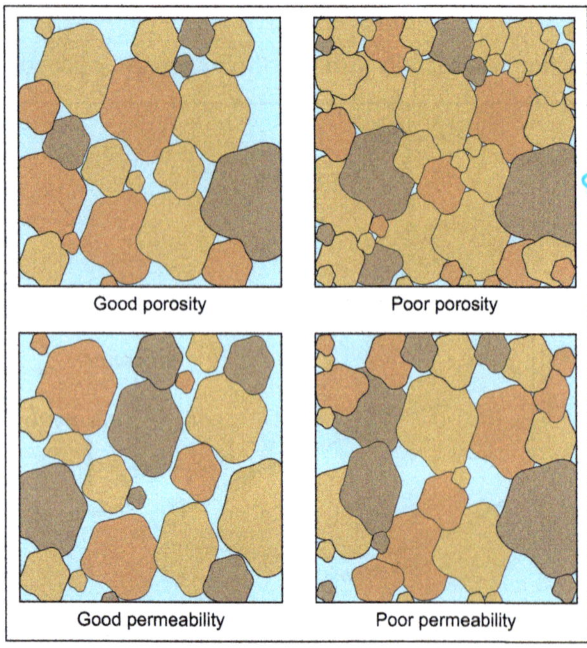

Figure 13.25: Porosity and permeability explained.

Swiss cheese and some are like sponges.

Bodies of rock with good porosity and good permeability ("sponges") are called **aquifers** while bodies of rock with poor porosity or poor permeability (or both) are called **aquitards**. Water passes fairly easily through an aquifer while water can barely pass through an aquitard, if at all. Sand, gravel, sandstone and highly fractured rocks of all kinds usually make good aquifers, while clay, shale, and poorly fractured rocks usually form aquitards. In some locations, the bedrock can be highly variable over small areas, and the juxtaposition of aquifers and aquitards in the subsurface can be quite complicated. Knowing where to drill for water, and how deep to drill, requires a thorough knowledge of the local geology and an understanding of how aquifers work.

An aquifer is kind of like a leaky bucket of water (Figure 13.26). The surface of the water in our "bucket" is called the **water table** because it is relatively flat. If you want the water table to go up in the bucket you have to add water. This process is known as **recharge**. As soon as you stop adding water to the bucket, the water table goes down as water is constantly leaking out of the bucket in a process known as **discharge**. How high the water table is in your bucket at any time depends on how fast the water is being discharged and how fast recharge is occurring. In the real world, precipitation is pretty much the sole source of water to recharge an aquifer. As water infiltrates down through the soil and into the bedrock, gravity pulls it down, filling the aquifer from the bottom up, just like a bucket.

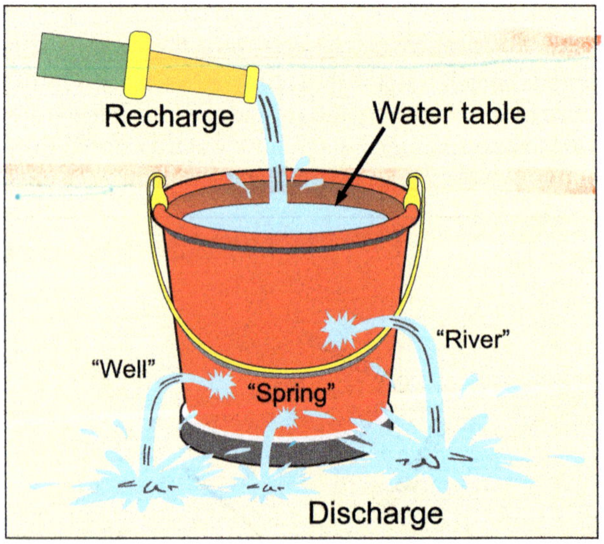

Figure 13.26: An aquifer behaves in a manner similar to a leaky bucket. When water is added to the bucket, the water table goes up. When you stop adding water, the water table goes down as water discharges from the bucket.

Any place where the water table intersects the Earth's surface, water will be discharged and surface waters will form. In fact, the surface of a lake or a stream is really just a place where the water table is exposed above the ground. Streams, rivers, lakes and springs are fed primarily from the groundwater system in this way. When there hasn't been any precipitation in a long time, the water table gets lower and lower until eventually surface waters begin to disappear. So in the case of our "bucket" in Figure 13.26, eventually all of the water would leak out of the bucket and water would stop flowing out of the "leaks." A common misconception is that surface waters "dry up." Although some of the water at the surface is indeed lost through evaporation, it is the lack of groundwater recharge that causes surface waters to go dry.

Wells also are a form of discharge. Just like with surface waters, when the water table drops below the bottom of the well casing, the well will stop producing. That's why it is important that wells are drilled deep enough so the bottom of the well is below the water table's lowest annual height. Unfortunately, sometimes we remove water from the aquifer faster than it can be recharged, dropping the water table to unnaturally low levels. This can lead to a couple problems. First, when aquifers are drained of water, they become compacted from the weight of overlying rock. This can cause the pores and fractures to close up, and once they are closed, they will never be able to open back up and hold water, so the capacity of the aquifer decreases. Second, the compaction of the earth can lead to subsidence of the land surface above. Thus, the groundwater is a valuable, finite resource that must be managed carefully.

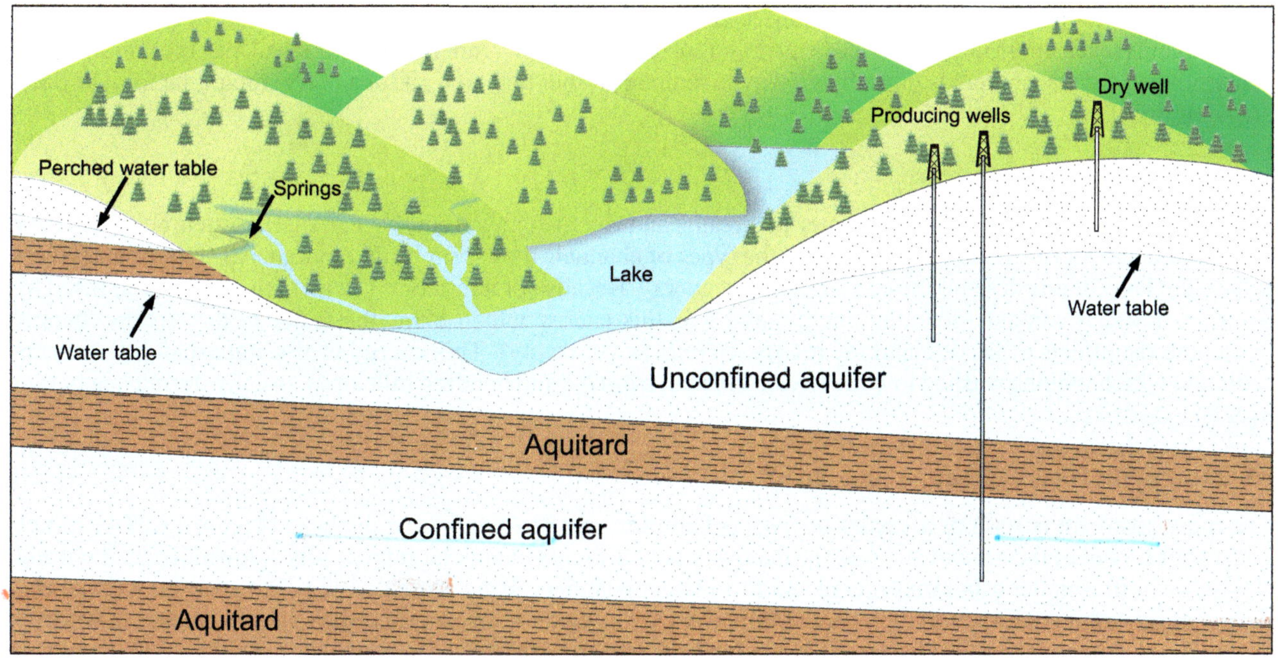

Figure 13.27: Schematic diagram showing a typical profile through the groundwater system.

There are two primary types of aquifers, unconfined aquifers and confined aquifers, and both are potential sources of water (Figure 13.27). **Unconfined aquifers** are the bodies of rock nearest the Earth's surface and are fed by water infiltrating in from directly above. Unconfined aquifers can be highly variable in both thickness and porosity. Sometimes small areas of clay or shale may produce a **perched water table** and as a result, **springs** may form (Figure 13.27). **Confined aquifers** are aquifers which are not recharged directly from the land surface above. Instead, water enters the aquifer in a recharge area at some other location, sometimes thousands of miles away. In many places, the subsurface is divided into multiple layers of aquifers separated by aquitards. The Delmarva Peninsula of Maryland and Delaware is a good example of this (Figure 13.28). In this case, the recharge area for the lower Delmarva Peninsula is in northern Delaware and the western shore of the Chesapeake Bay. Unfortunately, in many places the groundwater in the unconfined aquifer is polluted to the point where it can't be used by humans.

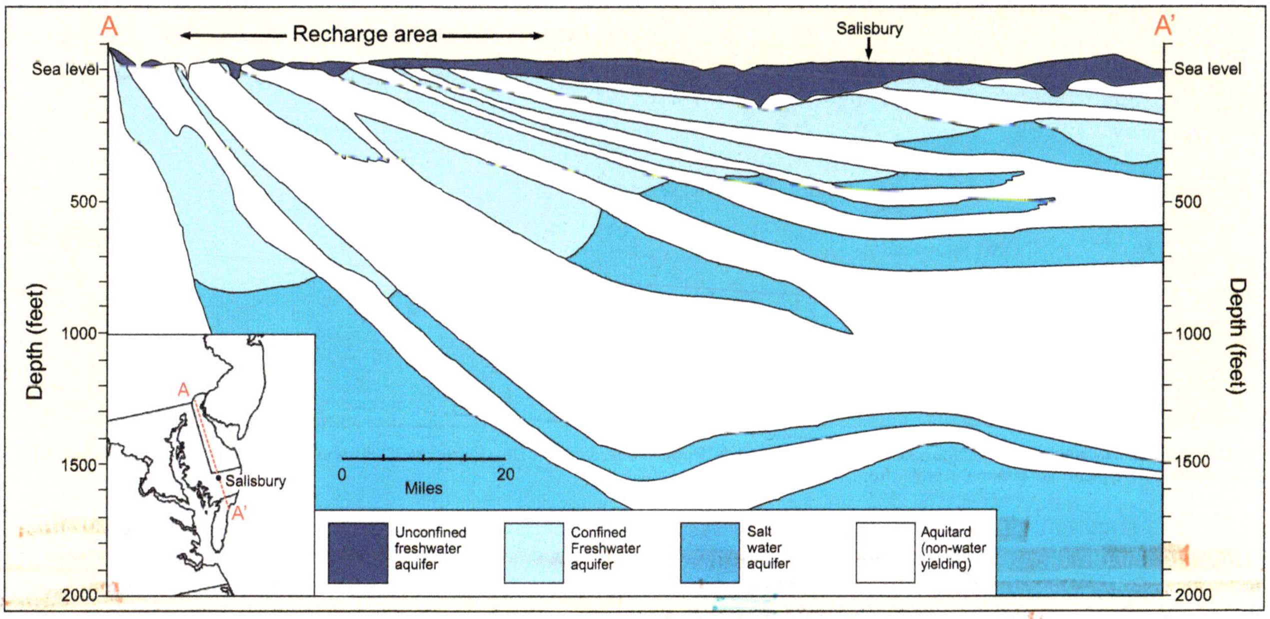

Figure 13.28: Cross-sectional profile of the Delmarva Peninsula and its groundwater resources. Most of the aquifers are recharged in the northern part of the peninsula and then flow southward. Because of its close proximity to the ocean, some aquifers are saline. Modified from U.S. Geological Survey Professional Paper 822.

When precipitation soaks into the ground, it can carry surface contaminates into the subsurface. If these contaminants reach the groundwater, they can pollute an entire aquifer, rendering it unusable. Once the pollution has contaminated the water, there is virtually no way to remove it other than to wait for a (very) long time for it to flush out naturally, if ever.

Section 13-5: River Dynamics

Now that we've discussed where the water in streams and rivers comes from, we can discuss their behavior over time. The behavior of streams and rivers and the types of channels they form is not random. All streams and rivers follow the laws of physics and the conservation of energy. Because precipitation is water which has been lifted up from sea level, it contains potential energy, and it is this energy which helps the water in the stream channel overcome the effects of friction and the resistance to flow (viscosity). This energy is also important for carrying sediment which is being washed into the stream channel from the surrounding hillslopes, usually during heavy rain events.

Anatomy of the River System

The most upstream portion of a stream or river is known as the **headwaters** or the head (Figure 13.29A). When water trickles out of the ground and into the headwaters of a stream channel, gravity pulls it downhill. As the water flows downhill, smaller tributaries merge and form larger channels. Thus, streams and rivers grow both wider and deeper as the volume of water in their channels increases. The water in streams and rivers also increases in velocity as you move downstream. This happens because there is less friction to overcome in a larger channel, so the water can flow faster. Geologists use the term **discharge** to describe the volume and speed of water passing by a specific point along a river channel (usually in terms of cubic feet per second). Eventually the stream or river reaches a large river, a lake or the ocean, forming the **mouth**. The elevation where a stream or river ceases to exist (that is, the elevation of the mouth) is known as **base level** (Figure 13.29B). The change in elevation along the channel from the head to the mouth is known as the channel's **gradient**, which is just another term for its slope. In general, stream channels are steeper towards their headwaters and less steep towards their mouths (Figure 13.29B). The entire land surface which contributes to the water flow out of the mouth of a given river is known as a **drainage basin** or sometimes as a **watershed** (Figure 13.29A). Drainage basins form when rivers and streams erode down into the landscape, and they are separated by **drainage divides**.

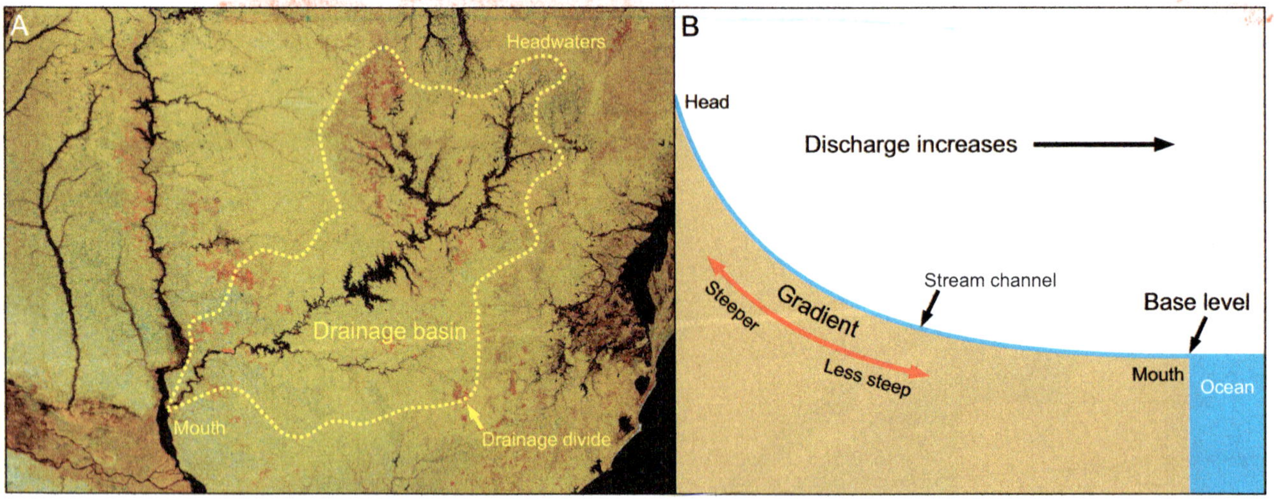

Figure 13.29: A. An example of a drainage basin and its different parts. Photo credit: Jacques Descloitres, MODIS Land Rapid Response Team, NASA/GSFC. B. The nature of a typical river channel in profile from its head to its mouth.

In an ideal situation, stream channels are perfectly balanced so that they are the right width, depth and gradient (slope) to handle the discharge of the river *and* the sediment supplied by erosion *while using the least amount of energy possible*. We refer to this condition as **grade**. When a stream is at grade, it is neither eroding downwards nor depositing sediment, although a river may erode laterally (side to side), widening the river valley and depositing sediments in the floodplain (Figure 13.30). *Most streams and rivers are rarely at grade at any given time but they*

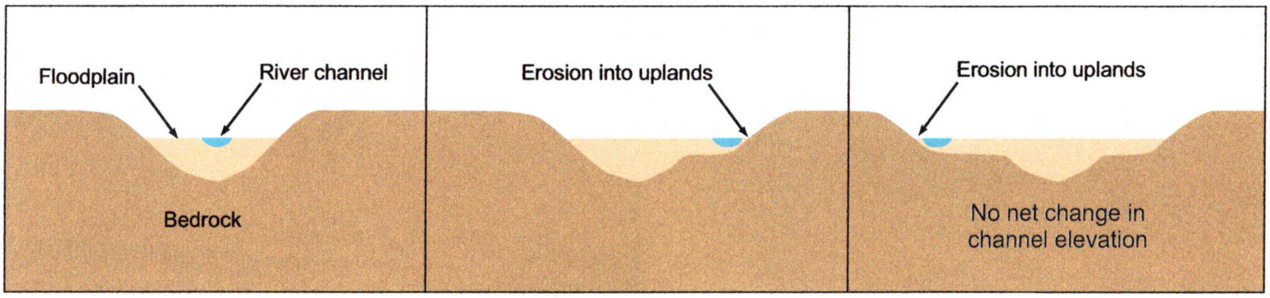

Figure 13.30: Lateral (side to side) erosion by a stream causes a valley to get wider over time but not necessarily deeper. Lateral erosion occurs when a stream approaches its graded state.

constantly seek to reach the graded state, and their behavior is directly related to this concept. When some part of a stream's condition changes, the stream will attempt to seek grade again by changing one or more of the stream's characteristics (i.e., width, depth, slope, etc.). It is important to note that the stream will not necessarily try to return to the same graded state as before but rather the stream will attempt to reach a new graded state which balances the current conditions imposed on the stream by the surrounding environment. So how can grade be disrupted? There are three ways:

1. A change in discharge: This could occur because precipitation patterns change or some of the water is diverted from the channel for human consumption or irrigation. For example, the lower parts of the Colorado have gone dry because huge amounts of water have been removed for irrigation (Figure 13.31A).

2. A change in sediment delivery to the channel: This could occur because of increased erosion due to farming, construction, logging, or forest fires; or it could occur because of a decrease in sediments due to upstream damming. For example, many of the world's rivers have gotten much shallower as sediments fill them in and in order for shipping lanes to be maintained, periodic dredging is required (Figure 13.31B). In other areas, rivers downstream from dams are experiencing sediment starvation. This has an impact on plants and animals that live near or in river channels.

Figure 13.31: Examples of how human activities impact river systems. A. intensive irrigation has reduced the Colorado River to a trickle. Photo credit: Robert Simmon/NASA. B. Deforestation in Madagascar is filling in and shallowing the Betsiboka River. In fact, ocean-going ships were once able to travel up the Betsiboka estuary, but must now berth at the coast. The channel pattern has also changed this braided pattern. Photo credit: NASA.

3. A change in base level: Streams and rivers are also graded to their base level. Changing the elevation of base level will cause a stream to fall out of its graded condition. A rise in base level will result in the deposition of sediment in the stream channel. A drop in the base level will produce a **knickpoint** (Figure 13.32A). A knickpoint is a distinct and sudden change in a stream's gradient. This knickpoint will then erode headward, changing the stream's elevation until it reaches a graded state again. In some cases, the knickpoint is very steep and forms a waterfall, like Niagara Falls. Niagara Falls exists because about 10,000 years ago, the lake level of Lake Ontario dropped down lower than the level of Lake Erie. The Niagara River is now eroding headwards towards Lake Erie in an attempt to reach a new graded state (Figure 13.32). In other cases, the knickpoint just forms inconspicuous rapids.

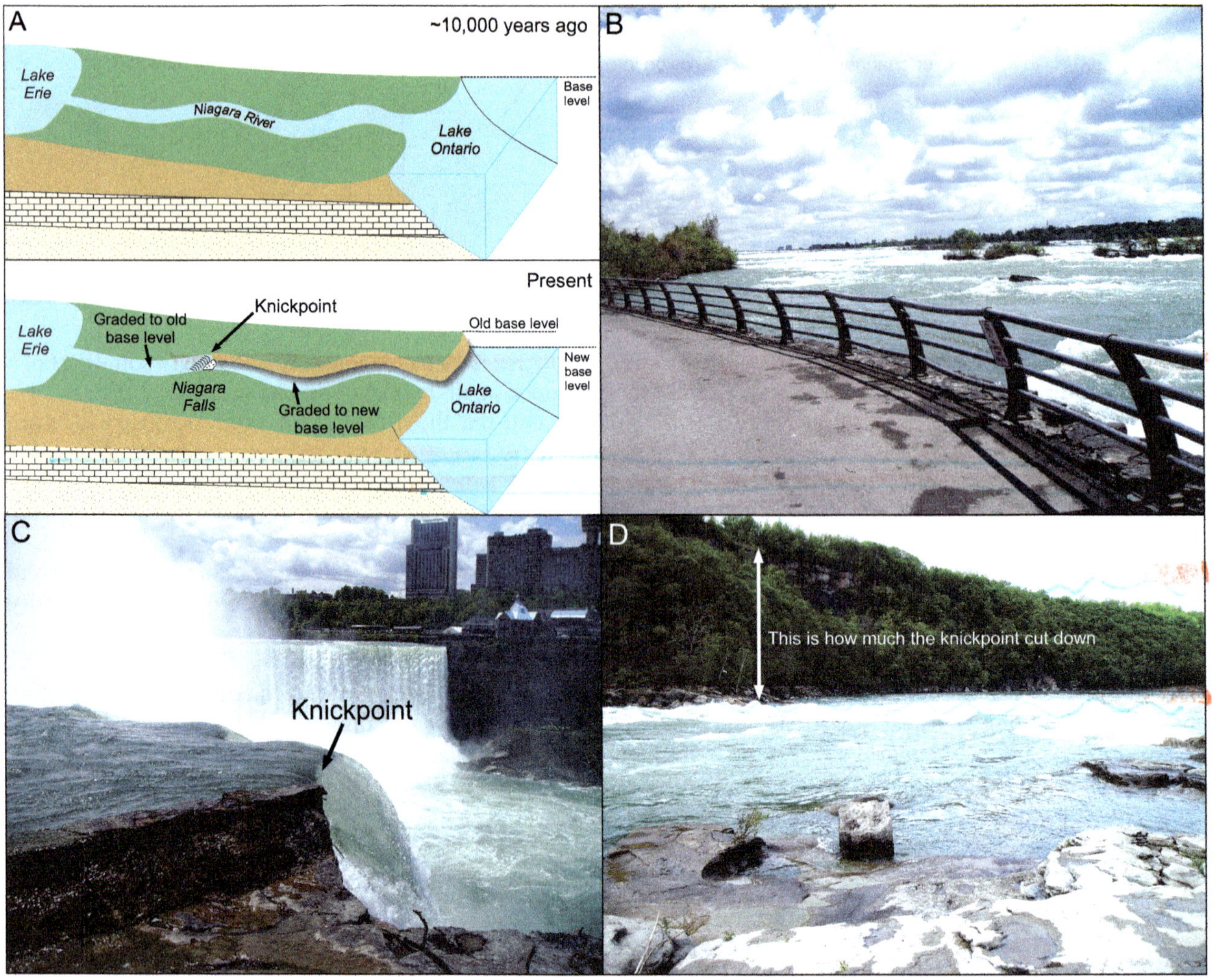

Figure13.32: A. A drop in base level results in erosion of the channel and the formation of a knickpoint. It is at the knickpoint where erosion and lowering of the channel occurs. A knickpoint can form a waterfall like Niagara Falls.B. Looking upstream from the lip of Niagara Falls. Notice how wide the river is - this portion of the river is not graded to the new base level yet. C. The lip of Niagara Falls at Terrapin Point. Here, the water falls 167 feet to the river below. D. The Niagara Gorge a few miles downstream from the falls. The walls of the gorge show how much downcutting needed to occur for the river to reach its new base level. Photo credits: B. Zaprowski.

Recurrence Intervals and the 100-year Flood

The highly variable nature of streams and rivers over time makes them a powerful and sometimes deadly force to be reckoned with. As streams and rivers cut down through the landscape, they produce a variety of different landforms worthy of discussion. The most obvious **fluvial** (fluvial means relates to rivers) landform is a **floodplain** (Figure 13.33). A floodplain is any area of land adjacent to a stream or river channel which is subject to flooding. On average, all floodplains are inundated with water *to some extent* twice every three years. Because they flood frequently, geologists study and measure streams and rivers to predict which areas are at high risk for flooding. Let's look closer at how they do that.

Figure 13.33: The Amazon River floodplain during a flood. Image credit: Image Science and Analysis Laboratory, NASA/JSC.

It is not uncommon to hear a meteorologist on TV discuss a "100-year flood" or a "100-year storm" when an especially large storm event is occurring. These terms can be misleading - a **100-year flood** does not occur once every 100 years; instead there is a 1 in 100 (or 1%) chance of a 100-year flood occurring *in any given year* (Table 13-1). Given the rather unpredictable nature of streams and rivers, there is no realistic way to predict with any certainty when a given stream or river will flood, but geologists can predict which areas of land will be flooded during such an event and how frequently. So if you live in the 25-year floodplain, there is a 4% chance that in any given year a flood would inundate your property. This knowledge helps to protect lives and property from future flood damage.

Recurrence interval (years)	Probability of occurring in any given year	% chance of occurring in any given year
500	1 in 500	0.1%
100	1 in 100	1%
50	1 in 50	2%
25	1 in 25	4%
10	1 in 10	10%
5	1 in 5	20%
2	1 in 2	50%

Table 13-1: This table shows how a stream's recurrence interval is related to the chances of a flood of a given size occurring in any given year.

Geologists collect data from thousands of **gaging stations** around the country in order to predict how frequently a flood of a given size may occur on a given river (Figure 13.34A). These data can then be used to make flood maps for the land areas adjacent to that river. For example, consider Figure 13.34B which is a plot of recurrence interval vs. water height for a river in Maryland. Based on 52 years worth of data, we can predict that there is a 1% chance that the Pocomoke River will flood ~16.2 feet above its normal level. Thus, people living at an elevation of 16 feet above the river have a 1 in 100 chance of being flooded in any given year. Of course there are many factors which dictate whether or not a 100-year flood will occur with any given storm. 100-year storms don't always produce a flood and 100-year floods can occur without a 100-year storm. Factors such as soil saturation, vegetation, and slope also play a role in how high a flood may get during a given storm event.

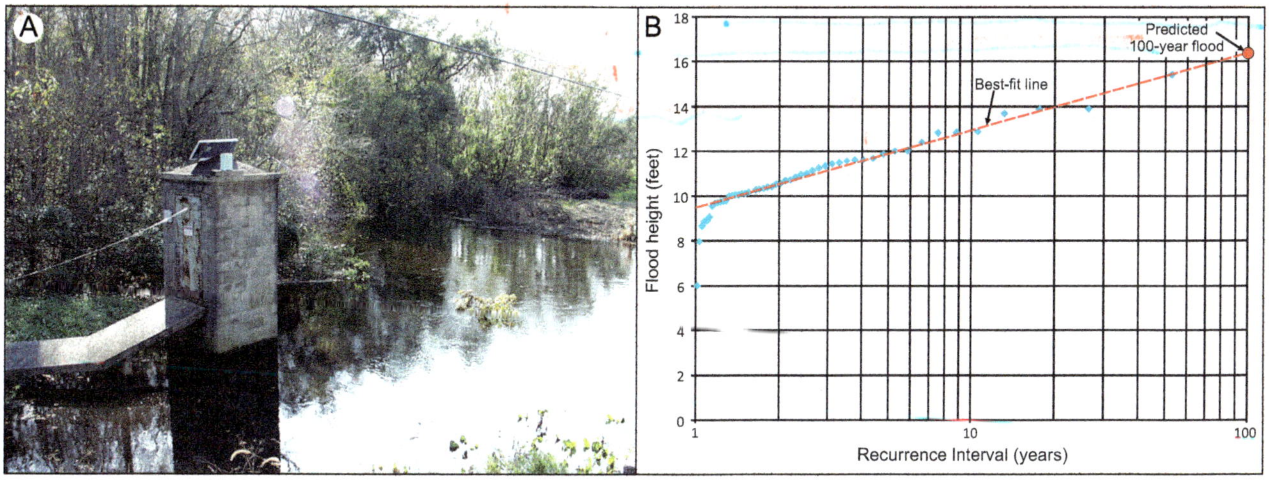

Figure 13.34: A. A gaging station on the Pocomoke River, MD. The river happens to be flooding in this picture. Photo credit: B. Zaprowski. B. A flood-frequency diagram for the Pocomoke River based on 52 years worth of flood data from the gaging station in A. Based on these data, we can predict that a flood of ~16.2 feet above the normal channel elevation has a 1% chance of occurring in any given year.

Floodplain Features

Flood events deposit sediments called **alluvium** on the floodplain surface, causing it to build up over time (Figure 13.35A). The coarsest sediments are deposited just adjacent to the channel, forming a natural levee, while finer-grained sediments like clay are deposited in quieter areas of the floodplain away from the channel. The size of the floodplain is a function of the size of the channel such that floodplains become wider and more well developed as you go downstream. We should point out that not all streams and rivers have floodplains. Floodplains take time to form and develop, so "young" rivers which are actively downcutting have not been stable long enough form a floodplain (e.g., the Colorado River in the Grand Canyon). It is important to note that *streams and rivers do most of their work during flood events* (Figure 13.35B). On a normal, day-to-day basis streams and rivers move very little sediment downstream.

Figure 13.35: A. Erosion has exposed the layers of alluvium which underlie this old floodplain deposit along the Belle Fourche River in South Dakota. B. This stream is carrying some very muddy water following an overnight rainstorm. Most of the time this stream is crystal-clear. Photo credits: B. Zaprowski.

Clearly, the most obvious feature on a floodplain is the stream or river channel. There are four principle types of stream channels, and each can produce their own unique landforms within a floodplain.

Most streams and rivers display a sinuous or snake-like patten as they flow across the land surface. These channels are said to have a meandering pattern (Figure 13.36). Meandering channels are the most common type of channel pattern because *meandering streams actually use less energy than straight channels*. As rivers meander across the floodplain, they redistribute sediment by depositing it in some places and eroding it from other places. Most of this sediment movement occurs in the **meander bends** of a river. On the inside bend of a meander, water velocity is low so sediments are deposited in the form of **point bars** (Figure 13.36). On the outside bend of the meander, the water velocity is higher, so erosion dominates forming **cut banks**. It is through erosion of the cutbank and deposition on the point bar that a river's channel can meander over time. Sometimes the meander bends become so exaggerated that they are cut-off and abandoned, forming **oxbow lakes** (Figure 13.36). Over time, these oxbows fill in with sediment, becoming **meander scars**.

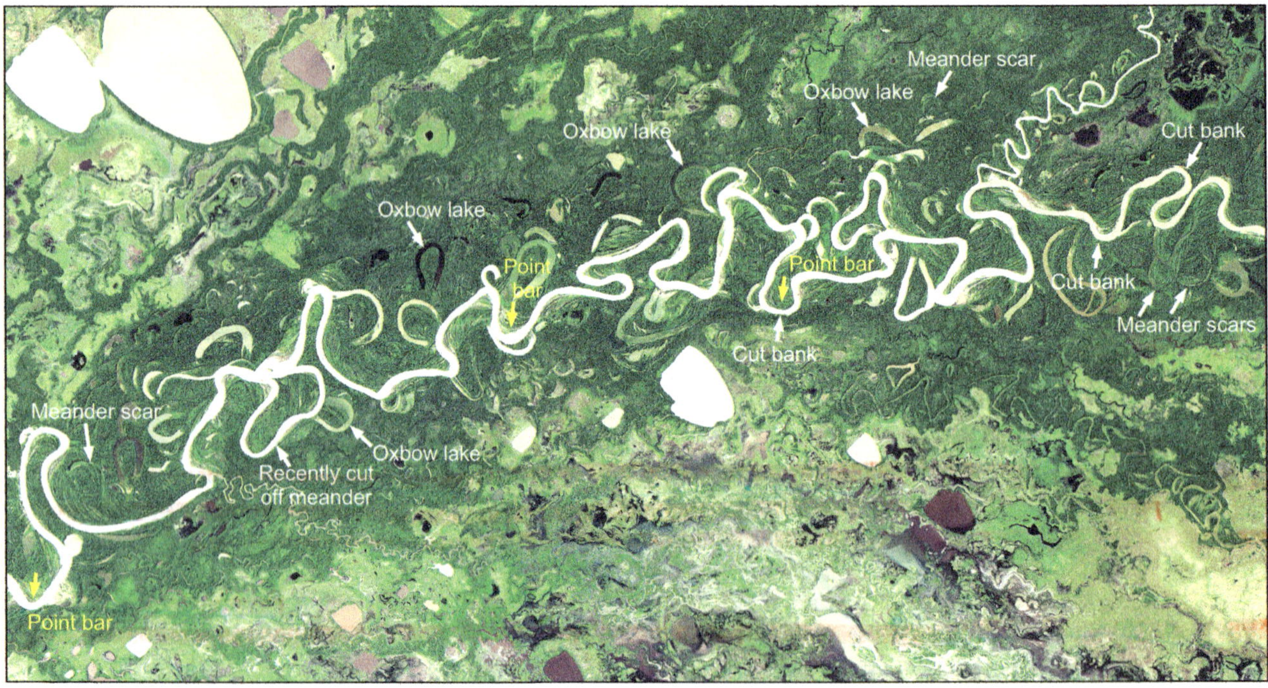

Figure 13.36: This image shows some of the common floodplain features found along a meandering river. Note that still bodies of water are black-colored or muddy and brown-colored, while flowing water appears white. Photo credit: Jesse Allen/NASA.

Figure 13.37: Examples of straight channels. A. The Zambezi River upstream from Victoria Falls has several straight channels. The channels are following weaknesses in the underlying rock. Photo credit: NASA. B. A newly forming alternate bar in the Grand Canyon. The dam upstream was opened up and sediment was allowed to flush out of the reservoir, replacing the alternate bars that had disappeared because of sediment starvation. Photo credit: Matt Kaplinski/Northern Arizona University

Straight channels can occur but they are rare and are not likely to have floodplains because most straight channels are found in narrow canyons or gorges where the bedrock forces them along a straight path (Figure 13.37A). Streams rarely stay on a straight course for more than a few miles before changing direction. Even though they are "straight" they still have a slight wiggle as they pass around **alternate bars** along the sides of the channel (Figure 13.37B). These alternate bars provide critical habitats for plants and animals that live along the river.

In some scenarios, streams and rivers are overwhelmed by sediment. They simply don't have enough water in their channels to carry all the sediment being washed into the channel from the adjoining hillslopes; they are literally "choking" on sediment. The water pushes its way through the piles of sediment any way it can, forming **braided channels** (Figure 13.38A). Braided channels have numerous **braid bars** which are constantly changing and shifting over time (Figure 13.38B). Braided channels are most common in mountainous areas and downstream from active glaciers. They typically display a "flashy" pattern - that is - they have one big flood each year (usually during the melt season in the spring) and then low-flow conditions the rest of the time.

In some cases, a river may form stable, permanent islands, and the river will display an **anabraching** pattern (Figure 13.39). These types of rivers form in a wide variety of settings, but typically they are most common in areas where frequent floods occur and the sediment in the river is mostly fine-grained.

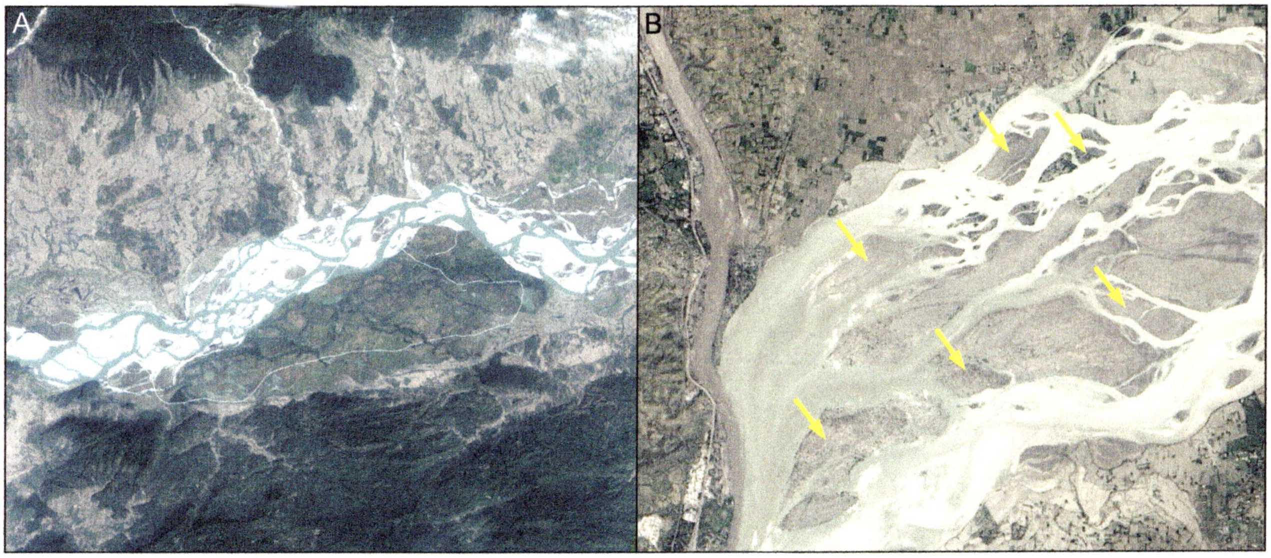

Figure 13.38: A. This image from the Himalayas is a good example of a braided stream channel. B. A closer view of braid bars (yellow arrows). Photo credits: Robert Simmon/NASA.

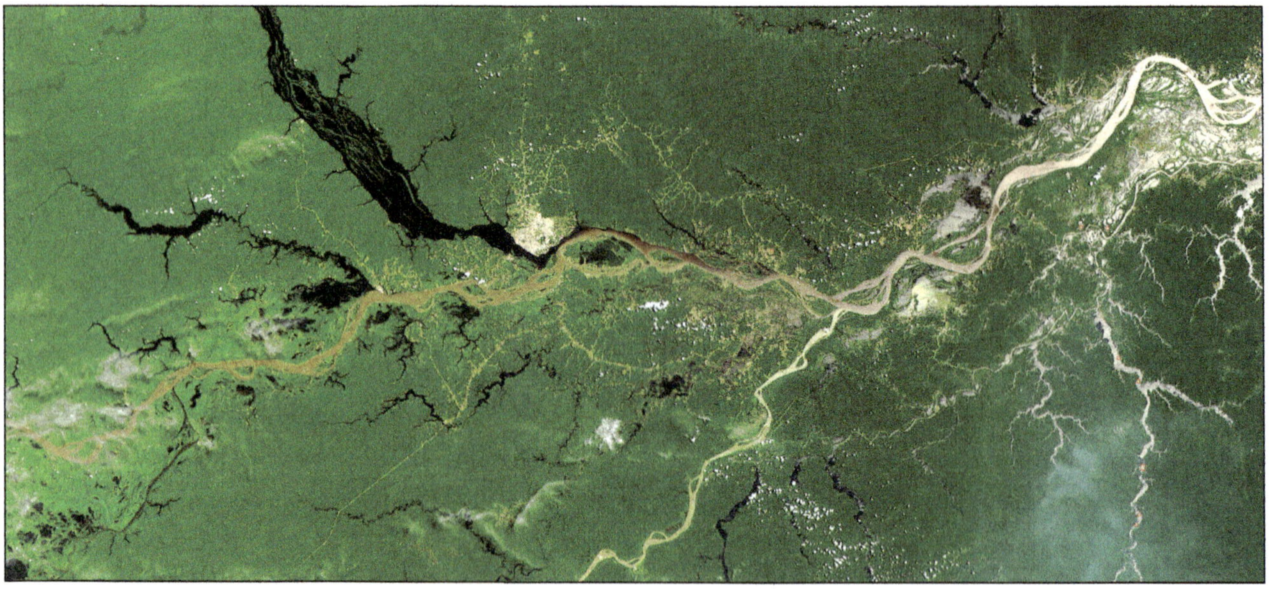

Figure 13.39: This stretch of the Amazon River displays an anabranching pattern with it stable islands. Photo credit: Jacques Descloitres, MODIS Land Rapid Response Team, NASA/GSFC.

One final note about stream channels. Not all stream channels have water in them at all times during the year. For instance, stream channels in arid climates like the desert southwest almost never have water in them. We called these streams **ephemeral** and ephemeral streams only have water in them after a large rainstorm (Figure 13.40A). Although they don't normally have water in them, ephemeral channels can be particularly dangerous because **flash floods** may occur without warning. If you travel in arid lands, be sure to never drive your car through standing water on the road, as even a few inches of fast-moving water can sweep your car away. You should also be very careful traveling in canyons because you can become trapped very quickly by rising waters in these settings (Figure 13.40B). In humid climates, **perennial** streams always have water flowing in them except in the case of extreme droughts. In between these two extremes are **intermittent** streams, that is, streams which flow seasonally. Semi-arid locations like the midwestern United States typically have streams which flow in the winter and spring and then go dry in the late summer and fall.

Section 13-6: Glaciers

Glaciers are commonly described as "giant dirt machines." When glaciers flow across the landscape, they break up large quantities of bedrock, producing immense quantities of sediment (Figure 13.41A). This sediment is then deposited, forming unique glacial landforms and landscapes. In fact, almost 10% of the Earth's surface is covered by

Figure 13.40: A. An ephemeral channel in the Mojave Desert. B. This sign warns hikers and drivers about the flash flood risk in this Utah canyon. Photo credits: B. Zaprowski.

Figure 13.41: A. Glacial sediments melting out of a glacier in Alaska. Photo credit: Sarah Burton. B. Glacial erratics scattered across the Lamar Valley in Wyoming. Photo credit: R.G. Johnsson/USGS.

glacial sediments. In the United States, old glacial deposits from the Ice Ages can be found as far south as Kansas. It is not unusual to see huge boulders called **glacial erratics** sitting out in the flat, open plains of the Midwest hundreds of miles away from their source in Canada (Figure 13.41B). So how can a mass of ice carry boulders for hundreds of miles? How can a supposedly "solid" block of ice move at all? Let's investigate.

Glacial Dynamics

Glacial ice begins its life as snow. However, one critical step is needed for snow to become glacial ice: At the end of each summer, there must be snow left over from the previous winter (Figure 13.42A). This unmelted snow is then buried by the following winter's snowfall. Over the course of centuries, layer upon layer of snow accumulates in the **zone of accumulation** (Figure 13.42B). It takes anywhere from 100-400 years for the snow to be slowly squeezed into glacial ice. Once the glacial ice grows to a thickness of about 50-100 meters, the crystalline structure of the ice can no longer support the weight of the snow above and it deforms plastically. The glacial ice is then able to start flowing under the influence of gravity. In this way, a glacier is kind of like a form of mass wasting. Once a critical mass is reached, the mass of the ice and the pull of gravity overcomes the resistence of the ice to flow. Slowly the ice will flow away from its source of origin as the snow continues to pile up over time. Eventually, the glacial ice will reach the **zone of ablation** (Figure 13.42B). Here is where the glacier meets its final demise as the ice melts away and deposits the sediment that it was carrying. This sediment is called **till** (Figures 13.42B and13.43A).

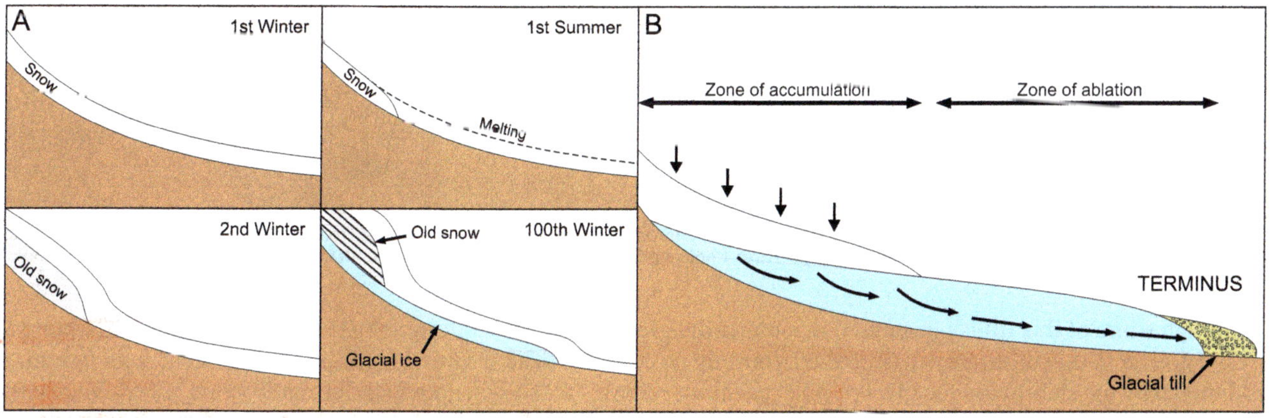

Figure 13.42: A. The formation of glacial ice from snow. It takes approximately 100-400 years for glacial ice to form this way. B. Basic glacial flow. Glacial ice forms in the zone of accumulation and melts away in the zone of ablation, leaving sediment behind.

The end of a glacier is called the **terminus** (Figure 13.42B) and the location of the terminus changes over time as climatic conditions change (Figure 13.43B-D). When more ice is created in the zone of accumulation than is being removed in the zone of ablation, the terminus moves forward and the glacier is said to be **advancing** (Figure 13.43C). When ice is melting faster in the zone of ablation than it is forming in the zone of accumulation, the glacier is said to be **retreating** and the terminus moves back towards the source of the ice (Figure 13.43D). It is important to

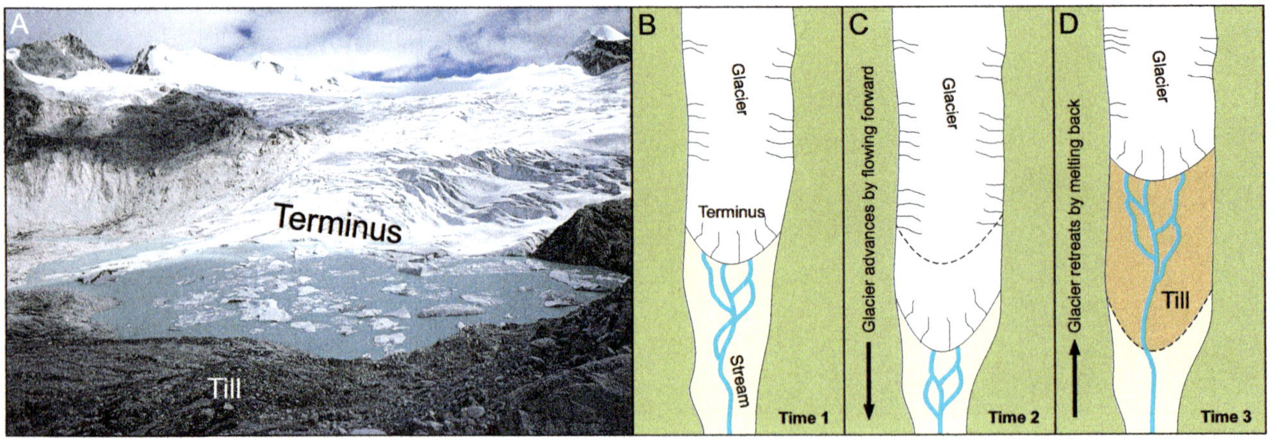

Figure 13.43: The movement of glaciers over time. A. The terminus of a small glacier at Gangrinchemzoe Pass, south of the main Himalayan divide, Bhutan. Note the dark-colored deposits of till in the foreground, which indicate this glacier is in retreat. Photo credit: Shuji Iwata, Tokyo Metropolitan University, Japan/USGS. B. Initial position of a terminus. C. Glacier advances by flowing forward. D. The glacier retreats by melting back, leaving behind a layer of till.

stress here that although the position of the terminus moves back over time, *the ice is always flowing downhill*. It is a common misconception that glaciers flow uphill when they are retreating, an idea that completely ignores gravity and scientific observations.

Because of its large mass, pieces of rock get embedded in the ice at the bottom of the glacier, grinding large **glacial grooves** into the bedrock as the glacier flows over it, further increasing the erosion process (Figure 13.44A). Smaller pieces of sediment trapped under the ice act like sandpaper and grind the rock down even further, producing a fine dust called **rock flour** (Figure 13.44B). All of the rocky debris produced and deposited by a glacier is collectively called till. Till is a unique sedimentary deposit because it has a wide variety of different-sized sediments; you can find clay, sand and boulders all mixed together in the same deposit (Figure 13.44C).

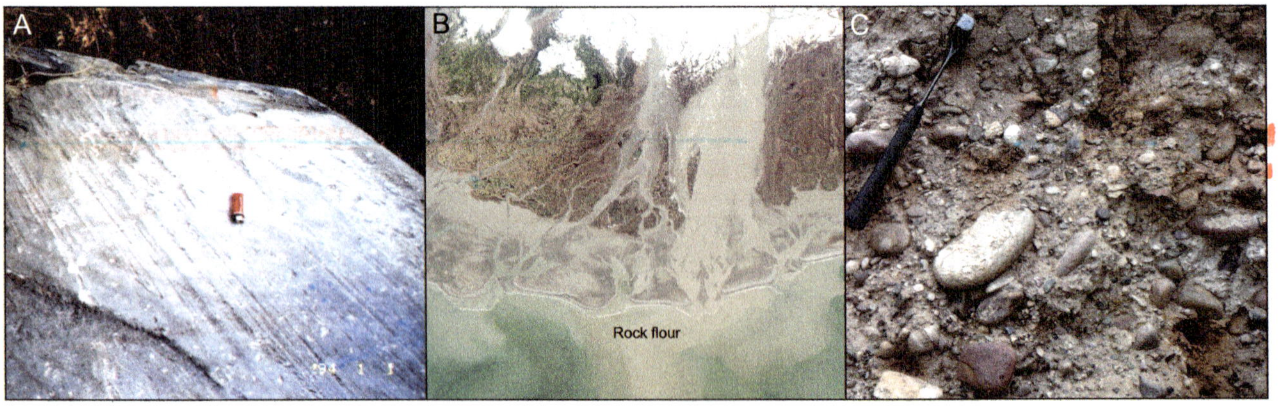

Figure 13.44: The effects of glaciers on bedrock. A. Glacial grooves on a bedrock outcrop in Massachusetts. Photo credit: B. Zaprowski. B. Plumes of rock flour are being washed into the ocean at the mouth of the Copper River. Photo credit: Robert Simmon and Jesse Allen/NASA. C. A close-up example of till. Photo credit: .

Although some of this till is plastered down under the glacier as it flows along, most of the sediment is transported by the glacier to the terminus and deposited, forming an **end moraine** (Figure 13.45A). The size of the end moraine is dictated by the stability of the glacier. If the terminus remains at the same position for many years, a very large end moraine can form (e.g. Long Island). Beyond the end moraine is the **outwash plain**, the place where melting glacial waters merge to form a river (Figure 13.45B). Outwash plains commonly have braided streams choked with coarse sediment. Sometimes glacial ice flows directly into a large lake or into the ocean, forming icebergs (Figure 13.45C).

Glaciers occur on a variety of spatial scales but generally speaking, we can place all glaciers into one of two groups: Glaciers **confined by topography** and glaciers **unconfined by topography**. Confined glaciers are generally small glaciers found in mountainous regions where the glacial ice is confined in valleys by high mountain peaks and ridges. Unconfined glaciers are large ice sheets that cover the entire land surface except perhaps the tops of the

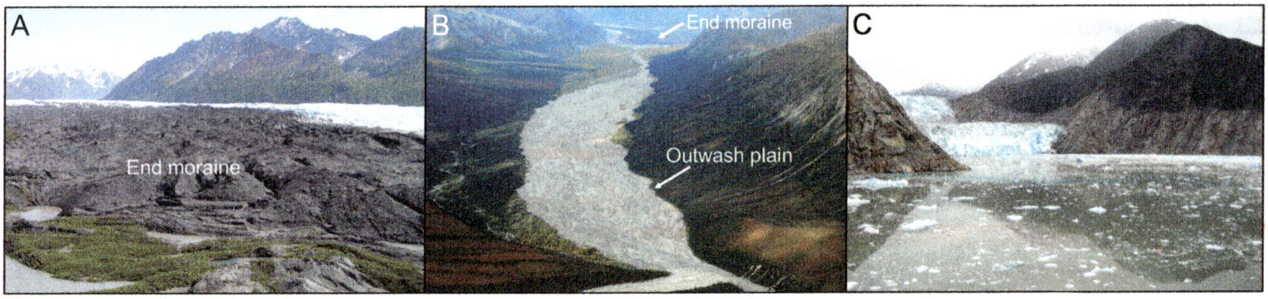

Figure 13.45: Glacial processes at the end of a glacier. A. This glacier has built up a sizable end moraine. Photo credit: Sarah Burton. B. An outwash plain in Alaska. Photo credit: Don Becker/USGS. C. Icebergs float away from the terminus of a glacier. Photo credit: Michelle Saunders.

highest peaks. The smallest glaciers, called **cirque glaciers**, are small and rarely travel more than a few miles (Figure 13.46A). Many times these cirque glaciers merge together forming **valley glaciers**. Valley glaciers may have ice streams hundreds of meters across and they may travel tens of kilometers (Figure 13.46B). Mountain tops may also have **ice fields** covering all but the tops of the mountain peaks (Figure 13.46C). Ice fields commonly have multiple valley glaciers draining them. The largest glaciers occur as continental **ice sheets**. These massive sheets of ice can be 3,000-4,000 m thick and cover thousands of square kilometers, like the sheets of ice that cover Antarctica and Greenland. All but the highest peaks are covered by the ice (Figure 13.46D). It was massive continental-sized ice sheets which deposited the glacial erratics in Figure 13.41B.

Figure 13.46: Types of glaciers. A. A small cirque glacier. Photo credit: USGS. B. In northern Tajikistan, several smaller cirque glaciers (arrows) are feeding a larger valley glacier. Photo credit: NASA. C. An ice field on Baffin Island feeds multiple valley glaciers. This false-color image shows vegetation in red. Photo credit: Jesse Allen/NASA. D. The southern tip of the Greenland ice sheet. Greenland is one of only two places on Earth where ice sheets are found (the other is Antarctica). Photo credit: Jacques Descloitres, MODIS Land Rapid Response Team, NASA/GSFC.

Glacial Landscapes

Although glaciers only cover a small portion of the landscape, they alter it in fundamentally different ways than other types of erosion, and produce a unique landscape with unique landforms. In this final section of the chapter, we are going to *briefly* examine those features. Confined glaciers like cirque glaciers and valley glaciers produce a very rugged landscape with drastic relief. Unlike streams, which can only erode at the narrow contact between the channel and the bedrock below, glacial ice is in contact with the entire bottom of the valley as it flows past. This accelerates the erosion process on the valley walls and produces a deep, U-shaped valley with a flat bottom and steep hillslopes (Figure 13.47A). Many mountainous areas owe their rugged nature to erosive power of glaciers. In areas where several glaciers sit adjacent to each other (like in Figures 13.46B and C), each smaller ice stream that contributes to a valley glacier also erodes the valley it flows in. When the erosional fronts from the widening, deepening valleys merge, a thin, jagged ridge called an **arête** forms between them (Figures 13.47B and 13.47C). Where three or more arêtes meet, a peak called a **horn** will form. Small lakes called **patternoster lakes** commonly form in glaciated valleys as the glacier retreats. The lakes are dammed by the terminal moraines left behind by the retreating ice (Figure 13.47D). Along coasts, these glacially-carved valleys may be flooded by the ocean to form

Figure 13.47: Examples of mountainous glacial landforms. A. This up-valley view of the Toklat River in Denali National Park shows the typical U-shape that glaciated valleys commonly display. Photo credit: Jacob W. Frank/NPS. B. An example of arêtes. Photo credit: NPS. C. The archipelago of Svalbard displays a classic example of a glaciated terrain. Photo credit: Joshua Stevens/NASA. D. A paternoster lake in Rocky Mountain National Park. Photo credit: NPS. E. Fjords off the coast of Norway (yellow arrows). Photo credit: Jeff Schmaltz/MODIS Land Rapid Response Team, NASA/GSFC.

embayments called **fjords** (Figure 13.47E).

Unlike confined glaciers, which only deepen and widen valleys, ice sheets modify the entire land surface, eroding down the bedrock in both the valleys and on the uplands. Just like valley glaciers, continental ice sheets can also gouge out very deep valleys as they flow over the land. The main difference is that the ice sheets round off the uplands, shaping them into streamlined hills which are orientated parallel to ice flow (Figure 13.48A). This results in a landscape with broad, flat valleys, steep hillslopes and rounded uplands (Figure 13.48B). Many of these valleys have large terminal moraine deposits which dam up rivers, causing the valleys to fill with water. The Finger Lakes in central New York State formed this way (Figure 13.48C). Glacial ice sheets also deposit vast quantities of till, resulting in areas with rolling topography and large, stream-lined hills called **drumlins** (Figure 13.48D). In some cases, the deposits of till are so large they form whole islands, like Long Island, NY.

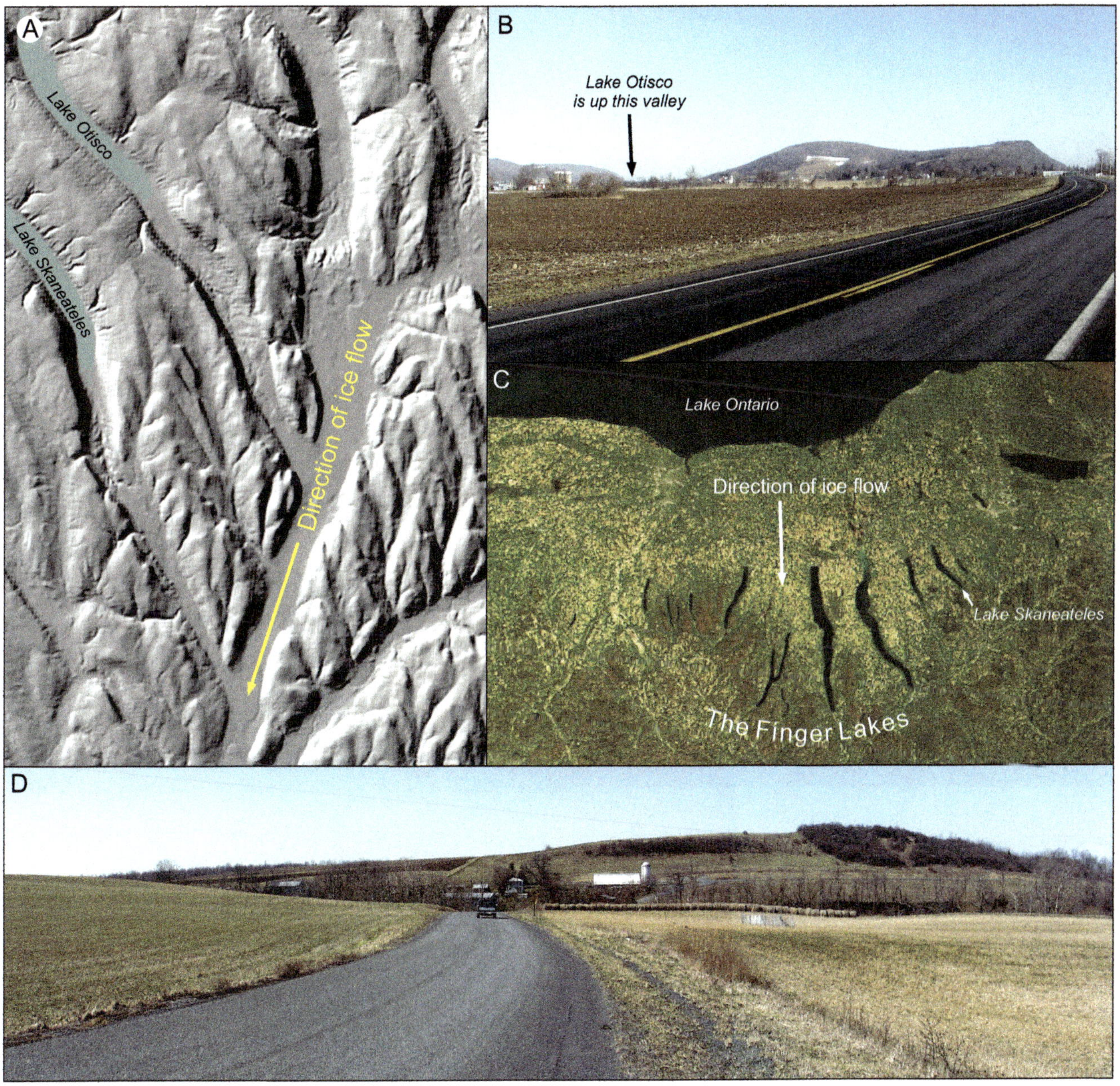

Figure 13.48: The effects of continental ice sheets on a landscape. A. This digital elevation model from upstate New York State shows the deep valleys called glacial troughs that are common in areas eroded by ice sheets. B. A ground-level photo of the valley visible in center of image A. Note the flat valley bottom and steep valley walls. Photo credit: B. Zaprowski. C. The Finger Lakes of New York formed in long, deep glacially-carved valleys that are dammed at their southern ends by massive till deposits. Photo credit: MODIS Land Rapid Response Team, NASA/GSFC. D. This large hill, called a drumlin, is made of sand and gravel that was deposited by a retreating ice sheet. Photo credit: B. Zaprowski.

Chapter 14 - Oceans and Coasts

Oceanography is a broad subject which encompasses all of the physical, chemical and geological aspects of the ocean waters and the basins which contain them, including the coastal zone. In this chapter, we will first look at the attributes of the ocean waters and how currents form. Then we'll look at waves and tides. Finally, we'll look at the world's coasts and the landforms commonly found there.

Section 14-1: The Nature of Seawater

In this section we are going to look at seawater temperature, salinity and density and discuss what causes them to vary both temporally and spatially.

Temperature

At the Equator, the sun's rays are the strongest and most intense throughout the year and the solar intensity decreases with increasing latitude towards the poles. It should come as no surprise that the oceans' **surface temperatures**, by and large, are warmest near the Equator and decrease towards the poles (Figure 14.1A). It should be noted that this pattern doesn't hold true everywhere because surface currents mix and redistribute ocean water (more on that in the next section). Even so, only the upper 500 m (~1600 ft.) or so of the ocean's surface contains warm water. Below 500m, water temperatures quickly decrease to near freezing (Figure 14.1B). In fact, most of the world's ocean waters are quite cold: 80% of the oceans are at a temperature of about 4°C (39°F), which is the maximum density point of liquid water. This rapid transition in water temperature with depth is called the **thermocline**. The thermocline is produced by two phenomenon. First, 500 m is the approximate depth to which the sun's rays can penetrate the water column and heat the water. Second, winds and the resulting waves cause mixing to occur - cooler waters are drawn up from below while warmer water is pulled down. Once you reach the polar regions, the thermocline becomes pretty much non-existant. In the polar regions, solar radiation is weak, so little heating of the water occurs. Furthermore, under the ice sheets, the lack of mixing by surface winds also suppresses water temperature.

Salinity

As most people know, the world's oceans are salty. So how much salt is there? If you were to take a pan with 1000 grams of average seawater in it and let it sit out in the sun until all the water is evaporated, you would have 35 grams (1.2 oz.) of salt in the bottom of the pan. Thus, the average salinity of the world's oceans is 35 parts salt per 1,000 parts seawater, or **35‰**. The **permil symbol (‰)** signifies parts per thousand, just like the percent symbol (%)

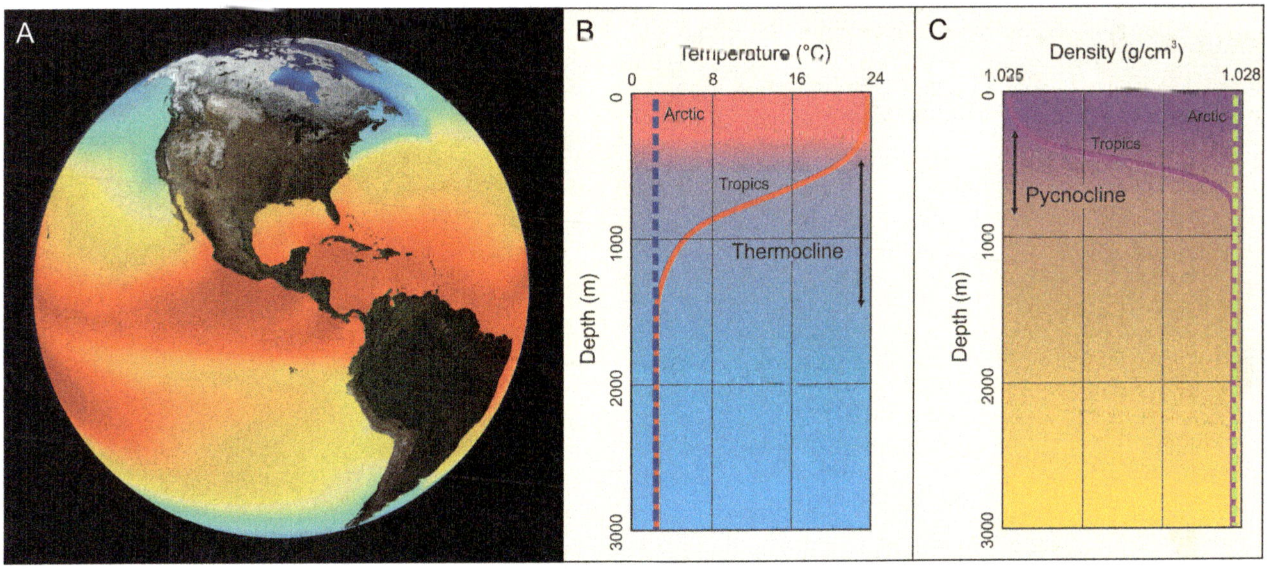

Figure 14.1: A. Mean sea surface temperatures in the Western Hemisphere. Image credit: NASA/ Goddard Space Flight Center Scientific Visualization Studio. B. The average temperature of the world's oceans with depth in the tropics and the arctic. C. The average density of the world's oceans with depth in the tropics and the arctic.

means parts per hundred. Notice that we did not say "per 1,000 grams of water." *Seawater mass is the combined mass of the dissolved solids plus the mass of the water.* Although the dissolved ions do not contribute significantly to the volume of the water, they do contribute mass. Thus, 1000 grams of average salinity seawater contains 35 grams of salt and 965 grams of pure water. Most of the dissolved ions delivered to the oceans by rivers are eventually removed by biological processes or through the precipitation of minerals like calcite. Some ions, like chlorine, don't have any significant physical or chemical mechanism to remove them from the water. Thus, over billions of years, the tiny amounts of chlorine released by weathered rocks have built up in the world's oceans to significant concentrations. In fact, chlorine makes up over 50% of the dissolved ions in average seawater (Figure 14.2). Second on the list is sodium, followed by sulfate, magnesium, calcium and potassium. Almost all of the remaining elements in the periodic table are found in seawater and yet collectively make up only 0.7% of all the ions in seawater. However, it is the dominance of chlorine and sodium ions in seawater (common salt) which makes the oceans taste salty.

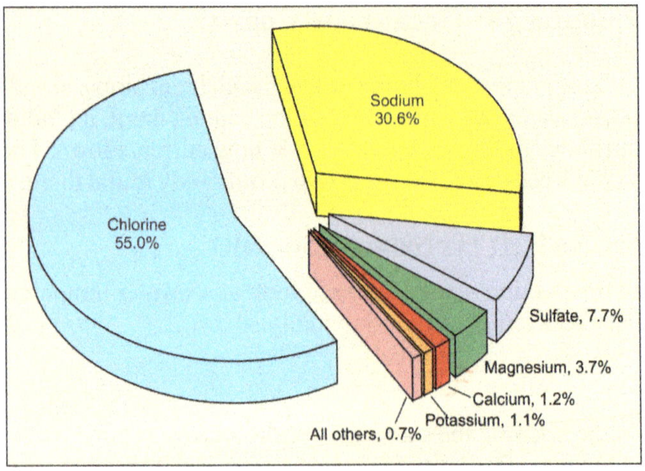

Figure 14.2: The percentage of different dissolved ions (by mass) in average seawater.

The salinity of the world's oceans is highly variable: It can reach as high as 50‰ in restricted inland seas like the Mediterranean Sea or as low as 1‰ in river-fed estuaries like the Chesapeake Bay. In the open ocean, most of the variation in the salinity of the surface waters is a function of climate (Figure 14.3). In humid equatorial regions, the salinity is lower than average as frequent rainfall dilutes the surface waters. In arid climates, particularly around the subtropical highs (30° latitude), evaporation dominates over precipitation. The evaporation of seawater leads to the concentration salts at the surface and makes the salinity higher than average as a result (Figure 14.3). Once you get towards the higher latitudes, the lack of intense solar radiation and the addition of precipitation (particularly around 60° latitude) causes the salinity to drop to below average levels. Just like with temperature, salinity doesn't vary much at depth. At ~4000 m (13,000 ft.) depth, the salinity of seawater only varies by 0.3‰ globally! Be warned however, under ice sheets water can become very salty. When sea ice forms, only fresh water is drawn up into the

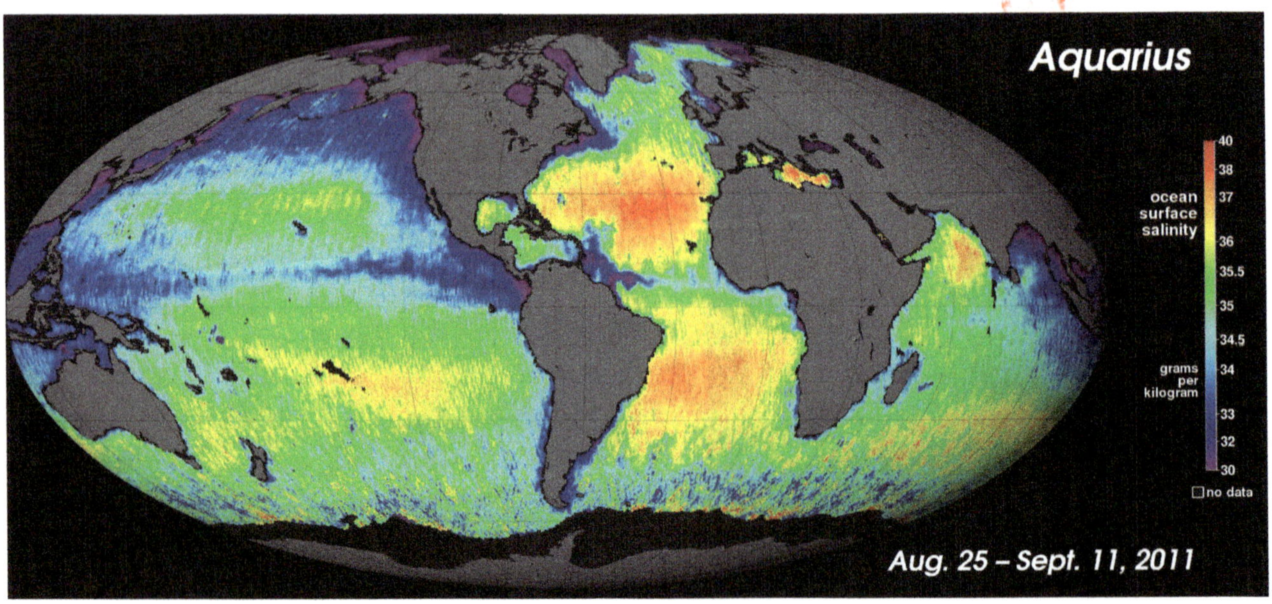

Figure 14.3: A global image of sea surface salinity (‰). The data was collected by the Aquarius/SAC-D (Satélite de Aplicaciones Científicas) observatory is a collaboration between NASA and Argentina's space agency, Comisión Nacional de Actividades Espaciales (CONAE). Image credit: NASA/GSFC/JPL-Caltech.

crystalline structure - the salts get left behind in the remaining unfrozen water below. We'll talk more about the importance of this phenomenon in the next section.

Density

Many processes in the ocean are driven by differences in water density. Two factors control ocean water densities: Salinity and temperature. First, salt water is denser than fresh water because the dissolved solids add mass to the water without increasing its volume. Second, cold water is denser than warm water because more molecules can be packed into a given volume of water at cold temperatures as compared to warm water. Thus, cold, salty water is very dense, and will quickly sink to the lowest parts of the water column. **Thus, the oceans are stratified by density**. As you can see in Figure 14.1C, the increase in density with depth is not constant. In the lower latitudes, there is a sharp transition between lower-density surface waters and the dense, deep-ocean waters called the **pycnocline**. The pycnocline is found at a depth of about 500 m (1,600 ft.) in most of the world's oceans, but the pycnocline is absent in the polar regions (Figure 14.1C). Since the salinity of the world's oceans doesn't vary much with depth, temperature differences are the key to the stratification process (although locally, salinity can be important). You can see this by comparing Figure 14.1B to Figure 14.1C.

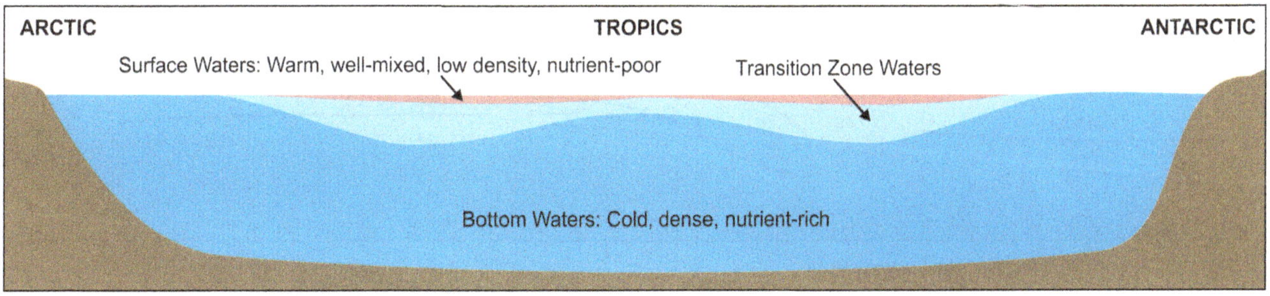

Figure 14.4: A schematic cross-section through the world's oceans showing the three main layers of ocean water.

Putting it all together: If you look at a profile through the world's oceans from north to south, you will see that there are essentially two thin layers of seawater floating on a thick layer of cold, dense nutrient-rich seawater called **bottom waters** (Figure 14.4). About 80% of ocean waters are bottom waters, another 18% are in the **transition zone waters** and the remaining 2% are **surface waters**. The surface waters are warm, low in density, well-mixed and generally depleted in nutrients. The transition zone waters, as the name implies, has water which rapidly transitions between surface water characteristics and deep water characteristics. Also note how the upper two layers both thin out and disappear completely towards the polar regions.

Section 14-2: Ocean Circulation Patterns

In this section we will review the two primary circulation patterns operating in the world's oceans, **surface currents** and **thermohaline currents**.

Surface currents are large, fast-moving streams of water operating within the upper layers of the ocean (Figure 14.5). Surface currents are driven by the wind. As the wind blows over the water's surface, energy is transferred and the mass transport of water ensues. There is a clear relationship between the location of the surface currents and the general circulation patterns of the atmosphere (refer back to Figure 4.10). There are three primary wind belts which influence surface currents: The Trade Winds, the Westerlies and the Polar Easterlies. For example, you can see that the Equatorial currents (which flow from east to west) line up with the Trade Winds, which blow from the northeast and southeast (Figure 14.6). However, the relationship is a little

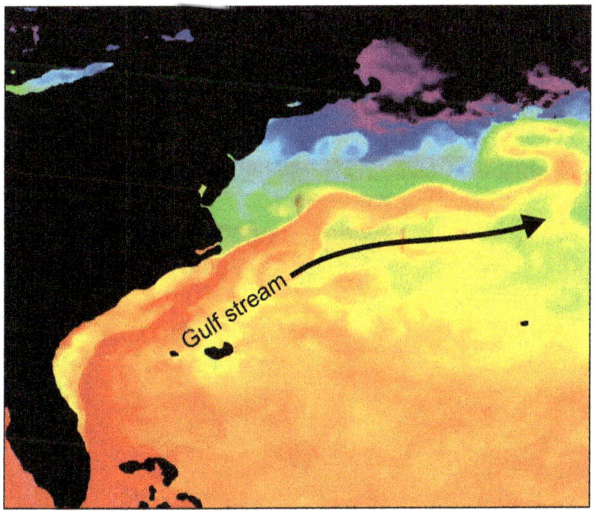

Figure 14.5: The Gulf Stream is a good example of a surface current. Image credit: Physical Oceanography Distributed Active Archive Center (PO.DAAC) at the NASA Jet Propulsion Laboratory, Pasadena, CA.

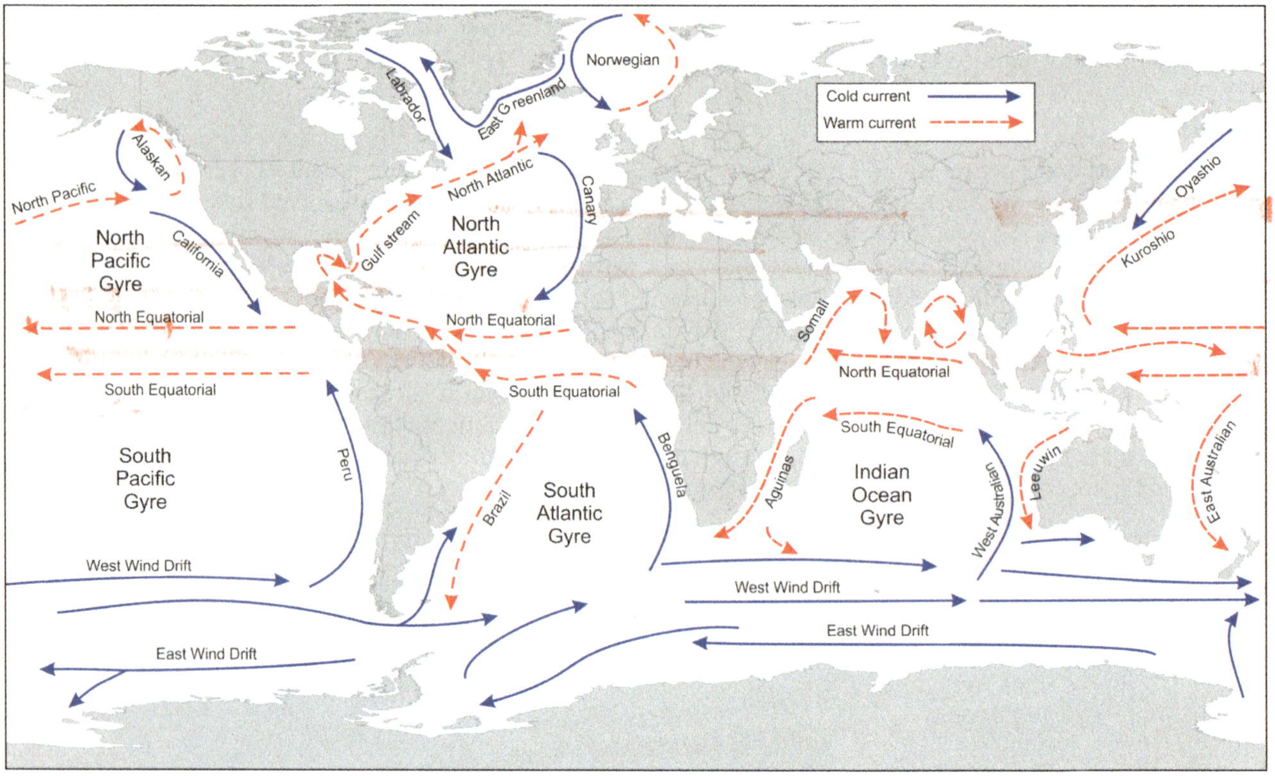

Figure 14.6: The world's primary surface ocean currents.

more complicated than that. Once the ocean currents are in motion, the Coriolis force causes them to be deflected. At the surface of the ocean, the Coriolis force modifies the direction of the moving water such that in the Northern Hemisphere, the water moves to the right of the prevailing winds, while in the Southern Hemisphere, the water moves to the left. For example, the North Equatorial current in the Atlantic Ocean is driven along the Equator by the Trade Winds. Once the current reaches the Americas, the Coriolis force causes it to be deflected to the right (or northward) forming the Gulf Stream (Figure 14.5). The Gulf Stream eventually merges with the North Atlantic Current. The Westerlies then push the North Atlantic Current back towards the European coast, where Coriolis then deflects the water towards the south, forming the Canary Current, which runs along the African Coast. Eventually, the Canary Current merges with the North Equatorial Current to start the cycle back over again. This looping circulation pattern is also known as a **gyre**. There are many gyres operating in the world's oceans (Figure 14.6). In the Northern Hemisphere, gyres rotate clockwise while in the Southern Hemisphere gyres rotate counterclockwise. These gyres cause cold water currents to travel from the polar regions towards the tropics, while warm water currents travel from the equator towards the poles. The Gulf Stream is a warm water current and as discussed in previous chapters, has a strong influence over weather and climate patterns in the Northern Atlantic.

Thermohaline currents also operate within the oceans, but unlike surface currents, these currents operate throughout the entire ocean from the surface to the bottom. Thermohaline currents are generated by differences in seawater density ***usually in areas of very high or very low salinity.*** For example, we noted earlier that under the polar ice sheets, salt is concentrated when seawater is converted into freshwater ice. This very cold, very salty, very

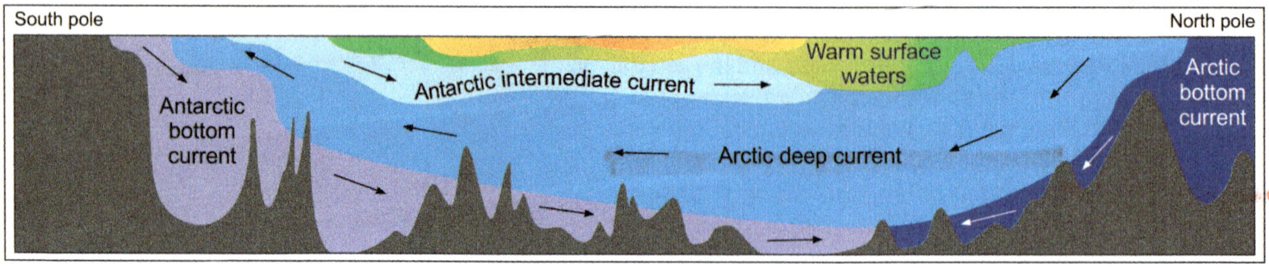

Figure 14.7: The principle thermohaline currents operating in the Altantic Ocean. The Arctic Deep Current is the most prominent of these currents.

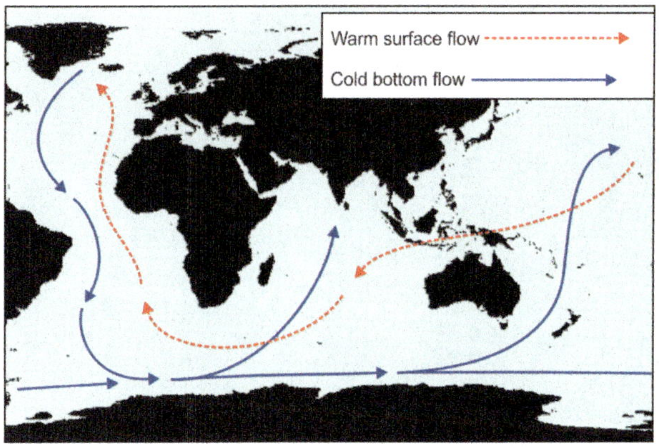

Figure 14.8: A map of the Deep Arctic current's circulation pattern.

dense water sinks to the bottom of the ocean and then slowly spreads outwards forming circulation cells (Figure 14.7). Currents are named based on both their source area (e.g., Arctic or Antarctic) and the general depth within the ocean in which they circulate (e.g., intermediate or deep). These currents move at much slower rates than surface currents, but they also play a role in moving heat around the globe. For example, consider the Arctic Deep Current, the largest and most dominant current of the various thermohaline circulation cells in the ocean (Figure 14.8). This dense body of water forms off the coast of Greenland, sinks to the bottom of the Atlantic Ocean and then slowly flows southward along the bottom of the Atlantic Ocean and works its way into the other major ocean basins. Slowly the water works its way back to the surface, warming and becoming less dense as it goes. Meanwhile, the dense sinking water produced near the edge of Greenland must be replaced by new water drawn in from elsewhere, so warm surface waters are pulled back towards the North Atlantic Ocean to complete the cycle all over again. The North Atlantic Ocean's temperature is actually considerably warmer than it would be if the current didn't operate this way. It takes about 1000 years for a water molecule to move through this thermohaline circulation pattern once.

Section 14-3: Waves

All ocean waves are some form of energy moving through the water; in the case of **wind waves**, the energy originates from the disturbing force of the wind. When the wind blows over the surface of the ocean, it displaces water, transferring mechanical energy from the air to the water (Figure 14.9). Once the energy has been transferred to the water, the energy disperses out in all directions from the wave generation area. It is again important to note that waves are caused by the displacement of water as the energy passes through. As you can see in Figure 14.9B, energy is transferred from water molecule to water molecule as the water molecules rotate in a circular fashion, displacing the water surface in the process. Thus, there is no ***significant*** net horizontal movement of water once a wave's energy has passed by a given point. This is why seagulls bobbing on the water never seem to move anywhere: The seagull moves up and down as the wave energy passes, but ultimately the seagull does not move very far. However, it would be incorrect to say there is ***no*** net transport of water. In most cases, there is a small amount of mass transport that does occur (which leads to the formation of surface currents).

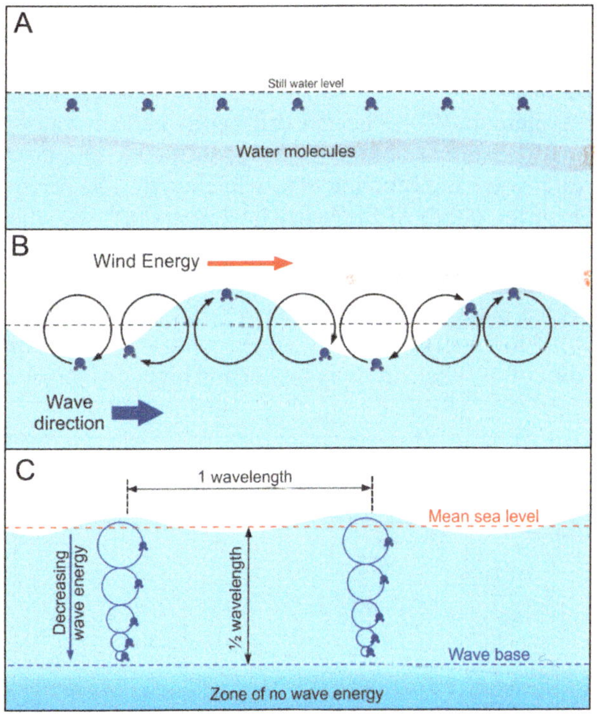

Figure 14.9: Understanding wave behavior. A. Water molecules in calm waters. B. When the wind blows over the water surface, energy is transferred to the water. In response, the water molecules rotate in an orbital motion and energy is transferred from molecule to molecule. This orbital motion deforms the water surface into the shape of a wave. C. Wave energy is strongest at the water surface and decreases with depth. Once the water reaches a depth of ½ the wavelength (the wave base), the wave energy reaches zero.

Wave energy is strongest at the surface and decreases with depth (Figure 14.9C). This should make sense given that wind waves are generated at the surface. Once you get to the **wave base**, wave energy decreases to zero. This decrease is due to friction which dissipates the energy. The location of the wave base is a function of the wavelength of the passing waves. The wave base is found at a water depth of ½ the wavelength of the wave. For example, a wave with a wavelength of 20 meters would have a wave base of 10 meters. Below that depth, no wave energy would be felt.

Although wind-generated waves can and do form near the shore, most of these waves are small in both wave length and amplitude. Waves generated in the open ocean are called

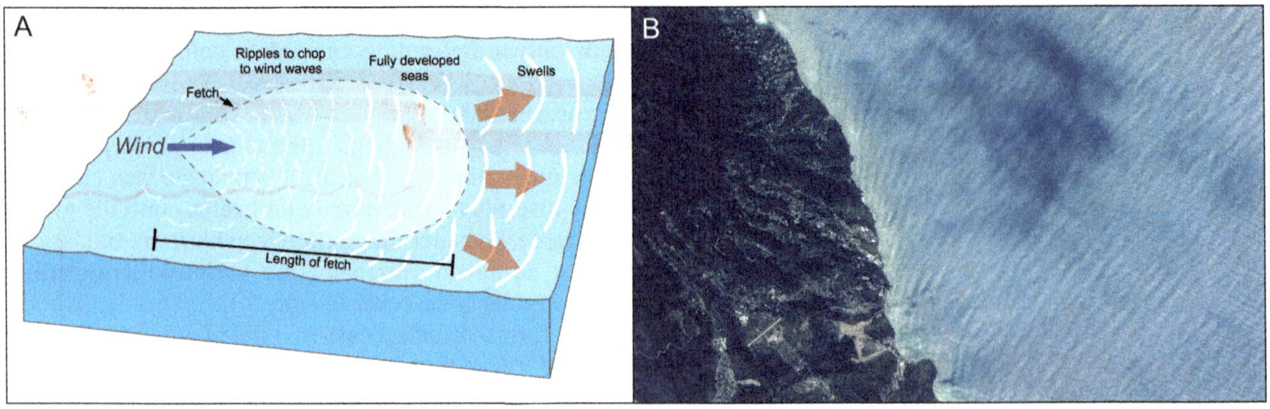

Figure 14.10: A. The generation of swells in the fetch area. How large a wave can ultimately grow is a function of both the wind speed and duration and the length of the fetch. B. Swells off the coast of Montserrat. Photo credit: Image Science and Analysis Laboratory, NASA.

swells and swells are the waves you see constantly breaking on the shore, even when the wind near the shore is calm (Figure 14.10A). Swells form hundreds of miles offshore in wave generation areas known as **fetch**. The size of a swell is controlled by three factors. The first two factors have to do with the wind: How hard is the wind blowing and how long has is been blowing for? An increase in wind speed will lead to the generation of larger swells. Likewise, the longer the wind blows, the bigger the swells will grow. The third factor is the size of the fetch. Ultimately, the size of the fetch limits how big waves can get. This is why waves generated near the shore or in small bodies of water (e.g., the Gulf of Mexico or the Great Lakes) tend to be smaller in nature as compared to open ocean swells. If the fetch is long enough and the wind remains relatively constant long enough, **a fully developed sea** will develop. At this point, the waves have reached their maximum possible size, and the waves should all be consistent in terms of wavelength and height. However, this is fairly rare condition. Once the waves leave the fetch, they are called swells. The swells begin to disperse outwards and although some waves may simply dissipate over time, the fate of most waves is to eventually reach some shoreline and break (Figure 14.10B).

Up to this point, we have been focusing on **deep water waves**, that is, waves operating in waters deeper than ½ their wavelength. Waves behave differently in deep water than they do in shallow water. In deep water, waves travel at a fairly constant speed as they move through the open ocean. And, as you are already quite aware, the speed of a deep water wave is a function of its wavelength. The wave's crest and trough are commonly broad and rounded in shape. Once the depth of the water becomes shallower than ½ wavelength, the wave becomes a **shallow water wave** (Figure 14.11). When the wave reaches shallow water, it begins to slow down because of friction between the wave base and the seafloor. This causes the wave to change in two significant ways. First, the wavelength begins to decrease and the waves begin to "bunch up" closer together much like a loose carpet might bunch up on a smooth wooden floor (Figure 14.12A). Second, the faster moving water at the back of the wave train pushes the waves at the front of the wave train up higher and higher as the waves approach the shore. The wave's shape begins to change as

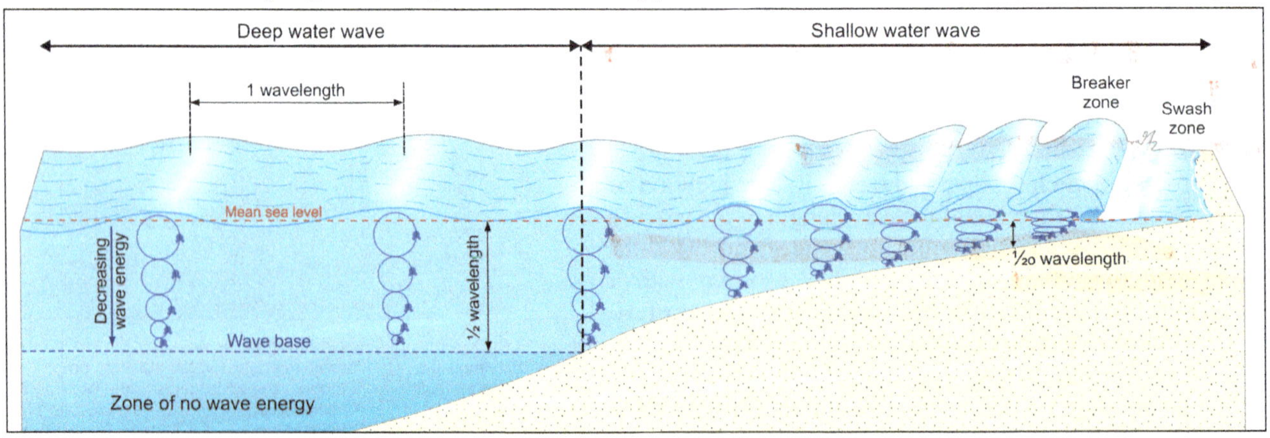

Figure 14.11: This diagram shows how waves behave in both deep and shallow water. When waves reach shallow water, friction causes the bottom of the wave to slow down. This causes the waves to grow in height while their wavelengths shorten. They continue to build upwards until they reach a water depth of 1/20 their deep-water wavelength. At that point, they become unstable and topple over, breaking on the shore.

Figure 14.12: A. This image shows how the wavelength of a wave decreases in shallow water. Also note how the waves are becoming "peaked" as well. Photo credit: Image Science and Analysis Laboratory, NASA/JSC. B. The author contemplates the breaking waves in Puerto Rico. Photo credit: A. Zaprowski.

well. Instead of being broad and rounded, the wave becomes steeper-sided and the crest will become "peaked." When the wave reaches water that is shallower than 1/20 its wavelength, the wave becomes unstable and topples over, breaking onto the shore and transferring energy from the water to the land (Figure 14.12B). Also note that the circular orbital paths the water molecules followed in deep water become progressively flatter or elliptical as the wave approaches the shore. Once the wave breaks, the water runs up the beach face in a process called uprush, and then returns back to the ocean through a process called backwash. The direction of uprush is a function of the direction the swell is coming from, while backwash always runs down the beach face perpendicular to the shoreline. The foamy, turbulent layer of water that forms on the beach face during uprush and backwash phases is known as swash and swash action in the swash zone can move sediment across the beach face.

If the beach slope is gentle, the returning water will be channeled out though narrow zones between the incoming waves. Because a large volume is being funneled through a relatively narrow area, strong currents called **rip currents** commonly form (Figure 14.13). Rip currents are part of a cell-like circulation of water in the nearshore regions as water from breaking waves return to deep waters. Rip currents can be quite strong - speeds of up to 2 m/sec (8 ft/sec) have been reported, which is faster than most people can swim. The strength of the rip current is a function of wave size. Rip currents can become quite dangerous when the surf is very rough and people are advised to stay out of the water during these times.

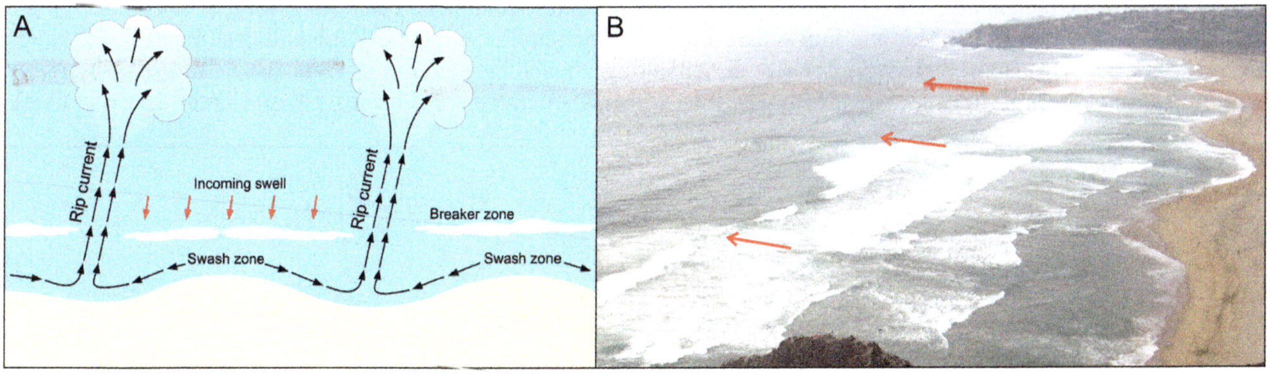

Figure 14.13: A. Rip currents form along gently sloped coasts as part of circulation cells in the nearshore zone. B. Rip currents along the coast of Chile (red arrows). Photo credit: Cecilia Lascody/NOAA.

How Waves Modify the Coast

Now that we have reviewed wave properties and behavior, we can now look at the many ways that waves shape the coast. In the most general sense, wave energy does two things along a coast. First, wave energy erodes bedrock and produces sediment and over time the waves continue to break the sediment down into small pieces. This happens primarily along rocky coasts with little to no beach. Second, wave energy is capable of transporting and depositing large quantities of sediment along the coast. These two processes work together to shape coasts and the myriad of coastal landforms seen around the world. Specifically, there are three wave-produced phenomena which impact the coast: **Onshore-offshore drift**, **Longshore transport** and **Wave refraction**.

Onshore-offshore transport: One effect that breaking waves have on a beach is that swash action entrains (moves) sediments across the beach face. Sand can be carried up the beach face during the uprush phase or carried back into the surf during the backwash phase. Over time, significant quantities of sediment can be moved from offshore to onshore and vice versa in a process called onshore-offshore transport (Figure 14.14). When wave energy is relatively high, the backwash phase dominates and sediment is transported offshore to the offshore bar. This is known as offshore transport (Figure 14.14B). When wave energy is lower, the uprush phase dominates and sediment is transported from the offshore bar to the beach (Figure 14.14C). This is known as onshore transport. When onshore transport dominates, the beach will build seaward, and when offshore transport dominates, erosion of the beach occurs. We'll explore the importance of this concept in more detail later in the chapter.

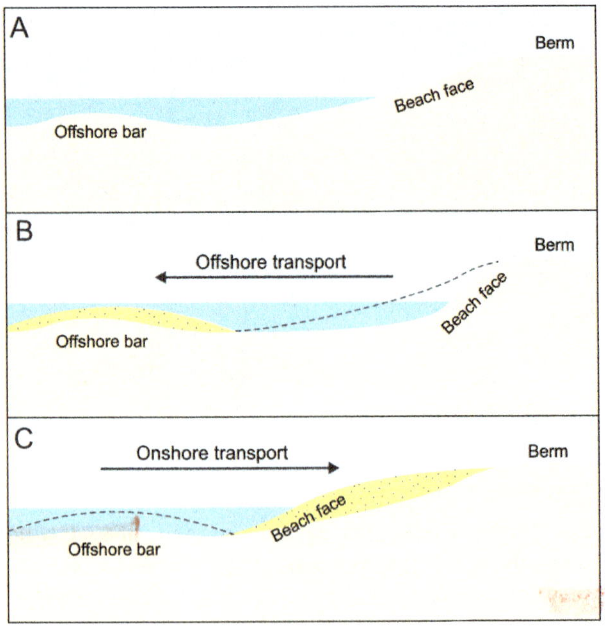

Figure 14.14: Onshore-offshore transport explained. A. The three main parts of a beach profile are the berm, the beach face and the offshore bar. B. During times of high wave energy, the backwash phase dominates and sand is eroded from the berm and transported to the offshore bar. This is called offshore transport. C. During times of lower wave energy, the upwash phase dominates and sand is transported from the offshore bar and deposited on the beach. This is called onshore transport.

Longshore transport: In addition to onshore-offshore transport, a coast may also experience longshore transport. Waves typically arrive at the beach at an oblique angle and thus the waves break at an angle (Figure 14.15A). This oblique motion is what causes longshore transport to occur. Longshore transport is actually a combination of two related phenomena: **Beach drift** and **longshore currents**.

Beach drift occurs because of swash action. When an oblique wave breaks, sediment is transported across the beach face at an angle relative to the coast (Figure 14.15B). However, when the wave goes into the backwash phase, the sediment travels back towards the sea *perpendicular* to the coast. This action leads to a net transport of the entrained sediment along the beach face. This mass transport of sand is what we call beach drift (or littoral drift). Although the amount of sediment entrained by one individual wave may be small, collectively the endless breaking of waves along the coast can transport huge volumes of sediment along a beach face over time. Sediments are also transported parallel to the shoreline in the nearshore regions by **longshore currents** (Figures 14.15B and 14.16A). These

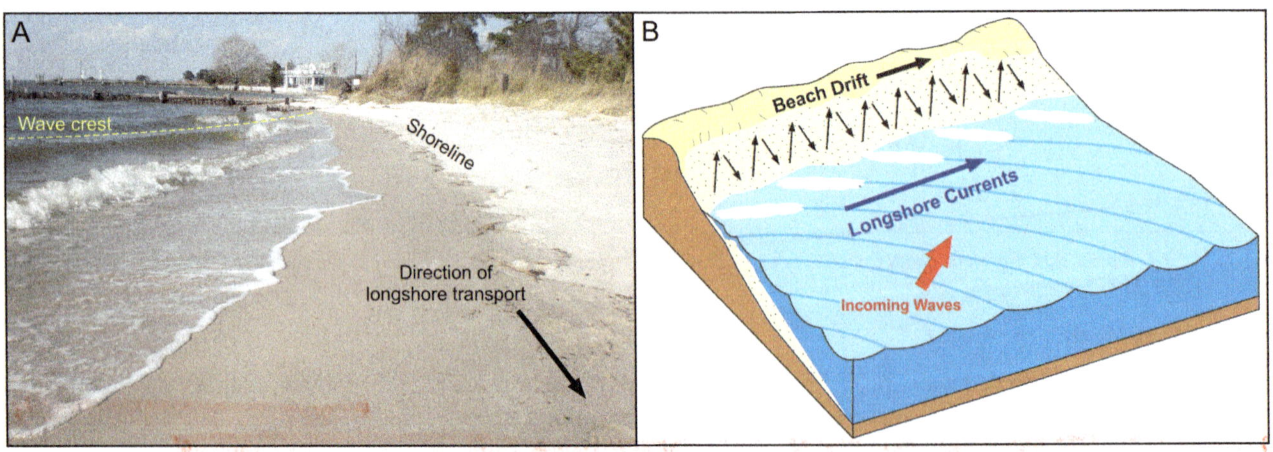

Figure 14.15: Longshore transport explained. A. An example of waves approaching the coast an at angle. Photo credit: B. Zaprowski. B. When waves approach the shore at an angle, they carry sediment across the beach at an angle but when the wave backwashes, the sediment is carried back down towards the coast perpendicularly. This results in the net flow of sediment along the beach face (beach drift).

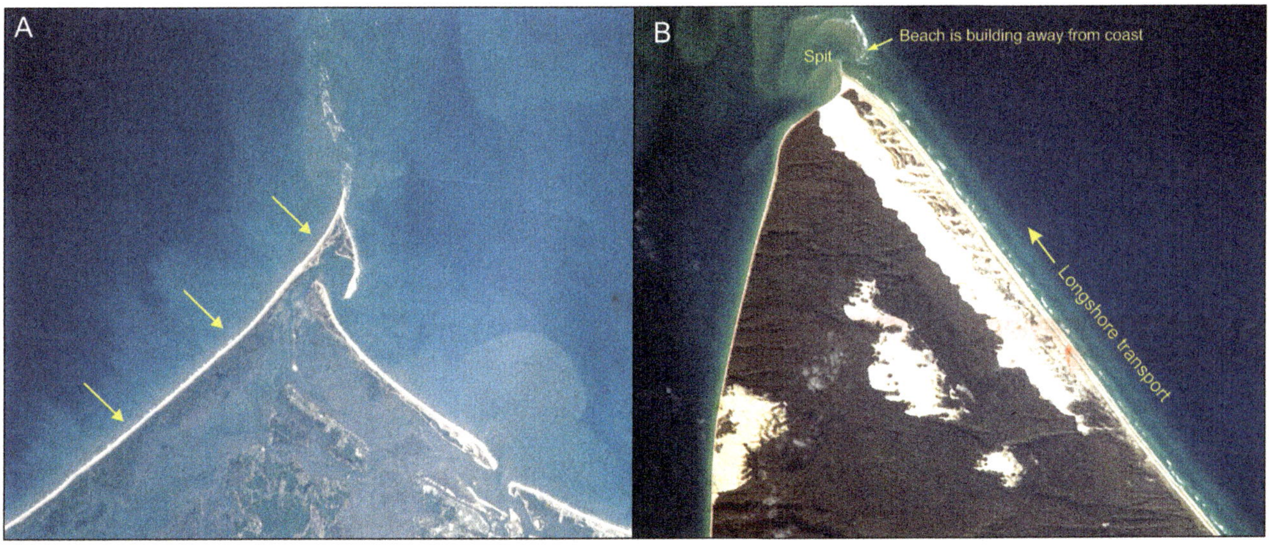

Figure 14.16: A. This image from the North Carolina coast shows longshore currents carrying brownish sediment along the coast (yellow arrows). B. Longshore transport along this stretch of the Australian coast is causing the beach to grow out away from the land, forming a simple spit. Photo credits: Image Science and Analysis Laboratory, NASA.

currents are generated out where the waves begin to set up and break and extend to the shoreline. Oblique waves generate a net transport of water parallel to the coast, and these currents can entrain and transport sediment, particularly during times of higher wave energy. Wave refraction (see below) can also aid in the formation of longshore currents. **Longshore transport** is the *combined* net transport of sand from *both* of these processes. Most shorelines have a dominant direction of longshore transport, although it can change seasonally. Longshore currents play an important role in the formation and migration of coastal features like spits and barrier islands (Figure 14.16B).

Wave refraction: Strictly speaking, wave refraction is a term which refers to the bending of any kind of wave. For our purposes, wave refraction refers to the bending of waves as they approach the shore. Wave refraction is best developed along coasts when the depth of the seafloor near the coast is highly variable, as is the case with jagged or irregular coastlines. Refraction occurs when one part of a wave crest reaches shallow water before the rest of the wave (Figure 14.17). This part of the wave starts to behave like a shallow water wave and slow down while the rest of the wave continues on at the faster deep water wave speed. In essence, one part of the wave "outruns" another part and the wave bends as a result. The significance of this bending can be see in Figure 14.17A. Since wave energy is

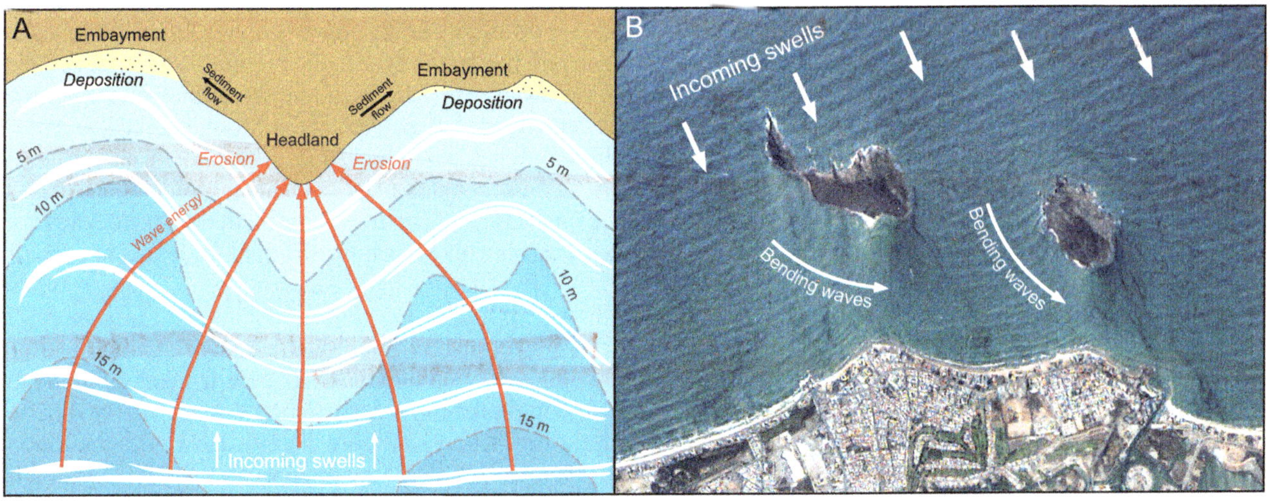

Figure 14.17: Wave refraction explained. A. When waves approach shore, part of each wave reaches shallow water first, slowing it down. This causes the wave to bend (or refract) and the wave mimics the shape of the coast as it breaks. Wave refraction focuses the energy towards the end and sides of the headland, causing intensified erosion, while the adjacent embayments receive reduced energy, allowing for the deposition of sediments. B. Waves refracting around a couple of islands near Mazatlan, Mexico. Photo credit: Image Science and Analysis Laboratory, NASA.

traveling in a direction which is perpendicular to the wave crest, any bending of the wave results in the wave energy moving in a new direction. In this case, the bending of the waves around a headland leads to the focusing of wave energy on the headland and away from the embayments. Because the energy is directed away from the embayments, sediment is able to accumulate, forming small pocket beaches. Wave refraction is particularly important in shaping rocky coasts like in Figure 14.17B, a topic we'll explore later in the chapter.

Section 14-4: Tides

Tides are a special type of wave that occur because of a complex web of gravitational forces between the Earth, the moon and the sun. Recall (hopefully) from physics that all objects exert a gravitational force on all other nearby objects. For example, the Sun's gravitational force holds the Earth in an orbit, and the Earth's gravitational force holds the Moon in orbit. These gravitational forces disturb the ocean surface and a very long wave forms as a result. Of course this gets complicated really quickly as the Earth, moon and sun are not static objects, they are constantly in motion and that has implications for the tidal patterns we observe around the world. Before we delve into the complicating factors, we will first look at the basic physics behind tides and then we can discuss how a variety factors modify them.

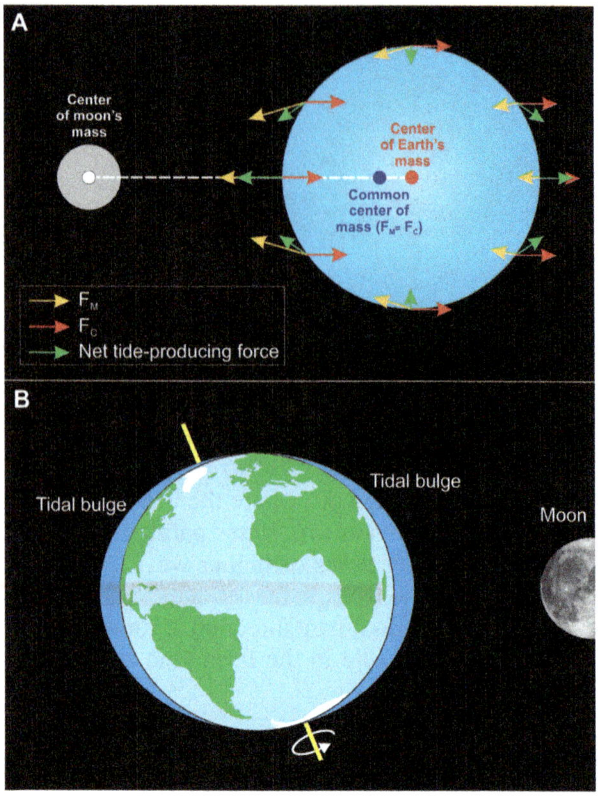

Figure 14.18: Explaining tidal forces. A. The forces of the Moon (F_M) and centrifugal forces (F_C) work together to produce a net tide-generating force. B. This net force causes the ocean waters to pile up into two "bulges" on opposite sides of the earth.

For now, let's ignore the influence of the sun and just focus on the Earth and the Moon, that is, the **lunar tides.** You'll see later on that the **solar tides** operate similarly. Next, we'll also have to assume a uniformly deep ocean with no continents in its way. Finally, let's define the two forces at work - the gravitational force of the Moon (F_M) and the offsetting centrifugal force generated as the Earth and the moon rotate around a common center of mass (F_C) (Figure 14.18A). At this common point, F_M equals F_C.

We can use arrows to define both the direction and the magnitude of the force each object exerts on any point on the Earth's surface. The key here is that the *combined* force or the **net tide-producing force** determines how the ocean will behave at that location (Figure 14.18A). On the side of the Earth facing the Moon, the moon's pull on the oceans exceeds the centrifugal forces, so the net tide-producing force is away from the Earth and the oceans pile up into a "bulge" (Figure 14.8B). On the opposite side of the Earth, the centrifugal forces exceed the Moon's forces, and net-tide producing forces cause the oceans to form another matching bulge. Thus, a high tide occurs when a position on Earth passes through the bulge and a low tide occurs when a portion of the Earth passes through the adjacent trough. We must stress at this point that the tides don't actually operate this way. The Earth has continents and the oceans are not uniformly deep so although this simple bulge model makes it easier to explain how different forces act in concert to produce the tides, it is just that - only a conceptual model.

One of the most basic tidal characteristics is their timing. The timing of the high and low tides each day is a function of the Earth's rotational speed and the speed of the moon's revolution around the Earth. The Earth takes 24 hours to rotate once on its axis, so if the moon didn't move, the time it would take to for a particular stretch of coastline to return back to the same spot under the same tidal bulge would be 24 hours (Figure 14.19). However, the moon revolves around the Earth along an orbital path that takes 27.3 days to complete (the lunar cycle). So during the 24-hour span of time when the Earth was rotating on its axis, the moon was moving approximately $1/27^{th}$ of the way along its orbital path around the Earth, pulling the bulges along with it. As a result, it takes an additional 50 minutes for the Earth to rotate into a position directly under the peak of the tidal bulge. Thus, in most places, the peak of the

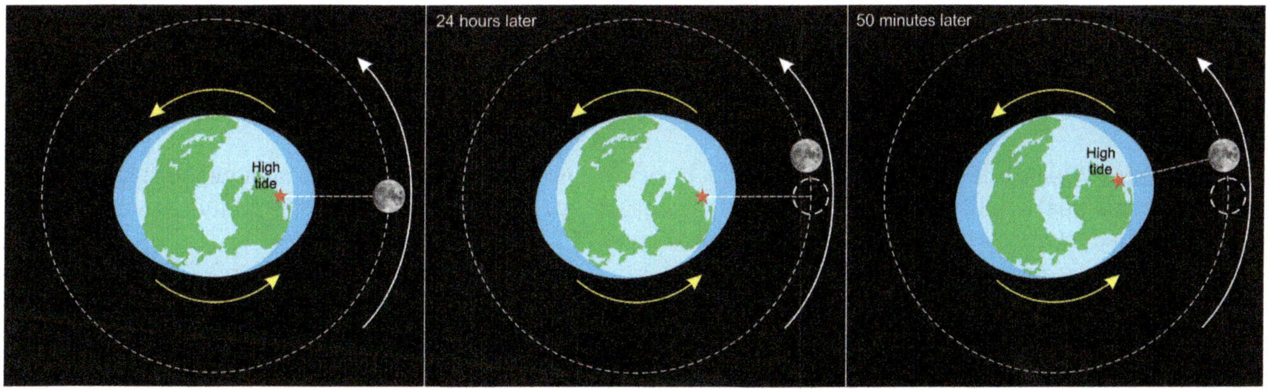

Figure 14.19: Explaining the timing of the tides. The Earth rotates counterclockwise on its axis as viewed from above once every 24 hours while the Moon rotates once around the Earth every 27.3 days. It is high tide along the coast and the moon is directly overhead. After 24 hours, the Earth is back exactly where is was but the Moon has now moved approximately 1/27th of the way around the Earth. Thus, it takes an additional 50 minutes for the same stretch of coast to move back under the moon.

tides occur approximately 50 minutes later each day (Figure 14.19). In places where there are two high tides each day, the timing between each high tide is approximately 12 hours and 25 minutes. In truth, the high tide doesn't always occur when the coast is directly under the Moon, in some places the high tide occurs hours before or after the Moon passes overhead.

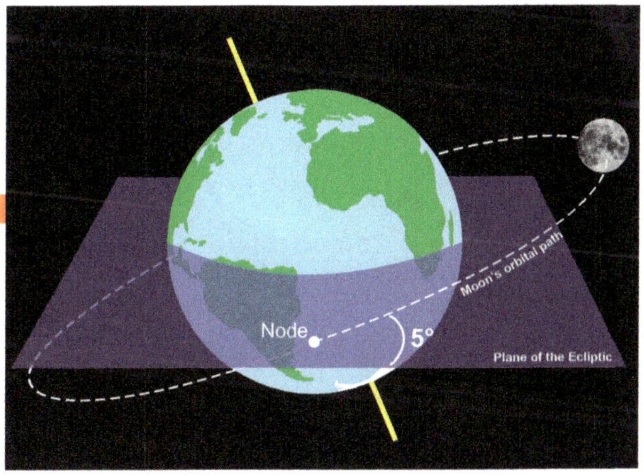

Figure 14.20: The Moon's declination relative to the plane of the ecliptic.

Now let us add a little more complexity to the story. First, let us establish that as the Earth orbits the Sun, its orbital path defines a two-dimensional surface called the plane of the ecliptic, and the Earth rotational axis is titled 23.5° relative to that plane (Figure 14.20). Next, the moon's orbital path is tilted by 5° relative to the plane of the ecliptic and the Moon's path intersects that plane at two nodes on opposite sides of the Earth. This phenomenon is known as **lunar declination**. The Moon's declination causes the position of the moon (relative to the Earth's surface) to vary between 28.5° north and 28.5° south latitude throughout the lunar cycle (because 23.5° + 5° = 28.5°). The Moon's declination produces three distinct tidal signatures during the course of the lunar cycle:

1. Semi-diurnal tides: A semi-diurnal tide has two highs and 2 lows per daily cycle (24 hours and 50 minutes). If the amplitude of the each of the tidal bulges is about the same, they are called **equal semi-dirunal** tides (Figure 14.21A). The East Coast of the United States is a place where these types of tide occur. However, in the case where the amplitude of the two tidal bulges are unequal, they are called **unequal semi-diurnal** tides (Figure 14.21B). Many parts of the West Coast experience these types of tides.

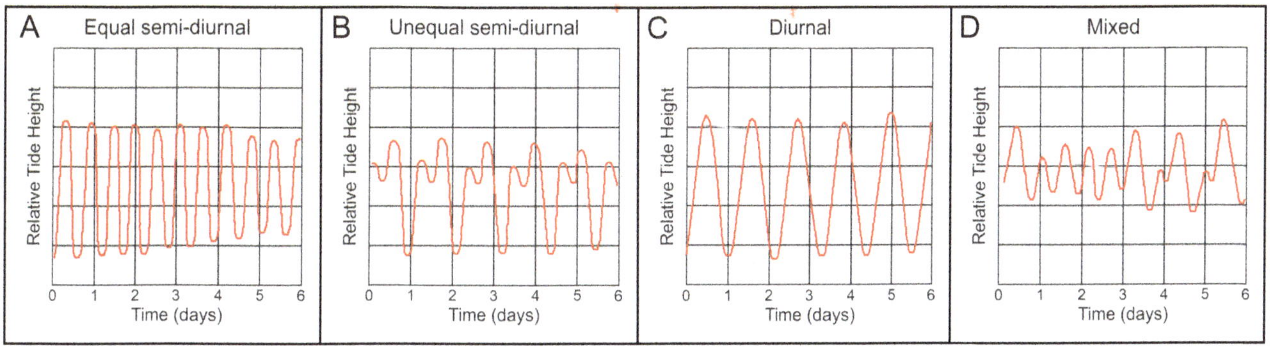

Figure 14.21: The different tidal patterns produced by the effects of the lunar declination. Note that some people consider unequal semi-diurnal tides to also be a type of mixed tide.

2. Diurnal tides: A diurnal tide has only one high and one low per daily cycle (Figure 14.21C). Relatively speaking, diurnal tides are fairly rare. They occur most frequently in restricted basins (e.g., the Gulf of Mexico) with a natural tidal oscillation of 24 hours and 50 minutes. In these basins, the tidal wave is distorted when it enters the basin from the ocean.

3. Mixed tides: Mixed tides are just that - tides which vary in amplitude significantly throughout the lunar cycle. In some cases, the tidal amplitudes may be extremely unequal (e.g., Figure 14.21B) or display a mix of semi-diurnal and diurnal characteristics throughout the lunar cycle (Figure 14.21D).

Now let's turn our attention back to the sun. The sun and the Earth also have a gravitational relationship with each other, much in the same way as the Earth and moon in Figure 14.18. In this case, the combined tide-producing forces between the Earth and the sun result in the formation of a pair of solar bulges (Figure 14.22). The amplitudes of these solar bulges are smaller than the lunar bulges because the net-tide generating force produced by the sun is about half as strong as that of the moon. The changing relative orientation of the solar and lunar bulges through the course of

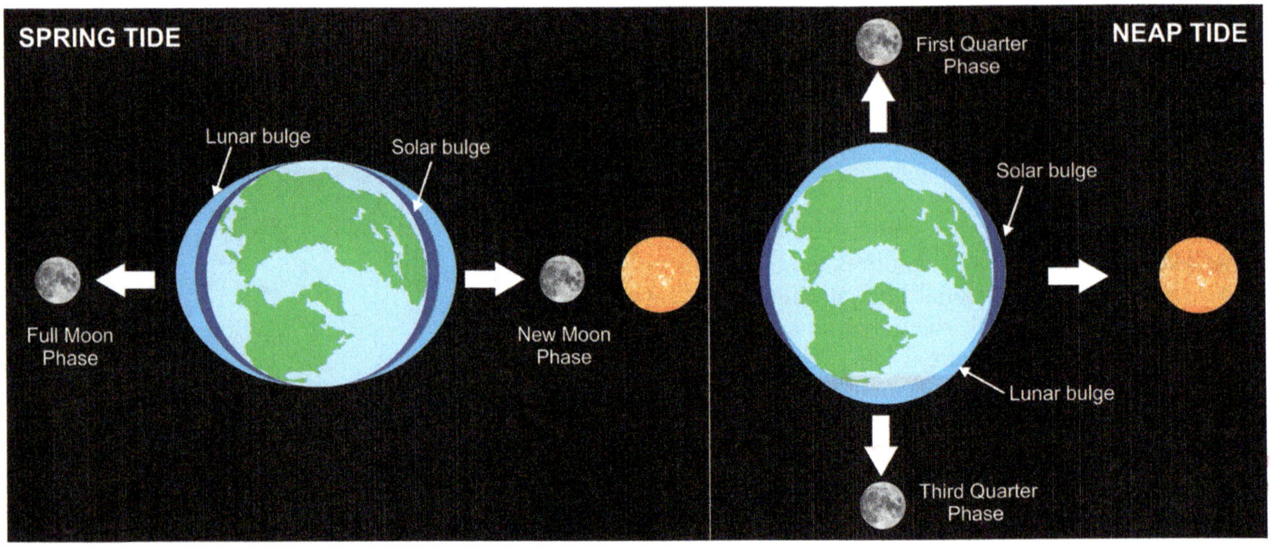

Figure 14.22: Spring and neap tides explained. When the lunar and solar bulges line up, spring tides occur. This happens during the new moon and full moon phases. When the moon and sun are oriented at a 90° angle (relative to the Earth), neap tides occur. This happens during the 1st quarter and 3rd quarter phases. Spring tides are the largest in the lunar month and the neap tides are the smallest.

the lunar cycle produces another cyclical variation in the amplitude of the tides called **spring tides** and **neap tides** (Figure 14.22). Spring tides occur during the new moon and full moon phases when the lunar and solar bulges are lined up, producing large tidal amplitudes (i.e., higher highs and lower lows). During the first quarter and third quarter moon phases, the lunar and solar bulges are oriented at 90° relative to the Earth and each other and neap tides occur. The two bulges partially cancel each other out and the result is small tidal amplitudes (i.e., lower highs and higher lows). Two spring-neap cycles occur every 29½ days which is the period of time it takes for the moon to go through one complete cycle of moon phases. The tides that occur between the spring and neap tides are called the mean tides. Spring-neap tide cycles are readily observable on tidal records like the one in Figure 14.23.

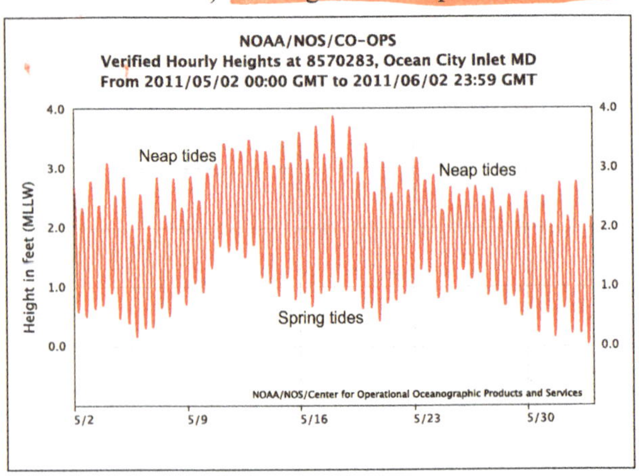

Figure 14.23: This tide gage station record from Ocean City, MD shows how the tides vary between spring tides to neap tides throughout the course of a month. Data courtesy of NOAA.

The irregular nature of the world's coastlines can also lead to an amplification of the tidal bulge. For example, at the Bay of Fundy in Nova Scotia, Canada, the tidal range can be over 6 meters (18+ feet!). In the case of the Bay of Fundy, the long, tapering shape of the bay forces the tidal bulge

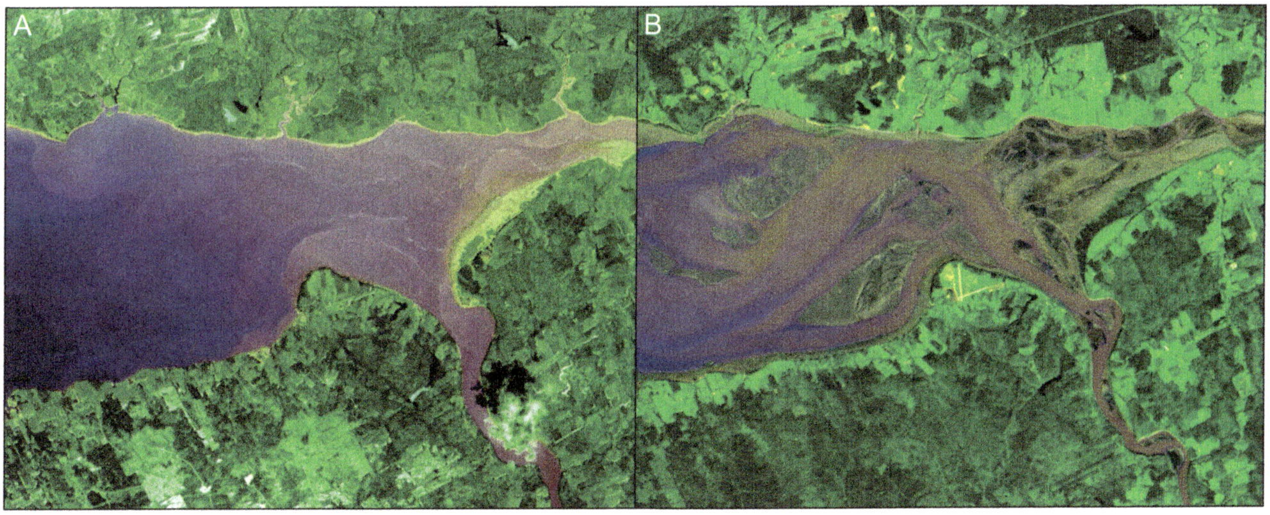

Figure 14.24: The Bay of Fundy has the world's largest tidal range. The long, funnel-shaped bay amplifies the tidal bulges as it moves up the bay. A. High tide. B. Low tide. Image credit: Visible Earth/NASA.

upwards as the bay narrows (Figure 14.24). Other parts of the world where large tidal ranges occur include the Amazon River, the English Channel and Indonesia.

Features of the Intertidal Zone

The intertidal zone encompasses all of the parts of the coast which experience the effects of the tides. When wave energy is high, beaches or wave-cut platforms occupy the intertidal zone. When the wave energy is low and the tidal range is moderate to large, depositional features form. These intertidal features occur between the mean spring high water and mean low water levels. **Tidal currents** are generated by incoming **flood tides** and outgoing **ebb tides**. Tidal currents also keep **tidal inlets** open and allow the tidal wave to enter and exit embayments (Figure 14.25A). Some embayments with really large tidal ranges, like the Bay of Fundy, also experience a phenomenon known as the **tidal bore**. The tidal bore is a large visible wave which forms along the edge of the incoming tide (Figure 14.25B). The tidal bore most commonly forms in narrow, tapering embayments and tidally-influenced rivers because the tidal wave height is amplified as the channel gets narrower and shallower (e.g., Figure 14.24).

These tidal currents are capable of entraining and redepositing large quantities of sediment along the coast and great quantities of sediment can be washed into embayments and deposited in areas where wave energy is low. **Tidal flats** are deposits of sediment found in the unvegetated part of the intertidal zone (Figures 14.24B and 14.26A). Most tidal flats occur in embayments, estuaries or along tidally-influenced rivers. Tidal flats are generally flat, have a slightly convex profile, and are dissected by a network of tidal channels (Figure 14.26B). During flood tide, seawater floods

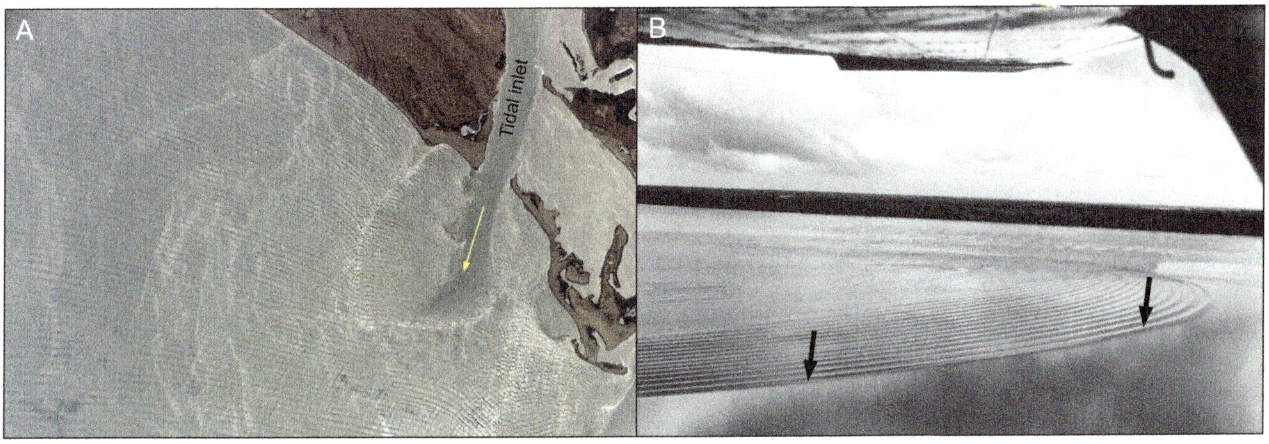

Figure 14.25: A. A strong ebb current (yellow arrow) pushes out through a tidal inlet. Photo credit: Image Science and Analysis Laboratory, NASA-Johnson Space Center. B. A tidal bore coming up the Araguari River in Brazil. Photo credit: S. Bartsch-Winkler and David K. Lynch/USGS.

Figure 14.26: Tidal flats. A. Muddy tidal flat exposed at low tide along the Washington coast. Photo credit: B. Zaprowski. B. A sandy tidal flat exposed along the California coast. Photo credit: Jessica Weinberg McClosky/NPS.

into the channels and then floods out onto the adjacent flats. During ebb tide, water first drains back into the channels and then moves out to the ocean. The channels may be lined with coarse sediments or shell debris as a direct result of the higher velocity currents that occur in the channels.

When the water gets shallow enough, salt marsh grasses and other types of plants will begin to colonize the tidal flats, forming the vegetated part of the intertidal zone. In the mid- to high latitudes, salt marshes form. Salt marshes are dominated by salt-tolerant grasses, shrubs and herbs (Figures 14.27A-B). In the tropics, the intertidal zone is mostly populated by forests of salt-tolerant shrubs and trees called **mangroves**. Mangroves can grow in the intertidal zone because their prop roots allow them to breathe (Figures 14.27C-D).

Figure 14.27: Examples of vegetation in the intertidal zone. A. Some of the extensive salt marshes behind Assateague Island, MD. B. A small tidal channel cuts through the salt marshes. C. A clump of red mangroves in Puerto Rico. Note the prop roots (inset) which allow the mangroves to "breathe" in this aqueous environment. D. A grove of black mangroves. Photo credits: B. Zaprowski.

Section 14-5: Coastal Landforms

In the last chapter, we explored how weathering, erosion and deposition shape the Earth's land surface. In this section, we'll take a similar look at the world's coastlines and explore how those same forces dictate the configuration of the coast and the types of landforms that develop there. In essence, there are two basic types of coasts: **Erosional coasts**, which are predominantly steep and rocky, and **depositional coasts**, which are dominated by sandy beaches and wide, flat coastal plains (Figure 14.28). Erosional coasts form in areas where the wave energy is high, the sediment supply is low, and/or along tectonically active coasts (e.g., the West Coast of the U.S.). Depositional coasts form in areas where wave energy is low and the sediment supply is high (e.g., the East Coast of the U.S.). Each of these types of coasts tend to have similar types of landforms as a result, although there are exceptions. For example, beaches, which are depositional landforms, may be found along rocky coasts, but they tend to be fairly small and poorly developed. In the next couple sections, we'll look at the most common types of landforms and how they form, starting with erosional landforms.

Figure 14.28: Examples of the two basic types of coasts. A. An erosional coast in Australia. Photo credit: D. Votta. B. The Gulf Coast is a good example of a depositional coast. Photo credit: Ben Mieremet/NOAA Photo Library.

Erosional Coastal Landforms

Erosional landforms develop when waves can directly attack the base of rocky cliffs exposed along a coast. Over time, this erosion at the base leads to the formation of a **wave-cut notch** in the bedrock (Figures 14.29A and C). As the notch grows, it undercuts the rock above, weakening it. This will eventually lead to the failure and collapse of the entire cliff face into the water below (Figures 14.29B). Over time, the sea cliffs retreat and leave behind a flat bench of bedrock called a **wave-cut platform** (Figure 14.30A). This platform is scoured down by the action of waves and

Figure14.29: A. Schematic diagram of a typical rocky coast. B. A schematic diagram of a coast which recently experienced cliff failure. C. A well-developed wave-cut notch along the coast of Thailand. Photo credit: Casey Condon.

Figure14.30: A. Kelp-covered wave-cut platform exposed at low tide. B. A tidal pool in a wave-cut platform with both a sea anemone and a starfish. Photo credits: B. Zaprowski.

sediment being tumbled over it, kind of like a giant polishing machine. In many areas, these platforms are exposed during low tide. These rocky features can host a wide array of animals, including sea anemones, starfish and mussels (Figure 14.30B)

Along tectonically-active coasts, things are a little different. Because these coasts are actively rising, sediments generally don't get the opportunity to build up to any great extent, although smaller beaches may still form in the quieter embayments. Sooner or later, wave-cut platform is lifted out of the water by tectonic uplift and the action of wave erosion on it ceases. The wave-cut platform has now become a **wave-cut terrace** (Figure 14.31). After an episode of uplift occurs, wave energy begins to attack the newly exposed bedrock at the base of the terrace. It is not unusual to have a whole series of wave-cut terraces "stepping up" away from the coast (Figure 14.32).

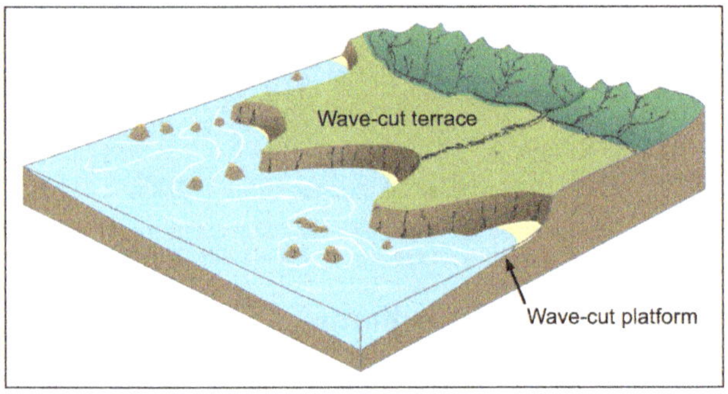

Figure 14.31: Wave-cut terraces along an actively rising coastline.

If the coastline has rocky headlands, wave refraction can erode the base of the headlands from the side, forming a pair of **sea caves** (Figure 14.33). The further action of wave erosion causes the caves to deepen and eventually connect through the headland, forming a **sea arch** (Figure 14.33A). As waves continue to eat away at the arch, the opening underneath the arch erodes higher and wider until eventually the arch collapses, leaving behind a **sea stack** (Figure 14.33B). Where they occur, it is not uncommon to have large numbers of these stacks in groups along rocky coasts (e.g., Figure 14.28A).

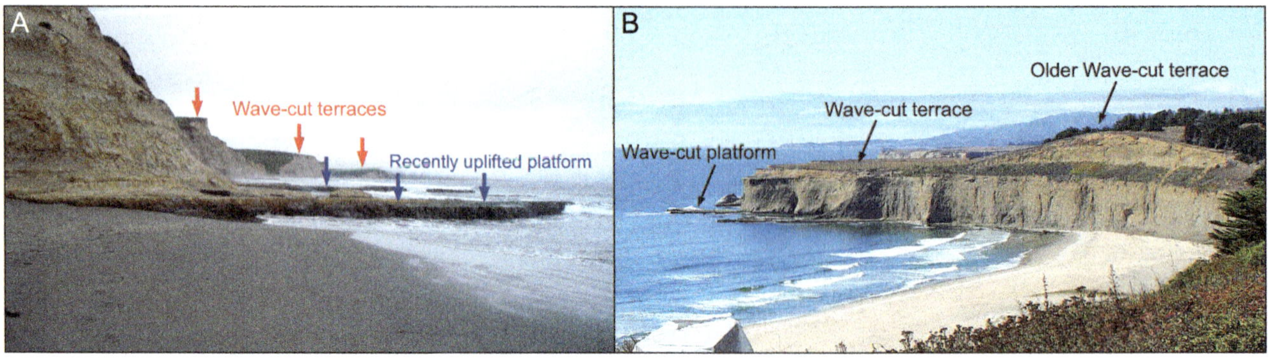

Figure 14.32: A. Wave-cut terraces and recently uplifted platforms near Point Reyes, CA. Photo credit: B. Zaprowski. B. A series of wave-cut terraces stepping up away from the coast. Photo credit: Casey Condon.

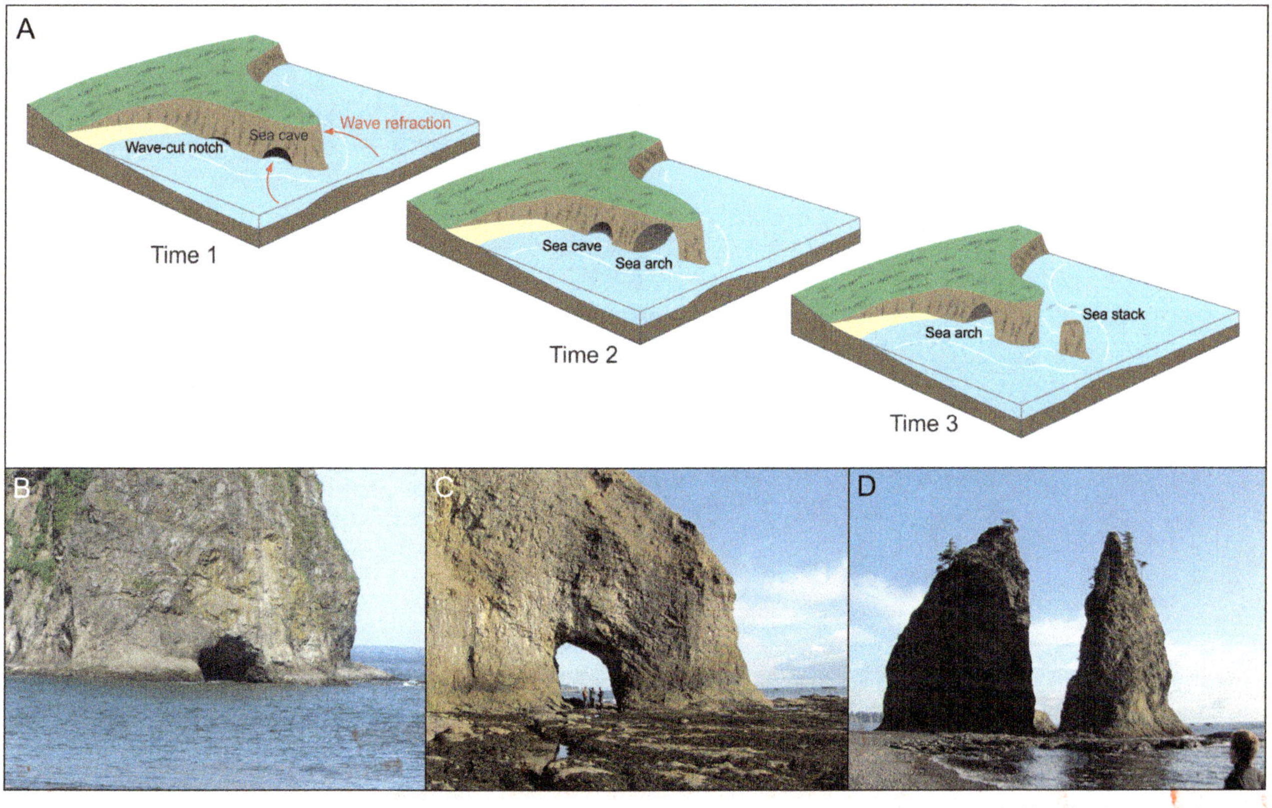

Figure 14.33: A. The evolution of rocky headlands and the formation of sea arches and sea stacks. On some coasts, eroded headlands may produce sea stacks but not sea arches or sea caves. B. An example of a sea cave. C. An example of a sea arch. D. An example of sea stacks. Photo credits: B. Zaprowski.

Depositional Coastal Landforms

Beaches: Beaches are nothing more than an accumulation of sediments along a shoreline. Although most people think of lush, tropical locales with wide, white sandy beaches when they think of beaches (Figure 14.34A), most "average" beaches are small in scale and composed of coarse sands and gravel (Figure 14.34B). The formation of a large, well-developed sandy beach only occurs when wave energy is relatively low and there is a good supply of sediment, as is common along depositional coasts. That's not to say erosional coasts are completely devoid of beaches - small pocket beaches may form as wave refraction deflects wave energy away from embayments (refer back to Figure 14.17). Nor is this to imply that beaches are static features; waves (through the process of onshore-offshore transport) and storms constantly reshape the beach over time. To understand how beaches evolve over time, we first need to consider the parts of a beach as viewed along a cross-sectional profile (Figure 14.35A).

Figure 14.34: A. The "typical" tropical resort beach isn't as common as you'd think. B. Many beaches, like this one along the Washington coast, are composed of coarse sands, pebbles and gravel. Photo credits: B. Zaprowski.

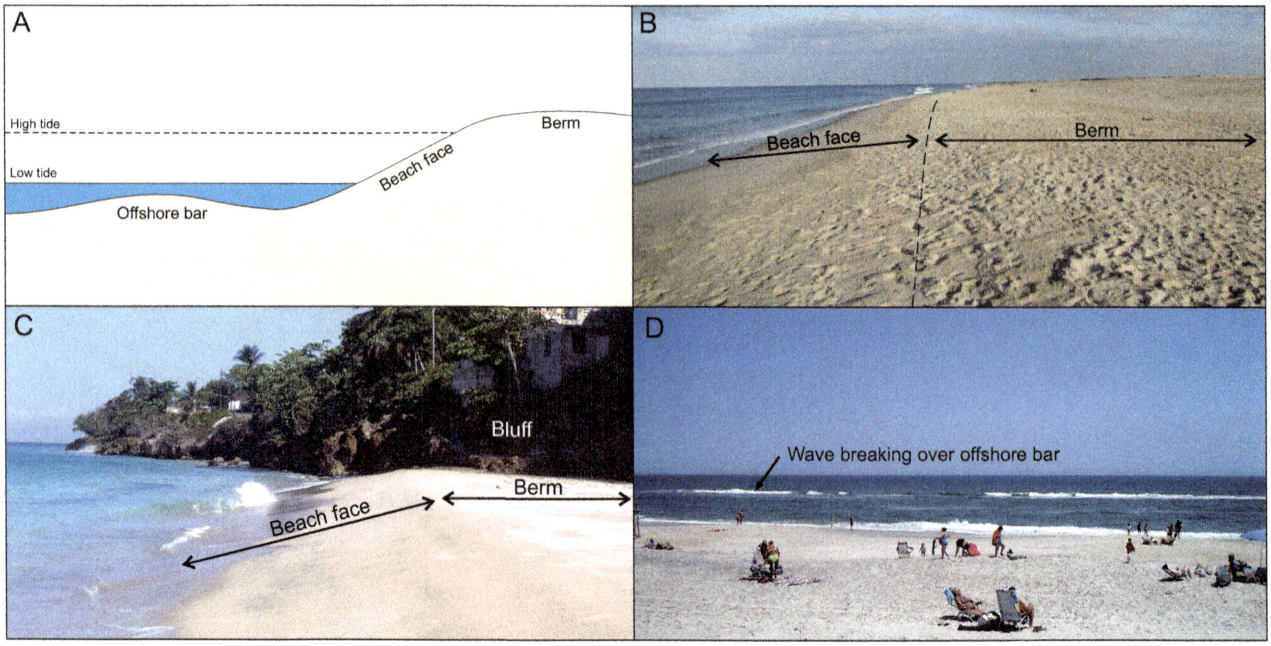

Figure 14.35: A. An idealized beach profile showing the main features of a beach. B. An example of the berm and the beach face along a barrier beach. C. An example of the berm and beach face on a small pocket beach abutting a bluff. D. A wave breaking over the offshore bar. Photo credits: B. Zaprowski.

An average beach has 3 basic parts: The berm, the beach face and the offshore bar (Figure 14.35). The berm is the part of the beach above the normal high tide line. It is relatively flat and only flooded by waves during spring tides and storms. The beach face is the part of the beach where the waves actually break and swash up onto the beach face. The beach face has a slope which starts at the edge of the berm and ends at the edge of the water. Beyond the water's edge, a long linear bar of sand called the offshore bar forms parallel to the beach. Although you can't see the offshore bar, sometimes you can see waves breaking as they pass over the bar (Figure 14.35D). When wave energy is low, onshore transport moves sediment onto the beach face, causing the berm to grow. When wave energy is high, offshore transport erodes sediment from the berm and carries it to the offshore bar, causing it to grow. During high-energy storm events, the erosion of the berm may be particularly severe, leading to the formation of a beach scarp (Figure 14.36).

Figure 14.36: This large scarp formed after a summer nor'easter passed by the coast. Photo credits: B. Zaprowski.

If you study a particular stretch of beach for a few years, patterns begin to emerge, that is, seasonal cycles of erosion and deposition occur. In many locations, there is a seasonal change in the average wave energy impacting a beach. In general, swells tend to have less energy in the summer as compared to the winter. This occurs because marine storms are stronger and more frequent in the winter. In the summer, onshore transport dominates and sand is transported landward and accreted to the berm, causing it to grow throughout the summer. As a result, a nice, wide, well-developed berm forms (Figures 14.37A-B). The sediments tend to be fine-grained as well. This process of beach growth can be quite rapid, with rates of centimeters per hour measured at some locales. In contrast, offshore transport dominates in the winter so sand is eroded from the berm and deposited on the offshore bar. This "flattens" the beach face and causes the berm to retreat landward (Figures 14.37C-D). The remaining sediment becomes much coarser-grained and dark, "heavy minerals" like hematite, magnetite and garnet may be concentrated in patches (Figure 14.37D). The sand will then spend the winter stored on the offshore bar until the following spring when wave conditions change. Of course at any time in the seasonal cycle a storm may come along and cause beach erosion. The resulting after-storm profile can be quite different from the before-storm beach profile, especially when it occurs in the summer (Figure 14.36).

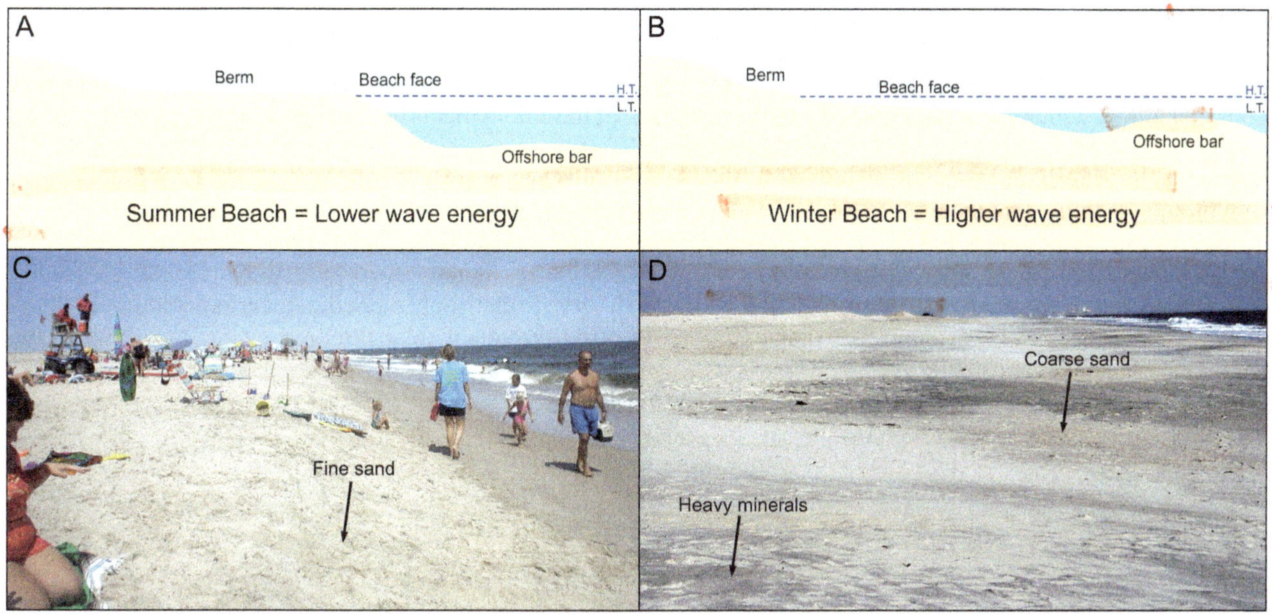

Figure 14.37: The evolution of beaches throughout the year. A. The typical summer beach profile. B. The typical winter beach profile. C. The beach at Assateague Island State Park in the summer. D. The same beach in the winter. Note the coarser orange-colored sand and dark heavy minerals. Photo credits: B. Zaprowski.

Although the resulting erosion may look catastrophic, storms are a natural part of the beach evolution process. The onshore transport of sand allows most beaches to repair themselves to pre-storm conditions within 6-8 weeks after a storm.

Barriers: Coastal barriers develop when beach material is deposited offshore, across river mouths or embayments, primarily by the action of waves and the wind. They most commonly form along coasts with strong longshore transport, and although barriers may be built out of gravel (especially in glaciated areas), they are dominantly built out of sand. 15% of the world's coastlines have some type of barrier fringing them. Barriers can form along almost any type of coast except in areas with a very large tidal range. They most commonly form along tectonically-quiet coasts with low wave energy and an abundant sediment supply. The two principle groups of barriers are **spits** and **barrier islands** (Figure 14.38). Spits are elongate deposits of sediment that grow out from a coast and are connected to the land while barrier islands are linear deposits of sediment which parallel the coast and are not connected to the

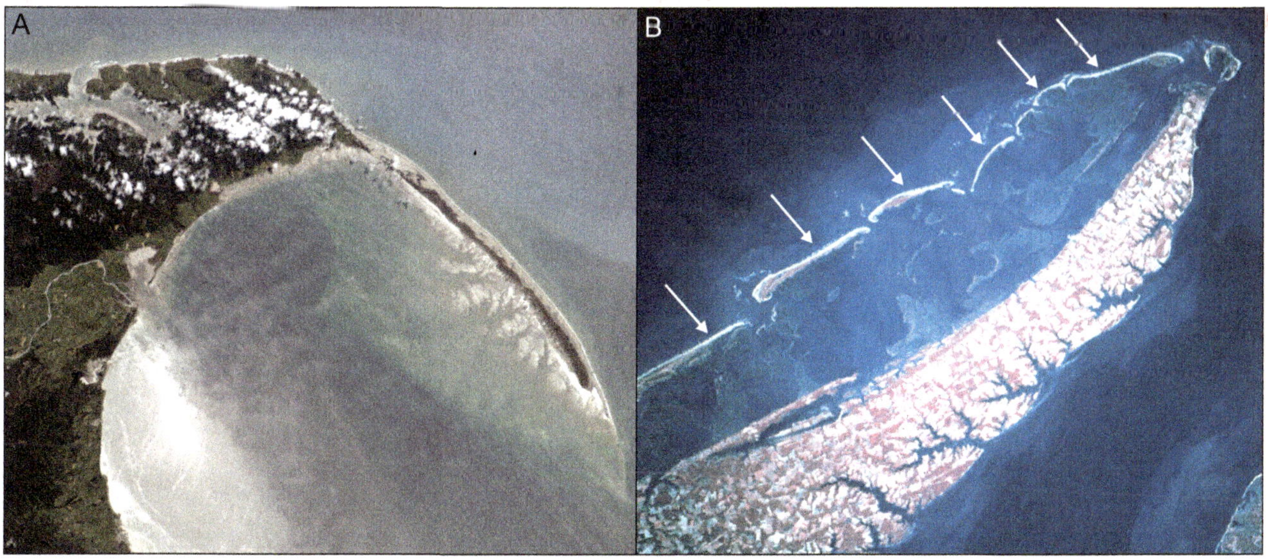

Figure 14.38: A. Farewell Spit in New Zealand. B. Multiple barrier islands fringe the southern tip of the Delmarva Peninsula in Virginia. Photo credits: Image Science and Analysis Laboratory, NASA-Johnson Space Center.

mainland. Closing off an embayment or a river mouth with barriers causes a significant decrease in wave energy. This decrease promotes the deposition of fine grained silts and clays in tidal flats and the subsequent colonization of those deposits by grasses or mangroves. Many barriers, especially barrier islands, are found in close association with tidal inlets, which allow tidal waters to move in and out of the embayments behind barriers.

Deltas: Deltas are large-scale deposits of sediment that form when a fast-moving current enters a large, slow moving body of water, either a lake or the ocean. The sediments settle out of the water column because velocity decreases at the river mouth and the water is no longer able to carry its sediment load (Figure 14.39). Over time, the delta grows upwards and outwards as the river continues to deliver sediment to the river mouth. Deltas are usually dissected by dozens of smaller **distributary channels** which serve to distribute the sediment around the edges of the delta, giving it a fan-like or cone-like shape (Figure 14.40A). Frequent tidal flooding events also lead to the formation of marshes and mangroves.

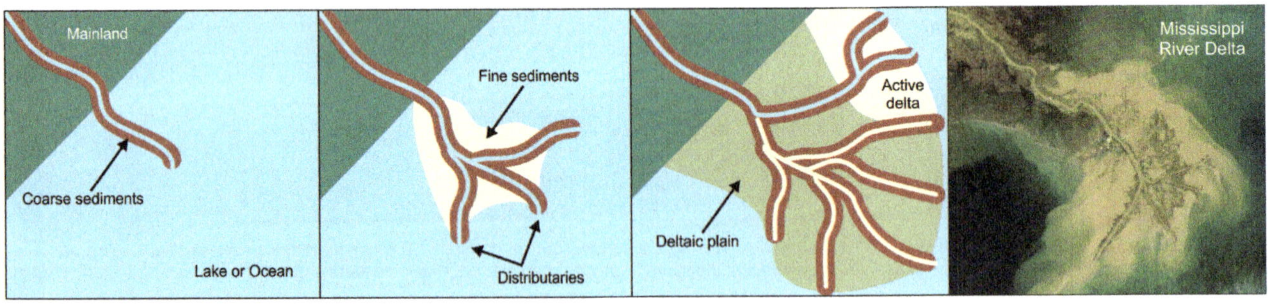

Figure 14.39: The evolution of a delta over time. Coarse sediments are deposited just adjacent to the tributaries and then fine grained sediments fill in the areas in between. Eventually the sediments can develop into a deltaic plain as vegetation colonizes the area. Photo credit: Image Science and Analysis Laboratory, NASA/JSC.

As the depositional process continues and the delta surface becomes vegetated, deltaic plains are created, essentially adding new land to the edge of the coast. These deltaic plains are some of the richest farmlands in the world and historically many cultures developed on deltaic plains like the Nile Delta (Figure 14.40A). The downside to living on a deltaic plain is the significant risk of flooding from heavy rains or the storm surge associated with hurricanes. Many countries in southeast Asia have large populations of people who live in deltaic plains like the Ganges Delta plain (Figure 14.40B) that get flooded every year during the monsoon season.

Coral Reefs: Coral reefs are quite common along most tropical coasts and are also places of great biological diversity (Figure 14.41A). Corals are distant relatives of jellyfish and share some of their basic characteristics like tentacles with stinging cells called nematocysts. Corals excrete a hard skeleton of calcium carbonate called a corallite (Figure 14.41) while the soft, living part of the coral in the corallite is called the polyp. Corals can grow in a solitary form (one corallite, one polyp) but most species of coral grow in colonial forms with hundreds or thousands

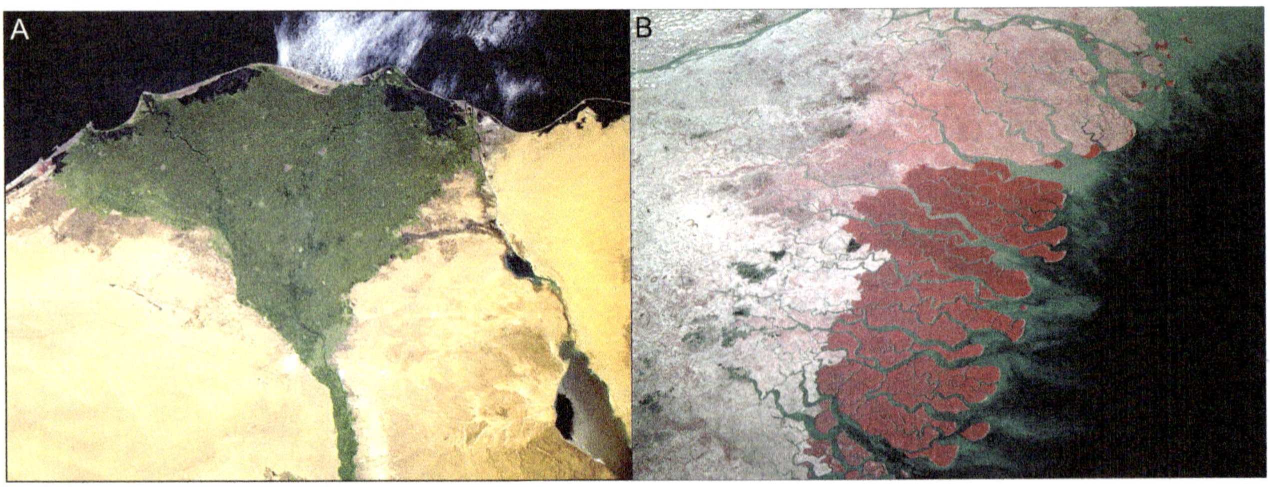

Figure 14.40: A. The Nile Delta with its rich, fertile deltaic plain. B. The Ganges Deltaic Plain stretches about 220 km (136 mi.) along the coast of the Indian Ocean and is home to one of the densest populations in the world. Photo credits: Image Science and Analysis Laboratory, NASA-Johnson Space Center.

Figure 14.41: A. Fish swimming through a patch of staghorn coral. Massive yellow mounded corals can also be seen in this image. Photo credit: David Burdick/NOAA Photo Library. B. This close up image of some corallum shows the tiny polyps that live in a coral colony. The polyp uses its tentacles to capture some of its food out of the water, but most corals rely on the zooxanthellae living in their epidermal tissues for most of their energy needs. Photo credit: G. P. SchmahlNOAA Photo Library.

of corallum covering the surface of the reef. Coral growth is confined to warm seas where the mean temperature of the coldest month does not fall below 18°C and the warmest month does not exceed 34°C. As a result, most coral reefs are found between 30°N and 30°S latitude. Corals require average salinity ocean waters which are relatively free of sediment. They also require active water circulation to keep temperature and salinity uniform, to supply them with nutrients and oxygen, and to wash off any fine-grained sediment that may have been deposited on them. Corals have no mechanism to remove sediment and if buried in thick layers of sediment, the polyps will be smothered and die. Although corals can and do get some of their energy from trapping food with their tentacles, most of their energy comes from symbiotic algae or zooxanthellae living in their epidermal tissues (Figure 14.41B). The algae convert sunlight into sugars which the coral then uses to survive and in return, the algae are protected from predators. As such, corals need an adequate supply of sunlight, thus most species of corals can rarely grow below 50 m (165 ft.) in depth, although some species can live up to 100 m (330 ft.) down in exceptionally clear waters.

Average rates of upward vertical growth for coral reefs are 0.5 cm to 2 cm per year and in ideal situations they can grow at rates of 4.5 cm to 10 cm per year. Some coral reefs have been in existence for millions of years and it is the corals' ability to grow vertically that allows them to grow into massive coastal features that can persist for extremely long periods of time. Corals will grow upwards until they reach a level where they are briefly exposed at low tide (Figure 14.42). Other reef organisms, however, can continue to build the reef up above the low tide line if they are able to tolerate longer exposure times out of the water. One particularly important group of organisms that contribute to reef growth are calcareous algae (Figure 14.43A). Calcareous algae help the coral reef community by cementing together sand, coral fragments, and other debris into a suitable hard substrate for sessile animals and by absorbing wave energy in the outer reef rim that would otherwise erode the reef and destroy the more fragile coral species. Other creatures who contribute significant skeletal material to the reef include echinoids (sea urchins and sand dollars), gastropods (snails), bivalves (clams and oysters), crinoids (sea-lillies), sponges and a wide variety of fish (Figure 14.43B). Because corals act like mangroves and reduce wave energy, calcareous sand, broken pieces of

Figure 14.42: The top of a fringing reef in the Mariana Islands, Guam, exposed at low tide. Photo credit: David Burdick/NOAA Photo Library.

Figure 14.43: A. Patches of pink algae covering a reef in the U.S. Virgin Islands. A brownish coral (bottom left), and some pink and orange sponges can also be seen in the image. Photo credit: Personnel of the NOAA Ship *Nancy Foster*/NOAA photo library. B. A colorful reef scene with multiple fish species including a squirrelfish in the foreground, two coneys, a French grunt, a surgeonfish, a triggerfish, and a rock beauty. Photo credit: Personnel of the NOAA Ship Nancy Foster/NOAA.

corals, shells and other organic debris fills in the spaces between the corals. Calcium carbonate precipitates out of seawater to cement the fragments together to form a solid carbonate reef.

There are three primary groups of coral reefs: Fringing reefs, barrier reefs and atolls. Sometimes these reefs are small and only locally-developed, but many reefs grow into very large, complex features with smaller reef features developed within them. Fringing reefs are reefs that have built upwards and outwards from the coast into the shallow seas that border continents or islands (Figure 14.44A). Fringing reefs are thinnest near shore and thicken seaward. They do not grow near river mouths as fresh water and excessive amounts of sediment can smoother and kill them.

Barrier reefs are long, linear reefs or chains of reefs that develop offshore and are roughly parallel to the coasts they fringe (Figure 14.44B). Most barrier reefs begin as fringing reefs and as relative sea-level has risen, the corals have kept pace by growing vertically. Meanwhile, the shoreline migrates landward and a large lagoon develops between the reefs and the coasts. Many reefs fail to keep up with rising sea level they may become submerged, and die off.

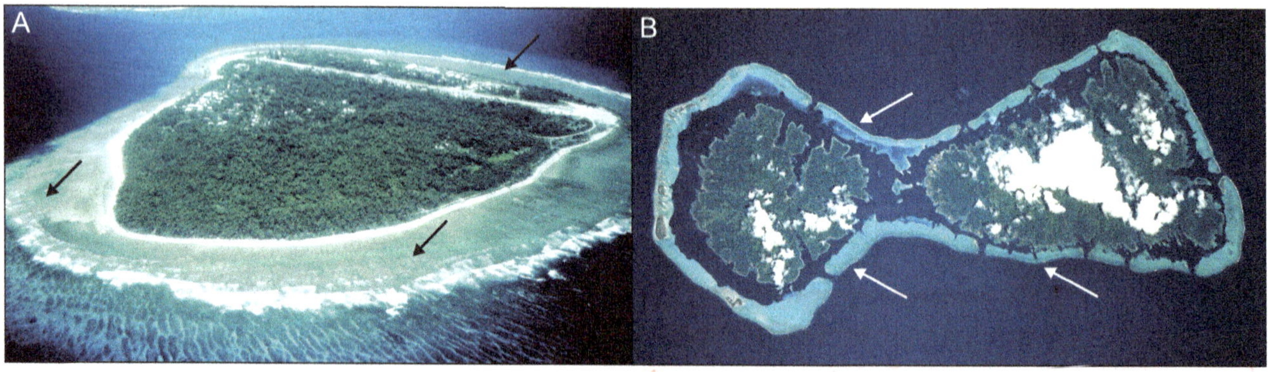

Figure 14.44: Basic types of reefs. A. This image is an excellent example of a fringing reef. Photo credit: Ben Mieremet/NOAA Photo Library. B. These two islands are surround by a "figure 8" of barrier reefs. Photo credit: Image Science and Analysis Laboratory, NASA-JSC.

Atolls are reefs which have built up into a more or less continuous circular, semi-circular or ovoid ring of reefs with a lagoon in the middle (Figure 14.45). There is no rock-based landmass exposed in or near an atoll. Most atolls begin their life as fringing and/or barrier reefs around an island (Figure 14.45A). Over time, the island slowly shrinks in both height and aerial extent as both wave erosion along the coast and stream erosion on land attack the island and weather it down. Rising sea level will also help to accelerate the destruction of the island. Eventually the central island disappears and the reef (which can continue to keep pace with changing sea level) is left behind (Figures 14.45B-C). Over time, the reef flat may become vegetated and develop into an island called a cay. Atolls are

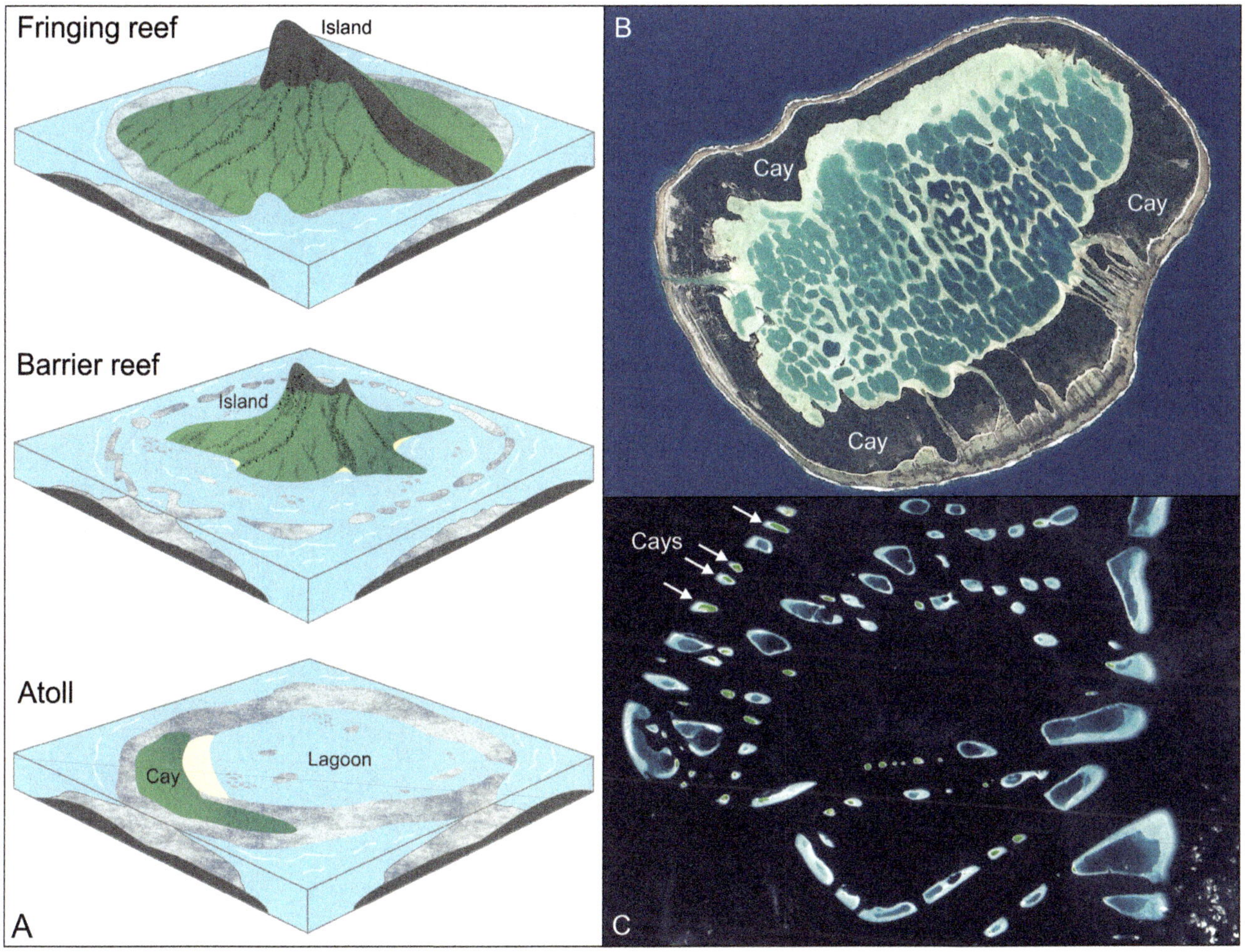

Figure 14.45: A. The evolution of atolls and cays. B. Mataiva Island in the Tuamotu Archipelago is a classic example of a well-developed oceanic atoll. Photo credit: Image Science and Analysis Laboratory, NASA-Johnson Space Center. C. A compound atoll in the Maldives, an island republic in the northern Indian Ocean, southwest of India. The Maldives are made up of a chain of 1,192 small coral islands that are grouped into clusters of atolls. Photo credit: NASA/GSFC/METI/ERSDAC/JAROS, and U.S./Japan ASTER Science Team.

particularly abundant in the Pacific and Indian Oceans where volcanic islands are common.

NASA

NASA

NASA

NASA

NASA/STSCI

Dana Berry/NASA

Chapter 15 - An Introduction to Astronomical Techniques

An experience that everyone should have is to look upward at a nighttime sky filled with stars. If you've ever looked into the expanse of space, you might wonder how the Earth fits into the larger picture- that of the solar system, the galaxy, and the universe. Humans have been doing so for thousands of years, and have used their observations to make scientific conclusions, but have also used them to develop myths and legends that have permeated many ancient cultures. The ongoing study of the universe, also known as astronomy, helps us to understand more about the Earth and its surroundings. In the next few sections, we will introduce the astronomical techniques that scientists use to help us understand Earth's place in the universe.

Section 15-1: What is a Universe?

What is a **universe**? A basic definition is: all of the matter, energy, and space existing in a certain area (or something to that effect). Our universe contains all of the stars, planets, moons, asteroids, nebulas, and all of the emptiness in between and beyond them. There are more galaxies and stars than we have ever been able to count, and astronomers are finding more every year. Figure 15.1 is known as the Hubble Deep Field. It is a small region of the sky (~1/15th of the Moon's diameter) near the bowl of the big dipper, representing only one-forty-millionth of the entire sky. If you looked at this region with your naked eye, you may only see one or two stars. But by collecting telescopic data from over 100 hours of observation and compiling it, we get the image below. If this image is typical of the rest of the universe, it means that there are some 125 *billion* galaxies in the universe. There are no doubt even more galaxies

Figure 15.1: This image was assembled by combining 10 years of NASA Hubble Space Telescope photographs taken of a patch of sky at the center of a small area of space in the constellation Fornax. Image credit: NASA; ESA; G. Illingworth, D. Magee, and P. Oesch, University of California, Santa Cruz; R. Bouwens, Leiden University; and the HUDF09 Team

since the universe is so vast and there is no way we can see everything contained within it.

You may ask, "If we can't see the entire universe, how can we study it?" We learn about the universe by studying those things we *can* see and those quantities we *can* measure, and then theorizing about what those measurements mean on the universal scale. But it also means that many things we think we understand about the universe are speculations without hard scientific evidence. For example, imagine that you walk to the edge of the universe. What would you find? A wall? Empty space? If there is an edge, what is beyond it? Since the universe contains everything, including empty space, it is believed that the universe has no edge, because the idea of a universal edge goes against common sense (modern astronomical observations indicate that the universe is infinite, and therefore has no edge or center). Up until the early 1800's, most astronomers looked at the universe as infinite in age, uniform, and unchanging. However, in 1826 an astronomer named Heinrich Olber argued that this view of the universe was incorrect.

Figure 15.2: Olber argued that within a dense forest, every line of sight ends at a tree. Similarly, every line of sight in the night sky should encounter a star if the sky was unchanging. Photo Credit: M. Bailey.

Olber's argument was this: if the universe is unchanging and full of an untold numbers of stars, then anywhere you look, your line of sight should run into a star. To understand this, look at Figure 15.2. If you are in a dense forest, everywhere you look, your line of sight should hit a tree. Olber argued that this should also be true for the universe, meaning that our nighttime sky should be much brighter than it is. Even though stars that are farther away should be dimmer, there should be many more stars farther away, and these two effects would cancel out. From this argument, astronomers began to consider the idea that the universe was different than they had previously supposed. Along with these early ideas, other theories about the universe are based solely on scientific measurements. For example, astronomers have used telescopes and spectroscopes to see that all galaxies not in our local galactic cluster are moving away from ours in all directions. A consequence of these observations is that we must live in a universe which is expanding, a topic we will discuss further in Chapter 17.

Figure 15.3: In its most basic form, a telescope is simply a device which gathers and focuses the light from distant objects. Photo credit: © SirinS/shutterstock.com.

Section 15-2: Telescopes

How do we know so much about the universe, galaxy, and stars? There are many scientific instruments that are used in the study of astronomy. One of the first instruments used to study the night sky, and one of the most important, is the **telescope** (Figure 15.3). Telescopes have been around for over 400 years. The basic idea of a telescope is to gather as much as light as possible and to concentrate it into one location to form an image. This can be done in a couple of different ways. One way is to focus all of the light by use of lenses. As light travels through a lens, the light is bent and can be concentrated at one point, known as the focal point (Figure 15.4A). Telescopes that use lenses only are known as **refracting telescopes**. Other types of telescopes use curved mirrors to reflect the light and concentrate it on a small area. Telescopes that use mirrors are known as **reflecting telescopes** (Figure 15.4B). In both types of telescopes, there is an objective lens or objective mirror (referred to as the *objective*), and then typically a second lens known as the *eyepiece*, which serves to magnify the image made by the objective.

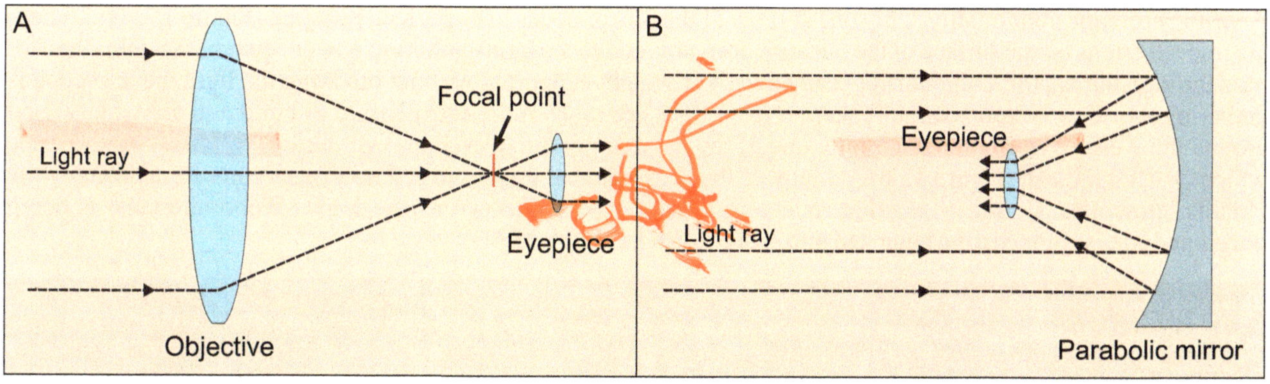

Figure 15.4: A. A refracting telescopes uses an objective (such as a lens) to focus light. B. In a reflecting telescope, a mirror is used to focus light.

Nearly all of the interesting objects in the sky are faint sources of light, so to study those objects astronomers need telescopes with large light-gathering power. In telescopes, the larger the objective (in diameter), the more light a telescope can gather. Light gathering power increases with the area of the objective, that is, it is proportional to the objective's diameter squared. So, a 2 m objective telescope would have 4 times as much light-gathering power as a 1m objective. For astronomers, the diameter of a telescope's objective is the single most important characteristic of a telescope. The world's largest operating single-mirror reflecting telescope is located on Mauna Kea in Hawaii and has a 10 m diameter objective mirror. This telescope can gather approximately 4 million times the amount of light the human eye can gather (Figure 15.5).

Almost all large telescopes are reflecting telescopes, because refracting telescopes have some drawbacks. If a large lens is used for a telescope's objective, it is very heavy and tends to change shape under its own weight. This will cause it to focus light at a different point than originally designed. If supports are built to help sustain the weight, these supports will block some of the light of the telescope. Another disadvantage of refractive telescopes it that different colors of light bend differently as they pass through a material such as glass. This means that in a simple telescope made of only two lenses, red and blue light will focus at different points. If you focus the eyepiece on the blue light, then all of the other colors will be out of focus and you will see a colored blur around the image. The same is true if you focus on the red light. This effect is known as a *chromatic aberration*. Chromatic aberrations can be corrected by making a compound lens with layers of several different materials, each with different shapes. However, such lenses are difficult to design, and at each layer's surface, some light is reflected, reducing the amount of light that passes through a lens, which in turn reduces the light-gathering power of the telescope. A third problem with refracting telescopes is that lens surfaces must be perfectly shaped; so if we have a compound lens that means every surface of every layer must be perfectly shaped.

Reflecting telescopes do not have these drawbacks. All colors reflect the same way off the mirrors' surface, so they

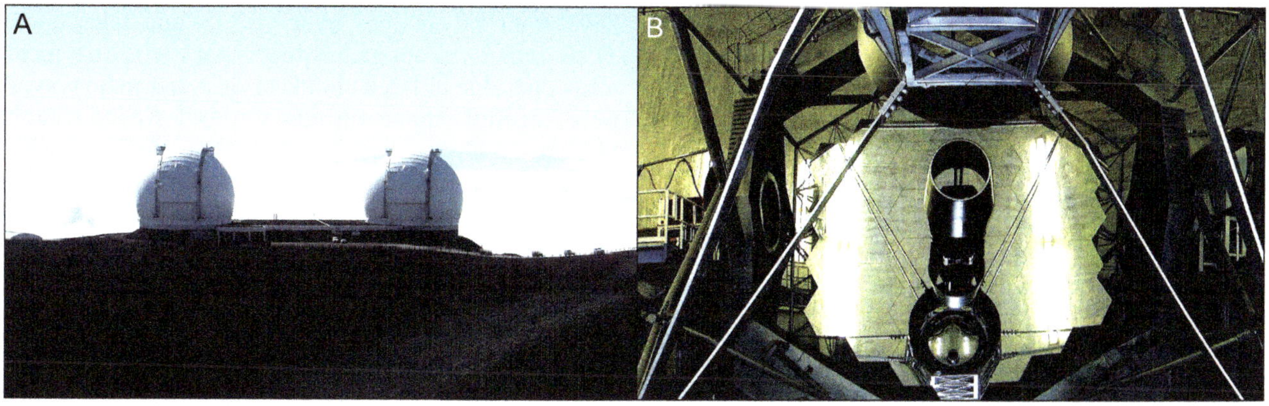

Figure 15.5: A. At the summit of Mauna Kea, Hawaii, NASA astronomers have linked the two 10 m (33 ft.) telescopes at the W. M. Keck Observatory. The linked telescopes, which together are called the Keck Interferometer, make up the world's most powerful optical telescope system. Photo credit: K. Baccala. B. A view of the Keck 2 primary mirror, made up of 36 1.8 m hexagonal mirror segments. The conical tube in the center of the mirror is a baffle, below which is reflected an image of the round secondary mirror. Photo credits: NASA/JPL.

have no chromatic abnormality. Only one surface of the mirror needs to be perfectly shaped (Figure 15.5B). If there are imperfections on the surface of the mirror it does not reduce the light-gathering power because the light does not pass through the mirror. Large mirrors can have their weight supported without blocking the light the telescope is gathering. For these reasons large reflecting telescopes are much more inexpensive to build and are widely used around the world. The Hubble telescope, one of the most important telescopes ever built, is a reflecting telescope which orbits the Earth (Figure 15.6A). Many of the astronomical images in scientific academia were taken by the Hubble telescopes. Currently, another reflecting space telescope known as the James-Webb telescope is being developed and is planned to be launched into space in 2018 (Figure 15.6B).

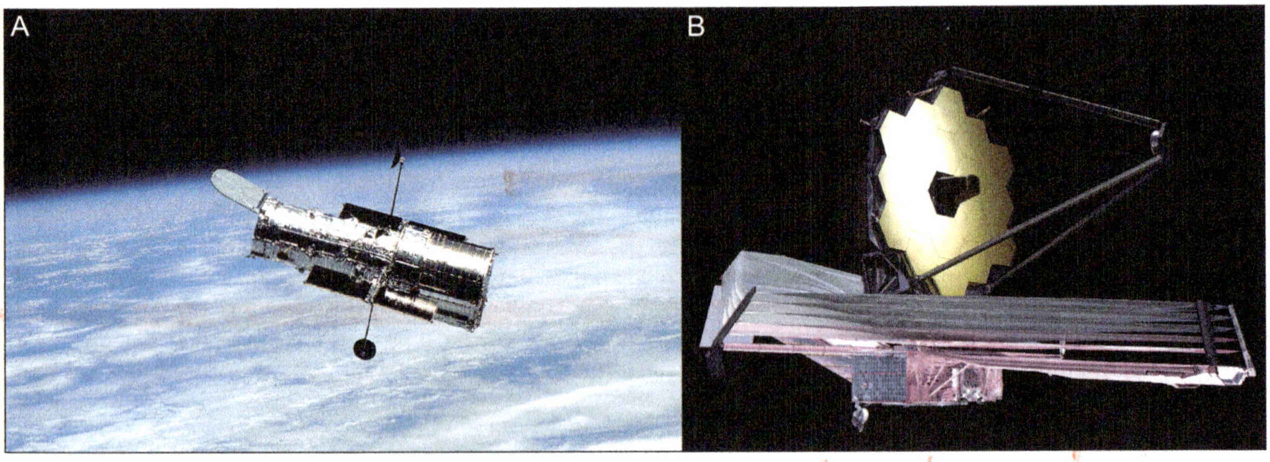

Figure 15.6: A. The Hubble telescope orbits the Earth, allowing it to avoid the distortion of the Earth's atmosphere. It has provided some of the most important images of our galaxy within the past two decades. Photo Credit: NASA/STScl. B. An artist's depiction of the James Webb Space Telescope (JWST), the Hubble telescope's eventual successor. JWST will be located in solar orbit 1 million miles from Earth. Image credit: NASA.

Section 15-3: Radio Astronomy

Astronomers have used telescopes to gather more than just the wavelengths of visible light. **Radio telescopes** (telescopes that gather radio waves) can see many things in space that are not very exciting when looking at them in typical telescopes. Examples of these objects are **quasars** and **pulsars**. Quasars are very luminous when viewed by radio telescopes, but look very faint in visible-light telescopes. Pulsars, which have given astronomers a lot of information about the final stages in the life of a star, would possibly never have been discovered using only visible-light telescopes, but are very prominent with radio telescopes (More on pulsars and quasars in Chapter 17).

Figure 15.7: A simple radio telescope antenna. Photo credit: Brian Day/NASA.

It is a common misconception to think of radio waves as related to sound, but in fact, they are electromagnetic waves that have fairly long wavelengths. Because of these long wavelengths (and smaller frequencies), they have some advantages over visible-light telescopes. There is much of our galaxy that is not visible to typical telescopes due to large clouds of dust, but radio waves pass through these clouds without losing much luminosity. It is technically very difficult to focus radio waves with a lens, so all radio telescopes are reflecting telescopes. An example of a radio telescope is shown in Figure 15.7. One problem with radio telescopes is that radio waves coming from space are typically of low intensity. Cosmic radio signals arriving on Earth are very weak, typically a million times or more, weaker than a typical radio station. To get detectable signals, astronomers must either build large telescopes, or combine multiple smaller telescopes in arrays. Examples of each of these can be seen in Figure 15.8.

Figure 15.8: A. The world's largest radio telescope is the Puerto Rican Arecibo Radio Observatory. The dish measures over 305 meters across. Photo credit: Z. Baccala. B. The Very Large Array (VLA), containing 27 receivers, is a radio telescope in New Mexico. Photo credit: ©IrinaK/shutterstock.com.

When arrays of telescopes are built, the signals from the telescopes are combined, helping to collect and resolve the radio signals. Even then, because the radio signals from space are so weak, the signals must be amplified before they can be measured and recorded.

Other wavelengths of electromagnetic radiation can also be used to study our galaxy and the universe. The James Webb Telescope (Figure 15.6B) will study the universe by collecting infrared radiation. Since almost all infrared radiation from space is absorbed by the Earth's atmosphere, telescopes with the most sensitive long-wavelength infrared detectors must be located above the Earth's atmosphere. The atmosphere also interferes with gamma rays, X-rays, and ultraviolet waves eminating from space, and astronomers have begun to use satellites to study these types of electromagnetic radiation. More on this in the next section.

Section 15-4: Spectroscopy

After collecting the electromagnetic radiation from telescopes, sometimes astronomers spread out the light according to the wavelength (color for visible light). This was first done by Isaac Newton in 1666, when he made a small hole in a window shutter, and then placed a prism over the incoming beam of light. The white light entered the prism, but a beautiful spectrum of light came out the other side (Figure 15.9). From this and other experiments, Newton showed that white light is made up of a mixture of all colors known as a **spectrum** (the plural of which is spectra). Other scientists built on Newton's studies to study the spectra of light (known as spectroscopy). One such scientist was Gustav Kirchhoff. Kirchhoff studied many aspects of light and came up with a few laws concerning all types of electromagnetic radiation. Kirchhoff's First Law is of interest here:

A hot solid, liquid, or a dense gas emits radiation at all wavelengths (a continuous spectrum of radiation). If the light from these objects were passed through a prism, you would see a continuous band of the whole rainbow of colors.

An incandescent light bulb (where electrical current flows through a tungsten filament) is an example of a solid giving off radiation; hot, molten steel is an example of a hot liquid; hot, dense gases are harder to find in everyday settings, but they exist in many places, such as the interior of stars. When these types of objects emit radiation, they do so across all the wavelengths.

Figure 15.9: A simple prism demonstrates how visible light is really just a combination of all the colors. Photo credit: © Mopic/ shutterstock.com.

Kirchhoff studied what types of spectra we would see when we look at objects in space. To understand his findings, let's use the following scenario. Let's say you have a friend on a different planet that has an incandescent bulb in a flashlight. If he aimed that light toward Earth, and you could collect the light in a telescope and pass it through a prism, you would see all colors of light. This is known as a **continuous spectrum** (Figure 15.10). However, if there was a cloud of gas between your friend's flashlight and your telescope, then some of the light would absorbed by the gas. Specifically, light of certain wavelengths would be absorbed, and the gas would become hot. Now, if you aimed your telescope so that light from the flashlight could not enter, but still looked at the areas of the gas near the flashlight, a second type of spectrum would be seen. The light you would see would be emitted by the hot gas, which initially absorbed the light of the flashlight. This type of spectrum is known as the **emission spectrum**, or a **bright-line spectrum**, and can be seen in Figure 15.10. This is the type of spectrum seen from an emission nebula. Now, a third type of spectrum can be seen if you looked directly at the flashlight through the cold gas with your telescope, and passed it through a prism. In this case, some of the colors would be missing. You would see their positions as dark absorption lines projected on a continuous spectrum. This is known as an **absorption spectrum**, or a **dark-line spectrum**, and can be seen in Figure 15.10. Looking at the types of spectra an object produces helps us to understand the object's properties.

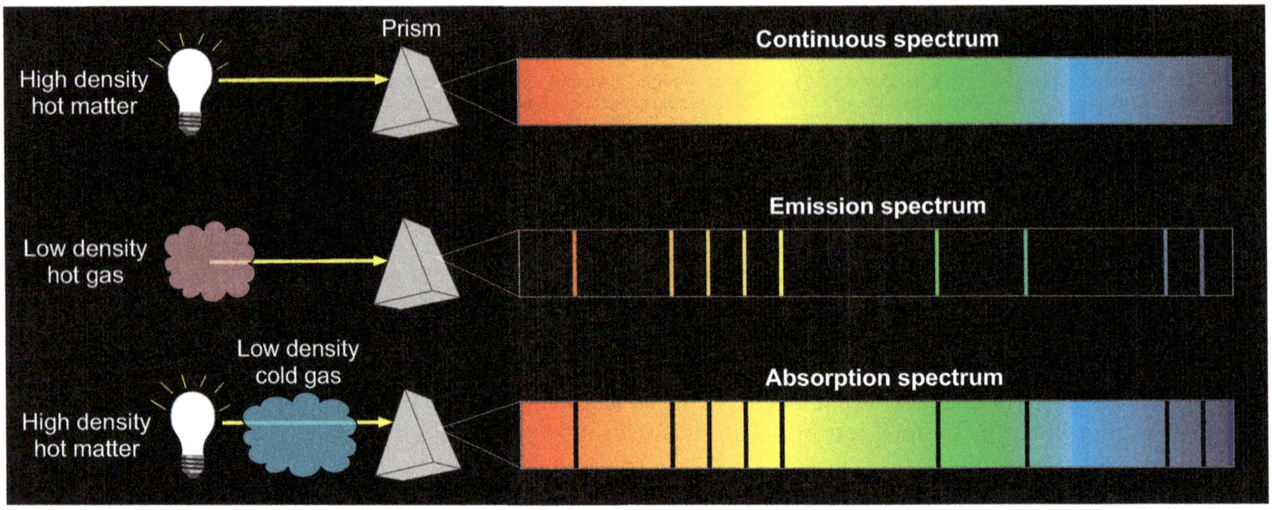

Figure 15.10: Three types of spectra. The type of spectrum seen gives us information about what type of object is emitting or absorbing the light.

Today, astronomers still use telescopes to gather the light, but use other instruments to separate the different wavelengths of the light. A spectrometer is used to look at the visible-light spectrum, while spectrographs use computers or other types of sensors to determine the wavelengths of other forms of electromagnetic radiation. If spectra are recorded digitally with a spectrograph, not only can their wavelengths be determined, but also the intensity of each wavelength. For example, even though both red and blue light will be seen in the spectrographs results, there may be more blue light than red light, and this difference in intensity can give us information about the object emitting the light.

Applications of Spectroscopy

If you place a fireplace poker into a fire, it will begin to glow, and the color of that glow changes from dark red, to light red, to white. According to Kirchhoff's first law, a hot solid will radiate all colors, so why do we see all the change in the glow? In fact, the poker is radiating all wavelengths, but not at the same intensity. As the poker gets hotter, it starts to emit more yellow, green, and blue light than red light, and so the combination of the colors of light changes, and it appears to glow a different color. This relationship was clarified by the German Physicist Wilhelm Wien, and is known as Wien's law:

Objects of different temperature emit spectra with intensities that peak at different wavelengths. Hotter objects emit most of their radiation at shorter wavelengths; hence they appear to be bluer. Cooler objects emit most of their radiation at longer wavelengths; hence they appear to be redder.

Using spectrographs, astronomers can determine the wavelengths of the electromagnetic radiation coming from stars or other objects, as well as the intensity of each wavelength. Chemists had long understood that different chemical elements exhibited different, and highly specific, emission spectra, allowing their identification. If a gas were made up of several elements, the resulting spectrum would contain features of each individual spectra, and each element could be identified, and the intensity of each individual spectrum could be used to determine the proportion of the elements within the gas. In essence, looking at the spectrum of an object, as well as the intensity of the wavelengths tells us the composition of the object.

Not only can spectra tell us what an object is made of, but it can also tell us about the object's motion. Imagine that you are standing on the side of the road and a firetruck is coming towards you with its sirens blaring. You will hear that not only does the firetruck get louder as it moves toward you, but the pitch of the sound gets higher. As the firetruck passes and moves away from you, you immediately notice that the pitch of the sound suddenly lowers – a sort of "weeeeeeee-oooooooh" sound. This change in frequency is known as the **Doppler Effect** (or Doppler shift) and is caused by an observer, or a source of waves, that are moving relative to each other. In the case described above, the source of the wave (sound) is the siren, and the observer is you.

Since sound is a wave, it has a specific speed in air (about 340 meters per second), as well as having a wavelength and a frequency. When a sound is emitted, these waves travel outwards in all directions. To understand why the Doppler Effect occurs, look at Figure 15.11. When the fire truck is stationary, the wavelength an observer hears is the wavelength that the siren is emitting (Figure 15.11A). However, when the fire truck moves toward an observer, the waves are crowded together between the observer and the siren, giving them a shorter wavelength and a higher frequency or pitch (Figure 15.11B). After the fire truck passes by, each wave is coming from a source further away from the observer, elongating the wavelengths and lowering the frequency.

The Doppler Effect also occurs in other types of waves, such as light. If an object is moving toward a stationary observer, the entire spectrum is shifted to shorter wavelengths, and the light from that object will look bluer than

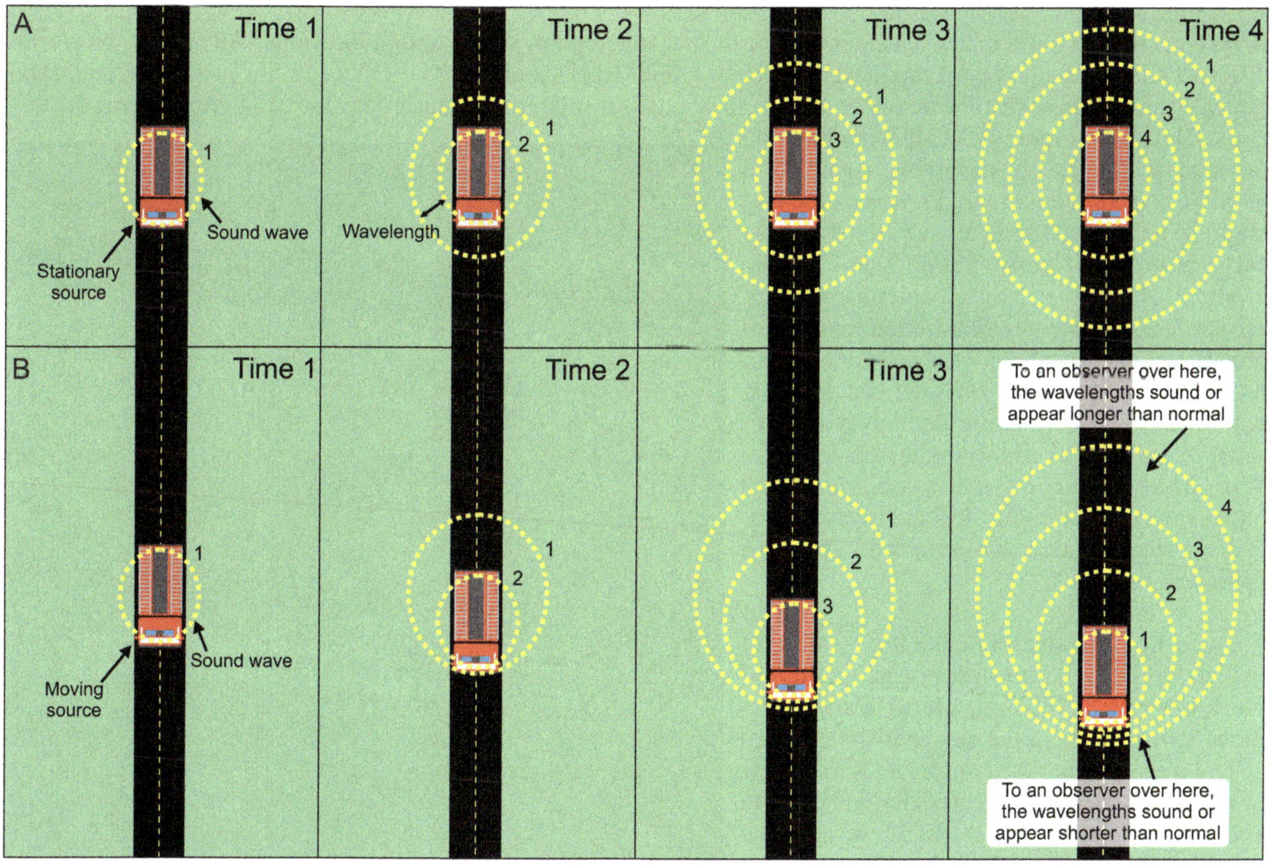

Figure 15.11: An explanation of the Doppler Effect. Each circle represents the wave crests of sound (or light) going in all directions from the source. A. When the source is stationary with respect to the observer, the wavelength emitted is the wavelength heard (or seen, in the case of light). B. The source moving with respect to the observer creates an observed wavelength that will be different than that emitted by the source.

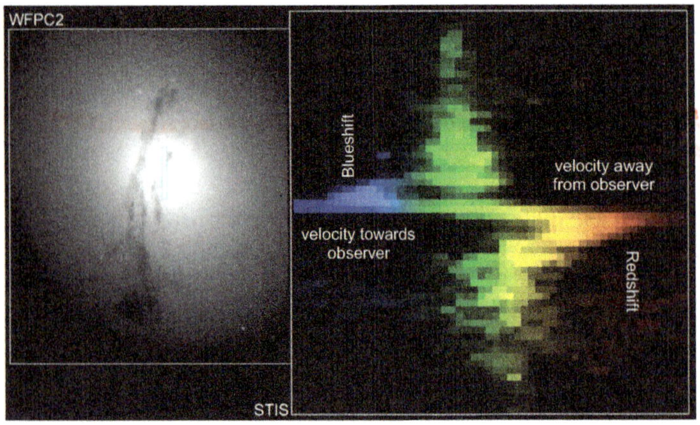

Figure 15.12: An example of how redshift and blueshift are used to measure the velocity of a galaxy. How far the light has shifted toward the red or the blue is a function of how fast the galaxy is moving towards or away from the observer. Image credit: B. Woodgate (GSFC), G. Bower(NOAO), STIS Instrument Definition Team, NASA.

normal. This is called a **blueshift**. When an object is moving away from the observer, the wavelengths will lengthen; the spectrum will shift to longer wavelengths, known as a **redshift**. The faster the object is moving, the larger the blue or red shifts, and the easier they are to detect. Using spectrographs, we can determine the wavelengths of the light coming in from the object, and compare that to spectral lines from a stationary object. These redshift and blueshift measurements are then used to calculate the speed of the distant objects relative to Earth. An example of these shifts is shown in Figure 15.12. It is from these types of measurements that we have learned that all other galaxies are moving away from ours, and the universe is expanding.

Section 15-5: Stellar Positions and the Celestial Sphere

Today we may take for granted that we are living on a small planet orbiting a typical star in one of the many galaxies in the universe. This conclusion is not obvious just by looking at the night sky, and has been determined through years of careful and precise observations and measurements. Ancient astronomers imagined that the sky was a great sphere surrounding the Earth, and the stars were stuck on the inside of the sphere, almost as if they were glued there. Today we understand that the idea of a celestial sphere is incorrect, but it is still a useful model, because it allows us to map the sky as seen from Earth. Assigning locations to objects in space helps us to study the objects.

If we consider the Earth to be the center of the celestial sphere, how do we assign locations? All around the world, different cultures such as the Assyrians, Babylonians, and Greeks looked into the sky and saw patterns of stars, and typically named these patterns based on their heroes, gods, or mythical beasts. Today we call these patterns of stars **constellations** (Figure 15.13). You may be familiar with some constellations, such as the great bear (more commonly known as the Big Dipper) or the constellation Orion. To these early cultures, a constellation was a loose grouping of stars that depicted a picture. Modern astronomers have defined constellations in more precise terms. A constellation is now defined as a region of the sky with well-defined borders; within these borders is the familiar pattern of stars which help us to locate the constellation. With this definition every constellation represents an area in the sky, and every star or every object in the sky is located within a constellation.

Figure 15.13: Humans have been using stars and constellations as a navagational guide for thousands of years. Image credit: © tatishdesign/shutterstock.com.

There are other ways to define the location of objects in the sky. The west-to-east rotation of the Earth causes the Sun, Moon, and stars to move east-to-west in the sky as if the celestial sphere were rotating westward. The stars appear to rotate about points directly above the North and South Poles of the earth. These apparent pivot points are the location where the extension of the Earth's axis intersects the celestial sphere,

Figure 15.14: A. Polaris, or the North Star, is a useful navigational guide because it never moves its position relative to the Earth. Photo credit: © Valerio Pardi/Shutterstock.com. B. A time-lapse photo shows how all the other stars move but Polaris does not. Photo credit: © shihina/shutterstock.com.

and are known as the **North Celestial Pole** and the **South Celestial Pole**. Coincidentally, there is a very bright star called *Polaris* or the *North Star* that is very close to the North Celestial Pole, so this star does not seem to move along as the other stars rotate westward (Figure 15.14). Located halfway between the celestial poles lies the **celestial equator**, which is an extension of the Earth's equator onto the celestial sphere. Another characteristic of the celestial sphere is the ecliptic, which is the apparent path of the Sun against the background of the stars. If our model of the celestial sphere was correct, the ecliptic would be the shadow cast by the Earth as it rotates around the Sun. Since the Earth's axis is tilted with respect to its orbit, the ecliptic and celestial equator are in different planes in space.

On Earth, we use coordinates known as latitude and longitude to describe a location on the Earth's surface. Using these standard lines, we can measure anything on Earth. Similarly, we can describe locations on the celestial sphere. We can measure the angular distance north or south of the celestial equator, known as the **declination**; this is equivalent to latitude on Earth. On the celestial sphere, **Right ascension** (RA) is similar to longitude. It gives the location of a star along the celestial equator. Right ascension has its zero point where the celestial equator and the ecliptic cross when the Sun moves from the Southern into the Northern Hemisphere. This point is also known as the vernal equinox (Chapter 2). Together, these quantities make up the **celestial coordinate system**, and are shown in Figure 15.15. With this type of coordinate system, we can catalog the location of any object in the sky, and anyone in the world can find the same object, because the location of the object does not change.

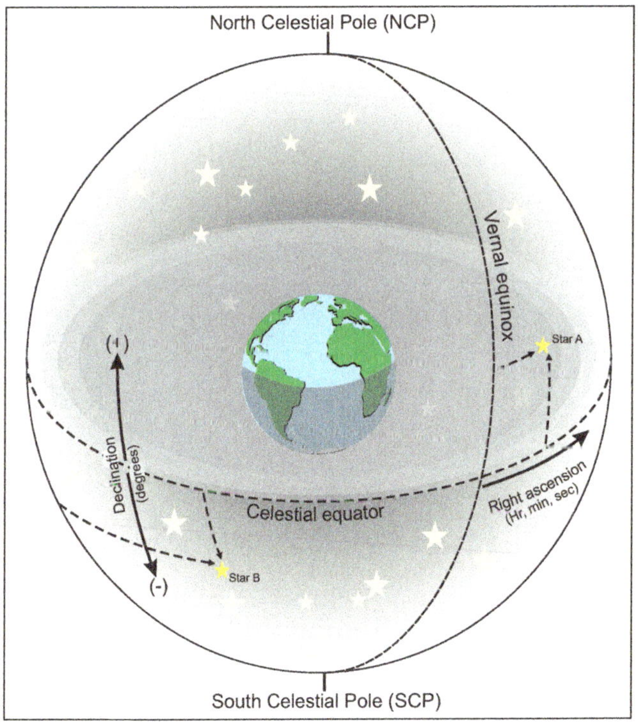

Figure 15.15: The celestial coordinate system. Declination is measured north and south of the celestial equator and measured in degrees, minutes, and seconds with stars north of the celestial equator having positive declination (Star A), and stars south of the celestial equator having negative declination (Star B). Right ascension is measured from the vernal equinox, and is measured in hours (0 – 24), minutes, and seconds.

NASA/JPL

NASA/Johns Hopkins University/Carnegie Institution of Washington

NASA

NASA

NASA/JPL

Chapter 16 – The Solar System

For thousands of years, men could only speculate on Earth's beginnings. They believed that the Earth was special, and for centuries they even believed that the Earth was the center of the solar system, as well as the universe. Discoveries over the last few hundred years have shown that the Earth is one of many planets in our solar system, and that our solar system is just one of many similar systems throughout the universe. In seeking to know about Earth's origin, we turn to our nearest heavenly neighbors, our own solar system. By studying the similarities and differences of these celestial bodies, we hope to understand our own cosmic origin. We'll start begin our journey through the solar system with our closest neighbor, the Moon.

Section 16-1: The Moon

The Moon is the Earth's constant partner, orbiting the planet as the Earth orbits the Sun (Figure 16.1). Because the Moon is the second brightest object in the sky, scientists have been studying it for thousands of years. Between 300-200 B.C., the Greek astronomer Aristarchus determined the distance between the Earth and the Moon (384,000 km/236,000 mi.), and consequently, determined the approximate size of the Moon. Since Aristarchus, other scientists, such as Galileo, have used telescopes to study the Moon. In the last 50-60 years, scientists have used satellites to study the Moon, and have even landed 12 astronauts upon its surface (Figure 16.1B). All of these studies help us to understand our celestial companion.

Figure 16.1: A. After Apollo 12 left lunar orbit this image of the Moon was taken from the command module on 11/24/69. Photo credit: NASA. B. View of Station 11 taken during the third Extravehicular Activity on the Apollo 16 mission. Note the lunar rover for scale. Photo credit: NASA.

The Moon orbits the Earth every 27 ⅓ days, and rotates on its own axis in 27 ⅓ days. Because these times are equivalent, the same side of the Moon is always pointing towards Earth. The only information we have about the far side of the Moon comes from satellite and spacecraft studies. Earth's rotation makes the moon appear to move in the the sky from east to west, just like all of the stars on the celestial sphere. However, because the Moon is orbiting the Earth, the Moon will move about 0.5 degree each hour relative to the background of the stars each night. This effect is quite noticeable through an entire nighttime period. If you were to look at the bright stars next to the Moon at dusk, and then check them again right before dawn, you will see that the Moon has moved considerably (Figure 16.2).

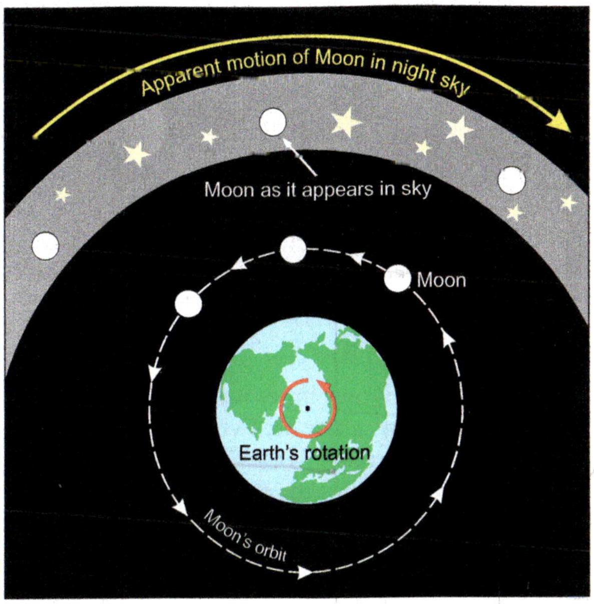

Figure 16.2: The Moon appears to move eastward compared to the stars behind it. This is due to the Moon's orbit around the Earth.

Phases of the Moon

As the Moon orbits the Earth, it will rise and set in the sky at different times each day. This orbit also causes a change in the appearance of the Moon, or its **lunar phase**. Half of the Moon is always lit by the Sun, just as half of the Earth is always illuminated. How much of the brightened side of the Moon we can see depends on the Moon's position relative to the Earth and the Sun. For example, when the Moon is on the opposite side of the Earth from the Sun, we will see the entire illuminated side (Figure 16.3). This is known as the **Full Moon**. When the Moon is on the same side of the Earth as the Sun, we see none of its illuminated side, and the Moon appears to be dark. This is known as a **New Moon**. The other phases of the moon range from full to new as the Moon moves around the Earth. If the Moon appears to be "growing", that is, we see more of its illuminated side than the day before, this is known as the **waxing** period; if the Moon appears to be "shrinking", this is known as its **waning** period. The "half-moon" phases are known as 1st quarter moon during the waxing period and 3rd quarter moon during the waning period. When less than half of the Moon is visible, we refer this to as a **crescent** phase, and when more than half of the Moon is visible (but less than a full moon) we refer to this as a **gibbous** phase.

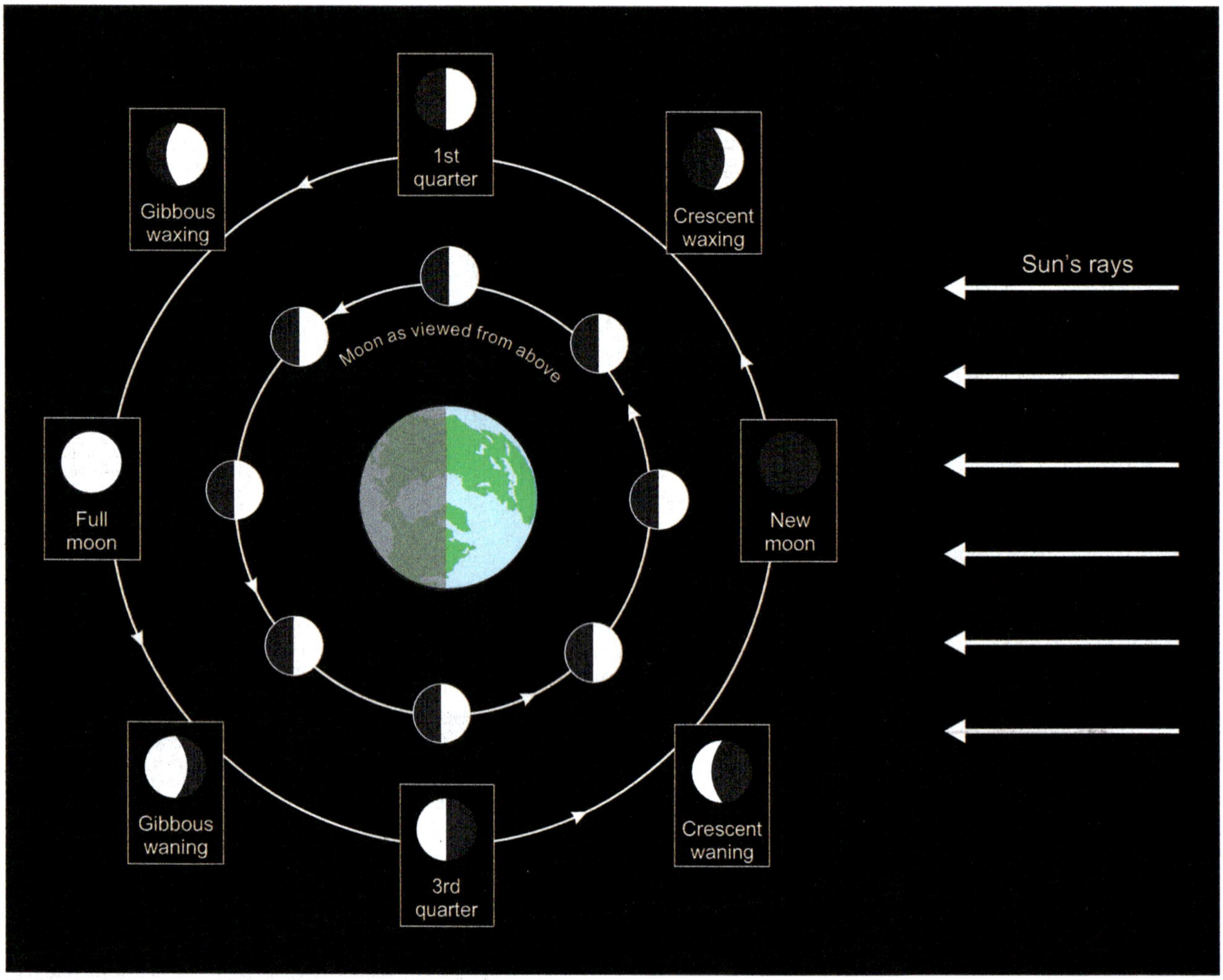

Figure 16.3: The phases of the Moon change as the Moon orbits the Earth.

Eclipses

There are two types of eclipses; **lunar eclipses** and **solar eclipses**. In a lunar eclipse, the Moon's orbit carries it into the Earth's shadow. There are two areas within the Earths' shadow; the region of complete shadow, known as the *umbra*, and an area of partial shadow, known as the *penumbra* (shown in Figure 16.4 for both types of eclipses). As the Moon moves through the penumbra, a partial shadow will be evident upon the Moon's surface. This is known as a *partial lunar eclipse*. Many eclipses are only partial lunar eclipses (Figure 16.4 A). If the Moon continues to move through the penumbra into the umbra (known as a total eclipse) the shadow of the Earth envelops the Moon completely. Even then, the Moon is still visible. The Earth's atmosphere can act like a lens, bending sunlight around Earth and giving a dim glow to the Moon. However, when the sunlight travels through Earth's atmosphere, blue

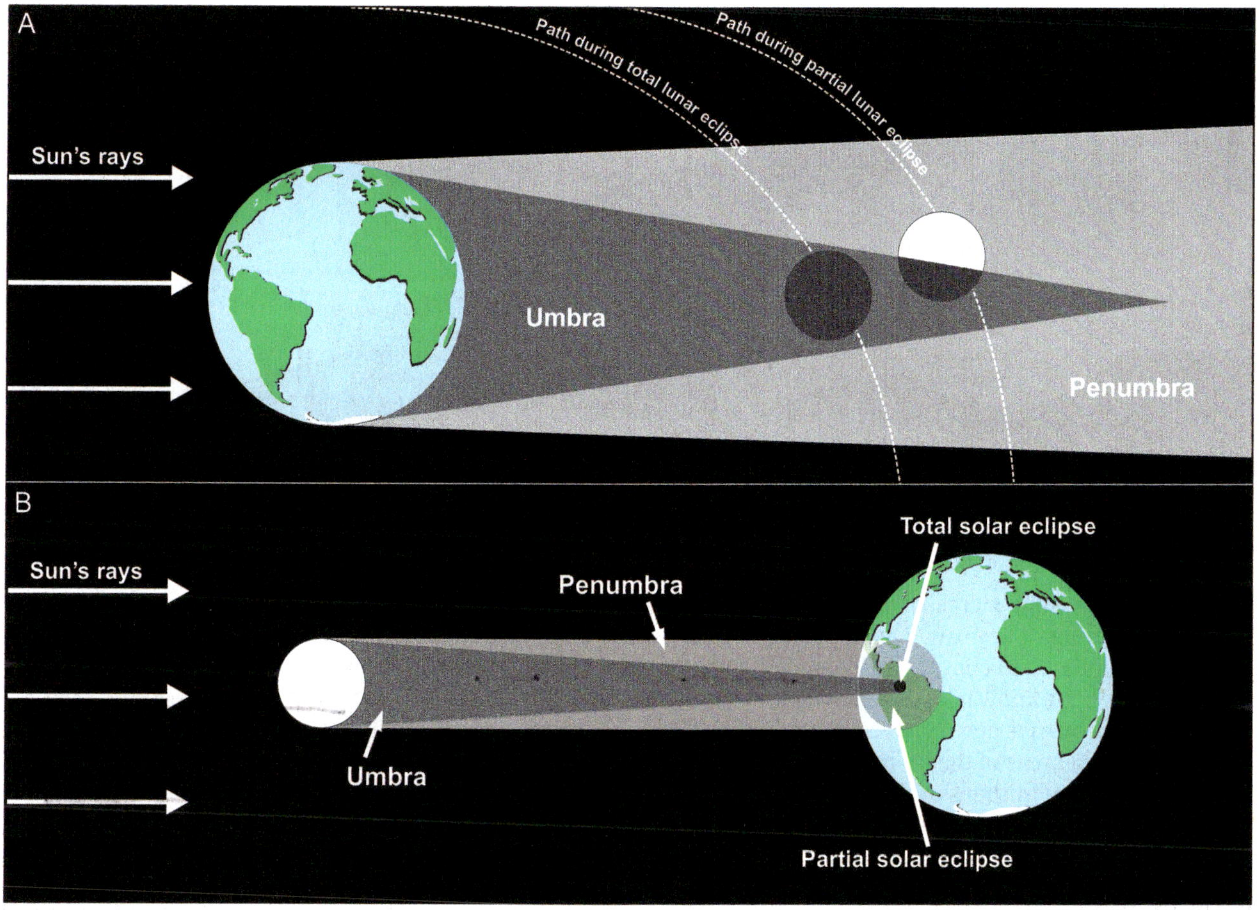

Figure 16.4: The two types of eclipses. A. During a lunar eclipse, the Moon moves into the Earth's shadow. B. During a solar eclipse, the Moon's shadow darkens part of the Earth.

light becomes somewhat scattered, giving the resultant light hitting the Moon a reddish hue. For a solar eclipse to occur, the Moon's orbit carries it between the Sun and Earth, causing the Moon's shadow to darken a section of the Earth's surface (Figure 16.4 B). Most observers of a solar eclipse, will only observe the penumbra, and thus see a *partial solar eclipse* (Figure 16.5). To see a *total solar eclipse*, an observer would need to be in the umbra's direct path, which is only ~240 km (150 mi.) wide.

Figure 16.5: A picture of a partial solar eclipse, taken by the Hinode spacecraft in 2009. Note that the visible part of the sun has been filtered to show the detail of the surface. Photo credit: NASA/JAXA.

When looking at the path of the Moon's orbit it is easy to see that lunar eclipses must occur during its full moon phase, while solar eclipses must occur during a new moon (**Figure 16.5**). However, it should be noted that every full or new moon does not necessarily mean there will be an eclipse. The Moon's orbital plane is tilted 5° from the ecliptic of the Earth (Figure 16.6A, see also Figure 2.27B). The angle of this plane causes the Moon to be higher or lower in the sky at different times. The difference in these rotational planes also means that in most cases, the Moon is above or below a line extending from the Sun to the Earth, not directly on it (Figure 16.6B). Thus, lunar eclipses occur between 2-5 times a year, while solar eclipses average 2.4 per year, with some years having none at all.

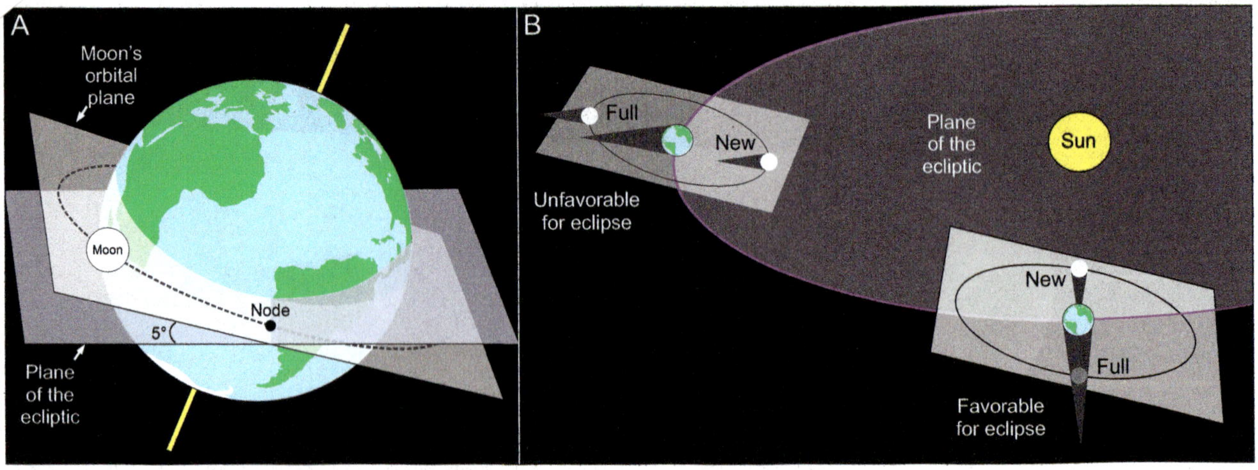

Figure 16.6: A. The plane of the Earth's rotation (the Ecliptic) is 5° different than the Moon's rotational plane. This angle controls the frequency of eclipses, as well as the height of the Moon in the sky. B. It's only when the moon intersects a node that conditions are favorable for an eclipse.

The Lunar Surface

As telescopes became increasingly more available, astronomers (including Galileo) pointed them at the Moon to make observations. They saw many unchanging features, including dark areas they compared to oceans and light areas they compared to continents. These dark areas they named *maria*, the Latin word for "seas" (the singular of which is *mare*). We know now that these areas contain no water, but are areas where lava flowed at some point in the Moon's evolution. It is now believed that billions of years ago, large meteors collided with the Moon, creating large craters in certain areas of the moon. These impacts were so powerful that they created large fissures in the Moon's crust. Hundreds of millions of years later, heat from radioactive decay with the Moon's interior caused molten rock to flow out of these cracks, filling the large craters with lava before cooling. These filled in craters are generally circular. There are 14 maria on the Moon, the largest with a diameter of 1,100 km (~680 mi.) (Figure 16.7).

On the far side of the Moon is a region with very few maria and many craters (Figure 16.7). This area has a higher average altitude than the near side of the Moon, and is known as the lunar highlands (Figure 16.8A). This area is filled with craters, and some of these are of the same magnitude as those of the maria, but they never filled with lava. Rocks brought back from both sides of the Moon show that the highland rocks are of lower density than those from the maria, and are close to 4.1 billion years old. Rocks sampled from the maria are 3.0 – 3.8 billion years old. These

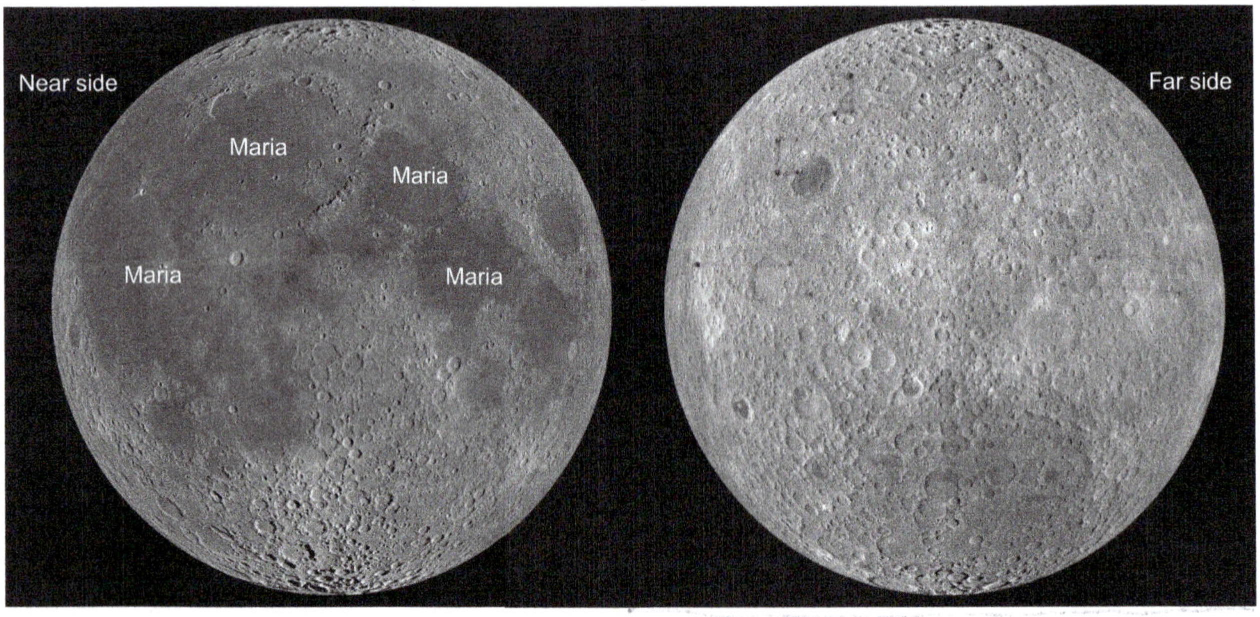

Figure 16.7: The two sides of the Moon. The left image is a view of the near side of the Moon that faces the Earth, while the right image is of the far side of the Moon, the side that faces away from the Earth. Image credit: NASA.

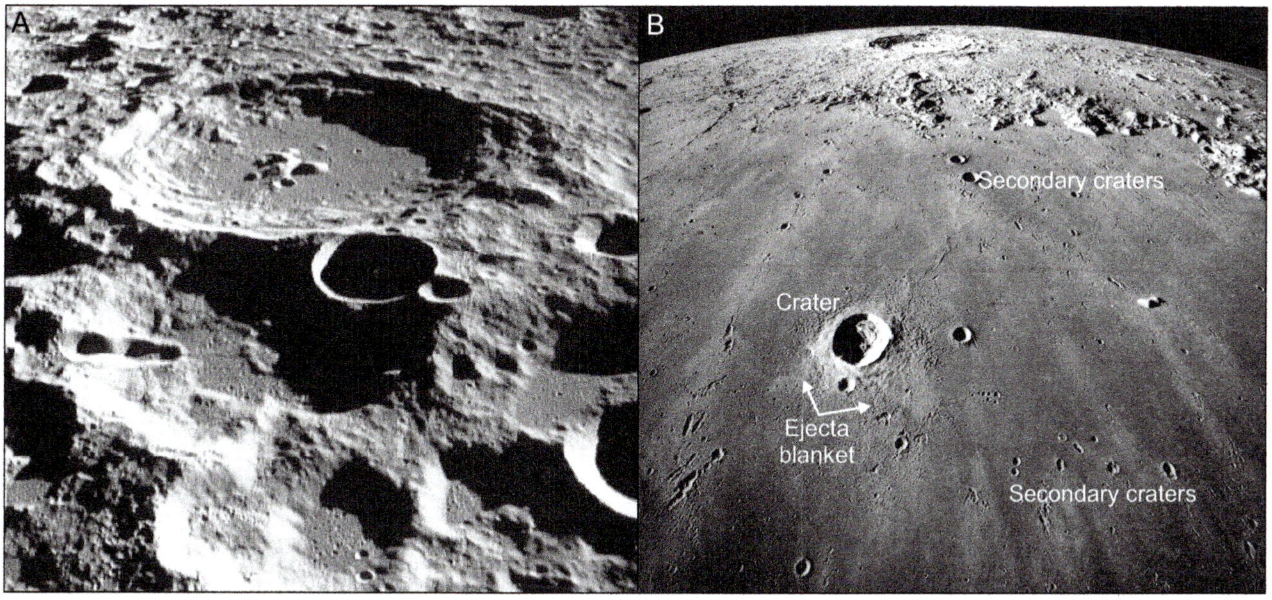

Figure 16.8: A. Craters are very prevalent in the lunar highlands. Photo credit: NASA/Johnson Space Center. B. Different types of meteoric craters on the moon. Photo credit: NASA.

findings provide evidence to collaborate scientist's theories about the maria's formation.

The Moon is riddled with craters, most of which are evidence of past collisions between the Moon and other celestial bodies. In old craters, many of the features of the craters appear to be leveled off, or worn down, due to billions of years of impacts from small meteors, changing the Moon's landscape. However, younger craters are apparent when looking in the near vicinity of the crater. In these younger craters, debris fields surrounding the crater are apparent. This field is known as an ejecta blanket (Figure 16.8B). The figure also shows secondary craters, created when there are large chunks of material in the ejecta blanket which, in turn, make their own craters. A small percentage of craters on the Moon are volcanic in origin, but these are typically easy to distinguish from impact craters due to their elongated shapes.

Due to the constant bombardment of the Moon's surface, there is a thick layer of dust and debris called **regolith** (Figure 16.9). Small meteorites of all sizes (including some as small as a grain of sand) crush rocks on the Moon's surface, leaving smaller and smaller particles with each collision. Some authors refer to this as the "sandblasting of the Moon". Temperature fluctuations also cause lunar rocks to expand and contract, many times leading to cracks within the rocks. In planning the lunar missions of the 1960's, scientists were concerned that the Apollo spacecrafts

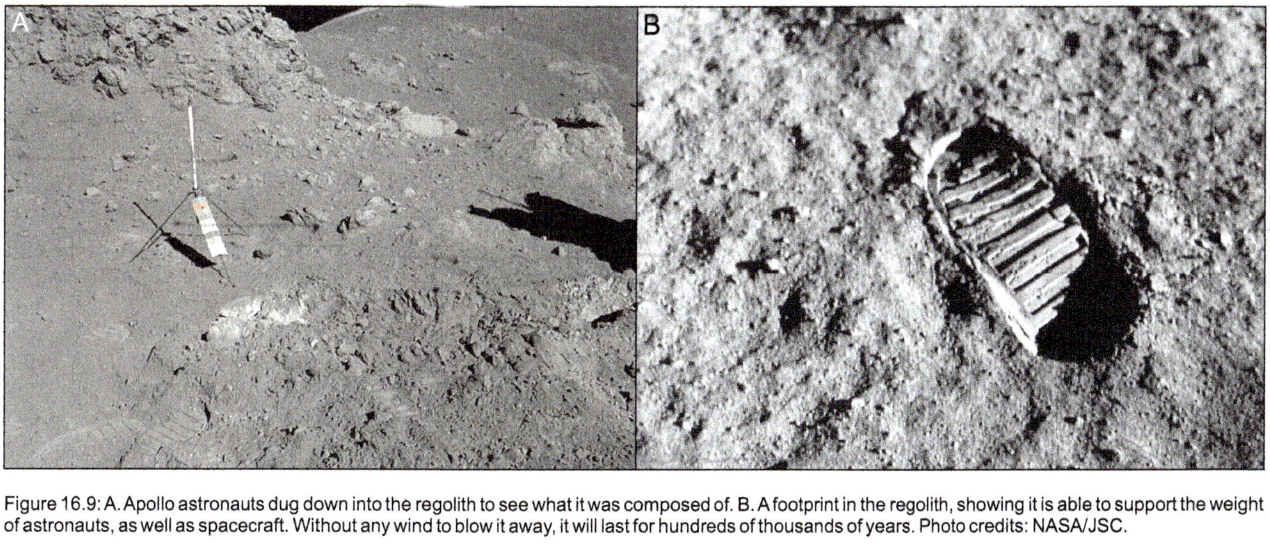

Figure 16.9: A. Apollo astronauts dug down into the regolith to see what it was composed of. B. A footprint in the regolith, showing it is able to support the weight of astronauts, as well as spacecraft. Without any wind to blow it away, it will last for hundreds of thousands of years. Photo credits: NASA/JSC.

would simply sink within this layer of regolith. Unmanned missions to the Moon showed that this dust was able to provide the support necessary for astronauts and spacecraft (Figure 16.9B).

History of the Moon

What do these results tell us about the formation of the Moon? The Moon formed approximately 4.5 billion years ago. Within the next half billion years or so, the surface melted and was entirely molten. During this time, **breccia**, a type of rock consisting of large fragments of minerals, with the spaces between them filled with smaller particles or some type of mineral cement were formed (Figure 16.10). This type of rock is found primarily in the lunar highlands. Around 4 billion years ago, larger craters were formed by an unusually large number of large collisions. Between 3.1 – 3.9 billion years ago, these craters filled up with lava from beneath the Moon's crust. For the last three billion years or so, the Moon has had no significant geological events, excepting collisions with meteorites.

Figure 16.10: A piece of breccia brought back from the Moon. Notice the white-colored fragments of rock and minerals cemented together. This sample is typical of ejecta from large basins. Photo credit: B. Zaprowski.

Section 16-2: Our Planetary System

If we look at other objects orbiting our Sun, we find that they have a personality all their own. They come in all compositions, shapes, sizes, rotational speeds and orbital paths; some have their own moons, and others only visit the inner solar system every few hundred years. However, a majority of objects in our solar system have many similarities; all orbital planes are very near our own ecliptic (Figure 16.11), all of the planets move in a counterclockwise direction as seen from the Earth's North Pole, and most planetary moons orbit in this same direction. The planets have elliptical paths that are almost perfect circles. Mercury has the most elliptical orbital path of the planets.

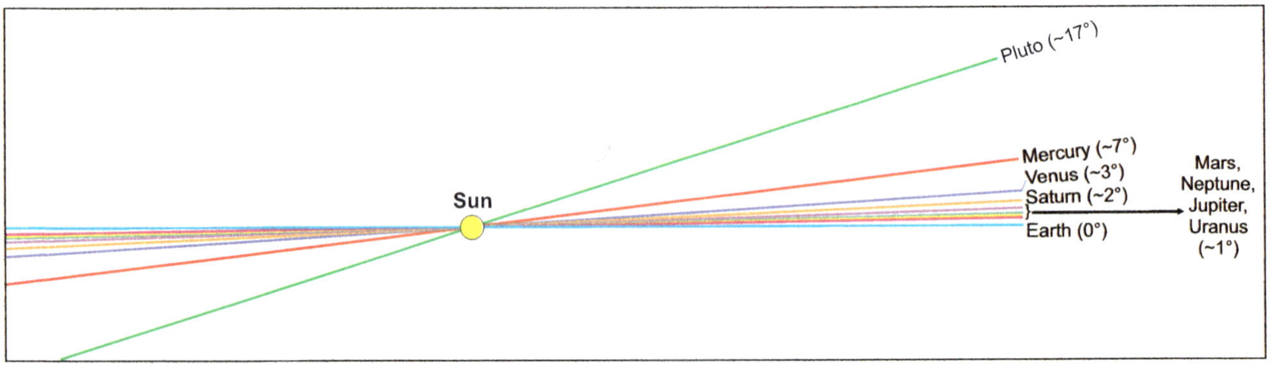

Figure 16.11: The orbital planes of the planets, as viewed from the side.

Moving outward from the Sun, the planets can be categorized in two groups. First come the terrestrial (or Earth-like) planets. These planets have rocky surfaces. In this group you will find the four planets closest to the Sun: Mercury, Venus, Earth, and Mars. Beyond Mars' orbit, lies an area known as the asteroid belt, consisting of thousands of objects. Beyond the asteroid belt lie the Jovian (or Jupiter-like) planets. These planets are also known as the gas giants (Jupiter and Saturn) and the ice giants (Uranus and Neptune). Beyond Neptune lies the dwarf planet Pluto and the other objects in the Kuiper Belt (Figure 16.12).

Mercury

Mercury has an average distance to the Sun of 58 million km (36 million mi.), or 0.39 astronomical units (AU). An astronomical unit is defined as the average distance from the Sun to the Earth, with 1 AU = 1.5 x 10^8 km (9.3 x 10^7 mi.), and is useful when comparing distances of planets from the Sun. As mentioned above, Mercury's orbit is the most elliptical, varying from 0.307 AU at its closest, to 0.467 AU at its farthest point. It is the smallest planet, with a

Figure 16.12: There are many objects that are orbiting our sun. Note: image is not to scale. All photos from NASA.

mass only 1/20 of Earth's. Mercury has no atmosphere, and its surface is covered with craters (Figure 16.13A), mainly due to collisions. The daytime temperature is 700 K at mid-day, while at nighttime it cools down to 100 K, mainly due to the lack of an atmosphere. It revolves very quickly around the Sun, with its year measuring 88 of our Earth days. Mercury has a density only slightly less than that of Earth's. This leads scientists to believe it is composed of many heavy metals within its core, surrounded by a rocky crust. Russian landers have landed on Mercury's surface (the only such landings to date), and readings taken from these spacecraft show that the rocky surface is made mostly of basalts and granites, types of rock formed through volcanic activity. Most surprisingly, Mercury exhibits a weak magnetic field, evidence for a molten interior of some sort. However, there is no evidence of plate tectonics on Mercury, although there is evidence that the planet has shrunk sometime in the last 4 million years, creating "wrinkles" on the surface (Figure 16.13B).

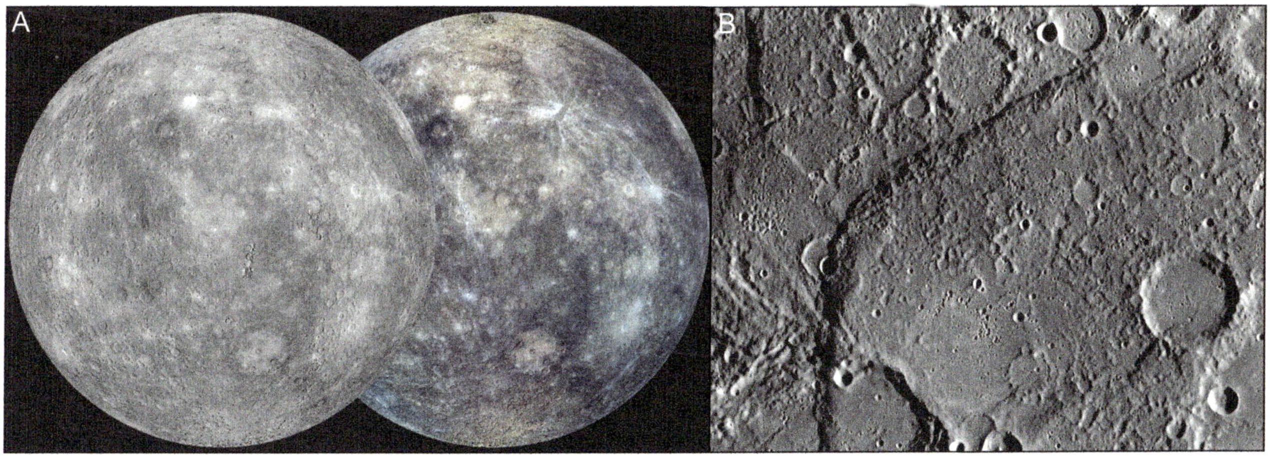

Figure 16.13: Images of Mercury. A. Mercury is pretty heavily cratered. B. A small area of Mercury's surface, showing its rocky, crater-like characteristics and one of the many "wrinkles" that cross the surface. Image credits: NASA/JHUAPL/CIW.

Venus

Venus is the planet closest to Earth (40 million km/29 million mi. at its nearest point), and the second planet from the Sun. Along with Mars, Venus is sometimes very bright near dawn or dusk, and has been called alternatively the "morning star" or the "evening star". Its orbit is nearly circular, and its diameter is only slightly smaller than the Earth's. A unique feature of Venus is its deep atmosphere (with an atmospheric pressure 90 times greater than Earth's), consisting mainly of carbon dioxide and a small percentage of nitrogen. Because of its similar size and proximity to Earth, as well as its atmosphere, scientists had hoped that perhaps life was present on Venus. This has turned out not to be the case.

Venus's surface, observed using radar techniques, shows a landscape almost completely free of erosion. There is no water or ice on the surface, and very little wind to cause any change in surface formations. Therefore, the surface has only been formed through geological activity, such as volcanic and tectonic activity (Figure 16.14). About 75% of Venus's surface consists of lava plains, similar to the Moon's maria. However, there are large "continents" of mountainous regions. Figure 16.14C shows the Aine Corona. The Corona is the larger circular area, approximately 200 km (124 mi.) across, where magma is causing the crust to bulge upward. The surface features here include small "pancake" shaped volcanos, whose lava did not spread far due to its high viscosity. These domes can be 25 km (15.5

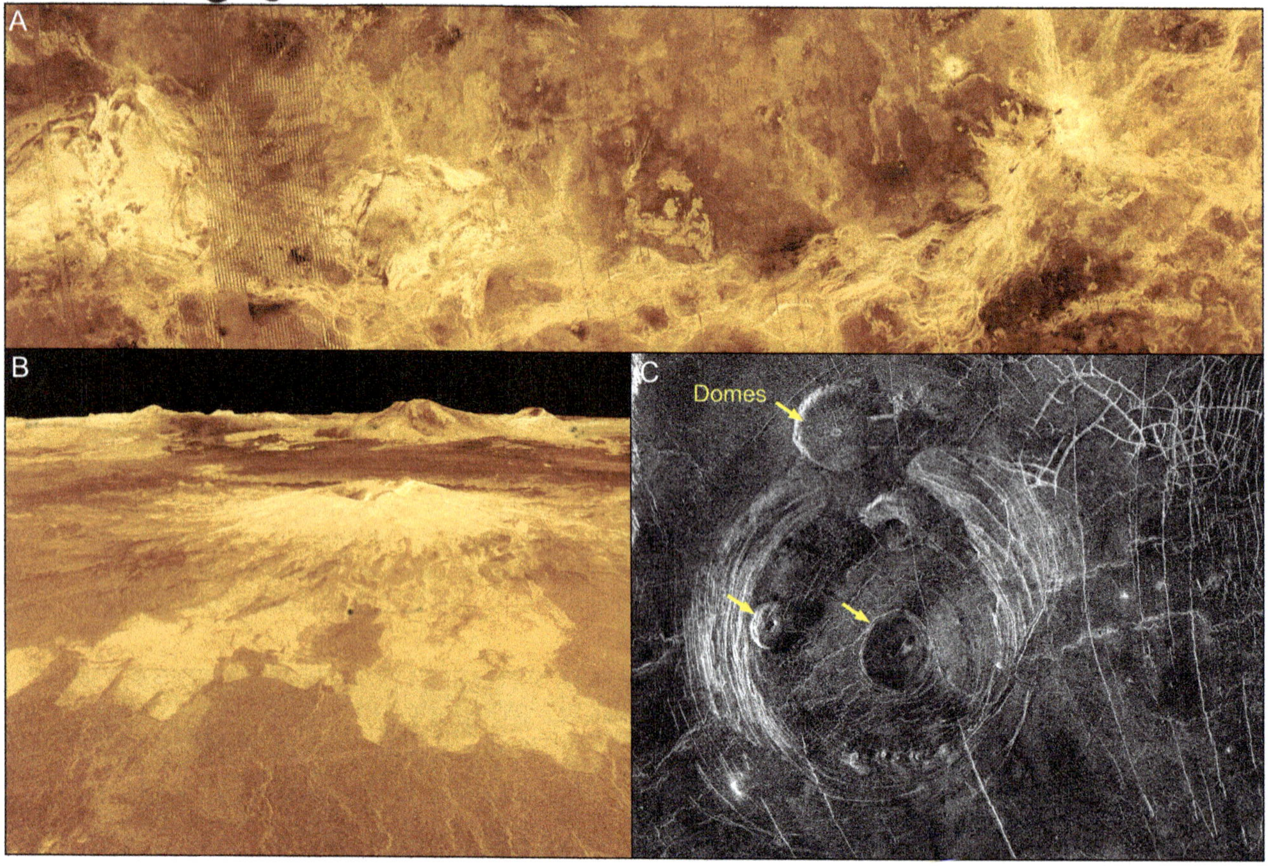

Figure 16.14: Views of Venus. A. This composite radar image show the volcanic nature of Venus. B. A close-up view of some of the volcanos. C. Image of the Aine Corona, taken with radar images through Venus' thick atmosphere. Some astronomers imagine they can see a famous puppet's face within the corona. Image credits: NASA/JPL.

mi.) across and nearly 2 km (1.25 mi.) high. Also note the cracks across the region. Such domes and cracks are very common on the surface of Venus. Scientists have not been able to determine if there is current volcanic activity on the surface of Venus, due to its incredibly thick and dense atmosphere. However, chemical traces in Venus's atmosphere point to volcanic activity in the not so distant past.

The temperature on Venus is very hot, with an average temperature of 730 K. Even though Venus is closer to the Sun than Earth, it is warmer than even Mercury. Scientists have determined that this high temperature is due to the greenhouse effect. This effect works the same on Venus as it does on Earth, but since Venus's atmosphere has almost a million times more carbon dioxide than the Earth's, the effect is much more profound. The surface temperature tends to stay fairly constant, night or day, due to a 20-30 km (~12-18 mi.) thick cloud made primarily out of sulfur dioxide particles. It is this cloud that makes visible light measurements of the surface impossible.

Mars

Known alternatively as the red planet, Mars is a medium-sized rocky planet, and the fourth planet from the Sun (Figure 16.15). It has a diameter of about half of Earth's, with a density 30% less than the average of the other terrestrial planets. Mars is very visible in the night sky, and astronomers have long wondered if life could exist on Mars. Although the atmosphere is made up primarily of carbon dioxide (similar to Venus), the atmospheric pressure of Mars was less than one percent of Earth's atmospheric pressure, and much smaller than that of Venus. Studies of Mars' surface show geological characteristics caused by water, such as canyons and valleys (Figure 16.15B). There are also large plains that could have once been oceans and seas. Mars has polar ice caps containing layers of frozen water that are covered by frozen carbon dioxide. From recent studies conducted at NASA, scientists believe that at one time up to a fifth of the planet could have been covered by water. On Earth, wherever there is water there are also a multitude of life forms. Because of this, scientists have speculated that there may have been life on Mars when there was water on the surface. Some even believe that life could exist in underground aquifers, although no life on Mars has been detected to date. Although the reason Mars lost most of its water is unknown, many scientists believe that Mars's size did not provide sufficient gravity to keep water vapor in its atmosphere, and most of the water escaped into space.

Figure 16.15: Views of Mars. A. A surface view from NASA's Curiosity Mars rover. Photo credit: NASA/JPL-Caltech/MSSS. B. Gullies on the Martian surface are evidence that water flowed on its surface in the past. Photo credit: NASA/JPL. C. Olympus Mons (Mount Olympus), the largest volcano in the solar system. The whitish areas are cliffs, some of which are over 6 km (3.7 mi.) tall. Photo credit: NASA/JPL.

Similar to the Earth and Venus, there are areas of higher altitude on Mars known as "continents" or highlands, as well as low-level plains caused by volcanic activity. These plains look similar to the Moon's maria, and geologists have determined that these areas formed at approximately the same time on both the Moon and Mars. This planar region is typically in the Southern Hemisphere, while highlands are mostly in the Northern Hemisphere of the planet. The largest volcano in the solar system, Olympus Mons, is located in the planet's Tharsis area. This volcano is massive, with the base as large as the state of Missouri, and its height of 27 km (18 mi.), over three times higher than Mt. Everest (Figure 16.15C). This volcano, as well as others on the Martian surface, is not active.

At times, it is useful to compare and contrast the physical properties of the terrestrial planets. In doing so, we begin to notice the similarities and differences between our home and the other rocky planets. Table 16.1 compares many physical properties of these planets.

	Mercury	Venus	Earth	Mars
Physical properties				
Orbital period (days)	88	224.7	365.25	687
Rotational period (hours)	58.65	243	23.9	24.6
Diameter (Earth units)	0.381	0.951	1	0.531
Mass (Earth units)	0.0558	0.815	1	0.107
Average density (g/cm^3)	5.42	5.25	5.52	3.94
Surface temperatures (°C)	-173 to 425	453	-13 to 37	-123 to 37
Satellites	None	None	1	2
Atmospheric composition				
Nitrogen	None	~3%	~78%	~3%
Oxygen	None	none	~21%	trace
Carbon dioxide	None	~96%	trace	~95%
Water Vapor	None	very little	<2%	very little
Clouds	None	Sulfur dioxide	Water	Water, dust

Table 16-1: A comparison of select physical properties of the terrestrial planets.

The Asteroid Belt

In the past astronomers were puzzled by the large gap between Mars and Jupiter. In fact, many believed there should be another planet at about 2.8 AU, based on the distances of the other planets from the Sun. In 1801 a monk named Giuseppe Piazzi discovered an object in this region of space. Named Ceres, many thought that this was the long lost planet (Figure 16.16). However, further studies showed that Ceres was only about ¼ the size of the Moon, much smaller than any other planet. In the following years, other objects similar in size to Ceres were discovered at about the same distance from the Sun. These objects were classified as **asteroids**, meaning small rocky objects which orbit the Sun.

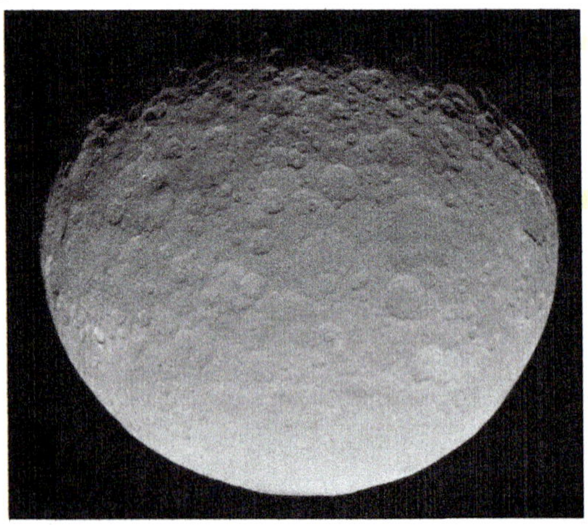

Figure 16.16: Ceres is a dwarf planet in the Asteroid Belt. Image credit: NASA/JPL-Caltech/UCLA/MPS/DLR/IDA

Today, astronomers have charted the orbits of over 300,000 asteroids. Many of these reside in the asteroid belt, located between Mars and Jupiter. Some of these objects, such as Ceres, are now classified as dwarf planets, because although their gravitational pull is strong enough to create a spherical shape, their gravitational pull is not strong enough to clear out other objects in their orbital paths. Most asteroids orbit the Sun in a fairly circular pattern, although there are some exceptions. Asteroids with a very elliptical orbit sometimes pass near Earth's orbit. There are close to 1000 asteroids in this class that have a diameter of over 1 km (0.6 mi.). It is possible that someday the Earth may collide with one of these asteroids. There are also asteroids located at various other locations throughout the galaxy. Many of these are near the Earth, and are known as **Near Earth Objects** (NEOs). Studies are underway to identify and catalog all of these NEO's, which someday could possibly hit the Earth.

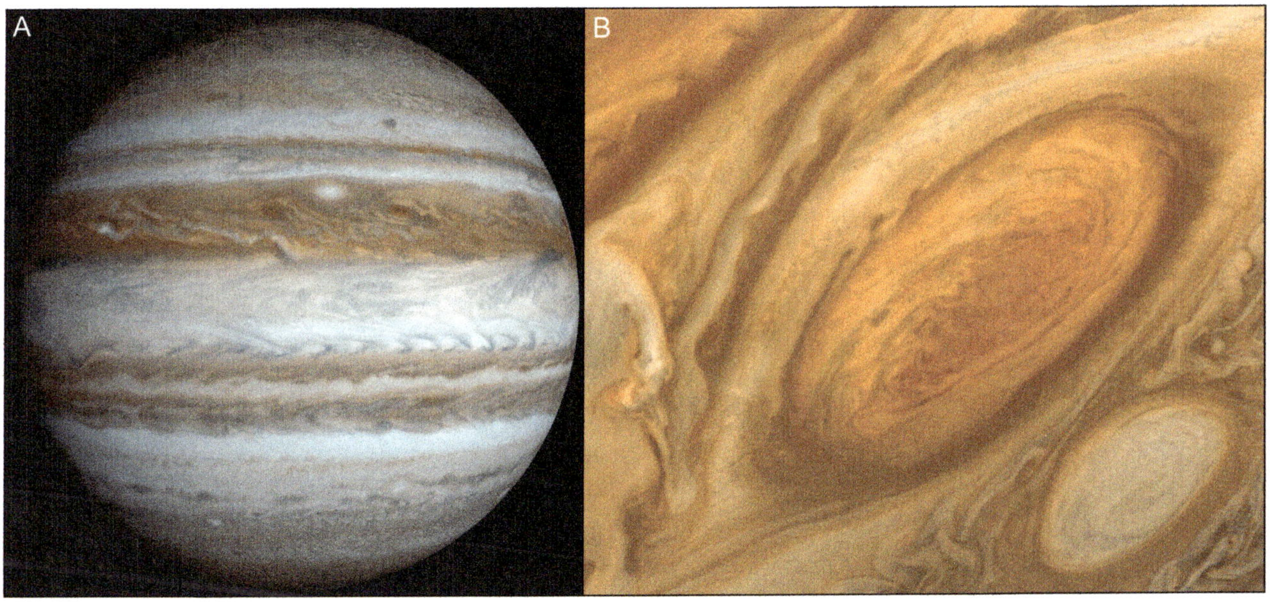

Figure 16.17: Views of Jupiter. A. Jupiter has regions of turbulent atmosphere, which appear as colored bands. B. The "Great Red Spot", a giant storm on Jupiter's surface. This storm has been active for at least 350 years. Photo credits: NASA/JPL.

Jupiter

The fifth planet from the Sun is also the largest, containing nearly 70% of all planetary mass in the solar system. Jupiter is located at 5.2 AU from the sun, and it could easily fit 1,300 Earths within it. Jupiter's density, however, is only slightly greater than that of water, and only 18% of Earth's density. Studies have also shown that Jupiter is a "flattened" sphere, meaning that the diameter of its equator is larger than the diameter around the poles. This type of flattening is not as evident with rocky planets, but unmistakable flattening occurs if the planet was mostly made of gas and/or liquid. Scientists now believe that all of the Jovian planets only have a small, solid core surrounding by gases and liquids. Because of this, scientists have defined the surface of the Jovian planets as that altitude at which we can no longer see through the gas. However, if you were to jump onto Jupiter's surface, you would fall through a gas that was gradually becoming denser until you realized that you were completely encased in a liquid. There is no definite boundary between the two.

Jupiter also has an amazingly complex atmosphere, primarily composed of hydrogen. In fact, most of the Jovian planets atmospheres are principally hydrogen. In Jupiter's atmosphere there are clouds filled with water, ammonia, and ammonium sulfide. These cloud layers are able to condense, forming bands that reflect different wavelengths of light based on their composition. Water and ammonia clouds reflect white light, but other types of clouds reflect brown and red light. Thus, Jupiter appears to have multiple circulating bands around it (Figure 16.17A). There are also storms on Jupiter's surface; these storms can last for tens of years (Figure 16.17B). In 1979, the Voyager spacecraft detected rings around the planet Jupiter (Figure 16.18). These rings are made out of rocky material and not icy material like found in Saturn's rings. As such, these rings are not nearly as luminous as Saturn's rings.

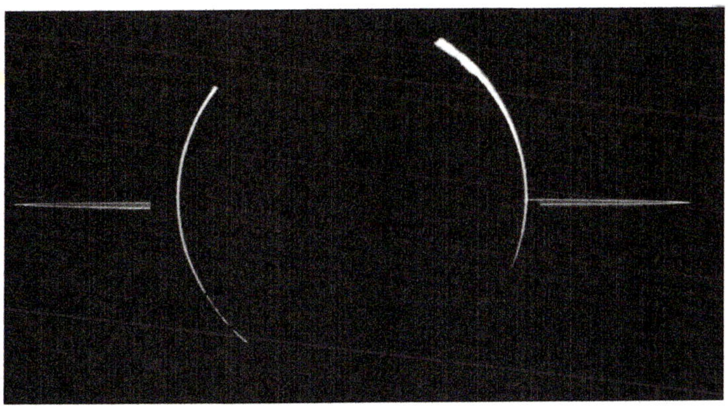

Figure 16.18: An eclipse of the Sun by Jupiter (from Galileo spacecraft). The thin dust particles within the rings are then visible as they reflect sunlight. Image credit: NASA, JPL, Galileo Project, (NOAO), J. Burns (Cornell) et al.

The scientist Galileo discovered four large moons around Jupiter. These are called the Galilean Moons: Io, Europa, Ganymede, and Callisto. There are many other small moons orbiting Jupiter, but these four are the largest (Figure 16.19). Io exhibits a lot of volcanic activity, with over 100 volcanic vents along its surface. These

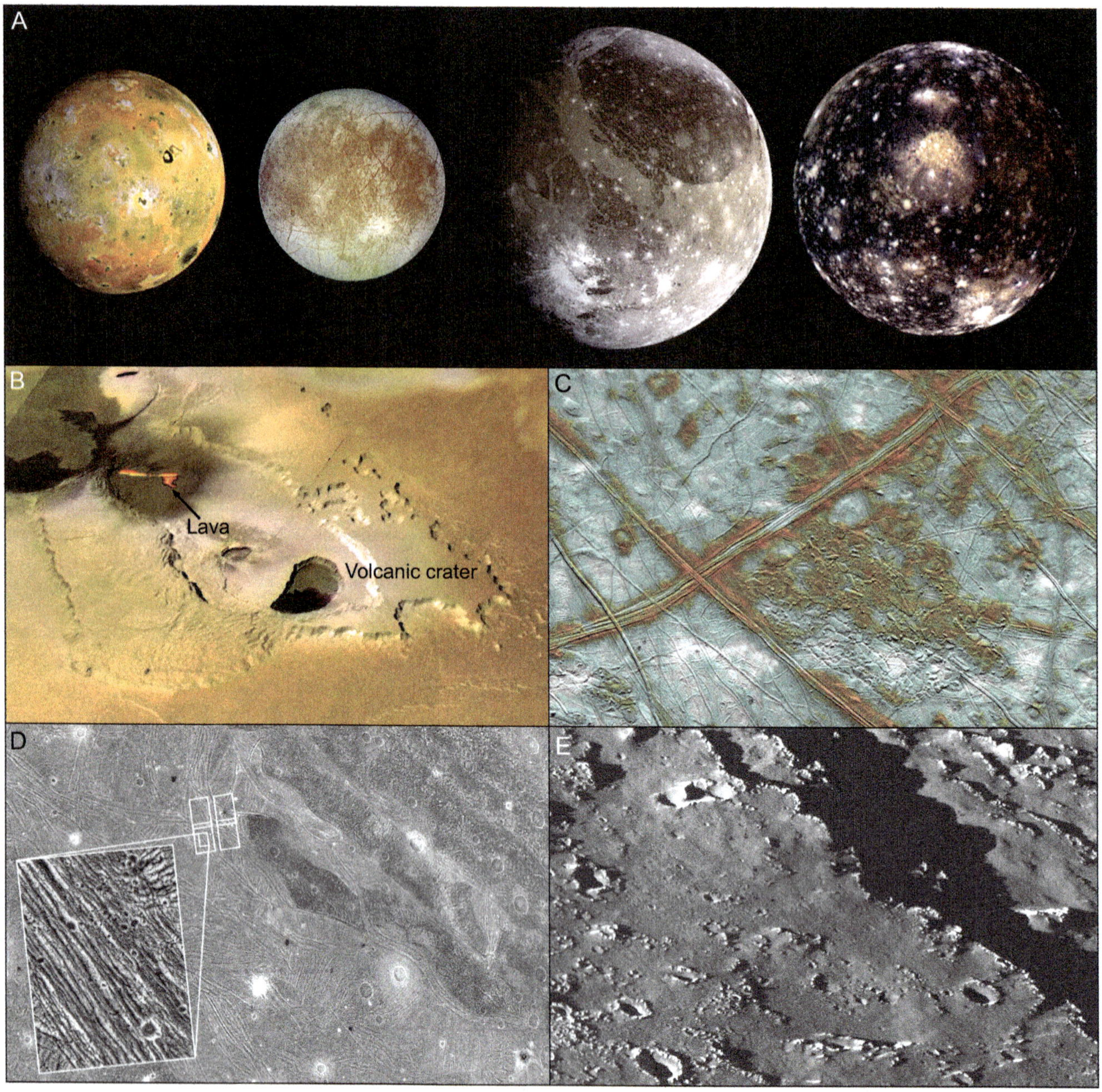

Figure 16.19: A. The Galilean moons; from left to right, Io, Europa, Ganymede, and Callisto. The Earth's moon is slightly larger than the moon Europa. Jupiter has over 60 moons. Image credit: NASA/JPL/DLR. B. A close-up view of volcanic activity on Io. Photo credit: NASA/JPL/University of Arizona. C. The surface of Europa is crisscrossed by thousands of ice cracks. Photo credit: NASA/JPL/University of Arizona. D. View of a portion of the Uruk Sulcus region on Ganymede showing the fine details of the grooved terrain that are the principal features in the brighter regions of this satellite. Photo credit: NASA/JPL. E. An area within the Valhalla region on Jupiter's moon, Callisto. Photo credit: NASA/JPL/ASU.

vents emit sulfur and oxygen atoms, as well as ash, into space. Many of these particles then fall to the ground and cover up almost all of the impact craters on its surface. Europa is made primarily of rock, with a thick icy layer. This icy layer may be affected by tides due to Jupiter's gravity, with the tidal action causing cracks within this layer. Studies also suggest that beneath Europa's icy exterior may reside an ocean of liquid water, perhaps as deep as 200 km (124 mi.). There are very few craters on Europa, pointing to a landscape that is constantly changing. Ganymede and Callisto are both roughly the size of Mercury, and both are believed to be composed of a rocky core surrounded by ice. It is also believed that there are layers of liquid water beneath their icy crusts. On Ganymede, there is evidence of fault lines, with some faults overlying others, indicating that there must have been eras of geological activity in the past. Callisto's surface is scarred with hundreds of craters, but the lack of any large highland-like areas may indicate that its surface cannot support these types of features.

Saturn

Figure 16.20: This approximate natural-color image shows Saturn, its rings, and four of its icy satellites. Three satellites (Tethys, Dione, and Rhea) are visible against the darkness of space, and another smaller satellite (Mimas) is visible against Saturn's cloud tops very near the left horizon and just below the rings.

Excluding Earth, possibly the most recognizable of the planets is Saturn. The magnificent rings of Saturn are visible even with backyard telescopes. These rings are made up of billions of ice particles ranging in size from miniscule to the size of tractor-trailers. Each of these particles has its own orbit around the planet, but also reflects sunlight toward Earth, making them extremely visible (Figure 16.20).

Saturn consists primarily of hydrogen and helium, with a density that is less than water, meaning it would float if you could somehow find a swimming pool big enough. Like Jupiter, Saturn has a flattened spherical shape, indicating that it has a small, solid core surrounded mainly by liquids and gases. Its orbit is fairly circular, with an average distance of 9.58 AU from the Sun. There are zones and belts of moving gases with Saturn's atmosphere, moving at higher rates of speed than even seen on Jupiter. However, the clouds of Saturn lie deeper in its atmosphere than Jupiter's. As a result, less light reaches the clouds, less light is reflected by the clouds, and this light is somewhat scattered by the atmosphere above them. This results in Saturn's hazy appearance. Electromagnetic readings from the Voyager spacecraft indicate that there may be large electrical storms within Saturn's atmosphere.

Saturn has over 60 moons, consisting of ice and rock mixtures. Due to their composition and orbital characteristics, scientists believe that many of Saturn's moons originated elsewhere but were captured by Saturn's large gravitational force. Although 50 of these moons have been named, and many of these moons have been known for over 400 years, two of are particular note. **Titan**, the largest of Saturn's moons, is slightly larger than the planet Mercury. Titan's temperature is low enough to prevent atmospheric molecules from escaping into space. In fact, Titan's dense atmosphere is the only one found on a natural satellite within the solar system. This thick atmosphere (primarily made of nitrogen) prevented telescopes and spacecraft from seeing the surface of this moon. In 2005, the *Cassini* spacecraft dropped the *Huygens* probe onto the surface of the moon. From the surface, the probe sent back images showing complex surfaces, including level plains, chunks of ice, and lakes believed to be filled with hydrocarbons, such as methane (Figure 16.21). This is the only place, other than Earth, where standing bodies of surface liquid have been observed. **Enceladus** is a small moon, less than 500 km (310 mi.) across. The small number of impact craters on its surface points to this moon being geologically active, erasing older impact craters. Scientists have discovered thermal vents near the Enceladus's south pole forming geysers which vent ice and water to the surface.

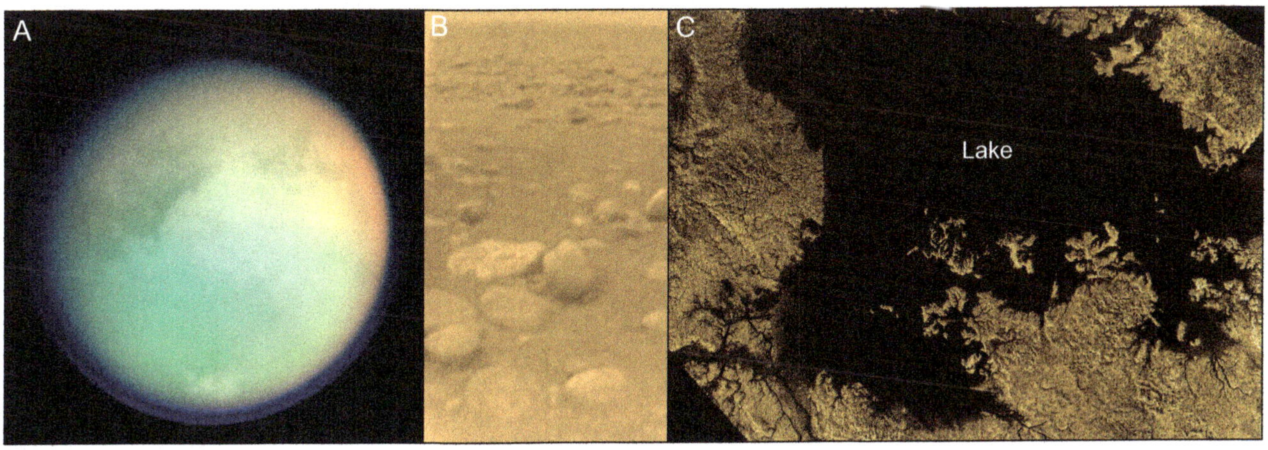

Figure 16.21: A. Image of Titan, taken from the Cassini spacecraft in 2004. This is a false-color image of both ultra-violet and infrared wavelengths. Photo credit: NASA/JPL/SSI. B. A ground-level view reveals pebble-sized pieces of rock and/or ice. Photo credit: NASA/JPL/ESA/University of Arizona. C. This view of Titan's surface shows a complex surface with lakes of liquid hydrocarbons. Photo credit: NASA/JPL/Space Science Institute.

This area is marked by blue "tiger stripes", which are cracks where the thermal geysers are located, and where new ice is being formed on the moon's surface (Figure 16.22). The rest of the surface is astonishingly white, reflecting almost all light. This observation indicates that the surface is covered by ice.

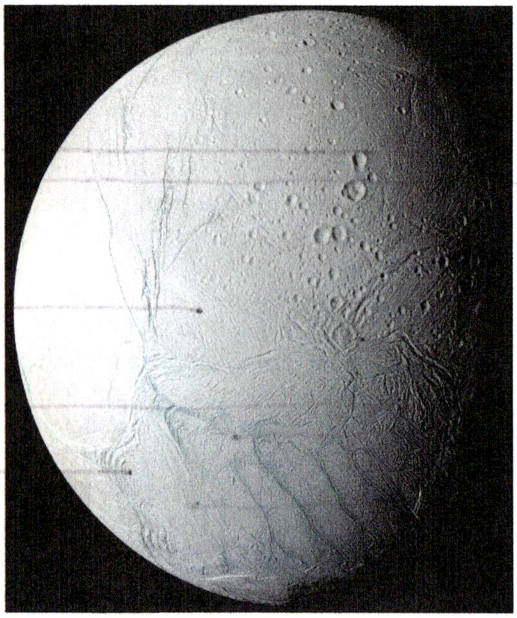

Figure 16.22: The moon Eneceladus, showing its blue "tiger stripes". Photo credit: NASA/JPL/ESA/SSI, Cassini Imaging Team.

Uranus

Located at approximately 19 AU from the Sun we find the planet Uranus. Uranus is unique among the planets in that it rotates sideways, with its south pole pointing near the direction of the Sun. Because of the orientation of its axis, Uranus has wide variations between its seasons, with each summer and winter lasting for 21 years. As satellites have studied the planet, Uranus seems to have none of the bands of clouds that Jupiter or even Saturn display. Uranus appears to be covered in a blue haze, made up mostly of hydrogen and helium. The blue color indicates methane in the atmosphere as well (Figure 16.23A). Methane is very effective in reflecting blue light while absorbing red light.

The rings of Uranus were not discovered until 1977 when an aircraft-based telescope detected the rings as they blocked out the light from a far-away star. Almost all of the information we know about the rings comes from the Voyager 2 spacecraft which imaged the rings in 1986 (Figure 16.23B). These rings are made primarily out of ice and frozen methane, appearing dark because of exposure to solar wind. Because of the particles dark color, they reflect only 3% of the light they are exposed to.

Uranus has over 20 moons, but until the Voyager 2 mission, only 5 of these were known. These five major moons are all cracked, heavily cratered, and dark. It is interesting to note that the **International Astronomical Union** (IAU), the organization responsible for naming celestial bodies, has specified that all of the moons of Uranus are to be named after William Shakespeare's characters. Thus, the five major moons are known as Ariel, Miranda, Titania, Oberon and Umbriel.

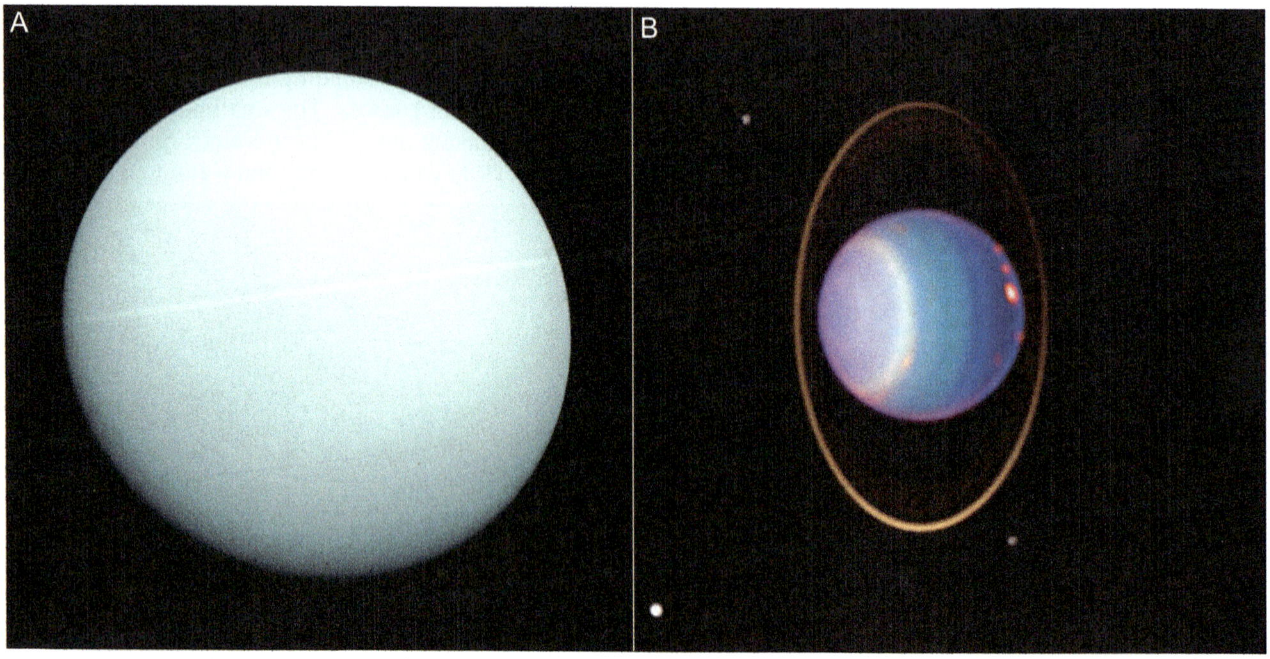

Figure 13.23: A. Image of Uranus taken from the Voyager 2 spacecraft in 1986. Notice the lack of cloud bands within the atmosphere. Photo credit: NASA/JPL-Caltech. B. Uranus' rings reveal its unusual orientation. Photo credit: NASA/JPL/STScI.

Neptune

Neptune is only slightly smaller than Uranus, and is the smallest of the Jovian planets. Located at 30.1 AU from the sun, Neptune appears somewhat different than Uranus. Its atmosphere is a darker blue than that of Uranus, indicating more methane in its atmosphere than its Jovian neighbor. Neptune also has a more active atmosphere with high clouds visible in the atmosphere. There are also large storms seen within its atmosphere (see Figure 16.24). The energy necessary to form these storms seems to come from within the planet itself, although scientists don't understand why this would be the case on Neptune and not on Uranus.

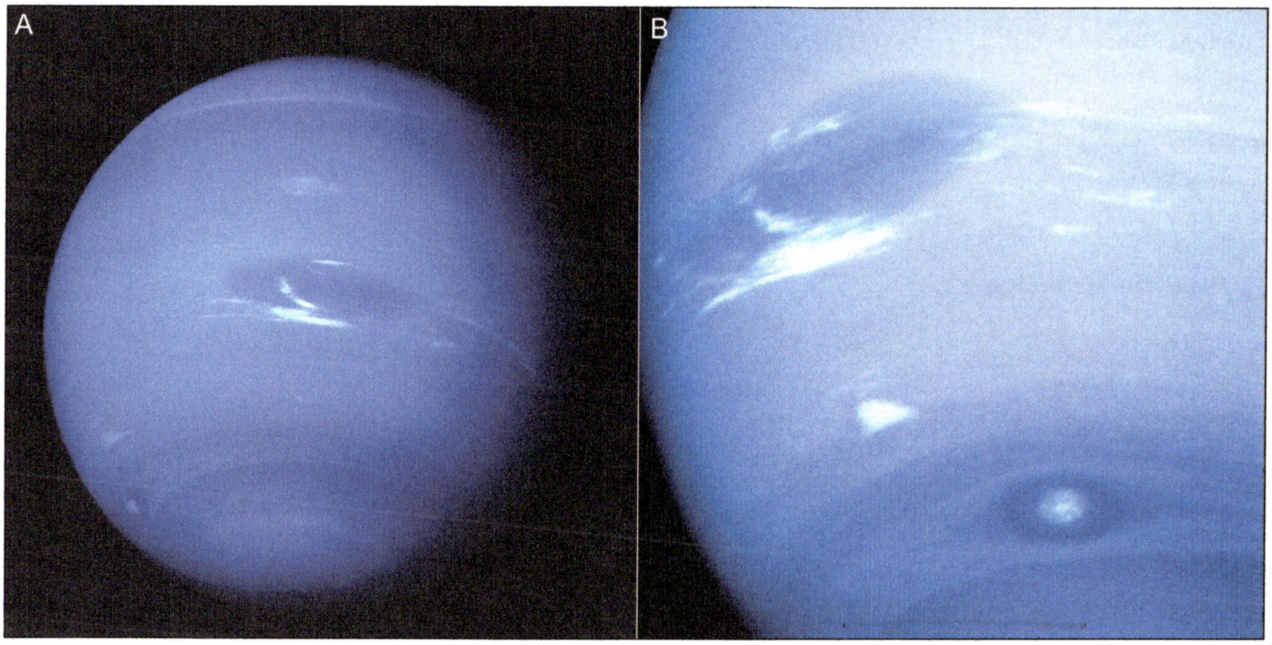

Figure 16.24: A. The planet Neptune. Note near the middle an area of darker blue. Known as the "Great Dark Spot", this is believed to be a large atmospheric storm, comparable to the Great Red Spot on Jupiter. Photo credit: NASA/JPL. B. A close-up of the Great Dark Spot and another smaller storm. Recent images have shown that this spot is no longer visible, indicating that the atmosphere on Neptune is dynamic and changing. Photo credit: NASA/JPL.

Earth-based telescopes had discovered two moons orbiting Neptune before Voyager 2 studied the planet in 1989. These two moons have unique characteristics. The moon Nereid follows a very elliptical path around the planet, completing a rotation approximately once a year. The other moon, Triton (not to be confused with Saturn's moon Titan) circles the planet in the opposite direction as both Neptune's orbit and Neptune's rotation. This unique rotation points to a large disturbance within the planetary system sometime in the past. Triton has an atmosphere made of nitrogen and methane, with its surface showing grooves, ridges, and evidence of tectonic processes (Figure 16.25). The Voyager 2 spacecraft discovered six more small moons, and since that time, many more have been

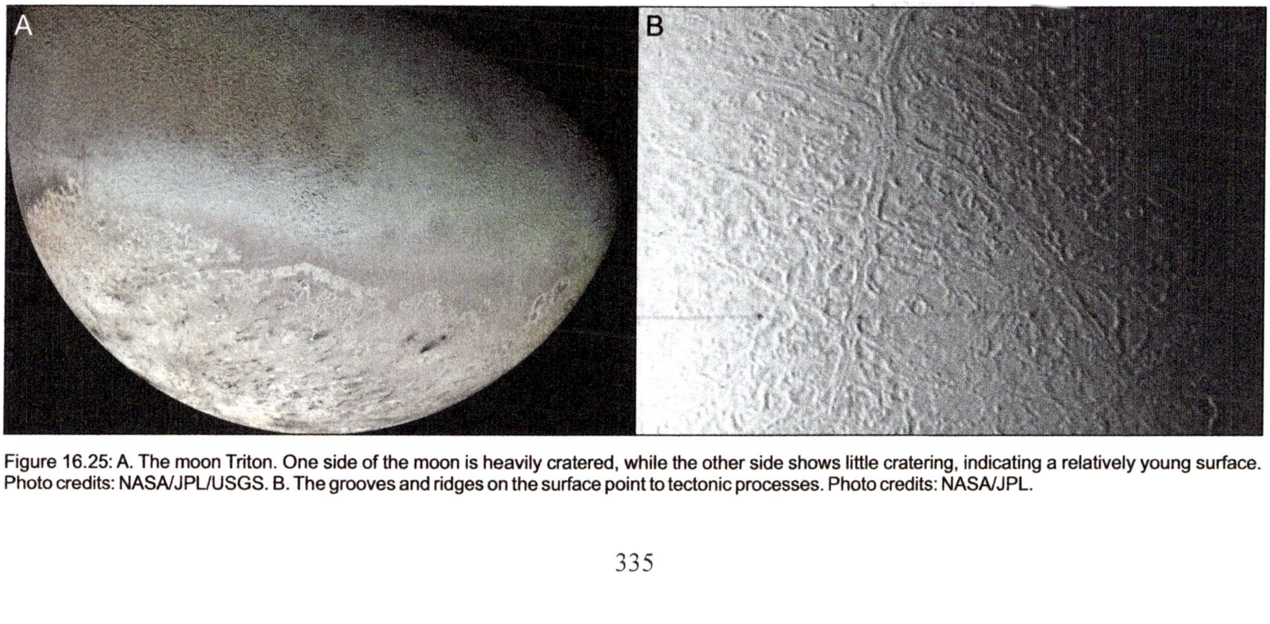

Figure 16.25: A. The moon Triton. One side of the moon is heavily cratered, while the other side shows little cratering, indicating a relatively young surface. Photo credits: NASA/JPL/USGS. B. The grooves and ridges on the surface point to tectonic processes. Photo credits: NASA/JPL.

discovered from more powerful Earth-based telescopes. Neptune also has rings, although they are very faint and very hard to detect from Earth.

Table 16-2 shows a comparison between the Jovian planets. Again, please notice the difference as well as the similarities between the planets.

	Jupiter	Saturn	Uranus	Neptune
Physical characteristics				
Orbital period (years)	11.4	29.46	84.01	164.79
Rotational period (hours)	9.92	10.66	17.24	16.11
Diameter (Earth units)	11.19	9.41	4.01	3.89
Mass (Earth units)	317.9	95.1	14.5	17.1
Density (g/cm^3)	1.31	0.69	1.29	1.64
Satellites	63	47	27	13
Atmospheric composition				
Hydrogen	90%	97%	83%	83%
Helium	10%	3%	15%	15%
Methane	None	None	2%	2%
Ammonia	trace	trace	None	None
Water Vapor	trace	trace	None	None

Table 16-2: A comparison of select physical properties of the Jovian planets.

Section 16-3: Pluto and the Kuiper Belt

Pluto sits on the edge of the solar system, part of a group of small, icy objects. First discovered in 1930, its location so far from the Earth gave us little information about the object itself. Some things were obvious from its discovery. It was not a large, gas planet like Jupiter or Saturn. Nor was it like Neptune and Uranus, as it was much smaller. It did not resemble the terrestrial planets either. In fact, Pluto is substantially smaller than Earth's Moon. Its orbit was highly elliptical, and is actually closer to the Sun than Neptune for large amounts of time. Its orbital plane is much more inclined than the other planets' (see Figure 16.2). Until 2015, scientists did not have an accurate picture of Pluto's surface. This changed in July 2015 when the New Horizon Spacecraft passed Pluto and sent back images of the planet to Earth (Figure 16.26). At the time of this writing, scientists are analyzing these images to learn more about Pluto's geologic activity.

Figure 16.26: The image sent to Earth by the New Horizon Spacecraft of Pluto. This image is taken from approximately 765,000 km (475,000 mi.) from Pluto's surface. Photo credit: NASA/ JHUAPL/SwRI.

Pluto is known to have five moons, four of which are very small. The exception is the satellite Charon, named after the mythical ferryman carrying the souls of the recently deceased to Hades. Charon was discovered in 1978, and although difficult to observe due to its distance from Earth, observations have shown that its diameter is nearly half that of Pluto's. Both Pluto and Charon are composed of approximately 65% rock and 35% ice.

When it was discovered in 1930, Pluto was considered a planet, and remained so until 2006, when the IAU changed its classification to a dwarf-planet. Pluto is no longer classified as a planet due to three main reasons: its size, its orbital plane, and its composition, which are all substantially different than any other planet. In fact, between 1930 and 2006, many objects unlike planets, yet similar to Pluto were discovered near Pluto's orbit. A sample of these objects is shown in Figure 16.27. Some of these objects are even larger than Pluto. Since the early 1990's, astronomers have discovered more

Figure 16.27: The largest known objects within the Kuiper belt. This is only a small sampling of the objects that astronomers believe reside within this region of space. The Earth is shown in the background for scale. Image Credit: NASA.

than a thousand of these small objects that reside farther from the Sun than Neptune's orbit. These objects are part of what is now referred to as the **Kuiper Belt**. Scientists have speculated that there may be over 100 million **Kuiper Belt Objects** (KBOs) which lie within the Kuiper Belt Figure 16.12). Since objects in the Kuiper Belt are very far from the Sun, astronomers believe that KBO's probably consist of ices and frozen gases. In 1992 the first Kuiper Belt object (other than Pluto) was observed, and data showed that it did indeed contain ice, but also contained carbon-rich material.

Section 16-4: Comets and the Oort Cloud

If we consider the Sun, the planets, the asteroid and Kuiper belt, is there anything else we will find in our solar system? The answer is a resounding yes. Some of these objects are known as **comets**. Comets are small, averaging only 10 km (6.2 mi.) in diameter, and having an average mass of 1×10^{14} kg (2.2×10^{14} lbs.). An expert on comets, Fred Whipple, once described a comet as a "dirty snowball" (Figure 16.28). Although this description is a good way to picture a comet, studies have shown that the densities of comets are only ¼ to 1/10th that of ice and scientists have concluded that they must be "fluffy" mixtures of water ice, rock, and frozen gases.

Figure 16.28: Comet Tempel really is just a big "dirty snowball." Photo credit: NASA/JPL/UMD.

Comets do not look very impressive most of the time. In fact, since they do not emit light and are so small, they are relatively invisible to telescopes for most of their orbit. However, as they approach the Sun, comets becomes very visible, some even visible to the naked eye (Figure 16.29). This is due to solar energy which begins to sublimate the frozen gases and ice, changing them directly into gases. During a comet's voyage to the Sun, the central nucleus appears as a bright spot encompassed by luminous gas. The head of the comet, or the coma, has a diameter of 105 km (65 mi.), although

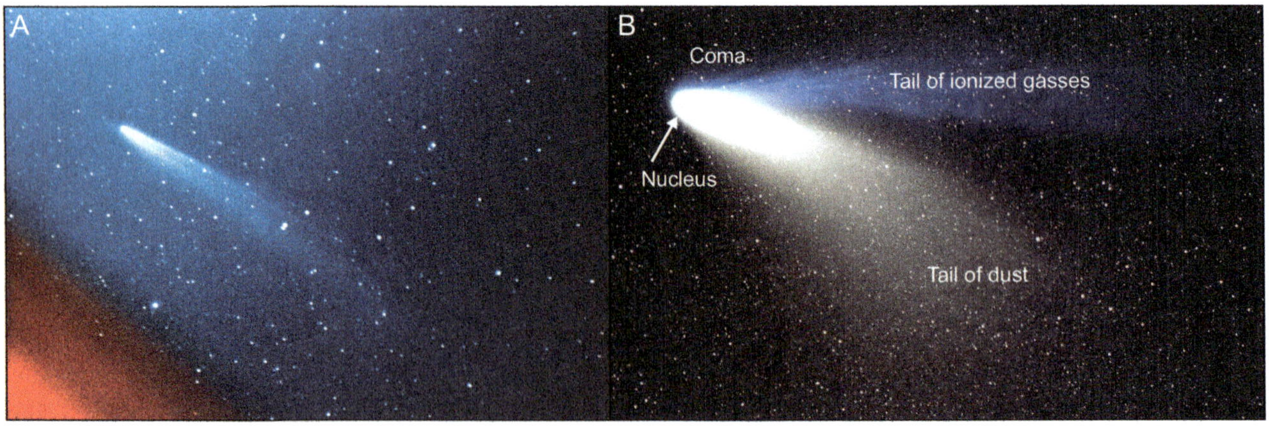

Figure 16.29: A. The Kohoutek Comet as it passes by Earth in 1974. Photo Credit: NASA/JSC. B. The parts of a comet. Photo credit: © Valerio Pardi/ shutterstock.com.

for some large comets, it can be one million km (620,000 mi.). Other parts of the comet include the two tails, one made of gas and the other comprised of dust. The gas tail is a result of ionized gas particles leaving the comet's nucleus due to the solar wind. The dust tail is composed of the particles that were released from the nucleus during the sublimation process, and are pushed away from the comet by the radiation of the sun. Because both the solar wind and the solar radiation are pointing away from the sun, both tails will extend away from the sun (Figure 16.30).

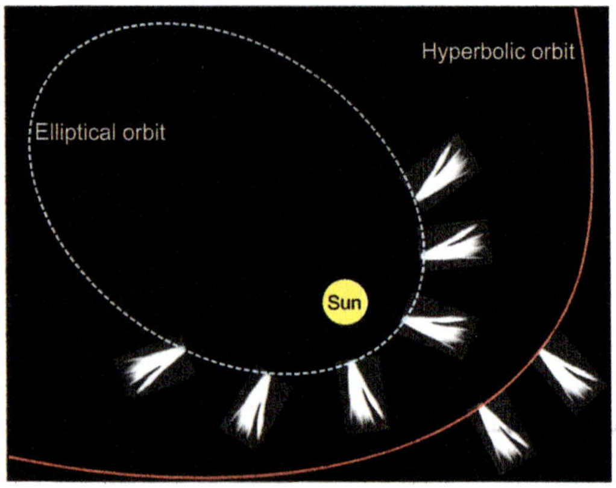

Figure 16.30: Different types orbital paths for comets. A comet in an elliptical orbit around the Sun repeats the pattern continuously. In contrast, hyperbolic orbits only take a comet past the Sun once. Note how the tails always point away from the sun.

Some comets have an orbital path that is elliptical in shape, while others follow a hyperbolic path. Those comets traveling in elliptical orbits will revolve around the Sun with a certain period. Hailey's comet has a elliptical orbit, revolving around the Sun once every 75 years. Hyperbolic comets do not travel on a closed path, and after passing by the Sun will be flung into space, perhaps never to return to the solar system (See Figure 16.30). Unlike planets, comets travel in both clockwise and counterclockwise directions around the Sun. Their orbits also tend to be oriented arbitrarily when compared to the elliptic plane of Earth. Because of these findings, a Dutch astronomer by the name of Jan Hendrik Oort hypothesized that comets must reside in a spherical cloud, somewhere between 10,000 and 100,000 AU from the Sun. This Oort Cloud, as it is now called, lies well beyond the Kuiper belt (Figure 16.31).

Due to the extreme distance from Earth, only a few objects within the Oort cloud have ever been observed, but astronomers believe that nearly 6 trillion comets exist within the Oort cloud. Each of these objects has a unique orbit around the Sun. Most comets within this region of space will never come near the Sun, but orbit the Sun at their extreme radii. In fact, the comets which do enter the inner solar system (the ones we are able to see) were probably affected by the gravity of a large object, such as another star, and are deflected toward the Sun.

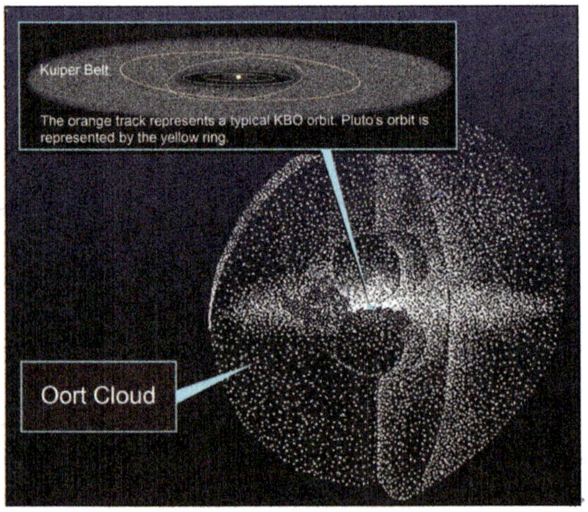

Figure 16.31: An illustration of the Oort Cloud, surrounding our solar system. Image credit: NASA.

Due to the spherical nature of the Oort cloud, many astronomers believed that the objects residing there must have been formed in the early history of the solar system, before it began to flatten. However, the density of the solar nebula that far from the Sun is so small that it would be almost impossible to form objects with the diameters we see in comets (<10 km/6 mi.). Another theory proposed, and more widely accepted, is that the comets were formed between the orbits of Saturn and Neptune. These objects were then flung into space due to the different gravitational pulls of the Jovian planets. Formation in this region would help to explain the icy-gas mixture which makes up most comets; however, recent observations of Sedna, the most distant object ever observed in the solar system have some scientists questioning this model as well. Sedna (Figure 16.32) was imaged in 2004 at a distance of 120 AU. Its orbit is very elliptical, positioning the object as close to 76 AU or as far away as 900 AU from the Sun. If estimates are correct, this object is nearly three quarters the size of Pluto. It also has a red surface, similar to Mars. These observations of Sedna do not fit well with the current theory of how Oort cloud objects may have been formed.

Figure 16.32: An illustration of Sedna, with the Sun in the upper right corner. Illustration credit: R. Hurt (SSC-Caltech), JPL-Caltech, NASA.

Section 16-5: Meteoroids, Meteors, and Meteorites

As the Earth revolves around the Sun, different objects in its orbital path are pulled toward the Earth due to gravity. Some of these are small particles known as **meteoroids**. Some of these particles are no bigger than dust grains, but others have considerable mass. Some of these particles will run into Earth's atmosphere at speeds anywhere between 10 to 70 km/sec (6.4 to 43 mi/sec). The force of friction between the atmosphere and these particles raises the temperature of the meteoroids, causing the surrounding air particles to glow. These "shooting stars" are also known as **meteors** as they burn within the atmosphere (Figure 16.33). Some meteors are completely vaporized, but others are large enough to make it all the way to the Earth's surface. These are known as **meteorites**.

Figure 16.33: The Perseids Meteor Shower. Photo credit: Fred Bruenjes/ NASA.

Meteorites can be grouped into three categories: *iron meteorites* are composed of 90% iron and 10% nickel; *stony meteorites* consist mainly of silicate material; *stony-iron meteorites* are a mixture of both iron and stone (Figure 16.34). An analysis of meteorites found on Earth shows that a "typical" meteoroid has a composition similar to Earth's crust, a mixture of iron, oxygen, silicon, magnesium, as well as smaller amounts of other elements. The similarity to Earth's surface composition, as well as the fact that most meteorites originate within the asteroid belt provide evidence that meteorites are fragments of parent bodies within the asteroid belt. These parent bodies must have been large enough when forming that their lighter elements were located in the crust while heavier elements migrated to their core. If one of these parent objects was broken apart (like in a collision), it would form smaller pieces, some of which were iron-laden, while others contained more stony material.

Figure 16.34: Examples of meteorites. A. A stony meteorite. B. A large iron meteorite at the Arecibo Radio Observatory. C. A cut and polished stony-iron meteorite, showing the peridot crystals inside. Photo credits: B. Zaprowski.

Figure 16.35: Some craters, like the basin which forms Quebec's Manicougan Reservoir, were only detected by taking images from high altitudes, or even from space. Photo credits: NASA.

Small meteoroids are entering the atmosphere constantly. Satellite data indicates that meteors with a diameter greater than 1 m (~3 ft.) enter the atmosphere at a rate of about once a week. Some of these larger meteors create "fireballs" in the sky, at times creating sonic booms with their high velocity. However, only more massive meteors (those with a weight close to a ton and over) can travel through the atmosphere and make it to the Earth's surface. Craters mark the location of impacts between the Earth and these large meteors, moving at speeds from 10 to 30 km/sec (~6 to 18 mi/sec) or more. The energy of motion (kinetic energy) an object has depends on its mass and its velocity; more mass and more speed mean more energy. The Barringer crater in Arizona (refer back to Figure 6.16) was caused by a meteoroid with a weight of 300,000 tons, creating a crater nearly a mile across. Since only about 25,000 tons of iron meteorites have been found within the crater, it is believed that much of the meteorite vaporized, along with much of the surrounding land. Scientists calculate that this collision was the equivalent to 2.5 million pounds of TNT. Satellites and high-altitude aircraft have discovered close to 100 craters that are at least 100 meters in diameter. Many of these were not recognized on the ground because of erosion and their great size (Figure 16.35). These large collisions do not occur frequently with the Earth, with astronomers estimating that they only occur every few hundred thousand years (Figure 16.36). In fact, only one person in the last hundred years has ever even been injured by a meteorite. Still, the idea of a collision between the Earth and a large meteor or small asteroid gives cause for concern. Scientists are using emerging technology to detect and track large objects that may come near Earth's orbit. These objects can have a major impact on our planet, and on our civilization.

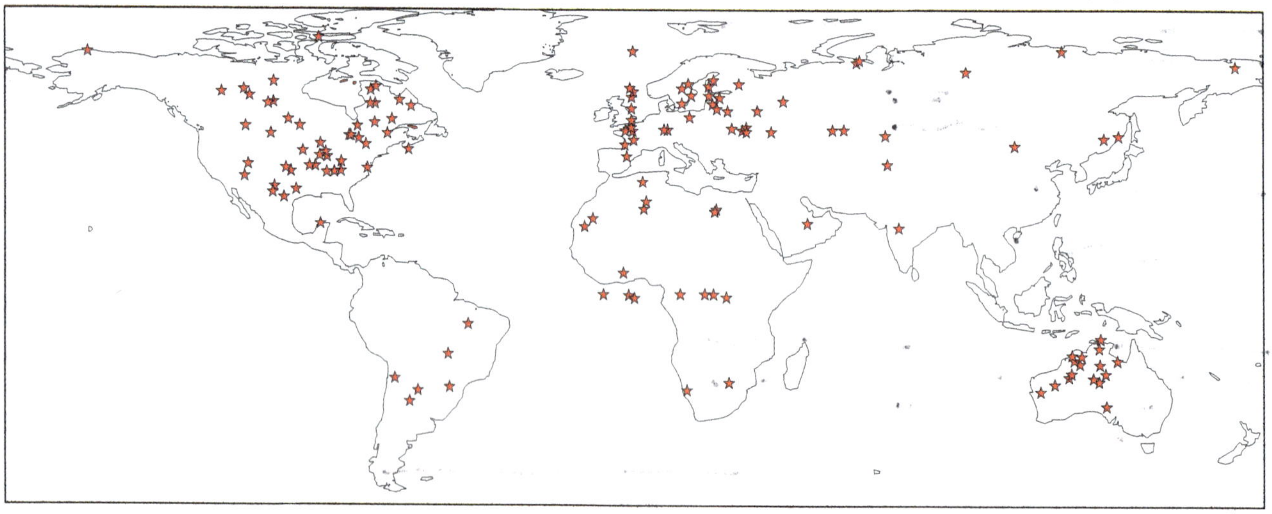

Figure 16.36: A map of the known meteorite craters around the world.

Section 13-6: The Sun

For generations, people have understood that the Sun is vital to human existence. Some worshiped the Sun as a god, or imagined that a chariot pulled by horses carried the Sun across the daytime sky. Only in the last 100 years have scientists learned about the structure of the Sun, as well as what gives it its power. Most of what we know about stars today comes from the study of our Sun.

The Sun's Atmosphere

The bright, visible surface of the Sun is known as the **photosphere**, and above that are the **chromosphere** and the **corona**. Together, these make up the solar atmosphere (Figure 16.37). The photosphere is the thin layer (less than 500 km/310 mi. thick) that emits most of the Sun's radiation. This layer is substantial enough to emit an abundance of radiation, but still thin enough that light can escape. The photosphere is known as an opaque gas, where opacity is a measurement of how much light will be absorbed when passing through a substance. Below the photosphere a lot of light is emitted, but most of that light is absorbed by gases above it. Since the corona and chromosphere are not opaque, they do not absorb much of the light emitted from the photosphere, and therefore it is the deepest level of the Sun that is visible to us. When the chromosphere and corona do absorb light, they do so only at certain wavelengths, and are responsible for the observed absorption spectrum of the sun (Chapter 15). Because of the opacity of the photosphere, it appears to be a solid surface. However, it is a gas thousands of times less dense than the air you breathe, and the boundary between the atmospheric layers is difficult to define.

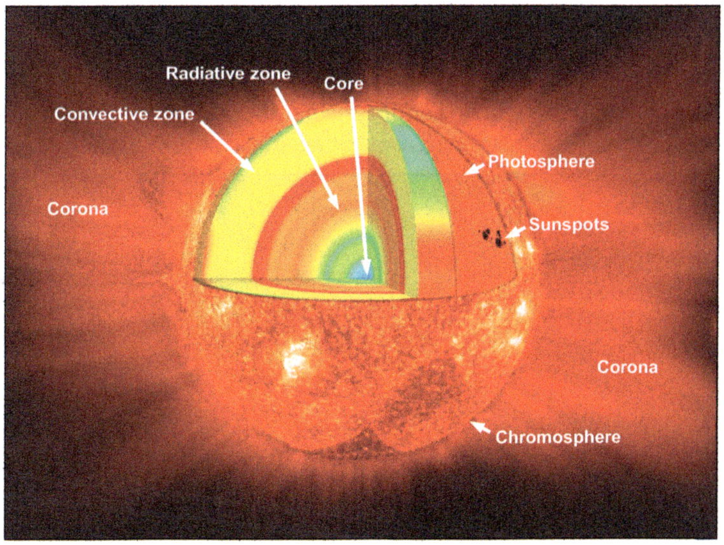

Figure 16.37: An illustration showing key parts of the Sun. Image credit: Goddard/NASA.

In good images (those taken above the Earth's atmosphere), the photosphere's surface has a blotchy appearance, with a pattern of light and dark areas known as **granulation** (Figure 16.38A). The bright areas are slightly hotter than the dark ones. Some of these granules can be a few thousand kilometers across (the size of Alaska), but last only for a few minutes before sinking back into the Sun. If we could watch a movie of the photosphere, it would look like a region of boiling gas, where the bubbles burst on the surface only to be replaced by other bubbles. Spectra of the granules show that within the granules the centers rise (blueshifted), while the outer edges of the granules move downward (red-shifted). From this evidence, astronomers have determined that energy is being transported from the interior of the Sun to the photosphere through convection.

Astronomers from Asia noticed dark spots on the Sun thousands of years ago (now known as sunspots), but only in the last few hundred years have they been studied in detail. In the 17th century, Galileo put a dark filter over his telescope, and saw spots on the Sun's surface. He used his observations to determine that the Sun was rotating around its own axis. Today, the easiest and safest way to see sunspots is to use a small telescope, point it at the Sun, and project the image on a screen (*never look directly at the Sun with a telescope*). Sunspots appear dark, and are

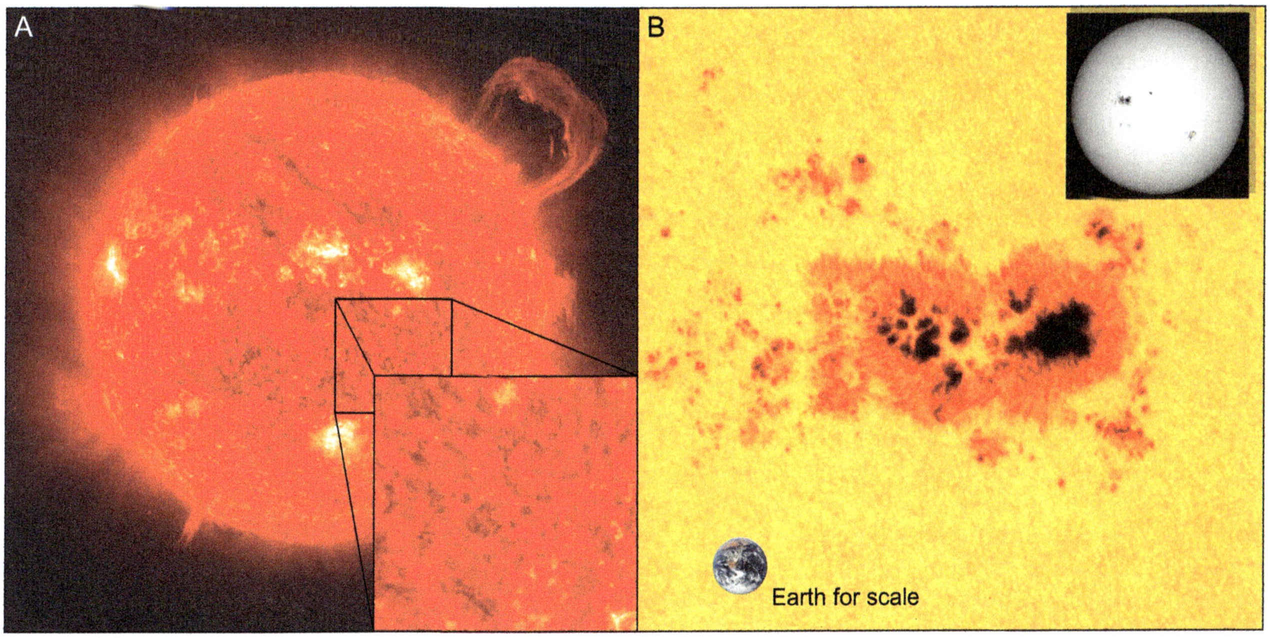

Figure 16.38: A. Detailed image showing the surface granulation of the photosphere. Darker areas are slightly cooler than the brighter areas. B. A very close-up image of a sunspot cluster. Photo credits: SOHO/ NASA/ESA.

cool spots on the Sun's photosphere. They usually appear in groups (Figure 16.38B), and can form and vanish within weeks or months. The number of sunspots on its surface at any given time varies.

As astronomers began to catalog the number of sunspots at any given time, a pattern began to emerge. The number of sunspots seemed to increase and decrease approximately every 11 years (Refer back to Figure 6.10). Between 1645 and 1715, records indicate that there were very few sunspots, an event known as the *Maunder minimum*. This lack of sunspots suggests a possible change within the Sun's activity. Corresponding with the Maunder minimum is the middle of a time of unusually cool weather in North America and Europe known as the "Little Ice Age". Some scientists point to this situation as a connection between sunspot activity and climate conditions on Earth. This idea is somewhat controversial, mainly because there have been other periods of cooler climate without a decrease in sunspots, but there is evidence suggesting a connection between the solar energy the received on Earth and the solar activity.

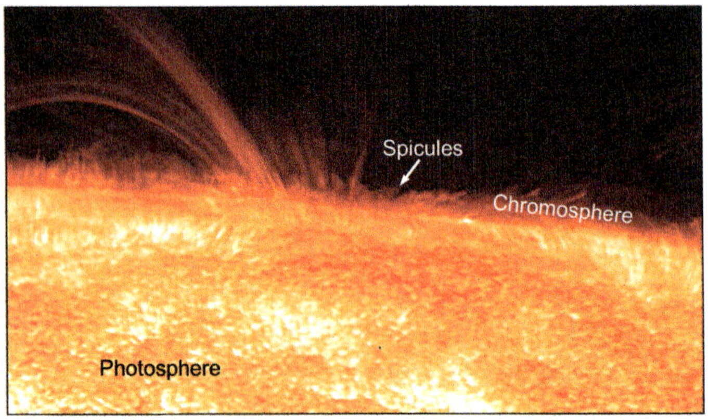

Figure 16.39: An edge-on view of the surface of the Sun. In the lower part of the image you can see the photosphere and its granules. Above the photosphere is the chromosphere where wisps of hot gas called spicules rise up. Photo credit: Hinode/JAXA/NASA.

The chromosphere is the central layer of the solar atmosphere, responsible for radiating most of the Sun's ultraviolet light. Astronomers define the lower edge of the chromosphere as the lowest visible surface of the Sun. If seen by itself, it appears as a thick pink layer, and its name comes from the Greek word "chroma" meaning color. The chromosphere's density is very low, ranging from approximately one ten-thousandth of the Earth's atmosphere near the photosphere to only one billionth of Earth's atmospheric density near the corona. Figure 16.39 is an image from the edge of the Sun, displaying finer details of the chromospheric region. In particular, notice the fine-structured spicules within the chromosphere. **Spicules** are flame-like jets of gas extending upward into the chromosphere. These spicules can extend upward to heights of 5,000 to 20,000 km (3,100 to 12,400 mi.) and can last anywhere from 5 to 15 minutes. The lowest temperature in the Sun's atmosphere is about 3,000 K at the base of the Sun's chromosphere, but increases with altitude, reaching a temperature of 30,000 K near the chromosphere's border with the corona.

The corona is the outermost part of the Sun's atmosphere. The word corona comes from the Latin word for crown. The corona is so faint that during the daytime, it only has the brightness of the full moon, and cannot be seen due to the brilliance of the photosphere below it. The density of the corona is very low, with lower-level densities 10 billion times lower than Earth's atmosphere, and upper-level densities lower than any vacuum ever created on Earth. The corona is constantly changing; sometimes it exhibits long streamers (Figure 16.40) that extend outward from the Sun, and at other times, the corona is essentially featureless. The corona's spectrum includes emission lines of highly ionized gases, similar to the upper chromosphere. Using spectroscopy techniques, scientists determined that in the lower corona the particles are not as highly ionized compared to particles at higher altitudes within the corona. This means that the temperature of the corona increases with altitude, which surprised many scientists. Further studies have shown that the heating of the corona (and in some part, the upper chromosphere) is not due to radiation, but due to turbulence of the Sun's magnetic field.

Figure 16.40: Image of the solar corona showing the inner streamer belt along the Sun's equator. Over the polar regions, one sees the polar plumes all the way out to the edge of the field of view. The field of view of this coronagraph encompasses 8.4 million km (5.25 million mi.) of the inner heliosphere. Photo credit: SOHO/ NASA/ESA.

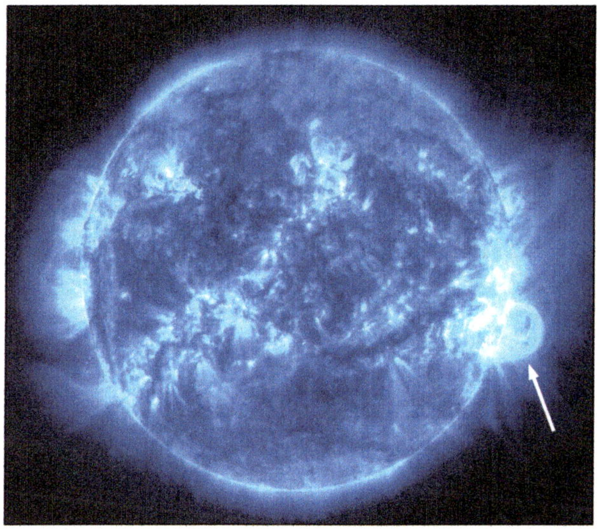

Figure 16.41: An X-ray image of the Sun showing a solar flare erupting from the Sun's surface (arrow). Photo credit: SOHO/NASA/ESA.

Solar Events

There are many types of phenomena that occur near the surface of the Sun due to the Sun's magnetic field. **Solar flares** (Figure 16.41) are sudden flashes on the surface of the Sun, occurring in active solar regions. These flares are believed to occur when two oppositely directed magnetic fields meet and cancel each other out. This is known as a reconnection event, and each such event releases a large amount of energy. Some flares are powerful enough to emit high-energy X-rays, as well as high energy particles such as protons and electrons. When these particles reach the Earth, they interact with a layer of the ionized particles that encircle the Earth's outer atmosphere (the **ionosphere**), which can then disrupt radio communications and navigation systems. The charged particles from solar flares can also cause disruptions in electrical power lines as well as damage satellites orbiting the Earth.

A **solar prominence** (Figure 16.42) is another surface feature that occurs on the Sun. It is caused by ionized gas particles confined within an arching magnetic field rising up through the solar atmosphere into the lower corona. They often are found in between active sunspot regions, and when seen at the edge of the Sun, can be quite spectacular. Some prominences can remain in the corona for many days, while some burst upward within hours.

The Sun not only emits particles during these types of events, but continuously radiates charged particles into space. Known as the **solar wind**, these particles primarily consist of protons, electrons, and helium nuclei. The solar wind can be enhanced by eruptions on the sun, such as solar flares and solar prominences. These particles travel at speeds of several hundred kilometers per hour, and can modify the Earth's magnetic environment. The particles in the solar wind constantly batter the Earth's magnetic field, compressing it on the daytime side of the Earth. On the nighttime side, the magnetic field lines are greatly distorted by the solar wind, with a long "tail" of magnetic field lines extending far into space, away from the Sun.

Some of these charged particles enter into the Earth's magnetic field near its poles. As these particles move toward the Earth, they spiral inward around Earths' magnetic field lines (Figure 16.43A). As they enter the Earth's atmosphere, they collide with atmospheric molecules, and transfer some of their energy. The atmospheric molecules, in turn, can re-radiate their extra energy by emitting visible light. If the solar wind is particularly strong, the amount of light given off can be quite dramatic (Figure 16.43B). These patterns of light are known as the *Aurora Borealis*, or Northern Lights, in the Northern Hemisphere (Figure 16.43C). In the Southern Hemisphere they are known as the *Aurora Australis*, or the Southern Lights (Figure 16.43D).

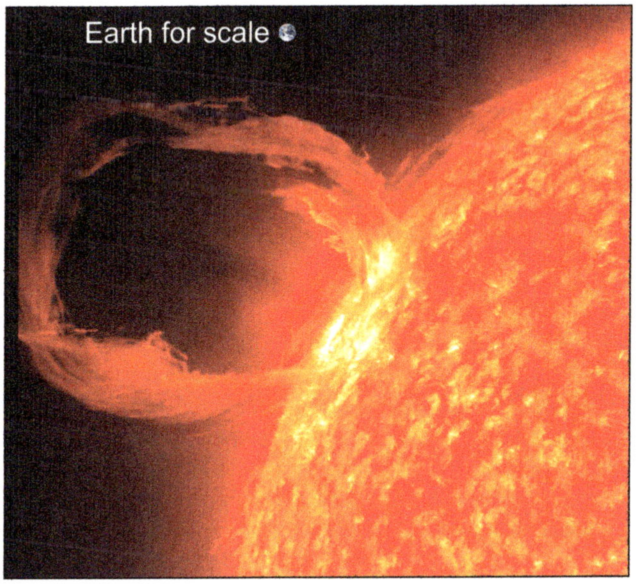

Figure 16.42: A solar prominence comprised of a very hot gas of ionized particles. These gases are trapped by magnetic fields anchored in the Sun's photosphere. Some prominences may continue for many months, and extend hundreds of thousands of miles above the Sun's surface. Image credit: NASA/SDO.

The Sun's Interior

Because we cannot see past the Sun's photosphere, it is difficult to know what lies beneath its surface. Scientists have used other types of techniques to study the Sun's interior. From these studies they have determined that directly beneath the photosphere is a convective zone,

Figure 16.43: A. An illustration showing how the auroras occur. Charged particles from the Sun follow the magnetic field lines of the Earth, spiraling toward the poles. These particles transfer energy to the atmospheric molecules, which then emit light. B. Space shuttle photograph (notice the wing on the left) showing the Aurora Australis. Photo Credit: Image Science and Analysis Laboratory, NASA-Johnson Space Center. C. A beautiful green and red aurora dancing over the Jokulsarlon lagoon, Iceland. Photo Credit: © Krissanapong Wongsawarng/shutterstock.com. D. The Aurora Australis and Milky Way at Cremorne in Tasmania Australia. Photo Credit: © Ricki Leigh Eaves/shutterstock.com.

where hot gas rises and cooler gas falls (see Figure 16.37). Traveling farther down, about a third of the way to the center, lays the radiative zone, the area where energy moves toward the surface predominantly in the form of high-energy electromagnetic radiation, such as X-rays and gamma rays. Below the radiative zone is the core of the sun, where nuclear fusion is taking place (Section 13-7). The core maintains the conditions necessary for fusion through a balance between gravity pulling everything toward the core and the pressure from the hot gases pushing outward.

Section 13-7: The Power Of The Sun

The Sun releases a massive amount of energy into space, not only in the form of visible light, but in other wavelengths as well, and by emitting mass through means such as the solar wind. But how much energy is it really radiating? Using detectors on Earth, scientists have found that the amount of energy reaching our atmosphere each second is 1,370 watts/m^2 (known as the solar constant). Since the Sun radiates in all directions, the amount reaching the Earth is only a small fraction of what the Sun is truly radiating. Using the distance from the Earth to the Sun, scientists have calculated the power output of the Sun, also known as the Sun's luminosity. It turns out the Sun's luminosity is 3.8×10^{26} watts. This is comparable to exploding 10 billion, 1-megaton nuclear bombs every second.

A majority of ancient scientists believed the Sun to be some type of fire, such as burning wood. However, it became obvious to astronomers in the 1800's that ordinary combustion simply could not supply the large amount of energy the Sun was radiating for a long period of time. Other ideas, such as chemical processes, were soon ruled out for similar reasons. A breakthrough came in 1905 with Einstein's theory of relativity. Einstein argued with his famous equation that mass and energy were interchangeable; that is, mass could be converted into energy. It took scientists until the 1920's to understand that the Sun's energy was being produced through nuclear reactions.

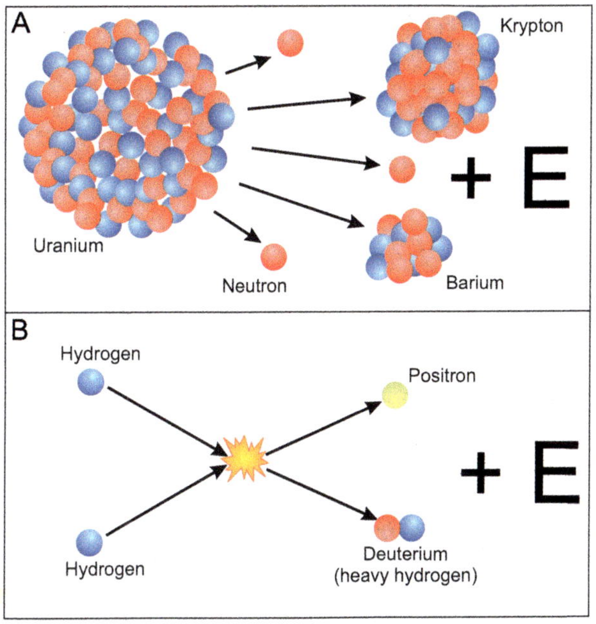

Figure 16.44: A. Fission occurs when a larger nucleus is broken into smaller nuclei. B. The first fusion step of the proton-proton chain. Two hydrogen nuclei are fused together to form the nucleus of deuterium. Energy is released in each of these processes.

There are two types of nuclear reactions. In both of these processes, energy is released. The process of splitting a nucleus into two smaller nuclei is known as **nuclear fission**. An example of a fission reaction is shown in Figure 16.44A, where a uranium nucleus is split into krypton and barium nuclei, as well as emitting neutrons. Combing smaller nuclei to make a nucleus with more protons or neutrons is called **nuclear fusion**. By studying spectra, astronomers knew that the Sun was made up almost entirely of light elements, such as hydrogen and helium, and they concluded that fusion was the source of the Sun's energy.

As fusion takes place within the Sun, the overall reaction actually proceeds through several steps called the **proton-proton chain**, involving just two nuclei at a time. In the first step, two hydrogen nuclei (which are protons) collide and form the nucleus of deuterium, an isotope of hydrogen (also known as heavy hydrogen). This process is shown in Figure 16.44B. In the second step, a deuterium nucleus collides with another hydrogen nucleus. The product of this collision is the nucleus of an isotope of helium known as helium-3 containing two protons and one neutron. The third and final step of nuclear fusion, the helium-3 nucleus gains another proton to create helium-4. This process can occur in many different ways within the Sun. Fusion occurs within the sun because collisions are occurring with an incredible frequency due to the Sun's high temperature, approximately 15 million Kelvin near its core.

If we count all of the protons that are used to create a helium nucleus, we see that it takes four protons to produce one helium-4 nucleus (Figure 16.45). However, if we were to measure the masses of the four protons before they fuse together, and compared it to the masses of the helium-4 nucleus (as well as the other small particles created in this process), we would see that they are not equivalent. In fact, the mass of the four protons is greater than the mass of all the particles after the fusion process. This disappearing mass confirms Einstein's theory, meaning that some of the mass on the Sun is being converted into energy. In fact, 596 million tons of helium are created through fusion within the Sun from 600 million tons of hydrogen *every second*. The missing 4 million tons is converted into energy following Einstein's famous equation $E = mc^2$ (where E = energy, m = mass and c = speed of light). This may seem like a large amount of gas, but because the Sun is so massive, it is only a very small portion of the Sun's mass, and its conversion into energy does not affect its mass in any meaningful way. Of course, the Sun won't last forever. The Sun will eventually use up all the hydrogen. When that happens, the Sun will change radically, as you will see in the next chapter.

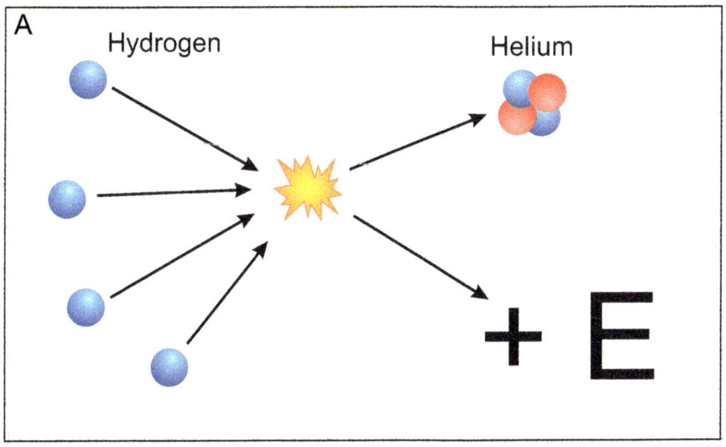

Figure 16.45: A. The simplified version of fusion in the Sun; four hydrogen nuclei combine to make one helium nucleus, releasing energy in the process.

NASA
NASA/ESA
NASA/ESA/STScl
ESA/NASA/Hubble
NPS
ESA/NASA/Hubble
R. Hurt/NASA/JPL-Caltech
NASA/ESA/M. Livio/STScl

Chapter 17– The Universe

It has often been said that we are stardust. In a way this is true; the carbon, iron, calcium, and all of the Earth's heavy elements were made within stars that have long since disappeared. These elements make up living objects here on Earth. Now consider: the Earth is one of eight planets surrounding our Sun, one of billions of stars. Our Sun lies within the Milky Way Galaxy, one of millions of galaxies. The more we know about how galaxies and stars form, evolve, and in the case of stars, how they die, the more it helps us to understand our past. This knowledge can also help us to understand what lies in store for our Sun, as well as for the Earth.

Section 17-1: The Beginning of the Universe

The universe is mostly empty space filled with **galaxies**, large gatherings of stars in space, each containing a few hundred million to more than a trillion stars. There are billions of galaxies located throughout the universe; some are fairly isolated, but many others are found in groups (groups of galaxies with more than a few dozen members are often called galaxy clusters). Using telescopes and spectroscopes, astronomers have observed that all galaxies not in our local group are moving away from our galaxy in all directions. A conclusion of these observations is that we must live in a universe which is expanding. The expansion of the universe does not imply that the Earth has a special location (the center of the universe). To see why, look at Figure 17.1, which shows an analogy of baking raisin bread, where we consider galaxies to be the raisins and the dough to be the universe. As the dough rises, the raisins are carried along by the expanding dough, but do not move through the dough. The loaf will get larger, but the raisins will remain the same size. While watching the dough rise, we see that it pushes the raisins away from each other at speeds that are proportional to distance. Two raisins that were originally close to each other are pushed apart slowly, but two raisins that were far apart from each other have more dough in between them and are pushed apart faster. No matter which raisin you examine, you will see that all other raisins have moved away from it, showing that it does not mean that the raisin is located at the center of the loaf.

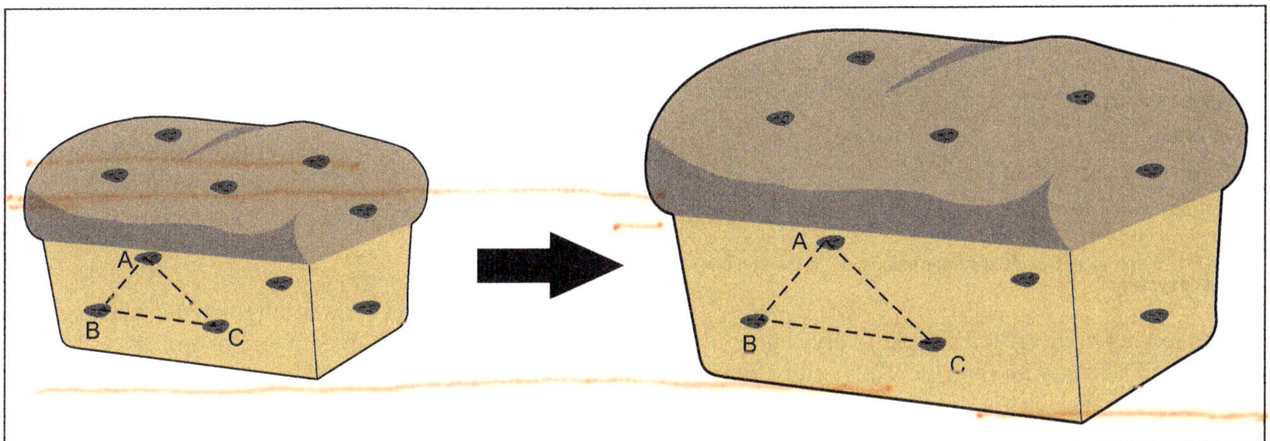

Figure 17.1: A loaf of raisin bread dough is a good analogy to explain the expanding universe. Each of the raisins represent a galaxy. As the bread dough expands, the raisins are carried outward with it. As a result, each raisin is moving away from each of its neighbors.

In the analogy above, the galaxies do not move through the universe, but are carried along by the expansion of the universe itself. Since the universe is expanding (and has been doing so for some time), this means that in the past the universe must have been much more compact, with galaxies closer together. Imagine that you have a video of the universe expanding, and then you watch this video backwards. Since there is no center of the universe, the galaxies would not converge at one single point, but the distance between the galaxies would decrease and the spaces between the galaxies would shrink. Eventually, the galaxies would begin to merge. Keep running the video back, and soon you would see all of the mass and energy of the universe packed into a high temperature, high pressure, and highly dense state within an infinitesimally-small volume of space. The expansion of our universe must have started at this moment of extreme conditions that astronomers call the **big bang**. The term "big bang" was initially invented by early critics of the theory. This label almost automatically leads to a misconception. The big bang was not like a bomb going off. It didn't cause a loud noise, and did not happen at one single location. It would be more accurate to call this event "the big expansion", or as some call it "the big stretch", because from that moment in time, our universe began to expand and continues to expand.

The big bang is a theory, a hypothesis based on scientific observations, and we use this theory to make predictions about what will happen to the universe in the future. But it is important to note that we know our current state of knowledge of the universe is limited, and that because of this, the big bang theory will see many changes as scientists collect more information. Some of these changes may be minor, and some may be major. Each of these changes will make the theory better, and help us make better predictions about the universe in the future.

Section 17-2: The Stars

Astronomers have learned much about our galaxy by studying individual stars. Stars are formed when giant clouds of gas in space begin to condense due to gravity. As this gas becomes dense, the pressure and temperature increase, and the atoms of gas begin to fuse together in a process known as nuclear fusion. During this process, large amounts of energy are released, and the stars emit light (as well as other forms of radiation). The colors of light emitted from a star depend on what types of atoms are fusing at any given time, and all stars are emitting more than one color of light at any given time. Each star is unique in its size, color and temperature, as well as how much energy it gives off every second (known as luminosity).

Object	Apparent Magnitude
Sun	-26.7
Full Moon	-12.7
Venus (at its brightest)	-4
Jupiter or Mars (at their brightest)	-2
Sirius (brightest star in the sky)	-1.5
Alpha Centuri (nearest star)	-0.1
Andromeda Galaxy	3.5
Naked-eye limit	6.5
Binocular limit	10
6-inch telescope limit	13
Hubble Space Telescope limit	28-29

Table 17-1: Some examples of apparent magnitude in the brightness magnitude system.

Over 2000 years ago, a Greek astronomer by the name of Hipparchus created a way to classify stars based on their brightness. The brightest stars he called the stars of "the first-magnitude". The next brightest group of stars was called the "second-magnitude". This classification continued down to the "sixth-magnitude" stars that were barely visible with the naked eye. This classification is still used today, although it has been extended. Since we have instruments which can see more than the naked eye, there are objects that have a magnitude greater than six. Also, if we consider that some very bright objects (e.g., Sun, Moon, and some planets) are actually brighter than many stars, their magnitudes have negative values. This type of classification is known as "**apparent magnitude**", because it is a number which describes how bright an object appears to be. A list of objects with their apparent brightness can found in Table 17.1.

If you are driving at night and see a light in the distance, it is hard to tell how powerful that light really is. It could be a powerful headlight of a distant tractor-trailer, or the dim headlight of a nearby motorcycle. How bright an object appears depends not only on its luminosity, but also how far away it is from the observer. A sixth-magnitude star looks faint to the naked eye, but this does not tell how luminous the star really is. Another scale was therefore developed to compare the true brightness of stars. This is known as the **absolute magnitude**, and tells us which stars are brighter than others. Absolute magnitude is a true measure of luminosity, and is related to the star's actual energy output. This scale is based on the idea that the absolute magnitude would be the same as a star's apparent magnitude if the star was approximately 33 light-years (ly) away. The Sun has an absolute magnitude of 4.38, dimmer than many of the stars in our sky; but the reason the Sun seems so bright to us, and why has a very large apparent magnitude, is because of its proximity to Earth.

Astronomers Ejnar Hertzsprung and Henry Russell independently began to study the luminosity of stars compared with their surface temperatures. The purpose of this was to try and classify the stars, and help to understand how stars are formed and how they change throughout their lifetimes. A diagram summarizing their work is shown in Figure 17.6, and is known as a **Hertzsprung-Russell diagram** (also known more commonly as an **H-R diagram**). This graph is one of the most important in all of astronomy, and helps astronomers sort and classify stars according to their size and luminosity. When we look at an H-R diagram we clearly see that some parts of the graph are empty, while other parts have high concentrations of stars.

A very prominent feature in the H-R diagram is a large group of stars running diagonally from the upper left to the

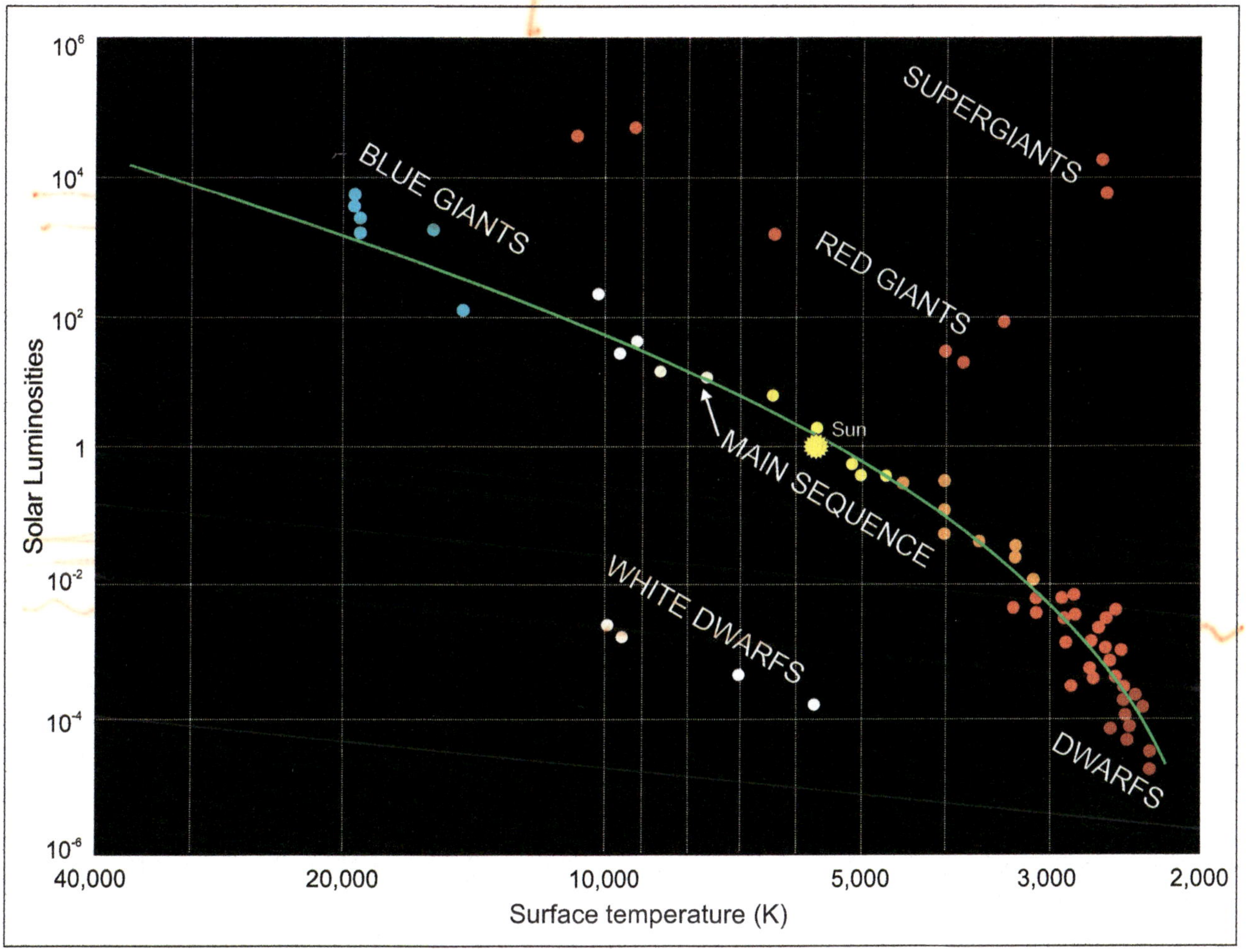

Figure 17.2: A Hertzsprung-Russell diagram showing the basic stellar groups. Blue giants and dwarfs are end members of the main sequence of stars. Data from Shawl and Lo Presto, 2000.

lower right - that is, from higher temperature and higher luminosity to lower temperature and lower luminosity. These are known as **main sequence stars**, and contain roughly 80 percent of all stars. Although stars will change their temperature and luminosities during their lifetime, this main sequence is where typical stars will spend most of their lifetime. For normal main sequence stars, the mass of the star determines its temperature and diameter. Stars that are larger in diameter have a larger surface area, and thus are typically more luminous. Therefore, the more massive the star, the more luminous it is. Stars on the upper part of the main sequence are hotter and more luminous than other stars on the main sequence because they are more massive.

Notice in the H-R diagram that some stars are very luminous, but are also cooler than main sequence stars, and thus lie above the main sequence. Since they are luminous but cool, they must be larger and have more surface area than main-sequence stars of the same temperature. These are known as giant stars, and are 10 to 100 times larger than our Sun. Notice also the supergiant stars in the upper right of the H-R diagram. These have diameters that are over 1,000 times that of our Sun. At the bottom of the H-R diagram are the very low-luminosity stars. They are not very luminous because they are very small. The red dwarfs are found at the far right on the main-sequence line, and the white dwarfs lie below the main sequence line. The white dwarfs are very dim when compared with their surface temperature, which means they must be very small and have very little surface area from which to radiate light.

Our Sun is a main sequence star, intermediate in size between the smallest and largest stars. However, studies have shown that only 3% of the stars in the Milky Way that are on the main sequence are more massive than our Sun, and 85% of the stars have half of the mass of the Sun or less. Thus, the Sun is more luminous and larger than a majority of the stars in the galaxy. Our Sun is known as a *typical* star, because it isn't the biggest or the smallest, and is located on

Figure 17.3: The Eagle Nebula is a good example of clouds which are the birthplace of stars. Image credit: NASA/ESA/ STScI/Arizona State University.

the main sequence. The only thing which makes it our Sun special is that it is close to us.

Star Formation

Within the interstellar medium of space lie areas where the density of gases is greater than others. Studies have shown that the matter within these regions has the same composition as fully formed stars. Within these nebula-type regions are the "nurseries of stars" where stars begin to form (Figure 17.8). However, just having the raw materials for the formation of a star is not sufficient to begin this process. Within the gas clouds are processes, such as thermal activity, magnetic fields, and rotation, which tend to keep the particles in the nebula from condensing to a localized area. However, if an outside mechanism, such as a passing shockwave, were to condense a local area of gas, gravity may be able to continue this condensation process. Sometimes such a shockwave will cause multiple areas to condense. As matter in one area becomes denser, the matter will exert a stronger gravitational attraction, pulling in other nearby surrounding particles. This process of **gravitational contraction** will cause the temperature of the gas to increase, creating a greater outward pressure to counteract the gravitational contraction. If the mass is large enough, the gravitational contraction will be stronger than this outward pressure, and the cloud will continue to condense and collapse.

The gas particles condense towards the center of a localized area, gaining kinetic energy as they move there. As these particles collide with similar particles, they transfer energy back and forth between them. The temperature of the gas increases, and when hot enough, the gas begins to emit short-wavelength radiation, thus reducing its energy. This object is now a **protostar**, an object that has enough mass to be a star, but radiates energy due to its high temperature, and *not* the fusion process (Figure 17.4). The protostar will continue to grow within the gas cloud, gathering more and more mass through gravity from the cloud around it. Because the gas cloud surrounds the protostar, it absorbs much of the short-wavelength radiation being emitted by the protostar. The gas, in turn, reradiates this energy, typically in the form of longer-wavelength infrared radiation. Thus, it is very difficult for us to see the protostar itself, but we can see evidence that one is forming.

Figure 17.4: A protostar forms in this expanding bubble of gas (see arrow). Image credit: NASA/JPL-Caltech/E. Churchwell (University of Wisconsin-Madison).

Because we can't easily observe the complete evolution of a protostar, astronomical models are typically relied on to understand what happens during formation. The gas is contracting due to gravity, increasing its temperature, and then radiating its energy away, cooling the gas body. As the gravitational collapse continues, the gas, once slowly rotating, will begin to spin faster and faster, causing a flattening of the gas cloud. Since gravity is stronger near the center of the protostar, particles near this area tend to fall to the center, reducing the outward pressure on the particles above them. This causes a domino effect, with more and more particles entering the protostar's interior. Eventually, what is left is a growing protostar in the center of a disk of dust and gas. The particles within this disk will either continue to fall into the protostar, be ejected into space by another force, or continue to orbit the protostar. In fact, it is believed that particles within this disk may condense to form planets, moons, asteroids, and other objects. As the gravitational collapse continues within the protostar, the energy transport mechanism changes from radiation to convection, which is a much more efficient way to move energy. When this change occurs, the surface of the protostar will increase in temperature (because it is receiving more energy). Still, the protostar as a whole is contracting, with the inner layers become more and more dense. When the temperature within the core is high enough (around 10 million K), the proton-proton chain will commence, and hydrogen fusion begins. This increases the amount of energy within the core, increasing the outward pressure, and counteracting the gravitational collapse. This state of balance between outward pressure and gravitational contraction is known as **hydrostatic equilibrium** (Figure 17.5). When a star has commenced nuclear fusion, and has reached hydrostatic equilibrium, it is now a main sequence star.

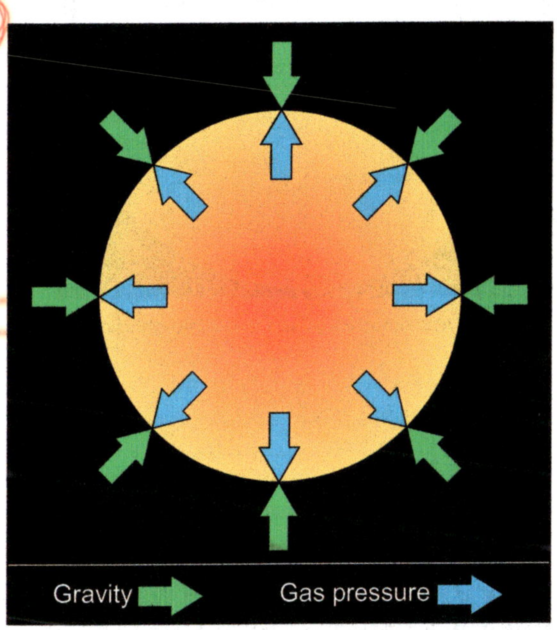

Figure 17.5: In hydrostatic equilibrium, the force from gas pressure due to high temperature within a star's core is equal to its gravitational contraction.

It is important to note here that some protostars do not have a large enough mass (0.08 solar masses) to reach the critical 10 million K threshold during its gravitational contraction; they will never be able to fuse hydrogen into helium. Such objects are known as **brown dwarfs**. These objects have contracted as far as they can, and only emit energy due to their contraction, typically in the infrared. Because they are not luminous, they are not very easy to find, yet astronomers have discovered hundreds of these objects in our galaxy.

Life on the Main Sequence

The main sequence is where a majority stars spend most of their lives. While on the main sequence, hydrostatic equilibrium is the rule: gravity and gas pressure are always working to try to equalize. However, as the star fuses hydrogen to helium within its core, the number of atoms within the core lessens. This reduces the outward pressure, and gravity causes more contraction with the core. This contraction leads to higher temperatures and an increased rate of fusion within the core itself. This releases more energy to the outer layers through convection and radiation, causing an expansion of the outer layers. This expansion causes a cooling in the outer layers, reducing the star's luminosity. These subtle changes are continually occurring within a star on the main sequence, and may lead to a slight shift in its position on the H-R diagram (Figure 17.2). Most stars will stay on the main sequence line for millions of years, until hydrostatic equilibrium can no longer occur.

How long a star stays in its main sequence stage depends on the initial mass of the star. More massive stars use up their fuel more rapidly than lower mass stars. Since larger masses produce a larger gravitational force, fusion takes place at a higher rate, resulting in the higher outward pressures necessary for them to maintain static equilibrium. Lower mass stars do not have large gravitational pressures, and consume their fuel at a much slower rate. In fact, red dwarf stars (the smallest type of star) are expected to continue converting hydrogen into helium for trillions of years (Table 17.2).

Solar Mass (Sun = 1)	Main sequence lifetime (years)
60	2 million
30	5 million
10	25 million
3	350 million
1.5	1.6 billion
1	9 billion
0.1	thousands of billions

Table 17-2: The expected lifetime of main-sequence stars with different solar masses. Data from Shawl and Lo Presto, 2000.

The helium core within a star is constantly growing due to the fusion process (Figure 17.6). Eventually, the amount of fusion needed to sustain hydrostatic equilibrium declines, and gravitational contraction begins to occur. Luminosity does not decrease immediately because stars are so massive that it takes thousands of years for energy at the core to reach the surface. This continual gravitational contraction increases the core's inner temperature. The star at this point is "over the hill", and big changes are about to occur. What types of changes occur depends on the star's mass.

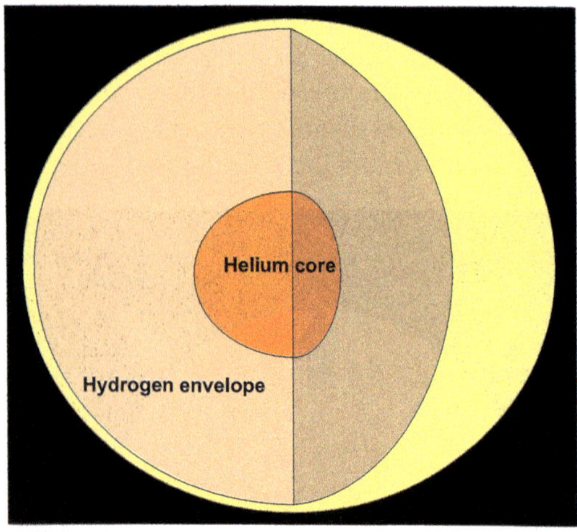

Figure 17.6: A helium core is being added to through fusion.

When discussing stars, it is very common to classify them in terms of low-mass, medium-mass, and high-mass stars; however, the boundaries between each of these categories depend on who you talk to. This difference stems from the purpose of the classification employed. The difference of the cut-off values also arises from our lack of data. In truth, we don't have enough information to give a definite cut-off for each category. The following information is based on one such set of definitions and cut-off values commonly used by many astronomers.

Low-Mass Stars (Sun-like Stars)

For low-mass stars similar in size to our Sun (0.3-8 solar masses), gravitational contraction causes heating within the helium core. This temperature increase is transported to the hydrogen directly in contact with the core, an area known as the *hydrogen burning shell* (Figure 17.7A). This increase in temperature re-starts the hydrogen-fusion process again, adding more helium to the core, increasing its density, and condensing it even further. The hydrogen burning shell radiates energy toward the outer layers of the star, which absorb the energy and expand (Figure 17.7B). This expansion causes the outer layers to cool, and emit a more red color. However, the expanded hydrogen envelope begins to transport heat through convection rather than radiation. Convection is a more efficient process to transport energy than is radiation. Thus, the cooling of the outer layers does not counteract the increase in surface area, and the overall luminosity increases. The star has become a **red giant**.

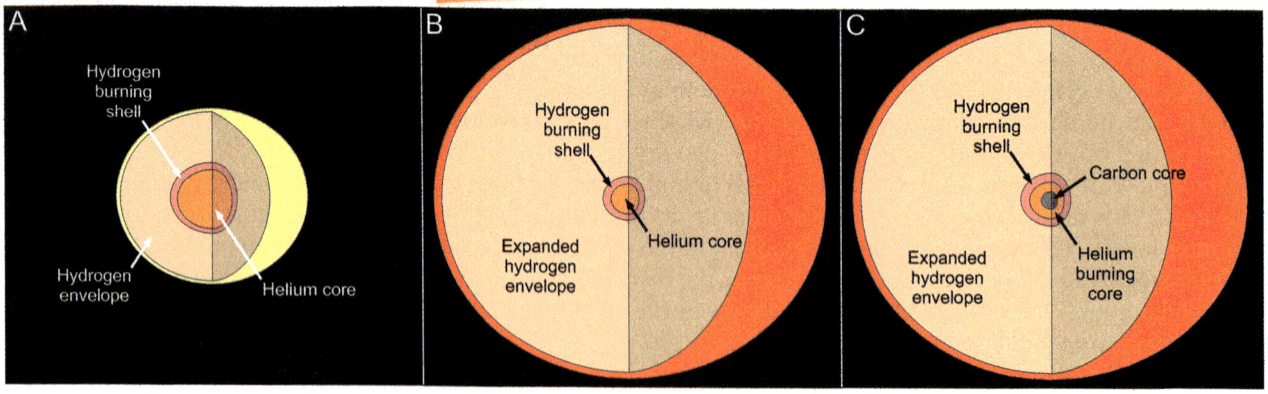

Figure 17.7: The making of a red giant. A. The high temperature of the helium core initiates a hydrogen burning shell. B. The hydrogen envelope expands due to energy emitted from the center of the star. C. The temperature within the core increases to the point where helium begins to fuse into carbon.

The process of becoming a red giant is slow, taking hundreds of millions of years. During this process, the hydrogen burning shell continues to produce helium, adding mass to the helium core. The increase in density causes larger gravitational contraction, increasing the core's temperature. As the core approaches a temperature of 100 million K, helium fusion will begin. Helium fusion is the process where three helium nuclei are fused into a single carbon nucleus. This process of making new elements within the core of a star is known as **nucleosynthesis**. This process continues, changing the helium core to a carbon core (Figure 17.7C). Because of their medium mass, when the helium fuel runs out within the core, temperatures are not sufficient to cause further nucleosynthesis, and the Sun moves toward its final stages. These stages will be discussed later in the chapter. Only a small portion of a star's life will be as a red giant. Because of this, there are relatively few red giants seen in the sky at any one time. However, because of their increased luminosity, they can be some of the most noticeable.

Very low-mass stars (<0.3 solar masses) generate energy and produce radiation similarly to their low-mass counterparts. One significant difference, however, is that when the hydrogen fuel runs out, there is not enough gravitational force to create the necessary pressure within the core to commence helium fusion. This means as the hydrogen fuel runs out, there will be a slow contraction, followed by the star simply "cooling off", radiating off its energy slowly.

Medium-Mass Stars

Stars that start with a mass between 5 and 25 solar masses are considered medium-mass stars. As these stars run out of hydrogen fuel, helium fusion occurs, creating carbon. Because of their larger mass, the temperature within the core can increase to a point where carbon and helium nuclei fuse, creating an oxygen nucleus. When the helium fuel runs out within the core, the core is composed of oxygen and carbon (Figure 17.8A). As gravitation contracts the core due to the lack of gas pressure, the temperature is increased which causes a helium-burning shell to form adjacent to the core. The energy from the helium-burning shell reignites a hydrogen-burning shell, and fusion continues. During this process, the outer layers expand and cool, creating a red giant. The increasing density within the core, along with increasing gravitational contraction, can produce temperatures high enough to cause nucleosynthesis of other heavier elements from the carbon-oxygen core.

High-Mass Stars

Stars that start with a mass over 25 solar masses are considered high-mass stars, or massive stars. They convert hydrogen to helium at an incredible rate, a necessary condition to produce the pressure needed to counteract their massive gravity. As the hydrogen begins to run out within the core, helium fusion occurs, creating oxygen and carbon. The gravitational force creates such high temperatures within the core that other elements are formed as well. Some of these elements are magnesium, silicon, sulfur, and even iron (Figure 17.8 B). For each element, a burning shell will form over the core, fusing smaller nuclei into larger elements. However, the core temperature never reaches a point where iron can be fused into larger nuclei; it is the most massive element that can be made within a star's core through fusion. The outer layers of the star expand to such a large distance from their core that they are known as red **supergiant stars**.

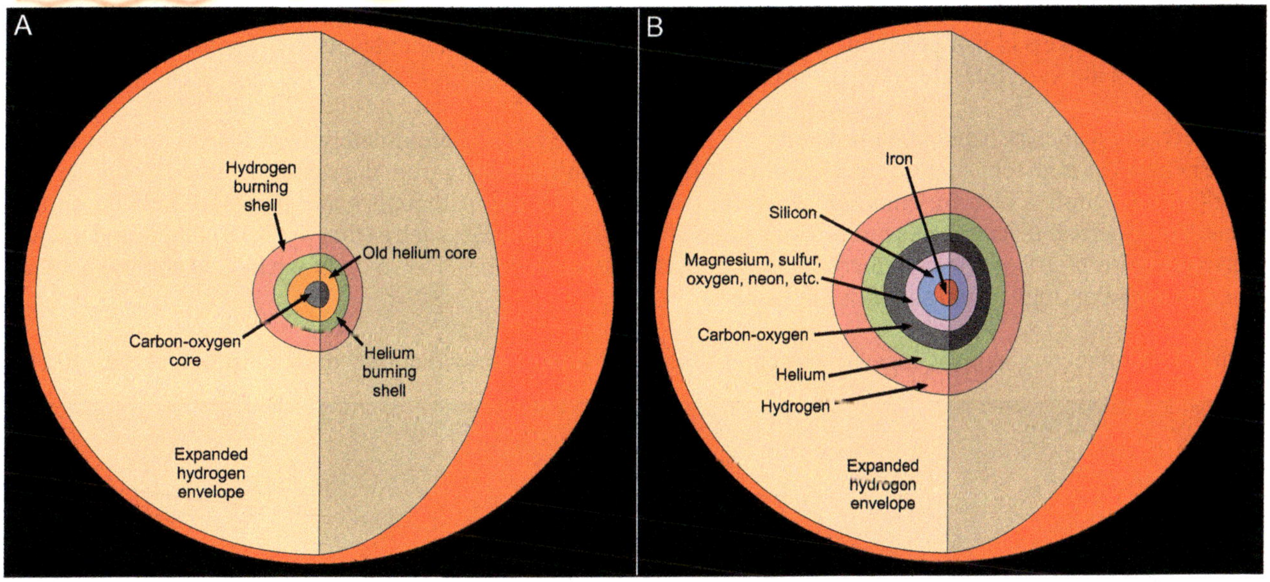

Figure 17.8 A. Two burning shells are present around the oxygen-carbon core, the hydrogen shell fusing into helium, and helium fusing into carbon and oxygen. B. A red supergiant has multiple burning shells, each fusing lighter nuclei into heavier nuclei.

The Death of Stars

As the fusion process ends within a star's core, the star enters its "death stage". The death of each type of star is unique, some of which are very spectacular, while others are not nearly as exciting.

Low-Mass Stars

Red giant stars contain a carbon core surrounded by a helium-burning shell. Their outer layers have expanded drastically; for our Sun, this expansion will put its surface near the Earth's orbit. The outer layers are cooling, and absorb more of the light coming from underneath them. This radiation pressure can cause some of the matter in the outer layers to leave the star completely. The fusion shells near the core are becoming unstable, and begin to pulsate, due to the fusion of the helium nuclei into carbon nuclei. This process is extremely sensitive to temperature, and the

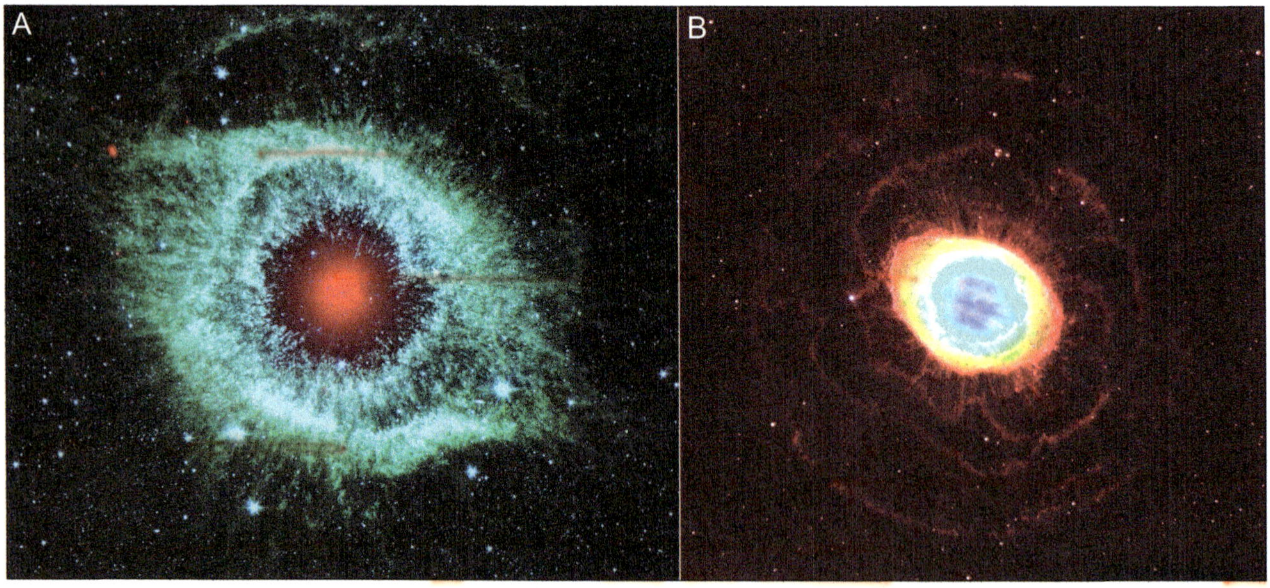

Figure 17.9: An Infrared image of the Helix Nebula, an example of a planetary nebula. Image credit: NASA/JPL-Caltech/Univ. of Ariz. B. Composite image of infrared and visible light images of the Ring Nebula, another planetary nebula. Image credit: NASA/ESA.

continual fluctuations within the core can cause large helium-shell eruptions, emitting massive amounts of energy toward the outer layers. These processes can eject much of the star's outer layers, creating a **planetary nebula** (Figure 17.9). This name was given to these clouds of ionized gas because when looking through an amateur telescope, they have a bluish-green hue similar to the color of the most distant planets in the solar system, Uranus and Neptune. However, they are nothing like planets.

The spectra of planetary nebulae show that the temperature of the gas is approximately 10,000 K. Within the center of the nebula is a high temperature star. The ultraviolet radiation emanating from the star is absorbed by the nebula, which in turn emits a discrete emission spectrum (see Figure 15.10). Although many planetary nebulas appear circular or spherical in nature, this is due to our perspective. Other nebulae, such as the MyCn18 Nebula and the Ant Nebula (Figure 17.10), show asymmetry, appearing as two separate lobes. However, if we looked at these nebulae along its long axis, they would also look circular.

Nearly 1500 planetary nebulae have been seen in space, and studies show that they do not last longer than 10,000

Figure 17.10: A. Image of a young planetary nebula, MyCn 18 located 8,000 light-years away from Earth. From this perspective, we can see two separate lobes, showing that all planetary nebulas are not spherical. Image credit: NASA/WFPC2/HST, R. Sahai and J. Trauger (JPL). B. The Ant Nebula is another good example of a lobed planetary nebula. Image credit: NASA/STSI.

years. Because of these findings, astronomers have concluded that they are a very common part of a star's life. Many scientists believe that our Sun is destined to become a planetary nebula, although some scientific models disagree with this presumption.

At the center of the planetary nebula is what is left of the star. As the star exhausts its hydrogen and helium fuel, it can no longer maintain hydrostatic equilibrium, and gravity takes over, condensing all of the matter that is left. This new object is known as a **white dwarf** (Figure 17.11). It is highly condensed, with a density over 2 million times that of water. This condensation packs the electrons more closely together than is typical, creating a **degenerate gas**. Unlike an ideal gas, in a degenerate gas, pressure and temperature are no longer related. This means that even as the temperature drops, the pressure remains the same. In other words, the white dwarf will not contract any more. White dwarfs are no longer true stars. Without fusion and without gravitational contraction, mass cannot be converted into energy. White dwarves simply radiate their remaining energy into space as they cool. Because of their small surface area, this cooling may take billions of years.

Figure 17.11: This star cluster contains dozens of white dwarf stars (circled). Image credit: NASA and H. Richer (University of British Columbia).

Very low-mass stars (red dwarfs) will also eventually burn through all their fuel, and become white dwarfs. It is important to point out, however, that since these stars burn their fuel so slowly, it will take billions of years for them to evolve into white dwarfs. In fact, no red dwarf star has yet gone through this metamorphosis as the universe is not old enough!

Medium-Mass Stars

Medium-mass stars have too much original mass to end up as white dwarfs. However, when such a star uses up all of its fusible material, it will begin to contract due to gravitation. This increases the temperature, which increases the outward pressure within the core, but is not sufficient to stop the contraction. Within a very short period of time, the core collapses on itself, creating a massive explosion that blasts the star's outer layers away. This is known as a **nova**. Large novas or novae (plurals of nova) are often known as **supernovae**. The word nova comes from the Latin *nova stella*, meaning "new star". It was not until the 1900's that astronomers understood that novae were not new stars, but the final stages of an old star's life. No one had seen a supernova since 1604, until Chilean astronomers witnessed one in 1987 (Figure 17.12). This event was bright enough to see with the naked eye.

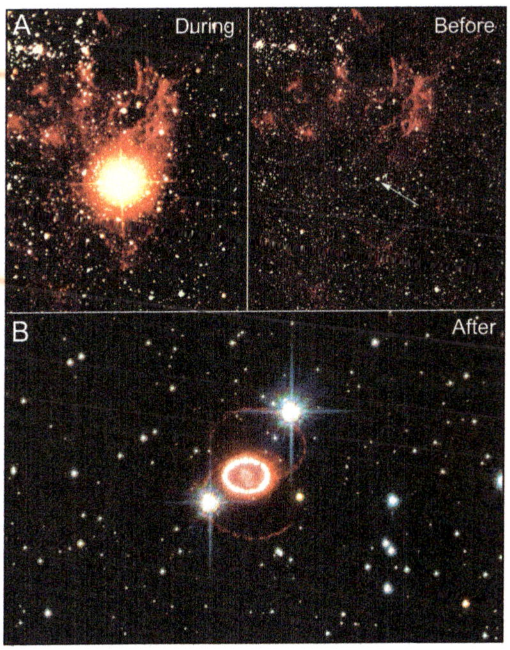

Figure 17.12: A. This pair of images shows supernova 1987a during (left image) and before (right image) the explosion. Photo credit: High Energy Astrophysics Science Archive Research Center (HEASARC), Dr. Nicholas White (Director), within the Exploration of the Universe Division (EUD) at NASA's Goddard Space Flight Center. B. The remnants of supernova 1987a.

As the star runs out of fuel, gravity begins to contract the core, causing a degenerate state. The electrons, moving at speeds near the speed of light, combine with iron nuclei, breaking the nuclei apart. Since the fast moving electrons helped to maintain the pressure within the core, a loss of any electrons will accelerate the gravitational contraction. This process raises the temperature further, increasing the speed of the electrons, which in turn combine more readily with the iron nuclei. This collapse and the disintegration of the iron nuclei creates massive numbers of neutrons which are soon absorbed by other nuclei, creating additional nuclei that are heavier than iron. In fact, it is only in this type of environment that elements heavier than iron can be created in a process known as **explosive nucleosynthesis**. These elements include gold, silver, and uranium. This release of neutrons and fusion of heavy nuclei release massive amounts of energy, and this energy causes the compete destruction of the star.

Supernovae

There are two types of supernovae. Type I supernovae are missing hydrogen lines within their spectra, while Type II supernovae spectra have hydrogen lines that are clearly visible. The description in the preceding paragraph describes a Type II supernova, where much hydrogen was still present before the explosion. When a supernova occurs, the enormous amount of energy released will increase the brightness of the star a billion-fold. There are a two different ways that Type I supernovae can occur. The first and most common way to form a Type I supernova is with a white dwarf that has a companion star (Figure 17.13). Within the white dwarf star, the degenerate electrons are moving at super high speeds, nearing the speed of light. Calculations have shown that if the white dwarf has a mass of over 1.4 solar masses, the electrons' speeds would exceed the speed of light. Einstein's theory of relativity states that nothing can travel faster than the speed of light. Therefore, if a white dwarf receives mass from a nearby star, and exceeds the limit of 1.4 solar masses, the core begins to collapse. The iron begins to disintegrate, and an implosion rapidly occurs. This implosion is immediately followed by an explosion, destroying the white dwarf. The second way for a Type I supernova to occur is if a massive star loses its outer layers (mainly hydrogen) due to the gravity of a second star. What's left of the first star can go through the same process as a Type II supernova, but since it no longer has its hydrogen layers, it will lack the hydrogen signature within its spectrum.

Figure 17.13: Artist's impression of a binary star system with a white dwarf and a low-mass companion that has become a red giant. The red giant "donates" mass to the white dwarf. If the white dwarf's mass exceeds 1.4 solar masses, it will collapse and explode as a Type Ia supernova. Image Credit: NASA/CXC/M.Weiss.

High-Mass Stars

High-mass stars evolve from their super-giant stage into supernovae, similar to their medium-mass counterparts. However, because their outer layers are so massive, it is more difficult for the inner core to blow them away; thus, the explosions of these stars tend to be smaller than for medium-mass stars. In fact, stellar models predict that for stars with more than 45 solar masses, the explosion would not even be visible from Earth.

Neutron Stars and Pulsars

After a Type II supernova occurs, and the outer layers of a star are blown away, the condensed core still exists in some form. For medium-mass stars, electrons will combine with the protons in the heavy-element nuclei, creating an object that only has neutrons. This result was predicted in 1934 by astronomers Fritz Zwicky and Walter Baade. This new object, called a **neutron star**, would have a mass of 3 solar masses or less, and a diameter of about 20 km (12 mi.). It would have no protons or electrons, only neutrons, and thus be super compact, with a density of approximately 1×10^{15} greater than water. For comparison, a teaspoon of a neutron star would have almost the same mass as 300 Empire State Buildings. The first evidence for neutron stars was found in 1967, when Jocclyn Bell, a 24-year-old Cambridge graduate student, discovered a very periodic radio signal coming from space. This signal repeated every 1.337301 seconds. This signal was so precise, that it rivaled the best man-made clocks of the era. Once astronomers knew what to look for, they began discovered many similar signals in other regions of the universe. They found that not all of the radio signals exhibited such a constant frequency like the first one discovered; some pulses were slowly decreasing in frequency. However, these increases in time between pulses were specific, unchanging amounts.

It is now known that these pulsing radio sources, called **pulsars**, are actually rotating neutron stars. Although the center of a neutron star is made entirely of neutrons, the outer layers may contain some ionized gas particles. As the neutron star spins, these charged particles can be thrown off the surface, and become subject to the magnetic field of

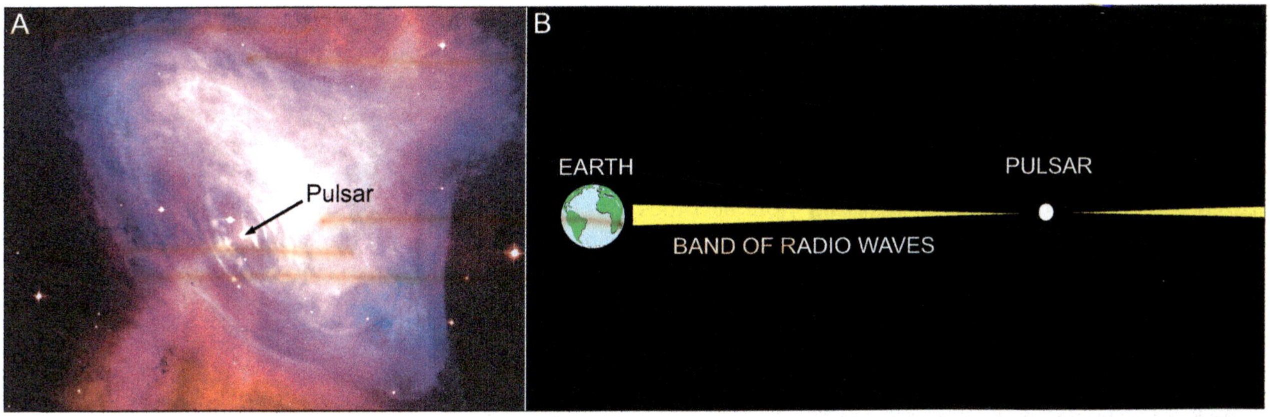

Figure 17.14: A. A composite image of both visible and X-ray images, showing the Crab Nebula. Image credit: NASA/HST/ASU/J. Hester et al. B. As streams of radio waves emitted by a neutron star sweep over the Earth, it appears as if the radio signals are "pulsing".

the neutron star, which is thousands of times greater than that of the Earth. These magnetic fields funnel the charged particles toward the magnetic poles, where the magnetic field is the strongest. Here, the charged particles can then be jettisoned, carried along by radiation emitted near the magnetic poles. If the magnetic poles are not aligned with its axial poles, the spinning will cause a "lighthouse effect", where every time the neutron star rotates its beam of radiation sweeps through an area of space. If Earth happens to be in this area, we will be able to observe the pulsar (Figure 17.14). We know that pulsars must be rotating neutron stars, because nothing else in the universe can rotate so quickly. Not all neutron stars become pulsars, and we have not discovered all the neutron stars which are pulsars, because we may not be in the correct location to observe them. Still, astronomers have discovered over 1300 to date.

Black Holes

In the late 1700's, an interesting idea was developed by French physicist Pierre Laplace and British philosopher John Mitchell. Scientists had used Newtonian physics to calculate the escape speed of an object; that is, the speed necessary for an object to escape the gravitational force of a larger body. If we want to send a rocket to Mars, we have to give it enough speed to escape Earth's gravity. This would be the Earth's escape speed. Laplace and Mitchell theorized that if an object was dense enough, the escape speed from that object would be greater than the speed of light, meaning not even light could escape. This theoretical object became known as a **black hole** (Figure 17.15).

As scientists understood more about supernovas and neutron stars, some began to wonder if the remnants of a star's core could collapse to a point where a black hole would occur. Models predict that if the core of a star contains more than 3 solar masses when it collapses, there is not a strong enough force in the universe to stop the contraction. The degenerative electrons will not be able to stop it as they do in a white dwarf, and the degenerative neutrons will not

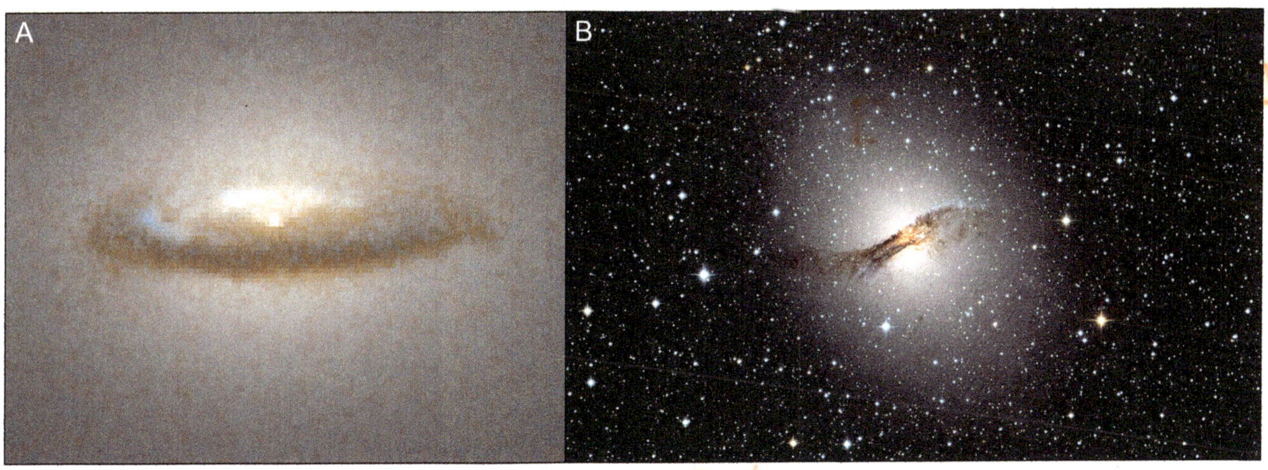

Figure 7.15: A. Because no light escapes a black hole, we can't technically "see it", but we can observe the disk of material which always surrounds it. Image credit: Roeland P. van der Marel (STScI), Frank C. van den Bosch (Univ. of Washington)/NASA. B. Galaxy NGC 5128, as it appears in visible light. This galaxy is one of the closest galaxies to the Milky Way that contains an active supermassive black hole. Image credit: NASA/Capella Observatory.

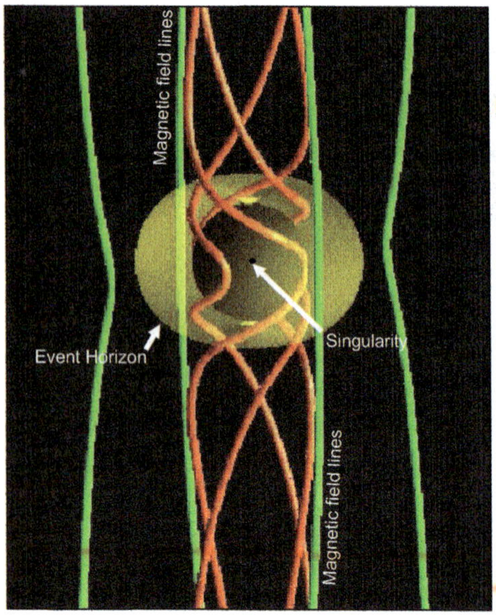

Figure 17.16: Diagram of a black hole. Anything that falls within the Event Horizon can never escape. This image was generated by supercomputers at Japan's National Institute for Fusion Science. Image credit: NASA.

be able to stop it as they do in a neutron star. It collapses to a single point with infinite density and gravity and zero radius. This is known as a **singularity**. The space around the singularity will feel immense gravitational force, with the force gradually lessening as you move away from the singularity. At a certain distance from the singularity, known as the *Schwartzchild radius*, the escape velocity for the black hole becomes less than the speed of light. This point is known as the **event horizon**; anything (even light) that approaches the singularity closer than the event horizon will not be able to escape its gravity, and will fall inward (Figure 17.16). The event horizon is the outer limit of the black hole.

Because light cannot escape a black hole, we cannot observe its interior (hence the term *black*). Within the black hole itself, scientists believe that there is only empty space between the singularity and the event horizon (hence the term *hole*). The only information that we can gather about black holes are their masses, how much electrical charge they contain, and their angular moment. All other information is lost, meaning we would have no information about the object(s) which formed the black hole. A common misconception about black holes is that once they form, they pull in everything else around them. Near or within the event horizon, this is a correct picture, but not so for more distant objects. If our Sun were to become a black hole, the planets would continue in their orbits, as the gravitational force they experience would not change.

Astronomical models predict that for high-mass stars, their endpoint will be as a black hole. For stars with 25 – 45 solar masses, their weak supernovae explosions will emit their outer layers into space, but not very far away. As the core becomes a neutron star, some of this mass can fall back onto the core. When the mass of the core exceeds 3 solar masses, the core will collapse into a black hole. For stars which begin with more than 45 solar masses, the explosion in the core is not powerful enough to jettison the outer layers, and the core becomes a black hole, pulling in the outer layers with it. Some black holes can contain millions of solar masses, and are known as **supermassive black holes** (Figure 17.15B).

Section 17-3: The Milky Way

Since ancient times, people have been aware of a hazy band of light around the nighttime sky. The Greeks called this band the "Milky Circle", while the Romans called the band the "Milky Road", or the "Milky Way". Our galaxy is known as the Milky Way Galaxy, which contains more than 100 billion stars. Astronomers have tried for hundreds of years to determine what our galaxy is like. Galileo used his telescope to reveal that the glowing Milky Way is made up of stars. Later astronomers realized that the Sun is part of a great wheel-shaped cloud of stars known as a star system. But until the early 20th century, astronomers did not understand we lived in a galaxy; in fact, they assumed that the Milky Way was the entire universe. This idea began to change in 1917 when a Harvard astronomer by the name of Harlow Shapely discovered how large our star system truly was. Shapely was studying **globular clusters** (Figure 17.17), giant spherical star groups

Figure 17.17: Messier-9 is a globular cluster that lies around 25,000 light-years from Earth, near the center of the Milky Way. There are 150 such clusters known in the Milky Way. Image credit: NASA/ESA.

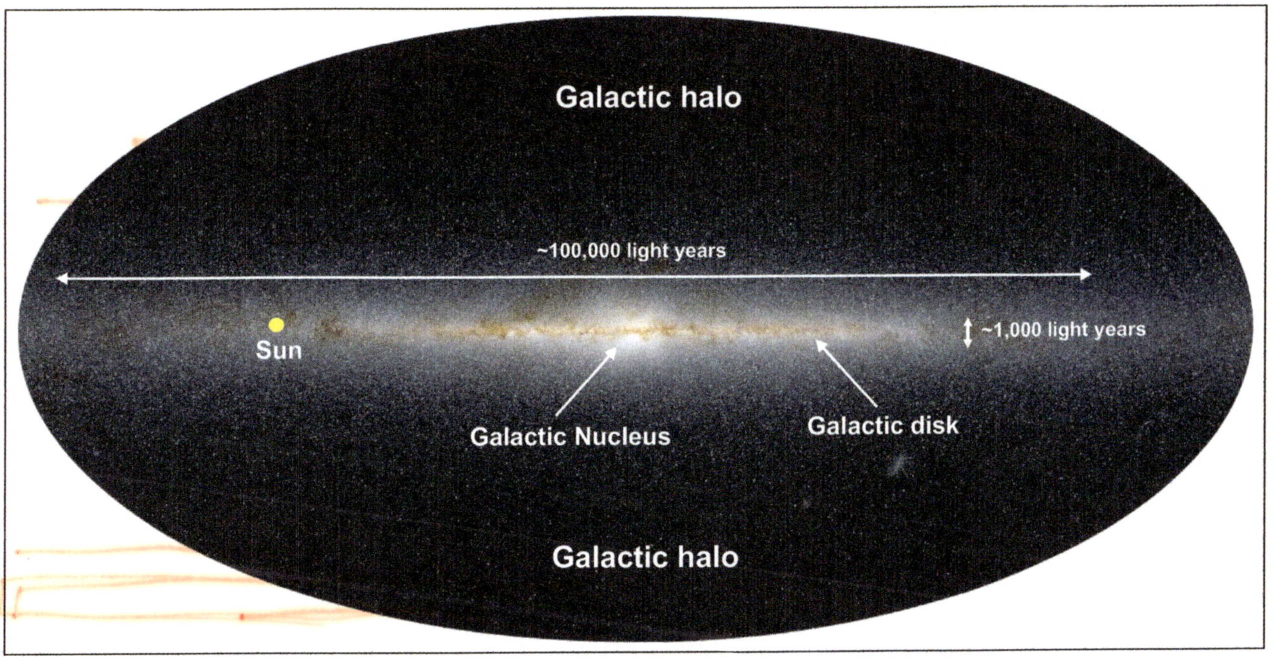

Figure 17.18: The Milky Way Galaxy, as seen from the side, using infrared imaging techniques. The Sun is approximately 30,000 ly from the center of the galaxy. Image credit: Atlas Image mosaic obtained as part of the Two Micron All Sky Survey (2MASS), a joint project of the University of Massachusetts and the Infrared Processing and Analysis Center/California Institute of Technology, funded by the National Aeronautics and Space Administration and the National Science Foundation.

containing approximately 100,000 stars. Shapely found that some of these globular clusters were very far from the Sun, and that our galaxy must be much larger than previously thought. He also found that nearly half of the globular clusters were centered in a great cloud with a center located in the direction of the constellation Sagittarius. Shapely used his understanding of gravity to reason that the center of the star system must be near the center of these globular clusters, because they were so massive and such an important component of the galaxy. From this he determined that the center of our galaxy is located somewhere toward Sagittarius. Is it now estimated that the Sun is between 1.6 x 10^{17} to 2.9 x 10^{17} km (1.5 x 10^{17} to 1.8 x 10^{17} mi.) away from the center of the galaxy (Figure 17.18). These distances are so large, that astronomers use a different measurement known as a **light-year** (ly) to describe them. A light-year is the distance that light could travel in one year, and is equivalent to 9.46 x 10^{12} km (5.88 x 10^{12} miles). Using this measurement, we can say that the Sun lies between 25,000 and 30,000 light-years from the galactic center. Our entire galaxy is about 100,000 ly across.

Using Shapley's work, as well as other astronomical observations, astronomers have theorized how a galaxy such as the Milky Way could begin to form (Please note that because we cannot see back to the birth of galaxies, models explaining their beginnings are mostly theoretical models and have some uncertainties). Sometime after the Big Bang, a massive cloud of gas would have begun to condense due to its gravity. As this gas condensed, gigantic areas of localized condensation, which eventually would become globular clusters, began to form within the cloud. Each of these localized areas contained at least 100,000 times the mass of our Sun, and quite possibly more. Each of these localized areas began to condense at a faster rate than the galaxy as a whole. The entire cloud of gas continued to collapse, with the most rapid condensation occurring near the center of the cloud, where the gas was denser. Astronomers believe that this nuclear region is where star formation would most likely occur, although the exact rate that stars would form is not known. The collapsing gas cloud would begin to spin more rapidly as it got smaller, just as an ice skater begins to spin

Figure 17.19: What the Milky Way Galaxy may have looked like as it formed. Image Credit: NASA/JPL-Caltech.

Figure 17.20: Artist's concept of the Milky Way Galaxy. Image Credit: NASA/ JPL-Caltech.

faster if she pulls in her arms and legs toward her body. In a very large region where particles are free to move, like in a galaxy, the system would begin to flatten into a disk as its rotation increased (Figure 17.19). Some of the local condensed regions began to break apart, and star formation would begin to proceed within the galactic disk. Since the rotation of this gas was not uniform, spiral shaped arms of dense gas would possibly begin to form, and in turn initiate more star formation. As the galaxy grew older, the gas cloud would have gained more and more heavy elements, remnants of stars which had already lived and died.

Since Shapley's time, other astronomers began to study other faint patches of light visible only through telescopes and determined that they were indeed other galaxies. Today, the largest telescopes can detect an estimated 100 billion galaxies similar to our own. These galaxies come in many shapes and sizes. From observations of the stars and dust in our own galaxy, as well as images of other galaxies, astronomers determined that the Milky Way is a spiral galaxy, which is a galaxy with spiral arms (Figure 17.20). Our Sun sits within one of these spiral arms, and orbits the center of the galaxy with a speed of 250 km/sec (155 mi/sec) relative to the galactic center. It takes the Sun about 200 to 250 million years to rotate once around the galactic center. Other stars similar to the Sun also rotate around the center of the Milky Way. The spiral arms are not distributed uniformly, but lie within the galactic disk (Figure 17.18), which contains most of the observable matter in the galaxy. This galactic disk is about 1,000 ly thick. Astronomers, however, have discovered many stars and globular clusters lying outside of this plane, with many of these further from the galactic center than the Sun. From these observations, astronomers have determined that the Milky Way Galaxy is surrounded by a **galactic halo,** a spherically-shaped cloud of stars and star clusters containing almost no gas and dust (Figure 17.18). The halo surrounding the Milky Way contains approximately 200 globular clusters, each of which contains 50,000 to one million stars.

Figure 17.21: This artist's concept illustrates the frenzied activity at the core of our Milky Way Galaxy. The galactic center hosts a supermassive black hole in the region known as Sagittarius A*, or Sgr A*, with a mass about four million times that of our Sun. Image Credit: NASA/ ESA/C. Carreau.

What lies at the center of the Milky Way? That question puzzled scientists for years because we cannot see all the way to the center of the galaxy. Using observations of the galaxy itself and new astronomical techniques, it is now known that the galactic center hosts a supermassive black hole called Sagittarius A* (Figure 17.21). In fact, it is highly likely that the Sagittarius A* was responsible for the formation of the Milky Way Galaxy.

Section 17-4: The Interstellar Medium

What else is there in the galaxy besides stars and planets? Of course, there are chunks of rock such as meteors, asteroids, and even moons. But what we think of as empty space truly isn't empty. It contains atoms, molecules, and dust grains, and is known as the **interstellar medium**. Stars can emit matter into the interstellar medium through many different processes, such as stellar winds or through explosions. The interstellar medium can also provide

Figure 17.22: A. An example of an emission nebula. NGC 3372, the Great Nebula in Carina, is home to massive stars and changing nebulas. Image credit: NASA/ESA/N. Smith (U. California, Berkeley) et al/STScI/AURA. B. The famous Horsehead Nebula is a dark nebula. Dark nebulas are dense enough to completely block out light from stars behind them, and thus appear "dark". Image credit: NASA, ESA, and the Hubble Heritage Team (STScI/AURA). C. Reflection nebula in the Pleiades Star Cluster. Here the blue light from the nearby stars illuminates the faint dust clouds. Image credit: NASA/ESA/AURA/Caltech.

matter for new star formation. The interstellar medium contains large amounts of gas, with most of the gas being hydrogen, although there are other elements present as well. Most of this gas is not distributed uniformly throughout the galaxy, but typically forms giant molecular clouds. There are many types of molecules in the interstellar medium; these molecules will be present in these clouds as well. These large clouds of gas are known as **nebulas** or **nebulae**. There are different classifications of nebulas, based on their appearance. Many nebulas are called **emission nebulas** (Figure 17.22A). These clouds of gas glow different colors, depending on the composition of the gas. The gases are excited by UV radiation emitted by nearby stars. Some nebulae have large amounts of dust along with the gases. This dust blocks light from passing through the gases and are known as **dark nebula** (Figure 17.22B). However, dust also reflects light coming from sources not behind the nebula. This type of nebula is known as **reflection nebula** (Figure 17.22C). You also saw examples of nebulas in Figures 7.1, 17.9,17.10 and 17.14.

Section 17-5: The Morphology of Galaxies

When Edwin Hubble realized that our galaxy was not alone in the universe (See Figure 15.1), it opened up a new area of study in astronomy. Since that time, astronomers have realized some galaxies are currently forming stars, while others have little star formation. Other differences became apparent as well. By studying galaxies, we learn much about our own galaxy and the universe in general.

Galactic Shapes

Galaxies have different shapes. As astronomers observed more and more galaxies, they began to classify them according to their shape, using a system developed by Hubble early in the 20th century. By placing objects into groups, comparisons can be made to try and understand each of the objects. This procedure is not unique to astronomers. A child may sort his toys according to ones that have feet and ones that do not; he may notice that everything that has feet also seems to have eyes, ears, and noses. Biologists sort plants and animals into kingdoms, phyla, classes, orders, etc., and realize that both chickens and dinosaurs have similar bone structure. Astronomers group objects such as galaxies, moons, planets, and other objects to help them see patterns and try to understand the universe.

Figure 17.23 shows a group of galaxies, each with a different shape. If we were to break these galaxies into smaller groups, we could call some elliptical (E subclass), and some spiral (S subclass). **Elliptical galaxies** have a round or elliptical shape and have little gas or dust. They are also missing any bright stars. Along with the designation of their subclass (E), they are assigned a numerical value to indicate how elliptical the galaxy is. The higher the numerical index, the flatter the galaxy is as well. An E0 galaxy is fairly spherical, an E3 may look like a football, and an E7 galaxy is very elliptical and flat (Figure 17.24).

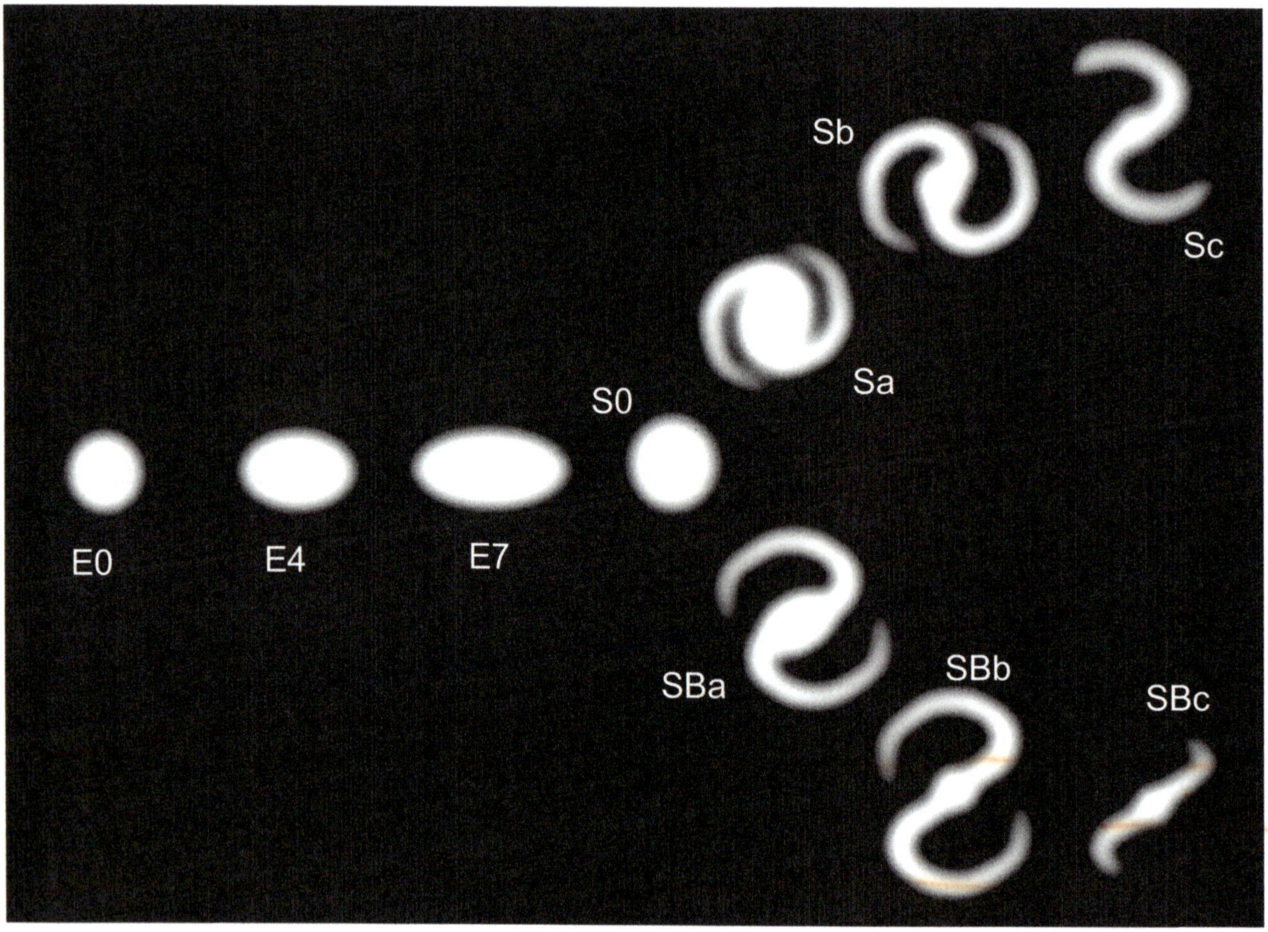

Figure 17.23: Edwin Hubble's classification of galaxies.

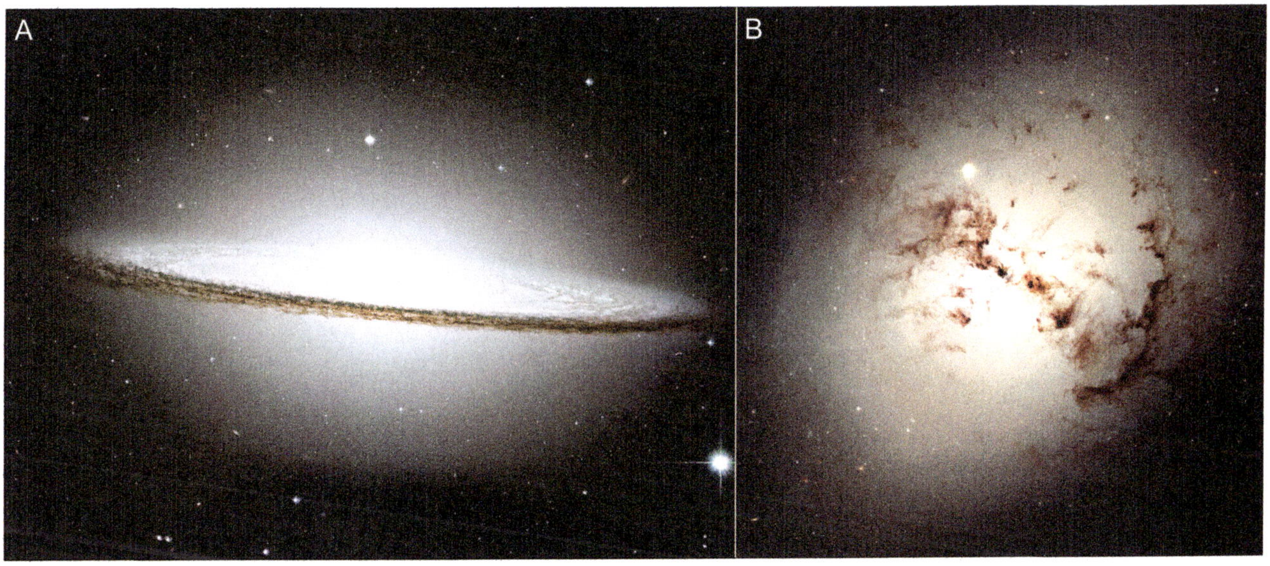

Figure 17.24: Some examples of elliptical galaxies. A. The Sombrero Galaxy. B. Surprisingly complex loops and blobs of cosmic dust lie hidden in the giant elliptical galaxy NGC 1316. Photo credits: NASA/STScI/AURA.

Spiral galaxies contain a visible disk, a halo, and spiral arms (like the Milky Way). Visible evidence shows us that star formation is occurring within these galaxies. Spiral galaxies can be further subdivided based on the relative size of their nuclei. Sa galaxies have a large nuclei, less gas and dust than other spiral galaxies, and have tightly-wound spiral arms (Figure 17.25A). Sc galaxies have smaller nuclei and more open spiral arms (Figure 17.25B). Approximately two thirds of spiral galaxies have a bar-like feature near the center, and are known as **barred spiral galaxies** (SB subclass). Within this subclass, astronomers have denoted SBa, SBb, and SBc categories, based on the same criteria used as for regular spiral galaxies (Figure 17.25C).

Figure 17.25: Different types of spiral galaxies. A. Spiral Galaxy NGC 3982 is a type Sa spiral galaxy. Photo credit: NASA, ESA, and the Hubble Heritage Team (HHT) (STScI/AURA) B. The Whirlpool Galaxy is an example of a Sc spiral galaxy. Photo credit: NASA and HHT (STScI/AURA) Acknowledgement: N. Scoville (Caltech) and T. Rector (NOAO). C. Galaxy NGC 1365 is an example of a SB spiral galaxy. Credit: FORS Team, 8.2-meter VLT Antu, ESO, NASA.

Spiral galaxies have hotter, brighter stars than their elliptical counterparts, and so they are much more visible in the sky. Because of this effect, astronomers initially believed that there were many more spiral galaxies than elliptical galaxies. This incorrect result was due to something scientists call the *selection effect*. Astronomers had concentrated on the galaxies that were brighter and clearly visible, and reached an erroneous conclusion (later studies have shown that there are more elliptical than spiral galaxies). Astronomers must be careful to take into account the selection effect when making conclusions about what they see. There also seems to be a galaxy with a structure halfway between elliptical and spiral galaxies. This "intermediate" galaxy is given the classification of S0. It is very flat (like spiral galaxies), and has a large nuclear budge, but no spiral arms (Figure 17.26A).

Some galaxies we see just don't fit into any of the categories originally defined by Hubble. These **irregular galaxies** have no nuclear bulge, have a varying mix of gas and dust, and have no spiral arms (Figure 17.26B). Irregular

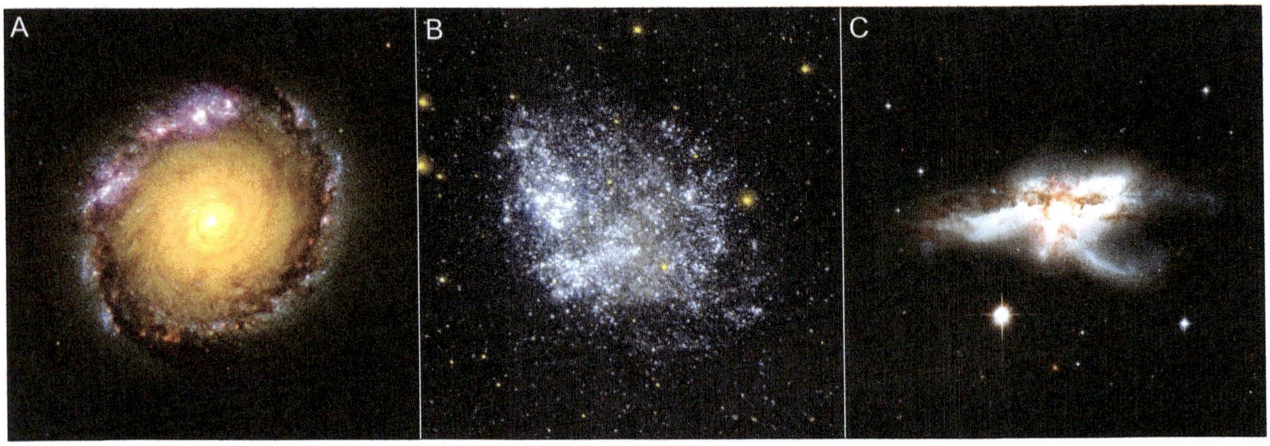

Figure 17.26: Examples of irregular galaxies. A. This image of Galaxy NGC1512 is a good example of a SO galaxy, one that is transitional between spiral and elliptical galaxies. Photo credit: NASA, Space Telescope Science Institute. B. Irregular Dwarf Galaxy IC 1613. Note the lack of any organized galactic nucleus or distinct shape. Photo credit: NASA/JPL-Caltech/SSC. C. Not all galaxies are neatly shaped, as this NASA/ESA Hubble Space Telescope image of a peculiar galaxy clearly demonstrates. Image credit: NASA, ESA, the Hubble Heritage (STScI/AURA)-ESA/Hubble Collaboration, and A. Evans (University of Virginia, Charlottesville/NRAO/Stony Brook University)

galaxies make up about 3% of all visible galaxies. Even within the irregular galaxy family, there are some galaxies that are so different and unique that they are often called **peculiar** galaxies (Figure 17.26C).

General Galaxy Attributes

Astronomers can measure the apparent brightness not only of stars, but also of galaxies as well. However, finding their true luminosity is much more difficult. This is due to a few reasons. First, galaxies have no designated outer edge, so how do we determine what objects to include in our calculations? Secondly, how do we account for light that may have been absorbed or reflected by dust or gas within the galaxy itself? Thirdly, to find true luminosity, we need to know how far away the galaxy is from us.

To calculate distance to galaxies, astronomers look for objects that we see and understand in galaxies closer to our own. These objects are known as *distance indicators*. One such indicator is a Cepheid variable star (Figure 17.27). This type of star gets brighter and dimmer over specific time periods. These periods are related to the object's true luminosity. Therefore, if astronomers can find such a star within the galaxy and measure its luminosity period, they can determine its true brightness. Comparing this brightness with its apparent brightness, they can determine the distance to the star, and thus the galaxy.

Figure 17.27: A Cepheid variable star is a star that varies in light intensity over regular, measurable periods. In this image, the star is at its brightest. Image credit: NASA.

To find the size of the galaxy, astronomers measure its angular diameter in the sky. Using this result, as well as the distance to the galaxy, astronomers calculate the object's diameter. Because there is uncertainty in the angular diameter (how do we know where the edge is?), as well as the distance calculation, there will always be uncertainty in this type of calculation. However, this type of study helps us paint a picture of what galaxies are truly like. For example, when compared with our own Milky Way Galaxy, irregular galaxies tend to be smaller, while spiral and elliptical galaxies can be either larger or smaller (Table 17-3).

Galactic Color

The color of a galaxy can help us understand what types of processes are occurring within the galaxy itself. When a galaxy forms, and the gas from the galactic disk begins to condense and form stars, it usually does so in

PROPERTY	Irregular	Spiral	Elliptical
Gas	Substantial	Substantial	Little
Dust	Yes	Substantial	Very little
Population I (old) stars	Yes	Yes	No
Population II (young) stars	No	Yes	Yes
Age of oldest stars (years)	10^{10}	10^{10}	10^{10}
Age of youngest stars (years)	Newly formed	Newly Formed	10^{10}
Rotation	Maybe	Yes	Low or none
Color	Bluish	Blue disk, red halo	Reddish
Range in Diameter (Milky Way units)	0.05 - 0.5	0.3 - 3	0.03 - 10
Range in Luminosity (Milky Way units)	0.0001 - 0.01	0.01 - 2	0.00005 - 10
Range in Mass (Milky Way units)	0.001 - 0.5	0.01 - 2	0.00001 - 50

Table 17-3: A comparison of common galaxy properties.

groups. Within these star clusters are stars of different masses, and thus different colors and luminosities. Bigger stars are hotter and brighter, emitting blue light, while smaller stars are not as bright and tend to emit more orange or reddish light (Refer back to Section 15-4). Thus, if we look at these clusters from a distance, the brighter or blue stars are more noticeable. Looking at images of spiral galaxies, we see a lot of blue light within the spiral arm region, indicating that star formation is currently taking place there (Figure 17.28A). The nuclear bulge of spiral galaxies seems to lack any blue color at all. This leads astronomers to believe that star formation took place in this region long ago. Since bigger, bluer stars tend to burn more quickly, all of the blue stars have used up their hydrogen, and are either in a red giant phase, or may have possibly already gone supernova. The smaller yellow, orange, and reddish stars, along with these red giants account for the color observed in this region.

Elliptical galaxies also seem to be lacking much of a blue color (Figure 17.28B). The reddish color indicates that most of the stars are of smaller mass, and new star formation is not occurring. Initially, astronomers believed that the lack of new star formation was due to lack of gas and dust in the region. Other studies have shown that there is, in fact, gas within this region. However, this gas has such a high temperature that it will not condense into a localized region, forming a new star. These high temperatures possibly occur because this gas has been ejected from existing stars in the form of solar wind or possibly as supernovae remnants.

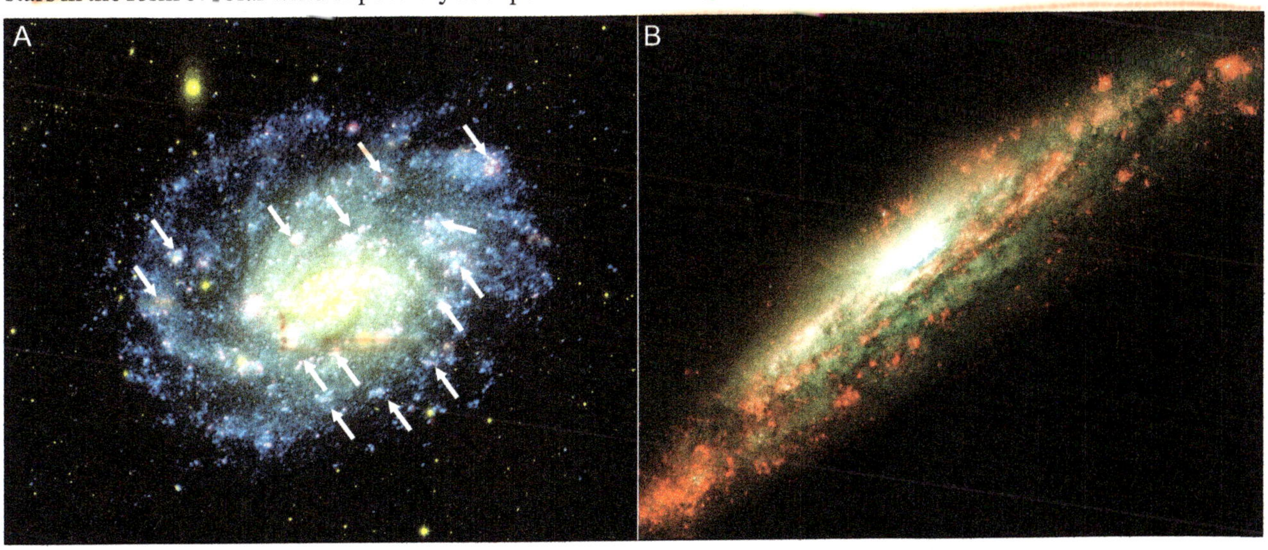

Figure 17.28: A. The spiral galaxy NGC300 shows that most of the newly-forming stars are in the arms. The white arrows show clusters where star formation is occurring. Image credit: NASA/JPL-Caltech/OCIW. B. Elliptical galaxies have more of a reddish color, indicating little active star formation. Image credit: NASA, Space Telescope Science Institute.

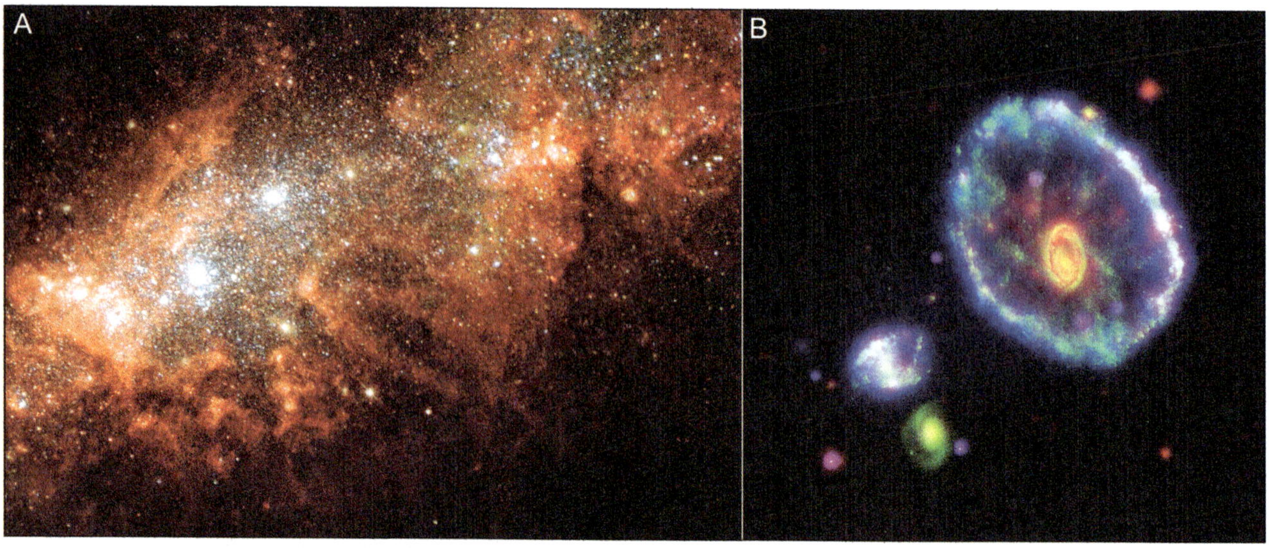

Figure 17.29: A. This irregular dwarf galaxy seems to contain many areas of new star formation (bluish areas). Image credit: NASA/JPL/Hubble. B. Known as the Cartwheel Galaxy, this irregular galaxy is believed to have received its unusual shape due to a collision with one of the two smaller galaxies in the lower left of the panel. Image credit: NASA/JPL-Caltech.

Irregular galaxies are unique in variety, each with its own color scheme. This indicates that different irregular galaxies have different types of star formation process occurring. Where blue stars are present (Figure 17.29A), star formation must be occurring, where as other galaxies may show no blue light whatsoever. Some irregular galaxies may have experienced a collision or an explosion in the past (Figure 17.29B), with their dust scattered in multiple directions.

Section 17.6: Galactic Centers

What lies at the center of a galaxy? This is a question that many astronomers have puzzled over for decades. By observing other stars within the Milky Way, scientists concluded that the center of the galaxy must have a very large mass there. If this was true, why couldn't we see it? A clue to this mystery was revealed in the early 1960's when Maarten Schmidt, a professor at the California Institute of Technology, was studying radio source emissions from the sky. Most radio emissions were known to come from distant galaxies. However, some radio sources did not appear to be galaxies, but looked like faint blue stars in telescopes. When analyzing their spectra, it looked like no element ever seen before (remember that each element has a unique spectrum). These new objects were given the name "quasi-stellar objects", also referred to as a **quasars**.

Figure 17.30: An image of Quasar 3C-273, the brightest quasar ever discovered. Image credit: ESA/Hubble/NASA.

One night, while studying an object known as 3C-273 (Figure 17.30) Dr. Schmidt realized that the spectrum from the quasar was not from a new element, but it was the hydrogen spectrum that was massively redshifted, indicating that the object was moving away from us very quickly. He then determined its distance from Earth, a staggering 2 billion light years away. Using this information, and using its apparent brightness, he determined that this one quasar had a brightness with an absolute magnitude that was hundreds of times greater than the entire Milky Way galaxy.

Since the light that left this object is 2 billion years old, it shows us what this object was like 2 billion years ago. These distant objects can help us understand how the universe formed, because we can see different ages of objects within the universe all at the same time. There is no question that 3C-273 looks different now than what we can see. But the question remains: What does a quasar become? And from where does its massive energy come? Since the 1960's, newer telescopes have shown that quasars are surrounded by other material with spectra similar to that of other galaxies we have seen. So it was determined that a quasar was part of a galaxy. These new telescopes also revealed that sometimes quasars would emit jets of material and radiation. Theoreticians had predicted that these types of jets would be associated with a supermassive black hole.

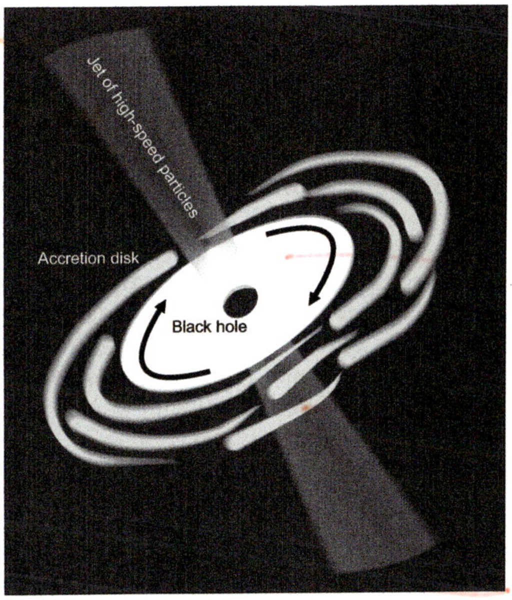

Figure 17.31: A basic diagram of a quasar. As material spins around a supermassive black hole, an accretion disk forms, as well two jets of particles.

If a black hole containing the mass of several million times our Sun was formed, what would it be like? Any matter outside of the event horizon would move toward the black hole, gaining kinetic energy and increasing its temperature. This matter would begin to spin around the black hole, forming what is known as an accretion disk (Figure 17.31). As the matter moved closer and closer to the event horizon, it would move faster and faster. The matter then released energy in form of radiation of all wavelengths and in all directions.

Black holes themselves seem to be able to jettison high-energy radiation and charged particles along its axis. This process is still not well understood, but scientists believe that it has to do with the magnetic field of the black hole. Particles trapped within these magnetic fields would be large emitters of radio-frequency radiation. This would mean that depending on our orientation compared to the black holes axis, we may be able to detect radio waves from one quasar, and not from another. The idea that quasars were massive black holes began to take shape, and helped explain some of the mystery behind their energy source, and why we cannot detect all of them easily.

Astronomers then began to wonder if there was a massive black hole at the center of every galaxy. If this theory were true, the center of some galaxies should be emitting large jets of charged particles, along with electromagnetic radiation, including x-rays. In 1994, the Hubble Space Telescope imaged the nucleus of M87, a giant elliptical galaxy. It appears that a cloud of gas is rotating around some invisible object (Figure 17.32A). Using the spectra and

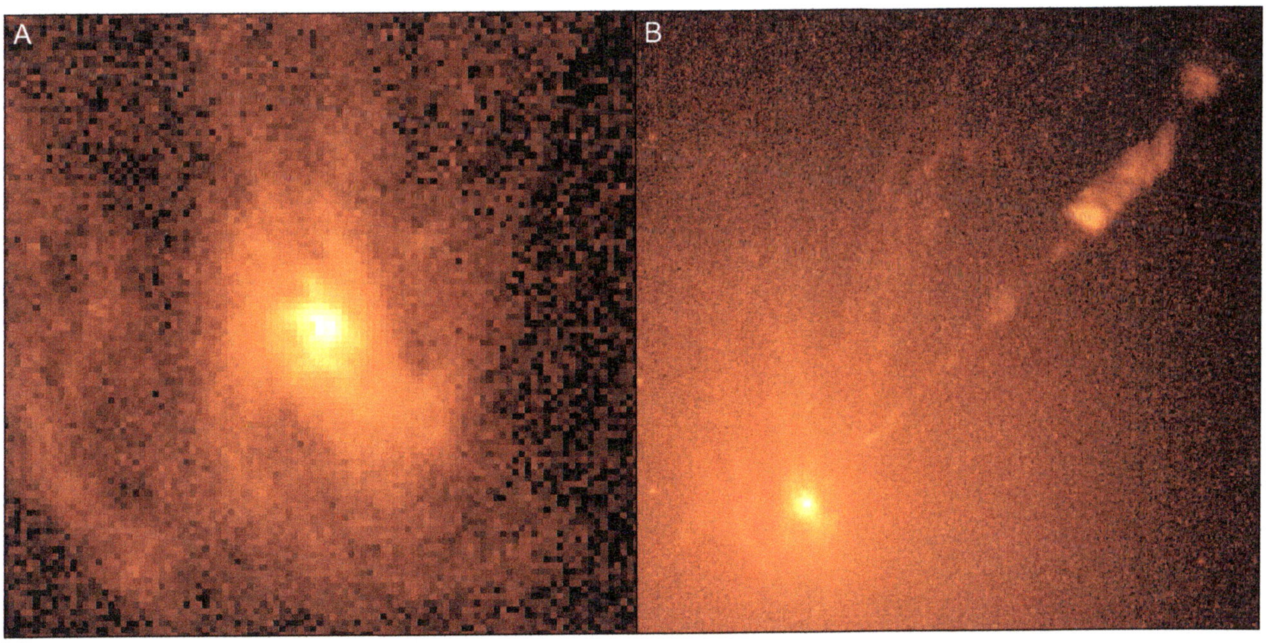

Figure 17.32: A. Astronomers using the Hubble telescope have found seemingly conclusive evidence for a massive black hole in the center of the giant elliptical galaxy M87, located 50 million light-years from Earth in the constellation Virgo. This image shows a spiral-shaped disk of hot gas in the core of M87. B. This image of M87 shows the high-energy jet the quasar emits. Image credits: Holland Ford, Space Telescope Science Institute/Johns Hopkins University; Richard Harms, Applied Research Corp.; Zlatan Tsvetanov, Arthur Davidsen, and Gerard Kriss at Johns Hopkins; Ralph Bohlin and George Hartig at Space Telescope Science Institute; Linda Dressel and Ajay K. Kochhar at Applied Research Corp. in Landover, Md.; and Bruce Margon from the University of Washington in Seattle.; NASA.

the Doppler Effect, scientists calculated that at 60 ly away from the center of this galaxy, gas is rotating with a speed of 450,000 m/s. This would mean that the black hole at the center of the galaxy contains close to 3 billion solar masses. This, along with the high energy jet being emitted provides evidence that there is indeed a black hole at the center of this galaxy (Figure 17.32B).

So if there are massive black holes at the center of galaxies closer to us, why don't we see radio pulses like we do from galaxies far away? Astronomers believe that as a galaxy matures, the black hole absorbs most of the material near its event horizon. Once no more mass enters the black hole, it stops emitting such large quantities of radiation and charged particles. Therefore, older massive black holes no longer have quasar-like properties. Then what are these distant quasars? Scientists now believe that quasars are the active nuclei of young galaxies. Although these galaxies may be much more mature today, having produced millions of stars and planets, we are seeing these galaxies in their youth, before star formation began. Matter is still falling into the black hole, and it is emitting large amounts of radiation. By studying quasars, we have the chance to understand what our galaxy could have been like billions of years ago.

Section 17-7: Galactic Clusters and the Local Group

Galaxies typically form in clusters, and the Milky Way has a few local neighbors. This cluster of galaxies is known as the **Local Group**. The number of galaxies within this cluster is uncertain, but there are probably close to 40 galaxies located within an area 3 million light years in diameter. Because some of these galaxies lie in the same plane as the Milky Way, they are very difficult to detect. Within the Local group, the Milky Way and the Andromeda Galaxy are the largest (Figure 17.33A). Each of these spiral galaxies seems to be associated with groups of smaller galaxies. For example, the Milky Way is associated with dwarf elliptical galaxies such as Fornax, Leo I, Sagittarius Dwarf and Canis Major Dwarf, among others (Figure 17.33B-C). In fact, these latter two galaxies are close enough that they may be actually colliding with the Milky Way's disk. As time passes, they may become part of the Milky Way. The Andromeda Galaxy also has several companion galaxies, including the NGC 205 and MC 32 galaxies.

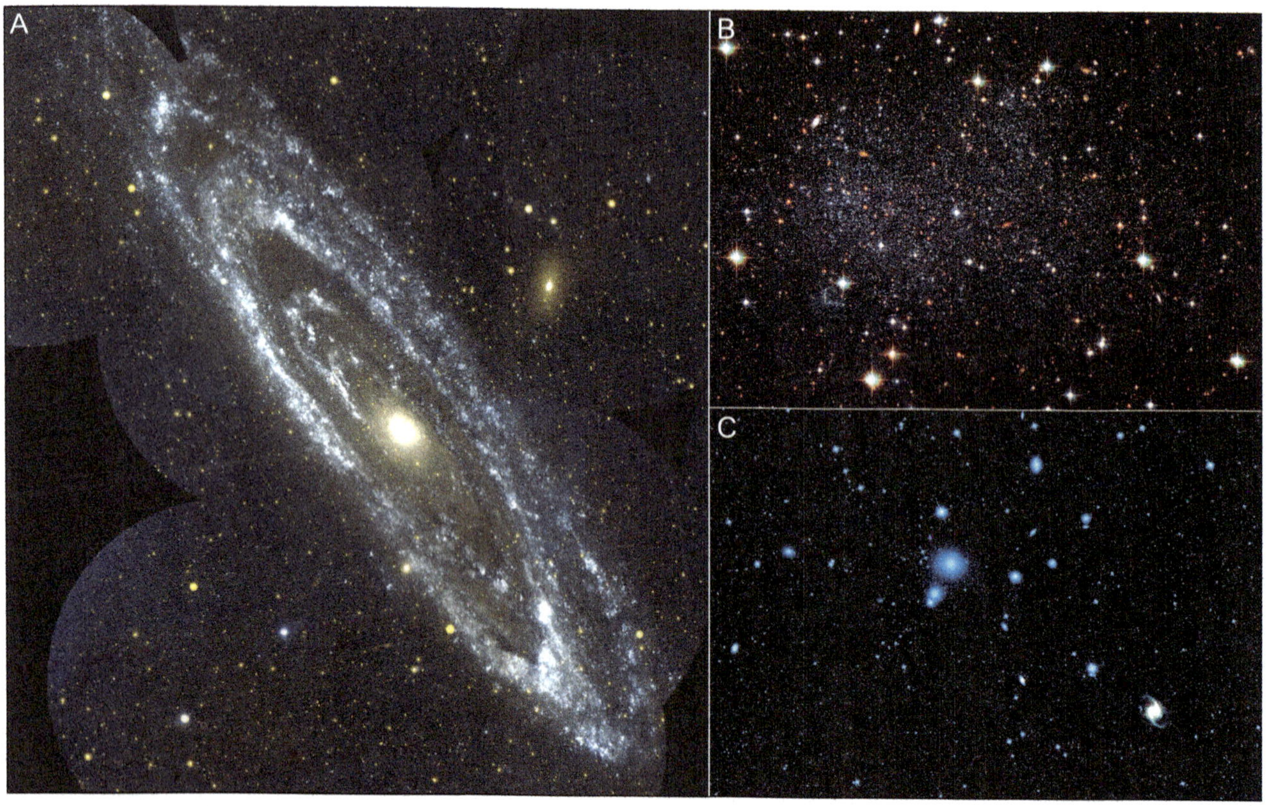

Figure 17.33: Some members of the Local Group. A. The Andromeda Galaxy, the closest large spiral galaxy near us. Image Credit: NASA/JPL/ California Institute of Technology. B. The Sagittarius dwarf irregular galaxy. Image credit: NASA, ESA, and the Hubble Heritage Team (STScI/AURA). C. The Fornax Galaxy Cluster. Image credit: NASA/JPL-Caltech/UCLA.

Other galaxies within the local group include several dwarf elliptical galaxies, such as Leo II and the Sculptor Galaxy (Figure 17.34A). The Large and Small Magellanic Clouds are irregular galaxies near the Milky Way (Figure 17.34B-C). These galaxies initially seem to have no structure. However, studies show that the Large Magellanic Cloud may be a barred spiral galaxy. In-depth images have indicated that the Small Magellanic Cloud is actually two separate galaxies, one overlaid on the other. Of all the galaxies within the Local Group, at least 4 are spiral, 15 are elliptical, and at least 13 are irregular.

It is important to keep in mind that even these smallest galaxies contain over a billion stars, and are only small when compared to the Milky Way. Astronomers believe that more galaxies will be discovered in our local cluster. By observing the interaction of these galaxies, as well as their evolution, astronomers hope to learn more about galaxies that are not as easy to observe.

Figure 17.34: Other members of the Local Group. A. The Scupltor Galaxy. Image credit: NASA/JPL-Caltech. B. The Large Magellanic Cloud in Infrared. Image credit: NASA/JPL-Caltech/M. Meixner (STScI) & the SAGE Legacy Team. C. The Small Magellanic Cloud. Photo credit: NASA, ESA and A. Nota (STScI/ ESA).

CPSIA information can be obtained
at www.ICGtesting.com
Printed in the USA
LVOW02s0832011215
463412LV00002BB/2/P